DEVELOPMENTAL MATHEMATICS

THE JOHNSTON/WILLIS DEVELOPMENTAL MATHEMATICS SERIES

Essential Arithmetic, **Sixth Edition** (paperbound, 1991)
Johnston/Willis/Lazaris

Essential Algebra, **Sixth Edition** (paperbound, 1991)
Johnston/Willis/Lazaris

Developmental Mathematics, **Third Edition** (paperbound, 1991)
Johnston/Willis/Hughes

Elementary Algebra, **Third Edition** (hardbound, 1991)
Johnston/Willis/Buhr

Intermediate Algebra, **Third Edition** (hardbound, 1991)
Johnston/Willis/Buhr

Intermediate Algebra, **Fifth Edition** (paperbound, 1991)
Johnston/Willis/Lazaris

DEVELOPMENTAL MATHEMATICS

THIRD EDITION

C. L. JOHNSTON
ALDEN T. WILLIS
Formerly of East Los Angeles College

GALE M. HUGHES
Cerritos College

Wadsworth Publishing Company
Belmont, California
A Division of Wadsworth, Inc.

This book is dedicated to our students, who inspired us to do our best to produce a book worthy of their time.

Mathematics Editor: Anne Scanlan-Rohrer
Assistant Editor: Tamiko Verkler
Special Projects Editor: Alan Venable
Editorial Assistant: Leslie With
Production: Ruth Cottrell
Print Buyer: Randy Hurst
Designer: Julia Scannell
Copy Editor: Marjory Simmons
Technical Illustration: Alexander Teshin Associates
Compositor: Polyglot Pte. Ltd.
Cover: Frank Miller
Signing Representative: Kenneth King

Printed in the United States of America

3 4 5 6 7 8 9 10—95 94 93 92

Library of Congress Cataloging-in-Publication Data

Johnston, C. L. (Carol Lee), 1911—
Developmental mathematics / C. L. Johnston, Alden T. Willis, Gale M. Hughes.—3rd ed.
p. cm.
Includes index.
ISBN 0-534-14208-7
1. Algebra. I. Willis, Alden T. II. Hughes, Gale M. III. Title.
QA152.2.J628 1990
512.9—dc20

90-41222
CIP

Contents

7 EVALUATING EXPRESSIONS

8 POLYNOMIALS

9 EQUATIONS AND INEQUALITIES

10 WORD PROBLEMS

11 FACTORING

12 ALGEBRAIC FRACTIONS

13 GRAPHING

14 SYSTEMS OF EQUATIONS

15 EXPONENTS AND RADICALS

Preface

In *Developmental Mathematics* we use simple, direct language to present the essential content of elementary algebra. An arithmetic review, Chapters 1–5, is provided for those students who need to improve their arithmetic skills. The instructor may choose to use all or part of the arithmetic review depending on the needs and abilities of his or her students. This third edition reflects various suggestions and ideas based on eight years of classroom use.

Features of This Book

Some of the major features of this book are:

1. The contents are arranged in small sections, each with its own examples and exercises. We use a one-step, one-concept-at-a-time approach.
2. Each topic is explained carefully and in detail, and many concrete examples lead up to general principles which can then be used to work the exercises.
3. Important concepts and algorithms are enclosed in boxes for easy identification and reference.
4. Common errors are clearly identified in special "Words of Caution," which will help reduce the number of mistakes commonly made by inexperienced students.
5. Liberal use is made of visual aids, such as number lines and shading.
6. The importance of checking solutions is stressed throughout the book.
7. Special attention is given to the operations with zero in a single section of Chapter 6; it includes a discussion of the common errors that students make with zero.
8. A detailed method is provided for changing a word statement into an algebraic equation for solving word problems. Many of the difficulties students have with percent problems are eliminated by the section on percent proportion, which has aids to help identify the numbers in percent problems.
9. Students are asked to do the following steps to help them set up word problems and organize their work:
   a. state what their variables represent,
   b. set up an equation and solve it, and
   c. answer the question.
10. More than 5,500 exercises are provided in this book. Answers to all odd-numbered exercises, as well as many selected solutions for these odd exercises, are provided in the back of this text. In most cases, the even-numbered exercises provide practice on problems analogous to the odd-numbered exercises.
11. A diagnostic test is included at the end of each chapter to help prepare students for taking the chapter test. Complete solutions to all problems in these diagnostic tests, together with section references, appear in the answer section.
12. A comprehensive summary and a set of review exercises are included at the end of each chapter.
13. Cumulative review exercises are provided after Chapters 5, 10, and 16.

14. Chapter 16 can be used to introduce students to geometry. The following topics are included: perimeter, area, volume, the Pythagorean Theorem, parallel and perpendicular lines, and similar and congruent triangles. In preparing this edition, we found that an increasing number of remedial mathematics courses include geometry in their syllabi. Geometry is also a topic that is included in competency tests and entry level examinations for schools in various states.
15. A comprehensive treatment of the metric and English systems is included in Appendix II.
16. The quadratic formula is covered in Appendix III.
17. For quick reference, English and metric tables, together with conversions and a list of geometric formulas, appear inside the front cover and on the page facing the back cover of this book.

Changes in the Third Edition

Some of the major changes in the third edition are:
The arithmetic chapters were expanded to include a wider variety of word problems. The new sections are:

Sect. 1.12	Average
Sect. 1.13	Perimeter and Area
Sect. 1.14	Estimation
Sect. 2.17	Comparing Fractions
Sect. 3.12	Comparing Decimals
Chap. 4	Ratio, Proportion, and Percent
Chap. 5	Graphs and Statistics

The algebra and geometry chapters were expanded to reflect the many helpful comments we received from our users. The major changes are:

Chap. 10	The word problems were expanded to include number problems, consecutive integer problems, geometry problems, mixture problems, solution problems, distance problems, and variation problems.
Chap. 12	Solving work problems was added to the section on word problems.
Chap. 13	Graphing lines by the slope-intercept method was added.
Chap. 16	A new section on congruent triangles was added.

Calculator problems with examples were added to Chapter 1 and to Chapters 3 through 7. They are labeled with a calculator symbol for easy identification. Cumulative review exercises were added after Chapter 5, 10, and 16.

This book can be used in three types of instructional programs:

1. *The conventional lecture course.* This book is particularly easy to fit into a program of regular assignments because it is divided into many small, self-contained units. A diagnostic test that students can use for review and diagnostic purposes is included at the end of each chapter.
2. *The learning laboratory course.* This book also fits easily into a laboratory learning class because of its format. Since explanations, examples, and exercises

are provided in each section of the book, there is a wide degree of latitude in the pace at which students can progress.

3. *Self-study.* This book lends itself to self study because (a) each new topic is short enough to be mastered before continuing, (b) each topic is introduced in an informal, readable manner, and (c) numerous examples and exercises are provided. In using the book for self-study, a student should begin by taking the diagnostic test for Chapter 1, and then checking the answers against the solutions at the back of the book. References given in the solutions will direct the student to specific sections of the book that explain the particular problems done incorrectly. The student can continue in this manner at his or her own pace throughout the book.

Ancillaries

The following ancillaries are available with this text:

1. The Instructor's Manual contains five different tests for each chapter, an arithmetic pretest, two forms of four unit examinations, and two final examinations that can be easily removed and duplicated for class use. These tests are prepared with adequate space for students to work the problems. Answer keys for these tests are provided in the manual, as are the answers to the even-numbered exercises. This manual also contains essays to help the instructor teach developmental mathematics students. Essays cover such topics as: writing in the mathematics classroom, running a lab, cooperative learning, and more.
2. The test bank is also available in a computerized format entitled EXP-Test. EXP-Test is a fast, highly flexible computerized testing system for the IBM PC and compatibles. Instructors can edit and scramble test items or create their own tests.
3. In addition, Wadsworth offers the *Johnston/Willis/Hughes Computerized Test Generator* (JeWeL TEST) software for Apple II and IBM PC or compatible machines. This software, written by Ron Staszkow of Ohlone College, allows instructors to produce many different forms of the same test for quizzes, work sheets, practice tests, and so on. Answers and cross-references to the text provide additional instructional support.
4. An "intelligent" tutoring software system is available for the IBM PC and compatibles. *Expert Tutor*, written by Sergei Ovchinnikov of San Francisco State University, uses a highly interactive format and sophisticated techniques to tailor lessons to the specific algebra and prealgebra learning problems of students. The result is individualized tutoring strategies with specific page references to problems, examples, and explanations in the textbook.
5. Twelve videotapes, created by John Jobe of Oklahoma State University, review the most essential and difficult topics from the textbook.

To obtain additional information about these supplements, contact your Wadsworth-Brooks/Cole representative.

Acknowledgments

We are grateful to the following people who reviewed the book and provided invaluable suggestions and comments: Kathleen Beisse, University of Oregon; Karen Sue Cain, Eastern Kentucky University; Audrey Douthit, The Pennsylvania State University; Miriam Keesey, San Diego State University; Susan McClory, San Jose State University; John Vangor, Housatonic Community College; and Alma McKinney

Wynn, North Florida Junior College. We'd also like to thank the following, who checked the text for mathematical accuracy: June L. Gaston, Borough of Manhattan Community College; Miriam Keesey, San Diego State University; Brad Lindemann, San Jose State University; Richard Spangler, Tacoma Community College; and Barbara Spanel, San Jose State University.

C. L. Johnston
Alden T. Willis
Gale M. Hughes

DEVELOPMENTAL MATHEMATICS

1 Whole Numbers

1.1 Basic Definitions

In this section we introduce some of the names and definitions of numbers and number relations.

Set

A **set** is a collection of objects or things. Sets are usually represented by listing their members, separated by commas, within braces { }.

Natural Numbers

The numbers

1, 2, 3, 4, 5, 6, 7, 8, 9, 10, 11, 12, and so on

are called the **natural numbers** (or **counting numbers**). These were probably the first numbers invented by man and used to count his possessions, such as sheep and goats. We call the set of natural numbers N, so that

$$N = \{1, 2, 3, 4, \ldots\}$$

This is read:

"N equals the set whose members are 1, 2, 3, 4, and so on."

The three dots to the right of the number 4 indicate that the remaining numbers are found by counting in the same way we have begun: namely, to add 1 to the preceding number to find the next number.

The smallest natural number is 1. The largest natural number can never be found because no matter how far we count there are always larger natural numbers.

Number Line

Natural numbers can be represented by numbered points equally spaced along a straight line (Figure 1.1.1). Such a line is called a **number line**. The arrow on the right of the line indicates that the number line continues forever.

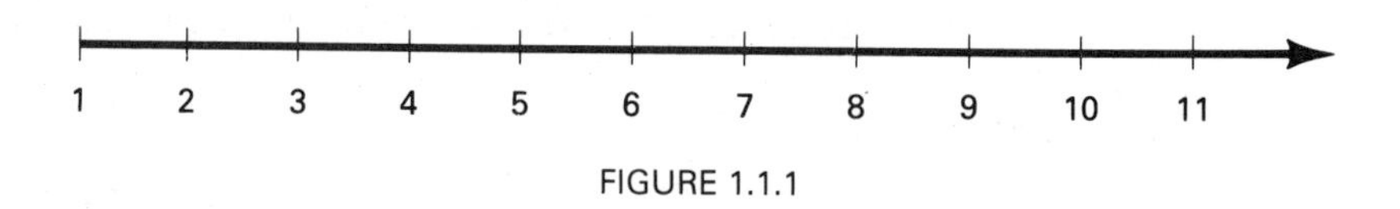

FIGURE 1.1.1

Whole Numbers

When 0 is included with the natural numbers, we have the set of numbers known as **whole numbers** (Figure 1.1.2).

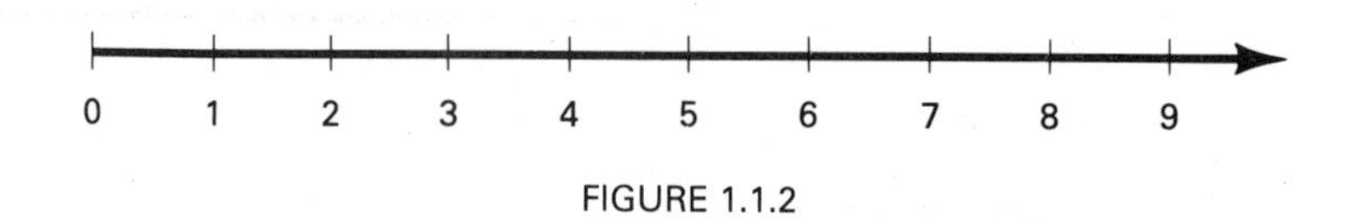

FIGURE 1.1.2

We call the set of whole numbers W. So that

$$W = \{0, 1, 2, 3, \ldots\}$$

This is read:

"W equals the set whose members are 0, 1, 2, 3, and so on."

Digits

In our number system a **digit** is any one of the first ten whole numbers {0, 1, 2, 3, 4, 5, 6, 7, 8, 9}. Any number can be written by using a combination of these digits. For this reason, the digits are sometimes called the building blocks of our number system. Numbers are often referred to as *one-digit* numbers, *two-digit* numbers, *three-digit* numbers, and so on.

Example 1

a. 35 is a two-digit number.

b. 7 is a one-digit number.

c. 275 is a three-digit number.

d. 100 is a three-digit number.

e. The first digit of 785 is 7.

f. The second digit of 785 is 8.

g. The third digit of 785 is 5. ■

We show the natural numbers, whole numbers, and digits on the number line in Figure 1.1.3.

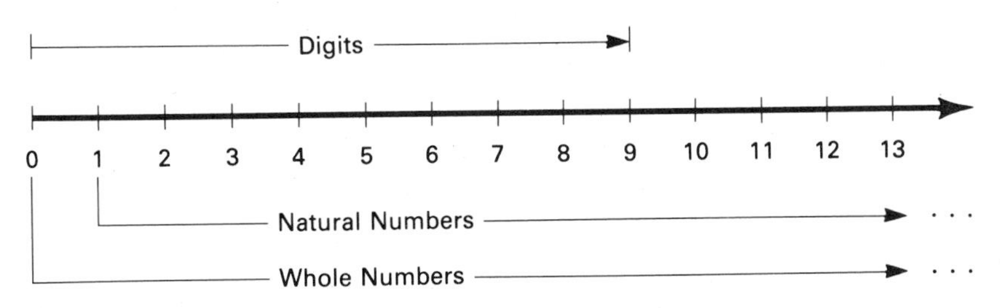

FIGURE 1.1.3

Inequality Symbols

The symbol <, read "less than" and the symbol >, read "greater than," are called **inequality symbols**. They are used to compare numbers that are not equal. Numbers get larger as we move to the right on the number line and smaller as we move to the left.

Example 2 $2 < 7$

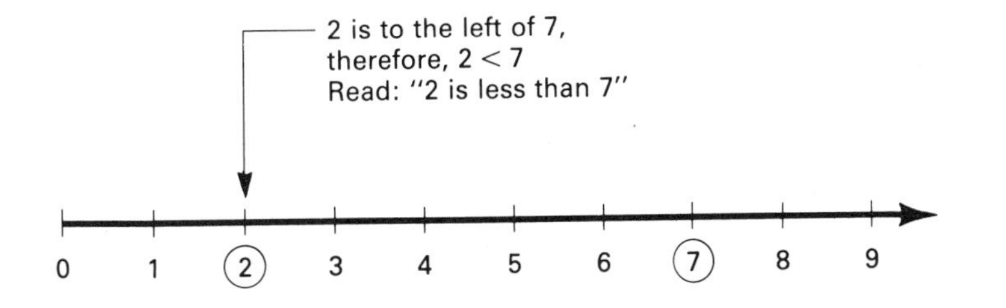

■

Example 3 $6 > 3$

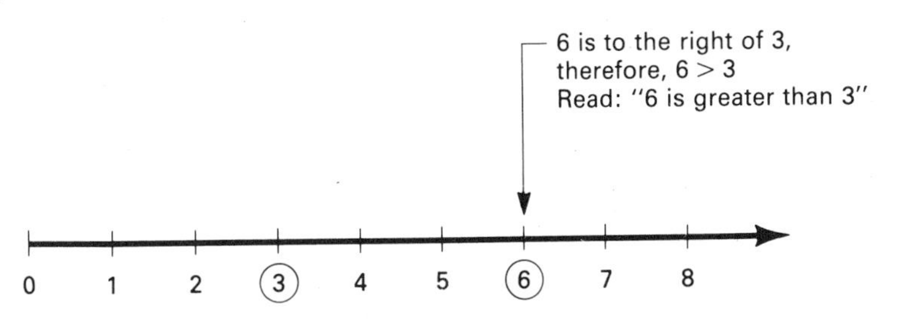

■

An easy way to remember the meaning of the symbol is to notice that the wide part of the symbol is next to the larger number.

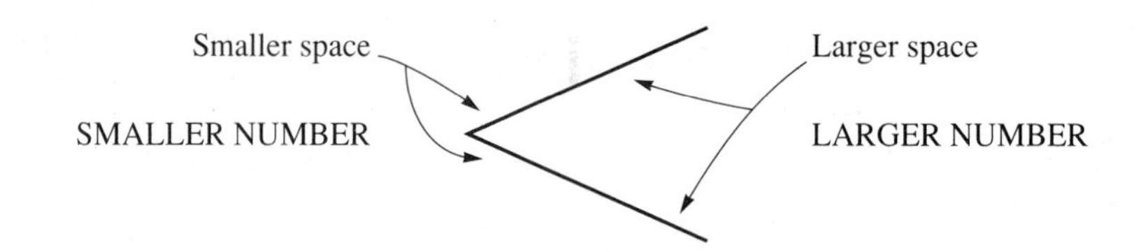

Some people like to think of the symbols $>$ and $<$ as arrowheads that point toward the smaller number.

Example 4

a. $7 > 6$ is read "7 is greater than 6."

b. $7 > 1$ is read "7 is greater than 1."

c. $5 < 10$ is read "5 is less than 10." ■

Note that $7 > 6$ and $6 < 7$ give the same information even though they are read differently.

Another inequality symbol is $\neq$. A slash line drawn through a symbol puts a "not" in the meaning of the symbol.

$=$ is read "is equal to."
$\neq$ is read "is *not* equal to."

$<$ is read "is less than."
$\nless$ is read "is *not* less than."

$>$ is read "is greater than."
$\ngtr$ is read "is *not* greater than."

Example 5 Examples showing the use of the slash line:

a. $4 \neq 5$ is read "4 is not equal to 5."

b. $3 \nless 2$ is read "3 is not less than 2."

c. $5 \ngtr 6$ is read "5 is not greater than 6." ■

EXERCISES 1.1

1. What is the second digit of the number 2,478?

2. What is the fourth digit of the number 1,975?

3. What is the smallest digit?

4. What is the smallest whole number?

5. What is the smallest natural number?

6. What is the largest digit?

7. What is the largest whole number?

8. What is the largest natural number?

9. What is the largest two-digit number?

10. What is the smallest two-digit number?

11. What is the smallest three-digit number?

12. What is the largest three-digit number?

13. Is 12 a digit?

14. Is 12 a whole number?

15. Is 12 a natural number?

16. Is 0 a natural number?

17. Write all the natural numbers < 6.

18. Write all the whole numbers < 6.

In Exercises 19–24, determine which of the two symbols, $>$ or $<$, should be used to make each statement true.

19. 5 ? 8 **20.** 0 ? 5 **21.** 7 ? 0 **22.** 2 ? 6 **23.** 13 ? 43 **24.** 27 ? 16

1.2 Reading and Writing Whole Numbers

Place-Value

Our number system is a **place-value** system. This means that the value of each digit in a written number is determined by its place in the number.

Example 1

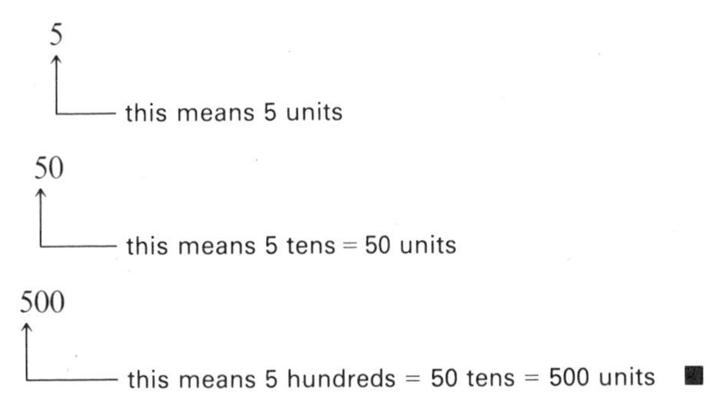

Notice in the above example that whenever a digit is moved one place to the left, its value becomes 10 times larger.

Example 2

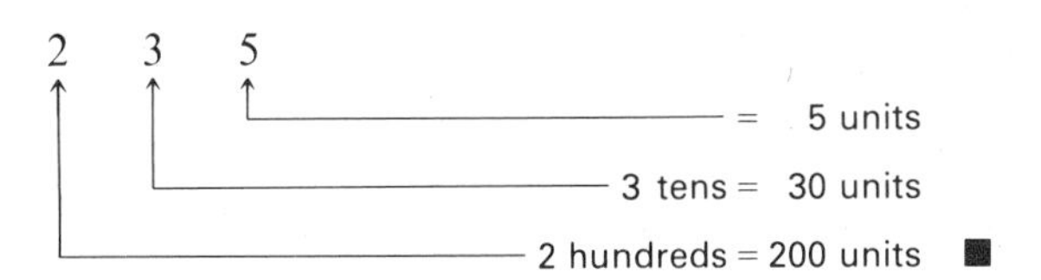

A more complete discussion of the idea of place value and our number system is given in Section 3.1.

Separating the Digits into Groups Commas are used to separate large numbers into smaller groups of three digits. In placing the commas, we count from the right as shown in Figure 1.2.1. Note that the first group on the left may have one, two, or three digits. Each group of three has a name: The first group on the right is the units group; the second group from the right is the thousands group; the third group from the right is the millions group; and so on (Figure 1.2.1).

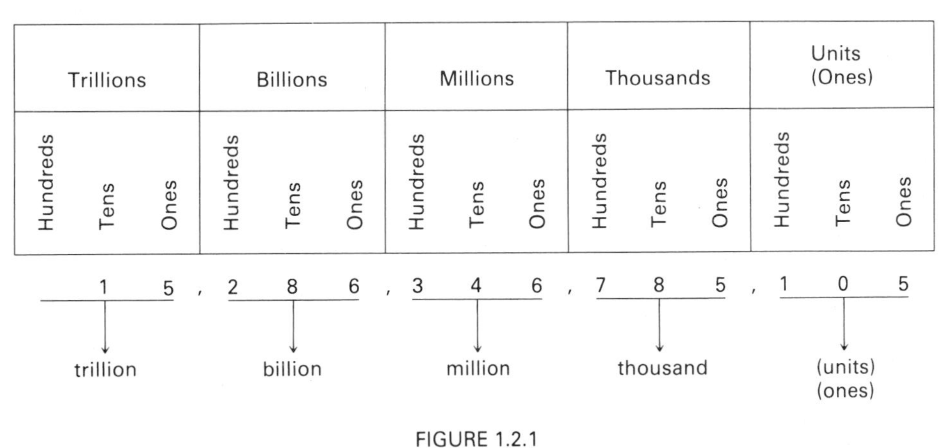

FIGURE 1.2.1

Reading a Number To read a number, we read the number formed by the digits in the left group, then say the name of that group. Then read the number formed by the digits in the next group and say the name of that group. We continue in this way until all groups in the number have been read. The name of the units group is usually omitted when reading the number. Study Figure 1.2.1.

Writing Numbers in Words In writing numbers in words, commas are used in the same places they are used when the number is written in digits. For example, we read in words the number shown in Figure 1.2.1 as follows: Fifteen trillion, two hundred eighty-six billion, three hundred forty-six million, seven hundred eighty-five thousand, one hundred five. The following examples demonstrate the reading and writing of numbers:

Example 3 Reading and writing numbers.

Notice in examples (a) and (b) that hyphens are used in writing two-digit numbers such as 25 (twenty-five), 37 (thirty-seven), 45 (forty-five).

a. 8,025 is read "eight thousand, twenty-five."

b. 37,045,200 is read "thirty-seven million, forty-five thousand, two hundred."

Notice in examples (c) and (d) that when a group is made up of all zeros it is not read.

c. 275,000,000 is read "two hundred seventy-five million."

d. 9,000,605,000 is read "nine billion, six hundred five thousand."

The word *and* is not used in writing whole numbers. ■

Numbers Larger Than Trillions Numbers larger than trillions have names. For those students who want to know the names of larger numbers, we include Figure 1.2.2. You will not be expected to know the names of numbers larger than trillions.

12,	315,	218,	557,	314,	708,	515,	900,	000,	304,	017,	708
Decillion	Nonillion	Octillion	Septillion	Sextillion	Quintillion	Quadrillion	Trillion	Billion	Million	Thousand	Units

FIGURE 1.2.2

Large numbers such as billions and trillions do not have much meaning for most of us. For example, one billion miles is farther than 40,000 times the distance around the earth. We often see large numbers written in newspapers and magazines. For example, the Gross National Product of the United States is over one trillion dollars. The distance light travels in one year is almost six trillion miles.

Using Digits to Write Numbers To write a number using digits remember that each group after the first group on the left must have three digits in it. Also, if a group is not mentioned, we will need three zeros as place holders for that group.

Example 4 Use digits to write the following numbers.

a. Forty-eight thousand, five hundred

48,500

b. Nine million, three hundred twenty

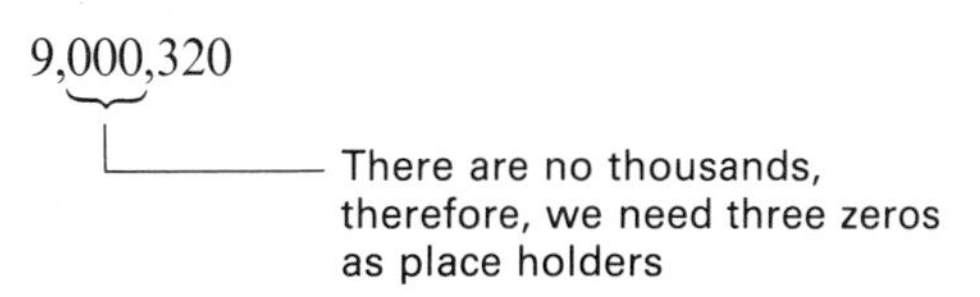

c. Seventeen million, six thousand, nine hundred

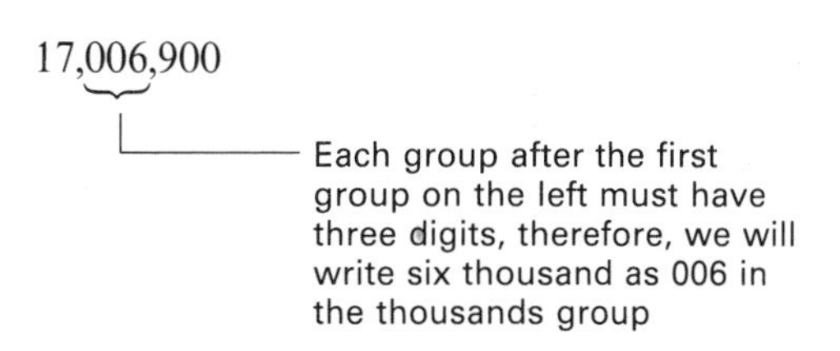

d. Five billion, fifty thousand

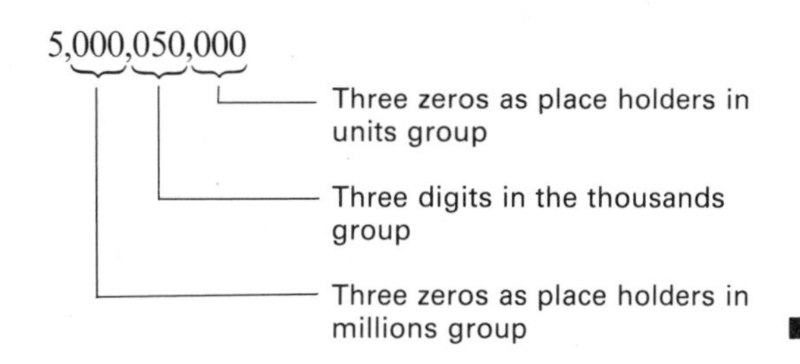

■

EXERCISES 1.2

1. In the number 576:
a. The 6 represents how many units?
b. The 7 represents how many units?
c. The 5 represents how many units?

2. In the number 914:
a. The 4 represents how many units?
b. The 1 represents how many units?
c. The 9 represents how many units?

3. In the number 348:
a. The 4 represents how many tens?
b. The 4 represents how many units?
c. The 3 represents how many hundreds?
d. The 3 represents how many tens?

4. In the number 862:
a. The 6 represents how many tens?
b. The 6 represents how many units?
c. The 8 represents how many hundreds?
d. The 8 represents how many tens?

In Exercises 5–12, write the numbers in words.

5. 16,346 **6.** 125,006 **7.** 1,070,200 **8.** 309,005,000

9. 49,000,070 **10.** 2,400,000,000 **11.** 8,000,725,000 **12.** 73,006,000,900

In Exercises 13–20, mark off with commas, then write the numbers in words.

13. 16601 **14.** 230023 **15.** 71040 **16.** 7100400

17. 710004000 **18.** 71000040000 **19.** 5206000000 **20.** 9000900909

In Exercises 21–32, use digits to write the numbers.

21. Twenty thousand, five

22. Two hundred thousand, fifty

23. Eight million, eight thousand, eight hundred eight

24. Twelve million, six hundred sixty thousand

25. Seven million, seven

26. Forty-five million, sixteen thousand

27. Ten billion, one hundred thousand, ten

28. Two billion, thirty million

29. One hundred million, twenty-nine thousand, six

30. Four hundred fifty million, nine hundred

31. One light-year is the distance light travels in one year. This is about five trillion, eight hundred eighty billion miles.

32. Sirius, the brightest star in our winter sky, is fifty trillion, five hundred sixty billion miles from the earth.

1.3 Rounding Off Whole Numbers

Numbers are often expressed to the nearest million, to the nearest thousand, or to the nearest hundred, and so on. When we say the earth is 93,000,000 miles from the sun, it is understood that 93,000,000 has been rounded off to the nearest million. The distance around the earth at the equator is 24,902 miles, but in speaking of this distance it is more common to say 25,000 miles. We say that 24,902 miles has been "rounded off to the nearest thousand" miles. When a whole number is rounded off, we must say what place it has been rounded off to.

The symbol $\approx$, read "is approximately equal to," is often used when numbers have been rounded off.

TO ROUND OFF A WHOLE NUMBER

1. Locate the digit in the round-off place. (We will circle the digit.)
2. If first digit to right of round-off place is less than 5: digit in round-off place is *unchanged.*

   If first digit to right of round-off place is 5 or more: digit in round-off place is *increased by 1.*
3. Digits to *left* of round-off place are unchanged. (Special case: Examples 3 and 4)
4. Digits to *right* of round-off place are replaced by zeros.

Example 1 Round off 243,280 to the nearest thousand.

Solution

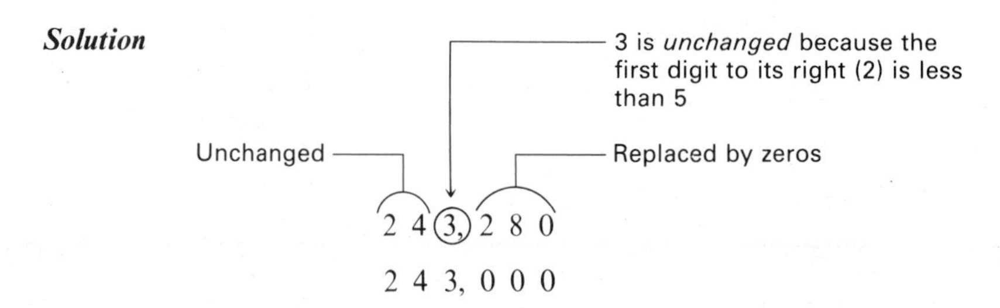

Therefore, 243,280 $\approx$ 243,000 rounded off to thousands. ■

Example 2 Round off 90,671 to the nearest hundred.

Solution

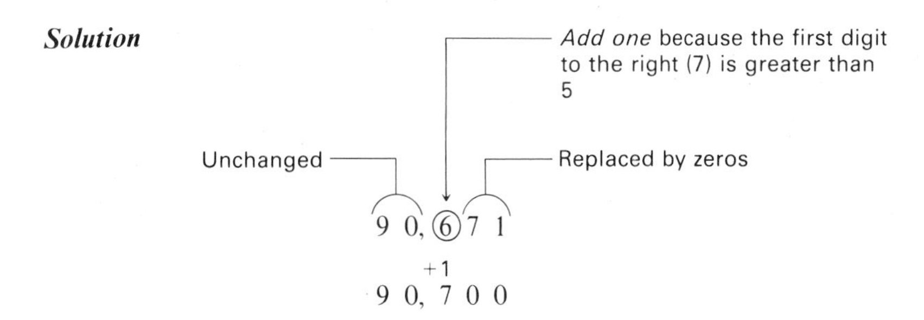

Therefore, 90,671 ≈ 90,700 rounded off to hundreds. ■

Example 3 Round off 31,962 to the nearest hundred.

Solution

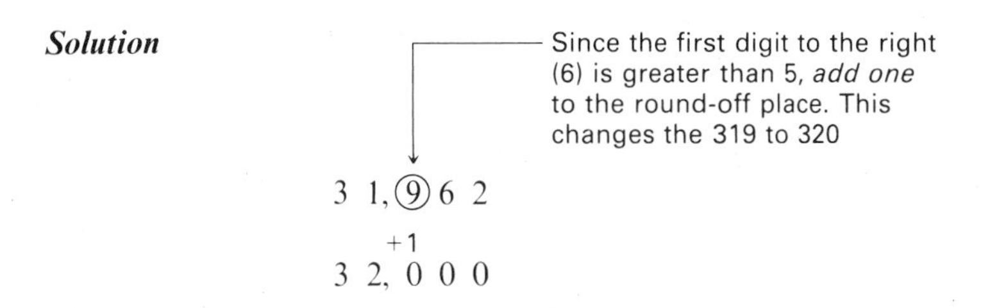

Therefore, 31,962 ≈ 32,000 rounded off to hundreds. ■

Example 4 Round off 999,507 to the nearest thousand.

Solution

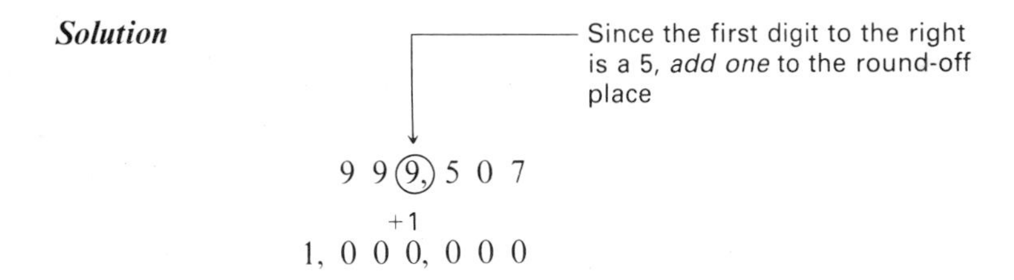

Therefore, 999,507 ≈ 1,000,000 rounded off to thousands. ■

EXERCISES 1.3

Round off the following numbers to the indicated place.

1. 4,728 Nearest hundred
2. 256,491 Nearest thousand
3. 926 Nearest ten
4. 28,619,000 Nearest million
5. 753 Nearest hundred
6. 485 Nearest ten
7. 63,195 Nearest thousand
8. 28,232 Nearest hundred
9. 798 Nearest ten
10. 629,653 Nearest thousand
11. 19,500,000 Nearest million
12. 26,500 Nearest thousand
13. 52,461,000 Nearest million
14. 853 Nearest ten

15. 3,472 Nearest hundred

16. 78,415 Nearest ten-thousand

17. 45,429,444 Nearest ten-thousand

18. 45,429,444 Nearest hundred-thousand

19. 1,648,723 Nearest hundred-thousand

20. 1,648,723 Nearest thousand

1.4 Addition of Whole Numbers

Terms Used in Addition

When whole numbers are added, the numbers being added are called the *addends*, and the answer is called the *sum*.

$$\begin{array}{rl} 2 & \text{addend} \\ +\,3 & \text{addend} \\ \hline 5 & \text{sum} \end{array}$$

Commutative Property of Addition

Example 1

a. $4 + 5 = 5 + 4 = 9$

b. $1 + 3 = 3 + 1 = 4$

c. $10 + 2 = 2 + 10 = 12$ ■

Example 1 suggests that reversing the order of two whole numbers in an addition problem does not change their sum. This important property of whole numbers is called the **commutative property of addition.**

> COMMUTATIVE PROPERTY OF ADDITION
>
> If a and b represent any numbers, then
>
> $$a + b = b + a$$

Associative Property of Addition

When adding three numbers, we can do either of the following:

1. Find the sum of the first two numbers, then add that sum to the third number.
2. Find the sum of the last two numbers, then add that sum to the first number.

We use parentheses ( ) to show which two numbers are being added first. Other grouping symbols can be used such as brackets [ ] or braces { }. When grouping symbols are used, the operations within those grouping symbols must be done first.

Example 2 Add: $2 + 3 + 4$

Solution

a. $(2 + 3) + 4 =$
$\quad 5 \quad + 4 = 9$

The parentheses mean that the 2 and 3 are to be added first

b. $2 + (3 + 4) =$
$\quad 2 + \quad 7 \quad = 9$

Here, the parentheses mean that the 3 and 4 are to be added first

Therefore, $(2 + 3) + 4 = 2 + (3 + 4)$ ■

In Example 2, the sum of three numbers was unchanged no matter how we grouped (or associated) the numbers. It is assumed that this property of addition is true when any three whole numbers are added. This is called the **associative property of addition**.

> ASSOCIATIVE PROPERTY OF ADDITION
>
> If a, b, and c represent any numbers, then
>
> $$(a + b) + c = a + (b + c)$$

Additive Identity

Adding zero to a number gives the identical number for the sum.

Example 3

a. $8 + 0 = 8$

b. $0 + 6 = 6$ ■

For this reason, zero is called the **additive identity**.

> ADDITIVE IDENTITY
>
> If a represents any number, then
>
> $$a + 0 = 0 + a = a$$

Addition of Whole Numbers

In adding numbers, arrange the numbers in vertical columns with all the units digits in the same vertical line, all the tens digits in the same vertical line, all the hundreds digits in the same vertical line, and so on. Then add the digits in each vertical line. Writing the numbers clearly and keeping the columns straight will help reduce the number of addition errors.

Example 4 Find the sum 587 + 265.

Explanation

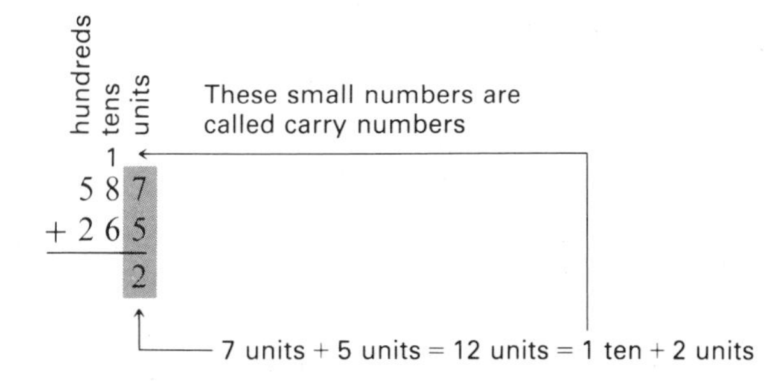

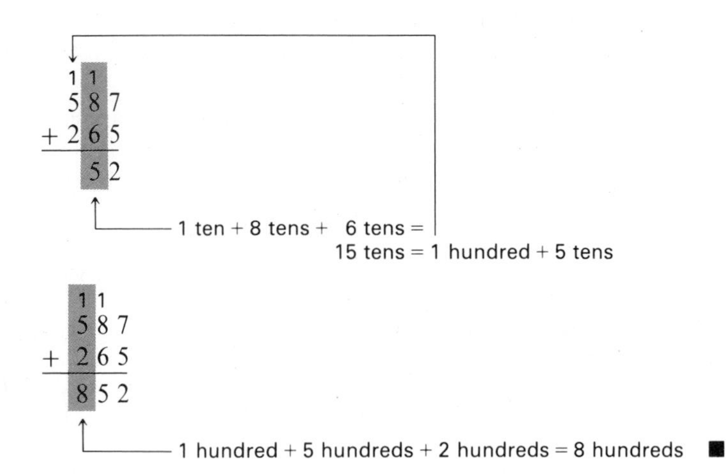

Example 5 Add: 85 + 354 + 6 + 5,871.

When adding a long column of numbers, the commutative and associative properties guarantee that we will obtain the same sum no matter what order or grouping of the numbers is used. Therefore, we can shorten the work by adding groups of 10.

Example 6 Add: 46 + 18 + 74 + 32.

Solution

$$\begin{array}{r} {}^{2}\\ 46 \\ 18 \\ 74 \\ +\ 32 \\ \hline 170 \end{array}$$

(46 and 74 make 10; 18 and 32 make 10; 6 and 4 make 10) ■

To improve your speed and accuracy with addition, see Appendix I for drill and practice with addition facts.

EXERCISES 1.4

In Exercises 1–18, find the sums.

1. 354 + 283

2. 678 + 254

3. 987 + 715

4. 817 + 209

5. 987 + 807

6. 483 + 967

7. 308 + 25 + 691

8. 755 + 9 + 307

9. 908 + 92 + 818

10. 562 + 267 + 48

11. 506 + 117 + 284

12. 274 + 638 + 752

13. $\begin{array}{r} 5{,}840 \\ 218 \\ 21 \\ +\ 3{,}002 \\ \hline \end{array}$

14. $\begin{array}{r} 75 \\ 3{,}594 \\ 315 \\ +\ 4{,}121 \\ \hline \end{array}$

15. $\begin{array}{r} 1{,}263 \\ 522 \\ 4{,}817 \\ +\ 63{,}548 \\ \hline \end{array}$

16. $\begin{array}{r} 26 \\ 47 \\ 64 \\ 13 \\ +\ 91 \\ \hline \end{array}$

17. $\begin{array}{r} 28 \\ 75 \\ 97 \\ 42 \\ +\ 15 \\ \hline \end{array}$

18. $\begin{array}{r} 203 \\ 576 \\ 524 \\ 837 \\ +\ 641 \\ \hline \end{array}$

In Exercises 19–26, arrange the following numbers in a vertical column and add.

19. $75{,}386 + 77 + 105{,}706{,}035 + 880{,}755{,}009 + 28{,}388{,}406$

20. $359 + 40{,}286 + 17 + 284{,}000{,}189 + 70{,}096{,}347$

21. $275 + 80 + 9 + 786{,}410{,}075 + 3{,}000{,}000 + 259{,}715{,}306$

22. $7{,}409 + 701{,}093{,}005 + 43 + 806{,}240 + 576$

23. $885{,}209{,}734 + 42{,}076 + 68 + 7{,}090{,}300 + 9{,}004$

24. $823 + 10{,}090 + 520{,}007{,}380 + 79 + 64{,}082$

25. $1{,}723 + 72 + 391{,}400{,}082 + 905 + 605{,}210 + 8$

26. $14 + 43{,}050{,}908 + 20{,}809 + 100{,}926 + 804$

In Exercises 27 and 28, use digits to write the following numbers, then arrange them in a vertical column and find their sum.

27. Thirty thousand, six. Seventy-five million, one hundred. Two billion, five hundred. Fifty million, one hundred thousand, ten.

28. Ten thousand, forty-seven. Twenty-three million, five thousand. Four billion, six million, seventy-three thousand, forty-two. Two hundred million, one hundred fifty-six thousand, six.

29. Find the sum of the digits.

30. Find the sum of the natural numbers less than 20.

31. The greatest known depth of our oceans is 36,198 feet, and the highest point on the earth is Mt. Everest, 29,028 feet. What is the vertical distance from the lowest point to the highest point on the earth?

32. The populations of some of the nations of the world are: China, over 958,230,000; India, 638,390,000; Russia, 258,930,000; United States, 219,500,000. Find the combined population of China, India, Russia, and the United States.

33. John has $75, Jim has $18, and Marie has $12 more than John and Jim together. Find the total amount of money the three have together.

34. Mary has $34, Jane has $15, and Helen has $27 more than Mary and Jane together. Find the total amount of money the three girls have together.

1.5 Subtraction of Whole Numbers

Subtraction is the *inverse* operation of addition. Every subtraction problem is related to an addition problem.

$$9 - 4 = 5 \qquad \text{and} \qquad 4 + 5 = 9$$

Terms Used in Subtraction

$$\begin{array}{rl} 9 & \text{minuend} \\ -\ 4 & \text{subtrahend} \\ \hline 5 & \text{difference} \end{array}$$

Subtraction Is Not Commutative

In Section 1.4 the operation of addition was commutative. That is,

$$a + b = b + a$$

This means that interchanging the order of the numbers does not change the sum. Can this be done with subtraction? We need only take a single example to see that **subtraction is not commutative**.

Example 1 $7 - 4 \neq 4 - 7$

The symbol $\neq$ is read "is not equal to." $7 - 4 = 3$, whereas $4 - 7$ does not represent a whole number. ■

Subtraction Is Not Associative

Addition is associative. That is,

$$(a + b) + c = a + (b + c)$$

This means that changing the way the numbers are grouped does not change the sum. Is this also true for subtraction? Again we need only a single example to show that **subtraction is not associative**.

Example 2 $(9 - 5) - 2 \neq 9 - (5 - 2)$

because $(9 - 5) - 2 = 4 - 2 = 2$
whereas $9 - (5 - 2) = 9 - 3 = 6$
$2 \neq 6$ ■

Subtraction of Whole Numbers

Example 3 Subtract 2,502 from 78,514.

Solution Always write units digits below units digits, tens digits below tens digits, hundreds digits below hundreds digits, and so on.

$$\begin{array}{rl} 78{,}514 & \text{minuend} \\ -\ 2{,}502 & \text{subtrahend} \\ \hline 76{,}012 & \text{difference} \end{array}$$ ■

Example 4 Find $\begin{array}{r} 692 \\ -456 \\ \hline \end{array}$

Solution

$$\begin{array}{r} {\scriptstyle 8\,12} \\ 6\not{9}\not{2} \\ -\ 456 \\ \hline 236 \end{array}$$

We borrow 1 ten from 9 tens, leaving 8 tens. The borrowed 1 ten (= 10 units) is added to the 2 units we already have, making 12 units
Then, $12 - 6 = 6$ units
and $8 - 5 = 3$ tens
and $6 - 4 = 2$ hundreds ■

Example 5 Find $\begin{array}{r} 58{,}067 \\ -\ 4{,}193 \\ \hline \end{array}$

Solution

$$\begin{array}{r} {\scriptstyle 9\ \ \ } \\ {\scriptstyle 7\ \not{10}\ 16} \\ 5\ \not{8},\ \not{0}\ \not{6}\ 7 \\ -\quad 4,\ 1\ 9\ 3 \\ \hline 5\ 3,\ 8\ 7\ 4 \end{array}$$

We borrow 1 thousand from 8 thousand, leaving 7 thousand. The borrowed 1 thousand (= 10 hundreds) is added to the 0 hundreds

$10 + 0 = 10$ hundreds

Then borrow 1 hundred from 10 hundreds, leaving 9 hundreds. The borrowed 1 hundred (= 10 tens) is added to the 6 tens

$10 + 6 = 16$ tens

Now we can subtract ■

Checking Subtraction Subtraction is checked using the related addition problem.

$$\textit{subtrahend} + \textit{difference} = \textit{minuend}$$

Example 6 Checking subtraction.

$$\begin{array}{rl} 6 & \text{minuend} \\ -2 & \leftarrow \text{subtrahend} \rightarrow \quad 2 \\ \hline 4 & \leftarrow \text{difference} \rightarrow +4 \\ & \qquad\qquad\qquad\quad 6 \quad \text{minuend} \end{array}$$

This same check can be done as follows:

$$\begin{array}{r} 6 \\ -2 \\ \hline 4 \\ \hline 6 \end{array} \quad \text{add } (2 + 4), \text{ check}$$

■

Example 7 Subtract 1,324 from 3,010 and check.

Solution

$$\begin{array}{r} 3{,}010 \\ -\ 1{,}324 \\ \hline 1{,}686 \\ \hline 3{,}010 \end{array} \quad \text{add } (1{,}324 + 1{,}686), \text{ check}$$

■

EXERCISES 1.5

In Exercises 1–20, find the differences.

1. $\begin{array}{r} 8{,}907 \\ -\ 702 \\ \hline \end{array}$ **2.** $\begin{array}{r} 2{,}806 \\ -\ 1{,}502 \\ \hline \end{array}$ **3.** $\begin{array}{r} 5{,}568 \\ -\ 2{,}563 \\ \hline \end{array}$ **4.** $\begin{array}{r} 7{,}186 \\ -\ 7{,}136 \\ \hline \end{array}$ **5.** $\begin{array}{r} 98{,}765 \\ -\ 4{,}565 \\ \hline \end{array}$

6. $\begin{array}{r} 642 \\ -\ 218 \\ \hline \end{array}$ **7.** $\begin{array}{r} 615 \\ -\ 287 \\ \hline \end{array}$ **8.** $\begin{array}{r} 833 \\ -\ 295 \\ \hline \end{array}$ **9.** $\begin{array}{r} 308 \\ -\ 179 \\ \hline \end{array}$ **10.** $\begin{array}{r} 740 \\ -\ 256 \\ \hline \end{array}$

11. $\begin{array}{r} 2{,}108 \\ -\ 1{,}896 \\ \hline \end{array}$ **12.** $\begin{array}{r} 6{,}914 \\ -\ 6{,}057 \\ \hline \end{array}$ **13.** $\begin{array}{r} 3{,}005 \\ -\ 684 \\ \hline \end{array}$ **14.** $\begin{array}{r} 2{,}804 \\ -\ 1{,}926 \\ \hline \end{array}$ **15.** $\begin{array}{r} 60{,}701 \\ -\ 10{,}808 \\ \hline \end{array}$

16. $\begin{array}{r} 90{,}084 \\ -\ 40{,}529 \\ \hline \end{array}$ **17.** $\begin{array}{r} 693{,}421 \\ -\ 355{,}818 \\ \hline \end{array}$ **18.** $\begin{array}{r} 173{,}041 \\ -\ 88{,}350 \\ \hline \end{array}$ **19.** $\begin{array}{r} 7{,}060{,}503 \\ -\ 6{,}829{,}711 \\ \hline \end{array}$ **20.** $\begin{array}{r} 5{,}200{,}365 \\ -\ 2{,}843{,}942 \\ \hline \end{array}$

21. Subtract 4,506 from 7,392.

22. Subtract 10,362 from 19,217.

23. The average distance to the sun is about 92,889,000 miles. The average distance to the moon is about 239,000 miles. How much farther is it to the sun than to the moon?

24. If the population of China is 958,230,000 and the population of the United States is 219,576,770, how many more people are there in China than in the United States?

25. If at the beginning of a trip your odometer* reading was 67,856 miles and at the end of the trip it read 71,304 miles, how many miles did you drive?

26. A used car had a list price of $4,120. Mr. Jones bought the car and was given a discount of $455. What did he pay for the car?

27. At the beginning of the month Mr. Hanson's checking account balance was $356. He made deposits of $225, $57, and $375. He wrote checks for $56, $135, $157, $38, and $417. What was his balance at the end of the month?

28. The height of Mt. Everest is twenty-nine thousand, twenty-eight feet. The height of Mt. Whitney is fourteen thousand, four hundred ninety-five feet. How much higher is Mt. Everest than Mt. Whitney?

29. Jim has $75 and Joe has $57. Jack has $17 more than Jim, and Mike has $13 less than Jim and Joe together. How much money do all four boys have together?

30. Nine years ago the Martinez family bought a house for $76,500. Today the value of the house is $125,000. How much did the value of the house increase in the last 9 years?

1.6 Multiplication of Whole Numbers

Multiplication can be thought of as repeated addition of the same number.

$$\left.\begin{array}{r} 5 \\ 5 \\ 5 \\ +\ 5 \\ \hline 20 \end{array}\right\} \text{four 5's} \quad \text{and} \quad 4 \times 5 = 20$$

Terms Used in Multiplication

$$\begin{array}{rl} 5 & \text{multiplicand} \\ \times\ 6 & \text{multiplier} \\ \hline 30 & \text{product} \end{array} \qquad \underset{\text{factor}}{6} \times \underset{\text{factor}}{5} = \underset{\text{product}}{30}$$

In the expression $6 \times 5 = 30$, the numbers 6 and 5 are said to be *factors* of 30. That is, the multiplier and multiplicand are factors of the product. The word factor is more commonly used for the numbers in a product than the words *multiplier* and *multiplicand.*

Example 1

$$\underset{\text{factor}}{3} \times \underset{\text{factor}}{4} = \underset{\text{product}}{12}$$

Therefore, 3 and 4 are factors of 12. ■

Symbols Used in Multiplication

Multiplication may be shown in several different ways.

Example 2

a. $3 \times 2 = 6$ — In algebra, we avoid using the times sign (×) because it may be confused with the letter *x*

b. $3 \cdot 2 = 6$ — The multiplication dot · is written a little higher than the decimal point

* The instrument on your car that is commonly called the "speedometer" is both a speedometer and an odometer. The part that tells how many miles the car has been driven is the odometer. The part that tells how fast you are going is the speedometer.

c. $3(2) = 6$

d. $(3)2 = 6$

e. $(3)(2) = 6$

When two numbers are written next to each other in this way with no symbol of operation, it is understood that they are to be multiplied ■

Commutative Property of Multiplication

Example 3

a. $7 \cdot 8 = 8 \cdot 7 = 56$

b. $4 \cdot 5 = 5 \cdot 4 = 20$

In Example 3, when we reversed the order of the factors, the product remained the same. This property of whole numbers is called the **commutative property of multiplication**.

> COMMUTATIVE PROPERTY OF MULTIPLICATION
>
> If a and b represent any numbers, then
>
> $$a \cdot b = b \cdot a$$

■

Associative Property of Multiplication

Example 4 Multiply: $3 \cdot 4 \cdot 2$

Solution

a. $(3 \cdot 4) \cdot 2 =$
$12 \cdot 2 = 24$

b. $3 \cdot (4 \cdot 2) =$
$3 \cdot 8 = 24$

Therefore, $(3 \cdot 4) \cdot 2 = 3 \cdot (4 \cdot 2)$ ■

In Example 4, the product of three numbers was unchanged no matter how we grouped the numbers. It is assumed that this property of multiplication is true when any three numbers are multiplied. This property of whole numbers is called the **associative property of multiplication**.

> ASSOCIATIVE PROPERTY OF MULTIPLICATION
>
> If a, b, and c represent any numbers, then
>
> $$(a \cdot b) \cdot c = a \cdot (b \cdot c)$$

The *commutative* and *associative* properties hold true for both addition and multiplication.

Multiplicative Identity

Multiplying any number by one gives the identical number for the product.

Example 5

a. $8 \cdot 1 = 8$

b. $1 \cdot 5 = 5$ ■

For this reason, one is called the **multiplicative identity**.

> MULTIPLICATIVE IDENTITY
>
> If a represents any number, then
>
> $$a \cdot 1 = 1 \cdot a = a$$

Multiplication By Zero

Since multiplication is repeated addition of the same number, multiplying a number by zero gives a product of zero. See Example 6.

Example 6

a. $3 \cdot 0 = 0 + 0 + 0 = 0$

b. $4 \cdot 0 = 0 + 0 + 0 + 0 = 0$ ■

Because of the commutative property of multiplication, it follows that

$$3 \cdot 0 = 0 \cdot 3 = 0$$
$$4 \cdot 0 = 0 \cdot 4 = 0$$

> MULTIPLICATION PROPERTY OF ZERO
>
> If a represents any number, then
>
> $$a \cdot 0 = 0 \cdot a = 0$$

Multiplication of Whole Numbers

Example 7 Find $4 \cdot 36$.
Explanation

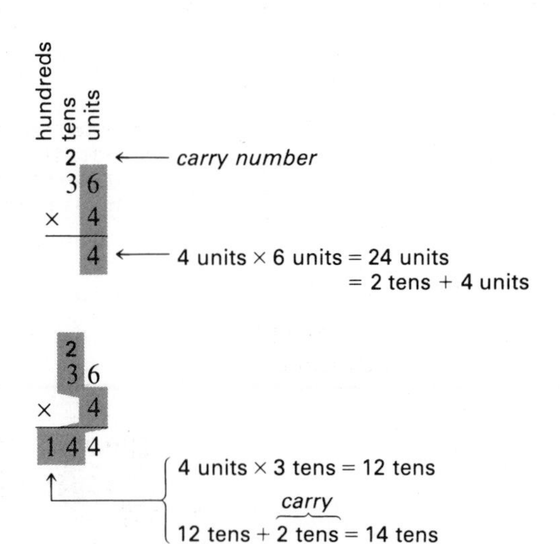

■

Example 8 Find 527 · 34.

Explanation

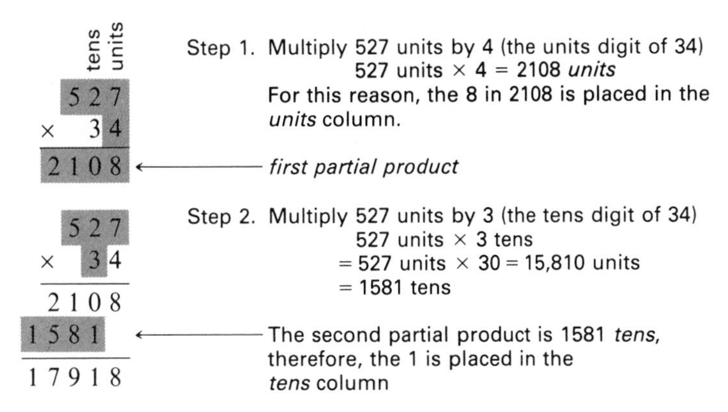

Step 1. Multiply 527 units by 4 (the units digit of 34)
527 units × 4 = 2108 *units*
For this reason, the 8 in 2108 is placed in the *units* column.

Step 2. Multiply 527 units by 3 (the tens digit of 34)
527 units × 3 tens
= 527 units × 30 = 15,810 units
= 1581 tens

Step 3. Add all partial products. This gives the final product (17,918). ■

Example 9 Find 4,385 · 739.

Explanation Notice that the right digit of each partial product is directly below the digit of the multiplier that was used to obtain it.

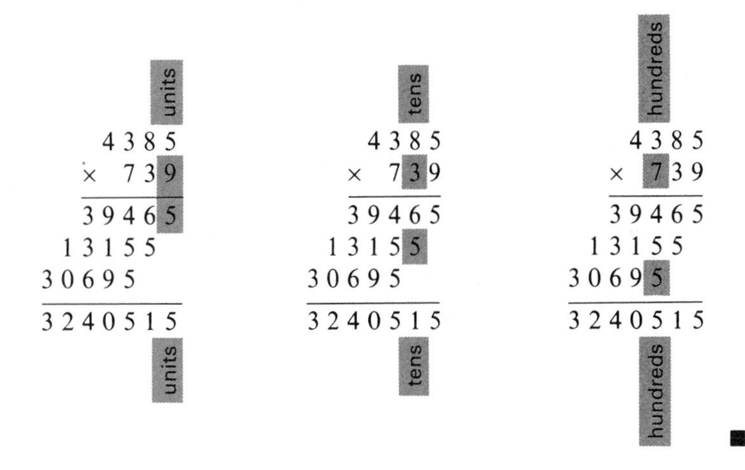

■

Example 10 Find 305 · 1,734.

```
  1734
×  305
  8670
 0000
5202
528870
```

Since the multiplication by zero results in a row of zeros, we shorten the writing by omitting this row of zeros

```
  1734
×  305
  8670
5202
528870 = 528,870
```

■

Example 11 Find 4,006 · 2,314.

Explanation

```
   2314
× 4006
  13884
9256
9269884
```

Notice that the right digit of each partial product is directly below the digit of the multiplier that was used to obtain it ■

Example 12 Find 18,000 · 2,341.

Solution Rather than write the problem in the form

```
  2341
×18000
```

it is better to move the multiplier to the right as shown here.

```
 2341
×18000
```

Draw a vertical line to separate the zeros from the 18. Multiply by 18, then attach the zeros later as shown below.

```
  2341|
×   18|000
 18728|
 2341 |
 42138|000 = 42,138,000
```

2,341 · 18,000
= 2,341 · (18 · 1,000)
= (2,341 · 18) · 1,000
= 42,138 · 1,000
= 42,138,000 ■

Example 13 Find 175,000 · 3,200.

Solution

```
   175|000      3 zeros } to the right
×   32|00      +2 zeros } of the line
   350|         5 zeros
  525 |
  5600|00000    5 zeros to the right of the line
= 560,000,000
```

■

To improve your speed and accuracy with multiplication, see Appendix I for drill and practice with multiplication facts.

EXERCISES 1.6

In Exercises 1–42, find the products.

1. 638 × 4 **2.** 794 × 6 **3.** 962 × 7 **4.** 548 × 9 **5.** 6,497 × 8

6. 8,697 × 5 **7.** 23,106 × 3 **8.** 68,075 × 9 **9.** 438 × 57 **10.** 897 × 98

11. 7,836 × 64 **12.** 9,267 × 38 **13.** 90,276 × 95 **14.** 43,807 × 76 **15.** 235 × 416

16. 487 × 395 **17.** 7,043 × 642 **18.** 8,106 × 387 **19.** 4,372 × 5,168 **20.** 7,864 × 8,794

21. 356 × 204 **22.** 725 × 306 **23.** 3,804 × 709 **24.** 9,067 × 508 **25.** 7,802 × 1,009

26. $\begin{array}{r} 3{,}085 \\ \times\ 4{,}007 \\ \hline \end{array}$

27. $\begin{array}{r} 60{,}058 \\ \times\ 9{,}005 \\ \hline \end{array}$

28. $\begin{array}{r} 20{,}109 \\ \times\ 6{,}008 \\ \hline \end{array}$

29. $\begin{array}{r} 75{,}009 \\ \times\ 30{,}007 \\ \hline \end{array}$

30. $\begin{array}{r} 820{,}040 \\ \times\ 90{,}007 \\ \hline \end{array}$

31. $2{,}500 \cdot 376$

32. $12{,}000 \cdot 507$

33. $9{,}200 \cdot 3{,}154$

34. $300 \cdot 7{,}855$

35. $500 \cdot 3{,}751$

36. $2{,}000 \cdot 799$

37. $6{,}600 \cdot 449$

38. $5{,}000 \cdot 7{,}008$

39. $9{,}500 \cdot 7{,}893$

40. $7{,}500 \cdot 3{,}500$

41. $8{,}960 \cdot 5{,}600$

42. $38{,}000 \cdot 7{,}800$

43. A ream of paper contains 500 sheets. A school uses 1,427 reams in a year. How many sheets of paper were used in that time?

44. A town has 7,500 inhabitants who pay property tax. If the average tax paid is $504, what is the town's income from property taxes?

45. If a man's car cost him 18¢ per mile to operate, how much does 16,000 miles of driving cost him?

46. Each tanker in an oil fleet can carry 503,024 gallons of oil. If the fleet has 207 tankers, how many gallons of oil can the fleet transport at once?

47. Debbie earns $850 a month. Her husband earns $275 a week. How much do they earn together in one year? (1 year = 12 months or 52 weeks)

48. If a machine can sort 1,650 checks in 1 minute, how many checks can it sort in an 8-hour day?

1.7 Powers of Whole Numbers

Now that we have learned to multiply whole numbers, it is possible to consider products in which the same number is repeated as a factor. For example:

$$3 \cdot 3 \cdot 3 \cdot 3 = 3^4 = 81$$

Base
Exponent
Power

In the symbol 3^4, the 3 is called the **base**. The 4 is called the **exponent** and is written above and to the right of the base 3. The entire symbol 3^4 is called the *fourth power of three* and is commonly read "three to the fourth power." See Figure 1.7.1.

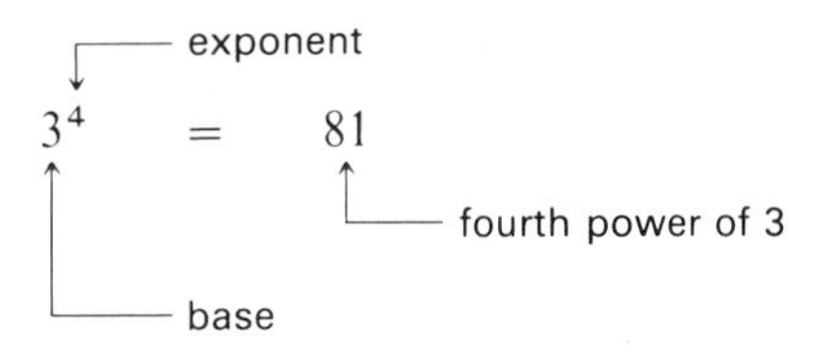

FIGURE 1.7.1

Note the importance of the position of the 4.

$$3^4 = 3 \cdot 3 \cdot 3 \cdot 3 = 81$$

the raised 4 indicates repeated *multiplication*

$$3 \cdot 4 = 3 + 3 + 3 + 3 = 12$$

this 4 indicates repeated *addition*

Example 1

a. $2^3 = 2 \cdot 2 \cdot 2 = 8$

b. $4^2 = 4 \cdot 4 = 16$

c. $1^4 = 1 \cdot 1 \cdot 1 \cdot 1 = 1$

d. $25^2 = 25 \cdot 25 = 625$

e. $10^1 = 10$

f. $10^3 = 10 \cdot 10 \cdot 10 = 1{,}000$ ■

Zero as a Base When 0 is raised to a power other than 0, we get 0 (Example 2).

Example 2

a. $0^2 = 0 \cdot 0 = 0$

b. $0^5 = 0 \cdot 0 \cdot 0 \cdot 0 \cdot 0 = 0$ ■

Zero as an Exponent When any whole number (other than 0) is raised to the 0 power, we get 1 (Example 3). The reason for this definition will be covered in Section 15.1.

Example 3

a. $2^0 = 1$

b. $5^0 = 1$

c. $10^0 = 1$ ■

In general,

> ZERO AS AN EXPONENT
>
> If a represents any number, except 0,
>
> $$a^0 = 1$$

Zero as Both Exponent and Base The symbol 0^0 is not defined or used in this book.

A calculator can be used to find powers of large numbers.

Example 4 Find 25^4 using a calculator.

Solution If your calculator has a power key that looks like [y^x] (or [x^y]), follow the steps below.

Key in: 25 [y^x] 4 [=]

Answer: 390,625

NOTE Your calculator may require you to use the [INV] or [2nd F] key before the [y^x] key. Consult your instruction manual for the correct order of keys for your calculator. ☑

If your calculator does not have a power key, you will need to use repeated multiplication.

Key in: 25 [×] 25 [×] 25 [×] 25 [=]

Answer: 390,625

Some calculators will allow you to repeat an operation with the same number by pressing the [=] repeatedly.

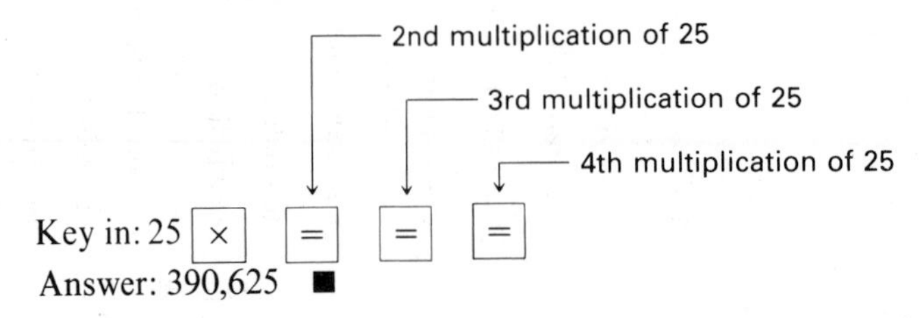

Answer: 390,625 ■

EXERCISES 1.7

In Exercises 1–24, find the value of each expression.

1. 2^3 **2.** 3^2 **3.** 5^2 **4.** 4^3 **5.** 3^3 **6.** 2^4 **7.** 6^2 **8.** 5^3

9. 10^0 **10.** 10^1 **11.** 10^2 **12.** 10^3 **13.** 10^4 **14.** 10^5 **15.** 10^6 **16.** 10^7

17. 0^3 **18.** 1^3 **19.** 3^0 **20.** 2^2 **21.** 4^2 **22.** 8^2 **23.** 1^5 **24.** 2^5

In Exercises 25–32, use a calculator to find the value of each expression.

25. 12^5 **26.** 36^3 **27.** 9^6 **28.** 8^5 **29.** 245^3 **30.** 765^2 **31.** 15^0 **32.** 0^{15}

1.8 Powers of Ten

Powers in which the base is 10 have many important uses in mathematics and science. We show some powers of 10.

	Power of 10	*Name*	
a.	$10^0 = 1$	$= 1$ unit	(Sec. 1.7)
b.	$10^1 = 10$	$= 1$ ten	
c.	$10^2 = 100$	$= 1$ hundred	
d.	$10^3 = 1{,}000$	$= 1$ thousand	
e.	$10^4 = 10{,}000$	$= 1$ ten-thousand	
f.	$10^5 = 100{,}000$	$= 1$ hundred-thousand	
g.	$10^6 = 1{,}000{,}000$	$= 1$ million	
	etc.		

In example (d), 10^{3} has a value of 1 followed by three zeros = 1,000 and is read "one thousand."

In example (f), 10^{5} has a value of 1 followed by five zeros = 100,000 and is read "one hundred thousand."

Notice also that the successive names of the powers of 10 correspond exactly to the names of the places when we read or write a number. See Figure 1.8.1.

Power of 10	10^{14}	10^{13}	10^{12}	10^{11}	10^{10}	10^9	10^8	10^7	10^6	10^5	10^4	10^3	10^2	10^1	10^0
Place Name	hundred trillion	ten trillion	one trillion	hundred billion	ten billion	one billion	hundred million	ten million	one million	hundred thousand	ten thousand	one thousand	hundred (unit)	ten (unit)	one (unit)

FIGURE 1.8.1

Multiplying a Whole Number by a Power of Ten Suppose you make \$5 a week for mowing and taking care of your neighbor's lawn. As you know, in 10 weeks you would make a total of \$50. In 100 weeks you would make a total of \$500. That is,

$$10 \cdot 5 = 50$$
$$100 \cdot 5 = 500$$
$$1{,}000 \cdot 5 = 5{,}000$$

Because $10 = 10^1$, $100 = 10^2$, $1{,}000 = 10^3$, and so on,

$$10^1 \cdot 5 = 50$$
$$10^2 \cdot 5 = 500$$
$$10^3 \cdot 5 = 5{,}000, \text{ and so on.}$$

> When a whole number is multiplied by a power of 10, follow the number by as many zeros as the exponent of 10.

Example 1

a. $12 \cdot 10^3 = 12{,}000$ or $12 \cdot 1{,}000 = 12{,}000$

b. $275 \cdot 10^4 = 2{,}750{,}000$ or $275 \cdot 10{,}000 = 2{,}750{,}000$

c. $4{,}806 \cdot 10^2 = 480{,}600$ or $4{,}806 \cdot 100 = 480{,}600$

d. $7 \cdot 10^0 = 7 \cdot 1 = 7$ ■

EXERCISES 1.8

Find the value of each expression without using written calculations.

1. $3 \cdot 100$
2. $75 \cdot 1{,}000$
3. $3 \cdot 10^2$
4. $75 \cdot 10^3$
5. $10 \cdot 16$
6. $100 \cdot 24$
7. $10^3 \cdot 8$
8. $10^4 \cdot 44$
9. $10 \cdot 10{,}000$
10. $100 \cdot 10$
11. $4 \cdot 10^0$
12. $10^0 \cdot 44$
13. $20 \cdot 100$
14. $1{,}000 \cdot 50$
15. $80 \cdot 10^3$
16. $400 \cdot 10^2$
17. $900 \cdot 10$
18. $12{,}000 \cdot 100$
19. $10^5 \cdot 103$
20. $250 \cdot 10^4$

1.9 Division of Whole Numbers

Division is the *inverse* operation of multiplication. Every division problem is related to a multiplication problem.

$$20 \div 5 = 4 \quad \text{and} \quad 4 \cdot 5 = 20$$

Just as multiplication is repeated addition of the same number, division can be thought of as repeated subtraction of the same number.

Multiplication as repeated addition	*Division as repeated subtraction*
When we think	When we think
$2 + 2 + 2 = 6$	$6 - 2 = 4$ ← 1st subtraction of 2 $4 - 2 = 2$ ← 2nd subtraction of 2 $2 - 2 = 0$ ← ③rd subtraction of 2
we say $3 \cdot 2 = 6$	we say $6 \div 2 = ③$

Terms Used in Division

The **dividend** is the number being divided.
The **divisor** is the number doing the dividing.
The **quotient** is the answer in a division problem.

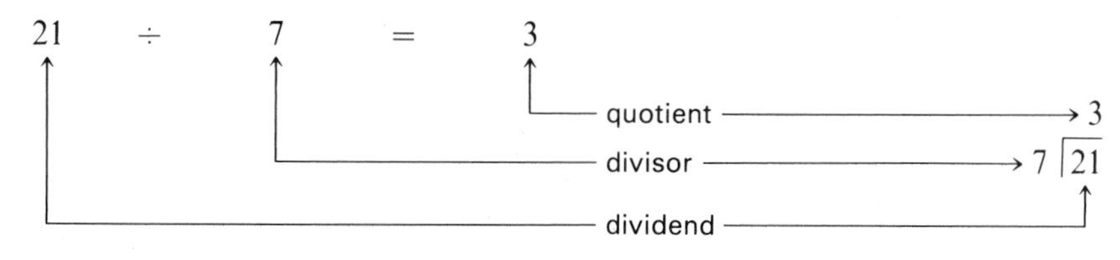

FIGURE 1.9.1

Other names are used for the various numbers in a division problem. Since $21 = 3 \cdot 7$, the numbers 3 and 7 are called *divisors* of 21. We say that 7 (and 3 as well) *divides 21 exactly* or that 3 (and 7 as well) *goes into 21 evenly*.

Example 1

a. Divisors of 15 are 1, 3, 5, 15 because $15 = 1 \cdot 15 = 3 \cdot 5$.

b. Divisors of 12 are 1, 2, 3, 4, 6, 12 because $12 = 1 \cdot 12 = 2 \cdot 6 = 3 \cdot 4$. ■

Checking Division Division is checked using the related multiplication problem.

$$\text{quotient} \cdot \text{divisor} = \text{dividend}$$

Example 2

a. $48 \div 6 = 8$ because $8 \cdot 6 = 48$

b. $63 \div 9 = 7$ because $7 \cdot 9 = 63$ ■

Division Is Not Commutative

Both addition and multiplication are commutative, but subtraction is not commutative. What is the case with division? A single example will show that **division is not commutative**.

$$6 \div 3 \neq 3 \div 6$$

because $6 \div 3 = 2$, whereas $3 \div 6$ does not represent a whole number.

Division Is Not Associative

Both addition and multiplication are associative, but subtraction is not associative. A single example will show that **division is not associative**.

$$(16 \div 4) \div 2 \neq 16 \div (4 \div 2)$$

because

$$(16 \div 4) \div 2 = 4 \div 2 = 2$$

whereas

$$16 \div (4 \div 2) = 16 \div 2 = 8$$

$2 \neq 8$

Division Involving Zero

Zero divided by a nonzero number is possible, and the quotient is zero.

$0 \div 2 = 0$ because $0 \cdot 2 = 0$

A nonzero number divided by zero is impossible.

$4 \div 0 = ?$ ← Suppose the quotient is some number x

Then

$4 \div 0 = x$ means $x \cdot 0 = 4$, which is certainly false

$x \cdot 0 \neq 4$ because any number multiplied by zero $= 0$. Therefore, dividing any nonzero number by zero is undefined.

Zero divided by zero cannot be determined.

$0 \div 0 = 0$ means $0 \cdot 0 = 0$ which is true

$0 \div 0 = 1$ means $1 \cdot 0 = 0$ which is true

$0 \div 0 = 5$ means $5 \cdot 0 = 0$ which is true

In other words, $0 \div 0 = 0$, 1, and also 5. In fact, it can be any number. Because there is *no unique* answer, we say that $0 \div 0$ is undefined.

DIVISION INVOLVING ZERO

If a represents any number, except 0, then

$0 \div a = 0$

$a \div 0$ is undefined } cannot divide by 0

$0 \div 0$ is undefined }

Long Division

Example 3 Find $73 \div 3$.

Solution We first work this division by showing that division is repeated subtraction.

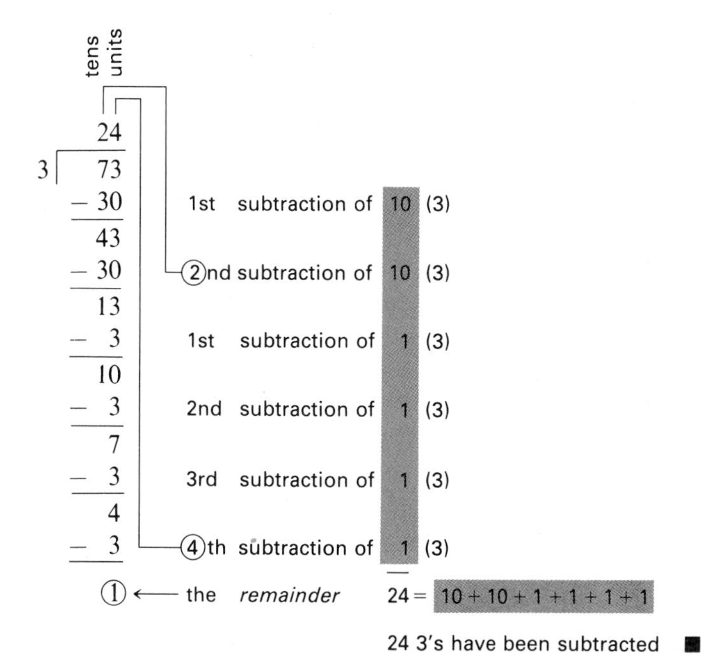

Remainder

The **remainder** is the number left over after the divisor has been subtracted as many times as possible.

This work is usually shortened as follows:

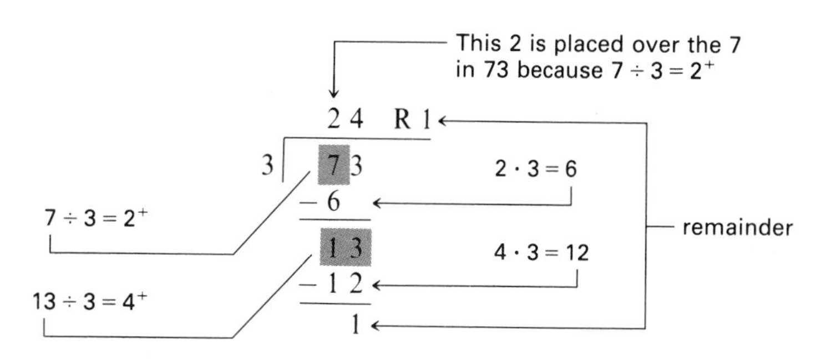

Checking Division

You can check division problems with remainders by using the following fact.

quotient · divisor + remainder = dividend

Example 4 Find $274 \div 6$.

Solution Because 2 cannot be divided by 6, we divide 27 by 6.

This 4 is placed over the 7 in 27 because $27 \div 6 = 4^+$

$$\begin{array}{r} 45 \text{ R } 4 \\ 6\overline{)274} \\ -24 \\ \hline 34 \\ -30 \\ \hline 4 \end{array}$$

$4 \cdot 6 = 24$

$5 \cdot 6 = 30$ ■

Check $45 \cdot 6 + 4 \stackrel{?}{=} 274$

$270 + 4 \stackrel{?}{=} 274$

$274 = 274$

Example 5 Find 9,035 ÷ 7.

Solution

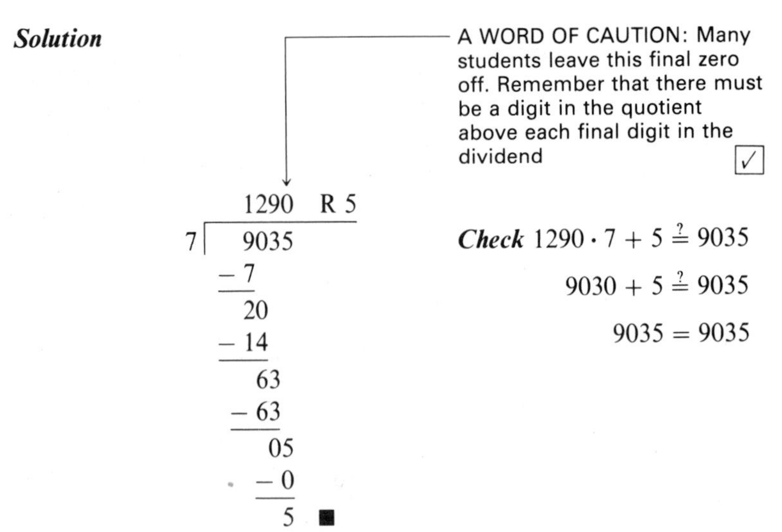

Check $1290 \cdot 7 + 5 \stackrel{?}{=} 9035$

$9030 + 5 \stackrel{?}{=} 9035$

$9035 = 9035$ ■

Trial Divisor

When the divisor has two or more digits, we will use the **trial divisor** method for long division. The *trial divisor* is usually the first digit of the divisor. We use examples to illustrate the method.

Example 6 Find 759 ÷ 31.
Solution Use 3 as a trial divisor.

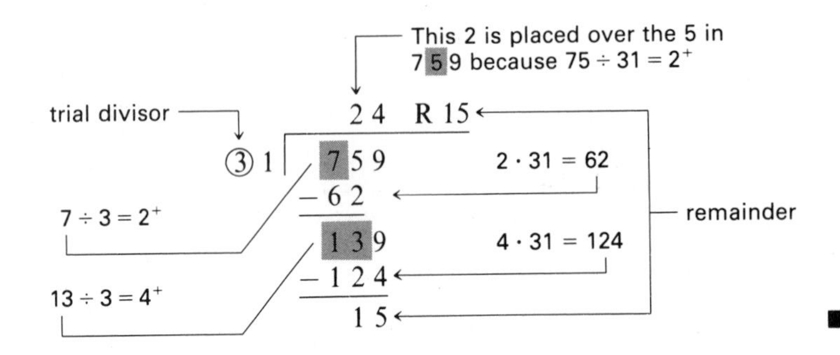

■

In the following example, instead of using the first digit, 3, of divisor 39 for our trial divisor, we use 4 because 39 is closer to four tens than it is to three tens.

Example 7 Find 6,842 ÷ 39.
Solution

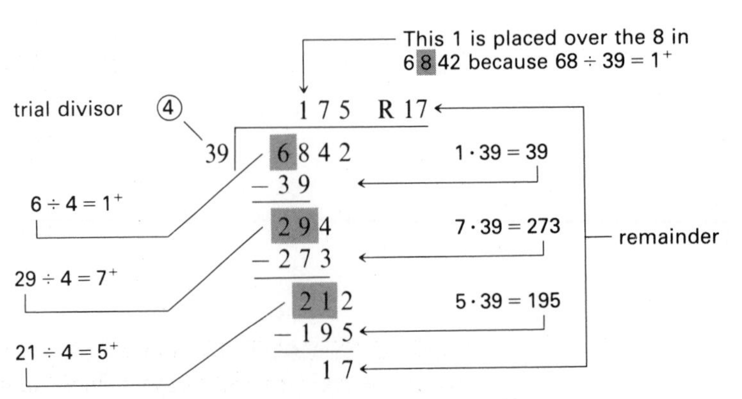

■

Example 8 Find 4,908 ÷ 67.

Solution

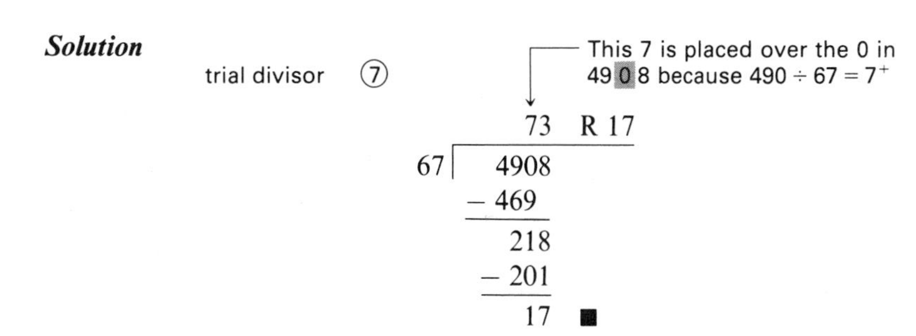

Example 9 Find 44,490 ÷ 218.

Solution

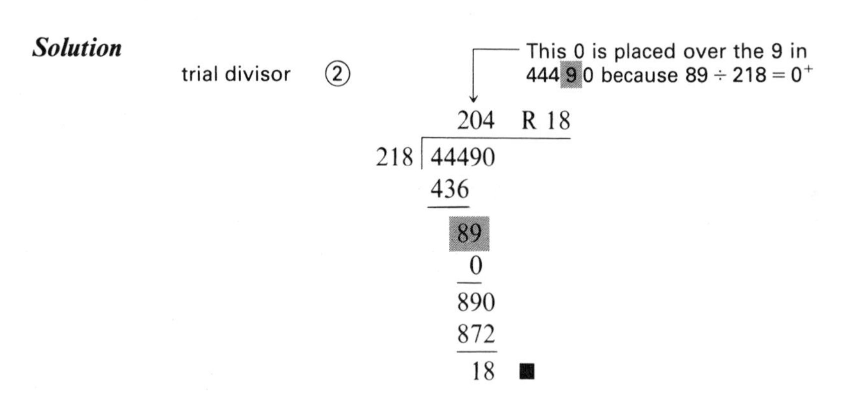

The trial divisor will help you estimate the quotient but sometimes you will need to adjust your estimate. (See Example 10.)

Example 10 Find 16,605 ÷ 27.
Solution Using a trial divisor of 3, 16 ÷ 3 = 5^+. We will try 5 in the quotient.

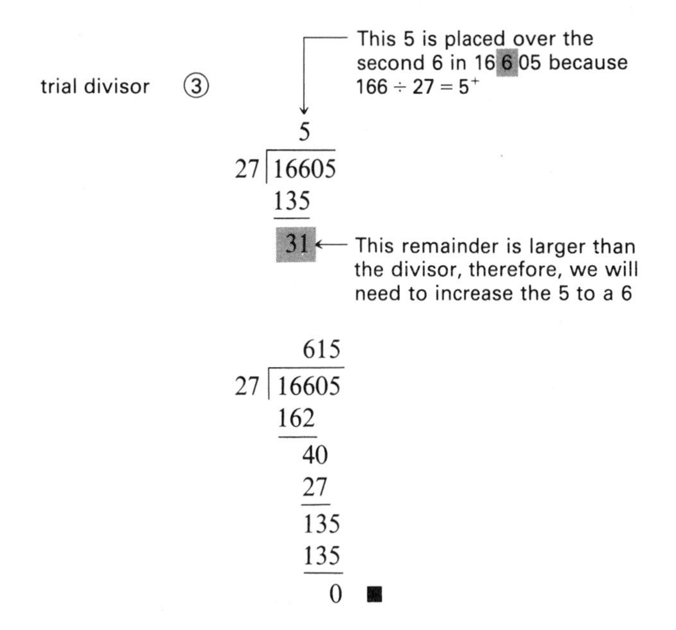

A WORD OF CAUTION Once the long division process has started, each remaining digit in the dividend must have a *single* digit above it in the quotient. ☑

EXERCISES 1.9

In Exercises 1–12, work each long division problem and check your answers.

1. 1,235 ÷ 8
2. 7,166 ÷ 9
3. 5,254 ÷ 5
4. 20,441 ÷ 6
5. 65,754 ÷ 9
6. 54,432 ÷ 8
7. 847 ÷ 31
8. 960 ÷ 15
9. 885 ÷ 43
10. 2,415 ÷ 69
11. 4,677 ÷ 87
12. 3,182 ÷ 63

In Exercises 13–36, work each long division problem.

13. 8,866 ÷ 31
14. 6,870 ÷ 15
15. 9,729 ÷ 47
16. 8,904 ÷ 84
17. 9,625 ÷ 26
18. 22,695 ÷ 27
19. 15,120 ÷ 56
20. 31,746 ÷ 78
21. 14,450 ÷ 24
22. 25,839 ÷ 41
23. 97,414 ÷ 91
24. 24,050 ÷ 12
25. 78,800 ÷ 56
26. 95,348 ÷ 73
27. 185,503 ÷ 89
28. 235,620 ÷ 63
29. 65,058 ÷ 185
30. 129,596 ÷ 206
31. 421,400 ÷ 301
32. 941,200 ÷ 362
33. 565,090 ÷ 715
34. 1,160,695 ÷ 144
35. 5,248,749 ÷ 583
36. 2,087,490 ÷ 298

37. A man can make 38 machine parts on a lathe in 1 hour. How many hours will he need to make 2,356 parts?

38. Frank figures a tire on his car lasts for 24,000 miles. How many sets of tires will he need to drive his car 96,000 miles?

39. A certain brand of soap sells for 21¢ a bar.
a. How many bars can be bought with $1.50?
b. How much change is left?

40. At a market, a 7-ounce bottle of shampoo costs 98¢. How much is this per ounce?

41. Rosali earns an annual salary of $18,720.
a. What is her monthly salary?
b. What is her weekly salary?

42. A family wants to save $5,000 to make a trip to Europe in 4 years. How much must they save each year? If they save $100 each month, will they meet their goal?

43. A soap company produces 96,400 pounds of detergent in one day. How many 16-pound boxes of detergent can be filled in one day?

44. A man pays off a debt of $3,048 in equal monthly payments over a period of two years. Find the amount he pays each month.

45. Eight boys going on a camping trip are trying to divide 100 small boxes of raisins up evenly among their packs. The leader (one of the eight) will take the remainder as well as his share. How many boxes did the leader have to carry?

46. Seven girls pooled their money for a lottery ticket and won $2,500. When they divide the money in dollars, there is a remainder. They draw straws to see who will get the remainder. What was the total amount received by the winner?

47. Form numbers by arranging the digits 1 through 9 in several different orders. Divide each number so obtained by 9. Compare the remainders in each division.

48. If it takes 6 minutes to saw a log into three pieces, how long will it take to saw the same log into four pieces?

1.10 Square Roots

Square Root

The symbol $\sqrt{}$ is called a radical sign, and it indicates the **square root** of the number under it.

$\sqrt{9}$ is read "square root of 9."

Finding the square root of a number is the *inverse* operation of squaring a number. Every square root problem has a related problem involving squaring a number.

$$\sqrt{9} = 3, \qquad \text{because } 3^2 = 9$$

Example 1 Find the square roots.

a. $\sqrt{4} = 2$ because $2^2 = 4$

b. $\sqrt{36} = 6$ because $6^2 = 36$

c. $\sqrt{100} = 10$ because $10^2 = 100$

d. $\sqrt{0} = 0$ because $0^2 = 0$

e. $\sqrt{1} = 1$ because $1^2 = 1$ ■

A calculator can be used to find the square roots of large numbers.

Example 2 Find $\sqrt{625}$ using the square-root key $\boxed{\sqrt{}}$.

Key in: 625 $\boxed{\sqrt{}}$

Answer: 25 ■

EXERCISES 1.10

In Exercises 1–15, find the indicated square roots.

1. $\sqrt{25}$ **2.** $\sqrt{49}$ **3.** $\sqrt{64}$ **4.** $\sqrt{4}$ **5.** $\sqrt{81}$

6. $\sqrt{9}$ **7.** $\sqrt{16}$ **8.** $\sqrt{36}$ **9.** $\sqrt{1}$ **10.** $\sqrt{0}$

11. $\sqrt{144}$ **12.** $\sqrt{121}$ **13.** $\sqrt{100}$ **14.** $\sqrt{400}$ **15.** $\sqrt{10{,}000}$

In Exercises 16–20, use a calculator to find the square roots.

16. $\sqrt{225}$ **17.** $\sqrt{529}$ **18.** $\sqrt{961}$ **19.** $\sqrt{5{,}625}$ **20.** $\sqrt{18{,}496}$

1.11 Order of Operations

What does $2 + 3 \cdot 4 = ?$
If we add first,

$$\underbrace{2 + 3}\cdot 4 =$$
$$5 \quad \cdot 4 = 20$$

But if we multiply first,

$$2 + \underbrace{3 \cdot 4} =$$
$$2 + \;12\; = 14$$

Which answer is correct, 20 or 14? To avoid this confusion, we perform the operations in the following order:

> **ORDER OF OPERATIONS**
>
> **1.** Any expressions in parentheses are evaluated first.
>
> **2.** Evaluations are done in this order:
>
> *First:* Powers and roots are done.
> *Second:* Multiplication and division are done in order from *left to right.*
> *Third:* Addition and subtraction are done in order from *left to right.*

Example 1

$2 + 3 \cdot 4 =$ Multiplication before addition

$2 + 12 = 14$ ■

Example 2

$16 \div 2 \cdot 4 =$ Multiplication and division are done *left to right*

$8 \cdot 4 = 32$ ■

Example 3

$4^2 - \sqrt{25} \cdot 2 =$ Powers and roots are done first

$16 - 5 \cdot 2 =$ Multiplication is next

$16 - 10 = 6$ Subtraction is last ■

Example 4

$2(20) + 2(15) =$ When two numbers are written next to each other with no symbol of operation, it is understood they are to be multiplied

$40 + 30 = 70$ ■

Example 5

$2(3)^2 - 4(3) + 1 =$ Powers are first

$2 \cdot 9 - 4(3) + 1 =$ Multiplication is next

$18 - 12 + 1 =$ Addition and subtraction are done *left to right*

$6 + 1 = 7$ ■

Example 6

$2 \cdot (5 + 6) =$ Operations in parentheses are done first

$2 \cdot 11 = 22$ ■

Example 7

$20 - (4 + 3 \cdot 2) =$ Multiply inside parentheses first

$20 - (4 + 6) =$ Add inside parentheses

$20 - 10 = 10$ Subtraction is last ■

Calculators use one of three types of logic: algebraic logic, arithmetic logic, and reverse Polish notation (RPN).

Algebraic Logic

For calculators using **algebraic logic**, numbers, symbols of operation, and parentheses can be entered into the calculator in the same order that they appear in the expression.

To find $2 + 3 \times 4$

Key in: 2 [+] 3 [×] 4 [=]

Answer: 14

Arithmetic Logic

For calculators using **arithmetic logic**, *all* operations are carried out from left to right. If you key in 2 [+] 3 [×] 4 [=], this calculator will display an incorrect answer of 20. Therefore, when using this type of calculator, you must be careful to use the correct order of operations.

Key in: 3 [×] 4 [+] 2 [=]

Answer: 14

RPN

For calculators using **RPN**, there is no [=] key. They have an [ENTER] or [SAVE] key, and the operations are keyed in after the numbers are entered.

Key in: 2 [ENTER] 3 [ENTER] 4 [×] [+]

Answer: 14

The remaining examples in this book will be performed on a scientific calculator using algebraic logic. Be sure to consult your instruction manual for the correct order of keys for your calculator.

Example 8 Find the value of each expression using a calculator.

a. $3^6 + 14 \cdot \sqrt{256}$

Key in: 3 [y^x] 6 [+] 14 [×] 256 [$\sqrt{\ }$] [=]

Answer: 953

b. $24 \cdot (75 - 18) \div 19$

Key in: 24 [×] [(] 75 [−] 18 [)] [÷] 19 [=]

Answer: 72 ■

EXERCISES 1.11

In Exercises 1–36, evaluate each expression using the correct order of operations.

1. $10 \div 2 \cdot 5$

2. $36 \div 6 \div 2$

3. $12 - 6 + 2$

4. $8 - 2 - 4$

5. $6 + 4 \cdot 5$

6. $48 - 18 \div 3$

7. $4 \cdot 5 - 15 \div 5$

8. $44 \div 11 + 8 \cdot 7$

9. $15 - 3 \cdot 4 + 6$

10. $16 - 8 \div 2 + 1$

11. $5^2 - 2^3 + \sqrt{9}$

12. $8^2 - \sqrt{16} - 3^3$

13. $4(5) + 3$

14. $4(5 + 3)$

15. $7(9 - 3)$

16. $7(9) - 7(3)$

17. $2(6) + 2(4)$

18. $2(6 + 4)$

19. $14 - (3 - 1)$

20. $14 - 3 - 1$

21. $100 \div 10 \div 2$

22. $100 \div (10 \div 2)$

23. $20 - (2 + 4 \cdot 3)$

24. $20 - 2 + 4 \cdot 3$

25. $12 + 9 - 6 \div 3$

26. $12 + (9 - 6 \div 3)$

27. $10^2 - 10^0 - 1^2$

28. $2^3 + 0^2 + 2^0$

29. $5 \cdot 8 + 2\sqrt{100}$

30. $4 \cdot 5 - 3\sqrt{36}$

31. $24 - 8 \div 4 \cdot 2 + 2^3$

32. $(24 - 8) \div 4 \cdot 2 + 2^3$

33. $24 - 8 \div (4 \cdot 2) + 2^3$

34. $24 - 8 \div 4 \cdot (2 + 2^3)$

35. $(24 - 8 \div 4) \cdot 2 + 2^3$

36. $24 - (8 \div 4 \cdot 2 + 2^3)$

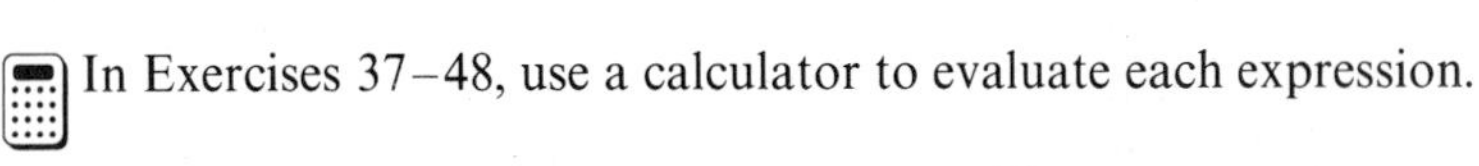

In Exercises 37–48, use a calculator to evaluate each expression.

37. $189 + 560 \div 35$ **38.** $2{,}400 - 18 \cdot 23$ **39.** $567 - 6 \cdot 43 + 297$

40. $46 \cdot 90 - 102 \div 17$ **41.** $560 \div (34 - 18) \cdot 100$ **42.** $7{,}040 - (9 \cdot 24 - 68)$

43. $(12 + 7) \cdot (48 - 12)$ **44.** $(234 + 416) \div (89 - 63)$ **45.** $909 + 8 \cdot 5^4$

46. $19^3 + 56^2$ **47.** $\sqrt{11{,}025} - 108 \div 12$ **48.** $\sqrt{9{,}216} - 5\sqrt{324}$

1.12 Average

If a student gets 70 on one test and 90 on another test, you probably know his average is 80. Why is 80 called his average?

It is because $70 + 90 = 160$ ← same total points

and $80 + 80 = 160$ ← same total points

↑ 80 on *every* test gives the same total points

In other words, the average is the score he would have to make on *every* test in order to get the same total points. To find the average, we take the sum of all the grades (160) and divide it by the number of grades (2).

We can find the average of any kinds of quantities such as grades, weights, speeds, and costs. The average is found by dividing the sum of the quantities by the number of quantities.

> TO FIND THE AVERAGE
>
> **1.** Find the sum of all the quantities.
>
> **2.** Divide this sum by the number of quantities.

Example 1 Find the average grade for 73, 84, 88, 92, and 68.

Solution

Step 1. $\underbrace{73 + 84 + 88 + 92 + 68}_{\text{5 grades}} = 405$

Step 2.

$$\begin{array}{r} 81 \leftarrow \text{Average} \\ 5\,\overline{)\,405} \\ \underline{40} \\ 05 \\ \underline{5} \end{array}$$ ■

Example 2 Four linemen on a football team weighed 270 lb, 241 lb, 265 lb, and 284 lb. What is their average weight?

Solution

Step 1. $\underbrace{270 + 241 + 265 + 284}_{\text{4 weights}} = 1060$

Step 2.

$$\begin{array}{r} 265 \text{ lb} \leftarrow \text{Average} \\ 4 \overline{)1060} \\ \underline{8} \\ 26 \\ \underline{24} \\ 20 \\ \underline{20} \end{array}$$ ■

Example 3 In Mrs. Martinez's quiz section two students scored 5, three students scored 4, and one student scored 2. Find the class average.

Solution Since *two* students scored 5 and *three* students scored 4, the number 5 must be added twice, and the number 4 must be added three times.

Step 1.

$$\underbrace{5 + 5 + 4 + 4 + 4 + 2}_{\text{6 students}} = 24$$

Step 2.

$$\begin{array}{r} 4 \leftarrow \text{Average} \\ 6 \overline{)24} \\ \underline{24} \end{array}$$ ■

Example 4 The sales figures for three months were \$54,230, \$86,045, and \$76,775. Find the average sales per month.

Solution For calculators using algebraic logic, the [=] key must be pressed at the end of the sum to get the total, then divide by 3.

Key in: 54230 [+] 86045 [+] 76775 [=] [÷] 3 [=]

Answer: \$72,350 ■

At this time we only consider problems in which the average is a whole number. Problems in which the average is not a whole number will be discussed in later chapters.

EXERCISES 1.12

In Exercises 1–8, find the average of each set of numbers.

1. {2, 7, 9}

2. {3, 5, 7}

3. {6, 8, 9, 5}

4. {9, 0, 6, 9}

5. {21, 24, 33}

6. {7, 10, 8, 5, 11, 7}

7. {74, 88, 85, 69}

8. {96, 92, 95, 89, 88}

9. Maria's examination scores during the semester were 75, 83, 74, 86, 95, and 61. What was her average score?

10. Mrs. Lindstrom recorded her weight each Monday morning for 6 weeks. The weights were 155 pounds, 150 pounds, 149 pounds, 148 pounds, 150 pounds, and 142 pounds. What is her average weight?

11. Five basketball players have heights of 76 inches, 78 inches, 84 inches, 72 inches, and 75 inches. What is the average height of the team?

12. Traveling across the country, a family stayed in motels. The cost per night for the motels was as follows: \$54, \$76, \$60, \$54, \$33, and \$47. Find the average cost per night for motels.

13. Find the class average on a history test if two students scored 90, one student scored 85, four students scored 70, and one student scored 55.

14. The manager earns \$25,000 a year, the assistant manager earns \$17,000 a year, and the four clerks each earn \$12,000 a year. What is the average income per year?

15. Two groups of five students took a test. The students in group A made scores of 78, 85, 97, 76, and 84. Those in group B made scores of 95, 87, 78, 80, and 55. Which group made the higher average, and by how much?

16. Ron's exam scores were 61, 73, 48, and 81. What grade will he need to get on his fifth test so that his average will be 70 for all five tests?

17. The monthly rainfall for a city was: 17 inches for January, 14 inches for February, 19 inches for March, 15 inches for April, 7 inches for May, 2 inches for June, 1 inch for July, 2 inches for August, 4 inches for September, 7 inches for October, 11 inches for November, and 9 inches for December. What is its average monthly rainfall?

18. The high temperatures for the week were 79°, 87°, 96°, 92°, 98°, 102°, and 97°. Find the average high temperature for that week.

1.13 Perimeter and Area

1.13A Perimeter

To measure the length of a line segment, we see how many times a unit of length divides into it.

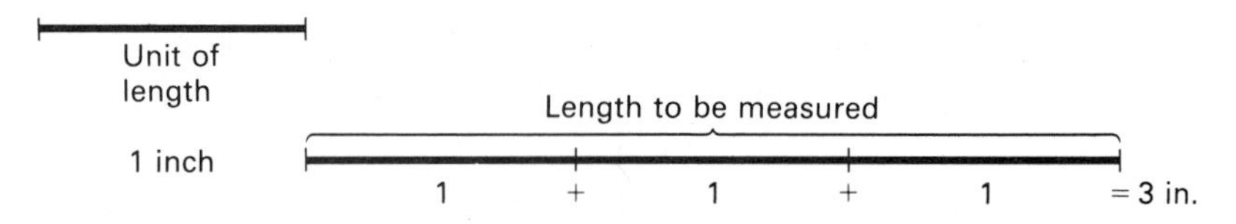

Since the unit of length (inch) fits three times into the length to be measured, we say the length is 3 inches.

Perimeter

The word **perimeter** means " the distance around a figure." To find the perimeter of a geometric figure, we sum the lengths of all its sides.

Example 1 Find the perimeter of the figure below.

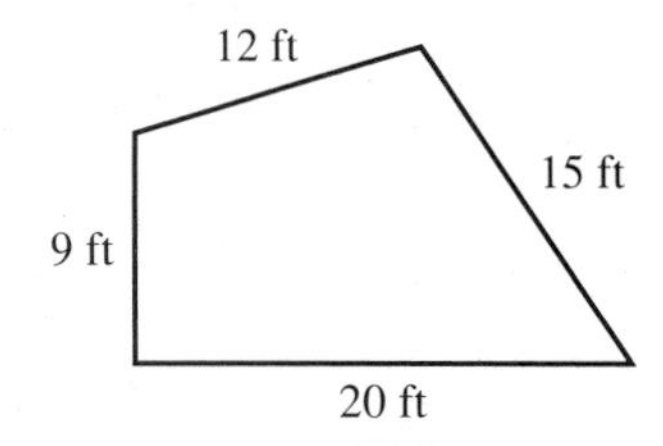

Solution To find the perimeter we must add the four sides.

$$\text{Perimeter} = 20 + 15 + 12 + 9$$
$$= 56 \text{ ft} \quad \blacksquare$$

Triangle

A **triangle** has three sides. The perimeter of a triangle is the sum of its three sides. Perimeter $= a + b + c$.

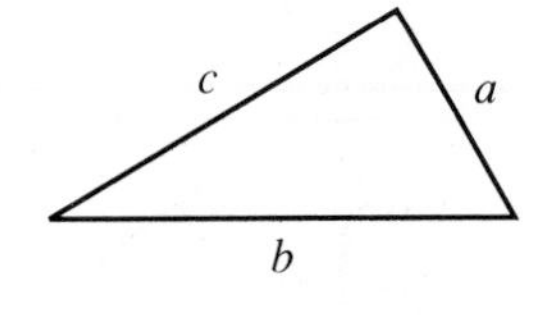

Triangle
$P = a + b + c$

Rectangle

A **rectangle** has four sides and four right angles (square corners). The opposite sides of a rectangle are equal. If we label the sides length (ℓ) and width (w), then the perimeter $= 2\ell + 2w$.

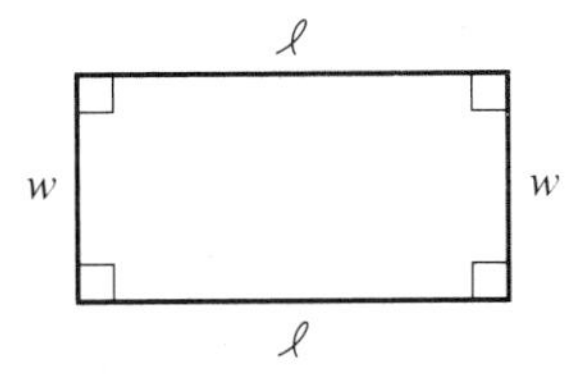

Rectangle
$P = 2\ell + 2w$

Square

A **square** is a rectangle with all sides equal. If we let s represent the length of one side, then the perimeter $= s + s + s + s = 4s$.

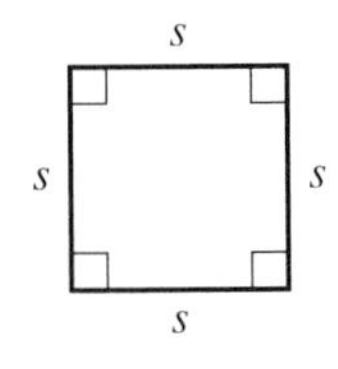

Square
$P = 4s$

Example 2 Find the perimeter of a rectangle with length 10 inches and width 6 inches.
Solution

$$\begin{aligned}\text{Perimeter} &= 2\ell + 2w \\ &= 2 \cdot 10 + 2 \cdot 6 \\ &= 20 + 12 \\ &= 32 \text{ in.}\end{aligned}$$ ■

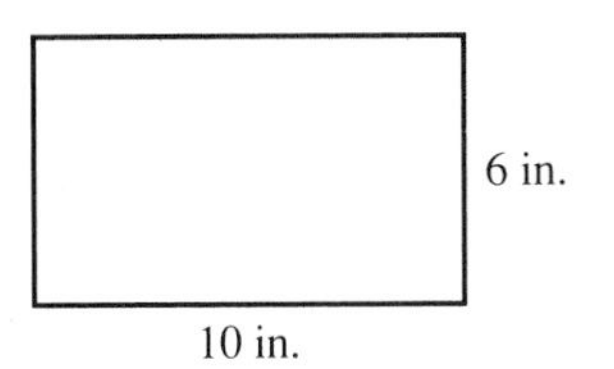

Example 3 Find the cost to fence in a square playground if the length of each side is 20 feet and the fencing costs \$12 per foot.
Solution

$$\begin{aligned}\text{Perimeter} &= 4s \\ &= 4 \cdot 20 \\ &= 80 \text{ ft}\end{aligned}$$

$$\text{Cost} = 80 \cdot 12 = \$960$$ ■

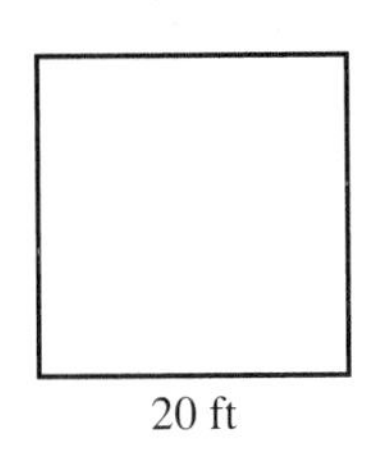

EXERCISES 1.13A

In Exercises 1–8, find the perimeter of each figure.

1.

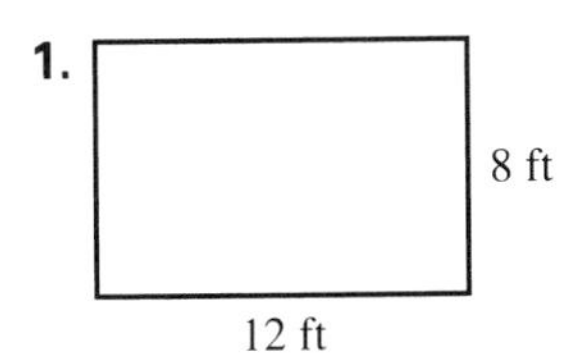

2.

8 cm

3.

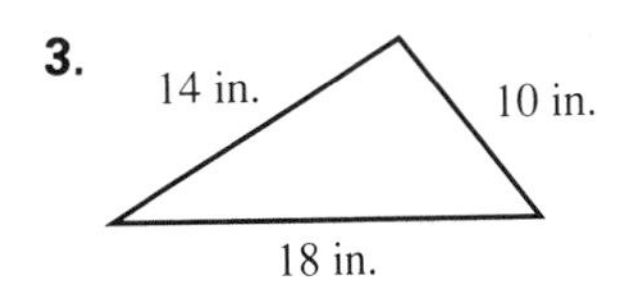

4.

6 in.
10 in.
8 in.

5.

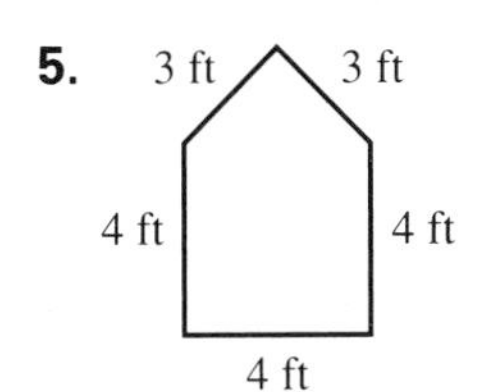

6.

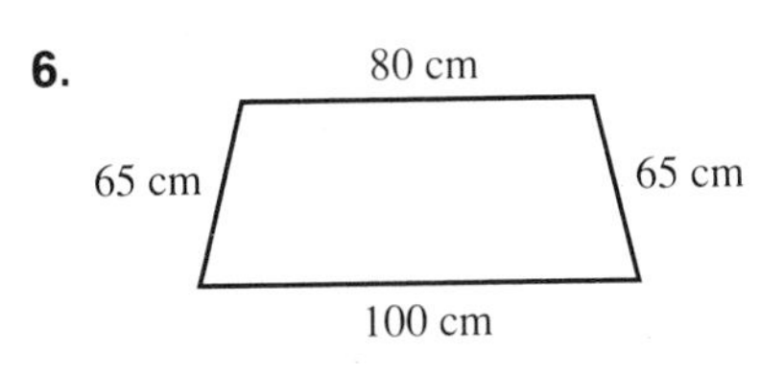

7.

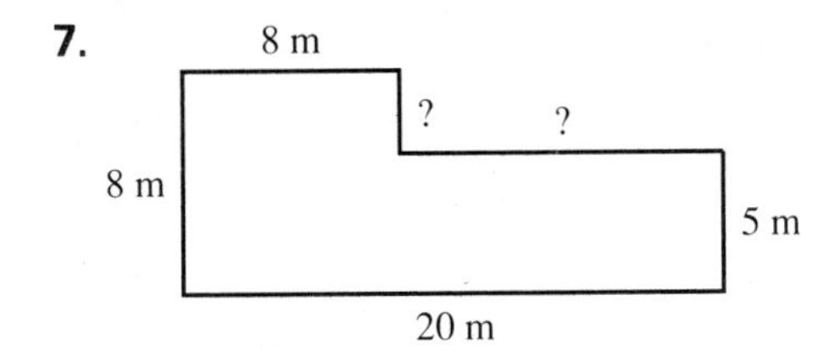

8.

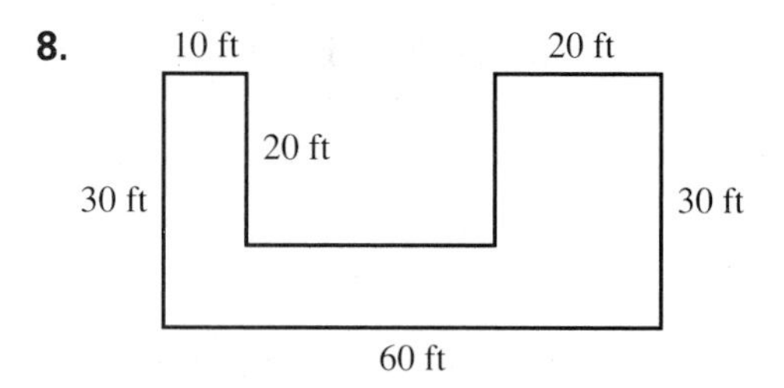

9. Find the perimeter of a square if the length of each side is 15 inches.

10. Find the perimeter of a rectangle with length 25 meters and width 18 meters.

11. What is the total cost to fence a rectangular garden that is 8 feet long and 4 feet wide if the wire fence costs \$3 per foot?

12. A door measures 3 feet wide and 6 feet high. Find the cost to replace the molding around the door if the molding costs \$2 per foot. (Note: There is no molding at the bottom of the door.)

1.13B Area

Area

The **area** of a geometric figure is the space inside the lines. Area is measured in square units: square feet, square inches, square meters, and so on.

To measure the area of the rectangle, we see how many times a unit of area fits into it.

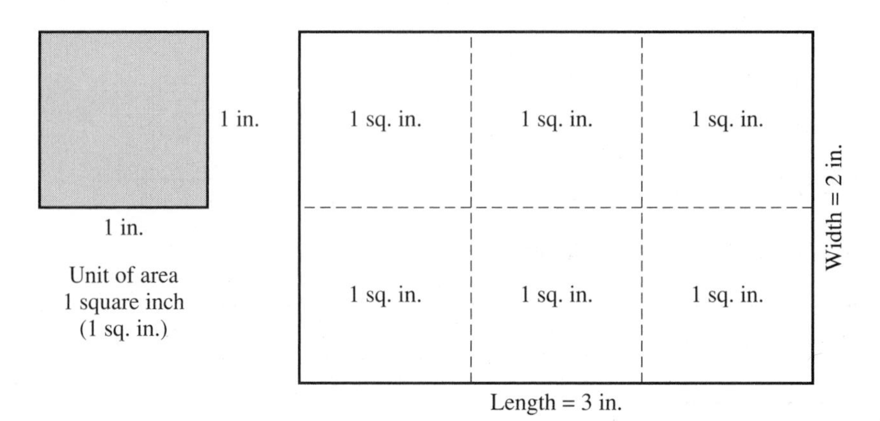

Since the unit of area (1 sq. in.) fits into the space six times, the area of the rectangle is 6 sq. in. We can find this area by multiplying the length times the width.

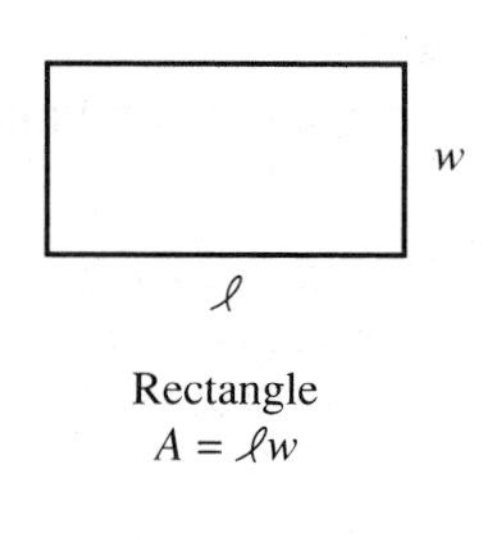

Rectangle
$A = \ell w$

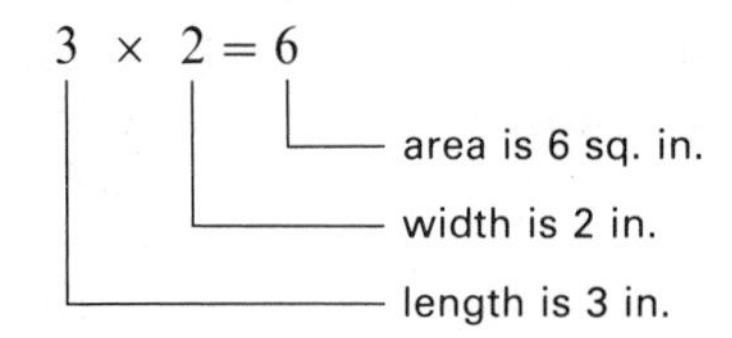

A square is a rectangle in which the length and width are equal. If we let s represent the length of a side, then the area of a square $= s \cdot s = s^2$.

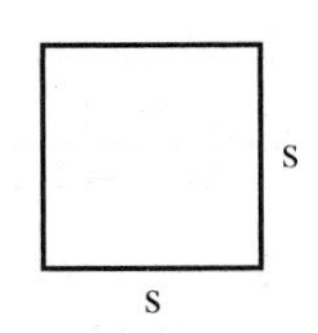

Square
$A = s^2$

Example 1 Find the area of a rectangle with length 9 inches and width 6 inches.
Solution

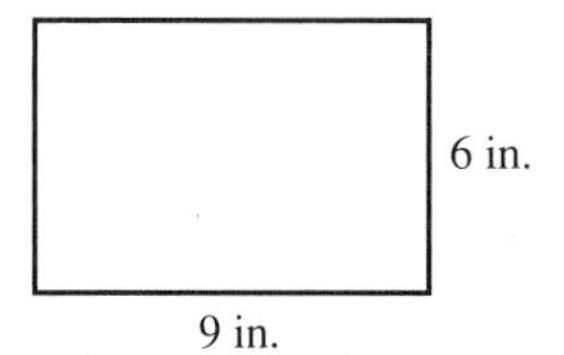

$$\text{Area} = \ell w$$
$$= 9 \cdot 6$$
$$= 54 \text{ sq. in.} \quad \blacksquare$$

Example 2 Find the area of a square playground if the length of each side is 20 feet.
Solution

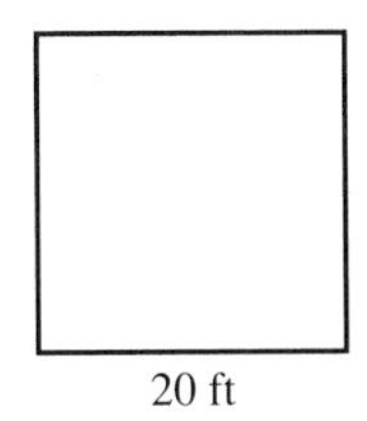

$$\text{Area} = s^2$$
$$= 20^2$$
$$= 400 \text{ sq. ft} \quad \blacksquare$$

Example 3 What is the cost to tile a kitchen that is 12 feet long and 8 feet wide if the ceramic tile costs \$3 per square foot?
Solution

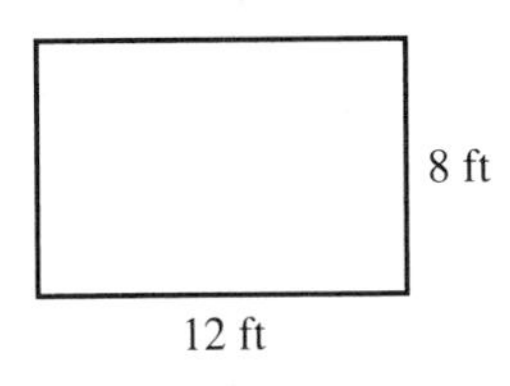

$$\text{Area} = \ell w$$
$$= 12 \cdot 8$$
$$= 96 \text{ sq. ft}$$
$$\text{Cost} = 96 \cdot 3 = \$288 \quad \blacksquare$$

EXERCISES 1.13B

In Exercises 1–3, find the area of each figure.

1.

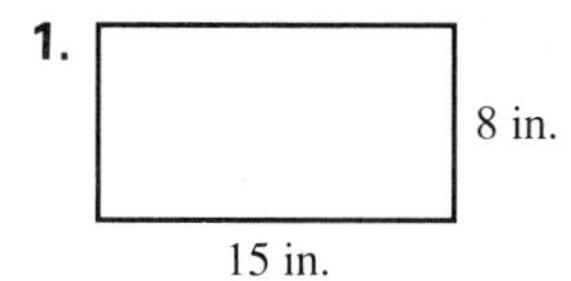

2.

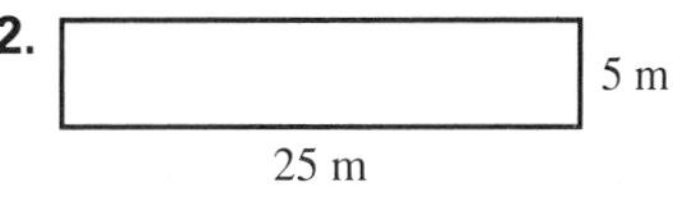

3.

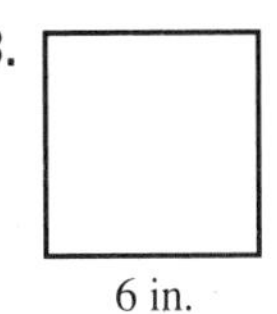

4. Find the area of a square if the length of a side is 15 centimeters.

5. The floor of a square balcony is 8 feet long. How many square feet of material are needed to cover this floor?

6. A rectangular window is 7 feet wide and 4 feet high. How many square feet of glass are needed for three windows of this size?

7. A rectangular room is 6 yards long and 5 yards wide.
a. Find the floor area of this room.
b. Find the cost to carpet this room if the carpet costs \$15 per square yard to install.

8. Find the cost to carpet a bedroom that is 4 yards long and 3 yards wide if the carpet costs \$18 per square yard.

9. What does it cost to replace a large bathroom mirror that measures 7 feet by 3 feet if glass costs \$2 per square foot?

10. Jan needs drapes that are 3 yards high and 5 yards long to cover a picture window in her home. If she pays \$7 a square yard for the material, what will the drapes cost?

1.14 Estimation

When we are not interested in an exact answer to a problem but only want a rough approximation, we can estimate the answer by rounding off each number in the problem and then performing the operations.

> TO ESTIMATE AN ANSWER
>
> **1.** Round off each number in the problem at the first digit. Replace all digits to the right of the round-off place with zeros.
>
> **2.** Perform the indicated operations.

Example 1 Estimate the answers to the following problems.

a. 5,306 + 1,775 + 9,588

Solution Round off 5,306 to 5,000, 1,775 to 2,000, and 9,588 to 10,000, then add.

$$5{,}000 + 2{,}000 + 10{,}000 = 17{,}000$$

b. $682 \cdot 43$

Solution Round off 682 to 700 and 43 to 40, then multiply.

$$700 \cdot 40 = 28{,}000$$

c. $78{,}361 \div 23$

Solution Round off 78,361 to 80,000 and 23 to 20, then divide.

$$80{,}000 \div 20 = \begin{array}{r} 4{,}000 \\ 20 \overline{)80{,}000} \end{array} \quad \blacksquare$$

Example 2 Joann is buying her textbooks for the fall semester. If three books cost \$27 each, one book costs \$44, and another book costs \$12, estimate the total cost of her textbooks.
Solution Round off \$27 to \$30, \$44 to \$40, and \$12 to \$10.

$$\begin{aligned} \text{Cost} &= 3(\$30) + \$40 + \$10 \\ &= \$140 \quad \blacksquare \end{aligned}$$

NOTE Estimating your answer to a problem before finding the exact answer may help you find some arithmetic errors. ☑

EXERCISES 1.14

In Exercises 1–10, estimate the answers to the following problems.

1. 6,723 + 8,908 + 3,215

2. 569 + 135 + 409 + 982

3. 92,750 – 18,060

4. 875,210 – 558,008

5. $6{,}623 \cdot 327$

6. $947 \cdot 86$

7. $8{,}274 \div 42$

8. $319{,}286 \div 631$

9. $54(296 + 676)$

10. $81 \cdot 45 - 2{,}765$

11. Dan had two hamburgers, an order of french fries, and a coke for lunch. Each hamburger contained 420 calories, the french fries contained 185 calories, and the coke contained 267 calories.
a. Estimate the number of calories in his lunch.
b. Find the exact number of calories in his lunch.

12. Sue earns $289 a week. She also receives a stock dividend of $1,225 once a year.
a. Estimate her total yearly income.
b. Find the exact amount of her yearly income.

13.

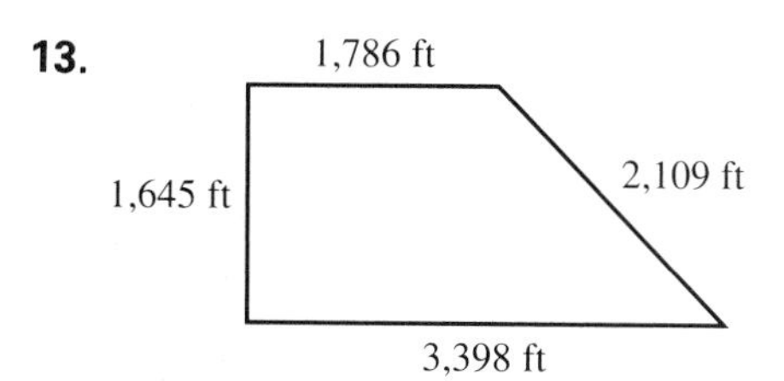

a. Estimate the perimeter of the figure.
b. Find the exact perimeter of the figure.

14.

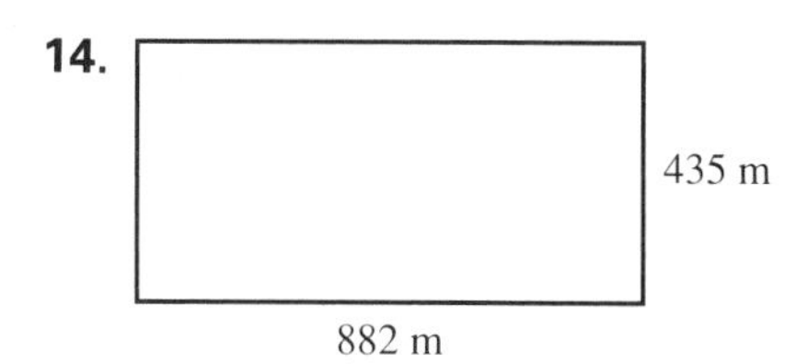

a. Estimate the area of the rectangle.
b. Find the exact area of the rectangle.

1.15 Chapter Summary

Numbers 1.1 *Natural numbers* are the numbers that start with 1 and continue on forever. When 0 is included with the natural numbers, we have the set of *whole numbers.* The *digits* are the first ten whole numbers. The digits are the building blocks used in writing all numbers. We show the natural numbers, whole numbers, and digits on the number line in Figure 1.15.1.

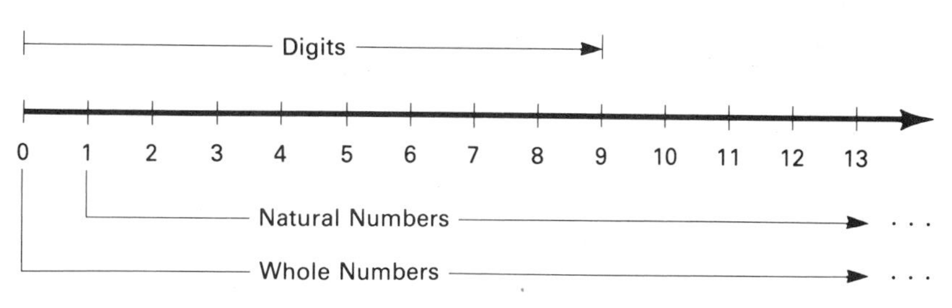

FIGURE 1.15.1

To Read Whole Numbers 1.2

1. Start at the right end of the number and separate the digits into groups of three by means of commas.
2. Start at the left end of the number and read the first group and follow it by that group's name. Continue in this way until all groups have been read.

To Write Whole Numbers in Words 1.2 Write the number in the same way it is read, putting commas in the same places they appear when the number is written in digits.

Addition 1.4 Addition is repeated counting; its inverse is subtraction.

$$\begin{array}{rl} 5 & \text{addend} \\ +\,4 & \text{addend} \\ \hline 9 & \text{sum} \end{array}$$

Commutative property of addition

$$a + b = b + a$$

Associative property of addition

$$(a + b) + c = a + (b + c)$$

Additive identity

$$a + 0 = 0 + a = a$$

Subtraction 1.5

Subtraction is the inverse of addition.

9	minuend
− 4	subtrahead
5	difference

Multiplication 1.6

Multiplication is repeated addition; its inverse is division.

4	·	5	=	20
factor		factor		product

Commutative property of multiplication

$$a \cdot b = b \cdot a$$

Associative property of multiplication

$$(a \cdot b) \cdot c = a \cdot (b \cdot c)$$

Multiplication identity

$$a \cdot 1 = 1 \cdot a = a$$

Multiplication property of zero

$$a \cdot 0 = 0 \cdot a = 0$$

Division 1.9

Division is repeated subtraction; its inverse is multiplication.

$$\begin{array}{r} 3 \\ 12\overline{\smash{)}40} \\ \underline{36} \\ 4 \end{array}$$

divisor: 12; quotient: 3; dividend: 40; remainder: 4

To check a division problem,

quotient	·	divisor	+	remainder	=	dividend
3	·	12	+	4	=	40

Raising a Number to a Power 1.7

Raising a number to a power is repeated multiplication.

$$2^3 = 2 \cdot 2 \cdot 2 = 8$$

exponent: 3; base: 2; power: 8

Square Roots 1.10 Finding the square root of a number is the inverse of squaring a number.

$$\sqrt{16} = 4 \qquad \text{because } 4^2 = 16$$

Order of Operations 1.11

1. Any expressions in parentheses are evaluated first.
2. Evaluations are done in this order:

   First: Powers and roots are done.
   Second: Multiplication and division are done in order from left to right.
   Third: Addition and subtraction are done in order from left to right.

To Find the Average 1.12

1. Find the sum of all the quantities.
2. Divide this sum by the number of quantities.

Perimeter and Area 1.13 Perimeter is the distance around a figure.

Area measures the space inside a figure and is measured in square units.

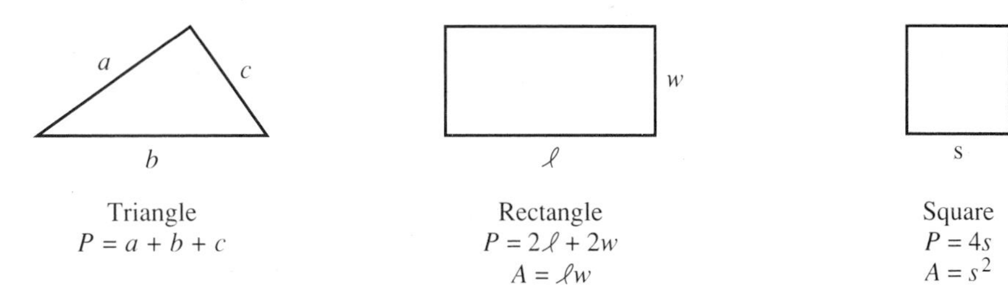

Triangle	Rectangle	Square
$P = a + b + c$	$P = 2\ell + 2w$	$P = 4s$
	$A = \ell w$	$A = s^2$

To Estimate an Answer 1.14

1. Round off each number in the problem at the first digit. Replace all digits to the right of the round-off place with zeros.
2. Perform the indicated operations.

Review Exercises 1.15

1. Write the number 3,075,600,008 in words.

2. Write "five million, seventy-two thousand, six" using digits.

3. Write all the digits less than 5.

4. Write the smallest two-digit natural number.

5. Which of the two symbols, $<$ or $>$, should be used to make each statement true?
a. 17 __?__ 12 b. 0 __?__ 19

6. Which of the operations—addition, subtraction, multiplication, and division—are neither associative nor commutative?

7. Since $5 \cdot 3 = 15$, 5 and 3 are ____________ of 15.

8. Division of whole numbers can be considered repeated ____________.

9. In many cases, the first digit of the divisor is used as the ____________ ____________.

10. Finding the square root of a number is the ____________ of squaring a number.

11. Write the term used for each part in the following subtraction example:

$$\begin{array}{r} 8 \leftarrow \underline{\qquad\qquad} \\ -2 \leftarrow \underline{\qquad\qquad} \\ \hline 6 \leftarrow \underline{\qquad\qquad} \end{array}$$

12. Write the term used for each part in the following division example:

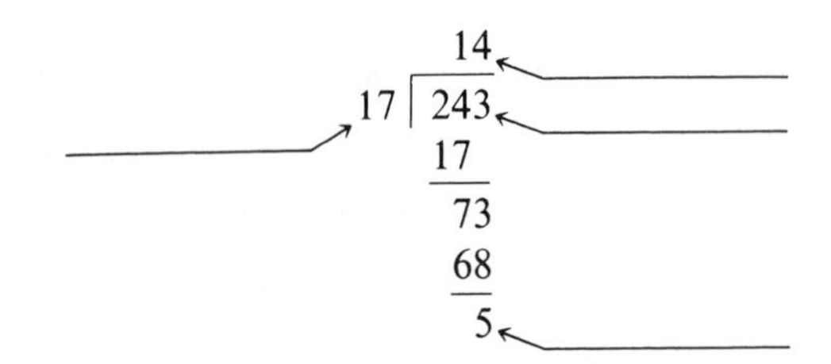

13. Round off the following numbers to the indicated place.

a. 284,791 Nearest ten-thousand
b. 9,685 Nearest hundred
c. 42,568,499 Nearest million

14. Find the following quotients. If division cannot be performed, write "undefined."

a. $6 \div 0$ b. $0 \div 3$
c. $4 \div 1$ d. $0 \div 0$

In Exercises 15–26, perform the indicated operations.

15. $\begin{array}{r} 1{,}045 \\ +\ 896 \\ \hline \end{array}$ **16.** $\begin{array}{r} 785 \\ 396 \\ +\ 29 \\ \hline \end{array}$ **17.** $\begin{array}{r} 78{,}619 \\ 788 \\ +\ 6{,}907 \\ \hline \end{array}$ **18.** $\begin{array}{r} 8{,}247 \\ -\ 358 \\ \hline \end{array}$

19. $\begin{array}{r} 7{,}906 \\ -\ 2{,}789 \\ \hline \end{array}$ **20.** $\begin{array}{r} 700{,}051 \\ -\ 20{,}893 \\ \hline \end{array}$ **21.** $\begin{array}{r} 786 \\ \times\ 35 \\ \hline \end{array}$ **22.** $\begin{array}{r} 9{,}207 \\ \times\ 704 \\ \hline \end{array}$

23. $\begin{array}{r} 3{,}967 \\ \times\ 867 \\ \hline \end{array}$ **24.** $7 \overline{)28{,}602}$ **25.** $38 \overline{)2{,}180}$ **26.** $609 \overline{)292{,}320}$

In Exercises 27–38, evaluate each expression.

27. 5^2 **28.** 2^4 **29.** 3^0 **30.** $\sqrt{36}$

31. $\sqrt{81}$ **32.** $75 \cdot 100$ **33.** $8 \cdot 10^4$ **34.** $30 \cdot 10^3$

35. $10 + 20 \div 2 \cdot 5$ **36.** $36 - 6 + 3 \cdot 2^3$

37. $2(3)^2 - 4(3) + 5$ **38.** $10 - (8 - 3 \cdot 2) - 6$

39. Mr. Jones earns an annual salary of $9,096. What is his monthly salary?

40. Fencing costs $7 per foot. What will 256 feet of fencing cost?

41. Arrange the following numbers in a vertical column, then add: $7{,}825 + 84 + 900 + 45{,}788 + 9 + 2{,}000{,}085$.

42. How many 3-ounce bottles can a druggist fill from a bottle that contains 16 ounces of peroxide?

43. A man pays off a debt of $3,048 in equal monthly payments over a period of two years. Find the amount he pays each month.

44. Mr. Smith pays the property tax on his home in equal monthly payments. If his yearly property tax is $576, how much does he pay each month?

45. The average page in a particular textbook has about 703 words. If there are 356 pages in the book, how many words would you estimate there are in the book?

46. John has $45. Harry has $23. Bill has $12 more than John and Harry together. Find the total amount of money the three have together.

47. Lee makes monthly payments to pay off a $1,950 debt. He makes 23 payments of $82 each and then one final payment to pay off the balance of the debt. How much was the final payment?

48. Two hundred seventy-five families are invited to a neighborhood picnic. The planning committee estimates the average family size to be four. How many people should they plan on providing refreshments for?

49. After trading his car in for a new car, Joe had a balance of $8,400 due. What equal monthly payments must he make to pay this off in five years?

50. In air, light travels 983,584,800 feet per second and sound travels 1,129 feet per second. The speed of light is how many times the speed of sound?

51. A man owed $2,365 on his car. After making 35 payments of $67 each, how much was left to be paid?

52. Suppose the gasoline tank of your car holds 22 gallons, and you average 16 miles to the gallon. How far can you drive on a tankful of gasoline?

53. The full price including tax and financing on a certain color television set was $456. Mr. White bought the set. He agreed to pay for it in 24 equal monthly payments.
a. Find the amount of his monthly payment.
b. After making 17 payments, how much does he still owe?

54. To get a gasoline mileage check, Mr. Perez filled his gasoline tank and wrote down the odometer reading, which was 53,408 miles. The next time he got gasoline his tank took 19 gallons and his odometer reading was 53,731 miles. How many miles did he get per gallon?

55. In the last four weeks a cashier worked 20 hours, 24 hours, 34 hours, and 26 hours. What was the average number of hours worked per week?

56. Find the class average on a test if one student scored 85, two students scored 80, one student scored 75, and two students scored 65.

57. A rectangle is 12 inches long and 8 inches wide.
a. Find the perimeter of the rectangle.
b. Find the area of the rectangle.

58. The length of a side of a square is 9 feet.
a. Find the perimeter of the square.
b. Find the area of the square.

59. Estimate the answers to the following.
a. $3{,}094 \cdot 575$ b. $9{,}880 \div 52$

60. Debbie wrote checks for the following amounts: $77, $12, $54, $48, and $192. Estimate the total amount of the checks.

Chapter 1 Diagnostic Test

Allow yourself about 50 minutes to do these problems. Complete solutions for every problem, together with section references, are given in the answer section at the end of the book.

1. Light travels about 5,879,200,000,000 miles in 1 year. Write this number in words.

2. Use digits to write the number fifty-four billion, seven million, five hundred six thousand, eighty.

In Problems 3–10, perform the indicated operations.

3. $\begin{array}{r} 5{,}843 \\ 209 \\ +6{,}027 \\ \hline \end{array}$

4. $\begin{array}{r} 946 \\ 7{,}328 \\ 407 \\ +\quad 24 \\ \hline \end{array}$

5. $\begin{array}{r} 3{,}564 \\ -\quad 782 \\ \hline \end{array}$

6. $\begin{array}{r} 50{,}406 \\ -35{,}008 \\ \hline \end{array}$

7. $\begin{array}{r} 576 \\ \times\ 89 \\ \hline \end{array}$

8. $\begin{array}{r} 3084 \\ \times\ 706 \\ \hline \end{array}$

9. $63\overline{)5{,}055}$

10. $495\overline{)349{,}470}$

11. Find the indicated powers.
a. 2^3 b. 8^2

12. Find the indicated square roots.
a. $\sqrt{16}$ b. $\sqrt{100}$

13. In the following, just look at the problem, then write your answer. Do not use written calculations.
a. $40 \cdot 100$ b. $16 \cdot 10^4$

14. Round off the following numbers to the indicated place.
a. 78,603 Nearest thousand
b. 3,749 Nearest hundred

15. If at the beginning of a trip your odometer reading was 67,856 miles and at the end of the trip it read 71,304 miles, how many miles did you drive?

16. After trading his car in on a new car, Joe had a balance of $2,016 due. What equal monthly payments must he make to pay this off in 3 years?

17. Enrique has a job paying $168 per week. How much does he earn in one year (52 weeks)?

18. Susan makes monthly payments to pay off a $1,350 debt. She makes 23 payments of $58 each, and then a final payment to pay off the balance. How much was the final payment?

19. Find the perimeter of a rectangle 18 inches long and 12 inches wide.

20. What is the area of a square whose side is 10 centimeters?

21. Estimate the answer.
$77{,}840 \cdot 216$

22. Peter's test scores during the semester were 76, 84, 92, 63, and 70. What was his average score?

In Problems 23–25, evaluate each expression.

23. $6 + 18 \div 3 \cdot 2$

24. $15 - 3^2 + 4$

25. $5 + 2(8 - 4)$

2 Fractions

So far we have been concerned mainly with whole numbers and the operations that can be performed on them. In this chapter, we introduce a new kind of number, *fractions*, and show how to perform the basic operations on fractions.

083193
2.1-2.4

2.1 The Meaning of Fraction

Definition of a Fraction

A **fraction** is part of a whole. It is written $\frac{a}{b}$. a and b are called the *terms* of the fraction. b is called the *denominator*. It tells us into how many parts the whole has been divided. a is called the *numerator*. It tells us how many of those parts we have.

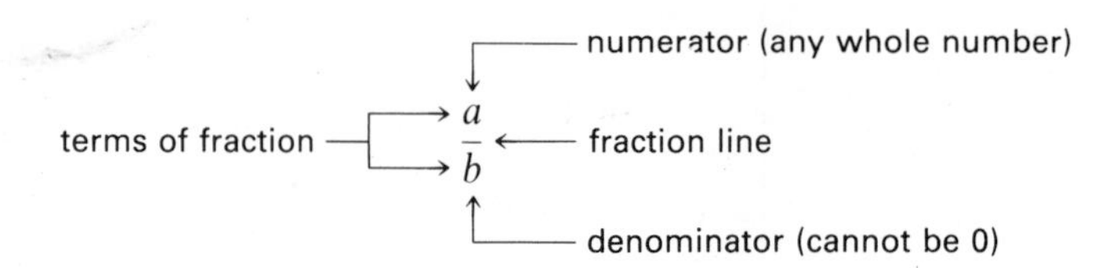

Example 1 Examples of fractions:

a. If we divide a whole rectangle into two equal parts, then the fraction $\frac{1}{2}$, read "one-half," means we have one of the two equal parts of the whole.

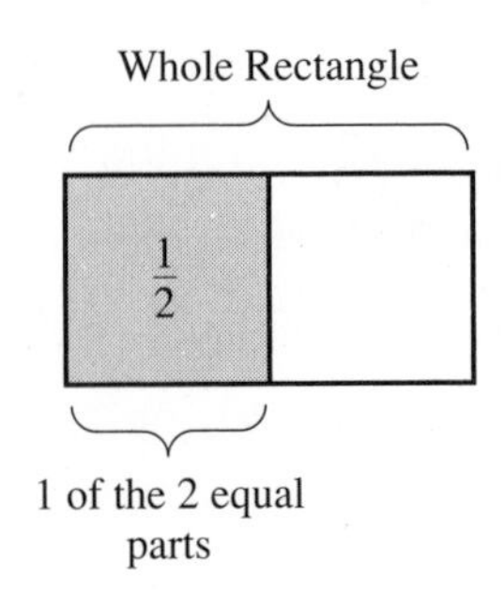

b. If we divide a whole circle into three equal parts, then the fraction $\frac{2}{3}$, read "two-thirds," means we have two of the three equal parts.

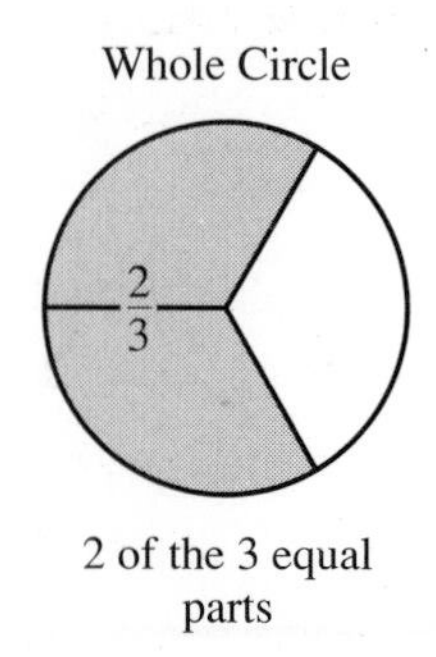

c. If we divide a whole square into four equal parts, then the fraction $\frac{4}{4}$, read "four-fourths," means we have four of the four equal parts.

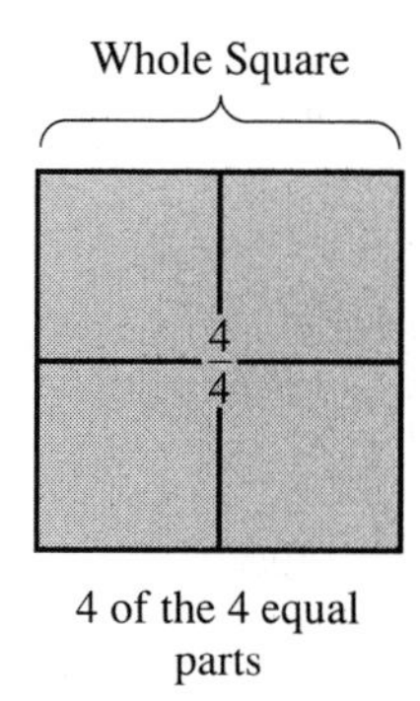

4 of the 4 equal parts

But $\frac{4}{4} = 1$ whole square. Similarly, $\frac{1}{1} = 1$, $\frac{2}{2} = 1$, $\frac{3}{3} = 1$, and so on. ■

In general,

> If a represents any number, except 0, then
>
> $$\frac{a}{a} = 1$$

Any Fraction is Equivalent to a Division

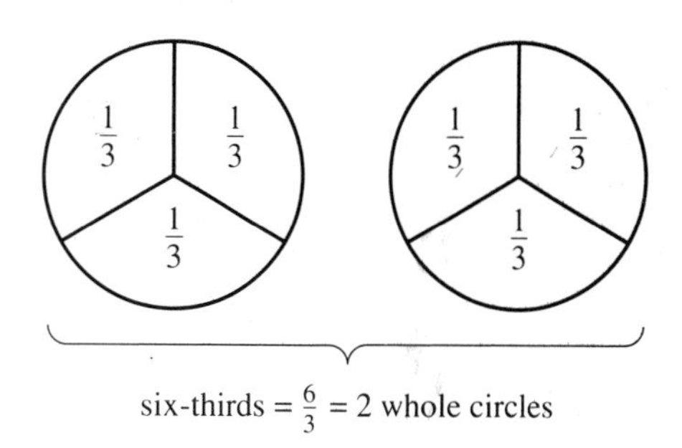

FIGURE 2.1.1

Figure 2.1.1 shows that $\frac{6}{3} = 2$. We also know that $6 \div 3 = 2$. Because both $\frac{6}{3}$ and $6 \div 3$ represent the same thing (in this case, 2), we can say:

$$\frac{6}{3} = 6 \div 3 = 2$$

> ANY FRACTION IS EQUIVALENT TO A DIVISION
>
> If a and b represent any numbers ($b \neq 0$), then
>
> $$\frac{a}{b} = a \div b$$

Example 2 Any fraction is equivalent to a division.

a. $\frac{4}{2} = 4 \div 2 = 2\overline{)4} = 2$

b. $\frac{36}{4} = 36 \div 4 = 4\overline{)36} = 9$

c. $\frac{39}{11} = 39 \div 11 = 11\overline{)39}$ ← Here the answer is not a whole number. Divisions of this type will be discussed in Section 2.3 ■

Proper Fraction

A **proper fraction** is a fraction whose numerator is less than its denominator. Any proper fraction has a value less than one unit. Examples of proper fractions are

$$\frac{1}{2}, \frac{2}{3}, \frac{3}{4}, \frac{4}{5}, \frac{3}{8}, \frac{18}{35}$$

Improper Fraction

An **improper fraction** is a fraction whose numerator is larger than or equal to its denominator. Any improper fraction has a value greater than or equal to one. Examples of improper fractions are

$$\frac{4}{3}, \frac{5}{4}, \frac{7}{2}, \frac{11}{11}, \frac{131}{17}$$

EXERCISES 2.1

1. If we divide a whole into eight equal parts and take three of them, what fraction would represent the part taken?
2. If you cut a pie into five equal pieces and serve two pieces, what fractional part of the pie is left?
3. If we divide a class into six equal groups and take five of the groups, what fractional part of the class was *not* taken?
4. In a football game, three of the eleven first-string players were injured. What fractional part of the team's first-string players were injured?
5. 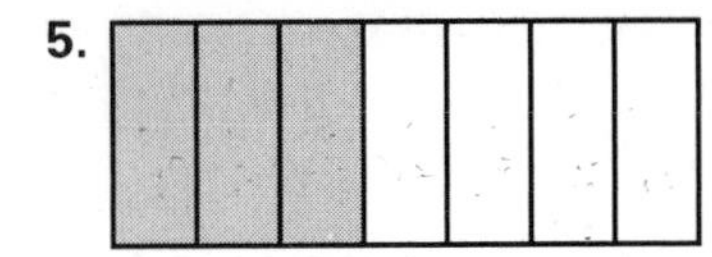 What fraction represents the shaded portion of the rectangle?
6. 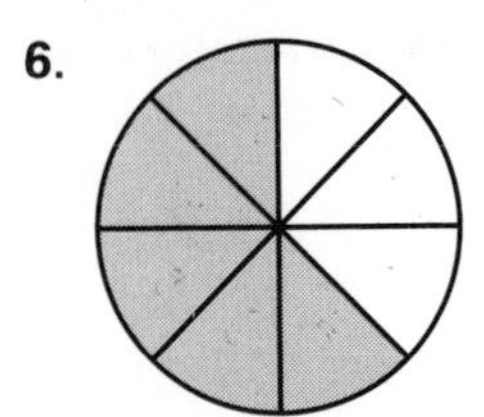 What fraction represents the shaded portion of the circle?

For Exercises 7–11, use the following list:

$$\frac{5}{11}, \frac{28}{13}, \frac{17}{22}, \frac{8}{8}, \frac{1}{2}, \frac{98}{107}, \frac{316}{219}, \frac{1}{31}, \frac{4}{4}$$

7. Which fractions in the list are proper fractions?
8. Which fractions in the list are improper fractions?
9. What is the numerator of the first fraction?
10. What is the denominator of the second fraction?
11. Name all of the fractions in the list that are equal to 1.
12. Is $\frac{4}{0}$ a fraction? Give a reason for your answer.

In Exercises 13–22, change the fractions to whole numbers.

13. $\frac{8}{2}$
14. $\frac{9}{3}$
15. $\frac{24}{6}$
16. $\frac{42}{7}$
17. $\frac{35}{35}$
18. $\frac{84}{7}$
19. $\frac{144}{16}$
20. $\frac{751}{751}$
21. $\frac{1,940}{97}$
22. $\frac{33,060}{551}$

2.2 Multiplication of Fractions

When dealing with whole numbers, addition was introduced first, and all other operations on whole numbers were based on an understanding of addition. With fractions, multiplication is introduced first, and all other operations follow.

TO MULTIPLY TWO FRACTIONS

$$\text{product of two fractions} = \frac{\text{product of their numerators}}{\text{product of their denominators}}$$

In symbols, this is

$$\frac{a}{b} \cdot \frac{c}{d} = \frac{a \cdot c}{b \cdot d}$$

Example 1 Product of two fractions.

a. $\frac{3}{5} \cdot \frac{2}{7} = \frac{3 \cdot 2}{5 \cdot 7} = \frac{6}{35}$

b. $\frac{13}{6} \cdot \frac{1}{8} = \frac{13 \cdot 1}{6 \cdot 8} = \frac{13}{48}$

c. $\frac{5}{12} \cdot \frac{7}{16} = \frac{5 \cdot 7}{12 \cdot 16} = \frac{35}{192}$ ■

Whole Numbers Written as Fractions Because $\frac{8}{1} = 8 \div 1 = 8$, in general, $\frac{a}{1} = a \div 1 = a$. This means that any whole number can be written as a fraction by writing it over 1. That is, the whole number becomes the numerator of a fraction whose denominator is 1.

If a is any whole number, then

$$\frac{a}{1} = a$$

Example 2 Writing whole numbers as fractions.

a. $5 = \frac{5}{1}$

b. $29 = \frac{29}{1}$

c. $117 = \frac{117}{1}$ ■

This makes it possible to multiply fractions by whole numbers.

Example 3 Multiplying fractions by whole numbers.

a. $2 \cdot \frac{4}{9} = \frac{2}{1} \cdot \frac{4}{9} = \frac{2 \cdot 4}{1 \cdot 9} = \frac{8}{9}$

b. $\frac{3}{4} \cdot 17 = \frac{3}{4} \cdot \frac{17}{1} = \frac{3 \cdot 17}{4 \cdot 1} = \frac{51}{4}$ ■

EXERCISES 2.2

Find the products.

1. $\frac{2}{3} \cdot \frac{5}{7}$ **2.** $\frac{1}{2} \cdot \frac{5}{3}$ **3.** $\frac{3}{4} \cdot \frac{7}{8}$ **4.** $\frac{5}{8} \cdot \frac{3}{2}$ **5.** $\frac{4}{5} \cdot \frac{3}{5}$

6. $\frac{5}{6} \cdot \frac{7}{8}$ **7.** $\frac{4}{9} \cdot \frac{2}{3}$ **8.** $\frac{7}{3} \cdot \frac{5}{9}$ **9.** $\frac{3}{4} \cdot \frac{3}{4}$ **10.** $\frac{6}{7} \cdot \frac{6}{7}$

11. $\frac{5}{12} \cdot \frac{5}{8}$ **12.** $\frac{3}{8} \cdot \frac{5}{16}$ **13.** $\frac{11}{32} \cdot \frac{3}{2}$ **14.** $\frac{7}{12} \cdot \frac{13}{15}$ **15.** $\frac{7}{8} \cdot \frac{11}{13}$

16. $\frac{37}{16} \cdot \frac{15}{43}$ **17.** $\frac{1}{2} \cdot \frac{3}{4} \cdot \frac{1}{5}$ **18.** $\frac{1}{3} \cdot \frac{2}{5} \cdot \frac{4}{3}$ **19.** $\frac{2}{5} \cdot \frac{3}{5} \cdot \frac{1}{7}$ **20.** $\frac{3}{4} \cdot \frac{5}{8} \cdot \frac{5}{7}$

21. $2 \cdot \frac{1}{3}$ **22.** $3 \cdot \frac{2}{7}$ **23.** $\frac{1}{5} \cdot 4$ **24.** $\frac{1}{8} \cdot 5$ **25.** $3 \cdot \frac{5}{2}$

26. $6 \cdot \frac{3}{5}$ **27.** $\frac{7}{12} \cdot 7$ **28.** $5 \cdot \frac{3}{16}$ **29.** $5 \cdot \frac{1}{12} \cdot \frac{5}{4}$ **30.** $\frac{3}{5} \cdot 7 \cdot \frac{1}{8}$

2.3 Changing an Improper Fraction to a Mixed Number

Mixed Number

A number such as $1\frac{2}{3}$ is called a **mixed number**. A mixed number is the *sum* of a whole number and a proper fraction. Although a mixed number is a *sum*, the + sign is omitted when writing mixed numbers.

A WORD OF CAUTION The mixed number $2\frac{1}{2}$ means $2 + \frac{1}{2}$. It does *not* mean $2 \cdot \frac{1}{2}$. ✓

Example 1 Examples of mixed numbers.

$$2\frac{1}{2},\quad 3\frac{5}{8},\quad 5\frac{1}{4},\quad 12\frac{3}{16} \quad \blacksquare$$

Consider the improper fraction $\frac{5}{3}$.

Since 5 thirds = 3 thirds + 2 thirds

then $\frac{5}{3} = \frac{3}{3} + \frac{2}{3}$

and $\frac{5}{3} = 1 + \frac{2}{3}$

so that $\frac{5}{3} = 1$ unit $+ \frac{2}{3}$ of another unit (Figure 2.3.1)

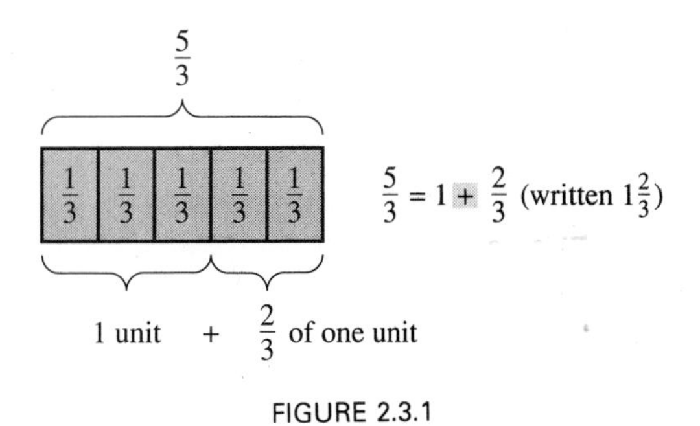

FIGURE 2.3.1

Since a fraction is equivalent to a division, we can change an improper fraction to a mixed number by dividing.

$$\frac{5}{3} = 5 \div 3 = 3\overline{)5}\ \ 1 \text{ R } 2 = 1\frac{2}{3}$$

TO CHANGE AN IMPROPER FRACTION TO A MIXED NUMBER

1. Divide the numerator by the denominator.
2. Write the quotient followed by the fraction: remainder over divisor.

$$\text{quotient}\,\frac{\text{remainder}}{\text{divisor}}$$

Example 2 Changing improper fractions to mixed numbers.

a. $\frac{3}{2} = 3 \div 2 = 2\overline{)3}\ \ 1 \text{ R } 1 = 1\frac{1}{2}$

b. $\frac{15}{11} = 15 \div 11 = 11\overline{)15}\ \ 1 \text{ R } 4 = 1\frac{4}{11}$

c. $\frac{37}{5} = 37 \div 5 = 5\overline{)37}\ \ 7 \text{ R } 2 = 7\frac{2}{5}$ ■

Example 3 Mark's test scores are 78, 82, 90, and 75. Find his test average.

Solution Recall from Section 1.12, to find the average we must find the sum of all the tests, then divide by the number of tests.

$$\text{Average} = \frac{78 + 82 + 90 + 75}{4} = \frac{325}{4} = 81\frac{1}{4}$$ ■

EXERCISES 2.3

In Exercises 1–18, change the following improper fractions to mixed numbers.

1. $\frac{5}{3}$ **2.** $\frac{7}{4}$ **3.** $\frac{9}{5}$ **4.** $\frac{11}{4}$ **5.** $\frac{13}{5}$ **6.** $\frac{11}{3}$

7. $\frac{15}{4}$ **8.** $\frac{35}{2}$ **9.** $\frac{23}{6}$ **10.** $\frac{86}{9}$ **11.** $\frac{16}{13}$ **12.** $\frac{32}{23}$

13. $\frac{20}{7}$ **14.** $\frac{56}{15}$ **15.** $\frac{207}{19}$ **16.** $\frac{37}{21}$ **17.** $\frac{54}{17}$ **18.** $\frac{87}{31}$

In Exercises 19–22, find the average of each set of numbers.

19. $\{7, 8, 3, 5\}$ **20** $\{75, 89, 81\}$ **21.** $\{12, 10, 15, 23, 12\}$ **22.** $\{9, 5, 2, 0, 8, 5\}$

23. On an English quiz, three students scored 5, five students scored 4, and two students scored 2. Find the class average.

24. Sue made five long distance calls. They were for 15 min, 20 min, 3 min, and two for 5 min each. Find the average time of the calls.

2.4 Changing a Mixed Number to an Improper Fraction

Consider the mixed number $2\frac{1}{3}$. This means $2 + \frac{1}{3} = 2$ units + 1 third. Each unit is 3 thirds; therefore, 2 units $= 2 \cdot 3 = 6$ thirds. So that $2\frac{1}{3} = 6$ thirds + 1 third = 7 thirds $= \frac{7}{3}$.

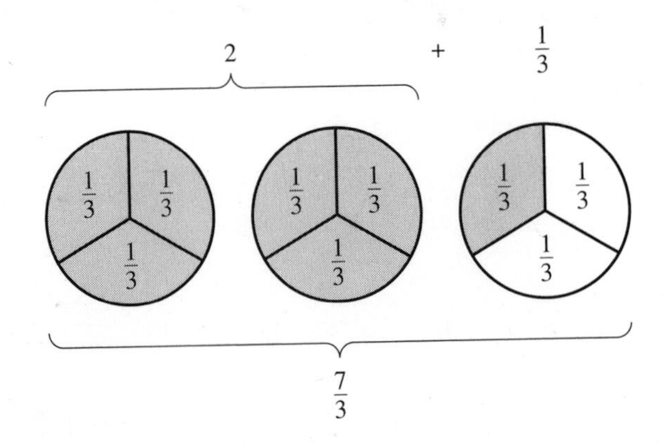

> TO CHANGE A MIXED NUMBER TO AN IMPROPER FRACTION
>
> **1.** Multiply the whole number by the denominator of the fraction.
>
> **2.** Add the numerator of the fraction to the product found in (1).
>
> **3.** Write that sum found in (2) over the denominator of the fraction.

Example 1 Changing a mixed number to an improper fraction.

a. $3\frac{1}{2} = \frac{3 \cdot 2 + 1}{2} = \frac{6 + 1}{2} = \frac{7}{2}$

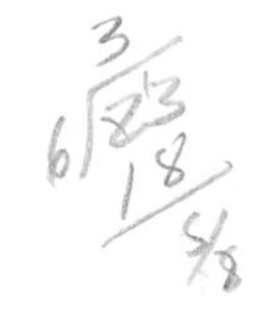

b. $7\frac{2}{5} = \frac{7 \cdot 5 + 2}{5} = \frac{35 + 2}{5} = \frac{37}{5}$

c. $4\frac{7}{13} = \frac{4 \cdot 13 + 7}{13} = \frac{52 + 7}{13} = \frac{59}{13}$ ■

EXERCISES 2.4

Change the mixed numbers to improper fractions.

1. $1\frac{1}{2}$ **2.** $2\frac{3}{5}$ **3.** $3\frac{1}{4}$ **4.** $2\frac{5}{8}$ **5.** $4\frac{5}{6}$

6. $4\frac{1}{2}$ **7.** $3\frac{7}{10}$ **8.** $3\frac{5}{16}$ **9.** $5\frac{7}{12}$ **10.** $6\frac{2}{7}$

11. $12\frac{2}{3}$ **12.** $3\frac{9}{13}$ **13.** $6\frac{3}{4}$ **14.** $4\frac{5}{11}$ **15.** $3\frac{7}{15}$

16. $1\frac{8}{17}$ **17.** $15\frac{3}{4}$ **18.** $21\frac{2}{3}$ **19.** $2\frac{7}{10}$ **20.** $8\frac{3}{100}$

2.5 Equivalent Fractions

Equivalent Fractions

Equivalent fractions are equal. They are different ways of writing the same number. A look at Figure 2.5.1 will convince you that $\frac{3}{4}$ and $\frac{9}{12}$ are equivalent fractions because they represent the same thing.

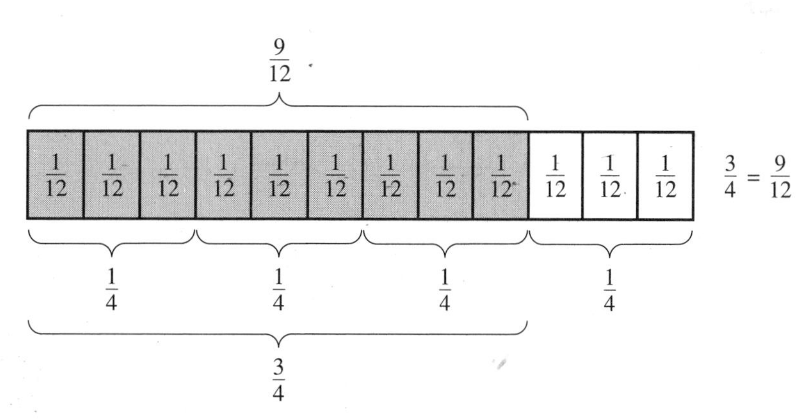

FIGURE 2.5.1

Another example of equivalent fractions:

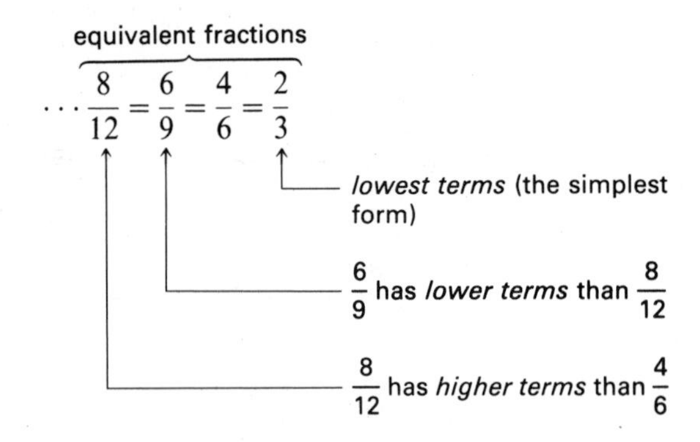

Forming Equivalent Fractions Recall from Section 1.6 that the number 1 is the multiplicative identity. That is, multiplying any number by 1 does not change its value. Since $1 = \frac{1}{1} = \frac{2}{2} = \frac{3}{3} = \frac{4}{4}$, and so on, multiplying any fraction by $\frac{2}{2}$, or $\frac{3}{3}$, or any fraction whose value is 1, will form an equivalent fraction.

Another way to see that $\frac{3}{4} = \frac{9}{12}$ is

$$\frac{3}{4} = \frac{3}{4} \cdot \boxed{1} = \frac{3}{4} \cdot \boxed{\frac{3}{3}} = \frac{3 \cdot 3}{4 \cdot 3} = \frac{9}{12}$$

The value of this fraction is 1

Hence, $\frac{3}{4}$ and $\frac{9}{12}$ are equivalent fractions $\left(\text{or simply, } \frac{3}{4} = \frac{9}{12}\right)$.

This leads to the following rule for forming equivalent fractions.

> Multiplying both numerator and denominator by the same nonzero number gives an equivalent fraction.
>
> $$\frac{a}{b} = \frac{a \cdot c}{b \cdot c}$$
>
> Equivalent fractions

Example 1 Finding a particular fraction equivalent to a given fraction.

a. $\frac{3}{4} = \frac{?}{8}$

Since the denominator 8 is twice the denominator 4, *multiply* the numerator and denominator by 2

$$\frac{3}{4} = \frac{3 \cdot \boxed{2}}{4 \cdot \boxed{2}} = \frac{6}{8}$$

The $\boxed{2}$ can be found by dividing 8 by 4. Since the denominator 4 must be multiplied by 2 to give 8, we must also multiply the numerator by 2

b. $\frac{7}{9} = \frac{?}{45}$

Since the denominator 45 is five times the denominator 9, multiply the numerator and denominator by 5

$$\frac{7}{9} = \frac{7 \cdot 5}{9 \cdot 5} = \frac{35}{45}$$

$45 \div 9 = 5$ ■

We have already shown that $\frac{3}{4}$ and $\frac{9}{12}$ are equivalent fractions.

$$\frac{3}{4} = \frac{3 \cdot 3}{4 \cdot 3} = \frac{9}{12}$$

We can look at these two fractions in another way.

$$\frac{9}{12} = \frac{9 \div 3}{12 \div 3} = \frac{3}{4}$$

In general,

> Dividing both numerator and denominator by the same nonzero number that is a divisor of both gives an equivalent fraction.
>
> $$\frac{a}{b} = \frac{a \div c}{b \div c}$$
>
> Equivalent fractions

Example 2 Finding a particular fraction equivalent to a given fraction.

a. $\frac{15}{20} = \frac{?}{4}$

Since the denominator 4 is less than the denominator 20, *divide* the numerator and denominator by 5

$$\frac{15}{20} = \frac{15 \div 5}{20 \div 5} = \frac{3}{4}$$

The 5 can be found by dividing 20 by 4. Since the denominator 20 must be divided by 5 to give 4, we must also divide the numerator by 5

b. $\frac{14}{35} = \frac{?}{5}$

Since the denominator 5 is less than the denominator 35, divide the numerator and denominator by 7

$$\frac{14}{35} = \frac{14 \div 7}{35 \div 7} = \frac{2}{5}$$

$35 \div 5 = 7$ ■

Forming equivalent fractions is necessary when reducing fractions (Section 2.7) and when adding unlike fractions (Section 2.10).

EXERCISES 2.5

Replace the question mark with a number that makes the fractions equivalent (have equal value).

1. $\frac{1}{2} = \frac{?}{10}$ **2.** $\frac{1}{3} = \frac{?}{6}$ **3.** $\frac{3}{6} = \frac{?}{2}$ **4.** $\frac{6}{8} = \frac{?}{4}$ **5.** $\frac{2}{3} = \frac{?}{6}$

6. $\frac{3}{4} = \frac{?}{40}$ **7.** $\frac{6}{10} = \frac{?}{5}$ **8.** $\frac{6}{16} = \frac{?}{8}$ **9.** $\frac{3}{5} = \frac{?}{15}$ **10.** $\frac{5}{6} = \frac{?}{12}$

11. $\frac{14}{20} = \frac{?}{10}$ **12.** $\frac{18}{45} = \frac{?}{15}$ **13.** $\frac{9}{13} = \frac{?}{52}$ **14.** $\frac{15}{18} = \frac{?}{36}$ **15.** $\frac{36}{54} = \frac{?}{9}$

16. $\frac{72}{24} = \frac{?}{3}$ **17.** $\frac{8}{11} = \frac{?}{55}$ **18.** $\frac{20}{26} = \frac{?}{52}$ **19.** $\frac{85}{34} = \frac{?}{2}$ **20.** $\frac{33}{77} = \frac{?}{7}$

2.6 Prime Factorization

Prime Number

A **prime number** is a natural number greater than 1 that can be divided evenly (the remainder is zero) only by itself and 1. A prime number has *no* factors other than itself and 1.

Composite Number

A **composite number** is a natural number that can be divided evenly by some natural number other than itself and 1. A composite number *has* factors other than itself and 1.

Note that 1 is neither prime nor composite.

Example 1 Prime and composite numbers.

a. 9 is a composite because $3 \cdot 3 = 9$, so that 9 has a factor other than itself or 1.

b. 17 is prime because 1 and 17 are the only factors of 17.

c. 45 is composite because it has factors 3, 5, 9, and 15 other than 1 and 45.

d. 31 is prime because 1 and 31 are the only factors of 31. ■

A partial list of prime numbers is 2, 3, 5, 7, 11, 13, 17, 19, 23, 29, ...

Prime Factorization

The **prime factorization** of a natural number is the product of all its factors that are themselves prime numbers. For example, the prime factorization of 6 is $2 \cdot 3$, because 2 and 3 are both prime numbers that are factors of 6. A factorization that is not a prime factorization is $1 \cdot 6$, because neither 1 nor 6 is a prime number.

Method for Finding the Prime Factorization The smallest prime is 2; the next smallest is 3; the next smallest is 5; and so on. To find the prime factorization of 24, first try to divide 24 by the smallest prime, 2. Two does divide 24 and gives a quotient of 12. Next try to divide 12 by the smallest prime, 2. Two does divide 12 and gives a quotient of 6. Next try to divide 6 by the smallest prime, 2. Two does divide 6 and gives a quotient of 3. This process ends here because the final quotient 3 is itself a prime. The work of finding the prime factorization of a number can be conveniently arranged as follows:

```
2 |24
2 |12
2 | 6
     3
```

The prime factorization is the product of these numbers →

Therefore, $24 = 2 \cdot 2 \cdot 2 \cdot 3 = 2^3 \cdot 3$, where $2^3 \cdot 3$ is the prime factorization of 24.

Example 2 Prime factorization of numbers.

a. Find the prime factorization of 30.

Solution

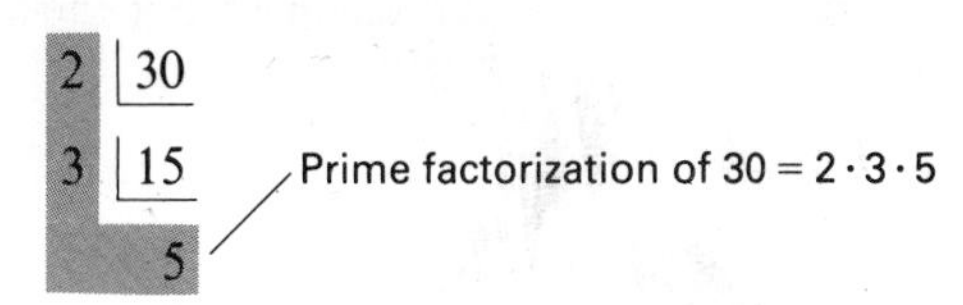

b. Find the prime factorization of 20.

Solution

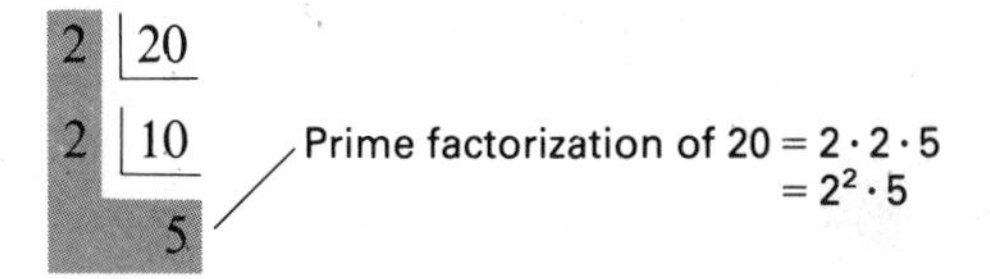

c. Find the prime factorization of 36.

Solution

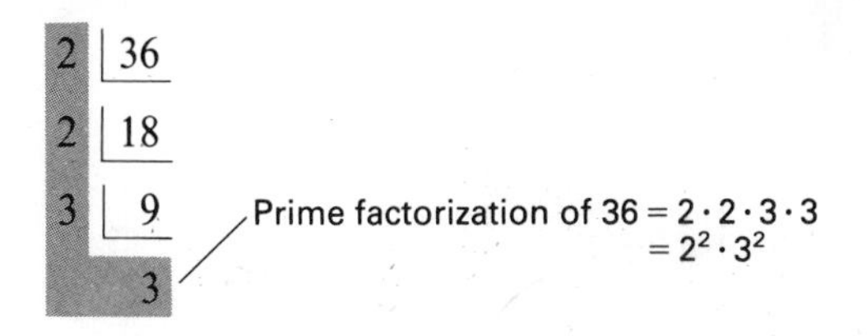

d. Find the prime factorization of 315.

Solution

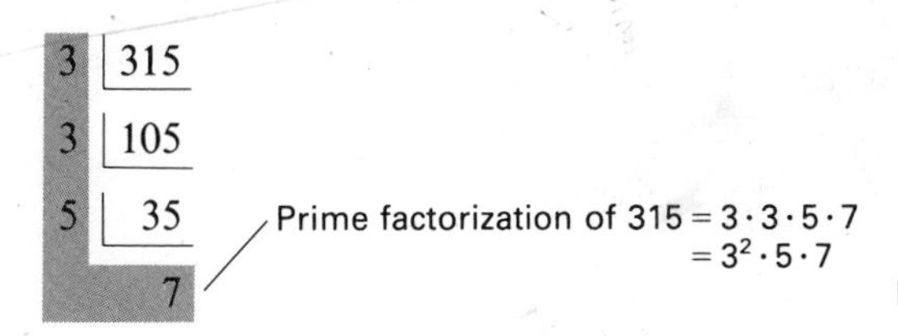

■

When trying to find a prime factor of a number, no prime whose square is greater than that number need be tried (see Example 3).

Example 3 Find the prime factorization of 97.

Solution primes in order of size

2 does not divide 97.
3 does not divide 97.
5 does not divide 97.
7 does not divide 97.
11 is not possible because $11^2 = 121$, which is greater than 97.

Therefore, 97 is prime. ■

Tests for Divisibility When finding the prime factorization, use the following rules to tell if a number is divisible by the primes 2, 3, or 5. We have listed only those rules that will be most useful.

Divisible by 2

A number is **divisible by 2** if its *last* digit is 0, 2, 4, 6, 8.

Example 4 1②; 30⓪; 203④; 57⑧ are divisible by 2. ■

Divisible by 3

A number is **divisible by 3** if the *sum* of its digits is divisible by 3.

Example 5 210 is divisible by 3 because $2 + 1 + 0 = 3$, which is divisible by 3.

5162 is not divisible by 3 because $5 + 1 + 6 + 2 = 14$, which is not divisible by 3. ■

Divisible by 5

A number is **divisible by 5** if its *last* digit is 0 or 5.

Example 6 25⓪ and 75⑤ are both divisible by 5, but 112⑦ is not. ■

We have not included tests of divisibility by larger primes because the tests are usually longer than the actual trial division.

EXERCISES 2.6

In Exercises 1–15, find the prime factorization of each number.

1. 14	**2.** 15	**3.** 21	**4.** 22	**5.** 16
6. 27	**7.** 29	**8.** 31	**9.** 32	**10.** 81
11. 84	**12.** 75	**13.** 144	**14.** 250	**15.** 360

In Exercises 16–30, state whether each of the numbers is prime or composite. To justify your answer, give the prime factorization for each number.

16. 6	**17.** 5	**18.** 8	**19.** 13	**20.** 15
21. 12	**22.** 11	**23.** 21	**24.** 23	**25.** 41
26. 55	**27.** 49	**28.** 31	**29.** 51	**30.** 42

2.7 Reducing Fractions to Lowest Terms

Lowest Terms

A fraction is reduced to **lowest terms** when there is no whole number (greater than 1) that divides both numerator and denominator exactly.

In Section 2.5 we divided numerator and denominator by a whole number (greater than 1) that is a divisor of both to obtain an equivalent fraction.

$$\frac{12}{18} = \frac{12 \div 2}{18 \div 2} = \frac{6}{9} \quad \text{and} \quad \frac{12}{18} = \frac{12 \div 3}{18 \div 3} = \frac{4}{6}$$

Since we have reduced the fraction $\frac{12}{18}$ to two different fractions having *lower* terms, the question arises: "How can we reduce $\frac{12}{18}$ to a fraction having the *lowest* terms?"

2.7A Reducing Fractions

TO REDUCE A FRACTION TO LOWEST TERMS

1. Divide numerator and denominator by any whole number (greater than 1) that you can see is a divisor of both.
2. When you cannot see any more common divisors, write the prime factorization of the numerator and the denominator.
3. Divide numerator and denominator by all factors common to both.

Example 1 Reduce $\frac{12}{18}$ to lowest terms.

Solution $\frac{12}{18} = \frac{12 \div 2}{18 \div 2} = \frac{6}{9}$ You can see that 2 is a common divisor

Then $\frac{6}{9} = \frac{6 \div 3}{9 \div 3} = \frac{2}{3}$ You can see that 3 is a common divisor

Therefore, $\frac{12}{18} = \frac{2}{3}$ Numerator and denominator are in prime-factored form and have no common factors

Dividing numerator and denominator by a common factor is often written as follows:

$$\frac{12}{18} = \frac{\overset{2}{\cancel{12}}}{\underset{3}{\cancel{18}}} = \frac{2}{3}$$ Both numerator and denominator were divided by 6 ■

Example 2 Reduce $\frac{30}{42}$ to lowest terms.

Solution $\frac{30}{42} = \frac{30 \div 2}{42 \div 2} = \frac{15}{21}$ You can see that 2 is a common divisor

Then $\frac{15}{21} = \frac{15 \div 3}{21 \div 3} = \frac{5}{7}$ You can see that 3 is a common divisor

Therefore, $\frac{30}{42} = \frac{5}{7}$ Numerator and denominator are in prime-factored form and have no common factor

Another way of writing this solution:

$$\frac{\overset{\overset{5}{\cancel{15}}}{\cancel{30}}}{\underset{\underset{7}{\cancel{21}}}{\cancel{42}}} = \frac{5}{7}$$ We first divide 30 and 42 by 2; then we divide 15 and 21 by 3 ■

Example 3 Reduce $\frac{150}{280}$ to lowest terms.

Solution

$$\frac{\overset{15}{\cancel{150}}}{\underset{28}{\cancel{280}}} = \frac{15}{28}$$ We first divide 150 and 280 by 10

$$\frac{15}{28} = \frac{3 \cdot 5}{2 \cdot 2 \cdot 7}$$ Write numerator and denominator in prime-factored form

Therefore, $$\frac{150}{280} = \frac{15}{28}$$ In lowest terms because numerator and denominator have no common factor ■

Sometimes numerator and denominator can be divided by the same number more than once. See Example 4.

Example 4 Reduce $\frac{18}{45}$ to lowest terms.

Solution

$$\frac{\overset{\overset{2}{\cancel{6}}}{\cancel{18}}}{\underset{\underset{5}{\cancel{15}}}{\cancel{45}}} = \frac{2}{5}$$ We first divide 18 and 45 by 3, then we divide 6 and 15 by 3 ■

After dividing numerator and denominator by a particular number, check to see if they can be divided again by the same number.

Another Way to Reduce Fractions to Lowest Terms *Start* with the prime factorization of numerator and denominator.

1. Write the prime factorization of both numerator and denominator (Sec. 2.6).
2. Divide numerator and denominator by all factors common to both.

Example 5 Reduce $\frac{10}{14}$ to lowest terms.

Solution The prime factorization of $10 = 2 \cdot 5$

The prime factorization of $14 = 2 \cdot 7$

Therefore, $$\frac{10}{14} = \frac{2 \cdot 5}{2 \cdot 7} = \frac{\cancel{2} \cdot 5}{\cancel{2} \cdot 7} = \frac{5}{7}$$

2 is a factor of the numerator

Both numerator and denominator were divided by 2

2 is a factor of the denominator ■

Example 6 Reduce $\frac{6}{20}$ to lowest terms.

Solution

$$\frac{6}{20} = \frac{\cancel{2} \cdot 3}{\cancel{2} \cdot 2 \cdot 5} = \frac{3}{10}$$ Both numerator and denominator were divided by 2

In reducing fractions, it is helpful to write every factor rather than using powers: $2 \cdot 2 \cdot 5$ instead of $2^2 \cdot 5$ ■

Example 7 Reduce $\frac{28}{56}$ to lowest terms.

Solution $\frac{28}{56} = \frac{\cancel{2} \cdot \cancel{2} \cdot \cancel{7}}{\cancel{2} \cdot \cancel{2} \cdot 2 \cdot \cancel{7}} = \frac{1}{2}$

Both numerator and denominator were divided by $2 \cdot 2 \cdot 7 = 28$

NOTE When the numerator is divided by $2 \cdot 2 \cdot 7$, the quotient is 1. ☑ ■

EXERCISES 2.7A

Reduce each fraction to lowest terms.

1. $\frac{6}{9}$	**2.** $\frac{15}{20}$	**3.** $\frac{12}{16}$	**4.** $\frac{14}{21}$	**5.** $\frac{30}{40}$
6. $\frac{12}{15}$	**7.** $\frac{24}{32}$	**8.** $\frac{36}{90}$	**9.** $\frac{10}{18}$	**10.** $\frac{16}{24}$
11. $\frac{32}{40}$	**12.** $\frac{48}{64}$	**13.** $\frac{54}{90}$	**14.** $\frac{135}{315}$	**15.** $\frac{84}{105}$
16. $\frac{42}{84}$	**17.** $\frac{33}{55}$	**18.** $\frac{44}{66}$	**19.** $\frac{36}{60}$	**20.** $\frac{12}{60}$
21. $\frac{39}{51}$	**22.** $\frac{33}{57}$	**23.** $\frac{80}{180}$	**24.** $\frac{210}{270}$	**25.** $\frac{3000}{4200}$

2.7B Shortening Multiplication of Fractions

Because of the definition of multiplication of fractions, we can reduce the fraction to lowest terms before multiplying. When we multiply fractions, a factor in any numerator can be canceled with the same factor in any denominator.

Example 8 Multiplying fractions.

a. $\frac{7}{12} \cdot \frac{4}{21} = \frac{\overset{1}{\cancel{7}} \cdot \overset{1}{\cancel{4}}}{\underset{3}{\cancel{12}} \cdot \underset{3}{\cancel{21}}} = \frac{1}{9}$ or $\frac{\overset{1}{\cancel{7}}}{\underset{3}{\cancel{12}}} \cdot \frac{\overset{1}{\cancel{4}}}{\underset{3}{\cancel{21}}} = \frac{1}{9}$

b. $\frac{4}{25} \cdot \frac{15}{8} = \frac{\overset{1}{\cancel{4}} \cdot \overset{3}{\cancel{15}}}{\underset{5}{\cancel{25}} \cdot \underset{2}{\cancel{8}}} = \frac{3}{10}$ or $\frac{\overset{1}{\cancel{4}}}{\underset{5}{\cancel{25}}} \cdot \frac{\overset{3}{\cancel{15}}}{\underset{2}{\cancel{8}}} = \frac{3}{10}$

c. $\frac{9}{2} \cdot \frac{1}{3} \cdot \frac{35}{14} = \frac{\overset{3}{\cancel{9}}}{2} \cdot \frac{1}{\underset{1}{\cancel{3}}} \cdot \frac{\overset{5}{\cancel{35}}}{\underset{2}{\cancel{14}}} = \frac{15}{4} = 3\frac{3}{4}$ ■

Fractional Part of a Number

Example 9 Find $\frac{2}{3}$ of 18.

Solution To find a fractional part of a number, we need to multiply the fraction times the number. In this case, the word *of* means to multiply.

$$\frac{2}{3} \text{ of } 18 = \frac{2}{3} \cdot \frac{18}{1} = \frac{2}{\cancel{3}_1} \cdot \frac{\cancel{18}^6}{1} = \frac{12}{1} = 12$$ ■

Example 10 $\frac{3}{5}$ of the class are women. How many of the 35 students are women?

Solution

$$\frac{3}{5} \text{ of } 35 = \frac{3}{5} \cdot \frac{35}{1} = \frac{3}{\cancel{5}_1} \cdot \frac{\cancel{35}^7}{1} = 21 \text{ women}$$ ■

EXERCISES 2.7B

In Exercises 1–16, find the products. Reduce your answers to lowest terms. Write any improper fraction as a mixed number.

1. $\frac{2}{3} \cdot \frac{6}{7}$ **2.** $\frac{2}{5} \cdot \frac{5}{8}$ **3.** $\frac{6}{8} \cdot \frac{4}{9}$ **4.** $\frac{7}{8} \cdot \frac{12}{7}$

5. $\frac{10}{18} \cdot \frac{9}{5}$ **6.** $\frac{25}{14} \cdot \frac{21}{10}$ **7.** $\frac{35}{72} \cdot \frac{18}{56}$ **8.** $\frac{80}{45} \cdot \frac{27}{60}$

9. $\frac{22}{75} \cdot \frac{20}{44}$ **10.** $\frac{3}{5} \cdot \frac{5}{8} \cdot \frac{1}{6}$ **11.** $\frac{5}{7} \cdot \frac{3}{10} \cdot \frac{14}{15}$ **12.** $\frac{4}{5} \cdot \frac{5}{6} \cdot \frac{10}{15}$

13. $\frac{16}{49} \cdot \frac{14}{8} \cdot \frac{9}{12}$ **14.** $\frac{40}{14} \cdot \frac{21}{5} \cdot \frac{2}{27}$ **15.** $\frac{4}{15} \cdot \frac{10}{28} \cdot \frac{21}{10}$ **16.** $\frac{4}{3} \cdot \frac{5}{14} \cdot \frac{7}{12} \cdot \frac{6}{15}$

17. Find $\frac{3}{4}$ of 12.

18. Find $\frac{5}{6}$ of 24.

19. Find $\frac{4}{5}$ of 300.

20. Find $\frac{5}{8}$ of 320.

21. An astronaut on the moon weighs $\frac{1}{6}$ of his weight on Earth. If he weighs 186 pounds on Earth, how much would he weigh on the moon?

22. If $\frac{2}{5}$ of the college's 25,000 students are over 25 years old, how many students are over 25 years old?

23. John filled his 24-gallon gas tank at the beginning of his business trip. At the end of the trip the tank was $\frac{3}{8}$ full. How many gallons of gas did he use on the trip?

24. On a math test of 20 problems, Marcia solved $\frac{3}{4}$ of the problems correctly. How many problems did she miss?

2.8 Adding and Subtracting Like Fractions

Like Fractions

Like fractions are fractions that have the same denominator.

Example 1 Like fractions. $\frac{2}{3}, \frac{5}{3}, \frac{1}{3}$ are like fractions

same denominator ■

Unlike Fractions

Unlike fractions are fractions that have different denominators.

Example 2 Unlike fractions.

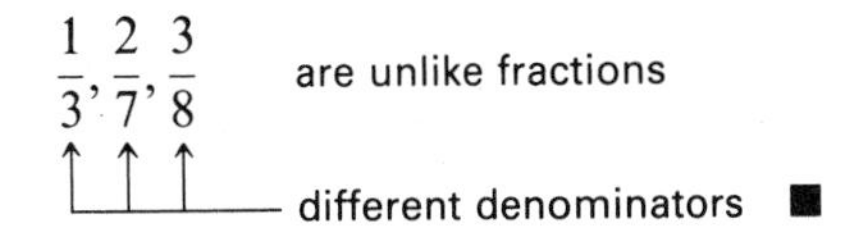

Adding Like Fractions Figure 2.8.1 illustrates how to add like fractions.

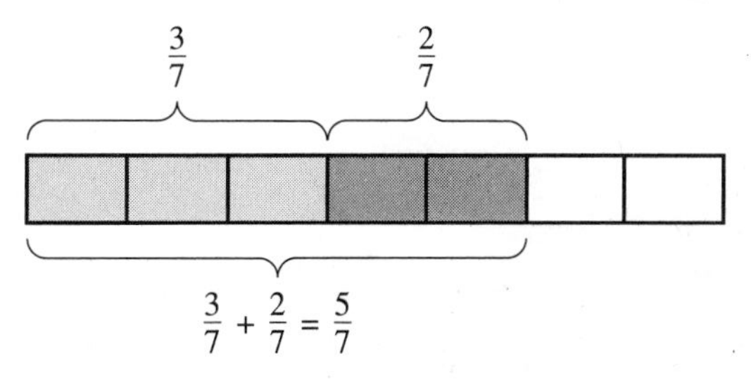

FIGURE 2.8.1

> TO ADD LIKE FRACTIONS
>
> 1. Add their numerators.
> 2. Write the sum found in (1) over the common denominator. $\frac{a}{c} + \frac{b}{c} = \frac{a+b}{c}$
> 3. Reduce the resulting fraction to lowest terms.
> 4. Any improper fraction found in (3) is usually changed to a mixed number.

Example 3 Adding like fractions.

a. Add: $\frac{11}{23} + \frac{5}{23} = \frac{11+5}{23} = \frac{16}{23}$ (already lowest terms)

b. Add: $\frac{1}{9} + \frac{4}{9} + \frac{7}{9} = \frac{1+4+7}{9} = \frac{12}{9} = \frac{\overset{4}{\cancel{12}}}{\underset{3}{\cancel{9}}} = \frac{4}{3} = 1\frac{1}{3}$ ■

Subtracting Like Fractions Subtracting like fractions is done in a similar manner.

TO SUBTRACT LIKE FRACTIONS

1. Subtract their numerators.
2. Write the difference found in (1) over the common denominator.

$$\frac{a}{c} - \frac{b}{c} = \frac{a-b}{c}$$

3. Reduce the resulting fraction to lowest terms.
4. Any improper fraction found in (3) is usually changed to a mixed number.

Example 4 Subtracting like fractions.

a. Subtract: $\frac{25}{8} - \frac{11}{8} = \frac{25-11}{8} = \frac{\overset{7}{\cancel{14}}}{\underset{4}{\cancel{8}}} = \frac{7}{4} = 1\frac{3}{4}$

b. Subtract: $\frac{23}{31} - \frac{17}{31} = \frac{23-17}{31} = \frac{6}{31}$ (already lowest terms) ■

EXERCISES 2.8

Perform the indicated operation. Reduce your answers to lowest terms. Write any improper fraction as a mixed number.

1. $\frac{1}{6}+\frac{3}{6}$ **2.** $\frac{3}{8}+\frac{1}{8}$ **3.** $\frac{3}{5}+\frac{2}{5}$ **4.** $\frac{3}{4}+\frac{1}{4}$ **5.** $\frac{7}{10}-\frac{5}{10}$

6. $\frac{11}{12}-\frac{3}{12}$ **7.** $\frac{5}{9}-\frac{2}{9}$ **8.** $\frac{9}{14}-\frac{5}{14}$ **9.** $\frac{5}{8}+\frac{7}{8}$ **10.** $\frac{5}{16}+\frac{13}{16}$

11. $\frac{11}{24}+\frac{4}{24}$ **12.** $\frac{35}{80}+\frac{27}{80}$ **13.** $\frac{27}{35}-\frac{6}{35}$ **14.** $\frac{45}{52}-\frac{6}{52}$ **15.** $\frac{70}{81}-\frac{19}{81}$

16. $\frac{123}{144}-\frac{43}{144}$ **17.** $\frac{1}{6}+\frac{5}{6}+\frac{3}{6}$ **18.** $\frac{5}{12}+\frac{2}{12}+\frac{9}{12}$ **19.** $\frac{3}{15}+\frac{1}{15}+\frac{6}{15}$ **20.** $\frac{29}{45}+\frac{16}{45}+\frac{3}{45}$

2.9 Lowest Common Denominator (LCD)

Before adding unlike fractions we ordinarily find their *lowest common denominator.*

LCD

The **lowest common denominator (LCD)** of two or more fractions is the smallest number that is exactly divisible by each of their denominators. In this section we show two methods for finding the LCD.

1. By inspection 2. By prime factorizaton

Finding the LCD by Inspection By inspection, try to find a number exactly divisible by each denominator. If you cannot find the LCD by inspection, then use the prime factorization method.

Example 1 Find the LCD for $\frac{1}{2} + \frac{3}{4}$.

Solution By inspection we see that the LCD = 4, because 4 is the smallest number exactly divisible by both denominators: 2 and 4. ■

Example 2 Find the LCD for $\frac{5}{2} + \frac{1}{3}$.

Solution By inspection we see that the LCD = 6, because 6 is the smallest number exactly divisible by both denominators: 2 and 3. ■

Finding the LCD by Prime Factorization Consider the problem $\frac{7}{12} + \frac{4}{15}$. Here the LCD is not easily found by inspection. To find the LCD in this case:

1. Find the prime factorization of each denominator.

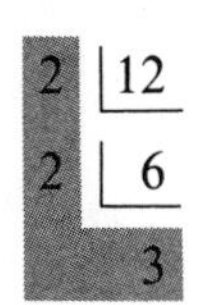

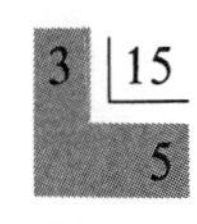

$12 = 2^2 \cdot 3$ $\quad$ $15 = 3 \cdot 5$

2. Write down each different factor that appears in the prime factorizations.

   2, 3, 5

3. Raise each factor to the highest power to which it occurs in any denominator.

   2^2, 3, 5

4. The LCD is the product of all the powers found in Step 3.

   $\text{LCD} = 2^2 \cdot 3 \cdot 5 = 60$

The prime factorization method for finding the LCD is summarized by the following box.

TO FIND THE LCD BY PRIME FACTORIZATION

1. Write the prime factorization of each denominator. Repeated factors should be expressed as powers.
2. Write down each different factor that appears.
3. Raise each factor to the highest power to which it occurs in any denominator.
4. The LCD is the product of all the powers found in Step 3.

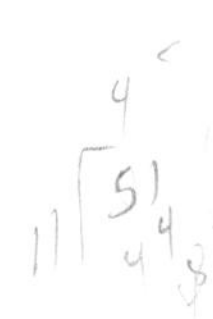

Example 3 Find the LCD for $\frac{3}{8} + \frac{7}{10}$.

Solution

Step 1. $8 = 2 \cdot 2 \cdot 2 = 2^3$ — Prime factorization of 8

$10 = 2 \cdot 5$ — Prime factorization of 10

Step 2. 2, 5 — 2 and 5 are the only different prime factors that appear in the prime factorizations

Step 3. $2^3, 5^1$ — The highest power of each factor

Step 4. $\text{LCD} = 2^3 \cdot 5^1 = 40$ — The product of all the powers found in Step 3

2^3 is a *power* of 2; 3 is only the *exponent* of 2 ■

Example 4 Find the LCD for $\frac{5}{14} + \frac{13}{21}$.

Solution

Step 1. $14 = 2 \cdot 7$, $21 = 3 \cdot 7$ — Prime factorization of each denominator

Step 2. 2, 3, 7 — The only different prime factors that appear

Step 3. $2^1, 3^1, 7^1$ — The highest power of each factor

Step 4. $\text{LCD} = 2^1 \cdot 3^1 \cdot 7^1 = 42$ — The product of all the powers found in Step 3 ■

Example 5 Find the LCD for $\frac{2}{9} + \frac{7}{15} + \frac{3}{20}$.

Solution

Step 1. $9 = 3 \cdot 3 = 3^2$, $15 = 3 \cdot 5$, $20 = 2 \cdot 2 \cdot 5 = 2^2 \cdot 5$ — Prime factorization of each denominator

Step 2. 2, 3, 5 — All the different prime factors

Step 3. $2^2, 3^2, 5^1$ — Highest power of each factor

Step 4. $\text{LCD} = 2^2 \cdot 3^2 \cdot 5^1$
$= 4 \cdot 9 \cdot 5 = 180$ ■

Example 6 Find the LCD for $\frac{3}{5} + \frac{1}{2} + \frac{2}{3}$.

Solution Since each denominator is a prime number, the LCD is the product of the three denominators.

$$\text{LCD} = 5 \cdot 2 \cdot 3 = 30$$ ■

EXERCISES 2.9

Assume the following sets to be denominators of fractions. In Exercises 1–6, find the LCD by inspection, then check it by prime factoring.

1. {2, 3, 4} **2.** {4, 5, 10} **3.** {2, 8, 4} **4.** {3, 6, 9} **5.** {3, 5, 15} **6** {7, 2, 14}

In Exercises 7–20, find the LCD by either method.

7. {14, 10} **8.** {16, 12} **9.** {7, 5} **10.** {4, 9}
11. {6, 8, 9} **12.** {4, 15, 18} **13.** {40, 15, 25} **14.** {3, 7, 5}
15. {4, 5, 21} **16.** {4, 6, 9, 12} **17.** {6, 13, 26} **18.** {45, 63, 98}
19. {66, 33, 132} **20.** {24, 40, 48, 56}

2.10 Adding and Subtracting Unlike Fractions

To add $\frac{1}{2}$ dollar and $\frac{1}{4}$ dollar (one quarter), we change the $\frac{1}{2}$ dollar to two $\frac{1}{4}$ dollars (two quarters). Then

$$\begin{aligned}\frac{1}{2}\text{ dollar} + \frac{1}{4}\text{ dollar} &= \frac{2}{4}\text{ dollar} + \frac{1}{4}\text{ dollar}\\ &= \left(\frac{2}{4} + \frac{1}{4}\right)\text{ dollar}\\ &= \frac{3}{4}\text{ dollar}\end{aligned}$$

We converted $\frac{1}{2}$ into $\frac{2}{4}$, which has the same denominator as $\frac{1}{4}$. The *lowest common denominator* (LCD) of $\frac{1}{2}$ and $\frac{1}{4}$ is 4. The method of adding unlike fractions is given in the following box.

> TO ADD OR SUBTRACT UNLIKE FRACTIONS
>
> **1.** Find the LCD.
>
> **2.** Convert all fractions to equivalent fractions having the LCD as denominator.
>
> **3.** Add or subtract the like fractions as before.
>
> **4.** Reduce the resulting fraction to lowest terms.
>
> **5.** Any improper fraction found in (4) is usually changed to a mixed number.

Example 1 Add $\frac{1}{2} + \frac{2}{3}$. ***Solution*** The LCD is 6, by inspection.

$$\left.\begin{array}{r} \frac{1}{2} = \frac{3}{6} \\ + \frac{2}{3} = \frac{4}{6} \end{array}\right\}$$

Convert each fraction to an equivalent fraction having the LCD 6 as a denominator. Then add the like fractions

$$\frac{7}{6} = 1\frac{1}{6}$$

The sum. ■

Example 2 Subtract $\frac{4}{5} - \frac{3}{10}$. ***Solution*** The LCD is 10 by inspection.

$$\begin{array}{r} \frac{4}{5} = \frac{8}{10} \\ - \frac{3}{10} = \frac{3}{10} \end{array}$$

$$\frac{5}{10} = \frac{\overset{1}{\cancel{5}}}{\underset{2}{\cancel{10}}} = \frac{1}{2}$$

Reduce ■

In Examples 3–6, the LCD is found by prime factorization.

Example 3 Add $\frac{5}{6} + \frac{3}{10} + \frac{4}{15}$.

Solution

$$\left.\begin{array}{r} \frac{5}{6} = \frac{25}{30} \\ \frac{3}{10} = \frac{9}{30} \\ + \frac{4}{15} = \frac{8}{30} \end{array}\right\}$$

Add like fractions

$$6 = 2 \cdot 3$$
$$10 = 2 \cdot 5$$
$$15 = 3 \cdot 5$$
$$\text{LCD} = 2 \cdot 3 \cdot 5 = 30$$

$$\frac{42}{30} = \frac{\overset{7}{\cancel{42}}}{\underset{5}{\cancel{30}}} = \frac{7}{5} = 1\frac{2}{5}$$ ■

Example 4 Add $\frac{3}{16} + \frac{5}{12} + \frac{5}{24}$.

Solution

$$\left.\begin{array}{r} \frac{3}{16} = \frac{9}{48} \\ \frac{5}{12} = \frac{20}{48} \\ + \frac{5}{24} = \frac{10}{48} \end{array}\right\}$$

Add like fractions

$$16 = 2^4$$
$$12 = 2^2 \cdot 3$$
$$24 = 2^3 \cdot 3$$
$$\text{LCD} = 2^4 \cdot 3 = 48$$

$$\frac{39}{48} = \frac{\overset{13}{\cancel{39}}}{\underset{16}{\cancel{48}}} = \frac{13}{16}$$ ■

Example 5 Subtract $\frac{7}{15} - \frac{5}{12}$.

Solution

$$\left.\begin{aligned} \frac{7}{15} &= \frac{28}{60} \\ -\frac{5}{12} &= \frac{25}{60} \end{aligned}\right\} \text{Subtract like fractions}$$

$15 = 3 \cdot 5$

$12 = 2^2 \cdot 3$

$\text{LCD} = 2^2 \cdot 3 \cdot 5$

$= 60$

$$\frac{3}{60} = \frac{\overset{1}{\cancel{3}}}{\underset{20}{\cancel{60}}} = \frac{1}{20}$$ ■

Example 6 Add $\frac{8}{48} + \frac{5}{12} + \frac{4}{18}$.

Solution

$$\left.\begin{aligned} \frac{8}{48} &= \frac{24}{144} \\ \frac{5}{12} &= \frac{60}{144} \\ +\frac{4}{18} &= \frac{32}{144} \end{aligned}\right\}$$

$48 = 2^4 \cdot 3$

$12 = 2^2 \cdot 3$

$18 = 2 \cdot 3^2$

$\text{LCD} = 2^4 \cdot 3^2$

$= 144$

$$\frac{116}{144} = \frac{\overset{29}{\cancel{116}}}{\underset{36}{\cancel{144}}} = \frac{29}{36}$$

If the original fractions had been reduced to lowest terms before adding them, the work could have been simplified. For example,

Add $\frac{\overset{1}{\cancel{8}}}{\underset{6}{\cancel{48}}} + \frac{5}{12} + \frac{\overset{2}{\cancel{4}}}{\underset{9}{\cancel{18}}}$ First, reduce the fraction to lowest terms

$$\frac{1}{6} + \frac{5}{12} + \frac{2}{9}$$

$$\left.\begin{aligned} \frac{1}{6} &= \frac{6}{36} \\ \frac{5}{12} &= \frac{15}{36} \\ +\frac{2}{9} &= \frac{8}{36} \end{aligned}\right\}$$

$6 = 2 \cdot 3$

$12 = 2^2 \cdot 3$

$9 = 3^2$

$\text{LCD} = 2^2 \cdot 3^2$

$= 36$

$$\frac{29}{36}$$ ■

NOTE It is not necessary to use the LCD when adding unlike fractions. *Any* common denominator will work. However, if the LCD is not used, larger numbers appear in the work. In this case, the resulting fraction must be reduced. ☑

Example 7 Add $\frac{1}{6} + \frac{2}{3} + \frac{3}{4}$.

Solution A *common denominator* (not the LCD) is 24, because each denominator divides into 24 exactly.

$$\frac{1}{6} = \frac{4}{24}$$

$$\frac{2}{3} = \frac{16}{24}$$

$$+\frac{3}{4} = \frac{18}{24}$$

$$\frac{38}{24} = \frac{\cancelto{19}{38}}{\cancelto{12}{24}} = \frac{19}{12} = 1\frac{7}{12}$$ ■

EXERCISES 2.10

First reduce fractions to lowest terms, then perform the indicated operation. Reduce your answers to lowest terms. Write any improper fraction as a mixed number.

1. $\frac{1}{2} + \frac{3}{4}$
2. $\frac{1}{3} + \frac{5}{6}$
3. $\frac{3}{5} + \frac{3}{10}$
4. $\frac{5}{8} + \frac{1}{2}$
5. $\frac{2}{3} - \frac{1}{6}$
6. $\frac{7}{10} - \frac{1}{2}$
7. $\frac{2}{3} - \frac{5}{12}$
8. $\frac{9}{15} - \frac{2}{5}$
9. $\frac{2}{3} + \frac{1}{4}$
10. $\frac{2}{5} + \frac{1}{3}$
11. $\frac{3}{6} + \frac{5}{15}$
12. $\frac{2}{6} + \frac{6}{8}$
13. $\frac{6}{7} - \frac{4}{12}$
14. $\frac{10}{16} - \frac{5}{12}$
15. $\frac{12}{30} - \frac{1}{5}$
16. $\frac{12}{16} - \frac{1}{6}$
17. $\frac{25}{32} - \frac{3}{4}$
18. $\frac{13}{16} - \frac{5}{8}$
19. $\frac{56}{64} - \frac{14}{24}$
20. $\frac{14}{35} - \frac{5}{20}$
21. $\frac{3}{20} + \frac{1}{8}$
22. $\frac{4}{15} + \frac{7}{12}$
23. $\frac{7}{9} + \frac{5}{6}$
24. $\frac{11}{12} + \frac{7}{8}$
25. $\frac{5}{12} - \frac{2}{15}$
26. $\frac{11}{18} - \frac{4}{15}$
27. $\frac{19}{35} - \frac{5}{14}$
28. $\frac{15}{16} - \frac{5}{24}$
29. $\frac{2}{3} + \frac{5}{6} + \frac{1}{2}$
30. $\frac{1}{2} + \frac{2}{3} + \frac{3}{4}$
31. $\frac{3}{5} + \frac{1}{2} + \frac{3}{10}$
32. $\frac{7}{10} + \frac{3}{5} + \frac{2}{3}$
33. $\frac{1}{4} + \frac{7}{12} + \frac{5}{8}$
34. $\frac{5}{6} + \frac{4}{12} + \frac{3}{8}$
35. $\frac{4}{6} + \frac{6}{14} + \frac{2}{3}$
36. $\frac{6}{9} + \frac{8}{12} + \frac{9}{10}$
37. $\frac{5}{6} + \frac{3}{8} + \frac{1}{12}$
38. $\frac{1}{3} + \frac{3}{5} + \frac{2}{11}$
39. $\frac{3}{4} + \frac{2}{14} + \frac{1}{2}$
40. $\frac{1}{12} + \frac{3}{16} + \frac{4}{10}$

2.11 Adding Mixed Numbers

To add mixed numbers, add the whole number parts and add the fractional parts.

Example 1 Find $12\frac{1}{2} + 14\frac{1}{3}$.

Solution

$$\begin{aligned} 12\frac{1}{2} &= 12 + \frac{3}{6} \\ + 14\frac{1}{3} &= 14 + \frac{2}{6} \\ \hline & 26 + \frac{5}{6} = 26\frac{5}{6} \end{aligned}$$

The fraction parts were changed to equivalent fractions having the LCD for denominator ■

Example 2 Find $12\frac{3}{4} + 21\frac{3}{8} + 45\frac{1}{2}$.

Solution

$$\begin{aligned} 12\frac{3}{4} &= 12 + \frac{6}{8} \\ 21\frac{3}{8} &= 21 + \frac{3}{8} \\ + 45\frac{1}{2} &= 45 + \frac{4}{8} \\ \hline & 78 + \frac{13}{8} = \\ & 78 + 1\frac{5}{8} = 79\frac{5}{8}. \end{aligned}$$

Replaced $\frac{13}{8}$ by $1\frac{5}{8}$ ■

Example 3 Find the perimeter of the triangle.

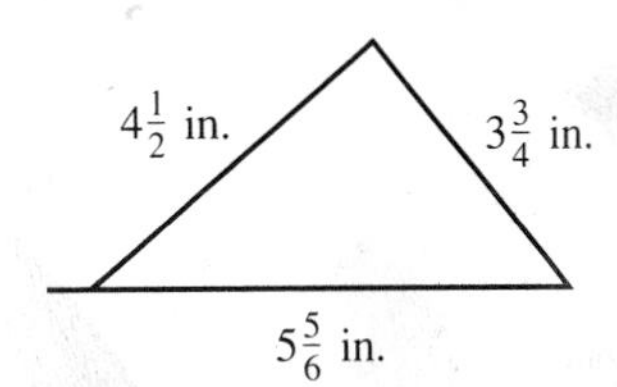

Solution To find the perimeter of a figure, we must add the lengths of all its sides.

$$\begin{aligned} 4\frac{1}{2} &= 4 + \frac{6}{12} \\ 3\frac{3}{4} &= 3 + \frac{9}{12} \\ + 5\frac{5}{6} &= 5 + \frac{10}{12} \\ \hline & 12 + \frac{25}{12} = \\ & 12 + 2\frac{1}{12} = 14\frac{1}{12} \text{ in.} \end{aligned}$$

■

EXERCISES 2.11

In Exercises 1–28, find the sums. Reduce answers to lowest terms. Write any improper fraction as a mixed number.

1. $2\frac{3}{5} + 1\frac{1}{10}$

2. $1\frac{3}{8} + 2\frac{1}{4}$

3. $3\frac{1}{3} + 2\frac{1}{2}$

4. $5\frac{3}{6} + 1\frac{1}{2}$

5. $4\frac{1}{6} + 3\frac{2}{3}$

6. $2\frac{1}{2} + 3\frac{3}{4}$

7. $1\frac{5}{8} + 2\frac{1}{2}$

8. $3\frac{1}{8} + 2\frac{1}{5}$

9. $7\frac{5}{6} + 3\frac{2}{3}$

10. $8\frac{1}{5} + 2\frac{7}{8}$

11. $2\frac{5}{8} + 3$

12. $4 + 2\frac{3}{5}$

13. $1\frac{1}{2} + 2\frac{1}{3} + 3\frac{1}{4}$

14. $2\frac{1}{3} + 1\frac{3}{4} + 3\frac{5}{6}$

15. $5\frac{1}{4} + 3 + 2\frac{3}{8}$

16. $6 + 2\frac{3}{5} + 1\frac{7}{10}$

17. $27\frac{5}{6} + 9\frac{3}{4}$

18. $46\frac{7}{9} + 21\frac{1}{6}$

19. $16\frac{3}{4} + 32\frac{7}{10}$

20. $58\frac{1}{6} + 43\frac{9}{10}$

21. $72\frac{2}{3}$, $81\frac{3}{4}$, $+\ 93\frac{1}{2}$

22. $68\frac{2}{9}$, $97\frac{1}{3}$, $+\ 55\frac{5}{6}$

23. $17\frac{1}{3}$, $28\frac{2}{5}$, $+\ 15\frac{4}{15}$

24. $32\frac{3}{14}$, $78\frac{19}{28}$, $+\ 21\frac{5}{7}$

25. $20\frac{5}{8}$, $46\frac{1}{16}$, $+\ 53\frac{3}{4}$

26. $56\frac{4}{5}$, $81\frac{7}{15}$, $+\ 22\frac{3}{10}$

27. $117\frac{5}{6}$, $28\frac{1}{2}$, $+\ 232\frac{7}{9}$

28. $168\frac{7}{16}$, $312\frac{19}{32}$, $+\ 406\frac{1}{2}$

29. Find the sum of $7\frac{3}{4}$ ounces, $5\frac{7}{8}$ ounces, $8\frac{5}{16}$ ounces, and $10\frac{2}{5}$ ounces.

30. Tony weighs $145\frac{1}{2}$ pounds. Mike weighs $157\frac{3}{4}$ pounds. Pat weighs $7\frac{3}{4}$ pounds more than Tony. Find the weight of the three men.

31. Find the perimeter of the figure.

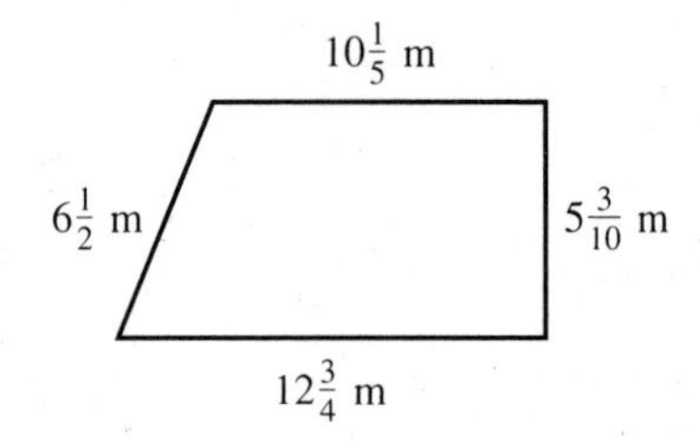

32. Find the perimeter of the triangle.

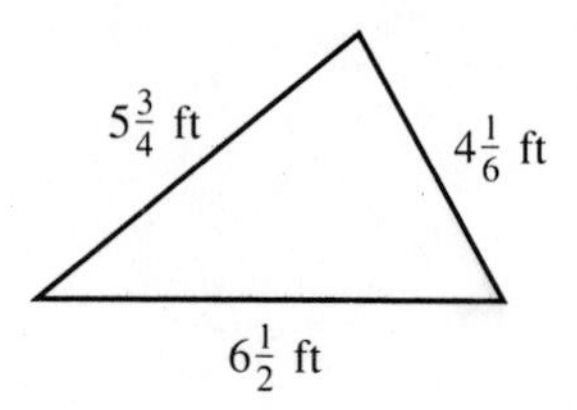

2.12 Subtracting Mixed Numbers

Example 1 Find $15\frac{3}{8} - 3\frac{1}{4}$.

Solution

$$\begin{array}{rr} 15\frac{3}{8} = & 15\frac{3}{8} \\ -\ 3\frac{1}{4} = & -\ 3\frac{2}{8} \\ \hline & 12\frac{1}{8} \end{array} \quad \blacksquare$$

Example 2 Find $12\frac{1}{6} - 4\frac{1}{2}$.

Solution

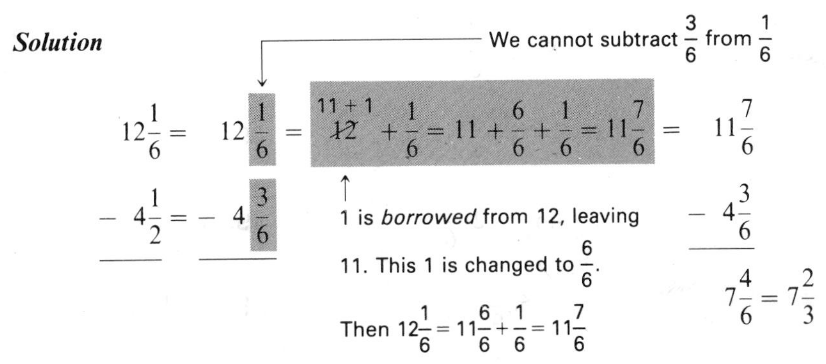

We shorten the work by writing the problem as follows:

$$\begin{array}{rr} & 11\frac{7}{6} \\ 12\frac{1}{6} = & \cancel{12}\frac{\cancel{1}}{6} \\ -\ 4\frac{1}{2} = & -\ 4\frac{3}{6} \\ \hline & 7\frac{4}{6} = 7\frac{2}{3} \end{array} \quad \blacksquare$$

Example 3 Find $58\frac{2}{7} - 25\frac{1}{3}$.

Solution

$$\begin{array}{rrl} & 57\frac{27}{21} & \longleftarrow \text{Borrow } 1 = \frac{21}{21} \\ 58\frac{2}{7} = & \cancel{58}\frac{\cancel{6}}{21} & \text{Then add } \frac{21}{21} + \frac{6}{21} = \frac{27}{21} \\ -\ 25\frac{1}{3} = & -\ 25\frac{7}{21} & \\ \hline & 32\frac{20}{21} & \blacksquare \end{array}$$

Example 4 Find $15 - 5\frac{3}{8}$.

Solution

$$\begin{array}{rl} 14\frac{8}{8} & \longleftarrow \text{Borrow } 1 = \frac{8}{8} \\ \cancel{15} & \\ -5\frac{3}{8} & \\ \hline 9\frac{5}{8} & \blacksquare \end{array}$$

EXERCISES 2.12

In Exercises 1–20, find the differences. Reduce answers to lowest terms. Write any improper fraction as a mixed number.

1. $14\frac{3}{4} - 10\frac{1}{4}$
2. $21\frac{5}{6} - 18\frac{1}{6}$
3. $17\frac{3}{4} - 5\frac{1}{8}$
4. $19\frac{3}{5} - 6\frac{1}{10}$
5. $8 - 4\frac{1}{2}$
6. $7 - 3\frac{1}{3}$
7. $6 - 2\frac{3}{5}$
8. $4 - 1\frac{5}{6}$
9. $4\frac{1}{4} - 1\frac{3}{4}$
10. $5\frac{1}{3} - 1\frac{2}{3}$
11. $12\frac{2}{5} - 7\frac{3}{10}$
12. $54\frac{2}{3} - 39\frac{1}{6}$
13. $3\frac{1}{12} - 1\frac{1}{6}$
14. $23\frac{5}{8} - 17\frac{3}{4}$
15. $45 - 38\frac{2}{3}$
16. $32 - 28\frac{7}{15}$
17. $68\frac{5}{16} - 53\frac{3}{4}$
18. $107\frac{2}{3} - 99\frac{1}{6}$
19. $234\frac{5}{14} - 157\frac{3}{7}$
20. $7{,}005\frac{2}{5} - 2{,}867\frac{2}{3}$

21. When a $1\frac{1}{4}$-pound steak was trimmed of fat, it weighed $\frac{7}{8}$ pound. How much did the fat weigh?

22. Mr. Angelini took $7\frac{1}{2}$ days to paint the interior of his house. A professional painter said he could do it in $2\frac{1}{3}$ days. How much time would be saved having the painter do it?

23. Mr. Segal has $5\frac{3}{4}$ square yards of carpet left after carpeting his living room. How much more will he need to carpet a bathroom that has a floor area of 7 square yards?

24. Jim wants three boards for shelves that measure $5\frac{3}{4}$ feet, $2\frac{5}{12}$ feet, and $3\frac{1}{2}$ feet. Can he cut all three shelves from a 12-foot board, allowing $\frac{1}{4}$ foot for waste?

2.13 Multiplying Mixed Numbers

To multiply mixed numbers, change each to an improper fraction, then multiply.

Example 1 Multiplying mixed numbers.

a. $2\frac{11}{12} \cdot 1\frac{3}{5} = \frac{\overset{7}{\cancel{35}}}{\underset{3}{\cancel{12}}} \cdot \frac{\overset{2}{\cancel{8}}}{\underset{1}{\cancel{5}}} = \frac{14}{3} = 4\frac{2}{3}$

b. $3\frac{1}{8} \cdot 16 = \frac{25}{\underset{1}{\cancel{8}}} \cdot \frac{\overset{2}{\cancel{16}}}{1} = 50$

c. $2\frac{1}{4} \cdot 4\frac{2}{3} \cdot 1\frac{1}{5} = \frac{\overset{3}{\cancel{9}}}{\underset{\underset{1}{\cancel{2}}}{\cancel{4}}} \cdot \frac{\overset{7}{\cancel{14}}}{\underset{1}{\cancel{3}}} \cdot \frac{\overset{3}{\cancel{6}}}{5} = \frac{63}{5} = 12\frac{3}{5}$ ■

Example 2 Find the area of a rectangle with length $2\frac{2}{3}$ yards and width $1\frac{3}{4}$ yards.

Solution Recall from Section 1.13B, to find the area of a rectangle we multiply the length times the width.

$$\text{Area} = \ell \cdot w$$

$$= 2\frac{2}{3} \cdot 1\frac{3}{4} = \frac{\overset{2}{\cancel{8}}}{3} \cdot \frac{7}{\underset{1}{\cancel{4}}} = \frac{14}{3} = 4\frac{2}{3} \text{ sq. yd}$$ ■

EXERCISES 2.13

In Exercises 1–16, find the products. Reduce answers to lowest terms. Write any improper fraction as a mixed number.

1. $1\frac{2}{3} \cdot 2\frac{1}{2}$

2. $1\frac{5}{6} \cdot 3\frac{1}{2}$

3. $2\frac{2}{3} \cdot 2\frac{1}{4}$

4. $2\frac{4}{5} \cdot 2\frac{1}{7}$

5. $8 \cdot 3\frac{3}{4}$

6. $4\frac{2}{3} \cdot 6$

7. $2\frac{5}{8} \cdot 4$

8. $6 \cdot 2\frac{5}{12}$

9. $1\frac{1}{12} \cdot 1\frac{4}{5}$

10. $2\frac{3}{10} \cdot 2\frac{1}{2}$

11. $2\frac{2}{15} \cdot 1\frac{1}{4}$

12. $1\frac{5}{16} \cdot 3\frac{1}{3}$

13. $3\frac{3}{10} \cdot \frac{6}{11} \cdot 1\frac{2}{3}$

14. $1\frac{1}{8} \cdot \frac{4}{9} \cdot 1\frac{5}{6}$

15. $3\frac{3}{4} \cdot 4\frac{2}{5} \cdot 3\frac{1}{2}$

16. $2\frac{1}{6} \cdot 1\frac{5}{7} \cdot 4\frac{2}{3}$

17. Each of eight hikers carries a food pack weighing $2\frac{7}{16}$ pounds. How much food are they carrying in all?

18. A roast is to be cooked for $\frac{1}{3}$ hour for each pound. If the roast weighs 7 pounds, how long will it take to cook?

19. A large truck can carry $10\frac{1}{2}$ tons of lettuce. If the truck is $\frac{4}{5}$ full, how many tons of lettuce is it carrying?

20. Steve's $7\frac{1}{2}$-gallon fish tank is only $\frac{2}{3}$ full. How many gallons does he need to add to fill the tank?

21. Find the area of the rectangle.

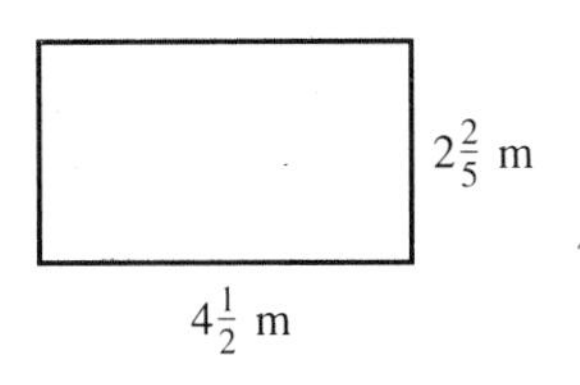

22. Find the area of the square.

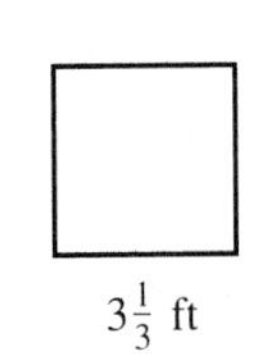

2.14 Division of Fractions

TO DIVIDE FRACTIONS

Invert the second fraction and multiply.

$$\frac{a}{b} \div \frac{c}{d} = \frac{a}{b} \cdot \frac{d}{c}$$

first fraction ($\frac{a}{b}$) — second fraction ($\frac{c}{d}$)

Example 1 Dividing fractions.

a. Find $\frac{3}{5} \div \frac{3}{7}$.

Solution $\frac{3}{5} \div \frac{3}{7} = \frac{3}{5} \cdot \frac{7}{3} = \frac{\cancel{3}^{1}}{5} \cdot \frac{7}{\cancel{3}_{1}} = \frac{7}{5} = 1\frac{2}{5}$

b. $\frac{5}{6} \div \frac{2}{3} = \frac{5}{6} \cdot \frac{3}{2} = \frac{5}{\cancel{6}_{2}} \cdot \frac{\cancel{3}^{1}}{2} = \frac{5}{4} = 1\frac{1}{4}$

c. $\frac{15}{32} \div 2 = \frac{15}{32} \cdot \frac{1}{2} = \frac{15}{64}$

d. $7 \div \frac{3}{8} = \frac{7}{1} \cdot \frac{8}{3} = \frac{56}{3} = 18\frac{2}{3}$ ■

To divide mixed numbers, change each mixed number to an improper fraction, then invert the second fraction and multiply.

Example 2 Dividing mixed numbers.

a. $6\frac{4}{5} \div 1\frac{7}{10} = \frac{34}{5} \div \frac{17}{10} = \frac{\cancel{34}^{2}}{\cancel{5}_{1}} \cdot \frac{\cancel{10}^{2}}{\cancel{17}_{1}} = 4$

b. $12 \div 2\frac{2}{3} = \frac{12}{1} \div \frac{8}{3} = \frac{\cancel{12}^{3}}{1} \cdot \frac{3}{\cancel{8}_{2}} = \frac{9}{2} = 4\frac{1}{2}$ ■

This method works because:

$$\frac{3}{5} \div \frac{4}{7} = \frac{\frac{3}{5}}{\frac{4}{7}} = \frac{\frac{3}{5}}{\frac{4}{7}} \cdot \frac{\frac{7}{4}}{\frac{7}{4}} = \frac{\frac{3}{5} \cdot \frac{7}{4}}{\frac{4}{7} \cdot \frac{7}{4}} = \frac{\frac{3}{5} \cdot \frac{7}{4}}{1} = \frac{3}{5} \cdot \frac{7}{4}$$

The value of this fraction is 1 ($\frac{\frac{7}{4}}{\frac{7}{4}}$)

Therefore, $\frac{3}{5} \div \frac{4}{7} = \frac{3}{5} \cdot \frac{7}{4}$

EXERCISES 2.14

In Exercises 1–35, find the quotients. Reduce answers to lowest terms. Write any improper fraction as a mixed number.

1. $\frac{3}{4} \div \frac{1}{2}$ **2.** $\frac{2}{5} \div \frac{1}{3}$ **3.** $\frac{1}{2} \div \frac{2}{3}$ **4.** $4 \div \frac{2}{5}$ **5.** $\frac{1}{2} \div 5$

6. $\frac{4}{3} \div \frac{8}{6}$ **7.** $\frac{5}{2} \div \frac{5}{8}$ **8.** $\frac{7}{8} \div 7$ **9.** $1 \div \frac{1}{2}$ **10.** $\frac{100}{150} \div \frac{4}{9}$

11. $\frac{3}{5} \div \frac{3}{10}$ **12.** $\frac{7}{12} \div \frac{1}{3}$ **13.** $\frac{3}{16} \div \frac{9}{20}$ **14.** $\frac{13}{28} \div 39$ **15.** $34 \div \frac{17}{56}$

16. $\frac{8}{15} \div \frac{24}{10}$ **17.** $\frac{35}{16} \div \frac{42}{22}$ **18.** $\frac{7}{18} \div \frac{21}{15}$ **19.** $36 \div \frac{4}{5}$ **20.** $\frac{4}{9} \div 36$

21. $\frac{6}{35} \div \frac{8}{15}$ **22.** $\frac{11}{84} \div \frac{22}{60}$ **23.** $\frac{14}{24} \div 210$ **24.** $22 \div \frac{11}{5}$ **25.** $\frac{56}{15} \div \frac{28}{90}$

26. $1\frac{3}{7} \div 1\frac{1}{4}$ **27.** $1\frac{7}{9} \div 2\frac{2}{3}$ **28.** $2\frac{3}{5} \div 1\frac{4}{35}$ **29.** $3\frac{2}{3} \div 1\frac{7}{15}$ **30.** $7 \div 4\frac{2}{3}$

31. $3\frac{4}{5} \div 19$ **32.** $3\frac{1}{3} \div 5$ **33.** $11 \div 3\frac{1}{7}$ **34.** $10\frac{1}{2} \div 5\frac{5}{6}$ **35.** $6\frac{2}{3} \div 3\frac{5}{9}$

36. A board $5\frac{1}{3}$ feet long is cut into 8 pieces. How long is each piece?

37. How many tablets, each containing 3 milligrams of a heart medicine, must be used to make up a $4\frac{1}{2}$-milligram dosage?

38. How many pins $2\frac{1}{2}$ inches long can be cut from a rod 30 inches long?

39. A bookshelf is 36 inches long. How many books that are $1\frac{1}{2}$ inches thick can stand on that shelf?

2.15 Complex Fractions

Simple Fraction

A **simple fraction** has only one fraction line.

Example 1 Examples of simple fractions.

$$\frac{2}{3}, \frac{3+5}{12}, \frac{7-4}{6}, \frac{18}{5}, \frac{2}{7-3}, \frac{13+6}{9-2} \quad \blacksquare$$

Complex Fraction

A **complex fraction** is a fraction having more than one fraction line.

Example 2 Examples of complex fractions.

$$\frac{\frac{2}{5}}{\frac{3}{2}}, \frac{8}{\frac{1}{9}}, \frac{\frac{4}{7}}{\frac{5}{6}}, \frac{\frac{3}{2}-\frac{2}{5}}{\frac{5}{6}+\frac{3}{4}} \quad \blacksquare$$

> **TO SIMPLIFY COMPLEX FRACTIONS**
>
> **1.** Simplify the numerator and the denominator when possible.
>
> **2.** Divide the simplified numerator by the simplified denominator.

Example 3 Simplifying complex fractions.

a. $\dfrac{\frac{5}{12}}{\frac{8}{15}} = \dfrac{5}{12} \div \dfrac{8}{15} = \dfrac{5}{\underset{4}{\cancel{12}}} \cdot \dfrac{\overset{5}{\cancel{15}}}{8} = \dfrac{25}{32}$

b. $\dfrac{6}{\frac{3}{5}} = 6 \div \dfrac{3}{5} = \dfrac{\overset{2}{\cancel{6}}}{1} \cdot \dfrac{5}{\underset{1}{\cancel{3}}} = \dfrac{10}{1} = 10$

c. $\dfrac{\frac{4}{5}}{12} = \dfrac{4}{5} \div \dfrac{12}{1} = \dfrac{\overset{1}{\cancel{4}}}{5} \cdot \dfrac{1}{\underset{3}{\cancel{12}}} = \dfrac{1}{15}$

d. $\dfrac{2\frac{3}{5}}{1\frac{5}{8}} = \dfrac{\frac{13}{5}}{\frac{13}{8}} = \dfrac{13}{5} \div \dfrac{13}{8} = \dfrac{\overset{1}{\cancel{13}}}{5} \cdot \dfrac{8}{\underset{1}{\cancel{13}}} = \dfrac{8}{5} = 1\dfrac{3}{5}$

e. $\dfrac{\frac{1}{6}+\frac{2}{3}}{\frac{5}{8}-\frac{1}{4}} = \dfrac{\frac{1}{6}+\frac{4}{6}}{\frac{5}{8}-\frac{2}{8}} = \dfrac{\frac{5}{6}}{\frac{3}{8}} = \dfrac{5}{6} \div \dfrac{3}{8} = \dfrac{5}{\underset{3}{\cancel{6}}} \cdot \dfrac{\overset{4}{\cancel{8}}}{3} = \dfrac{20}{9} = 2\dfrac{2}{9}$ ■

NOTE It is necessary to know which fraction line is the *main* fraction line of a complex fraction. The main fraction line will be indicated by making it heavier and longer than the other fraction lines. ☑

Example 4 Example 4 shows that

$$\dfrac{2}{\frac{3}{4}} \neq \dfrac{\frac{2}{3}}{4}$$

Solution If the upper fraction line is the main fraction line, then

$$\dfrac{2}{\frac{3}{4}} = \dfrac{2}{1} \div \dfrac{3}{4} = \dfrac{2}{1} \cdot \dfrac{4}{3} = \boxed{\dfrac{8}{3}}$$

If the lower fraction line is the main fraction line, then

$$\frac{\frac{2}{3}}{4} = \frac{2}{3} \div \frac{4}{1} = \frac{\overset{1}{\cancel{2}}}{3} \cdot \frac{1}{\underset{2}{\cancel{4}}} = \boxed{\frac{1}{6}}$$

which is different from the first value, $\frac{8}{3}$. ■

EXERCISES 2.15

Simplify the complex fractions.

1. $\dfrac{\frac{3}{4}}{\frac{1}{6}}$ **2.** $\dfrac{\frac{15}{16}}{\frac{12}{5}}$ **3.** $\dfrac{\frac{2}{3}}{\frac{1}{2}}$ **4.** $\dfrac{\frac{3}{4}}{\frac{7}{8}}$ **5.** $\dfrac{\frac{3}{5}}{\frac{3}{10}}$

6. $\dfrac{\frac{7}{16}}{\frac{7}{24}}$ **7.** $\dfrac{\frac{3}{8}}{\frac{5}{12}}$ **8.** $\dfrac{\frac{5}{7}}{\frac{10}{21}}$ **9.** $\dfrac{6}{\frac{2}{3}}$ **10.** $\dfrac{\frac{15}{6}}{9}$

11. $\dfrac{14}{\frac{8}{5}}$ **12.** $\dfrac{\frac{3}{4}}{8}$ **13.** $\dfrac{1\frac{2}{3}}{10}$ **14.** $\dfrac{8\frac{1}{3}}{100}$ **15.** $\dfrac{6\frac{2}{3}}{100}$

16. $\dfrac{2\frac{1}{2}}{10}$ **17.** $\dfrac{1\frac{1}{2}}{3\frac{3}{4}}$ **18.** $\dfrac{3\frac{5}{9}}{2\frac{2}{3}}$ **19.** $\dfrac{\frac{1}{4}+\frac{2}{5}}{\frac{1}{6}}$ **20.** $\dfrac{\frac{3}{8}}{\frac{9}{10}-\frac{2}{5}}$

21. $\dfrac{\frac{1}{8}+\frac{3}{4}}{\frac{1}{2}-\frac{1}{3}}$ **22.** $\dfrac{\frac{11}{4}-\frac{5}{9}}{\frac{7}{18}+\frac{13}{36}}$ **23.** $\dfrac{\frac{1}{7}+\frac{9}{28}}{\frac{13}{14}-\frac{3}{7}}$ **24.** $\dfrac{\frac{16}{5}-\frac{7}{15}}{\frac{9}{30}+\frac{3}{10}}$ **25.** $\dfrac{\frac{13}{18}-\frac{11}{24}}{\frac{5}{12}-\frac{7}{36}}$

26. The monthly rainfall for a city was $2\frac{1}{2}$ inches in January, $6\frac{3}{4}$ inches in February, 10 inches in March, $7\frac{1}{2}$ inches in April, 3 inches in May, and $1\frac{3}{4}$ inches in June. What was the average monthly rainfall for the first six months?

27. ABC's stock prices for the week were $11\frac{1}{4}$, $12\frac{1}{2}$, $11\frac{5}{8}$, $9\frac{7}{8}$, and 11. Find the average price of the stock for that week.

2.16 Combined Operations

In evaluating expressions with more than one operation, the same order of operations is used with fractions that was used with whole numbers.

ORDER OF OPERATIONS

1. Any expressions in parentheses are evaluated first.

2. Evaluations are done in this order:
First: Powers and roots
Second: Multiplication and division in order from *left to right*
Third: Addition and subtraction in order from *left to right*

Example 1

$$2 - \frac{3}{4} + 3\frac{1}{2} =$$

$$\frac{2}{1} - \frac{3}{4} + \frac{7}{2} =$$

$$\underbrace{\frac{8}{4} - \frac{3}{4}} + \frac{14}{4} =$$

$$\frac{5}{4} + \frac{14}{4} =$$

$$\frac{19}{4} = 4\frac{3}{4} \quad \blacksquare$$

Example 2

$$2\frac{4}{5} \div 7 \cdot 1\frac{2}{3} =$$

$$\underbrace{\frac{14}{5} \div \frac{7}{1}} \cdot \frac{5}{3} =$$

$$\frac{\overset{2}{\cancel{14}}}{\underset{1}{\cancel{5}}} \cdot \frac{1}{\underset{1}{\cancel{7}}} \cdot \frac{\overset{1}{\cancel{5}}}{3} = \frac{2}{3} \quad \blacksquare$$

Example 3

$$2\frac{1}{3} + \frac{5}{6} \div 1\frac{3}{4} =$$

$$\frac{7}{3} + \frac{5}{6} \div \frac{7}{4} =$$

$$\frac{7}{3} + \underbrace{\frac{5}{\underset{3}{\cancel{6}}} \cdot \frac{\overset{2}{\cancel{4}}}{7}} =$$

$$\frac{7}{3} + \frac{10}{21} =$$

$$\frac{49}{21} + \frac{10}{21} =$$

$$\frac{59}{21} = 2\frac{17}{21} \quad \blacksquare$$

Example 4

$$8 - \frac{2}{3} \cdot 2\frac{1}{2} =$$

$$\frac{8}{1} - \frac{\cancel{2}^{1}}{3} \cdot \frac{5}{\cancel{2}_{1}} =$$

$$\frac{8}{1} - \frac{5}{3} =$$

$$\frac{24}{3} - \frac{5}{3} =$$

$$\frac{19}{3} = 6\frac{1}{3} \quad \blacksquare$$

Example 5

$$\left(\frac{3}{4}\right)^2 + 2\frac{4}{5} \cdot 1\frac{1}{4} =$$

$$\frac{3}{4} \cdot \frac{3}{4} + \frac{\cancel{14}^{7}}{\cancel{5}_{1}} \cdot \frac{\cancel{5}^{1}}{\cancel{4}_{2}} =$$

$$\frac{9}{16} + \frac{7}{2} =$$

$$\frac{9}{16} + \frac{56}{16} =$$

$$\frac{65}{16} = 4\frac{1}{16} \quad \blacksquare$$

Example 6

$$\left(1\frac{3}{8} - \frac{1}{2}\right) \div 1\frac{5}{16} =$$

$$\left(\frac{11}{8} - \frac{1}{2}\right) \div \frac{21}{16} =$$

$$\left(\frac{11}{8} - \frac{4}{8}\right) \div \frac{21}{16} =$$

$$\frac{7}{8} \div \frac{21}{16} =$$

$$\frac{\cancel{7}^{1}}{\cancel{8}_{1}} \cdot \frac{\cancel{16}^{2}}{\cancel{21}_{3}} = \frac{2}{3} \quad \blacksquare$$

Example 7 Find the perimeter of a rectangle with length $3\frac{1}{4}$ feet and width $2\frac{1}{2}$ feet.

Solution

$$\text{Perimeter} = 2\ell + 2w$$
$$= 2 \cdot 3\frac{1}{4} + 2 \cdot 2\frac{1}{2}$$
$$= \frac{\overset{1}{\cancel{2}}}{1} \cdot \frac{13}{\underset{2}{\cancel{4}}} + \frac{\overset{1}{\cancel{2}}}{1} \cdot \frac{5}{\underset{1}{\cancel{2}}}$$
$$= \frac{13}{2} + \frac{5}{1}$$
$$= \frac{13}{2} + \frac{10}{2} = \frac{23}{2}$$
$$= 11\frac{1}{2} \text{ ft}$$

$2\frac{1}{2}$ ft

$3\frac{1}{4}$ ft

■

EXERCISES 2.16

In Exercises 1–18, evaluate each expression using the correct order of operations.

1. $3\frac{3}{4} \div \frac{5}{8} \cdot \frac{5}{9}$
2. $\frac{4}{5} \cdot 1\frac{2}{3} \div 1\frac{7}{9}$
3. $4 - \frac{2}{3} + 1\frac{1}{2}$
4. $2\frac{1}{8} + \frac{3}{4} - 1\frac{1}{2}$
5. $\frac{1}{2} \div \frac{1}{3} \div \frac{5}{6}$
6. $4\frac{1}{2} \div \frac{1}{4} \div \frac{3}{5}$
7. $\frac{7}{8} + \frac{3}{4} \cdot \frac{5}{6}$
8. $\frac{3}{4} + \frac{2}{3} \div \frac{4}{9}$
9. $\frac{3}{5} - \frac{1}{6} \div 1\frac{1}{4}$
10. $2\frac{1}{2} - 1\frac{1}{5} \cdot \frac{3}{4}$
11. $\left(\frac{2}{3}\right)^2 + 1\frac{2}{3} \cdot \frac{1}{10}$
12. $\left(\frac{1}{5}\right)^2 + \frac{4}{5} \div 1\frac{1}{3}$
13. $\left(8\frac{1}{4} - 1\frac{2}{3}\right) \cdot 3$
14. $\left(\frac{2}{3} + \frac{4}{5}\right) \div 2\frac{1}{5}$
15. $2\frac{2}{5} \div \left(3 - \frac{3}{10}\right)$
16. $1\frac{1}{5} \cdot \left(\frac{5}{6} + \frac{1}{4}\right)$
17. $7\frac{1}{2} \div \frac{3}{5} + 1\frac{7}{8} \cdot 2\frac{2}{5}$
18. $1\frac{1}{4} \cdot \frac{2}{3} - 1\frac{1}{6} \div 1\frac{1}{2}$

19. A room measures $5\frac{1}{3}$ yards long and $3\frac{1}{4}$ yards wide.
    a. Find the perimeter of the room.
    b. What would it cost to put ceiling molding around the room if the molding costs $12 per yard?

20. What is the total length of weather stripping needed to go around all sides of three windows if one window measures $5\frac{1}{2}$ feet by $2\frac{1}{4}$ feet and the other two square windows measure $2\frac{1}{4}$ feet on each side?

2.17 Comparing Fractions

Sometimes we need to recognize which of a group of fractions is largest. We can compare the size of fractions by converting all the given fractions to equivalent fractions having the same denominator. It is convenient (but not necessary) to use the LCD for the denominator of all the equivalent fractions.

Example 1 Arrange the following fractions in order of size—the largest first: $\frac{3}{8}, \frac{1}{3}, \frac{5}{12}$.

Solution

Finding the LCD:

$$8 = 2 \cdot 2 \cdot 2 = 2^3$$
$$3 = 3$$
$$12 = 2 \cdot 2 \cdot 3 = 2^2 \cdot 3$$
$$\text{LCD} = 2^3 \cdot 3 = 8 \cdot 3 = 24$$

Finding the equivalent fractions:

$$\frac{3}{8} = \frac{9}{24}$$
$$\frac{1}{3} = \frac{8}{24}$$
$$\frac{5}{12} = \frac{10}{24}$$

Arranging the *equivalent* fractions in order of size, we have

$$\frac{10}{24} > \frac{9}{24} > \frac{8}{24}$$

Therefore, the *original* fractions arranged in order of size are

$$\frac{5}{12} > \frac{3}{8} > \frac{1}{3} \quad ■$$

EXERCISES 2.17

In Exercises 1–8, arrange the fractions in order of size—the largest first.

1. $\frac{3}{4}, \frac{5}{6}, \frac{2}{3}$ **2.** $\frac{7}{9}, \frac{2}{3}, \frac{5}{6}$ **3.** $\frac{5}{8}, \frac{3}{4}, \frac{11}{16}$ **4.** $\frac{4}{9}, \frac{5}{12}, \frac{1}{3}$

5. $\frac{9}{14}, \frac{5}{7}, \frac{3}{4}$ **6.** $\frac{7}{10}, \frac{3}{4}, \frac{4}{5}$ **7.** $\frac{2}{15}, \frac{3}{10}, \frac{1}{6}$ **8.** $\frac{19}{32}, \frac{5}{8}, \frac{9}{16}$

In Exercises 9–14, determine which of the two symbols > or < should be used to make each statement true.

9. $5\frac{1}{2} ? 5\frac{13}{24}$ **10.** $3\frac{7}{18} ? 3\frac{1}{2}$ **11.** $10\frac{5}{8} ? 10\frac{7}{12}$ **12.** $25\frac{6}{25} ? 25\frac{4}{15}$

13. $\frac{1}{20}$ of 15 ? $\frac{5}{16}$ of 2 **14.** $\frac{1}{4}$ of 10 ? $\frac{2}{3}$ of 4

15. A stock price went from $12\frac{5}{8}$ to $12\frac{3}{4}$. Did it go up or down? By how much?

16. A stock price went from $16\frac{3}{8}$ to $16\frac{1}{4}$. Did it go up or down? By how much?

17. A salesperson measured the length of a drape to be $45\frac{3}{8}$ inches. Is the length closer to 45 or 46 inches?

18. A machinist measured the length of a steel rod to be $9\frac{17}{32}$ inches. Is the length closer to 9 or 10 inches?

2.18 Chapter Summary

Definitions 2.1 A **fraction** is equivalent to a division.

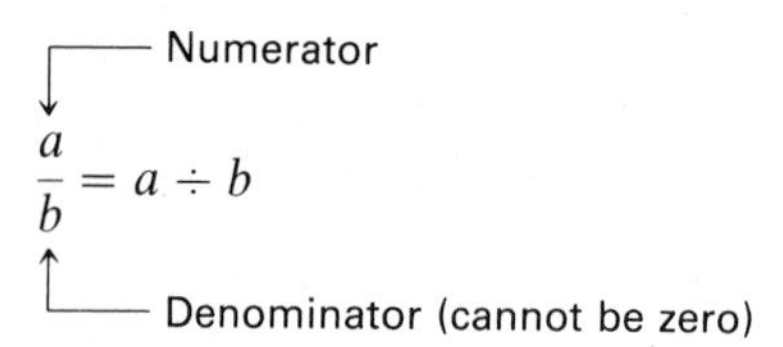

2.1 A **proper fraction** is a fraction whose numerator is less than its denominator.

2.1 An **improper fraction** is a fraction whose numerator is larger than or equal to its denominator. Any improper fraction has a value greater than or equal to 1.

2.3 A **mixed number** is the sum of a whole number and a proper fraction.

2.6 A **prime number** is a natural number greater than 1 that can be divided evenly only by itself and 1. A prime number has no factors other than itself and 1.

2.6 A **composite number** is a natural number that can be divided evenly by some natural number other than itself and 1. A composite number has factors other than itself and 1.

2.15 A **simple fraction** is a fraction having only one fraction line.

2.15 A **complex fraction** is a fraction having more than one fraction line.

Changing Improper Fractions to Mixed Numbers 2.3 To change an improper fraction to a mixed number divide the numerator by the denominator. Write the quotient, followed by the fraction: remainder over divisor.

Changing Mixed Numbers to Improper Fractions 2.4 To change a mixed number to an improper fraction:

1. Multiply the whole number by the denominator of the fraction.
2. Add the numerator of the fraction to the product found in (1).
3. Write the sum found in (2) over the denominator of the fraction.

Equivalent Fractions 2.5 To form an equivalent (equal) fraction either

1. Multiply both numerator and denominator by the same nonzero number, *or*
2. Divide the numerator and denominator by the same nonzero number that is a divisor of both.

Reducing Fractions 2.7A To reduce a fraction to lowest terms:

1. Divide numerator and denominator by any whole number greater than 1 that you can see is a divisor of both.
2. When you cannot see any more common divisors, write the prime factorization of the numerator and the denominator.
3. Divide numerator and denominator by all factors common to both.

Finding a Fractional Part of a Number 2.7B

To find a fractional part of a number, multiply the fraction times the number.

Finding the LCD 2.9

To find the LCD by prime factorization:

1. Write the prime factorization of each denominator. Repeated factors should be expressed as powers.
2. List each different factor that appears.
3. Raise each factor to the highest power to which it occurs in any denominator.
4. The LCD is the product of all the powers found in Step 3.

Adding and Subtracting Fractions 2.10

To add or subtract fractions:

1. Find the LCD.
2. Convert each fraction to an equivalent fraction having the LCD as denominator.
3. Add or subtract the numerators and write the result over the common denominator.
4. Reduce the fraction to lowest terms.

$$\frac{a}{c} + \frac{b}{c} = \frac{a+b}{c} \qquad \frac{a}{c} - \frac{b}{c} = \frac{a-b}{c}$$

Multiplying Fractions 2.2 and 2.13

To multiply fractions, write the product of the numerators over the product of the denominators.

$$\frac{a}{b} \cdot \frac{c}{d} = \frac{a \cdot c}{b \cdot d}$$

Dividing Fractions 2.14

To divide fractions, invert the second fraction and multiply.

$$\frac{a}{b} \div \frac{c}{d} = \frac{a}{b} \cdot \frac{d}{c}$$

Simplifying Complex Fractions 2.15

To simplify a complex fraction, simplify its numerator and denominator, then divide the simplified numerator by the simplified denominator.

Order of Operations 2.16

Order of Operations

1. Any expressions in parentheses are evaluated first.
2. Evaluations are done in this order:
   First: Powers and roots
   Second: Multiplication and division in order from left to right
   Third: Addition and subtraction in order from left to right

Comparing Fractions 2.17

To compare fractions, convert the given fractions to equivalent fractions having the same denominators.

Review Exercises 2.18

In Exercises 1–4, change the improper fractions to mixed numbers.

1. $\frac{7}{3}$ **2.** $\frac{11}{8}$ **3.** $\frac{28}{5}$ **4.** $\frac{29}{12}$

In Exercises 5–8, change the mixed numbers to improper fractions.

5. $2\frac{3}{5}$ **6.** $3\frac{7}{8}$ **7.** $9\frac{2}{11}$ **8.** $13\frac{5}{6}$

In Exercises 9–12, reduce to lowest terms.

9. $\frac{42}{54}$ **10.** $\frac{45}{105}$ **11.** $\frac{28}{57}$ **12.** $\frac{200}{250}$

In Exercises 13–30, perform the indicated operations. Reduce answers to lowest terms. Write any improper fraction as a mixed number.

13. $\frac{1}{2}+\frac{5}{6}+\frac{4}{9}$ **14.** $2\frac{2}{3}+1\frac{3}{5}$ **15.** $4\frac{5}{16}+\frac{5}{8}$

16. $153\frac{2}{5}+135\frac{3}{4}$ **17.** $5\frac{4}{5}-3\frac{7}{10}$ **18.** $5\frac{1}{3}-2\frac{3}{4}$

19. $16-5\frac{7}{8}$ **20.** $351\frac{2}{3}-272\frac{7}{9}$ **21.** $\frac{4}{5}\cdot\frac{7}{8}\cdot 15$

22. $3\frac{2}{3}\cdot 6\frac{3}{5}$ **23.** $1\frac{3}{5}\cdot 3\frac{3}{4}$ **24.** $9\cdot 1\frac{5}{12}$

25. $2\frac{2}{5}\div 1\frac{1}{15}$ **26.** $1\frac{1}{6}\div 4\frac{2}{3}$ **27.** $16\div\frac{8}{13}$

28. $1\frac{9}{16}\div 10$ **29.** $\frac{7}{12}+2\frac{2}{3}\cdot\frac{1}{4}$ **30.** $\frac{2}{3}\div 1\frac{1}{4}\cdot\left(\frac{3}{4}\right)^2$

In Exercises 31–34, simplify the complex fractions.

31. $\dfrac{\frac{5}{8}}{\frac{5}{6}}$ **32.** $\dfrac{12}{\frac{6}{11}}$ **33.** $\dfrac{1\frac{2}{3}}{\frac{3}{4}+\frac{1}{2}}$ **34.** $\dfrac{\frac{2}{3}+\frac{1}{2}}{\frac{8}{9}-\frac{1}{3}}$

In Exercises 35 and 36, arrange the fractions in order of size, largest to smallest.

35. $\frac{2}{5}, \frac{1}{4}, \frac{3}{10}$ **36.** $\frac{5}{8}, \frac{2}{3}, \frac{5}{6}$

37. If you sleep 8 hours out of every 24, what fraction of the time do you sleep?

38. A bookshelf is 42 inches long. How many books that are $\frac{3}{4}$ inch thick can stand on that shelf?

39. In driving across the country, a man drove $4\frac{1}{2}$ hours on Monday, $12\frac{3}{4}$ hours on Tuesday, $8\frac{1}{3}$ hours on Wednesday, and $15\frac{1}{6}$ hours on Thursday. What was his total driving time for the trip?

40. A family used $10\frac{1}{2}$ square yards of carpet for the living room, and $8\frac{3}{4}$ square yards for the hall and bedroom combined.
a. What was the total amount of carpet used?
b. If carpet cost $8 per square yard, how much did it cost them to carpet the hall and bedroom?

41. A satellite travels around the earth in $2\frac{1}{3}$ hours. How many trips does it make in 1 week?

42. A board that is 73 inches long is cut into three pieces of equal length. If $\frac{1}{8}$ inch is wasted each time the board is sawed, how long is each of the three finished pieces?

43. A man has a 16-foot board he is going to use for shelves. If the shelves are to be $3\frac{1}{2}$ feet long:
a. How many shelves can be cut from the board? $\left(\frac{1}{8}\right.$ inch is wasted each time the board is sawed.$\left.\right)$
b. What length will be left from the original board after he cuts as many shelves from it as he can?

44. A seamstress is making the costumes for the high school drill team, which has 26 members. If $1\frac{1}{4}$ yards of material are needed for each skirt, and $\frac{5}{8}$ yard of material is needed for each vest, how many yards of material should she order for the entire team?

45. A freight car can carry $10\frac{2}{3}$ tons of wheat. If the car is $\frac{3}{4}$ full, how many tons of wheat is it carrying?

46. A man weighing 200 pounds went on a diet and lost $3\frac{1}{2}$ pounds the first week, $2\frac{3}{4}$ pounds the second week, and $1\frac{5}{8}$ pounds the third week. How much did he weigh at the end of three weeks?

47. Princess had five puppies that weighed $2\frac{1}{4}$ pounds, $2\frac{1}{2}$ pounds, $1\frac{7}{8}$ pounds, $1\frac{3}{4}$ pounds, and $2\frac{1}{4}$ pounds. Find the average weight of her puppies.

48. Jack spends $\frac{1}{3}$ of his income on rent, $\frac{1}{4}$ of his income on food, and $\frac{1}{3}$ of his income on car expenses. If he earns $18,000 a year, how much does he have left over after paying for rent, food, and his car?

49. The length of the side of a square is $1\frac{1}{4}$ feet.

a. Find the perimeter of the square.
b. Find the area of the square.

50. A rectangle is $4\frac{1}{2}$ meters long and $2\frac{3}{4}$ meters wide.

a. Find the perimeter of the rectangle.
b. Find the area of the rectangle.

Chapter 2 Diagnostic Test

Allow yourself about 50 minutes to do these problems. Complete solutions for every problem, together with section references, are given in the answer section at the end of the book.

1. Change $\frac{69}{8}$ to a mixed number.

2. Change $5\frac{7}{12}$ to an improper fraction.

3. Reduce $\frac{180}{540}$ to lowest terms.

4. Find $\frac{7}{8}$ of 120.

In Problems 5–20, perform the indicated operations. Reduce all answers to lowest terms. Write any improper fractions as mixed numbers.

5. $\frac{5}{6}+\frac{7}{8}$

6. $4\frac{2}{3}+3\frac{1}{2}$

7. $\frac{5}{6}\cdot\frac{3}{20}$

8. $1\frac{1}{3}\cdot 42$

9. $\frac{3}{8}\div\frac{9}{16}$

10. $\frac{7}{10}-\frac{1}{6}$

11. $8-3\frac{4}{9}$

12. $2\frac{2}{9}\div 3\frac{1}{3}$

13. $4\frac{1}{5}\cdot 2\frac{1}{7}$

14. $5\frac{1}{4}-2\frac{5}{6}$

15. $124\frac{2}{3}-17\frac{4}{5}$

16. $\frac{5}{12}+\frac{3}{8}+\frac{5}{6}$

17. $\dfrac{\frac{5}{8}}{\frac{15}{16}}$

18. $\dfrac{\frac{3}{8}+\frac{3}{4}}{7\frac{1}{2}}$

19. $\frac{4}{5}\div 2\frac{2}{3}\cdot\frac{5}{6}$

20. $\frac{3}{4}+\frac{1}{4}\cdot 1\frac{3}{5}$

21. When a $1\frac{1}{8}$-pound steak was trimmed of fat, it weighed $\frac{3}{4}$ pound. How much of the steak was fat?

22. How many tablets, each containing 3 milligrams of medicine, must be used to make up a $7\frac{1}{2}$-milligram dosage?

23. Find the area of a rectangle with length $5\frac{1}{4}$ feet and width $3\frac{1}{3}$ feet.

24. Kevin received three shipments that weighed $23\frac{1}{2}$ pounds, $16\frac{1}{4}$ pounds, and $37\frac{7}{8}$ pounds. What was the total weight of the shipments?

25. Arrange in order, largest to smallest: $\frac{8}{15}, \frac{7}{10}, \frac{3}{5}$.

3 Decimal Fractions

3.1 Reading and Writing Decimal Numbers

Decimal Fraction

A **decimal fraction** is a fraction whose denominator is a power of 10.

Example 1 Decimal fractions.

a. $\frac{3}{10^1} = \frac{3}{10}$ is read "three tenths"

b. $\frac{3}{10^2} = \frac{3}{100}$ is read "three hundredths"

c. $\frac{25}{10^3} = \frac{25}{1{,}000}$ is read "twenty-five thousandths"

d. $\frac{5}{10^0} = \frac{5}{1} = 5$

e. $\frac{0}{10^0} = \frac{0}{1} = 0$ ■

Examples d and e show that whole numbers are also decimal fractions.

A decimal fraction can be written in two ways: in fraction form or in decimal form using a decimal point.

Example 2

	Fraction form		Decimal form	
a.	$\frac{4}{10}$	=	0.4	is read "four tenths"
b.	$\frac{5}{100}$	=	0.05	is read "five hundredths"
c.	$\frac{27}{1000}$	=	0.027	is read "twenty-seven thousandths" ■

Whole numbers are decimal fractions. When the decimal point is not written in a whole number, it is understood to be to the right of the units digit.

Example 3

	Whole number		Decimal form
a.	4	=	4.
b.	130	=	130. ■

Although you should never lose sight of the fact that decimal fractions are *fractions*, it is common practice to shorten the term *decimal fraction* to *decimal*. In most cases, we will use the term *fraction* for fractions having denominators other than powers of 10.

In Figure 3.1.1, we show place values of decimals. Note that the decimal point is written between the units place and the tenths place.

NOTE In Figure 3.1.1, you can see that the value of each place is one-tenth the value of the first place to its left. In addition, you can see that the place names to the right of the decimal point all end in "ths." The place names to the left of the decimal point do

Place	Value
millions	$10^6 = 1,000,000$
hundred-thousands	$10^5 = 100,000$
ten-thousands	$10^4 = 10,000$
thousands	$10^3 = 1,000$
hundreds	$10^2 = 100$
tens	$10^1 = 10$
units (ones)	$10^0 = 1$
decimal point →	.
tenths	0.1 = 1/10
hundredths	0.01 = 1/100
thousandths	0.001 = 1/1,000
ten-thousandths	0.0001 = 1/10,000
hundred-thousandths	0.00001 = 1/100,000
millionths	0.000001 = 1/1,000,000

Place Values of Decimals

FIGURE 3.1.1

not end in "ths." For example, the second place to the left of the units place is the "hundreds" place. The second place to the right of the units place is the "hundred**ths**" place. ☑

> TO READ A DECIMAL
>
> 1. Read the number to the left of the decimal point as you read a whole number.
> 2. Say "and" for the decimal point.
> 3. Read the number to the right of the decimal point as a whole number, then say the name of the place occupied by the right-hand digit of the number.

Reading and Writing Decimals Less Than One

Example 4

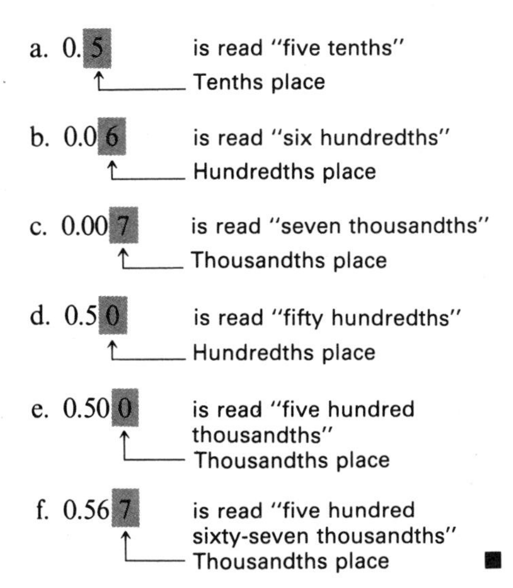

In writing decimals less than 1, we usually write a 0 to the left of the decimal point to call attention to the decimal point so that it is not overlooked. For example, seventy-five hundredths is written 0.75. However, both 0.75 and .75 are correct ways of writing seventy-five hundredths.

Reading and Writing Decimals Greater Than One

Example 5

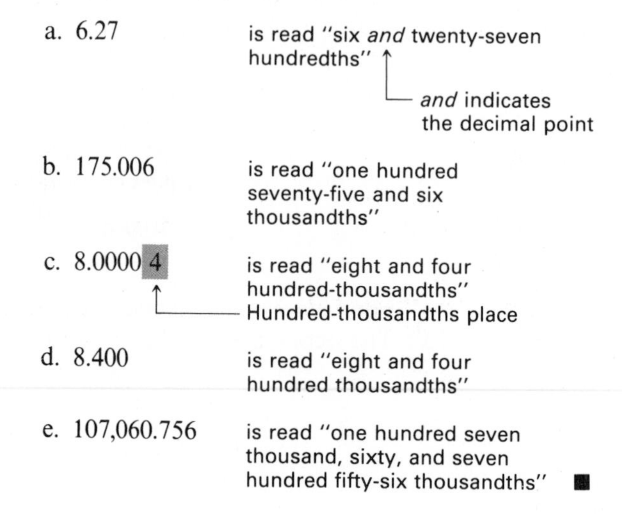

Writing Numbers in Decimal Notation

Example 6 Write the following numbers in decimal notation.

a. Twenty and nine tenths

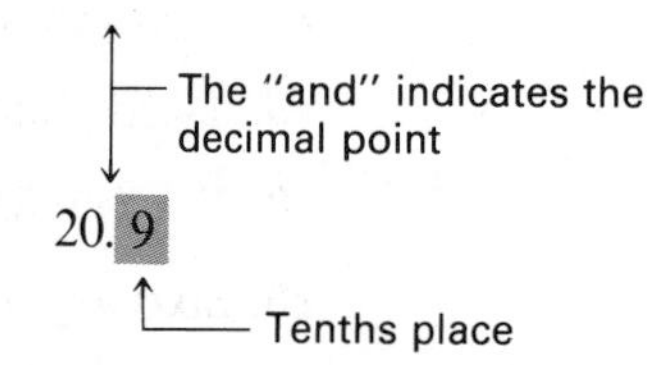

b. Five hundredths

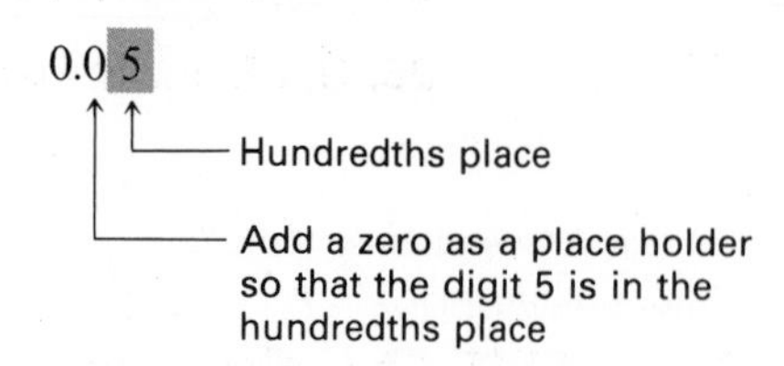

c. Four hundred six thousandths

d. Four hundred and six thousandths

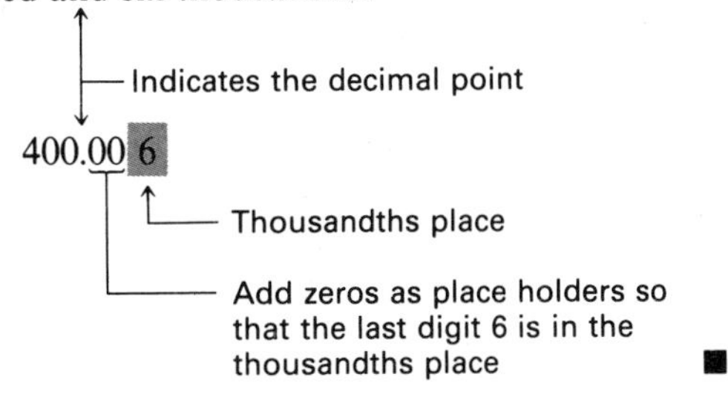

An Easy Way to Remember the Name of a Decimal Place To find the name of a decimal place, write a 0 for each decimal place in the number, then precede them by a 1.

Example 7 Find the name of the place marked with the X.

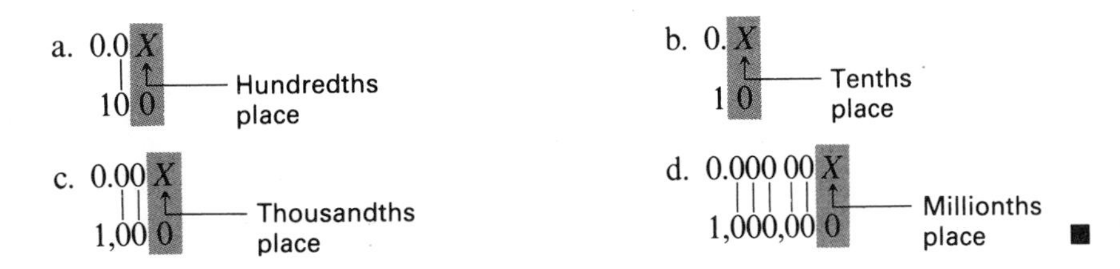

EXERCISES 3.1

In Exercises 1–12, write the numbers in words.

1. 0.35
2. 0.054
3. 3.016
4. 7.08
5. 0.0004
6. 0.070
7. 20.900
8. 860.03
9. 9,000.50
10. 5,006
11. 0.5006
12. 9.00275

In Exercises 13–26, write the numbers in decimal notation.

13. Nine hundredths
14. Eighteen thousandths
15. Two and three thousandths
16. Twelve and six ten-thousandths
17. Four hundred ten-thousandths
18. Five hundred thousandths
19. Sixty and eight hundredths
20. Sixty-eight hundredths
21. Seven hundred twenty thousandths
22. Seven hundred and twenty thousandths
23. Three thousand and fifty-five hundredths
24. Three thousand fifty and five hundredths
25. One hundred and four ten-thousandths
26. One hundred four ten-thousandths

3.2 Rounding Off Decimals

In Section 1.3 we explained rounding off whole numbers and why it is done. The same need for rounding off decimals exists. For example, since the smallest coin we have is the 1¢ piece, any calculations done with numbers representing money are usually rounded off to the nearest cent. When figuring federal income taxes, you are permitted to round off your calculations to the nearest dollar to make it easier to figure and pay your taxes. Some measuring instruments are accurate to thousandths of an inch or tenths of a foot, and so on. For this reason, we usually round off measurements to the accuracy of the instruments used.

> TO ROUND OFF A DECIMAL
>
> 1. Locate the digit in the round-off place. (We will circle the digit.)
> 2. If first digit to right of round-off place is less than 5: digit in round-off place is *unchanged.*
>
>   If first digit to right of round-off place is 5 or more: digit in round-off place is *increased by 1.*
> 3. Digits to *left* of round-off place are unchanged. (Special case: Examples 5 and 6)
> 4. Digits to *right* of *both* round-off place and decimal point are dropped.
> 5. Digits to *right* of round-off place and *left* of decimal point are replaced by zeros.

Example 1 Round off 3.249 to the nearest tenth.
Solution

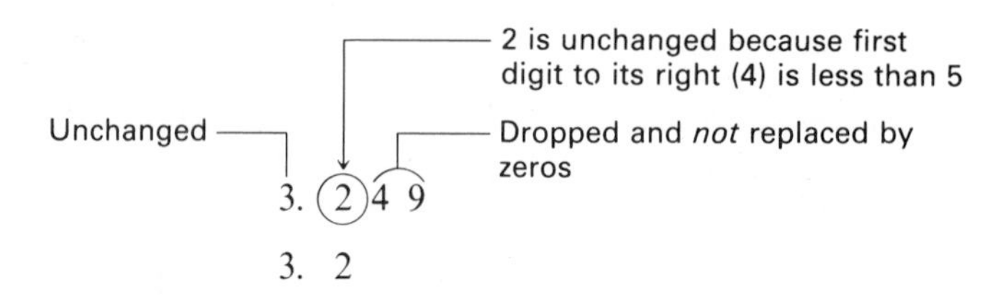

Therefore, 3.249 ≈ 3.2 round off to tenths. ■

NOTE The symbol ≈ means "is approximately equal to." We use this symbol to show that we are rounding off the number. ☑

Example 2 Round off 473.28 to the nearest ten.
Solution

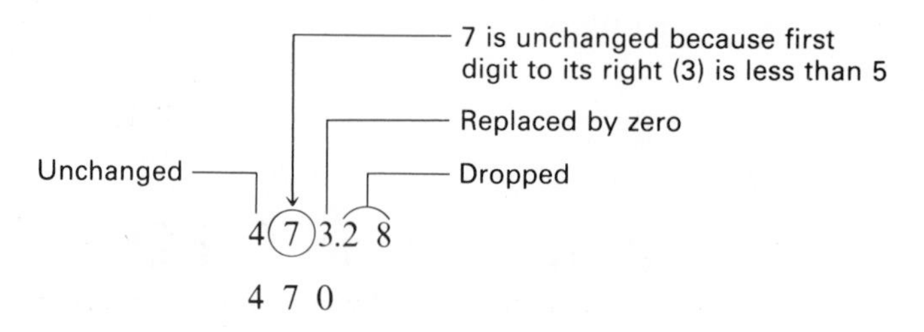

Therefore, 473.28 ≈ 470 rounded off to tens. ■

Example 3 Round off 2.4856 to the nearest hundredth.

Solution

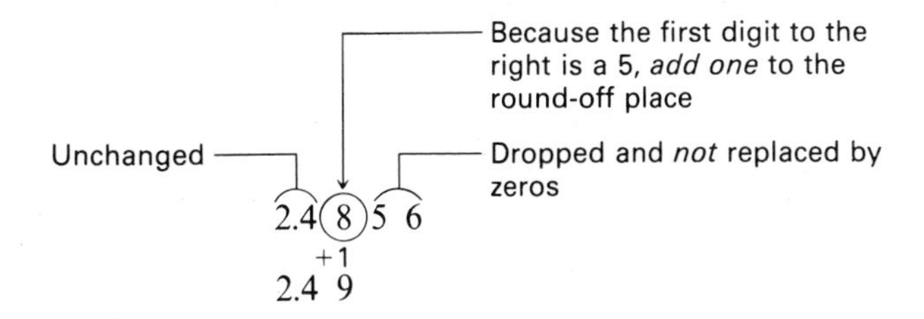

Therefore, 2.4856 ≈ 2.49 rounded off to hundredths. ■

Example 4 Round off 82,674.153 to the nearest hundred.
Solution

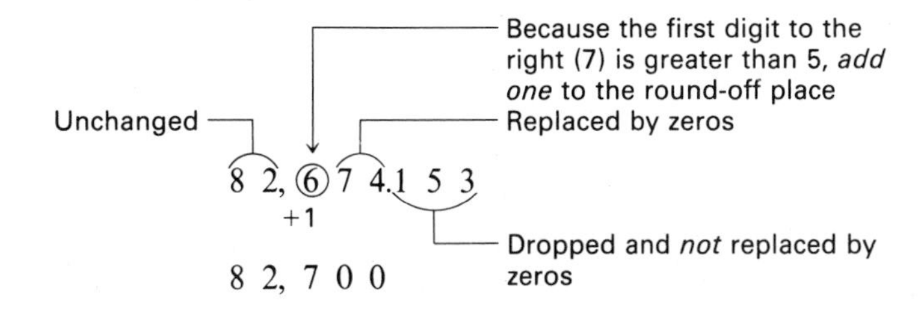

Therefore, 82,674.153 ≈ 82,700 rounded off to hundreds. ■

Example 5 Round off 64.982 to the nearest tenth.
Solution

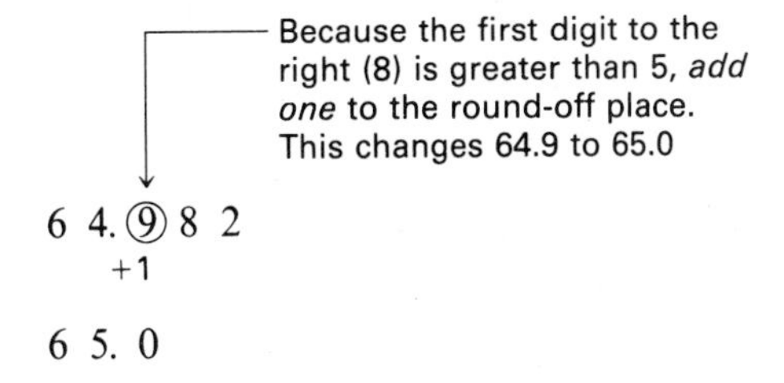

Therefore, 64.982 ≈ 65.0 to the nearest tenth. The zero in the tenths place shows that the number has been rounded to the nearest tenth. ■

Example 6 Round off 3.99964 to the nearest thousandth.
Solution

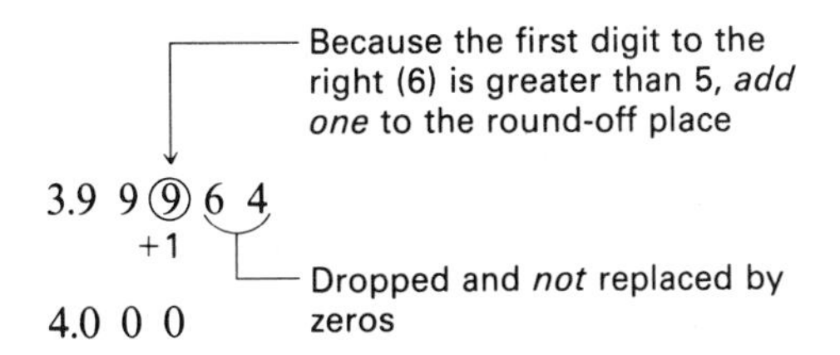

Therefore, 3.99964 ≈ 4.000 rounded off to thousandths. ■

A WORD OF CAUTION Do not accumulate rounding offs. For example, to round off 1.7149 to the nearest hundredths, do not round off 1.7149 to 1.715, then round off 1.715 to 1.72. Actually, 1.7149 ≈ 1.71, rounded off to hundredths. ☑

EXERCISES 3.2

Round off each number to the indicated place.

1. 7.16	tenths	**2.** 0.324	hundredths
3. 9.028	hundredths	**4.** 0.06372	thousandths
5. 427.301	hundreds	**6.** 804.49	tens
7. 0.06034	ten-thousandths	**8.** 0.8265	hundredths
9. 3.2096	thousandths	**10.** 56.97	tenths
11. 106.63	units	**12.** 0.704	hundredths
13. 494.949	tens	**14.** 494.949	tenths
15. 0.0098234	hundred-thousandths	**16.** 0.167761	ten-thousandths
17. 0.0975	hundredths	**18.** 0.06925	thousandths

3.3 Adding Decimals

> TO ADD DECIMALS
>
> **1.** Write the numbers under one another with their decimal points in the same vertical line.
>
> **2.** Add the numbers like you add whole numbers.
>
> **3.** Place the decimal point in your answer (sum) in the same vertical line as the other decimal points.

Writing the numbers clearly and keeping the columns straight helps to reduce the number of addition errors.

Example 1 Add 75.4 + 186 + 0.056 + 1.207 + 2,350.
Solution

	Thousands	Hundreds	Tens	Units	↓ *Decimal point*	Tenths	Hundredths	Thousandths
			7	5	.	4		
		1	8	6	.			
					.	0	5	6
				1	.	2	0	7
+	2,	3	5	0	.			
	2,	6	1	2	.	6	6	3

You may find it easier to keep the columns straight when zeros are written in the open spaces, as shown.

$$
\begin{array}{r}
75.400 \\
186.000 \\
000.056 \\
001.207 \\
+\,2{,}350.000 \\
\hline
2{,}612.663
\end{array}
\quad \blacksquare
$$

The method for adding decimals shown in Example 1 works because:

Only like things can be added directly.

Example 2 Add 1 apple + 2 pears. These cannot be added because they are *not* like things. ■

Example 3 Add 3 apples + 5 apples. These can be added because they *are* like things. The sum is 8 apples. ■

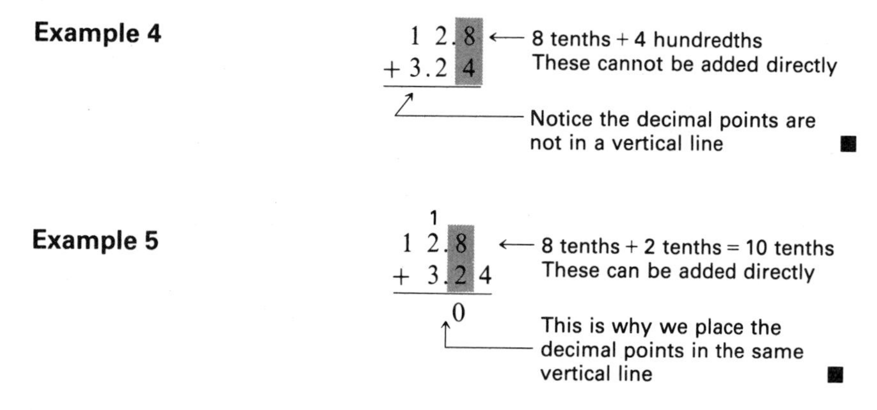

EXERCISES 3.3

In Exercises 1–8, find the indicated sums.

1. 6.5 + 0.66 + 80.75 + 287 + 0.078

2. 100 + 20 + 7 + 0.6 + 0.09 + 0.008

3. \$0.35 + \$24.79 + \$127.50 + \$18.84 + \$96

4. \$0.85 + \$286.83 + \$7.89 + \$46 + \$19.95

5. 75.5 + 3.45 + 180 + 0.0056

6. 185 + 35.06 + 0.186 + 0.0007

7. 987.46 + 35.778 + 1,750.46 + 706.188 + 7,556.189

8. 75,000 + 398.46 + 79.06 + 5.0789 + 186,300 + 35.45

9. Mrs. Ramirez spent the following for lunch: Monday \$5.33, Tuesday \$7.47, Wednesday \$3.89, Thursday \$6.28, Friday \$4.65.
a. Estimate the total spent for lunch.
b. How much did she spend for lunch that week?

10. Frank checked his gasoline credit slips after making a short trip and found that he had used the following amounts of gasoline: 11.2 gallons, 10.8 gallons, 14.1 gallons, 6.7 gallons, 9.4 gallons. How many gallons did he use for the trip?

11. Find the sum of the following numbers:
three thousand, fifty, and thirty-seven hundredths;
five and two hundred-thousandths;
seventy and one hundred fifty ten-thousandths.

12. a. Estimate the perimeter of the triangle.
b. Find the exact perimeter of the triangle.

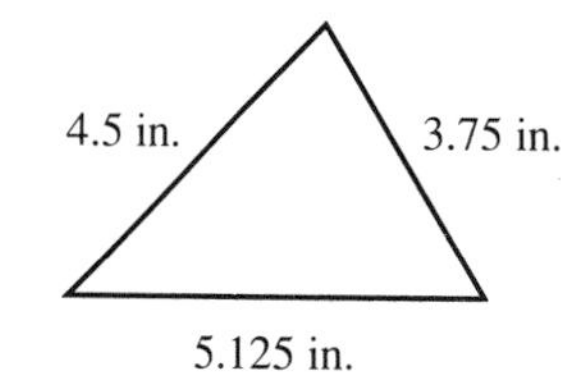

3.4 Subtracting Decimals

> TO SUBTRACT DECIMALS
>
> **1.** Write the number being subtracted (*subtrahend*) under the number it is being subtracted from (*minuend*) with their decimal points in the same vertical line.
>
> **2.** Subtract the numbers as whole numbers.
>
> **3.** Place the decimal point in your answer (*difference*) in the same vertical line as the other decimal points.

Example 1 Subtract 6.07 from 75.14.

Solution

```
 6 15 0 14
 7  5.1  4   (crossed out)
-    6.0 7
----------
 6  9.0 7  ■
```

Example 2 Subtract 121.6 from 304.178.

Solution

```
 2 10 3 11
 3  0 4.1 7 8   (3, 0, 4, 1 crossed out)
-1  2 1.6
------------
 1  8 2.5 7 8  ■
```

Example 3 Subtract 1.654 from 38.6.

Solution

```
    15 9
  7  5 10 10
3 8.6  0  0  ← Add zeros to the right of 6 in
-  1.6  5  4    the hundredths and
-----------     thousandths places
3  6.9  4  6                                ■
```

The method for subtracting decimals shown in Examples 1, 2, and 3 works because:

Only like things can be subtracted directly.

Example 4 Subtract 1 apple from 2 pears. These cannot be subtracted directly because they are *not* like things. ■

Example 5 Subtract 2 apples from 5 apples. These can be subtracted directly because they *are* like things. The difference is 3 apples. ■

Example 6 Subtract 3.24 from 12.8.

```
 0 12 7 10
 1  2.8  0   (crossed out)
-   3.2  4  ← 10 hundredths − 4 hundredths = 6 hundredths
----------
    9.5  6  ■
```

EXERCISES 3.4

In Exercises 1–10, find the indicated differences.

1. 7.85 – 3.44

2. 84.07 – 0.66

3. 208.5 – 7.16

4. 715.75 – 28.19

5. 300 – 0.145

6. 7,000 – 3.68

7. 81,284.56 – 2,784.8

8. 2,000,046 – 30,015.8

9. 5.785 – 0.9665

10. 6.005 – 0.8476

11. Subtract 46.8 from 224.

12. Subtract 2.093 from 187.5

13. Mrs. Geller's bank statement showed a balance of $254.39 at the beginning of the month. During the month she made the following deposits: $183.50, $233.75, and $78.86. During the month she wrote the following checks: $27.15, $86.94, $123.47, $167.66, $122.20, and $38.67. Find her balance at the end of the month.

14. When two dragsters raced, the first car's time was 6.05 seconds and the second car's time was 6.375 seconds. Find the difference in their times.

3.5 Multiplication of Decimals

The number of decimal places in a number is the number of digits written to the right of the decimal point. A whole number has no decimal places.

Example 1 The number of decimal places in a number.

a. 0.25 has two decimal places

b. 0.054 has three decimal places

c. 14.5 has one decimal place

d. 0.5000 has four decimal places

e. 167. has no decimal places ■

TO MULTIPLY DECIMALS

1. Multiply the numbers as whole numbers.
2. Add the number of decimal places in the two numbers being multiplied.
3. Place the decimal point in your answer so that your answer has as many decimal places as the *sum* found in (2).

In Example 2, we show why the number of decimal places in the product is equal to the sum of the decimal places of the numbers being multiplied.

Example 2 Multiply 0.2 by 0.003.

Solution

$$0.2 \times 0.003 = \frac{2}{10} \times \frac{3}{1{,}000} = \frac{6}{10{,}000} = 0.0006$$

1 decimal place + 3 decimal places = 4 decimal places ■

Example 3 Multiply 0.035 by 0.25.

Solution

```
      0.2 5        2 decimal places
  ×   0.0 3 5    + 3 decimal places
  -----------    ---
        1 2 5      5
        7 5
  -----------
  0.0 0 8 7 5      5 decimal places  ■
```

Since changing the order of the numbers being multiplied does not change the product, it is usually easier to use the number with fewer nonzero digits as the multiplier. We show this by working Example 4 two different ways.

Example 4 Multiply 4.6 by 3.749.

First Solution

```
     3.7 4 9
  ×      4.6
  ----------
   2 2 4 9 4
 1 4 9 9 6
  ----------
 1 7.2 4 5 4
```

Second Solution

```
         4.6
  ×  3.7 4 9
  ----------
       4 1 4
     1 8 4
   3 2 2
 1 3 8
  ----------
 1 7.2 4 5 4  ■
```

Example 5 Multiply 0.86 by 18,000.

We handle final zeros the same way we did in Section 1.6.

Solution

```
      0.8 6|            2 decimal places
  ×     1 8|0 0 0     + 0 decimal places
  ---------------     ---
      6 8 8|            2
      8 6  |
  ---------------
  1 5,4 8|0.0 0         2 decimal places  ■
```

Example 6 Gasoline costs 92.9 cents per gallon.

a. Estimate the cost of 17.5 gallons of gasoline.

b. Find the exact cost of 17.5 gallons of gasoline.

Solution First, because the answer to this problem will be in dollars and cents, we will rewrite 92.9 cents as \$0.929.

a. To estimate the answer, round off each number at the first nonzero digit, then perform the operation.

Round off \$0.929 to \$0.9 and 17.5 to 20, then multiply.

$$\$0.9 \times 20 = \$18.00$$

b.

```
      0.9 2 9        3 decimal places
  ×     1 7.5      + 1 decimal place
  -----------      ---
      4 6 4 5        4
    6 5 0 3
    9 2 9
  -----------
  1 6.2 5 7 5        4 decimal places
≈ $16.26             Rounded to the nearest cent  ■
```

EXERCISES 3.5

In Exercises 1–16, find the products.

1. 0.03 × 0.8
2. 0.9 × 0.007
3. 0.04 × 0.06
4. 0.005 × 0.03
5. 42 × 0.3
6. 0.06 × 94
7. 0.17 × 0.8
8. 0.029 × 0.5
9. 3.6 × 0.75
10. 8.4 × 9.5
11. 1.06 × 0.37
12. 0.41 × 20.4
13. 500 × 0.79
14. 0.26 × 800
15. 0.19 × 43,000
16. 6,800 × 6.7

In Exercises 17–20, estimate the products.

17. 0.875 × 0.39
18. 0.0718 × 0.56
19. 0.685 × 42.5
20. 0.00934 × 206

21. Marcia earns $6.15 an hour.
    a. Estimate her earnings if she worked 26 hours.
    b. How much did she earn if she worked 26 hours?

22. Gasoline sells for 87.9 cents a gallon.
    a. Estimate the cost of 13.2 gallons of gasoline.
    b. Find the cost of 13.2 gallons of gasoline.

23. A rectangular playground measures 12.5 meters long and 8.6 meters wide.
    a. Estimate the area of the playground.
    b. What is the area of the playground?
    c. Find the cost to blacktop the playground if the blacktop costs $20.50 per square meter.

24. A square acre measures approximately 69.6 yards on each side.
    a. Estimate the number of square yards in an acre.
    b. Find the number of square yards in an acre. Round off your answer to the nearest tens.

25. Ann makes a $65 downpayment and pays $16.25 a month for a stereo set. If she takes 18 months to pay for it, what is the total cost?

26. Irv makes $12.68 an hour.
    a. What is his salary for a regular 40-hour week?
    b. Irv receives time-and-a-half for each hour over 40 that he works in 1 week. How much does he make for working a 50-hour week?

27. A car rental agency charges $12.00 a day and 15 cents per mile to rent their cars. How much would it cost to rent the car for a week if you drove 875 miles?

28. The Cliffton County Telephone Company charges $8.25 a month plus 18 cents for each call over 30 during the month. What is the monthly phone bill for 57 calls made that month?

3.6 Division of Decimals

3.6A Dividing a Decimal by a Whole Number

TO DIVIDE A DECIMAL BY A WHOLE NUMBER

1. Place a decimal point above the quotient line directly above the decimal point in the dividend.
2. Divide the numbers as whole numbers.

Example 1 Divide 150.4 by 47.

Solution

$$\begin{array}{r} 3.2 \\ 47\overline{)150.4} \\ \underline{141} \\ 9\,4 \\ \underline{9\,4} \end{array}$$ ■

Example 2 Divide 48.4 by 85. (Round off your answer to two decimal places.)

Solution

```
        .5 6 9 ≈ .5 7
  8 5 ) 4 8.4 0 0
        4 2 5
        -----
          5 9 0
          5 1 0
          -----
            8 0 0
            7 6 5
            -----
              3 5  ■
```

In this example, the division has been carried out to 3 decimal places, then rounded off to two decimal places

When rounding off, carry out the division to *one more place* than required, then round off.

Example 3 Scott swam the 100-meter freestyle in 54.23 seconds, 59.60 seconds, and 57.26 seconds. What was his average time?
Solution

$$\text{Average} = \frac{54.23 + 59.60 + 57.26}{3}$$

$$= \frac{171.09}{3} = 3\overline{)1\,7\,1.0\,9}\quad \overset{5\,7.0\,3\text{ sec}}{} \quad ■$$

EXERCISES 3.6A

In Exercises 1–10, the quotients are exact. Do not round off the quotients.

1. 86.96 ÷ 8 **2.** 249.2 ÷ 7 **3.** 93.6 ÷ 6 **4.** 673.2 ÷ 9 **5.** 33.6 ÷ 32

6. 43.26 ÷ 21 **7.** 6.825 ÷ 39 **8.** 28.13 ÷ 58 **9.** 4.977 ÷ 63 **10.** 311.1 ÷ 85

In Exercises 11–20, divide and round off the quotient to the indicated place.

11. 8.56 ÷ 7 2 decimal places

12. 456.7 ÷ 9 1 decimal place

13. 376.3 ÷ 8 3 decimal places

14. 514.7 ÷ 6 3 decimal places

15. 58.6 ÷ 42 tenths

16. 75.4 ÷ 51 hundredths

17. 3.86 ÷ 76 thousandths

18. 5.77 ÷ 84 ten-thousandths

19. 76.5 ÷ 208 hundredths

20. 90.6 ÷ 555 thousandths

21. Find the average of 6.3, 8.4, 10.2, and 9.7.

22. Find the average of 0.12, 0.21, 0.34, 0.15, and 0.22.

23. The rainfall in Smalltown for five weeks was 2 inches, 1.6 inches, 3.1 inches, 4.7 inches, and 0.5 inches. What was the average rainfall for the five weeks? Round off answer to the nearest tenth of an inch.

24. William ran the 400-meter hurdles in 53.02 seconds, 56.28 seconds, 52.93 seconds, and 54.67 seconds. Find his average time. Round off answer to the nearest hundredth of a second.

25. Terry bought three tapes for $8.50 each and two tapes for $6.95. Find the average cost of the tapes.

26. Virginia bought four birthday cards for $1.75, one card for $2.25, and one card for $3.65. What was the average cost of the cards?

3.6B Dividing a Decimal by a Decimal

In our study of fractions, we learned that the value of a fraction is not changed when the numerator and denominator are *multiplied* by the same number.

$$\frac{5}{8} = \frac{5 \times 10}{8 \times 10} = \frac{50}{80}$$

Example 4 A fraction is equivalent to a division, therefore,

$$0.7\overline{)16.8} = 16.8 \div 0.7 = \frac{16.8 \times 10}{0.7 \times 10} = \frac{168}{7} = 7\overline{)168.} \quad \blacksquare$$

This shows that $0.7\overline{)16.8} = 7.\overline{)168.}$ Therefore, *the quotient is unchanged when the decimal point in the divisor and dividend are both moved the same number of places to the right.* Using carets (∧), we would rewrite the last example:

$$0.7\overline{)1\ 6.8} = 0.7_\wedge\overline{)1\ 6.8_\wedge}$$

These carets (∧) are used to indicate the new positions of the decimal points

$$\begin{array}{r} 2\ 4. \\ 0.7_\wedge\overline{)1\ 6.8_\wedge} \\ 1\ 4 \\ \hline 2\ 8 \\ 2\ 8 \\ \hline \end{array}$$

The divisor has become 7 instead of .7

It is also true that the value of a fraction is not changed when the numerator and denominator are *divided* by the same number. Therefore, *the quotient is unchanged when the decimal point in the divisor and dividend are both moved the same number of places to the left.*

Example 5 Divide 166.4 by 40.
Solution

$$166.4 \div 40 = 40\overline{)166.4} = \frac{166.4}{40} = \frac{166.4 \div 10}{40 \div 10} = \frac{16.64}{4} = 4\overline{)16.64}$$

Both numerator and denominator *divided* by 10

We will write $\quad 40\overline{)166.4} = 4_\wedge0\overline{)16_\wedge6.4}$

These carets are used to indicate the new positions of the decimal points

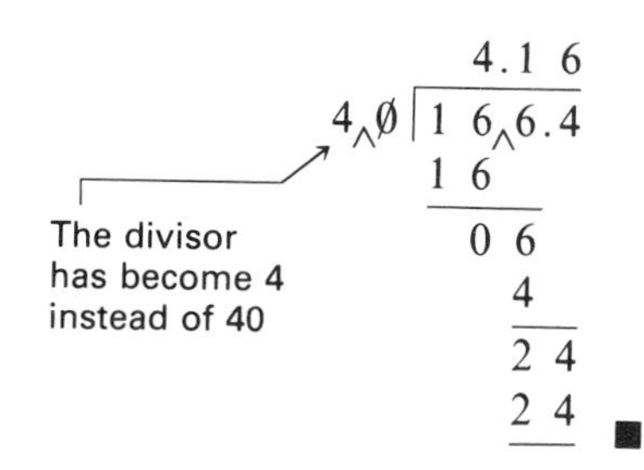

■

> **TO DIVIDE A DECIMAL BY A DECIMAL**
>
> 1. Place a caret ($\wedge$) to the right of the last nonzero digit of the divisor.
> 2. Count the number of places between the decimal point and the caret in the divisor.
> 3. Place a caret in the dividend the same number of places to the right (or left) of its decimal point as in (2). Add zeros to the dividend when needed.
> 4. Place a decimal point in the quotient directly above the caret in the dividend.
> 5. Divide the numbers as whole numbers.

Example 6 Divide 2.368 by 0.32.

Solution

```
            7.4
0.3 2^ ) 2.3 6^8
         2 2 4
         -----
           1 2 8
           1 2 8  ■
           -----
```

Example 7 Divide 0.144 by 1.20.

Solution

```
                   .1 2
1.2^0 ) 0.1^4 4
            1 2
            ---
            2 4
            2 4  ■
            ---
```

The divisor has become 12

Example 7 shows that *we place the caret so that we make the divisor the smallest possible whole number.*

Example 8 Divide 3.51 by 0.065.

Solution

```
                5 4.
0.0 6 5^ ) 3.5 1 0^  ←— This zero was added so that
           3 2 5          three places come between
           -----          caret and decimal point
             2 6 0
             2 6 0  ■
             -----
```

Example 9 Divide 197.2 by 0.29.

Solution

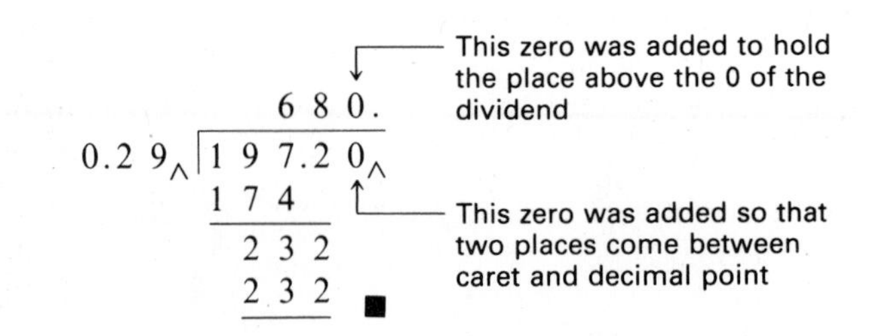

Example 10 Divide 0.2429 by 5.7, and round off your answer to three decimal places.

Solution

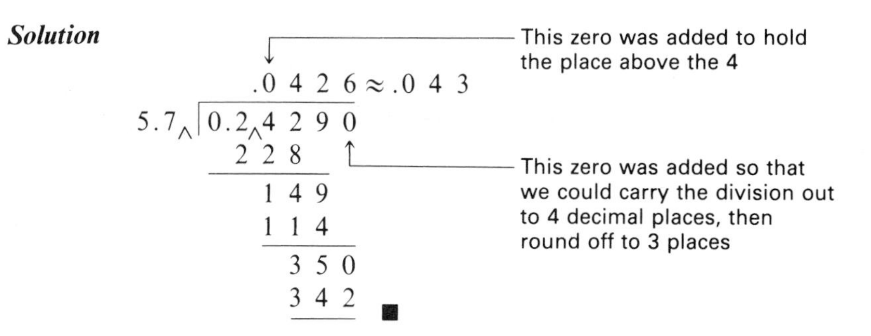

Example 11 Divide 7 by 4.1, and round off your answer to the nearest hundredth.

Solution

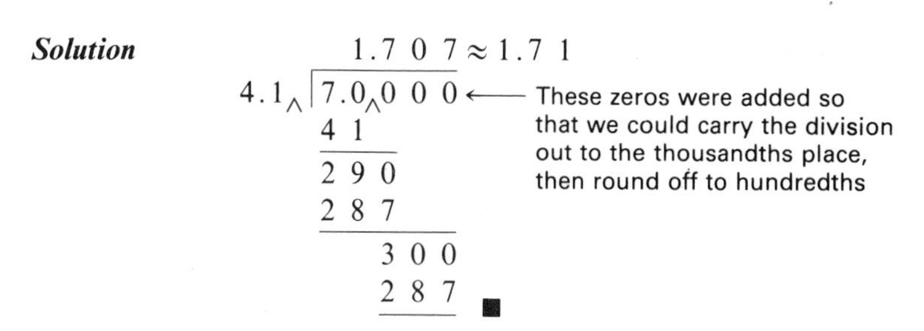

EXERCISES 3.6B

In Exercises 1–16, the quotients are exact. Do not round off the quotients.

1. $9.612 \div 2.7$ **2.** $17.898 \div 3.8$ **3.** $478.24 \div 6.1$ **4.** $12.42 \div 9.2$

5. $51.2 \div 800$ **6.** $40.5 \div 500$ **7.** $0.0196 \div 1.40$ **8.** $0.242 \div 1.10$

9. $0.4077 \div 0.45$ **10.** $5.226 \div 0.65$ **11.** $2.0349 \div 0.057$ **12.** $1.6472 \div 0.29$

13. $2173 \div 2.65$ **14.** $6120 \div 8.16$ **15.** $1.0488 \div 0.368$ **16.** $7.3143 \div 0.774$

In Exercises 17–26, divide and round off the quotient to the indicated place.

17. $7.68 \div 2.90$ hundredths

18. $0.0089 \div 0.056$ thousandths

19. $12 \div 0.39$ tenths

20. $29 \div 1.4$ hundredths

21. $0.041 \div 0.0064$ thousandths

22. $0.0023 \div 0.093$ thousandths

23. $0.25 \div 8.07$ 4 decimal places

24. $0.2 \div 35.9$ 5 decimal places

25. $66 \div 0.073$ 1 decimal place

26. $246 \div 0.816$ 2 decimal places

27. The balance owed on a car amounted to \$2,467.14. What would the monthly payments be in order to pay it off in 3 years? Round off the payment to the nearest cent (two decimal places).

28. At the beginning of a trip, Raul's odometer reading was 65,479 miles. At the end of the trip the reading was 67,784 miles. He used 147 gallons of gasoline. How many miles did he get to the gallon of gasoline? Round off the answer to one decimal place.

29. How many pieces of ribbon 3.5 feet long can be cut from a spool 50 feet long?

30. If Al paid \$13.92 a month on his finance company loan, how long would it take him to pay off a balance of \$250.56?

3.7 Multiplying and Dividing Decimals by Powers of Ten

Here are some powers of 10:

$$10^1 = 10, \quad 10^2 = 100, \quad 10^3 = 1{,}000$$

Refer to Section 1.8 for a more complete explanation of powers of 10.

Multiplying by Powers of 10

Example 1 Multiplying 5.3 by 10 and 100.

```
  5.3          5.3
 ×1 0        ×1 0 0
 ----        ------
  0 0          0 0
 5 3          0 0
 ----        5 3
 5 3.0       ------
             5 3 0.0  ■
```

Notice that when we multiplied 5.3 by 10 or 100, the decimal point in the product moved to the right the same number of places as zeros in the power of ten.

> TO MULTIPLY A DECIMAL BY A POWER OF TEN
>
> Move the the decimal point to the *right* as many places as the number of zeros in the power of 10.

Example 2 Multiplying a decimal by a power of ten.

a. $24.7 \times 10 = 24_{\wedge}7. = 247$

1 zero → 1 place

b. $24.7 \times 100 = 2{,}4_{\wedge}70. = 2{,}470$

2 zeros → 2 places

c. $1.0567 \times 1{,}000 = 1_{\wedge}056.7 = 1{,}056.7$

3 zeros → 3 places

d. $0.0973 \times 10^2 = 0_{\wedge}09.73 = 9.73$

exponent 2 → 2 places

e. $34 \times 10^4 = 34_{\wedge}0000. = 340{,}000$

exponent 4 → 4 places ■

Dividing by Powers of 10

Example 3 Dividing 5.3 by 10 and 100.

$$5.3 \div 10 = 10\overline{)5.30} = .53$$

```
        .5 3
10 ) 5.3 0
     5 0
     ---
       3 0
       3 0
```

$$5.3 \div 100 = 100\overline{)5.300} = .053$$

```
         .0 5 3
100 ) 5.3 0 0
      5 0 0
      -----
        3 0 0
        3 0 0  ■
```

When we divided 5.3 by 10 or 100, the decimal point in the quotient moved to the left the same number of places as zeros in the power of ten.

> TO DIVIDE A DECIMAL BY A POWER OF TEN
>
> Move the decimal point to the *left* as many places as the number of zeros in the power of 10.

Example 4 Dividing a decimal by a power of ten.

a. $395 \div 100 = 3.95_{\wedge} = 3.95$

2 zeros ← 2 places

b. $75.6 \div 10 = 7.5_{\wedge}6 = 7.56$

1 zero ← 1 place

c. $\frac{0.315}{100} = 0.00_{\wedge}315 = 0.00315$

2 zeros ← 2 places

d. $\frac{4{,}165.2}{10^3} = 4.165_{\wedge}2 = 4.1652$

exponent 3 ← 3 places

e. $75.6 \div 10^4 = 0.0075_{\wedge}6 = 0.00756$

exponent 4 ← 4 places ■

EXERCISES 3.7

In Exercises 1–16, perform the indicated operations.

1. $\frac{95.6}{10}$

2. $7.98 \div 10$

3. $573 \div 100$

4. $\frac{64.8}{100}$

5. 27.8×100

6. 8.95×10

7. $1000(0.2094)$

8. $1000(3.097)$

9. $\frac{750.2}{10^2}$

10. $\frac{98.47}{10^2}$

11. 9.846×10^2

12. 0.0837×10^3

13. $\frac{100}{10^3}$

14. 200×10^3

15. $10^4 \times 27.4$

16. $0.48 \div 10^2$

17. The $146.35 cost of a party was shared by 10 people. How much did each person have to pay? (Be sure to round your answer off to the nearest cent.)

18. A club charged $1.50 for each lottery ticket. If 10,000 tickets were sold, how much money did the club raise from this lottery?

19. 537 people attended a $100-a-plate fund-raising dinner given for a political candidate. How much campaign money did this dinner raise?

20. The $23,758 cost of putting in electric service along a rural road was shared equally by 100 home owners. How much did each owner have to pay?

3.8 Combined Operations

In evaluating expressions with more than one operation, the same order of operations is used with decimals that was used with whole numbers and fractions.

> ORDER OF OPERATIONS
>
> **1.** Any expressions in parentheses are evaluated first.
>
> **2.** Evaluations are done in this order
> First: Powers and roots
> Second: Multiplication and division in order from *left to right*
> Third: Addition and subtraction in order from *left to right*

Example 1

$4.6 + 2.3 \times 5.4 - 8.6 =$ Multiply first

$4.6 + 12.42 - 8.6 =$ Addition and subtraction are done left to right

$17.02 - 8.6 = 8.42$ ■

Example 2

$54 \div 10 - 0.07 \times 2^3 =$ $2^3 = 2 \times 2 \times 2 = 8$

$54 \div 10 - 0.07 \times 8 =$ Multiplication and division are done left to right

$5.4 - 0.07 \times 8 =$

$5.4 - 0.56 = 4.84$ Subtraction is last ■

Example 3 Estimate the answer to $\frac{0.38 \times 63}{0.51}$.

Solution Round off 0.38 to 0.4, 63 to 60, and 0.51 to 0.5.

$$\frac{0.4 \times 60}{0.5} = \frac{24}{0.5}$$

Now we can round off 24 to 25 (because 25 is a *convenient* number to divide by 0.5), or we can round off 24 to 20 and then divide.

If we round off 24 to 25: $\frac{25}{0.5} = 0.5_{\wedge}\overline{)25.0_{\wedge}}$ = 50.

If we round off 24 to 20: $\frac{20}{0.5} = 0.5_{\wedge}\overline{)20.0_{\wedge}}$ = 40. ■

Both 50 and 40 are considered good estimates

Example 4 Find $\frac{0.38 \times 63}{0.51}$ using a calculator. Round off answer to one decimal place.

Key in: .38 [×] 63 [÷] .51 [=]
Answer: 46.941176 ≈ 46.9 ■

See Section 1.11 for a discussion on calculators and their use.

Example 5 The Southern Electric Co. charges 7.95 cents per kWh for the first 250 kWh used and 12.2 cents for each kWh used over the first 250. Mr. Carson's electric meter read 22,607 on September 1 and 23,039 on October 1. (Note that kWh means kilowatt-hour and is a measure of electrical energy. One kWh of electricity will light a 100-watt bulb for 10 hours.)

a. Find the amount of Mr. Carson's electric bill for September.
b. Find his daily average of kWh used for September (30 days).

Solution First, find the number of kWh used by subtracting the meter readings.
Key in: 23039 [−] 22607 [=] 432 ← kWh used
And: [−] 250 [=] 182 ← kWh at higher rate

a. To find the cost of his electric bill,
Key in: 250 [×] .0795 [+] 182 [×] .122 [=]
Answer: 42.079 ≈ \$42.08

b. To find the daily average, divide the total number of kWh used by the number of days.
Key in: 432 [÷] 30 [=]
Answer: 14.4 ■

EXERCISES 3.8

In Exercises 1–12, find the value of each expression using the correct order of operations.

1. $2.5 + 4.3 \times 0.8 - 1.5$

2. $12 \div 0.2 - 1.8 \times 4$

3. $7.06 - 0.4 \times 0.6 + 3^2$

4. $18 + 2.4 \div 6 \times 0.4$

5. $0.09 \cdot \sqrt{25} + 75 \div 10$

6. $1.42 \times 10^3 - 4.65 \times 10^2$

7. $42 \div 1{,}000 + 0.057 \times 100$

8. $0.5(2.8 + 12)$

9. $46 - (5.6 - 0.34)$

10. $46 - 5.6 - 0.34$

11. $2.3 + 1.6 + 1.2 \div 3$

12. $(2.3 + 1.6 + 1.2) \div 3$

In Exercises 13–18, estimate the answer (estimates may vary).

13. $3.8 + 7.1 - 4.3$

14. $0.72 \times 0.394 \times 0.53$

15. $2(8.9 + 4.7)$

16. $\dfrac{0.405}{7.69}$

17. $\dfrac{2.1 \times 6.3}{0.292}$

18. $\dfrac{7.2 \times 4.8}{0.0613}$

In Exercises 19–24, use a calculator to evaluate each expression. Round off to two decimal places when necessary.

19. $6.89 + 0.53 \times 7.34$

20. $\sqrt{39.69} - 3.51 \div 0.78$

21. $1{,}722(1 + 0.055)$

22. $750 \times 0.085 \times \dfrac{30}{365}$

23. $\dfrac{4.1 \times 1.9}{0.625}$

24. $5{,}000\left(1 + \dfrac{0.8}{12}\right)$

25. A rectangle measures 8.65 meters long and 5.3 meters wide. What is the perimeter of the rectangle?

26. A door measures 2.5 feet wide and 6.5 feet high. Find the cost to replace the molding around the door if the molding costs \$1.80 per foot. (Note that there is no molding at the bottom of the door.)

27. A taxi charges \$2 for the first mile and \$0.75 for each additional mile. What is the taxi fare for 6 miles?

28. A delivery company charges \$12 plus \$0.35 per pound. How much would it cost to send a 12-pound package?

29. The gas company charges \$0.365 per therm (a unit of heat) for the first 65 therms used and \$0.785 for each therm used over the first 65. The gas meter read 9,876 on November 1 and 9,963 on December 1.
a. What was the amount of the gas bill for November?
b. Find the daily average of therms used for November (30 days).
c. Find the average daily cost of gas for November.

30. The electric company charges 6.78 cents per kWh for the first 280 kWh used and 10.9 cents for each kWh used over the first 280. The electric meter read 12,865 on May 1 and 13,237 on June 1.
a. Find the amount of the electric bill for May.
b. Find the daily average of kWh used for May (31 days).
c. Find the average daily cost of electricity for May.

3.9 Changing a Fraction to a Decimal

Because

$$\frac{a}{b} = a \div b$$

we can say,

> TO CHANGE A FRACTION TO A DECIMAL
> Divide the numerator by the denominator.

Terminating Decimal
Repeating Decimal

Any fraction can be changed either to a **terminating decimal** for which the division comes to an end (see Example 1), or to a **repeating decimal** in which a digit or a group of digits will repeat indefinitely (see Examples 2 and 3).

Example 1 Change $\frac{5}{8}$ to a decimal.

Solution

$$\frac{5}{8} = 5 \div 8 = 8\,\overline{)\,5.000}$$

$$\begin{array}{r} .625 \\ 8\,\overline{)\,5.000} \\ \underline{4\,8} \\ 20 \\ \underline{16} \\ 40 \\ \underline{40} \end{array}$$ ■

Example 2 Change $\frac{2}{11}$ to a decimal.

Solution

$$\frac{2}{11} = \begin{array}{r} .1818\ldots \\ 11\,\overline{)\,2.0000} \\ \underline{1\,1} \\ 90 \\ \underline{88} \\ 20 \\ \underline{11} \\ 90 \\ \underline{88} \\ 2 \end{array}$$

← The three dots indicate that the decimal continues forever

The digits 1 and 8 repeat forever, and we may indicate this with a bar above the 1 and 8.

Therefore, $\frac{2}{11} = 0.1818\ldots$

or $\frac{2}{11} = 0.\overline{18}$ ■

Example 3 Change $\frac{1}{12}$ to a decimal.

Solution

$$\frac{1}{12} = \begin{array}{r} .08333\ldots = 0.08\bar{3} \\ 12\,\overline{)\,1.00000} \\ \underline{9\,6} \\ 40 \\ \underline{36} \\ 40 \\ \underline{36} \\ 40 \\ \underline{36} \\ 4 \end{array}$$ ■

← The bar is over the 3 because only the digit 3 is repeating

Sometimes we are interested in an answer that has been rounded off. For example, numbers representing money are usually rounded off to the nearest cent. When rounding off, carry out the division to one more place than required, then round off as in Section 3.2.

Example 4 One share of stock in a computer company is selling for $5\frac{3}{8}$ dollars. Find the cost of one share of stock rounded off to the nearest cent.

First solution Change $\frac{3}{8}$ to a decimal, then add its value to 5.

$$\begin{array}{r} .375 \\ 8\overline{)3.000} \\ \underline{24} \\ 60 \\ \underline{56} \\ 40 \\ \underline{40} \end{array}$$

Therefore, $\frac{3}{8} \approx 0.38$. Then $5\frac{3}{8} \approx 5 + 0.38 = \$5.38$.

Second Solution Change $5\frac{3}{8}$ to the improper fraction $\frac{43}{8}$. Then change $\frac{43}{8}$ to a decimal.

$$5\frac{3}{8} = \frac{43}{8} = \begin{array}{r} 5.375 \\ 8\overline{)43.000} \\ \underline{40} \\ 30 \\ \underline{24} \\ 60 \\ \underline{56} \\ 40 \\ \underline{40} \end{array}$$

Therefore, $5\frac{3}{8} \approx \$5.38$. ■

Example 5 Change $3\frac{7}{24}$ to a decimal using a calculator. Round off to the nearest thousandths.

Solution $3\frac{7}{24} = 3 + \frac{7}{24}$

Key in: 3 [+] 7 [÷] 24 [=]

Answer: $3.2916666 \approx 3.292$ ■

EXERCISES 3.9

In Exercises 1–18, change the fractions and mixed numbers to exact decimals.

1. $\frac{3}{4}$	**2.** $\frac{5}{8}$	**3.** $2\frac{1}{2}$	**4.** $5\frac{1}{4}$	**5.** $\frac{1}{8}$	**6.** $\frac{3}{5}$
7. $\frac{2}{3}$	**8.** $\frac{7}{9}$	**9.** $\frac{3}{11}$	**10.** $\frac{5}{18}$	**11.** $3\frac{1}{15}$	**12.** $2\frac{5}{12}$
13. $\frac{1}{25}$	**14.** $\frac{1}{40}$	**15.** $\frac{5}{16}$	**16.** $\frac{7}{8}$	**17.** $6\frac{1}{20}$	**18.** $3\frac{3}{250}$

19. Change $\frac{7}{11}$ to a decimal rounded off to two decimal places.

20. Change $4\frac{5}{7}$ to a decimal rounded off to three decimal places.

21. Change $\frac{2}{9}$ to a decimal rounded off to two decimal places.

22. Change $2\frac{7}{12}$ to a decimal rounded off to three decimal places.

23. Change $\frac{2}{23}$ to a decimal rounded off to the nearest thousandths.

24. Change $\frac{3}{11}$ to a decimal rounded off to the nearest thousandths.

25. Change $6\frac{7}{15}$ to a decimal rounded off to the nearest thousandths.

26. Change $9\frac{1}{14}$ to a decimal rounded off to the nearest ten-thousandths.

In Exercises 27–32, use a calculator to change the fractions and mixed numbers to decimals. Round off to four decimal places when necessary.

27. $\frac{9}{16}$ **28.** $\frac{17}{40}$ **29.** $2\frac{7}{24}$ **30.** $3\frac{8}{27}$ **31.** $9\frac{5}{13}$ **32.** $6\frac{13}{21}$

3.10 Changing a Decimal to a Fraction

3.10A Simple Decimals

Because a decimal is actually a fraction, in this section we write the decimals in fraction form (with numerator and denominator), then reduce the fraction to lowest terms.

TO CHANGE A DECIMAL TO A FRACTION

1. *Read* the decimal, then *write* it in fraction form with numerator and denominator.
2. Reduce the fraction to lowest terms.

Example 1 Change 0.4 to a fraction in lowest terms.

(4: tenths place)

Solution *Read* the decimal "four tenths."

Write $\frac{4}{10} = \frac{2}{5}$ reduced to lowest terms ■

Example 2 Change 0.25 to a fraction in lowest terms.

(5: hundredths place)

Solution *Read* the decimal "twenty-five hundredths."

Write $\frac{25}{100} = \frac{1}{4}$ reduced to lowest terms ■

Example 3 Change 7.04 to a mixed number in lowest terms.

(4: hundredths place)

Solution *Read* the decimal "seven and four hundredths."

Write $$7\frac{4}{100} = 7\frac{1}{25}$$ ■

Example 4 Change 0.00785 to a fraction. We show *an easy way to write a decimal as a fraction*:

Solution

$$0.00785 \longrightarrow \frac{785}{100{,}000} = \frac{157}{20{,}000}$$

Write 1 to the left of the zeros → 1 0 0,0 0 0

Under each digit to the right of the decimal point, write a zero ■

EXERCISES 3.10A

In Exercises 1–18, change the decimals to fractions or mixed numbers reduced to lowest terms.

1. 0.6 **2.** 0.8 **3.** 0.05 **4.** 0.65 **5.** 0.075 **6.** 0.750

7. 0.875 **8.** 1.8 **9.** 2.5 **10.** 3.7 **11.** 5.9 **12.** 4.3

13. 0.0625 **14.** 2.125 **15.** 37.5 **16.** 2.1875 **17.** 65.625 **18.** 0.000875

19. A machinist must drill a 0.875-inch hole through a metal brace. What fractional size drill should he use?

20. A carpenter must drill a 0.625-inch hole through a rafter. What fractional size drill should he use?

3.10B Complex Decimals

Decimals such as $0.33\frac{1}{3}$ and $0.67\frac{1}{2}$ are called *complex decimals.*

> TO CHANGE A COMPLEX DECIMAL TO A FRACTION
>
> **1.** Write the complex decimal as a complex fraction.
>
> **2.** Simplify the complex fraction as in Section 2.15.

Example 5 Change $0.12\frac{1}{2}$ to a fraction in lowest terms.

Solution (2nd decimal place: $2\frac{1}{2}$)

$$0.12\frac{1}{2} \longrightarrow \frac{12\frac{1}{2}}{100} = \frac{\frac{25}{2}}{100} = \frac{25}{2} \div \frac{100}{1} = \frac{\overset{1}{\cancel{25}}}{2} \cdot \frac{1}{\underset{4}{\cancel{100}}} = \frac{1}{8}$$

(1 0 0 written under 0.12) ■

Example 6 Change $0.2\frac{2}{9}$ to a fraction in lowest terms.

Solution

1st decimal place

$$0.2\frac{2}{9} \longrightarrow \frac{2\frac{2}{9}}{10} = \frac{\frac{20}{9}}{10} = \frac{20}{9} \div \frac{10}{1} = \frac{\overset{2}{\cancel{20}}}{9} \cdot \frac{1}{\underset{1}{\cancel{10}}} = \frac{2}{9}$$

1 0 ■

Example 7 Change $2.16\frac{2}{3}$ to a fraction in lowest terms.

Solution

$$2.16\frac{2}{3} = 2 + .16\frac{2}{3}$$

$$0.16\frac{2}{3} \longrightarrow \frac{16\frac{2}{3}}{100} = \frac{\frac{50}{3}}{100} = \frac{50}{3} \div \frac{100}{1} = \frac{\overset{1}{\cancel{50}}}{3} \cdot \frac{1}{\underset{2}{\cancel{100}}} = \frac{1}{6}$$

1 0 0

Therefore, $2.16\frac{2}{3} = 2 + \frac{1}{6} = 2\frac{1}{6}$. ■

EXERCISES 3.10B

Change the complex decimals to fractions or mixed numbers reduced to lowest terms.

1. $0.37\frac{1}{2}$ **2.** $0.62\frac{1}{2}$ **3.** $0.33\frac{1}{3}$ **4.** $0.1\frac{1}{4}$ **5.** $0.5\frac{3}{4}$ **6.** $2.062\frac{1}{2}$

7. $1.0\frac{1}{5}$ **8.** $0.0\frac{2}{3}$ **9.** $0.00\frac{5}{12}$ **10.** $2.00\frac{1}{4}$ **11.** $0.001\frac{1}{6}$ **12.** $1.05\frac{3}{8}$

13. $0.8\frac{1}{3}$ **14.** $0.2\frac{6}{7}$ **15.** $0.13\frac{1}{3}$ **16.** $0.16\frac{2}{3}$ **17.** $2.10\frac{5}{7}$ **18.** $3.08\frac{8}{9}$

3.11 Operations with Decimals and Fractions

When a problem involves both decimals and fractions, either change the decimal to its fractional form or change the fraction to its decimal form, then perform the indicated operation.

Example 1 Find $3.2 + \frac{3}{4}$.

First solution Change the fraction $\frac{3}{4}$ to a decimal, then add.

$$\frac{3}{4} = 4\overline{)3.00} = .75$$

```
     .7 5
 4 | 3.0 0
     2 8
     ---
       2 0
       2 0
```

Therefore, $3.2 + \frac{3}{4} = 3.2 + .75 = \begin{array}{r} 3.2 \\ +\ .75 \\ \hline 3.95 \end{array}$

Second solution Change the decimal 3.2 to a fraction, then add.

$$3.2 = 3\frac{2}{10} = 3\frac{1}{5} \qquad \text{Reduce by 2}$$

Therefore,

$$3.2 + \frac{3}{4} = 3\frac{1}{5} + \frac{3}{4} = \begin{array}{r} 3\frac{1}{5} = 3\frac{4}{20} \\ +\ \frac{3}{4} = \frac{15}{20} \\ \hline 3\frac{19}{20} \end{array} \qquad \blacksquare$$

Example 2 Find $\frac{5}{12} \times 0.45$. Write your answer in fractional form.

Solution Because we want the answer in fractional form, we will change the decimal 0.45 to a fraction.

$$0.45 = \frac{45}{100} = \frac{9}{20} \qquad \text{Reduced by 5}$$

Then,

$$\frac{5}{12} \times 0.45 = \frac{\overset{1}{\cancel{5}}}{\underset{4}{\cancel{12}}} \times \frac{\overset{3}{\cancel{9}}}{\underset{4}{\cancel{20}}} = \frac{3}{16} \qquad \blacksquare$$

Example 3 Find $\dfrac{0.76}{\frac{4}{5}}$. Write your answer in decimal form.

Solution Because we want the answer in decimal form, we will change the fraction $\frac{4}{5}$ to a decimal.

$$\frac{4}{5} = 5\overline{)4.0} = .8$$

Then,

$$\frac{0.76}{\frac{4}{5}} = \frac{0.76}{0.8} = 0.8_\wedge\overline{)0.7_\wedge 60} = .95$$

$$\begin{array}{r} 72 \\ \hline 40 \\ 40 \\ \hline \end{array} \qquad \blacksquare$$

Example 4 Find the cost of $9\frac{1}{4}$ feet of copper pipe selling for 98 cents a foot. Round off your answer to the nearest cent.

Solution Because Example 4 deals with money rounded to the nearest cent, we will change the fraction $9\frac{1}{4}$ to a decimal, then multiply.

$$9\frac{1}{4} = \frac{37}{4} = \begin{array}{r} 9.25 \\ 4\overline{)37.00} \\ \underline{36} \\ 1\,0 \\ \underline{8} \\ 20 \\ \underline{20} \end{array}$$

Therefore,

$$\begin{array}{r} 9.25 \\ \times\ .98 \\ \hline 7400 \\ 8325 \\ \hline 9.0650 \end{array} \approx \$9.07$$

rounded to the nearest cent ■

EXERCISES 3.11

In Exercises 1–8, perform the indicated operations. Write your answers in decimal form.

1. $4.62 + \frac{7}{10}$

2. $9.6 - \frac{3}{100}$

3. $2\frac{3}{20} - 1.6$

4. $1\frac{6}{25} + 5.9$

5. $\frac{5}{8} \times 0.6$

6. $0.84 \times \frac{5}{16}$

7. $\dfrac{\frac{7}{20}}{0.014}$

8. $\dfrac{7.6}{\frac{19}{50}}$

In Exercises 9–16, perform the indicated operations. Write your answers in fractional form.

9. $\frac{7}{8} + 0.35$

10. $3.6 + 4\frac{2}{3}$

11. $6.75 - 2\frac{5}{6}$

12. $8\frac{1}{12} - 4.125$

13. $0.56 \times \frac{5}{6}$

14. $1\frac{2}{3} \times 2.7$

15. $\dfrac{2\frac{5}{6}}{0.85}$

16. $\dfrac{1.25}{1\frac{9}{16}}$

In Exercises 17–20, (a) estimate the answers, (b) use a calculator to find the answers rounded off to the nearest thousandths.

17. $8\frac{5}{6} + 4.797$

18. $9.0836 - 6\frac{1}{3}$

19. $7\frac{2}{11} \times 0.1238$

20. $\dfrac{19\frac{1}{6}}{41.6}$

In Exercises 21–26, round off your answers to the nearest cent when necessary.

21. Debbie needs $2\frac{3}{8}$ yards of material to make a dress. If she buys a fabric costing \$3.95 a yard, what is the total cost of the material?

22. Find the cost of $2\frac{3}{4}$ yards of silk at \$4.59 a yard.

23. Carmen makes $6.45 an hour. How much does she earn in a $7\frac{1}{2}$-hour day?

24. Kevin makes $7.80 an hour on Saturday. How much does he earn for $5\frac{3}{4}$ hours of work at that rate?

25. Suppose you drive 18,000 miles a year, and your car gets $22\frac{1}{2}$ miles per gallon. If the average cost of gasoline is $1.05 per gallon, how much do you spend on gasoline in one year?

26. Larry paid $60.50 for fabric to reupholster his chair. If he bought $3\frac{2}{3}$ yards of fabric, find the cost per yard of the fabric.

3.12 Comparing Decimals

Sometimes we need to recognize which of a group of decimals is largest. *We can compare the sizes of decimals by writing them all with the same number of decimal places.*

Example 1 Arrange the following decimals in order of size—the largest first: 0.27, 0.205, 0.2, 0.250.

Solution Because the largest number of decimal places in any of the given decimals is three, we add final zeros wherever necessary to change all the given decimals to three decimal places.

$$0.27\boxed{0},\quad 0.205,\quad 0.2\boxed{00},\quad 0.250$$

Now arrange the 3-decimal-place numbers in order of size—the largest first.

$$0.270,\quad 0.250,\quad 0.205,\quad 0.200$$

Therefore, the *original* decimals arranged according to size are

$$0.27,\quad 0.250,\quad 0.205,\quad 0.2 \quad ■$$

Example 2 Which is larger, $\frac{5}{8}$ or 0.65?

Solution Change $\frac{5}{8}$ to a decimal $\quad \frac{5}{8} = 8\overline{)5.000}$ = .625

And $\quad 0.65 = 0.65\boxed{0}$

We see $\quad 0.650 > 0.625$

therefore, $\quad 0.65 > \frac{5}{8}$ ■

EXERCISES 3.12

In Exercises 1–8, arrange the decimals in order of size—the largest first.

1. 0.409, 0.49, 0.41, 0.4

2. 0.35, 0.3, 0.305, 0.335

3. 3.075, 3.1, 3.05, 3.009

4. 7.0, 7.1, 7.08, 7.099

5. 0.075, 0.07501, 0.0749, 0.07

6. 0.06, 0.1998, 0.6, 0.059

7. 5.05, 5.5, 5.0501, 5, 5.0496

8. 8.0505, 8.051, 8.0695, 8.199, 8

In Exercises 9–12, determine which of the two symbols > or < should be used to make each statement true.

9. $\frac{3}{4}$? 0.8

10. $\frac{3}{8}$? 0.35

11. 0.5 ? $\frac{9}{20}$

12. 0.075 ? $\frac{2}{25}$

13. A drill press operator uses a $\frac{5}{16}$-inch drill to put a hole through a steel bracket. Would a 0.325-inch pin fit in the hole?

14. Will a $\frac{7}{8}$-inch pin fit in a 0.8215-inch hole?

15. Star Fish tuna costs $1.02 for 7 ounces, and Whale of the Sea tuna costs $1.89 for 12 ounces. Which is the best buy, and by how much? (Round off to the nearest tenth of a cent.)

16. Which brand is the better buy?
Brand A: $1.98 for 24 ounces
Brand B: $1.20 for 16 ounces
Brand C: $0.69 for 7.5 ounces

3.13 Chapter Summary

Decimal 3.1

A **decimal** is a fraction whose denominator is a power of 10.

To Read a Decimal 3.1

1. Read the number to the left of the decimal point as a whole number.
2. Say "and" for the decimal point.
3. Read the number to the right of the decimal point as a whole number, then say the name of the place occupied by the right-hand digit of the number.

To Round Off a Decimal 3.2

See the box on page 100.

The Number of Decimal Places 3.3

The number of decimal places in a number is the number of digits written to the right of the decimal point.

To Add Decimals 3.3

Write the numbers under one another with their decimal points in the same vertical line, then add the numbers like whole numbers. Place the decimal point in your answer in the same vertical line as the other decimal points.

To Subtract Decimals 3.4

Write the number being subtracted under the number it is being subtracted from, with their decimal points in the same vertical line. Subtract the numbers as whole numbers. Place the decimal point in the answer in the same vertical line as the other decimal points.

To Multiply Decimals 3.5

Multiply the numbers as whole numbers. Your answer has as many decimal places as the sum of the decimal places in the numbers being multiplied.

To Divide Decimals 3.6

1. Place a caret ($\wedge$) in the divisor to make it the smallest possible whole number.
2. Place a caret in the dividend the same number of places to the right (or left) of its decimal point as was done in the divisor.

3. Place the decimal point in the quotient directly above the caret in the dividend.
4. Divide the numbers as whole numbers.

To Multiply a Decimal by a Power of 10 3.7 To multiply a decimal by a power of 10, move the decimal point to the right as many places as the number of zeros in the power of 10.

To Divide a Decimal by a Power of 10 3.7 To divide a decimal by a power of 10, move the decimal point to the left as many places as the number of zeros in the power of 10.

To Change a Fraction to a Decimal 3.9 To change a fraction to a decimal, divide the numerator by the denominator.

To Change a Decimal to a Fraction 3.10

1. Read the decimal, then write it in fraction form (with numerator and denominator).
2. Reduce the fraction to lowest terms.

Order of Operations 3.8 The order of operations for decimals is the same as the order of operations for whole numbers and fractions.

Comparing Decimals 3.12 To compare decimals write all of them with the same number of decimal places.

Review Exercises 3.13

In Exercises 1 and 2, write the numbers in words.

1. 0.145

2. 250.06

In Exercises 3 and 4, write the numbers using digits.

3. Sixteen ten-thousandths

4. Five hundred and seventy-five hundredths

In Exercises 5–8, round off at the indicated place.

5. 0.83671 thousandths

6. 0.5967 hundredths

7. 24.74 tenths

8. 185.75 tens

In Exercises 9–20, perform the indicated operations.

9. $75.23 + 186.6 + 34{,}932 + 8.0205$

10. $23 + 5.6 + 0.875 + 0.0016 + 7.29$

11. 7.8×0.64

12. $0.0817 \div 0.95$

13. $509.6 - 41.345$

14. $7.85 \times 1{,}000$

15. $0.64 \div 10^2$

16. 0.014×0.59

17. $57.4 \div 28$

18. $10^4 \times 0.0056$

19. $476 - 39.4$

20. $9.6 \div 100$

In Exercises 21 and 22, divide and round off your answer to two decimal places.

21. $50 \div 4.8$

22. $0.262 \div 0.086$

In Exercises 23–26, perform the indicated operations in the correct order.

23. $12.5 + 5.6 \times 10^2$

24. $67 - (9.6 - 1.8)$

25. $48 \div 0.8 \times 0.3$

26. $2(2.7) + 2(1.9)$

In Exercises 27 and 28, change the fraction and mixed number to exact decimals.

27. $\frac{7}{8}$

28. $5\frac{1}{18}$

In Exercises 29 and 30, change the fraction and mixed number to decimals rounded off to two decimal places.

29. $\frac{5}{6}$

30. $3\frac{7}{11}$

In Exercises 31–34, change the decimals to fractions or mixed numbers in lowest terms.

31. 0.68

32. 4.025

33. $0.16\frac{2}{3}$

34. $0.4\frac{1}{6}$

In Exercises 35 and 36, perform the indicated operations and write your answer in decimal form.

35. $7.06 - \frac{5}{8}$

36. $\dfrac{4\frac{1}{20}}{0.9}$

In Exercises 37 and 38, perform the indicated operations and write your answer in fraction form.

37. $\frac{9}{10} + 0.24$

38. $0.025 \times 3\frac{1}{3}$

In Exercises 39 and 40, estimate the answers to the following problems.

39. $0.32(6.87 + 4.32)$

40. $\dfrac{0.52 \times 0.47}{7.9}$

In Exercises 41 and 42, arrange the decimals in order of size, largest to smallest.

41. 0.6, 0.603, 0.063, 0.06

42. 3.89, 3.9, 3.098, 3.908, 3

43. Find the average of 8.7, 6.2, and 7.6.

44. Find the average of 0.62, 0.57, 0.48, and 0.53.

45. A rectangle has a length of 6.25 meters and a width of 4.3 meters.
  a. Find the perimeter of the rectangle.
  b. Find the area of the rectangle.

46. The length of the side of a square is 3.4 centimeters.
  a. Find the perimeter of the square.
  b. Find the area of the square.

47. How many pieces of wire 1.5 meters long can be cut from a spool 60 meters long?

48. A machinist uses a $\frac{11}{16}$-inch drill to put a hole through a metal plate. Would a 0.625-inch pin fit in the hole?

49. Carlos bought one pair of shoes for \$19.95, two neckties for \$3.95 each, three pairs of socks for \$1.25 a pair, and one suit for \$89.95. What was his total bill?

50. At the beginning of the month, Jim's bank balance was \$275.38. During the month he wrote the following checks: \$15.98, \$46.75, \$87.45, \$135.46, and \$68. He made deposits of \$250 and \$350. Find his bank balance at the end of the month.

51. Nora made 18 equal monthly payments on her new stereo set. If the total cost of the set was \$355, what was her monthly payment? (Round off to the nearest cent.)

52. Find the cost of $15\frac{1}{2}$ yards of electrical wire at 29 cents a yard. (Round off to the nearest cent.)

53. Rudy drove his car 9,600 miles last year. His total car expenses were \$625 for the year. Find the average cost per mile. Round off your answer to the nearest tenth of a cent.

54. A car travels 196 miles. If it gets 14 miles per gallon of gas, how many gallons did it use for the trip? If gas is \$1.329 per gallon, how much was spent for gas?

Chapter 3 Diagnostic Test

Allow yourself about 50 minutes to do these problems. Complete solutions for every problem, together with section references, are given in the answer section at the end of the book.

1. Write 9.015 in words.

2. Use digits to write four hundred twenty and five hundredths.

In Exercises 3–10, perform the indicated operations.

3. $7.8 + 56 + 0.017 + 500.94$

4. $40.6 - 3.54$

5. $9.073 - 0.87$

6. 3.7×0.058

7. $62.79 \div 0.078$

8. $26 + 1.3 \times 0.8$

9. 0.46×10^3

10. $6.9 \div 100$

11. Divide and round off to the nearest tenth.

$7.2 \div 0.35$

12. Change $2\frac{3}{16}$ to an exact decimal.

13. Change 0.78 to a fraction in lowest terms.

14. Change $0.1\frac{9}{11}$ to a fraction in lowest terms.

15. Perform the indicated operation and write your answer in decimal form.

$1\frac{4}{5} \times 0.65$

16. Estimate the answer.

$$\frac{9.9 \times 2.3}{0.42}$$

17. Arrange the decimals in order, largest to smallest.

7.48, 7.408, 7.084, 7.4

18. Sheila skated the 500-meters in 45.60 seconds, 42.76 seconds, and 44.02 seconds. What was her average time rounded to the nearest hundredth of a second?

19. The balance owed on a car amounts to $2,870, including interest. What would the monthly payment be in order to pay it off in 36 months? Round off the payment to the nearest cent (two decimal places).

20. At the beginning of the month, Jeff's bank balance was $346.52. During the month he wrote the following checks: $17.75, $64.57, $91.35, $135.46, and $186.40. He made a deposit of $325. Find his bank balance at the end of the month.

4 Ratio, Proportion, and Percent

4.1 Ratio Problems

Ratio

A **ratio** is a comparison of two quantities usually written in fraction form. In algebra, fractions are called *ratio*nal numbers.

The ratio of a to b is written $\frac{a}{b}$

The ratio of b to a is written $\frac{b}{a}$

where a and b are the *terms of the ratio.*

The terms of a ratio can be any kind of number; the only restriction is that the denominator cannot be zero.

A WORD OF CAUTION The order of the terms of a ratio is important. The number that is before the *to* is in the numerator, and the number that is after the *to* is in the denominator. Therefore, the ratio of 5 to 6 is $\frac{5}{6}$, not $\frac{6}{5}$. ☑

Example 1 In an arithmetic class there are 13 men and 17 women.

a. Find the ratio of men to women.

b. Find the ratio of women to men.

c. Find the ratio of men to the total number of students in the class.

Solution

a. The ratio of men to women is $\frac{13}{17}$.

b. The ratio of women to men is $\frac{17}{13}$.

c. The total number of students in the class is $13 + 17 = 30$.

The ratio of men to the total number of students in the class is $\frac{13}{30}$. ■

Reducing a Ratio to Lowest Terms

Because a ratio is a fraction, a ratio can be reduced to lowest terms using the same methods we used in Chapter 2 for reducing fractions to lowest terms.

TO REDUCE A RATIO TO LOWEST TERMS

1. Divide numerator and denominator by any number that you can see is a divisor of both.
2. When you cannot see any more common divisors, use prime factorization.

Example 2 Reduce the ratio of 48 to 30 to lowest terms.

Solution

$$\frac{48}{30} = \frac{\overset{\overset{8}{\cancel{24}}}{\cancel{48}}}{\underset{\underset{5}{\cancel{15}}}{\cancel{30}}} = \frac{8}{5}$$ ■

Example 3 Reduce the ratio of 42 to 156 to lowest terms.

Solution Write the prime factorization of numerator and denominator, then divide both by the common factors.

The prime factorization of $42 = 2 \cdot 3 \cdot 7$

The prime factorization of $156 = 2 \cdot 2 \cdot 3 \cdot 13$

$$\frac{42}{156} = \frac{\overset{1}{\cancel{2}} \cdot \overset{1}{\cancel{3}} \cdot 7}{\underset{1}{\cancel{2}} \cdot 2 \cdot \underset{1}{\cancel{3}} \cdot 13} = \frac{7}{26}$$ ■

When we are using a ratio to compare like quantities, the numerator and the denominator should be written in the *same units.* (See Example 4.)

Example 4 Reduce the ratio of 6 inches to 1 foot to lowest terms.

Solution

$$\frac{6 \text{ inches}}{1 \text{ foot}} = \frac{6 \cancel{\text{ inches}}}{12 \cancel{\text{ inches}}} = \frac{\overset{1}{\cancel{6}}}{\underset{2}{\cancel{12}}} = \frac{1}{2}$$

because 1 foot = 12 inches ■

Example 5 Reduce the ratio of 5 cents to one dollar to lowest terms.

Solution

$$\frac{5 \text{ cents}}{1 \text{ dollar}} = \frac{5 \cancel{\text{ cents}}}{100 \cancel{\text{ cents}}} = \frac{\overset{1}{\cancel{5}}}{\underset{20}{\cancel{100}}} = \frac{1}{20}$$

because 1 dollar = 100 cents ■

Example 6 Greg gets a 15-minute break for every 2 hours he works. Find the ratio of work time to break time.

Solution

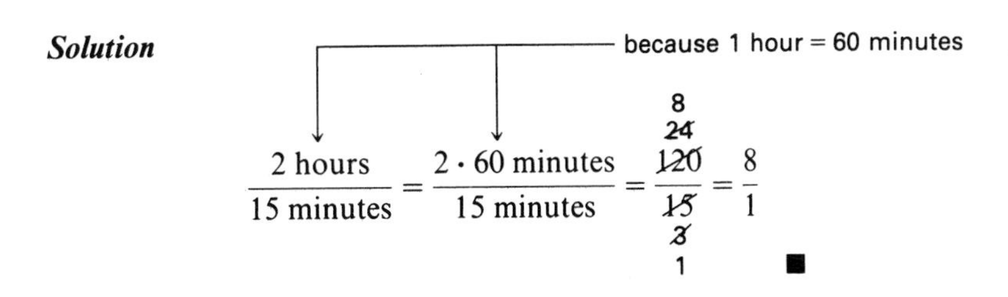

$$\frac{2 \text{ hours}}{15 \text{ minutes}} = \frac{2 \cdot 60 \text{ minutes}}{15 \text{ minutes}} = \frac{\overset{\overset{8}{\cancel{24}}}{\cancel{120}}}{\underset{\underset{1}{\cancel{3}}}{\cancel{15}}} = \frac{8}{1}$$ ■

NOTE Since a ratio is a comparison of two quantities, we will leave the 1 in the denominator in Example 6. ☑

Reducing a Ratio with Terms That Are Not Whole Numbers

In all the ratios studied so far, the terms have been whole numbers. This is not always the case. The terms of a ratio can be any kind of number, the only restriction being that the denominator cannot be 0.

Example 7 Simplify the ratio of $1\frac{3}{8}$ to 11.

Solution

$$\frac{1\frac{3}{8}}{11} = 1\frac{3}{8} \div 11 = \frac{11}{8} \div \frac{11}{1} = \frac{\overset{1}{\cancel{11}}}{8} \cdot \frac{1}{\underset{1}{\cancel{11}}} = \frac{1}{8}$$ ■

Example 8 Simplify the ratio of $1\frac{3}{4}$ to $3\frac{1}{2}$.

Solution

$$\frac{1\frac{3}{4}}{3\frac{1}{2}} = 1\frac{3}{4} \div 3\frac{1}{2} = \frac{7}{4} \div \frac{7}{2} = \frac{\cancel{7}}{\underset{2}{\cancel{4}}} \cdot \frac{\overset{1}{\cancel{2}}}{\cancel{7}} = \frac{1}{2}$$ ■

Example 9 Simplify the ratio of 3.6 to 2.4.

Solution

$$\frac{3.6}{2.4} = \frac{36}{24} = \frac{\overset{3}{\cancel{36}}}{\underset{2}{\cancel{24}}} = \frac{3}{2}$$ ■

Example 10 A 6-foot man casts a $7\frac{1}{2}$-foot shadow. Find the ratio of the man's height to the length of his shadow.

Solution

$$\frac{\text{height of man}}{\text{length of shadow}} = \frac{6\ \cancel{\text{feet}}}{7\frac{1}{2}\ \cancel{\text{feet}}} = 6 \div 7\frac{1}{2} = \frac{6}{1} \div \frac{15}{2} = \frac{\overset{2}{\cancel{6}}}{1} \cdot \frac{2}{\underset{5}{\cancel{15}}} = \frac{4}{5}$$

This means that the man's height is $\frac{4}{5}$ of the length of his shadow. ■

EXERCISES 4.1

In Exercises 1–24, reduce the ratios to lowest terms.

1. 18 to 27
2. 32 to 48
3. 135 to 60
4. 40 to 16
5. 60 to 12
6. 75 to 15
7. 85 cents to 15 cans
8. 42 yards to 12 dresses
9. 20 cents to 1 dollar
10. 1 hour to 15 minutes
11. 9 inches to 1 foot
12. 45 cents to 3 dollars
13. 2 hours to 40 minutes
14. 18 inches to 2 feet
15. 6 to $\frac{2}{3}$
16. 15 to $\frac{3}{5}$
17. $1\frac{1}{2}$ to 12
18. $2\frac{2}{3}$ to 24

19. $1\frac{1}{6}$ to $2\frac{1}{3}$

20. $4\frac{1}{2}$ to $3\frac{3}{4}$

21. 1.8 to 2.4

22. 7.5 to 4.5

23. 1.5 to 6

24. 2.4 to 12

25. A college basketball team won 24 out of 30 games played. There were no tie games.
a. Find the ratio of wins to games played.
b. Find the ratio of wins to losses.
c. Find the ratio of losses to wins.

26. In a history class 30 of the 48 students are women.
a. Find the ratio of women to students in the class.
b. Find the ratio of women to men.
c. Find the ratio of men to women.

27. A coffee shop received its order of pies. The order consisted of 12 apple pies, 8 cherry pies, and 4 lemon pies.
a. What is the ratio of apple pies to cherry pies?
b. What is the ratio of apple pies to lemon pies?
c. What is the ratio of cherry pies to the total number of pies?

28. A restaurant received its order of steaks. The order contained 20 top sirloin steaks, 15 porterhouse, and 10 filet mignon.
a. What is the ratio of porterhouse to filet mignon?
b. What is the ratio of top sirloin to porterhouse?
c. What is the ratio of filet mignon to the total number of steaks?

29. Bob can do a job in 2 hours 40 minutes. Hal takes 4 hours to do the same job. Find the ratio of Bob's time to Hal's time.

30. Machine A can produce 100 bushings in 2 hours. Machine B takes 1 hour 15 minutes to produce 100 bushings. What is the ratio of machine A's time to machine B's time?

31. A 15-foot pole casts a $4\frac{1}{2}$-foot shadow. What is the ratio of the length of the shadow to the height of the pole?

32. A room is $4\frac{2}{3}$-yards long and $3\frac{1}{2}$-yards wide. What is the ratio of the length to the width?

4.2 Proportion Problems

Proportion

A **proportion** is a statement that two ratios are equal. Common notation for a proportion is

$$\frac{a}{b} = \frac{c}{d}$$

Read: "*a* is to *b* as *c* is to *d*"
or: "*a* over *b* equals *c* over *d*"

Terms of a Proportion

The **terms of a proportion:**

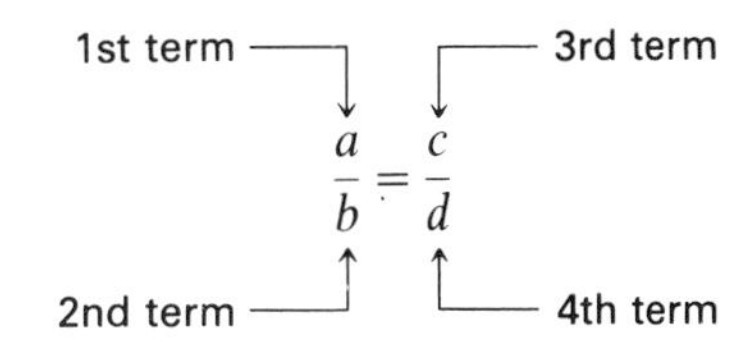

Means and Extremes

The **means and extremes** of a proportion:

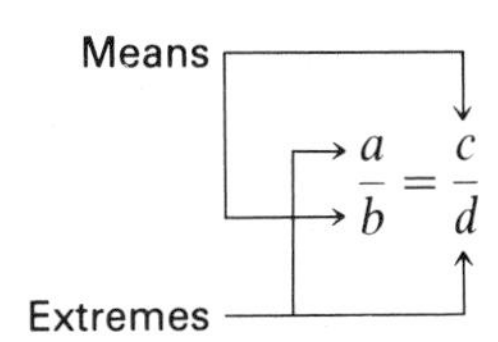

The *means* of this proportion are b and c.

The *extremes* of this proportion are a and d.

Example 1 In the proportion $\frac{2}{7} = \frac{6}{21}$, identify each term, the means, and the extremes.

Solution Read: "2 is to 7 as 6 is to 21."

The first term is 2. The second term is 7.
The third term is 6. The fourth term is 21.
The means are 7 and 6. The extremes are 2 and 21. ■

Product of Means Equals Product of Extremes In any proportion, *the product of the means equals the product of the extremes.* This is sometimes called the *cross-multiplication rule* and these products are called *cross products.*

In the proportion $\frac{2}{3} = \frac{4}{6}$,

$$\underbrace{3 \times 4}_{\text{Product of means}} = \underbrace{2 \times 6}_{\text{Product of extremes}}$$

> In the proportion
>
> $$\frac{a}{b} = \frac{c}{d}$$
>
> $$\underbrace{b \cdot c}_{\text{product of means}} = \underbrace{a \cdot d}_{\text{product of extremes}}$$
>
> (the cross products are equal)

We can use the cross-multiplication rule to see if two ratios form a proportion.

If the product of the means equals the product of the extremes, then the ratios form a proportion. If the product of the means does not equal the product of the extremes, then the ratios do not form a proportion.

Example 2 Do the given ratios form a proportion?

a. $\frac{5}{12} \stackrel{?}{=} \frac{3}{7}$. No, because $5 \cdot 7 \neq 12 \cdot 3$ This is not a proportion.

$$35 \neq 36$$

b. $\frac{16}{6} \stackrel{?}{=} \frac{8}{3}$. Yes, because $16 \cdot 3 = 6 \cdot 8$ This is a proportion.

$$48 = 48$$

c. $\frac{\frac{1}{2}}{3} \stackrel{?}{=} \frac{1\frac{1}{3}}{8}$. Yes, because $\frac{1}{2} \cdot 8 = 3 \cdot 1\frac{1}{3}$ This is a proportion.

$$\frac{1}{\cancel{2}_1} \cdot \frac{\cancel{8}^4}{1} = \frac{\cancel{3}^1}{1} \cdot \frac{4}{\cancel{3}_1}$$

$$4 = 4 \quad ■$$

Solving a Proportion

When three of the four terms of a proportion are known, it is always possible to find the value of the unknown term.

Example 3 In the proportion $\frac{2}{3} = \frac{x}{51}$, the third term is unknown and is represented by the letter x. Since the product of the means equals the product of the extremes,

$$\frac{2}{3} = \frac{x}{51}$$

$$3 \cdot x = 2 \cdot 51 \qquad \text{Cross products are equal}$$

$$3 \cdot x = 102$$

The = sign tells us that $3 \cdot x$ and 102 are the same number. Therefore, if $3 \cdot x$ and 102 are both divided by 3, the resulting numbers will be equal. That is,

$$\frac{\overset{1}{\cancel{3}} \cdot x}{\underset{1}{\cancel{3}}} = \frac{\overset{34}{\cancel{102}}}{\underset{1}{\cancel{3}}}$$

$$x = 34 \quad \blacksquare$$

METHOD OF SOLVING A PROPORTION FOR AN UNKNOWN LETTER

1. Set the product of the means equal to the product of the extremes. (The cross products are equal.)
2. Divide both sides by the number that is multiplying the unknown letter.

To Check Your Solution

1. Replace the unknown letter in the proportion by the value you obtained for it.
2. Cross-multiply, and verify that the cross products are equal.

Example 4 Solve $\frac{x}{25} = \frac{6}{5}$ for x.

Solution

$$\frac{x}{25} = \frac{6}{5}$$

$$5 \cdot x = 25 \cdot 6 \qquad \text{Cross products are equal}$$

$$5 \cdot x = 150 \qquad \text{Because 5 is multiplying } x \text{, divide both sides by 5}$$

$$\frac{\overset{1}{\cancel{5}} \cdot x}{\underset{1}{\cancel{5}}} = \frac{\overset{30}{\cancel{150}}}{\underset{1}{\cancel{5}}}$$

$$x = 30$$

Check

$$\frac{30}{25} \stackrel{?}{=} \frac{6}{5} \qquad \text{Replace } x \text{ by 30}$$

$$30 \cdot 5 \stackrel{?}{=} 25 \cdot 6$$

$$150 = 150 \quad \blacksquare$$

Example 5 Solve $\frac{2}{3} = \frac{x}{150}$ for x.

Solution

$$\frac{2}{3} = \frac{x}{150}$$

$$3 \cdot x = 2 \cdot 150 \qquad \text{Cross products are equal}$$

$$\frac{\overset{1}{\cancel{3}} \cdot x}{\underset{1}{\cancel{3}}} = \frac{\overset{100}{\cancel{300}}}{\underset{1}{\cancel{3}}} \qquad \text{Divide both sides by 3}$$

$$x = 100$$

Check

$$\frac{2}{3} \stackrel{?}{=} \frac{100}{150} \qquad \text{Replace } x \text{ by 100}$$

$$2 \cdot 150 \stackrel{?}{=} 3 \cdot 100$$

$$300 = 300 \quad \blacksquare$$

Example 6 Solve $\frac{42}{x} = \frac{28}{12}$ for x.

Solution We can simplify our work by reducing the ratio $\frac{28}{12}$ to lowest terms before cross-multiplying.

$$\frac{42}{x} = \frac{28}{12}$$

$$\frac{42}{x} = \frac{7}{3} \qquad \text{Reduce } \frac{28}{12} \text{ to } \frac{7}{3}$$

$$7 \cdot x = 42 \cdot 3 \qquad \text{Cross-multiply}$$

$$\frac{\overset{1}{\cancel{7}} \cdot x}{\underset{1}{\cancel{7}}} = \frac{\overset{18}{\cancel{126}}}{\underset{1}{\cancel{7}}} \qquad \text{Divide both sides by 7}$$

$$x = 18$$

Check

$$\frac{42}{18} \stackrel{?}{=} \frac{28}{12} \qquad \text{Replace } x \text{ by 18}$$

$$42 \cdot 12 \stackrel{?}{=} 18 \cdot 28$$

$$504 = 504 \quad \blacksquare$$

Example 7 Solve $\frac{x}{2} = \frac{15}{25}$ for x.

Solution

$$\frac{x}{2} = \frac{15}{25}$$

$$\frac{x}{2} = \frac{3}{5} \qquad \text{Reduce } \frac{15}{25} \text{ to } \frac{3}{5}$$

$$5 \cdot x = 2 \cdot 3 \qquad \text{Cross-multiply}$$

$$\frac{\overset{1}{\cancel{5}} \cdot x}{\underset{1}{\cancel{5}}} = \frac{6}{5} \qquad \text{Divide both sides by 5}$$

$$x = 1\frac{1}{5}$$

Check

$$\frac{1\frac{1}{5}}{2} \stackrel{?}{=} \frac{15}{25} \qquad \text{Replace } x \text{ by } 1\frac{1}{5}$$

$$1\frac{1}{5} \cdot 25 \stackrel{?}{=} 2 \cdot 15$$

$$\frac{6}{\underset{1}{\cancel{5}}} \cdot \frac{\overset{5}{\cancel{25}}}{1} \stackrel{?}{=} 30$$

$$30 = 30 \quad \blacksquare$$

Solving a Proportion with Terms That Are Not Whole Numbers

In all the ratios studied so far, the terms have been whole numbers. This is not always the case. The terms of a proportion can be any kind of number, the only restriction being that the denominator in either ratio cannot be 0.

Letters other than x are often used to represent the unknown term.

Example 8 Solve $\frac{P}{3} = \frac{\frac{5}{6}}{5}$ for P.

Solution

$$\frac{P}{3} = \frac{\frac{5}{6}}{5}$$

$$5 \cdot P = \frac{\overset{1}{\cancel{3}}}{1} \cdot \frac{5}{\underset{2}{\cancel{6}}}$$

$$5 \cdot P = \frac{5}{2}$$

$$\frac{\cancel{5} \cdot P}{\cancel{5}} = \frac{\frac{5}{2}}{5}$$

$$P = \frac{5}{2} \div 5 = \frac{\overset{1}{\cancel{5}}}{2} \cdot \frac{1}{\underset{1}{\cancel{5}}} = \frac{1}{2} \quad \blacksquare$$

Example 9 Solve $\frac{3\frac{1}{2}}{5\frac{1}{4}} = \frac{x}{4}$ for x.

Solution

$$\frac{3\frac{1}{2}}{5\frac{1}{4}} = \frac{x}{4}$$

$$5\frac{1}{4} \cdot x = 3\frac{1}{2} \cdot 4$$

$$\frac{21}{4} \cdot x = \frac{7}{\cancel{2}_1} \cdot \frac{\cancel{4}^2}{1}$$

$$\frac{\frac{\cancel{21}}{\cancel{4}} \cdot x}{\frac{\cancel{21}}{\cancel{4}}} = \frac{14}{\frac{21}{4}}$$

$$x = 14 \div \frac{21}{4}$$

$$x = \frac{\cancel{14}^2}{1} \cdot \frac{4}{\cancel{21}_3}$$

$$x = \frac{8}{3} = 2\frac{2}{3} \quad \blacksquare$$

Example 10 Solve $\frac{1.5}{100} = \frac{A}{2.4}$ for A.

Solution

$$\frac{1.5}{100} = \frac{A}{2.4}$$

$$100 \cdot A = 1.5 \times 2.4$$

$$\frac{\cancel{100}^1 \cdot A}{\cancel{100}_1} = \frac{3.6}{100}$$

$$A = 0.036 \quad \blacksquare$$

EXERCISES 4.2

1. In the proportion $\frac{8}{14} = \frac{16}{28}$, find the following:

a. the first term
b. the second term
c. the third term
d. the fourth term
e. the means
f. the extremes

2. In the proportion $\frac{3}{5} = \frac{x}{20}$, find the following:

a. the first term
b. the second term
c. the third term
d. the fourth term
e. the means
f. the extremes

In Exercises 3–8, do the given ratios form a proportion?

3. $\frac{9}{12} \stackrel{?}{=} \frac{6}{8}$ **4.** $\frac{3}{7} \stackrel{?}{=} \frac{4}{9}$ **5.** $\frac{6}{9} \stackrel{?}{=} \frac{10}{16}$ **6.** $\frac{\frac{2}{3}}{8} \stackrel{?}{=} \frac{\frac{3}{4}}{9}$ **7.** $\frac{2\frac{1}{4}}{27} \stackrel{?}{=} \frac{3\frac{1}{3}}{40}$ **8.** $\frac{60}{100} \stackrel{?}{=} \frac{4.2}{0.7}$

In Exercises 9–32, solve for the letter in each proportion.

9. $\frac{4}{7} = \frac{x}{21}$ **10.** $\frac{5}{12} = \frac{25}{x}$ **11.** $\frac{15}{12} = \frac{10}{x}$ **12.** $\frac{24}{30} = \frac{16}{x}$ **13.** $\frac{P}{100} = \frac{30}{40}$ **14.** $\frac{P}{100} = \frac{75}{125}$

15. $\frac{2}{3} = \frac{x}{4}$ **16.** $\frac{4}{5} = \frac{3}{x}$ **17.** $\frac{10}{x} = \frac{15}{4}$ **18.** $\frac{6}{5} = \frac{9}{x}$ **19.** $\frac{30}{45} = \frac{x}{12}$ **20.** $\frac{12}{21} = \frac{x}{56}$

21. $\frac{7}{100} = \frac{A}{5}$ **22.** $\frac{9}{100} = \frac{36}{B}$ **23.** $\frac{\frac{3}{4}}{6} = \frac{x}{16}$ **24.** $\frac{\frac{4}{5}}{2} = \frac{x}{25}$ **25.** $\frac{x}{9} = \frac{3\frac{1}{3}}{5}$ **26.** $\frac{x}{8} = \frac{2\frac{1}{4}}{18}$

27. $\frac{\frac{5}{6}}{1\frac{1}{4}} = \frac{24}{x}$ **28.** $\frac{x}{2\frac{1}{2}} = \frac{6}{1\frac{2}{3}}$ **29.** $\frac{6.5}{100} = \frac{A}{3.2}$ **30.** $\frac{4}{8.24} = \frac{0.5}{y}$ **31.** $\frac{P}{100} = \frac{4.8}{1.5}$ **32.** $\frac{7.6}{y} = \frac{1.9}{5}$

4.3 Word Problems Using Proportions

SOLVING WORD PROBLEMS THAT LEAD TO PROPORTIONS

1. Represent the unknown quantity by a letter.
2. Set up a proportion with the unit of measure next to each number in your proportion.
3. The same units must occupy corresponding positions in the two ratios of your proportion.

Correct Arrangements	*Incorrect Arrangements*
$\frac{\text{miles}}{\text{hours}} = \frac{\text{miles}}{\text{hours}}$	$\frac{\text{dollars}}{\text{weeks}} = \frac{\text{weeks}}{\text{dollars}}$
$\frac{\text{hours}}{\text{miles}} = \frac{\text{hours}}{\text{miles}}$	$\frac{\text{dollars}}{\text{weeks}} = \frac{\text{dollars}}{\text{days}}$
$\frac{\text{miles}}{\text{miles}} = \frac{\text{hours}}{\text{hours}}$	

4. Once the numbers have been correctly entered in the proportion by using the units as a guide, drop the units when cross-multiplying to solve for the unknown.

Example 1 A man used 10 gallons of gas on a 180-mile trip. How many gallons of gas can he expect to use on a 300-mile trip?

Solution Let x = number of gallons of gas for the 300-mile trip.

10 gallons of gas used for a 180-mile trip.	How many gallons of gas for a 300-mile trip?

$$\frac{10 \text{ gallons}}{180 \text{ miles}} = \frac{x \text{ gallons}}{300 \text{ miles}}$$

$$180 \cdot x = 10 \cdot 300$$

$$\frac{\overset{1}{\cancel{180}} \cdot x}{\underset{1}{\cancel{180}}} = \frac{\overset{50}{\cancel{3000}}}{\underset{3}{\cancel{180}}}$$

$$x = \frac{50}{3} = 16\frac{2}{3} \text{ gallons}$$ ■

NOTE The ratios used on each side have gallons in the numerator and miles in the denominator. ☑

Example 2 At a soda fountain 8 quarts of ice cream were used to make 100 milk shakes. How many quarts are needed to make 550 milk shakes?
Solution Let x = number of quarts of ice cream.

8 quarts of ice cream were used to make 100 milk shakes.	How many quarts are needed to make 550 milk shakes?

$$\frac{8 \text{ quarts}}{100 \text{ milk shakes}} = \frac{x \text{ quarts}}{550 \text{ milk shakes}}$$

$$100 \cdot x = 8 \cdot 550$$

$$\frac{100 \cdot x}{100} = \frac{4400}{100}$$

$$x = 44 \text{ quarts}$$ ■

Example 3 If a 6-foot man casts a $4\frac{1}{2}$-foot shadow, how tall is a tree that casts a 30-foot shadow?

Solution Let x = the height of the tree.

A six-foot man casts a $4\frac{1}{2}$-foot shadow	How tall is a tree that casts a 30-foot shadow?

$$\frac{6 \text{ feet high (man)}}{4\frac{1}{2} \text{ feet high (shadow)}} = \frac{x \text{ feet high (tree)}}{30 \text{ feet high (shadow)}}$$

$$\frac{6}{4\frac{1}{2}} = \frac{x}{30}$$

$$4\frac{1}{2} \cdot x = 6 \cdot 30$$

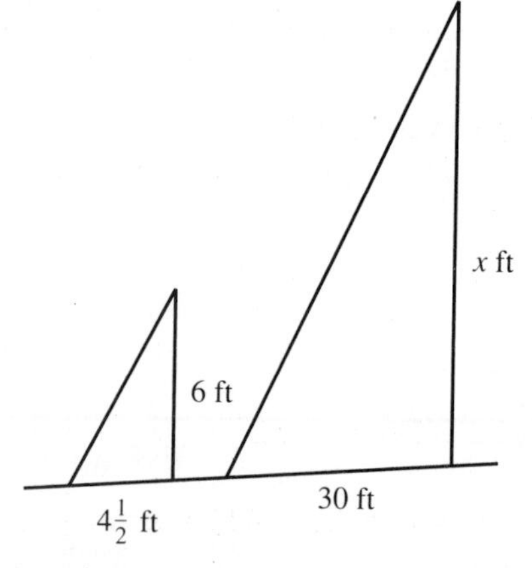

NOTE The ground does not need to be horizontal. ☑

$$\frac{9}{2} \cdot x = 180$$

$$\frac{\frac{9}{2} \cdot x}{\frac{9}{2}} = \frac{180}{\frac{9}{2}}$$

$$x = \frac{180}{1} \div \frac{9}{2}$$

$$x = \frac{\overset{20}{\cancel{180}}}{1} \cdot \frac{2}{\underset{1}{\cancel{9}}} = 40 \text{ feet}$$

■

EXERCISES 4.3

Solve the following problems using a proportion.

1. A painter uses about 3 gallons of paint in doing 2 rooms. How many gallons would he need to paint 20 rooms?
2. A person drives 600 miles in $1\frac{1}{2}$ days. How long would it take to drive 3,000 miles?
3. A 6-foot man has a 4-foot shadow when a tree casts a 20-foot shadow. How tall is the tree?
4. Seven men finish 10 houses in a month. How many houses could 35 men finish in the same time?
5. An investment of $3,000 earned $180 for a year. How much would have to be invested to earn $540 in the same time?
6. A store has a bargain price of 85¢ for three jars of grape jelly. How many jars could someone buy for $5.95?
7. The property tax on a $15,000 home is $450. What would be the tax on a $25,000 home?
8. A man used 12 gallons of gas on a 200-mile trip. How many gallons of gas can he expect to use on a 750-mile trip?
9. The ratio of a woman's weight on earth compared to her weight on the moon is 6 to 1. How much would a 150-pound woman weigh on the moon?
10. The ratio of a man's weight on Mars compared to his weight on earth is 2 to 5. How much would a 180-pound man weigh on Mars?
11. The ratio of the weight of lead to the weight of an equal volume of aluminum is 21 to 5. If an aluminum bar weighs 150 pounds, what would a lead bar of the same size weigh?
12. The ratio of the weight of platinum to the weight of an equal volume of copper is 12 to 5. If a platinum bar weighs 18 pounds, what would a copper bar of the same size weigh?
13. A market is selling three cans of beets for 99 cents. How much will 12 cans cost at the same rate?
14. A baseball team wins seven of its first 12 games. How many would you expect it to win out of its first 36 games if the team continues to play with the same degree of success?
15. On a map $\frac{1}{2}$ inch represents 12 miles. How many miles would $2\frac{1}{4}$ inches represent?
16. An apartment house manager spent 22 hours painting 3 apartments. How long can he expect to take painting the remaining 15 apartments?
17. Ralph drove 420 miles in $\frac{3}{4}$ of a day. About how far can he drive in $2\frac{1}{2}$ days?
18. Amalia noticed that her shadow was 4 feet long when that of a 16-foot flagpole was 12 feet long. What is her height in feet and inches?

19. A crew of 10 men take a week to overhaul 25 trucks in a fleet of 100 trucks. How many men would it take to complete the fleet overhaul in one week?

20. A car burns $2\frac{1}{2}$ quarts of oil on an 1800-mile trip. How many quarts of oil can the owner expect to use on a 12,000-mile trip?

21. The scale in an architectural drawing is 1 inch equals 8 feet. What are the dimensions of a room that measures $2\frac{1}{2}$ by 3 inches on the drawing? (Hint: Find the length and width of the room separately.)

22. Fifteen defective axles were found in 100,000 cars of a particular model. How many defective axles would you expect to find in the 2 million cars made of that same model?

4.4 Percent

Consider a 5-gallon paint can containing 2 gallons of paint. We can say, "The can is $\frac{2}{5}$ full."

5 gal

2 gal

Because $\frac{2}{5} = \frac{2 \cdot 20}{5 \cdot 20} = \frac{40}{100}$, we can also say, "The can is $\frac{40}{100}$ full." The decimal representation for $\frac{40}{100}$ is .40.

Percent

The word **percent** means per hundred. (*Cent*um means 100 in Latin.) The symbol for percent is %. Since $\frac{40}{100}$ is 40 per hundred, we say, "The can is 40% full."

We have described the amount of paint in the can in three different ways:

1. The can is $\frac{2}{5}$ full. *Fraction* representation
2. The can is 0.40 full. *Decimal* representation
3. The can is 40% full. *Percent* representation

Example 1 On a test a student answered 17 out of 20 problems correctly. We can describe his score three ways.

a. $\frac{17}{20}$ correct *fraction form*

b. $\frac{17}{20} =$ .85 correct (20 ⟌17.00) *decimal form*

c. $\frac{17}{20} = \frac{17 \cdot 5}{20 \cdot 5} = \frac{85}{100} = 85\%$ *percent form* ■

$\frac{17}{20}$, 0.85, and 85% are three different ways of saying the same thing. We need to know how to change from one form to another.

4.4A Changing a Decimal to a Percent

Because percent means the number of hundredths, to change a decimal to a percent, determine the number of hundredths in the decimal and then write the percent (%) symbol.

Example 2 Changing a decimal to a percent.

a. 0.25 = 25 hundredths = 25%

b. 0.03 = 3 hundredths = 3%

c. 0.5 = 0.50 = 50 hundredths = 50% ■

Notice in Example 2 that:

> TO CHANGE A DECIMAL TO A PERCENT
>
> Move the decimal point two places to the *right* and write the percent symbol (%).

Example 3 Changing a decimal to a percent.

a. 0.136 = 0.13‸6 = 13.6%

b. 0.07 = 0.07‸ = 7%

c. 0.3 = 0.30‸ = 30%

d. 2 = 2.00‸ = 200% ■

EXERCISES 4.4A

Change each decimal to a percent.

1. 0.27	**2.** 0.35	**3.** 0.06	**4.** 0.125	**5.** 1.4	**6.** 2.05
7. 0.186	**8.** 0.015	**9.** 0.075	**10.** 0.175	**11.** 2.9	**12.** 3.8
13. 2.005	**14.** 3.015	**15.** 1.36	**16.** 2.11	**17.** 4	**18.** 3

4.4B Changing a Percent to a Decimal

Because percent means the number of hundredths, to change a percent to a decimal, write the percent in the hundredths column and remove the percent symbol.

Example 4 Changing a percent to a decimal.

a. 75% = 75 hundredths = 0.75

b. 8% = 8 hundredths = 0.08

c. 20% = 20 hundredths = 0.20 = 0.2 ■

Notice in Example 4 that:

> TO CHANGE A PERCENT TO A DECIMAL
>
> Move the decimal point two places to the *left* and remove the percent symbol (%).

Example 5 Changing a percent to a decimal.

a. $38.5\% = {}_\wedge 38.5\% = 0.385$

b. $2\% = {}_\wedge 02.\% = 0.02$

c. $80\% = {}_\wedge 80.\% = 0.8$

d. $400\% = 4{}_\wedge 00.\% = 4$

e. $5\frac{1}{2}\% = 5.5\% = {}_\wedge 05.5\% = 0.055$

First change the mixed number $5\frac{1}{2}$ into the decimal 5.5 ■

EXERCISES 4.4B

Change each percent to a decimal. If the decimal is not exact, round off to four decimal places.

1. 45% **2.** 78% **3.** 125% **4.** 150% **5.** 6.5% **6.** 8.6%

7. 9% **8.** 7% **9.** $2\frac{1}{2}\%$ **10.** $4\frac{3}{4}\%$ **11.** $3\frac{1}{4}\%$ **12.** $5\frac{2}{5}\%$

13. 10.05% **14.** 2.08% **15.** $\frac{3}{4}\%$ **16.** $\frac{1}{2}\%$ **17.** $66\frac{2}{3}\%$ **18.** $33\frac{1}{3}\%$

4.4C Changing a Fraction to a Percent

> TO CHANGE A FRACTION TO A PERCENT
>
> **1.** Change the fraction to a decimal by dividing.
>
> **2.** Change the decimal to a percent by moving the decimal point two places to the right and writing the percent symbol.

Example 6 Changing a fraction to a percent.

fraction → decimal → percent

a. $\frac{3}{4} = 4\overline{)3.00}$ giving $.75 = 0.75{}_\wedge = 75\%$

b. $\frac{3}{5} = \begin{array}{r} .6 \\ 5\overline{)3.0} \end{array} = 0.60_{\wedge} = 60\%$

c. $1\frac{1}{8} = \frac{9}{8} = \begin{array}{r} 1.125 \\ 8\overline{)9.000} \end{array} = 1.12_{\wedge}5 = 112.5\%$ ■

EXERCISES 4.4C

Change each fraction to a percent. If it is not an exact answer, round off to two decimal places.

1. $\frac{1}{2}$ **2.** $\frac{1}{4}$ **3.** $\frac{2}{5}$ **4.** $\frac{4}{5}$ **5.** $\frac{3}{8}$ **6.** $\frac{7}{10}$

7. $\frac{9}{10}$ **8.** $\frac{3}{20}$ **9.** $\frac{4}{25}$ **10.** $\frac{17}{25}$ **11.** $\frac{1}{3}$ **12.** $\frac{1}{6}$

13. $\frac{5}{6}$ **14.** $\frac{5}{3}$ **15.** $\frac{7}{16}$ **16.** $1\frac{3}{4}$ **17.** $2\frac{5}{8}$ **18.** $1\frac{2}{5}$

4.4D Changing a Percent to a Fraction

Because percent means number of hundredths, to change a percent to a fraction, write the percent as a fraction over 100 and reduce the fraction to lowest terms.

$$25\% = 25 \text{ hundredths} = \frac{25}{100} = \frac{1}{4}$$

↑ Reduced form

> TO CHANGE A PERCENT TO A FRACTION
>
> **1.** Write the percent as the numerator of a fraction with a denominator of 100.
>
> **2.** Reduce the fraction to lowest terms.

Example 7 Changing a percent to a fraction.

a. $6\% = \frac{6}{100} = \frac{3}{50}$

b. $20\% = \frac{20}{100} = \frac{1}{5}$

c. $150\% = \frac{150}{100} = \frac{3}{2} = 1\frac{1}{2}$

d. $12\frac{1}{2}\% = \frac{12\frac{1}{2}}{100} = \frac{25}{2} \div \frac{100}{1} = \frac{\overset{1}{\cancel{25}}}{2} \cdot \frac{1}{\underset{4}{\cancel{100}}} = \frac{1}{8}$ ■

EXERCISES 4.4D

In Exercises 1–18, change each percent to a proper fraction or mixed number.

1. 75% **2.** 50% **3.** 10% **4.** 30% **5.** 35% **6.** 65%

7. 80% **8.** 60% **9.** 5% **10.** 4% **11.** 250% **12.** 300%

13. $\frac{1}{2}\%$ **14.** $\frac{3}{4}\%$ **15.** $2\frac{1}{2}\%$ **16.** $3\frac{1}{3}\%$ **17.** $33\frac{1}{3}\%$ **18.** $66\frac{2}{3}\%$

In Exercises 19–44, change the given form into the missing forms.

	Fraction	Decimal	Percent
19.	$\frac{1}{2}$		
21.		0.6	
23.			10%
25.	$\frac{3}{4}$		
27.			6%
29.		0.48	
31.	$1\frac{1}{8}$		
33.			44%
35.		0.025	
37.			350%
39.		6.25	
41.			$5\frac{1}{4}\%$
43.			$\frac{3}{4}\%$

	Fraction	Decimal	Percent
20.	$\frac{1}{4}$		
22.		0.4	
24.			20%
26.		0.36	
28.	$\frac{4}{5}$		
30.			8%
32.		0.075	
34.	$2\frac{3}{8}$		
36.			28%
38.			275%
40.			$4\frac{1}{2}\%$
42.		8.75	
44.			$\frac{1}{2}\%$

45. Due to illness a student was absent from school 10 days out of 40 days.
a. What fraction of the time was he absent?
b. What decimal part of the time was he absent?
c. What percent of the time was he absent?

46. A student takes an examination in mathematics and solves 9 problems correctly out of 12.
a. What fraction of the problems did he solve correctly?
b. What decimal part of the problems did he solve correctly?
c. What was his percent grade?

47. The price of a car is discounted 20%. What fraction of the original price must the buyer *pay*?

48. If $\frac{1}{3}$ of a man's salary is deducted from his paycheck, what percent of his check does he get?

49. Wayne spends 25% of his income on rent and 20% of his income on his car. What fraction of his income does he spend on rent and his car?

50. On an exam, 12% of the students got an A, 20% got a B, and 33% got a C. What fraction of the class passed the exam with an A, B, or C?

4.5 Finding a Fractional Part of a Number

In Section 2.7B we showed that to find a fractional part of a number, we multiply the fraction times the number. When the fractional part is expressed as a decimal or as a percent, we may multiply using the following rules.

TO FIND A FRACTIONAL PART OF A NUMBER

1. If the fractional part is expressed as a *fraction*, multiply the fraction times the number.
2. If the fractional part is expressed as a *decimal*, multiply the decimal times the number.
3. If the fractional part is expressed as a *percent*, change the percent to a decimal or fraction, then multiply.

Example 1 On a 20-problem test, Kathy answered $\frac{4}{5}$ of the problems correctly. How many problems did she have correct?

Solution $\frac{4}{5}$ of $20 = \frac{4}{\cancel{5}_1} \cdot \frac{\cancel{20}^4}{1} = 16$ problems

of means to multiply in problems of this type ■

Example 2 Find 0.3 of 25.

Solution 0.3 of 25 =

$$\begin{array}{r} 25 \\ \times\ .3 \\ \hline 7.5 \end{array}$$ ■

Example 3 Find 75% of 24.

Solution Change 75% to a fraction.

$$75\% = \frac{75}{100} = \frac{3}{4}$$

Then,

$$75\% \text{ of } 24 = \frac{3}{\cancel{4}_{1}} \cdot \frac{\cancel{24}^{6}}{1} = 18$$

■

Example 4 Find $3\frac{1}{3}\%$ of 150.

Solution Change $3\frac{1}{3}\%$ to a fraction.

$$3\frac{1}{3}\% = \frac{3\frac{1}{3}}{100} = 3\frac{1}{3} \div 100 = \frac{\cancel{10}^{1}}{3} \cdot \frac{1}{\cancel{100}_{10}} = \frac{1}{30}$$

Then,

$$3\frac{1}{3}\% \text{ of } 150 = \frac{1}{\cancel{30}_{1}} \cdot \frac{\cancel{150}^{5}}{1} = 5$$

■

Example 5 A man's weekly salary is $725. His total deductions amount to 23% of his check.

a. How much is deducted from his check?

b. What is his take-home pay?

Solution Change 23% to a decimal.

$$23\% = 0.23$$

a. Then,

$$\begin{array}{lr} 23\% \text{ of } \$725 = & \$\,7\,2\,5 \\ & \times \quad .2\,3 \\ \hline & 2\,1\,7\,5 \\ & 1\,4\,5\,0 \\ \hline & \$\,1\,6\,6.7\,5 \end{array} = \text{amount deducted from his paycheck}$$

b. To find his take-home pay, subtract the deductions from his salary.

$$\begin{array}{r} \$725.00 \\ -\$166.75 \\ \hline \$558.25 \end{array} = \text{his take-home pay}$$

■

Example 6 Use a calculator to find the following:

a. $\frac{17}{32}$ of 36

Key in: 17 [÷] 32 [×] 36 [=]

Answer: 19.125

b. 0.0375 of 1,250
Key in: .0375 [×] 1250 [=]
Answer: 46.875

c. 6.5% of 342
Key in: 342 [×] 6.5 [%]
Answer: 22.23 ■

EXERCISES 4.5

1. Find $\frac{5}{6}$ of 48.

2. Find $\frac{7}{8}$ of 32.

3. Find 0.8 of 15.

4. Find 0.35 of 36.

5. Find 15% of 60.

6. Find 8% of 75.

7. Find 200% of 15.

8. Find 300% of 7.

9. Find $\frac{4}{15}$ of 9.

10. Find $\frac{5}{12}$ of 32.

11. Find 0.06 of 25.

12. Find 0.05 of 98.

13. Find $13\frac{1}{3}\%$ of 600.

14. Find $6\frac{2}{3}\%$ of 504.

15. Find $\frac{1}{2}\%$ of 300.

16. Find $\frac{1}{4}\%$ of 200.

17. A man's weekly salary is $375. His total deductions amount to 27% of his check.
a. How much is deducted from his check?
b. What is his take-home pay?

18. On a mathematics examination of 20 problems, John solved 85% of the problems correctly.
a. How many problems did he have correct?
b. How many problems did he miss?

19. During an 88-day spring semester, a student was absent $\frac{1}{11}$ of the time. How many days was he absent?

20. A man's weekly salary is $165. His deductions amount to $\frac{1}{5}$ of his check. Find his take-home pay.

21. In purchasing a car, a 15% down payment is required. Find the down payment on a car that sells for $6,150.

22. The present value of a house is 250% of its 1965 cost. If the house sold for $15,500 in 1965, what is its present value?

23. If a student sleeps $\frac{1}{3}$ of each 24-hour day, how many hours does he sleep in a week?

24. If $\frac{3}{4}$ of the total weight of a steer will produce usable products, how many pounds of usable products can be taken from a 1,250-pound steer?

25. An automobile listed at $2,490 was sold at a 15% discount. What was the selling price?

26. Mr. Miller, with a salary of $12,500 a year, was given an 8.5% raise. What was his salary after the raise?

In Exercises 27–36, use a calculator to work the following problems.

27. Find $\frac{5}{6}$ of 8.7.

28. Find $\frac{7}{12}$ of 40.5.

29. Find 0.18 of 565.

30. Find 2.4 of 9.6.

31. Find 29% of 824.

32. Find 6% of 905.

33. Find 4.5% of 3,750.

34. Find $6\frac{1}{2}\%$ of 78.

35. Manuel is ordering three items that cost $12.95, $19.50, and $7.95. He must add 4.5% sales tax and $3.50 for postage and handling. What is the total cost of his order? (Round off to the nearest cent.)

36. The total surface area of the earth is 196,940,000 square miles, of which 29% is land and 71% is water. Find the land area and the water area of the earth.

4.6 Percent Problems

Take a second look at the example of the paint can discussed in Section 4.4. We described that situation in three different ways.

1. The can is $\frac{2}{5}$ full.
2. The can is 0.40 full.
3. The can is 40 percent full.

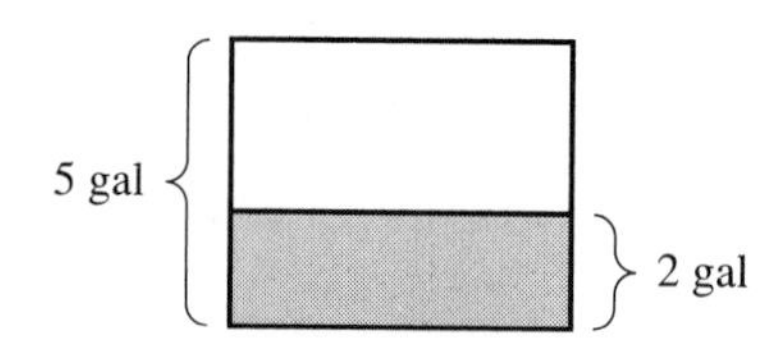

We have talked about three quantities:

1. 40 *percent* (P)
2. 5, which represents the *whole thing*, and is called the *base* (B)
3. 2, which is called the *amount* (A)

These three numbers are related by the proportion:

$$\frac{2}{5} = \frac{40}{100}$$

When the numbers are replaced by the corresponding letters, A, B, and P, we have:

The Percent Proportion

THE PERCENT PROPORTION

$$\frac{A}{B} = \frac{P}{100}$$

where A = Amount, B = Base, P = Percent

There are three letters in the percent proportion. If any two are known, the proportion can be solved for the remaining letter by the method given in Section 4.2.

1. The easiest term to identify is the percent P. "P" is the number written with the word *percent* or the symbol %.
2. Next, identify the base B. "B" is the number that represents 100 percent or the *total.* It is what we are finding the percent of, and therefore, it follows the words *percent of*.
3. "A" is the remaining number after P and B have been identified. "A" is the *amount*.

TO IDENTIFY THE NUMBERS IN A PERCENT PROBLEM

1. P is the number followed by the word *percent* (or %).

2. B is the number that follows the words *percent of*.

3. A is the number left after P and B have been identified.

Example 1 Identifying the numbers in a percent problem:
What number is 8 percent of 40?
Solution

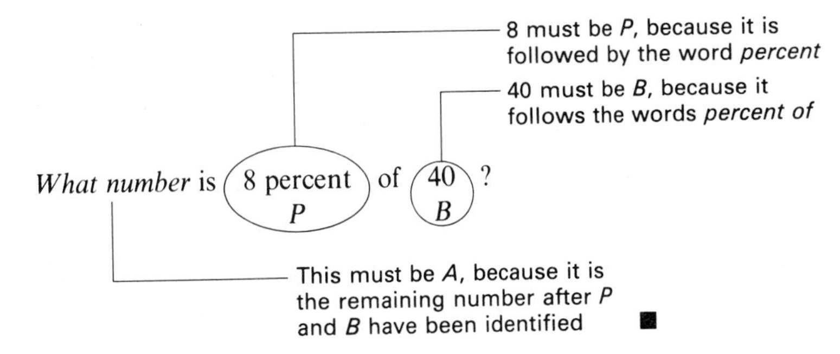

■

TO SOLVE A PERCENT PROBLEM

1. Identify the given numbers as A, B, or P.
2. Substitute the given numbers in the percent proportion.
3. Solve the percent proportion for the unknown.

Example 2 What is 15% of 72?
Solution Identify the numbers A, B, and P.

What (A) is 15% (P) of 72 (B)?

$$\frac{A}{B} = \frac{P}{100}$$ The percent proportion

$$\frac{A}{72} = \frac{15}{100}$$

$$100 \cdot A = 72 \cdot 15$$

$$\frac{\cancel{100} \cdot A}{\cancel{100}} = \frac{1080}{100}$$

$$A = 10.8$$

Therefore, 10.8 is 15% of 72.

Alternate Solution $A = 15\%$ of 72

$$= 0.15 \times 72 = 10.8$$

$15\% = \frac{15}{100} = 0.15$

"of" means to multiply in problems of this type ■

Example 3 5 is what percent of 20?

Solution Identify the numbers A, B, and P.

$$\underset{A}{5} \text{ is } \underset{P}{\text{what percent}} \text{ of } \underset{B}{20}\text{ ?}$$

$$\frac{A}{B} = \frac{P}{100} \qquad \text{The percent proportion}$$

$$\frac{5}{20} = \frac{P}{100} \qquad \text{Reduce } \frac{5}{20} = \frac{1}{4}$$

$$\frac{1}{4} = \frac{P}{100}$$

$$4 \cdot P = 100$$

$$P = 25$$

Therefore, 5 is 25% of 20. ■

Example 4 30% of what number is 12?

Solution

$$\underset{P}{30\%} \text{ of } \underset{B}{\text{what number}} \text{ is } \underset{A}{12}\text{ ?}$$

$$\frac{A}{B} = \frac{P}{100}$$

$$\frac{12}{B} = \frac{30}{100} \qquad \text{Reduce } \frac{30}{100} = \frac{3}{10}$$

$$\frac{12}{B} = \frac{3}{10}$$

$$3 \cdot B = 120$$

$$B = 40$$

Therefore, 30% of 40 is 12. ■

EXERCISES 4.6

Solve the following percent problems.

1. 15 is 30% of what number?

2. 16 is 20% of what number?

3. 115 is what percent of 250?

4. 90 is what percent of 225?

5. What is 25% of 40?

6. What is 45% of 65?

7. 15% of what number is 127.5?

8. 32% of what number is 256?

9. What percent of 8 is 18?

10. What percent of 6 is 12?

11. 65% of 48 is what number?

12. 87% of 49 is what number?

13. 750 is 125% of what number?

14. 325 is 130% of what number?

15. 20 is what percent of 16?

16. 30 is what percent of 25?

17. What is 200% of 12?

18. What is 300% of 9?

19. 15% of what number is 37.5?

20. What is 80% of $135?

21. 42 is $66\frac{2}{3}\%$ of what number?

22. 36 is $16\frac{2}{3}\%$ of what number?

4.7 Word Problems Using Percent

In the percent proportion

$$\frac{A}{B} = \frac{P}{100}$$

the base B represents 100 percent, or the whole thing. Therefore, B represents the *total* number or the *original* number.

Example 1 In an examination a student worked 15 problems correctly. This was 75% of the problems. Find the total number of problems on the examination.
Solution In this problem we are saying:

15 (A) is 75% (P) of some number (B) ← *total* number of problems on exam

$$\frac{A}{B} = \frac{P}{100}$$

$$\frac{15}{B} = \frac{75}{100} \qquad \text{Reduce } \frac{\overset{3}{\cancel{75}}}{\underset{4}{\cancel{100}}} = \frac{3}{4}$$

$$\frac{15}{B} = \frac{3}{4}$$

$$3 \cdot B = 4 \cdot 15$$

$$\frac{\cancel{3} \cdot B}{\cancel{3}} = \frac{60}{3}$$

$$B = 20$$

Therefore, 20 problems were on the examination. ■

Example 2 30 grams (about 1 cup) of a popular breakfast cereal contains 12 grams of sugar. What percent of the cereal is sugar?

Solution

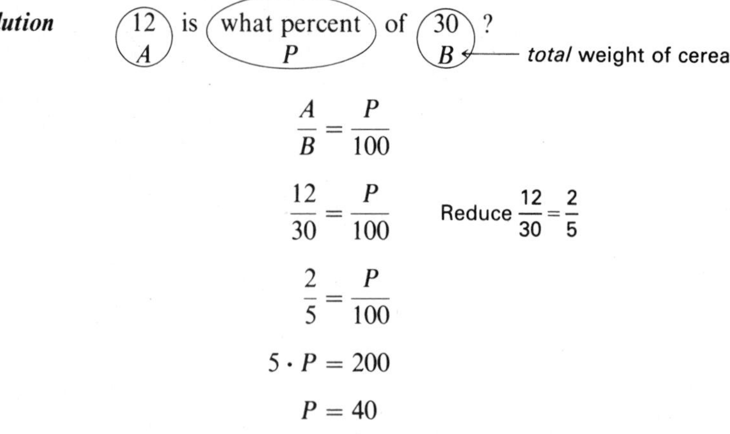

$$\frac{A}{B} = \frac{P}{100}$$

$$\frac{12}{30} = \frac{P}{100} \qquad \text{Reduce } \frac{12}{30} = \frac{2}{5}$$

$$\frac{2}{5} = \frac{P}{100}$$

$$5 \cdot P = 200$$

$$P = 40$$

Therefore, the cereal is 40% sugar. ■

Example 3 A record player that originally sold for \$300 is on sale at 25% off.

a. What is the amount of discount?

b. What is the sale price?

Solution

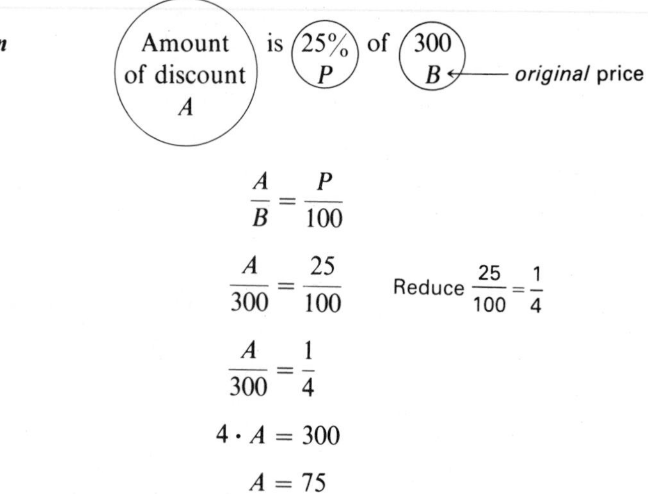

$$\frac{A}{B} = \frac{P}{100}$$

$$\frac{A}{300} = \frac{25}{100} \qquad \text{Reduce } \frac{25}{100} = \frac{1}{4}$$

$$\frac{A}{300} = \frac{1}{4}$$

$$4 \cdot A = 300$$

$$A = 75$$

Therefore, the discount is \$75.

To find the sale price, subtract the discount from the original price.

sale price = original price − discount

Sale price = \$300 − \$75 = \$225. ■

Percent Increase or Decrease Many changes are stated using the percent of increase or decrease.

Food prices increased 8%.

Enrollment has decreased 12%.

Unemployment has increased 25% this year.

In the percent proportion

$$\frac{A}{B} = \frac{P}{100}$$

P is the *percent of increase* (or decrease).

A is the *amount of increase* (or decrease).

B is the *original* amount.

Example 4 Mr. Lee's salary last year was \$15,000. This year his salary has increased 5%.

a. Find the amount of increase.

b. Find his new salary.

Solution

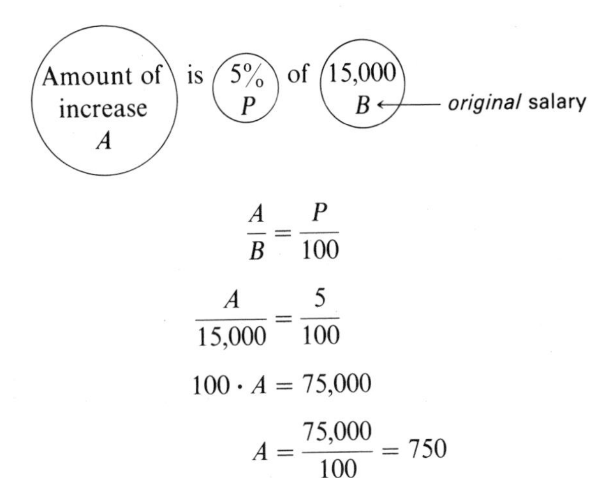

$$\frac{A}{B} = \frac{P}{100}$$

$$\frac{A}{15{,}000} = \frac{5}{100}$$

$$100 \cdot A = 75{,}000$$

$$A = \frac{75{,}000}{100} = 750$$

Therefore, the amount of increase is \$750.

To find his new salary, add the amount of increase to his original salary.

new amount = original amount + amount of increase

New salary = \$15,000 + \$750 = \$15,750. ■

Example 5 The population of a town increased from 20,000 to 25,000.

a. Find the amount of increase.

b. Find the percent of increase.

Solution To find the amount of increase, subtract the original population from the new population.

amount of increase = new amount − original amount

Amount of increase = 25,000 − 20,000 = 5,000.

amount of increase → 5,000 (A) is what percent (P) of 20,000 (B ← *original* population)?

$$\frac{A}{B} = \frac{P}{100}$$

$$\frac{5{,}000}{20{,}000} = \frac{P}{100} \qquad \text{Reduce } \frac{5{,}000}{20{,}000} = \frac{1}{4}$$

$$\frac{1}{4} = \frac{P}{100}$$

$$4 \cdot P = 100$$

$$P = 25$$

Therefore, the population increased 25%. ■

Example 6 Maria weighed 150 pounds. After going on a diet, she weighed 120 pounds. What was the percent of decrease?

Solution First, find the amount of decrease by subtracting the new weight from the original weight.

amount of decrease = original − new amount

Amount of decrease = 150 − 120 = 30 pounds.

amount of decrease → 30 (A) is what percent (P) of 150 (B ← *original* weight)?

$$\frac{A}{B} = \frac{P}{100}$$

$$\frac{30}{150} = \frac{P}{100} \qquad \text{Reduce } \frac{30}{150} = \frac{1}{5}$$

$$\frac{1}{5} = \frac{P}{100}$$

$$5 \cdot P = 100$$

$$P = 20$$

Therefore, the percent of decrease is 20%. ■

EXERCISES 4.7

1. A team wins 80% of its games. If it wins 68 games, how many games has it played?

2. On a Spanish test, Sally answered 17 problems correctly and scored 68%. How many problems were on the test?

3. In a class of 42 students, 7 students received a grade of B. What percent of the class received a grade of B?

4. John's weekly gross pay is \$110, but 23% of his check is withheld. How much is withheld?

5. Fifty-four out of 216 civil service applicants pass their exams. What percent of the applicants pass?

6. A 4200-pound automobile contains 462 pounds of rubber. What percent of the car's total weight is rubber?

7. Mr. Delgado, a salesman, makes a 6% commission on all items he sells. One week he made \$390. What were his gross sales for the week?

8. Mrs. Clark makes a 9% commission on all items she sells. One week she made \$369.81. What were her gross sales for the week?

9. An affirmative action committee demands that 24% of the 1800 entering freshmen be minority students. How many more than the actual 256 minority freshmen entering would satisfy the committee?

10. Sergio got a \$696 raise on his \$11,600 yearly salary. There is a 9% rate of inflation for the year. How much more (or less) should his raise have been in order for him to keep up with the inflation?

11. The evening enrollment at a certain college is 6% less than the day enrollment. If the day enrollment is 2450, find the evening enrollment.

12. Mr. Edmonson's new contract calls for a 12% raise in salary. His present salary is \$11,250. What will his new salary be?

13. Gloria thinks that at least 40% of the 50 teachers at her school should be women. The school board decides to increase the number of women teachers at her school from 16 to 22 (without changing the total number). Will this satisfy Gloria?

14. Seventy percent of the 46,000 burglaries in a city were committed by persons that had previously been convicted at least 3 times for the same crime. If all of these criminals had been kept in jail after the third conviction, how many burglaries could have been prevented?

15. There are 43 grams of sulfuric acid in 500 grams of solution. Find the percent of acid in the solution.

16. A team wins 105 games. This is 70% of the games played. How many games were played?

17. Rosie's weekly salary is \$225. If her deductions amount to \$63, what percent of her salary is take-home pay?

18. 500 grams of a solution contain 27 grams of a drug. Find the percent of drug strength.

19. Mrs. McMahan's salary was \$12,000 a year. At the end of one year, her salary was increased 8%.
    a. Find the amount of increase.
    b. Find her new salary.

20. Last year the Johnsons paid \$847.50 for property taxes. This year their property taxes will increase 6%. Find their property tax for this year.

21. Smalltown's population of 13,000 decreased 6%. Find the new population.

22. Sidney Sabot sold \$214,000 worth of sailboats in the summer. The company expects its sales to decrease 60% in the winter. Find the expected sales for the winter months.

23. Last year, the average monthly utility bill for a family was \$46.00. Today the average utility bill is \$80.50.
    a. Find the amount of increase.
    b. Find the percent of increase.

24. When Mary began selling real estate, the average family home in her city cost \$21,000. Today the average family home in her city costs \$45,150. Find the percent of increase in the cost of a family home in Mary's city.

25. When a man bought his new car in 1981, it cost \$8,750. The resale value of the car today is \$2,100. Find the percent of decrease in the value of the car.

26. After lowering its thermostat from 75°F to 68°F, a family found that its heating bill decreased from \$70.60 to \$48.10. Find the percent of decrease in the heating bill. (Round off to the nearest tenth of a percent.)

4.8 Chapter Summary

Ratio 4.1

A **ratio** is a comparison of two quantities written in fraction form.

Proportion 4.2

A **proportion** is a statement that two ratios are equal.

To Solve a Proportion for an Unknown Letter 4.2

1. Set the product of the means equal to the product of the extremes. (The cross products are equal.)
2. Divide both sides by the number that is multiplying the unknown letter.

To Check Your Solution of a Proportion 4.2

1. Replace the unknown letter in the original proportion by the value you obtained for it.
2. Cross-multiply, and verify that the cross products are equal.

To Solve a Word Problem by Using a Proportion 4.3

1. Read the word problem completely, then use x to represent the unknown quantity.
2. Set up a proportion with the same units occupying corresponding positions in both ratios of your proportion.
3. Solve your proportion for the unknown x.

Percent 4.4

Percent means per hundred or the number of hundredths.

To Change a Decimal to a Percent 4.4A

Move the decimal point two places to the right and write the percent symbol (%).

To Change a Percent to a Decimal 4.4B

Move the decimal point two places to the left and remove the percent symbol (%).

To Change a Fraction to a Percent 4.4C

1. Change the fraction to a decimal by dividing.
2. Change the decimal to a percent by moving the decimal point two places to the right and writing the percent symbol.

To Change a Percent to a Fraction 4.4D

1. Write the percent as the numerator of a fraction with a denominator of 100.
2. Reduce the fraction to lowest terms.

To Find a Fractional Part of a Number 4.5

1. If the fractional part is expressed as a *fraction*, multiply the fraction times the number.
2. If the fractional part is expressed as a *decimal*, multiply the decimal times the number.
3. If the fractional part is expressed as a *percent*, change the percent to a decimal or fraction, then multiply.

To Solve a Percent Problem 4.6

1. Identify the given numbers as A, B, or P.
   P is the number followed by the word *percent* (or %).
   B is the number that follows the words *percent of*.
   A is the number left after P and B have been identified.

2. Substitute the given numbers in the percent proportion:

$$\frac{A}{B} = \frac{P}{100}$$

3. Solve the percent proportion for the unknown.

Percent Increase or Decrease 4.7

In the percent proportion

$$\frac{A}{B} = \frac{P}{100}$$

P is the *percent of increase* (or decrease).

A is the *amount of increase* (or decrease).

B is the *original* amount.

Review Exercises 4.8

In Exercises 1–3, reduce the ratios to lowest terms.

1. 27 to 12

2. $1\frac{1}{3}$ to 12

3. 2 hours to 20 minutes

In Exercises 4–6, do the given ratios form a proportion?

4. $\frac{5}{12} \stackrel{?}{=} \frac{125}{300}$

5. $\frac{46}{32} \stackrel{?}{=} \frac{3}{2}$

6. $\frac{1\frac{2}{3}}{3} \stackrel{?}{=} \frac{5}{9}$

In Exercises 7–12, solve for the unknown letter.

7. $\frac{45}{70} = \frac{x}{56}$

8. $\frac{39}{x} = \frac{130}{210}$

9. $\frac{24}{15} = \frac{18}{x}$

10. $\frac{x}{\frac{2}{3}} = \frac{\frac{9}{16}}{\frac{5}{6}}$

11. $\frac{2.8}{5.2} = \frac{A}{6.5}$

12. $\frac{2\frac{1}{2}}{B} = \frac{9}{2\frac{7}{10}}$

13. If 10 pounds of apples cost $6, how much does $2\frac{1}{2}$ pounds cost?

14. Fred knows his car needs a quart of oil every 1,500 miles. How many quarts will he need for a 12,000-mile trip?

15. Four students finish a class for every five students who begin. For 252 students to finish, how many must have begun the class?

16. If a 6-foot man casts an 8-foot shadow, find the height of a tree that casts a 96-foot shadow.

17. On a map $\frac{1}{2}$ inch represents 10 miles. How many miles would $3\frac{1}{2}$ inches represent?

18. A machine can finish 80 parts in 2 hours 30 minutes. How many parts can it finish in an 8-hour day?

19. Change to a percent.
a. 0.7 b. 1.25

20. Change to a decimal.
a. 4% b. 65%

21. Change to a percent
a. $\frac{3}{5}$ b. $\frac{7}{8}$

22. Change to a reduced fraction.
a. 24% b. $37\frac{1}{2}\%$

23. Find $\frac{5}{12}$ of 48.

24. Find 0.4 of 120.

25. Find 45% of 300.

26. What is 6% of 40?

27. What percent of 20 is 25?

28. 360 is 150% of what number?

29. At a certain college the fall enrollment was 5% more than the spring enrollment. The spring enrollment was 3,560. Find the fall enrollment.

30. If you work 9 problems correctly on an examination of 12 problems, what is your percent grade?

31. A $95 suit of clothes is marked down 20%. Find the selling price of the suit.

32. The rent on a $400-a-month apartment is increased $12\frac{1}{2}\%$. Find the new rent.

33. 500 grams of a solution contain 75 grams of a drug. Find the percent of drug.

34. One week a salesman working on a 15% commission made $255. Find his total sales for the week.

35. Mr. Garcia, with a salary of $9,500, was given a 6% raise. What was his salary after the raise?

36. Last year, the Carters paid property taxes of $875. This year their property taxes are $1,015. Find the percent of increase in taxes.

37. A credit card company charges 1.75% interest each month on the unpaid balance. Find the interest on an unpaid balance of $369.

38. A lawn mower that originally sold for $250 is on sale at 25% off. Since Jack works at the store, he gets a 15% employee discount. How much will Jack pay for the lawn mower?

Chapter 4 Diagnostic Test

Allow yourself about 50 minutes to do these problems. Complete solutions for every problem, together with section references, are given in the answer section at the end of the book.

1. A college basketball team won 15 out of 24 games played. There were no ties.
a. What is the ratio of wins to games played?
b. What is the ratio of wins to losses?

2. Reduce the ratio of 9 inches to 2 feet to lowest terms.

In Exercises 3–5, solve the proportions.

3. $\frac{x}{12} = \frac{40}{60}$

4. $\frac{20}{75} = \frac{x}{300}$

5. $\frac{3\frac{1}{2}}{x} = \frac{21}{40}$

6. Change 0.8 to a percent.

7. Change 12.5% to a decimal.

8. Change $\frac{3}{40}$ to a percent.

9. Change 76% to a reduced fraction.

10. Find $\frac{5}{8}$ of 72.

11. Find 0.7 of 35.

12. What is 36% of 250?

13. 15 is 125% of what number?

14. What percent of 30 is 12?

15. If 5 pounds of strawberries cost $4, what will $7\frac{1}{2}$ pounds cost?

16. If a 6-foot man casts a 4-foot shadow, find the height of a flagpole that casts an 18-foot shadow.

17. Bonny's car used 2 quarts of oil on a 1,500-mile trip. How many quarts can she expect to use on a 6,000-mile trip?

18. On a test Bill answered 34 problems correctly and scored 85%. How many problems were on the test?

19. A bike that originally sold for $96 is on sale at 25% off. What is the sale price?

20. The population of a town increased from 4,000 to 4,600. What was the percent of increase?

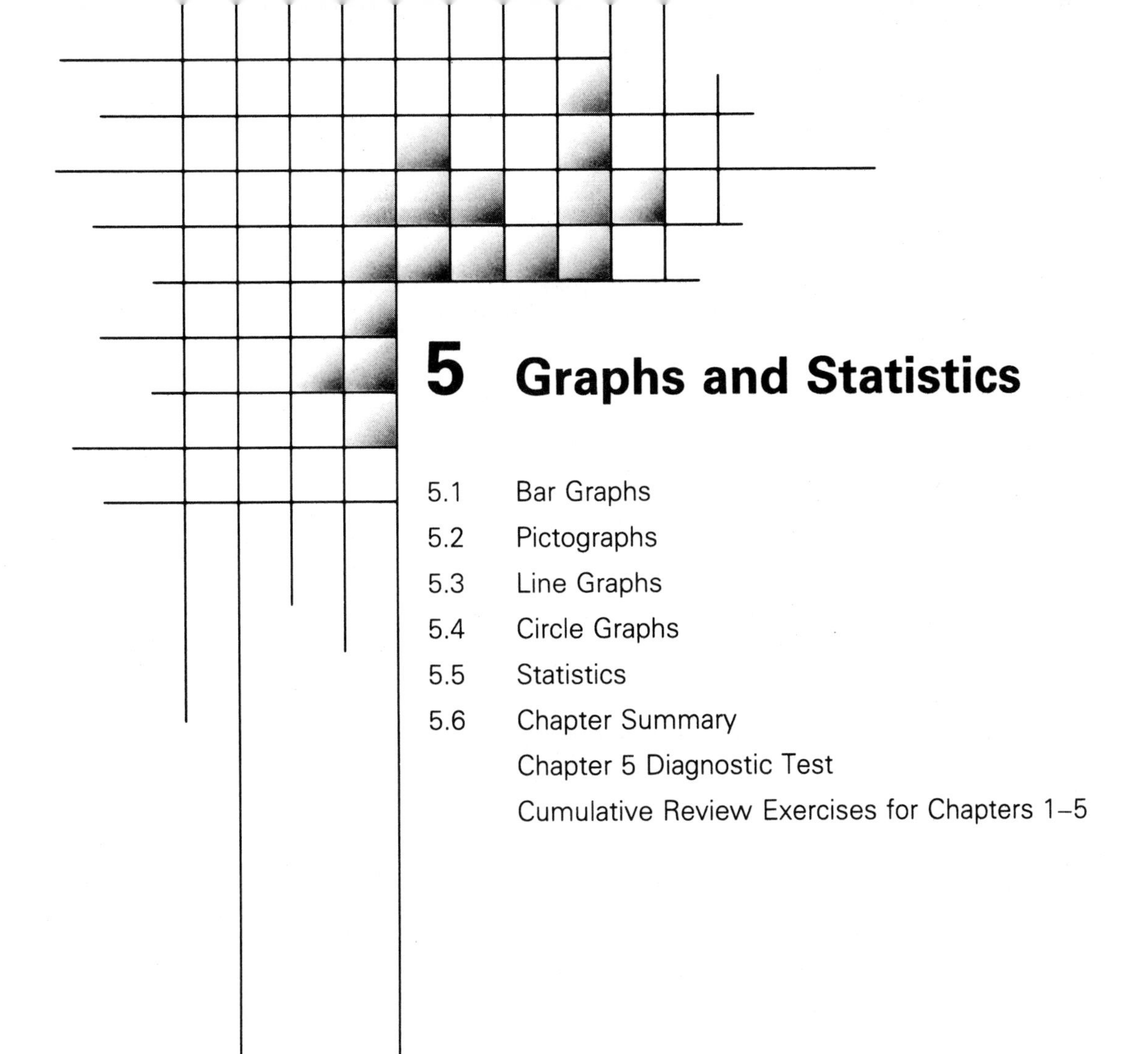

5 Graphs and Statistics

Graphs are used to present numerical data in an organized way. It is often easier to compare amounts or to see a change if the information is in graph form. Graphs are used by scientists, economists, technicians, business organizations, government agencies, and advertising firms.

5.1 Bar Graphs

Bar graphs are used to compare amounts. The bar graph in Figure 5.1.1 shows land area for the seven continents. The vertical scale on the left is labeled "Area in Millions of Square Miles," and each interval represents 1 million square miles. The horizontal scale on the bottom is labeled with the names of the continents. The height of each bar corresponds to the land area of the continent.

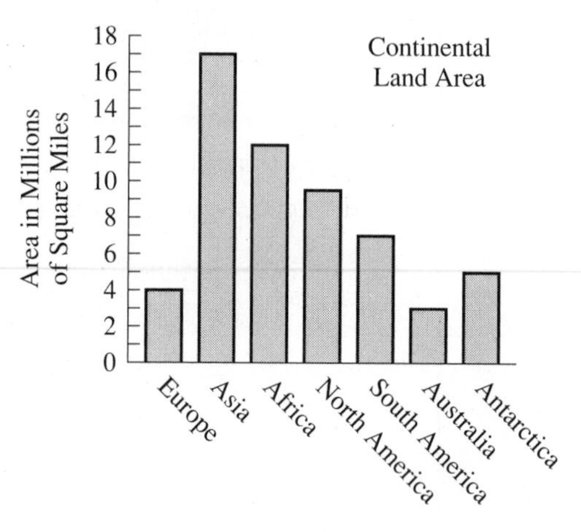

FIGURE 5.1.1

Notice that you can estimate the area of a continent from the graph, but you cannot read the area exactly. For example, the area of Europe is approximately 4 million square miles (4,000,000 sq. mi.)

It is easy to see that the largest continent is Asia (because its bar is the tallest) and that the smallest continent is Australia. We can also compare the areas. Africa is how many times larger than Europe? Because the bar representing Africa is about 3 times taller than the bar representing Europe, Africa is about 3 times larger than Europe.

Example 1 Examine the bar graph in Figure 5.1.2 and answer the following questions.

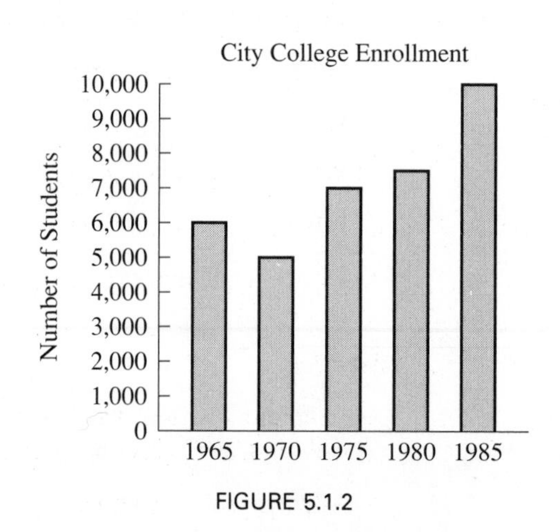

FIGURE 5.1.2

a. What was the enrollment in 1975?

b. In what year was the enrollment about 6,000?

c. What year had the least enrollment?

d. During which years did the number of students increase?

e. Which 5-year period had the largest increase?

f. How much greater was the enrollment in 1985 than in 1975?

Solution

a. About 7,000 students

b. 1965

c. 1970

d. Comparing the heights of the bars in order, we find the enrollment had *decreased* from 1965 to 1970, but it had continually *increased* from 1970 to 1985.

e. The largest difference between the heights of the bars is from 1980 to 1985.

f. The enrollment was about 10,000 in 1985 and about 7,000 in 1975. Therefore, the difference is $10,000 - 7,000 = 3,000$ more students. ■

Figures 5.1.1 and 5.1.2 used vertical bars. Figure 5.1.3 in Example 2 uses horizontal bars.

Example 2 Examine the bar graph in Figure 5.1.3 and answer the following questions.

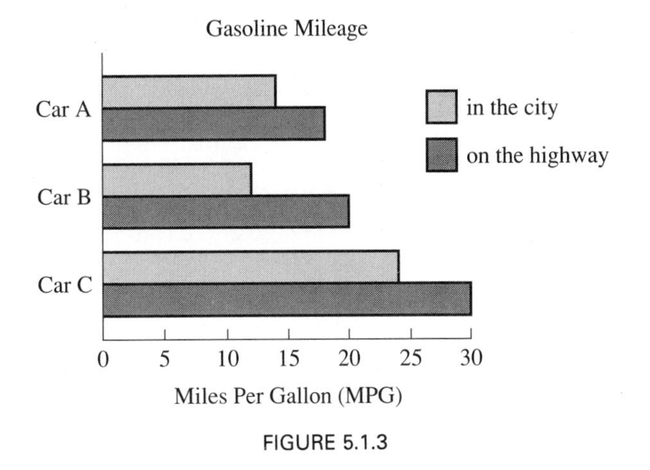

FIGURE 5.1.3

a. What gasoline mileage does Car C get in the city?

b. What gasoline mileage does Car C get on the highway?

c. Which car gets the best gasoline mileage in the city?

d. Which car gets the least gasoline mileage in the city?

e. Which car gets the least gasoline mileage on the highway?

f. How many times greater is the gasoline mileage for Car C compared to Car B in city driving?

Solution

a. Different shadings have been used for city driving and highway driving. For city driving Car C gets approximately 24 mpg.

b. Approximately 30 mpg

c. Car C

d. Car B

e. Car A

f. About 2 times ■

Slanting the Picture

When we read information from a graph, it is important to note the vertical and the horizontal scales. See Figure 5.1.4 in Example 3.

Example 3

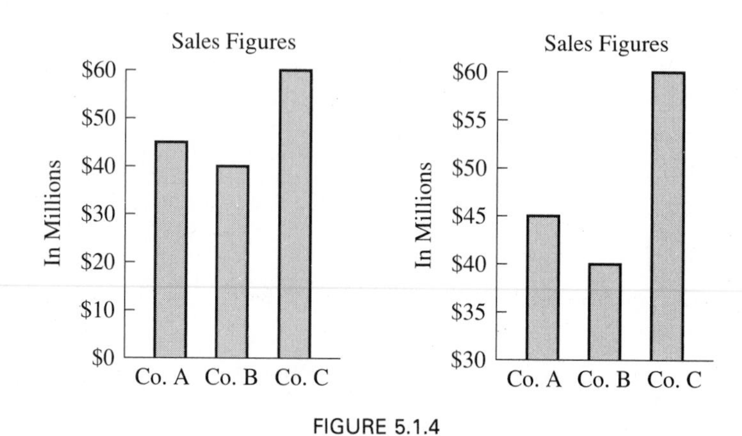

FIGURE 5.1.4

The two graphs in Figure 5.1.4 show the same sales figures for the three companies, but the graph on the right distorts the differences between the companies because the vertical scale does not start at 0. ■

EXERCISES 5.1

In Exercises 1–9, use the bar graph in Figure 5.1.5.

1. What is the population of California?

2. What is the population of Texas?

3. Which state has a population of 18,000,000?

4. Which state has a population of 10,000,000?

5. Which state has the greatest population?

6. Which state has the least population?

7. How many more people live in California than in New York?

8. How many more people live in California than in Illinois?

9. How many times greater is the population of California compared to the population of Ohio?

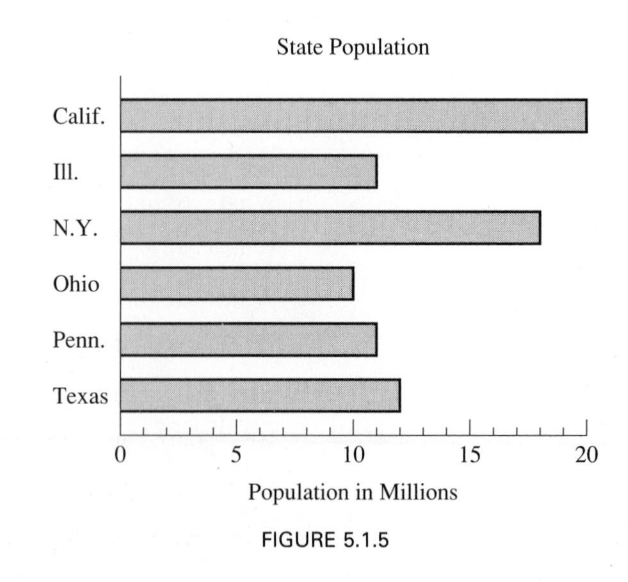

FIGURE 5.1.5

In Exercises 10–18, use the bar graph in Figure 5.1.6.

10. What were the store sales for January?

11. What were the store sales for April?

12. During which month did the store sell $45,000?

13. Which month had the least sales?

14. Which month had the best sales?

15. How much more was sold in May than in March?

16. During which months did the sales increase?

17. During which months did the sales decrease?

18. Find the total sales for January to June.

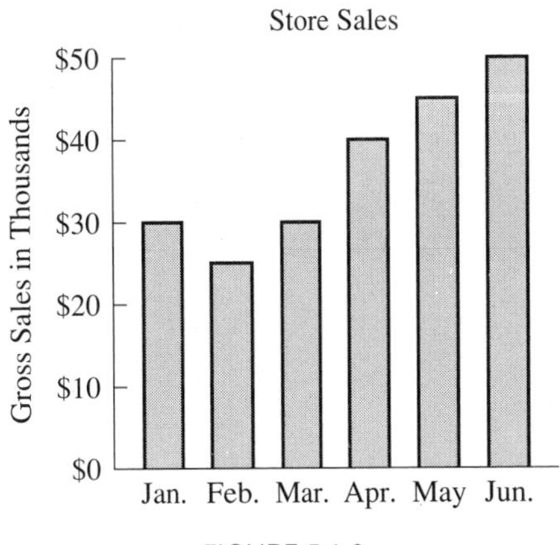

FIGURE 5.1.6

In Exercises 19–24, use the bar graph in Figure 5.1.1 on page 166.

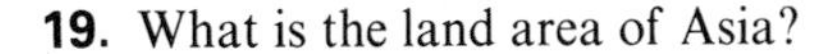

19. What is the land area of Asia?

20. What is the land area of Africa?

21. What is the total area of North America and South America?

22. What is the total area of Europe, Asia, and Africa?

23. How much larger is North America compared to South America?

24. How much larger is Asia compared to Europe?

5.2 Pictographs

Pictographs are basically bar graphs that use rows of picture symbols to replace the bars. Each picture symbol represents a certain number of units. Partial symbols represent fractional parts of the given unit.

Example 1 Examine the pictograph in Figure 5.2.1 and answer the following questions.

FIGURE 5.2.1

a. How many voters are 18–29 years old?

b. How many voters are 45–59 years old?

c. How many voters are over 44 years old?

d. How many registered voters are in Small County?

e. What percent of the voters are under 30 years old?

Solution

a. $6 \times 100 = 600$ voters

b. There are $11\frac{1}{2}$ or 11.5 symbols. Therefore, $11.5 \times 100 = 1{,}150$ voters.

c. $(11.5 + 3.5) \times 100 = 15 \times 100 = 1{,}500$ voters.

d. Total number of symbols $= 3.5 + 11.5 + 9 + 6 = 30$. Therefore, $30 \times 100 = 3{,}000$ voters.

e. 600 is what percent of 3,000?

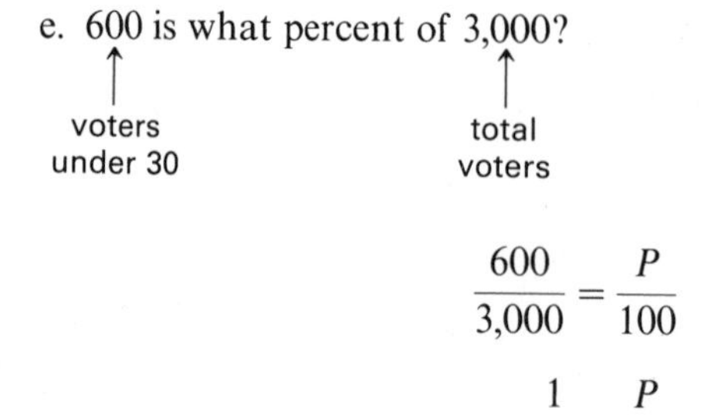

$$\frac{600}{3{,}000} = \frac{P}{100}$$

$$\frac{1}{5} = \frac{P}{100} \qquad \text{Reduce } \frac{600}{3000} = \frac{1}{5}$$

$$5P = 100$$

$$P = 20$$

Therefore, 20% of the voters are under 30 years old. ■

Slanting the Picture

Example 2

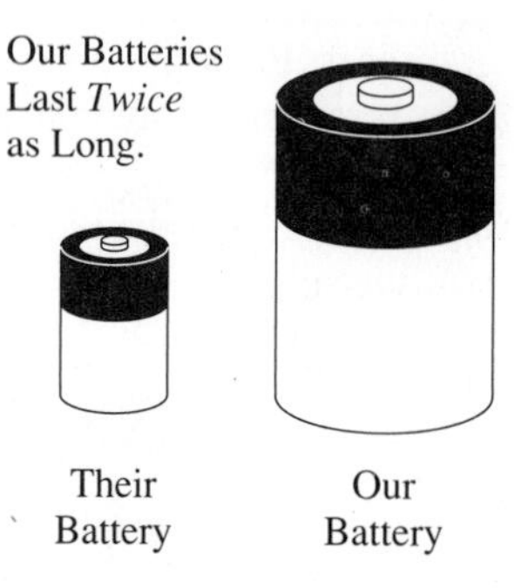

FIGURE 5.2.2

The height of the large battery is twice the height of the small battery, but to keep the figure proportional, the width is also twice as long. What the reader sees is an area that is 4 times larger.

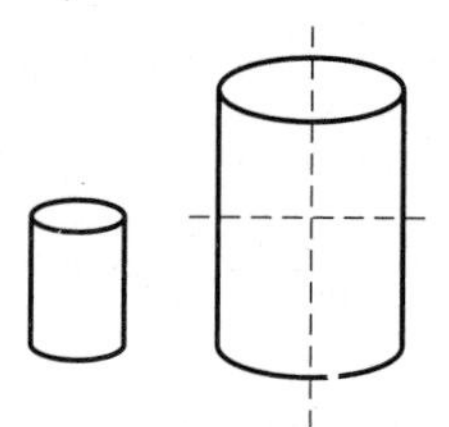

■

EXERCISES 5.2

In Exercises 1–6, use the pictograph in Figure 5.2.3.

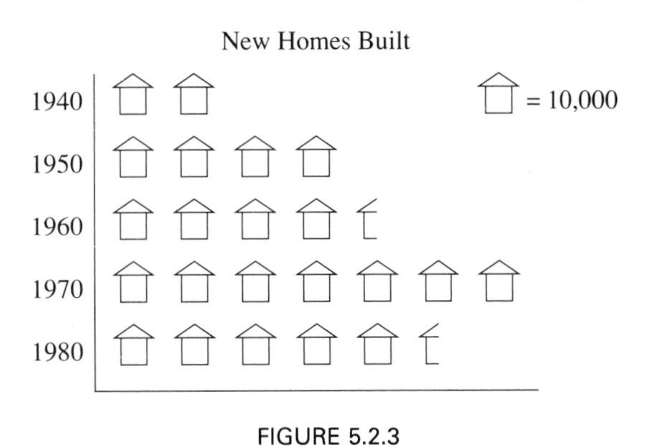

FIGURE 5.2.3

1. How many homes were built in 1940?

2. How many homes were built in 1960?

3. Did the number of new homes built increase or decrease from 1960 to 1970?

4. Did the number of new homes built increase or decrease from 1970 to 1980?

5. How many more homes were built in 1950 than in 1940?

6. How many more homes were built in 1970 than in 1960?

In Exercises 7–16, use the pictograph in Figure 5.2.4.

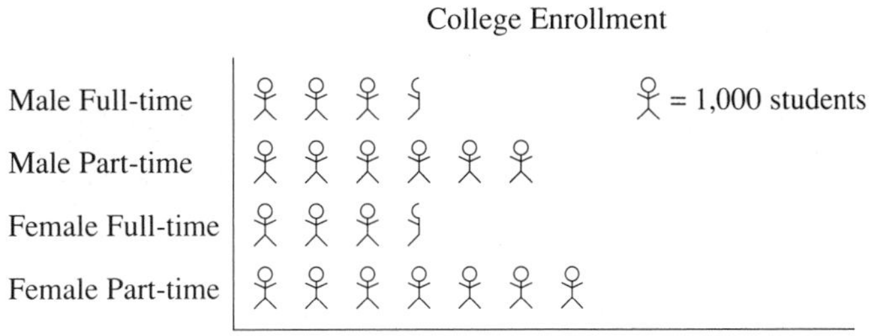

College Enrollment

Male Full-time

Male Part-time

Female Full-time

Female Part-time

= 1,000 students

FIGURE 5.2.4

7. How many male students attend the college full-time?

8. How many female students attend the college part-time?

9. How many students attend the college full-time?

10. How many students attend the college part-time?

11. Estimate the number of male students.

12. Estimate the number of female students.

13. What is the college's total enrollment?

14. What percent of the total enrollment are part-time students?

15. What percent of the total enrollment are full-time students?

16. What percent of the total enrollment are female?

5.3 Line Graphs

Line (or broken-line) graphs are similar to bar graphs except the data are represented by points instead of bars. Then the points are connected in order by line segments. Line graphs are used to show change over a period of time.

Example 1 Examine the line graph in Figure 5.3.1 and answer the following questions.

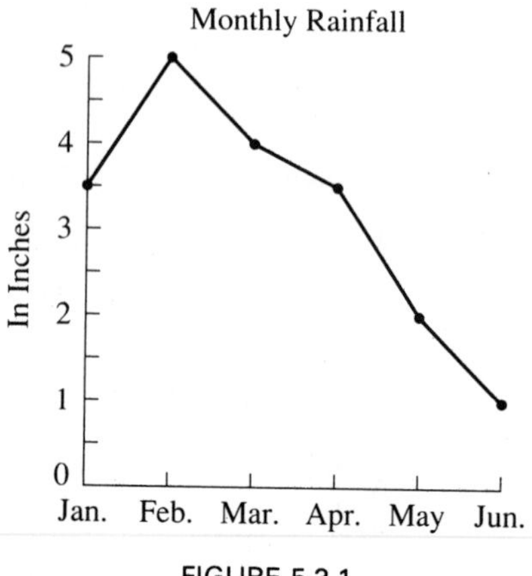

FIGURE 5.3.1

a. What was the monthly rainfall for January?

b. What was the monthly rainfall for May?

c. Which was the wettest month from January to June?

d. During which months did the rainfall increase?

e. During which months did the rainfall decrease?

f. Between which two months was the decrease in rainfall the greatest?

Solution

a. About 3.5 in.

b. About 2 in.

c. February

d. The line segment between January and February is rising, therefore, the rainfall increased from January to February.

e. All of the line segments between February and June are dropping. This means the rainfall decreased from February to June.

f. The greatest decrease in rainfall can be found by looking for the steepest line segment. This occurs between April and May. ■

Example 2 Examine the line graph in Figure 5.3.2 and answer the following questions.

a. What was the amount of sales for Store A in 1987?

b. Which store had higher sales in 1986?

c. Which store had higher sales in 1989?

d. How much more did Store B sell than Store A in 1988?

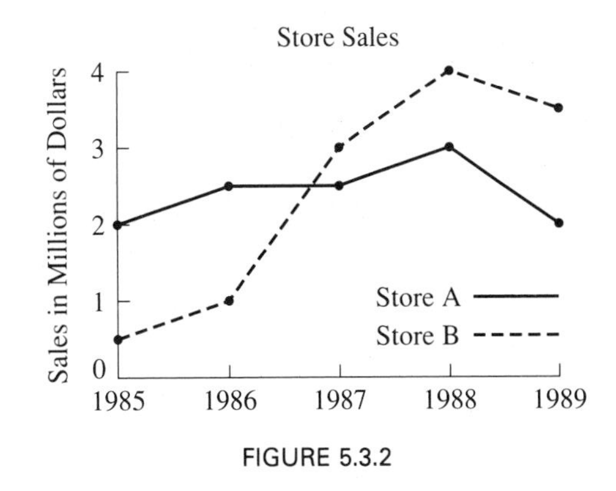

FIGURE 5.3.2

e. Estimate the total sales for Store A from 1985 to 1989.

f. Between which two years did the sales increase the most for Store B?

Solution

a. The sales for Store A is represented by the solid line. In 1987 Store A sold about 2.5 million dollars $= \$2.5 \times 1{,}000{,}000$
$= \$2{,}500{,}000$

b. Store A

c. Store B

d. In 1988 Store A sold \$3,000,000, and Store B sold \$4,000,000. The difference is \$4,000,000 − \$3,000,000 = \$1,000,000 more.

e. In

1985	\$ 2,000,000
1986	2,500,000
1987	2,500,000
1988	3,000,000
1989	+ 2,000,000
Total =	\$12,000,000

f. From 1986 to 1987 ■

Slanting the Picture

Example 3

FIGURE 5.3.3

Both graphs in Figure 5.3.3 show the same stock prices, but the graph on the right has a vertical scale that was stretched. This makes the line segments rise faster. ■

EXERCISES 5.3

In Exercises 1–8, use the line graph in Figure 5.3.4.

1. What was the population of California in 1950?

2. What was the population of California in 1970?

3. In what year was the population of California 16,000,000?

4. In what year was the population of California 22,000,000?

5. How many more people lived in California in 1980 than in 1940?

6. How many more people lived in California in 1970 than in 1950?

7. During which years did the population of California increase?

8. Between which 10-year period did the population increase the most?

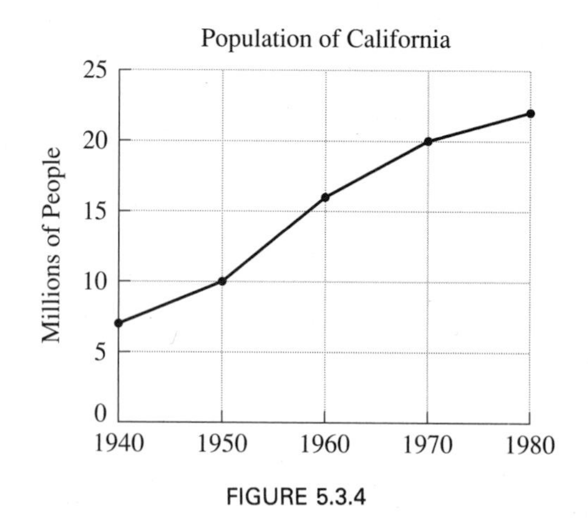

FIGURE 5.3.4

In Exercises 9–18, use the line graph in Figure 5.3.5.

9. What was the price of Stock A in 1987?

10. What was the price of Stock B in 1988?

11. What was the highest price for Stock A, and when did it occur?

12. What was the highest price for Stock B, and when did it occur?

13. Which stock had a higher price in 1986?

14. Which stock had a higher price in 1987?

15. How much more was the price of Stock A than Stock B in 1986?

16. How much more was the price of Stock A than Stock B in 1989?

17. Between which two years did the price of Stock A increase the most?

18. Between which two years did the price of Stock B increase the most?

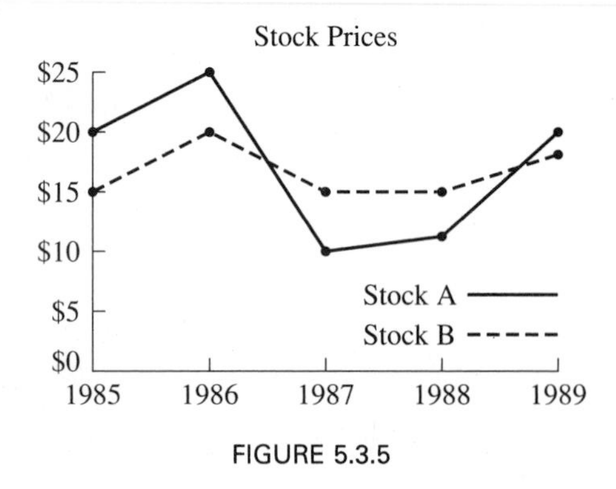

FIGURE 5.3.5

5.4 Circle Graphs

A circle (or pie) graph is a graph in the form of a circle that has been divided into wedges or slices. The size of each wedge is proportional to the size of the data it represents. Circle graphs are used to compare parts of a whole. The whole circle represents 100%.

Example 1 Examine the circle graph in Figure 5.4.1 and answer the following questions.

a. What percent of the course grade is based on a student's test scores?

b. What percent of the course grade is based on a student's quiz scores?

c. What percent of the course grade is based on a student's test and quiz scores?

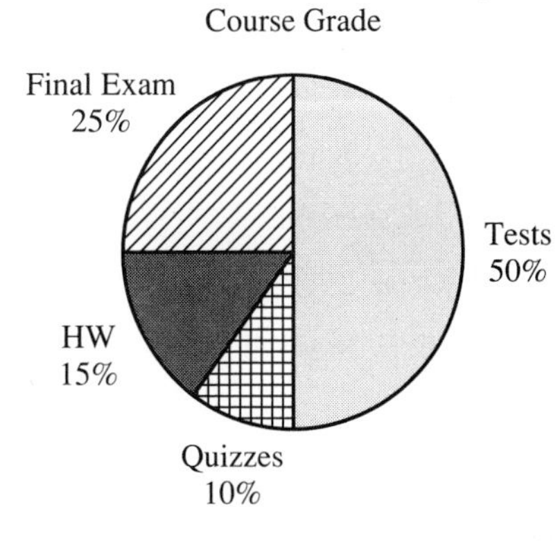

FIGURE 5.4.1

d. Which score will count as $\frac{1}{4}$ of the course grade?

e. Which score will count the least?

Solution

a. 50% (Note: $50\% = \frac{50}{100} = \frac{1}{2}$ of the whole circle.)

b. 10% (Note: $10\% = \frac{10}{100} = \frac{1}{10}$ of the whole circle.)

c. $50\% + 10\% = 60\%$

d. $\frac{1}{4} = 0.25 = 25\%$, therefore, the final exam will count as $\frac{1}{4}$ of the course grade.

e. Since the smallest wedge represents the quizzes, the quiz scores will count the least. ■

Example 2 Examine the circle graph in Figure 5.4.2 and answer the following questions.

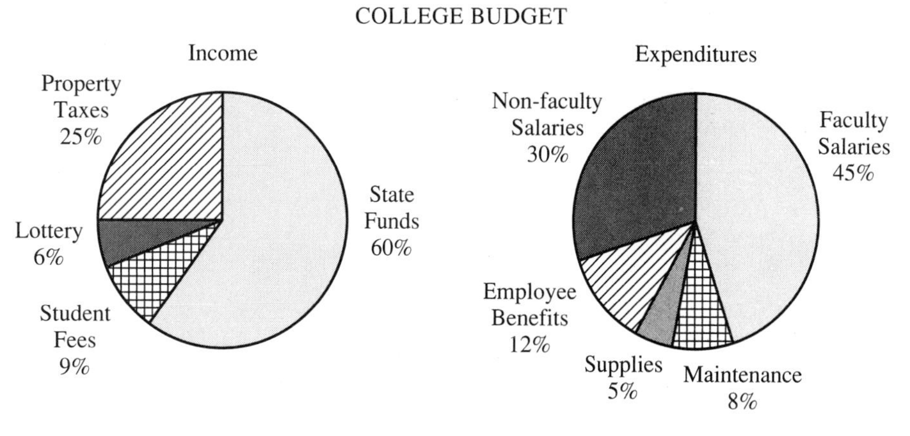

FIGURE 5.4.2

a. What percent of the college's income comes from property taxes?

b. What is the largest source of income for the college?

c. What is the smallest source of income for the college?

d. What percent of the college's budget is spent on faculty and non-faculty salaries?

e. If the college's total budget is \$50,000,000, how much is spent on supplies?

f. If the college's total budget is \$50,000,000, how much is spent on faculty salaries?

Solution

a. 25%

b. State funds

c. Lottery

d. 45% + 30% = 75%

e. What is 5% of 50,000,000?

(arrow under "What": amount for supplies; arrow under "50,000,000": total budget)

$$\frac{A}{50{,}000{,}000} = \frac{5}{100}$$

$$100A = 250{,}000{,}000$$

$$\frac{100A}{100} = \frac{250{,}000{,}0\not{0}\not{0}}{1\not{0}\not{0}}$$

$$A = \$2{,}500{,}000 \text{ for supplies}$$

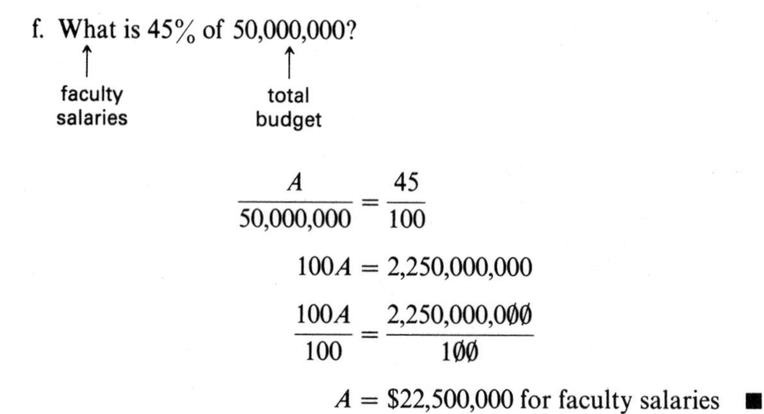

f. What is 45% of 50,000,000?

$$\frac{A}{50{,}000{,}000} = \frac{45}{100}$$

$$100A = 2{,}250{,}000{,}000$$

$$\frac{100A}{100} = \frac{2{,}250{,}000{,}0\not{0}\not{0}}{1\not{0}\not{0}}$$

$$A = \$22{,}500{,}000 \text{ for faculty salaries}$$ ■

EXERCISES 5.4

In Exercises 1–12, use the circle graph in Figure 5.4.3.

1. What percent of the students are 21 to 25 years old?

2. What percent of the students are 26 to 30 years old?

3. What percent of the students are over 30 years old?

4. What percent of the students are under 26 years old?

5. What is the largest age group?

6. What is the smallest age group?

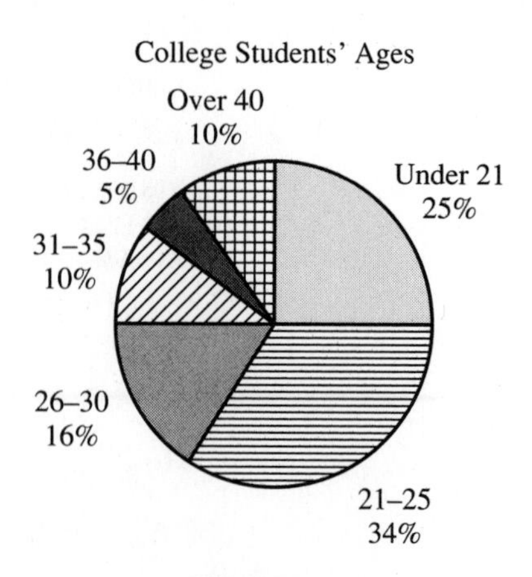

FIGURE 5.4.3

7. $\frac{1}{4}$ of the students are in which age group?

8. Approximately $\frac{1}{3}$ of the students are in which age group?

9. If the total number of students is 20,000, how many students are under 21 years old?

10. If the total number of students is 20,000, how many students are 21 to 25 years old?

11. If the college's total enrollment is 30,000 students, how many students are under 21 years old?

12. If the college's total enrollment is 30,000 students, how many students are 21 to 25 years old?

In Exercises 13–24, use the circle graph in Figure 5.4.4.

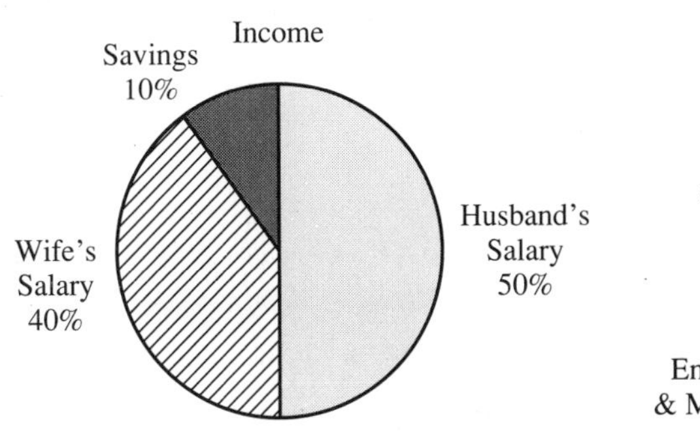

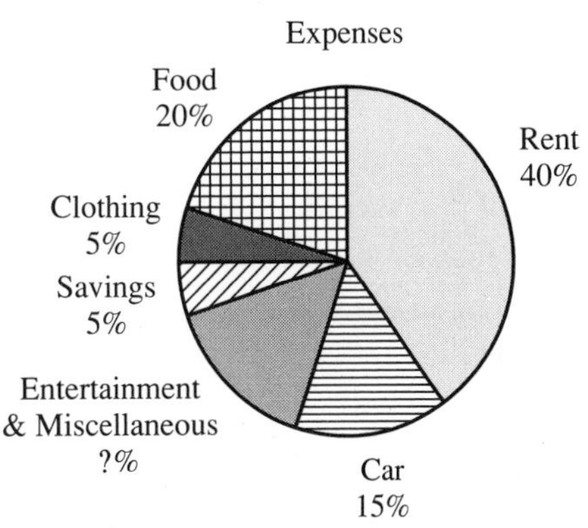

FIGURE 5.4.4

13. What percent of the family income comes from the wife's salary?

14. What percent of the family income comes from their savings?

15. What percent of the family budget is spent on food?

16. What percent of the family budget is spent on rent?

17. What is the largest source of income?

18. What is the biggest expense for the family?

19. What percent of the family budget is left over for entertainment and miscellaneous expenses?

20. $\frac{1}{5}$ of the family budget is spent for what item?

21. If the total family income after taxes is \$30,000 a year, what is the husband's salary?

22. If the total family income after taxes is \$30,000 a year, what is the wife's salary?

23. If the total family income after taxes is \$30,000 a year, how much is spent on rent each *month*?

24. If the total family income after taxes is \$30,000 a year, how much is spent on clothing each *month*?

5.5 Statistics

Statistics helps us to organize and interpret large amounts of data.

Mean

The **mean** of a set of numbers is found by adding all of the numbers and then dividing this sum by the number of numbers. The mean is what we usually refer to as the "average."

$$\text{Mean} = \frac{\text{sum of numbers}}{\text{number of numbers}}$$

Example 1 A bowler scored the following in four games: 175, 194, 150, and 165. What is the mean?

Solution $\text{Mean} = \dfrac{175 + 194 + 150 + 165}{4} = \dfrac{684}{4} = 171$ ■

Example 2 There are 10 students in a chemistry lab class. On a quiz 1 student scored 10 points, 4 students scored 9 points, 3 students scored 8 points, and 2 students scored 5 points. Find the class mean.

Method 1 Listing all of the scores

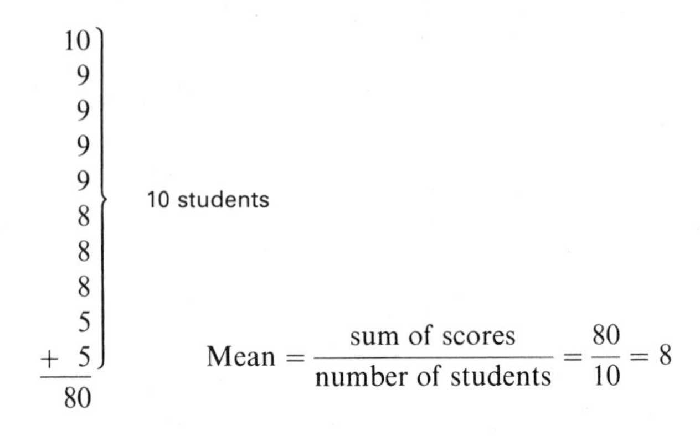

Method 2 Grouping the scores

Students		*Points*		
1	×	10	=	10
4	×	9	=	36
3	×	8	=	24
+ 2	×	5	=	+ 10
10				80 ← Sum of scores

↑ Number of students

$$\text{Mean} = \frac{\text{sum of scores}}{\text{number of students}} = \frac{80}{10} = 8 \quad ■$$

Median

The **median** of a set of numbers is the *middle* number *after the numbers have been arranged in order of size.* If there is an even numbers of items, there will be no middle number. In that case we take the average of the two numbers in the middle.

Example 3 The daily high temperatures for a week were 75°F, 82°F, 83°F, 85°F, 75°F, 77°F, and 70°F. Find the median.

Solution To find the median, arrange the temperatures in order of size.

85, 83, 82, 77, 75, 75, 70

↑

The middle number, 77°F, is the median. ■

Example 4 Find the median of the set of numbers $\{19, 16, 12, 15, 13, 9\}$.

Solution To find the median, arrange the numbers in order of size.

$$19, \quad 16, \quad \underbrace{15, \quad 13,} \quad 12, \quad 9$$

$$\text{The median} = \frac{15 + 13}{2} = \frac{28}{2} = 14. \quad ■$$

Mode

The **mode** of a set of numbers is the number that occurs the most often. There may be more than one mode, or there may be no mode.

Example 5 Find the mode of the set of numbers $\{3, 3, 4, 6, 6, 6, 7, 9\}$.

Solution $3,\ 3,\ 4,\ \underbrace{6,\ 6,\ 6,}_{\text{Mode}}\ 7,\ 9$

The number 6 occurs the most often, therefore, the mode is 6. ■

Example 6 Find the mode of the set of numbers $\{24, 17, 19, 24, 26, 17\}$.

Solution $\underbrace{17,\ 17,}_{\text{Mode}}\ 19,\ \underbrace{24,\ 24,}_{\text{Mode}}\ 26$

Both 17 and 24 occur twice. There are two modes, 17 and 24. ■

Range

The **range** of a set of numbers is the difference between the highest and the lowest number.

Example 7 Find the range in the following test scores: 78, 95, 86, 64, 72, 52, and 85.

Solution $\text{Range} = \underset{\text{Highest}}{\underset{\uparrow}{95}} - \underset{\text{Lowest}}{\underset{\uparrow}{52}} = 43$ ■

Example 8 Mrs. Furness paid the following heating bills last winter: \$12 in Oct., \$20 in Nov., \$30 in Dec., \$78 in Jan., \$65 in Feb., and \$35 in Mar. Find the mean, median, mode, and range.

Solution

$$\text{Mean} = \frac{\$12 + \$20 + \$30 + \$78 + \$65 + \$35}{6} = \frac{\$240}{6} = \$40$$

Arrange in order of size: \$12, \$20, \$30, \$35, \$65, \$78

$$\text{Median} = \frac{\$30 + \$35}{2} = \frac{\$65}{2} = \$32.50$$

There is no mode.

Range = \$78 − \$12 = \$66 ■

Example 9 Compare the salaries paid by Company A with Company B by finding the mean, median, and range of salaries for each company.

Company A	
President	\$120,000
Vice President	60,000
Manager #1	20,000
Manager #2	20,000
Manager #3	20,000
Manager #4	20,000
Manager #5	20,000
Total =	\$280,000

Company B	
President	\$ 80,000
Vice President	50,000
Manager #1	30,000
Manager #2	30,000
Manager #3	30,000
Manager #4	30,000
Manager #5	30,000
Total =	\$280,000

Solution

Company A	*Company B*
Mean $= \dfrac{\$280{,}000}{7} = \$40{,}000$	Mean $= \dfrac{\$280{,}000}{7} = \$40{,}000$
Median = \$20,000	Median = \$30,000
Range = \$120,000 − \$20,000 = \$100,000	Range = \$80,000 − \$30,000 = \$50,000

Although the mean salary for each company is \$40,000, the medians are very different. This is because one or two very large numbers, such as the president's salary for Company A, will have a big effect on the mean. We can see the big difference in the salaries for Company A by noting its large salary range. Sometimes the median gives a better idea of the "average." ■

EXERCISES 5.5

1. Use the following numbers:
{6, 4, 3, 4, 7, 9, 2}
a. Find the mean.
b. Find the median.
c. Find the mode.
d. Find the range.

2. Use the following numbers:
{3, 7, 9, 3, 4, 6, 9, 1, 3}
a. Find the mean.
b. Find the median.
c. Find the mode.
d. Find the range.

3. Use the following numbers:
{22, 26, 29, 21, 20, 32}
a. Find the mean.
b. Find the median.
c. Find the mode.
d. Find the range.

4. Use the following numbers:
{98, 74, 83, 81}
a. Find the mean.
b. Find the median.
c. Find the mode.
d. Find the range.

5. Five people drove a car on the highway to measure the car's fuel consumption in miles per gallon. The results were 24.7 mpg, 18.9 mpg, 20.0 mpg, 16.6 mpg, and 19.8 mpg.
a. Find the range.
b. Find the median.
c. Find the mean.

6. The rainfall for the month was 2.6 inches the first week, 1.7 inches the second week, 0.8 inches the third week, and 1.3 inches the fourth week.
a. Find the range.
b. Find the median.
c. Find the mean.

7. A bowler scored the following in six games: 180, 156, 164, 210, 176, and 170.
a. Find the median score.
b. Find the mean score.
c. Find the range in scores.

8. A golfer scored the following in seven games: 74, 86, 72, 76, 95, 70, and 80.
a. Find the median score.
b. Find the mean score.
c. Find the range in scores.

9. On a math quiz, 2 students scored 10, 1 student scored 9, 2 students scored 8, 3 students 7, and 4 students scored 6.
a. Find the class mean.
b. Find the class median.
c. Find the class mode.
d. Find the class range.

10. On an English quiz, 1 student scored 20, 3 students scored 14, 4 students scored 12, and 2 students scored 10.
a. Find the class mean.
b. Find the class median.
c. Find the class mode.
d. Find the class range.

11. Nine light bulbs were tested to measure the life of a light bulb in hours. The results are listed here. Round off answers to the nearest unit when necessary.)

12. Mr. Lumus' monthly electric bills for last year are listed here. (Round off answers to the nearest cent when necessary.)

726 hr	764 hr	780 hr
825 hr	801 hr	678 hr
715 hr	690 hr	679 hr

a. Find the mean.
b. Find the median.
c. Find the range.

\$52.50	\$43.43	\$51.36
\$47.00	\$38.00	\$45.00
\$48.50	\$43.95	\$61.75
\$36.25	\$48.27	\$70.50

a. Find the mean.
b. Find the median.
c. Find the range.

5.6 Chapter Summary

Mean 5.5 The **mean** is the sum of the numbers divided by the number of numbers.

$$\text{mean} = \frac{\text{sum of the numbers}}{\text{number of numbers}}$$

Median 5.5 The **median** is the middle number after the numbers have been arranged in order of size. If there is an even number of numbers, the median is the average of the two numbers in the middle.

Mode 5.5 The **mode** is the number that occurs the most often. There may be more than one mode or there may be no mode.

Range 5.5 The **range** is the difference between the highest and the lowest number.

Review Exercises 5.6

In Exercises 1–8, use the line graph in Figure 5.6.1.

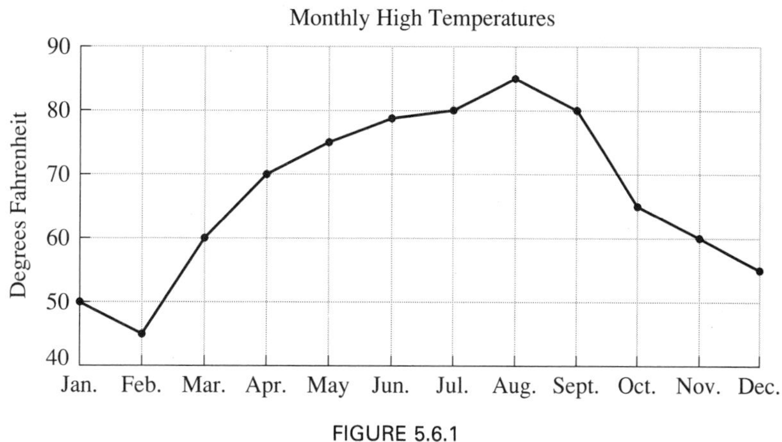

FIGURE 5.6.1

1. What was the high temperature for March?

2. What was the high temperature for April?

3. Which month was the coldest?

4. Which month was the hottest?

5. During which months did the temperature increase?

6. During which months did the temperature decrease?

7. Between which two months did the temperature increase the most?

8. Between which two months did the temperature decrease the most?

In Exercises 9–14, use the bar graph in Figure 5.6.2.

9. How many feet does it take to stop a car at 60 mph?

10. How many feet does it take to stop a car at 50 mph?

11. At what speed does it take approximately 40 feet to stop?

12. At what speed does it take approximately 120 feet to stop?

13. If you increase your speed from 20 mph to 40 mph, how many more feet will it take to stop the car?

14. If you increase your speed from 40 mph to 60 mph, how many more feet will it take to stop the car?

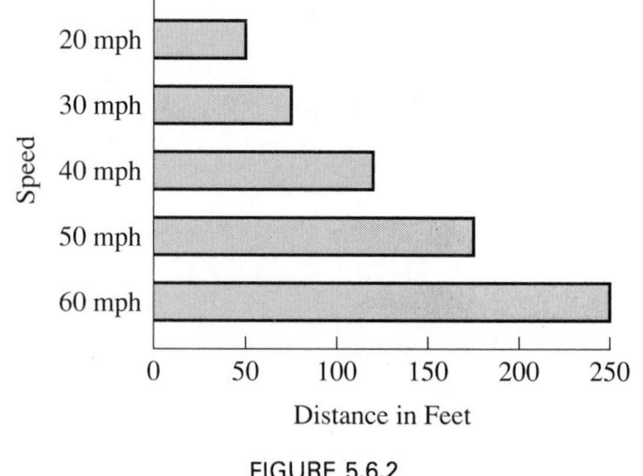

FIGURE 5.6.2

In Exercises 15–20, use the circle graph in Figure 5.6.3.

15. What percent of the people have black hair?

16. What percent of the people have blond hair?

17. What percent of the people do not have red hair?

18. What percent of the people do not have brown hair?

19. Out of 1,000 people, how many would have brown hair?

20. Out of 100,000 people, how many would have brown hair?

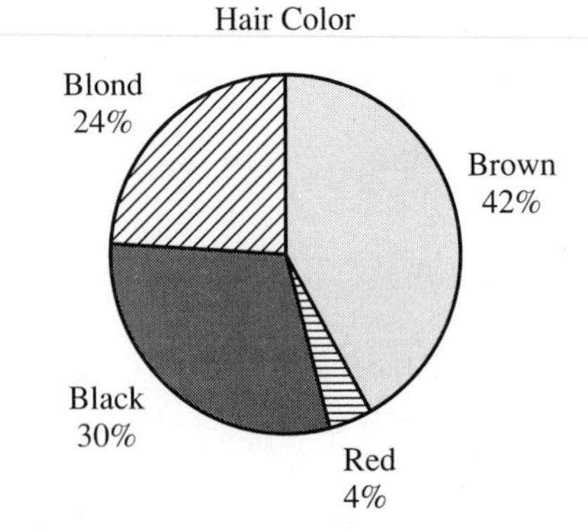

FIGURE 5.6.3

In Exercises 21–30, use the pictograph in Figure 5.6.4.

Daily Sales

Sun.	\$ \$ \$ \$ \$ \$ \$ \$ \$
Mon.	\$ \$ \$ \$ \$ \$
Tue.	\$ \$ \$ \$ \$
Wed.	\$ \$ \$ \$ \$
Thur.	\$ \$ \$ \$ \$ \$ \$ \$
Fri.	\$ \$ \$ \$ \$ \$ \$ \$
Sat.	\$ \$ \$ \$ \$ \$ \$ \$ \$ \$

\$ = \$1,000

FIGURE 5.6.4

21. What day of the week had the highest sales?

22. What day of the week had the lowest sales?

23. What was the amount of sales on Friday?

24. What was the amount of sales on Thursday?

25. On what day of the week did the sales equal $6,000?

26. On what day of the week did the sales equal $4,500?

27. How much more was sold on Saturday compared to Wednesday?

28. What was the total sales for the week?

29. What percent of the total week's sales were sold on Saturday?

30. What percent of the total week's sales were sold on Wednesday?

In Exercises 31–34, use the following set of numbers: {8, 6, 5, 3, 6, 2}

31. Find the mean.

32. Find the median.

33. Find the mode.

34. Find the range.

In Exercises 35–38, use the monthly temperatures on the line graph in Figure 5.6.1 to find the following.

35. Find the mean temperature for the year.

36. Find the median temperature for the year.

37. Find the temperature mode.

38. Find the range of the temperatures.

Chapter 5 Diagnostic Test

Allow yourself about 30 minutes to do these problems. Complete solutions for every problem, together with section references, are given in the answer section at the end of the book.

In Problems 1–5, use the line graph in Figure 5.DT.1.

1. What was the temperature at 8 A.M.?
2. At what time was the temperature 75°F?
3. What was the difference in temperature from 4 P.M. to 6 P.M.?
4. During what hours did the temperature increase?
5. At what time of the day was it the hottest?

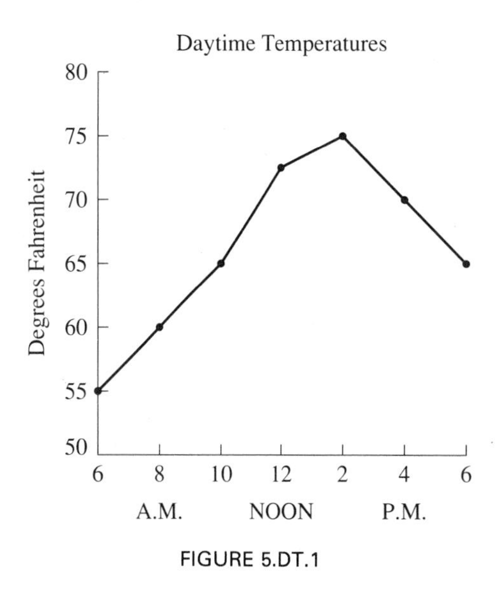

FIGURE 5.DT.1

In Problems 6–10, use the bar graph in Figure 5.DT.2.

6. Which cereal has the least sugar?
7. How many grams of sugar are in one ounce of Rice Krispies?
8. Which cereal has 8 grams of sugar in 1 ounce of cereal?
9. How many times more sugar is in Raisin Bran compared to Corn Flakes?
10. How much more sugar is in one ounce of Frosted Flakes than in one ounce of Rice Krispies?

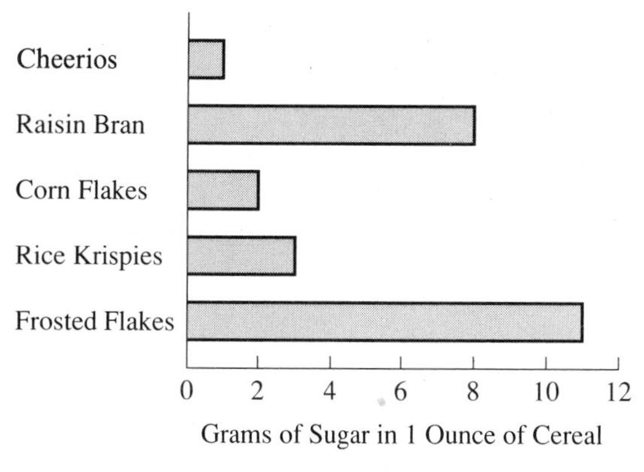

FIGURE 5.DT.2

In Problems 11–15, use the pictograph in Figure 5.DT.3.

11. How many cars did Mr. Brown sell?
12. Which salesman sold the most cars?
13. How many more cars did Mr. Carson sell compared to Mr. Brown?
14. What was the total sales for all three salesmen?
15. What percent of the total sales did Mr. Able sell?

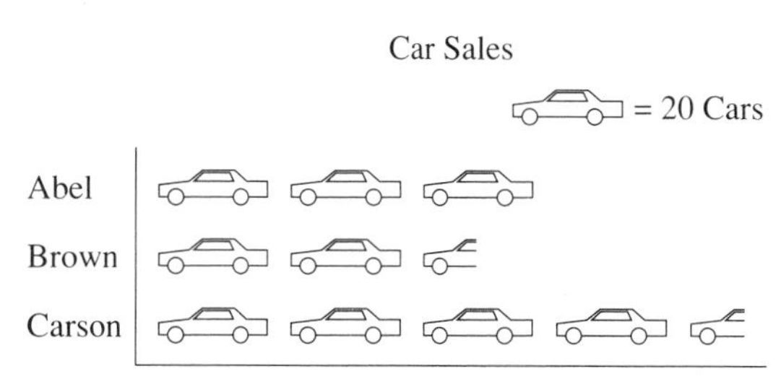

FIGURE 5.DT.3

In Problems 16–19, use the circle graph in Figure 5.DT.4.

16. What percent of the class got an A?

17. What percent of the class got a C or better?

18. What grade did approximately $\frac{1}{3}$ of the class receive?

19. If there were 48 students in the class, how many students got a B?

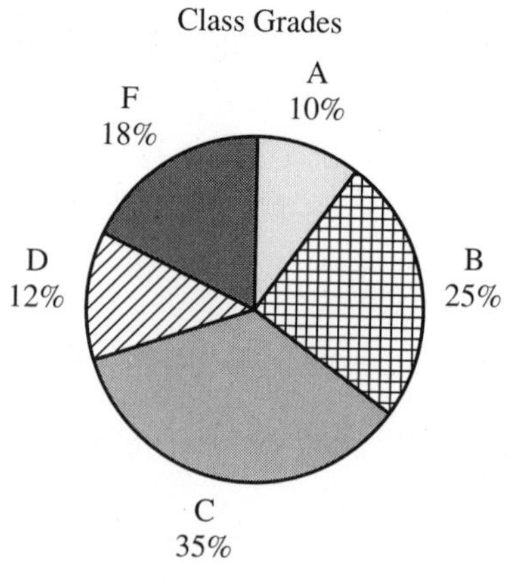

FIGURE 5.DT.4

In Problems 20 and 21, use the following test scores: {87, 72, 82, 93, 71}

20. Find the mean test score.

21. Find the median test score.

In Problems 22 and 23, use the following temperatures: {53°F, 66°F, 71°F, 62°F}

22. Find the temperature range.

23. Find the mean temperature.

In Problems 24 and 25, use the following numbers: {9, 6, 4, 2, 6, 2, 5, 6}

24. Find the median.

25. Find the mode.

Cumulative Review Exercises for Chapters 1–5

1. Write using digits: Five thousand, seven hundred and nine hundredths.

In Exercises 2–9, perform the indicated operations. Reduce all fractions to lowest terms, and change any improper fractions to mixed numbers.

2. $\frac{2}{9}+\frac{5}{6}+\frac{1}{2}$

3. $23\frac{1}{3}-10\frac{7}{12}$

4. $1\frac{1}{6}\cdot 3\frac{3}{4}$

5. $\dfrac{\frac{1}{2}+\frac{2}{3}}{\frac{3}{4}}$

6. $54.3-5.18$

7. 9.6×0.48

8. $630\div 1.8$

9. $15-12\div 3\cdot 2+3^2$

10. Arrange in order, largest to smallest.
$\frac{3}{4}$, $\frac{5}{6}$, $\frac{7}{12}$

11. Solve for the unknown letter.
$\frac{x}{9}=\frac{8}{12}$

12. Change 0.024 to a fraction in lowest terms.

13. Change $\frac{2}{5}$ to a percent.

14. Change 45% to a fraction in lowest terms.

15. 12 is 15% of what number?

16. On a map $\frac{1}{4}$ inch represents 6 miles. How many miles would $2\frac{1}{2}$ inches represent?

17. Find the perimeter of a rectangle with length 16 centimeters and width 12 centimeters.

18. A machinist uses a $\frac{5}{8}$-inch drill to put a hole through a metal plate. Would a 0.65-inch pin fit through the hole?

19. A woman bought a refrigerator for \$665. After making a downpayment of \$125, she paid the balance in 12 monthly payments. Find the amount of each monthly payment.

20. The rainfall for the first six months was 9.4 inches in January, 10.4 inches in February, 7.8 inches in March, 5.5 inches in April, 2.3 inches in May, and 1.2 inches in June. Find the average rainfall for the first six months.

21. A seamstress needs $1\frac{1}{4}$ yards of material for the top and $3\frac{1}{2}$ yards of material for the skirt of a bridesmaid's dress. If she plans to make 4 dresses, how many yards of material will she need?

22. A coat that originally sold for \$120 is on sale at 20% off. What is the sale price?

23 The college enrollment increased from 8,000 to 9,200. What was the percent of increase?

In Exercises 24 and 25, use the graph in Figure 5.CR.1.

24. How many IRA accounts were there in 1986?

25. Between which two years did the number of IRA accounts increase the most?

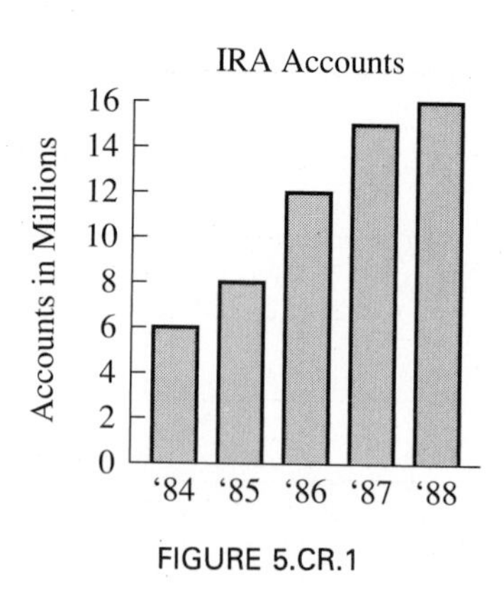

FIGURE 5.CR.1

6 Signed Numbers

Arithmetic is calculation with numbers using fundamental operations such as addition, subtraction, multiplication, and division. *Algebra* deals with the same fundamental operations with numbers but uses letters to represent some of the numbers.

Before beginning the study of algebra we review for your benefit a few basic definitions relating to numbers.

6.1 Signed Numbers

In Section 1.1 we introduced the natural numbers and the whole numbers.

Natural Numbers

The set of **natural numbers** (or *counting numbers*) is

$$N = \{1, 2, 3, 4, 5, \ldots\}$$

↑ Read "and so on"

The smallest natural number is 1. The largest natural number can never be found because no matter how far we count there are always larger natural numbers.

Number Line

Natural numbers can be represented by numbered points equally spaced along a straight line (Fig. 6.1.1). Such a line is called a **number line**. The arrow on the right of the line indicates that the number line continues forever.

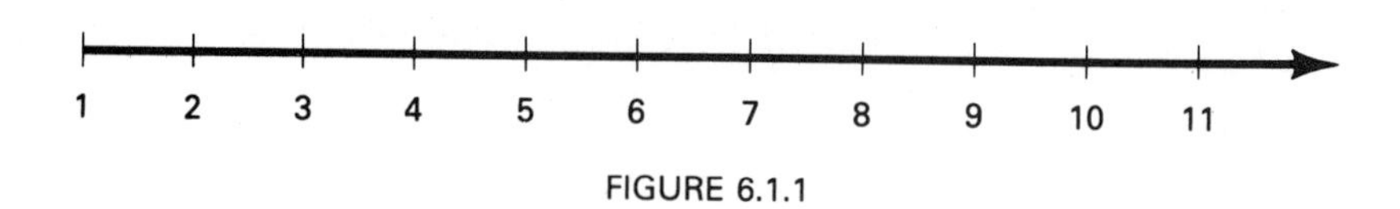

FIGURE 6.1.1

Whole Numbers

When 0 is included with the natural numbers, we have the set of numbers known as **whole numbers** (Fig. 6.1.2).

$$W = \{0, 1, 2, 3, 4, \ldots\}$$

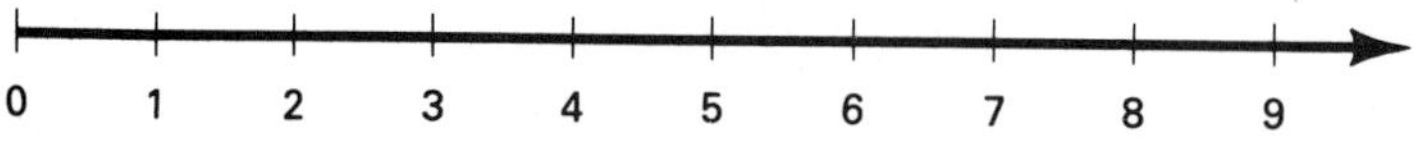

0 1 2 3 4 5 6 7 8 9

FIGURE 6.1.2

Digits

In our number system a **digit** is any one of the first ten whole numbers {0, 1, 2, 3, 4, 5, 6, 7, 8, 9}. Any number can be written by using a combination of these digits.

Signed Numbers

We now extend the number line to the left and continue with the set of equally spaced points.

Numbers used to name the points to the left of 0 on the number line are called *negative numbers.* Numbers used to name the points to the right of 0 on the number line are called *positive numbers.* Zero itself is neither positive nor negative. The positive and negative numbers are referred to as **signed numbers** (Fig. 6.1.3).

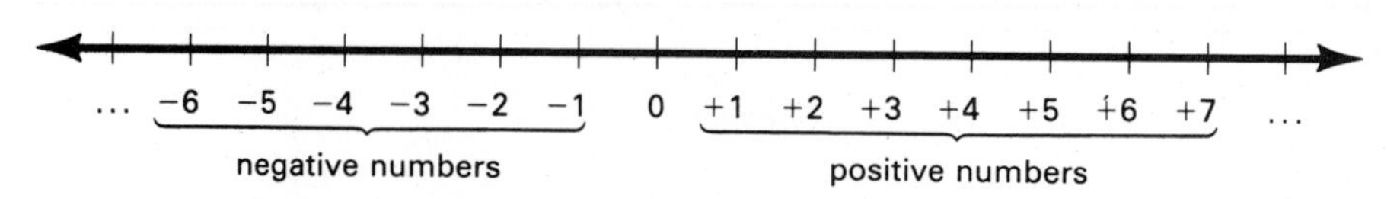

FIGURE 6.1.3

Integers

The numbers used to name the points shown in Fig. 6.1.3 are called **integers**. The set of integers can be represented in the following way:

$$\{\ldots, -3, -2, -1, 0, +1, +2, +3, \ldots\}$$

We have stated that a largest natural number could never be found because no matter how far we count there are always larger natural numbers. Similarly, no matter how far we count along the number line to the left of 0 we never reach a smallest integer.

Example 1 Reading positive and negative integers.

a. -1 Read "negative one."

b. -575 Read "negative five hundred seventy-five."

c. 25 Read "twenty-five" or "positive twenty-five." ■

When reading or writing positive numbers, we usually omit the word *positive* and the + sign. Therefore, when there is no sign in front of a number, it is understood to be positive.

Example 2 We show the use of positive and negative integers.

a. On a cold day in Minnesota, the temperature was $-20°$F. This means that the temperature was 20°F *below* 0.

b. The altitude of Mt. Everest is 29,028 feet. This means that the peak of Mt. Everest is 29,028 ft *above* sea level.

c. The lowest point in Death Valley, California, is -282 feet. This means that the lowest point in Death Valley is 282 ft *below* sea level. ■

Graphing Points on the Number Line

Many points exist on the number line that are not integers. Decimals, fractions, and mixed numbers are part of the real number system and can be represented by points on the number line (Fig. 6.1.4).

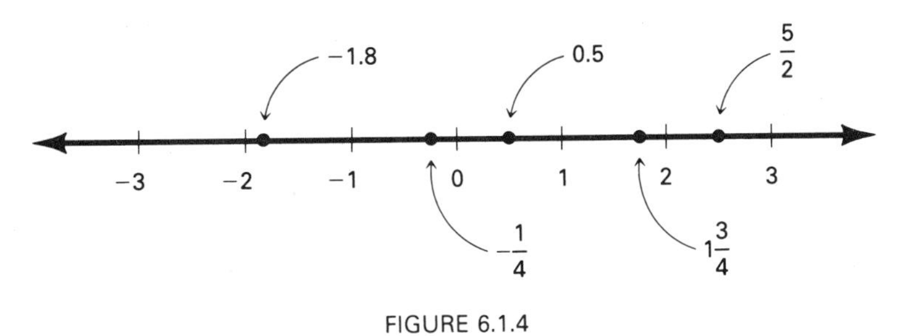

FIGURE 6.1.4

Example 3 On the number line below, which letter best locates the following numbers?

a. $1\frac{1}{2}$ b. $-\frac{1}{2}$ c. $-2\frac{1}{3}$ d. 0.75

A B C D E F G

−3 −2 −1 0 1 2 3

Solution

a. $1\frac{1}{2}$ is $1\frac{1}{2}$ units to the right of zero; therefore, $1\frac{1}{2} = G$.

b. $-\frac{1}{2}$ is $\frac{1}{2}$ unit to the left of zero; therefore, $-\frac{1}{2} = C$.

c. $-2\frac{1}{3}$ is $2\frac{1}{3}$ units to the left of zero; therefore, $-2\frac{1}{3} = A$.

d. 0.75 is $\frac{75}{100} = \frac{3}{4}$ unit to the right of zero; therefore, $0.75 = F$. ■

Using Inequality Symbols Recall that numbers get larger as we move to the right on the number line and smaller as we move to the left.

Example 4 $-3 > -7$

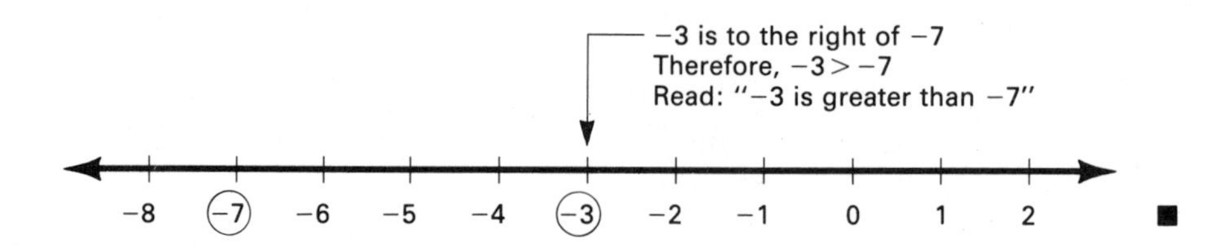

■

Example 5 $-5 < -2$

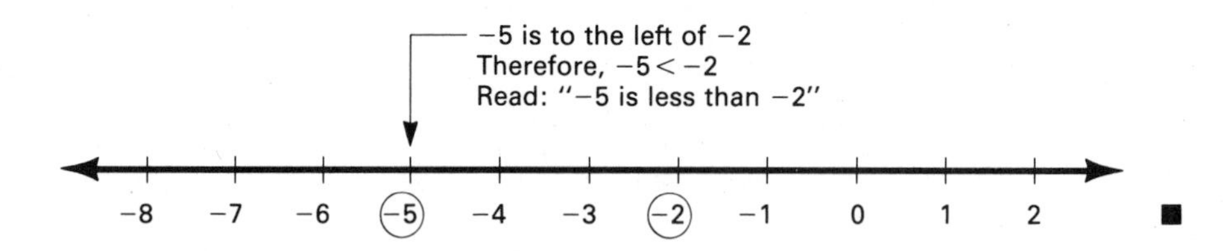

■

An easy way to remember the meaning of the inequality symbol is to notice that the wide part of the symbol is next to the larger number.

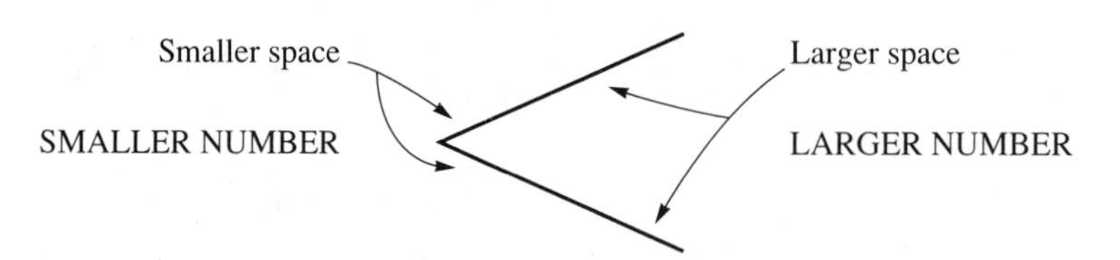

Some people like to think of the symbols $>$ and $<$ as arrowheads that point toward the smaller number.

Example 6 Verify the following inequalities by using a number line to determine whether the first number of each pair is to the right or left of the second number of that pair.

a. $6 > 4$ Read "6 is greater than 4."

b. $-2 > -7$ Read "-2 is greater than -7."

c. $-1 < -\frac{1}{2}$ Read "-1 is less than $-\frac{1}{2}$."

d. $-5 < 3$ Read "-5 is less than 3."

e. $3 > -5$ Read "3 is greater than -5."

Note that in Example 6 (d) and (e), $-5 < 3$ and $3 > -5$ give the same information even though they are read differently. ■

EXERCISES 6.1

1. Write -75 in words.
2. Write -49 in words.
3. Use digits to write negative fifty-four.
4. Use digits to write negative one hundred nine.
5. A scuba diver descends to a depth of sixty-two feet. Represent this number by an integer.
6. The temperature at Fairbanks, Alaska, was forty-five degrees Fahrenheit below zero. Represent this number by an integer.
7. Which is larger, -2 or -4?
8. Which is larger, 0 or -10?
9. Which is smaller, -5 or -10?
10. Which is smaller, -1 or -15?
11. What is the largest negative integer?
12. What is the largest integer?
13. What is the smallest integer?
14. What is the smallest whole number?

In Exercises 15–22, which letter best locates the following numbers on the number line below?

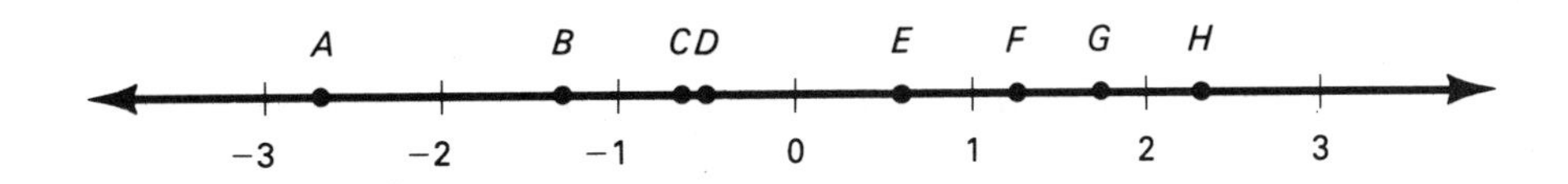

15. $1\frac{1}{4}$ **16.** 1.75 **17.** $-\frac{1}{2}$ **18.** $-\frac{2}{3}$ **19.** $-2\frac{2}{3}$ **20.** -1.3 **21.** 0.6 **22.** $\frac{9}{4}$

In Exercises 23–30, determine which of the two symbols > or < should be used to make each statement true.

23. $0 \mathbin{?} -3$ **24.** $-2 \mathbin{?} -6$ **25.** $-5 \mathbin{?} 2$ **26.** $-7 \mathbin{?} -4$

27. $-\frac{1}{2} \mathbin{?} -\frac{1}{4}$ **28.** $-2\frac{1}{3} \mathbin{?} -2\frac{2}{3}$ **29.** $-0.2 \mathbin{?} -0.7$ **30.** $\frac{1}{4} \mathbin{?} -0.5$

6.2 Adding Signed Numbers

In this section we show how to add signed numbers. We can represent a signed number by an arrow beginning at the point representing 0 and ending at the point representing that particular number on the number line.

Example 1 Represent 4 by an arrow.

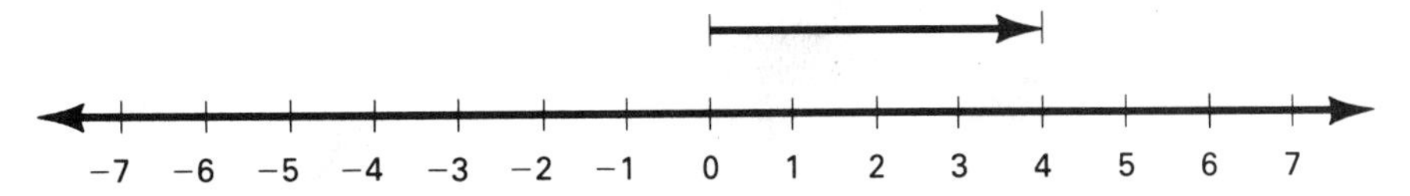

This arrow represents a movement of 4 units to the *right*. Any *positive* number is represented by an arrow directed to the *right*. The arrow need not start at zero so long as it has a length equal to the number it represents. ■

Example 2 Represent -5 by an arrow.

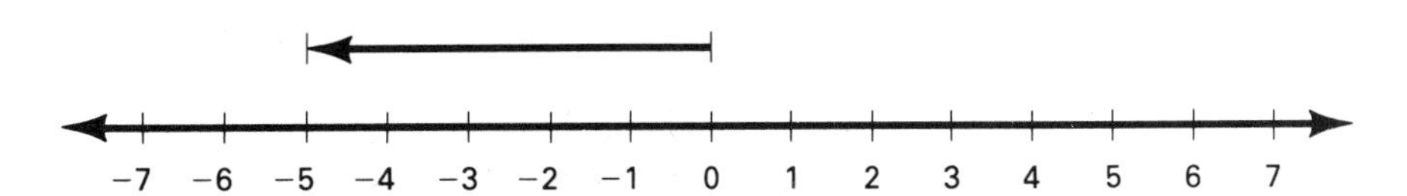

This arrow represents a movement of 5 units to the *left*. Any *negative* number is represented by an arrow directed to the *left*. ■

We can also represent the addition of signed numbers by means of arrows.

Example 3 Add 3 to 2 by means of arrows.

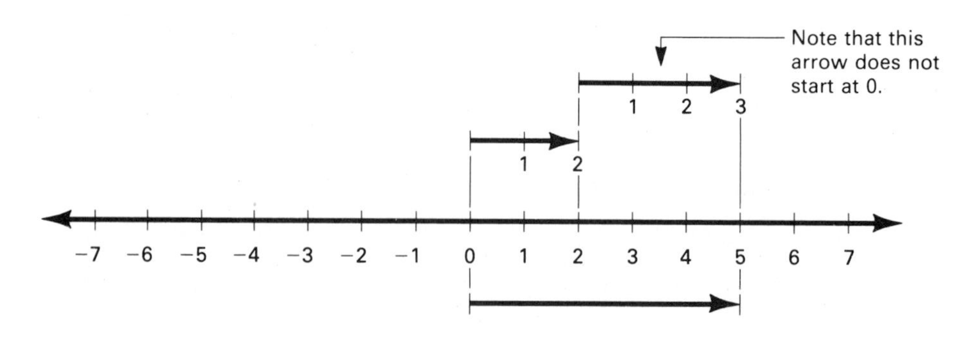

To add 3 to 2 on the number line, begin by drawing the arrow representing 2. Draw the arrow representing 3 starting at the arrowhead end of the arrow representing 2. These two movements represent a net movement to the right of 5 units. Therefore, $2 + 3 = 5$. ■

Example 4 Add -7 to 5 by means of arrows.

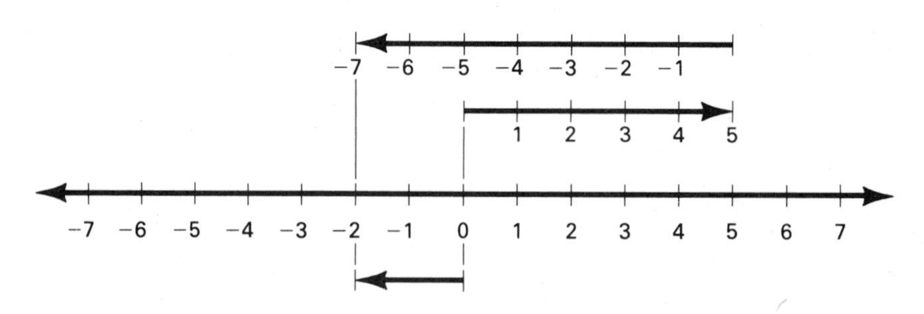

Begin by drawing the arrow representing 5. Draw the arrow representing -7 starting at the arrowhead end of the arrow representing 5. These two movements represent a net movement to the left of 2 units. Therefore, $5 + (-7) = -2$. ■

Example 5 Add -4 to -3 by means of arrows.

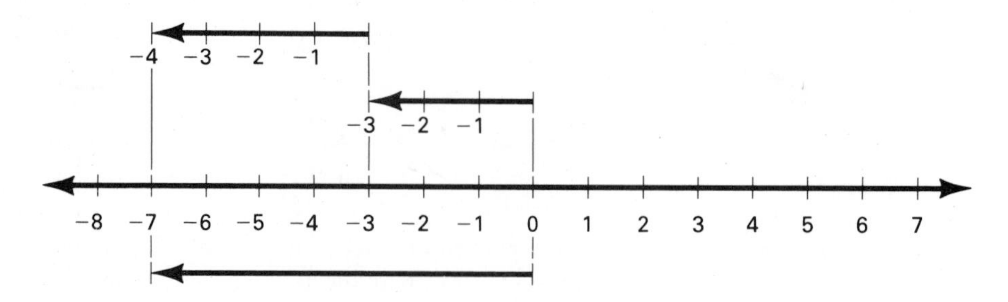

Begin by drawing the arrow representing -3. Draw the arrow representing -4 starting at the arrowhead end of the arrow representing -3. These two movements represent a net movement to the left of 7 units. Therefore, $-3 + (-4) = -7$. ■

Absolute Value

The **absolute value** of a number is the distance between that number and 0 on the number line *with no regard for direction*. See Figure 6.2.1. The absolute value of a real number x is written $|x|$.

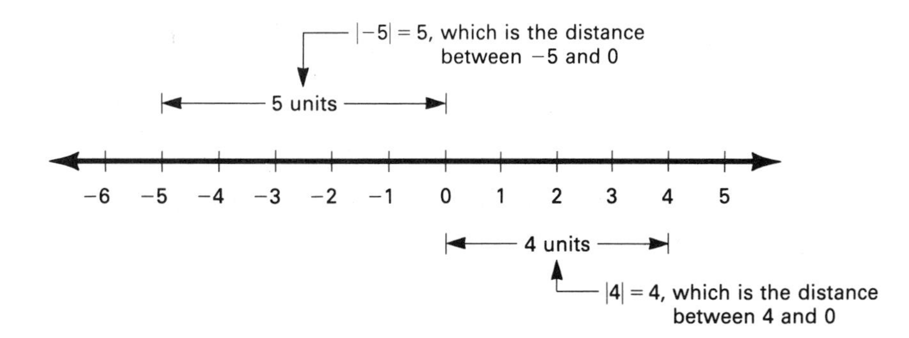

FIGURE 6.2.1

Example 6 Absolute value of numbers

a. $|9| = 9$ a positive number

b. $|0| = 0$ zero

c. $|-4| = 4$ a positive number

The absolute value of a number can never be negative ■

A signed number has two distinct parts: its absolute value and its sign.

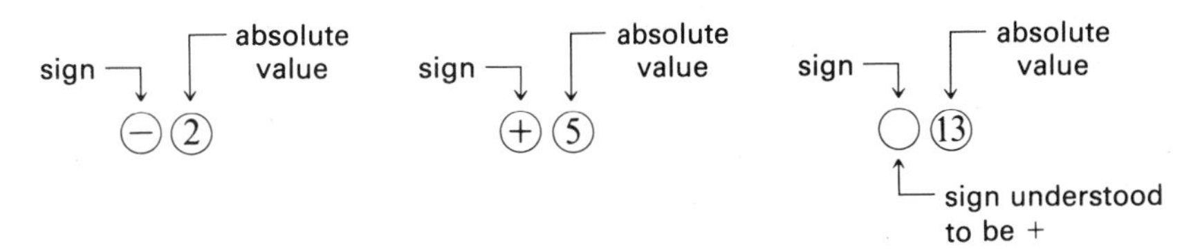

NOTE The absolute value of a signed number is the number written without its sign. ☑

Adding signed numbers by means of arrows is easy to understand, but it is a very slow process. The following rules give an easier and faster method for adding signed numbers. The previous four examples can be used to show why the following rules are true.

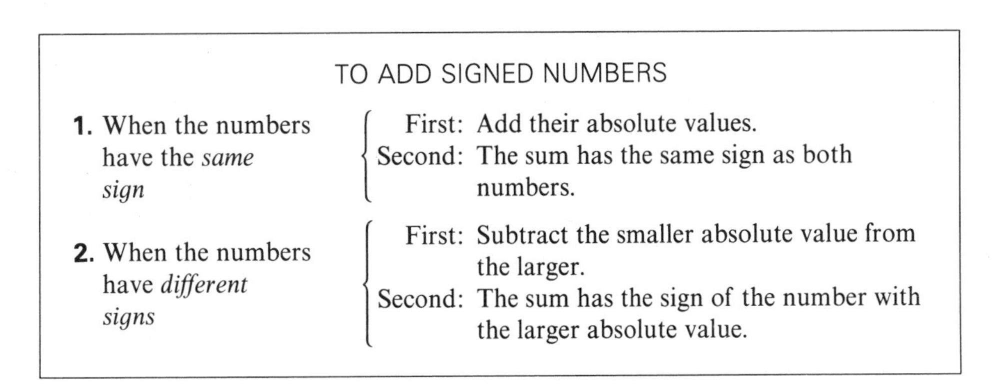

TO ADD SIGNED NUMBERS

1. When the numbers have the *same sign*
- First: Add their absolute values.
- Second: The sum has the same sign as both numbers.

2. When the numbers have *different signs*
- First: Subtract the smaller absolute value from the larger.
- Second: The sum has the sign of the number with the larger absolute value.

We show how to add signed numbers by means of the rules.

Example 7 Find $(-7) + (-11)$.

Solution Because -7 and -11 have the *same* sign, *add* their absolute values: $7 + 11 = 18$.

Sum has the same sign: $(-)$

Therefore, $(-7) + (-11) = -18$. ■

Example 8 Find $(-24) + (17)$.

Solution Because -24 and 17 have *different* signs, *subtract* their absolute values: $24 - 17 = 7$.

Sum has sign of number with larger absolute value $(-)$ since -24 has the larger absolute value.

Therefore, $(-24) + (17) = -7$. ■

Example 9 Find $(-18) + (32)$.

Solution Because -18 and 32 have *different* signs, *subtract* their absolute values: $32 - 18 = 14$.

Sum has sign of number with larger absolute value $(+)$ because 32 has the larger absolute value.

Therefore, $(-18) + (32) = 14$. ■

Example10 Find $(-29) + (-35)$.

Solution

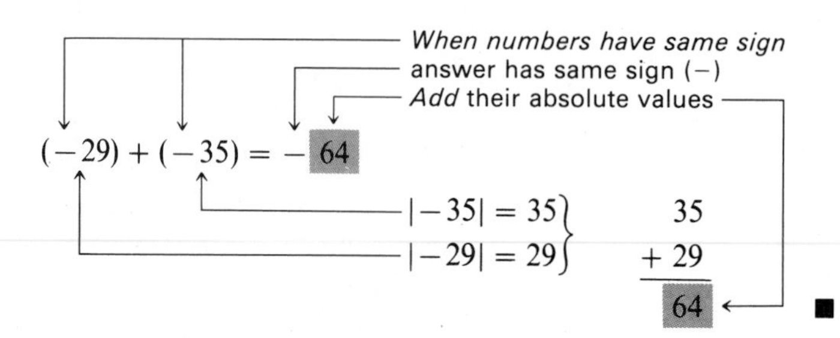

■

Example 11 Find $(-9) + (+23)$.

Solution

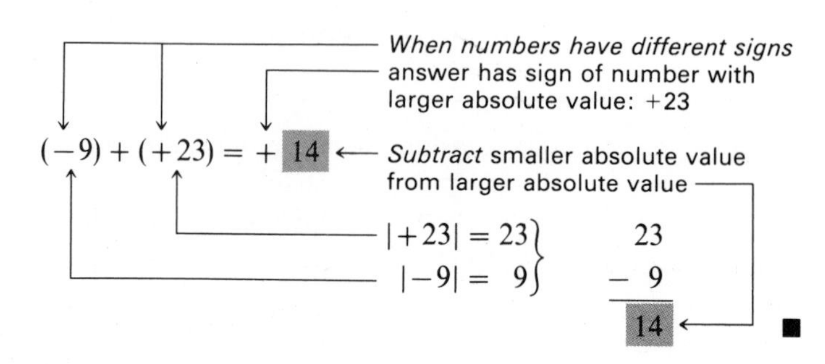

■

Example 12 Find $\left(-2\frac{1}{2}\right) + \left(-4\frac{1}{3}\right)$.

Solution LCD = 6

$$\left(-2\frac{1}{2}\right) = \left(-2\frac{3}{6}\right)$$

Because signs are the *same*, *add* their absolute values

$$+\left(-4\frac{1}{3}\right) = \left(-4\frac{2}{6}\right)$$

$$-6\frac{5}{6}$$

Sum has same sign: (−)

Therefore, $\left(-2\frac{1}{2}\right) + \left(-4\frac{1}{3}\right) = -6\frac{5}{6}$. ■

Example 13 Find the following using a calculator.

a. $(-6) + (3)$

b. $(-2) + (-5)$

Solution To enter a negative number, press the [+/−] after the numeral.

a. Key in: 6 [+/−] [+] 3 [=]
Answer: -3

b. Key in: 2 [+/−] [+] 5 [+/−] [=]
Answer: -7 ■

EXERCISES 6.2

In Exercises 1–56, find the sums.

1. $(4) + (5)$
2. $(6) + (2)$
3. $(-3) + (-4)$
4. $(-7) + (-1)$
5. $(-6) + (5)$
6. $(-8) + (3)$
7. $(7) + (-3)$
8. $(9) + (-4)$
9. $(-2) + (-9)$
10. $(-4) + (-8)$
11. $(-5) + (8)$
12. $(-3) + (6)$
13. $(5) + (6)$
14. $(4) + (2)$
15. $(2) + (-5)$
16. $(6) + (-8)$
17. $(-1) + (-8)$
18. $(-5) + (-5)$
19. $(-9) + (2)$
20. $(-7) + (4)$
21. $(3) + (-3)$
22. $(-9) + (9)$
23. $(8) + (-1)$
24. $(6) + (-4)$
25. $(-3) + (-9)$
26. $(-5) + (-7)$
27. $(6) + (-15)$
28. $(4) + (-12)$
29. $(-8) + (2)$
30. $(-6) + (-5)$
31. $(4) + (-1)$
32. $(7) + (-5)$
33. $(-3) + (-6)$
34. $(-9) + (4)$
35. $(-7) + (9)$
36. $(-8) + (-5)$
37. $(-18) + (-3)$
38. $(-16) + (9)$
39. $(-8) + (15)$
40. $(-3) + (-17)$
41. $(15) + (-5)$
42. $(18) + (-6)$
43. $(-27) + (-13)$
44. $(-42) + (-12)$
45. $(-80) + (121)$
46. $(69) + (-134)$
47. $(-105) + (73)$
48. $(218) + (-113)$
49. $\left(-1\frac{1}{2}\right) + \left(-3\frac{2}{5}\right)$
50. $\left(-2\frac{1}{2}\right) + \left(-5\frac{1}{4}\right)$
51. $\left(4\frac{5}{6}\right) + \left(-1\frac{1}{3}\right)$
52. $\left(6\frac{3}{4}\right) + \left(-2\frac{1}{8}\right)$
53. $(-7.3) + (-5.48)$
54. $(-12.67) + (-4.092)$
55. $(1.03) + (-0.946)$
56. $(-43.2) + (9.85)$

57. At 6 A.M. the temperature in Hibbing, Minnesota, was -35°F. If the temperature had risen 53°F by 2 P.M., what was the temperature at that time?

58. At midnight in Billings, Montana, the temperature was -50°F. By noon the temperature had risen 67°F. What was the temperature at noon?

In Exercises 59–62, use a calculator to find the sums.

59. $(-8) + 3$
60. $(-9) + (-6)$
61. $(-12) + (-26)$
62. $(14) + (-36)$

6.3 Subtracting Signed Numbers

The Negative of a Number

The idea of **negative** suggests the opposite of something. For example, the negative of taking two steps to the right would be taking two steps to the left. Because $+2$ can be thought of as a movement of two units to the right, then the negative of $+2$ would be thought of as a movement of 2 units to the left. This means the negative of $+2$ is -2.

Example 1 The negative of a *positive* number

a. The negative of 5 is -5.

b. The negative of 12 is -12, and so on. ■

Because -3 can be thought of as a movement of 3 units to the left, the negative of -3 can be thought of as an opposite movement of 3 units to the right. This means that the negative of -3 is $+3$. This can be written: $-(-3) = +3 = 3$.

Example 2 The negative of a *negative* number

a. The negative of -10 is $+10$ written $-(-10) = 10$.

b. The negative of -14 is $+14$ written $-(-14) = 14$. ■

Examples 1 and 2 lead to the following rules for finding the negative of a number.

TO FIND THE NEGATIVE OF A NUMBER

Change the sign of the number.
The negative of $b = -b$
The negative of $-b = -(-b) = b$
The negative of $0 = 0$

The negative of a number is used in the definition of subtraction.

DEFINITION OF SUBTRACTION

$$a - b = a + (-b)$$

In words: To subtract b from a, add the negative of b to a.

This definition leads to the following rule for subtracting signed numbers.

TO SUBTRACT ONE SIGNED NUMBER FROM ANOTHER

1. Change the subtraction symbol to an addition symbol, and change the sign of the number being subtracted.
2. Add the resulting signed numbers as shown in Section 6.2.

For example,

$$(8) - (+5)$$
$$= (8) + (-5)$$
$$= 3$$

Change the sign of the number being subtracted

Change the subtraction symbol to an addition symbol

Example 3 Find $(-3) - (-7)$.

Solution

Change subtraction to addition

Change the sign of the number being subtracted

$$(-3) - (-7) =$$
$$(-3) + (+7) = 4$$ ■

Example 4 Find $(-5)-(2)$.

Solution

Change subtraction to addition

Change the sign of the number being subtracted

$$(-5) - (+2) =$$
$$(-5) + (-2) = -7 \quad \blacksquare$$

Example 5 Find $(9)-(-8)$.

Solution

$$(+9)-(-8)=$$
$$(+9)+(+8)=17 \quad \blacksquare$$

Example 6 Subtract 87 from 25.

Solution To subtract 87 from 25 means

$$(+25)-(+87)=$$
$$(+25)+(-87)=-62 \quad \blacksquare$$

Example 7 Find $\left(-3\frac{1}{2}\right)-\left(-2\frac{1}{4}\right)$.

Solution LCD = 4

$$\left(-3\frac{1}{2}\right)-\left(-2\frac{1}{4}\right)=$$
$$\left(-3\frac{1}{2}\right)+\left(+2\frac{1}{4}\right)=$$
$$\left(-3\frac{2}{4}\right)+\left(+2\frac{1}{4}\right)=-1\frac{1}{4}$$

or,

$$\left(-3\frac{1}{2}\right)=\left(-3\frac{2}{4}\right)$$
$$-\left(-2\frac{1}{4}\right)=+\left(+2\frac{1}{4}\right)$$
$$-1\frac{1}{4}$$

Because signs are *different, subtract* their absolute values

Sum has sign of number with larger absolute value (−) ■

Example 8 Find $(-16.5)-(9.83)$.

Solution

$$(-16.5)-(+9.83)=$$
$$(-16.5)+(-9.83)=-26.33$$

or,

$$(-16.5)=(-16.50)$$
$$-(+9.83)=+(-9.83)$$
$$-26.33$$

Because signs are *same, add* their absolute values

Sum has same sign (−) ■

EXERCISES 6.3

In Exercises 1–56, find the differences.

1. $(10)-(4)$

2. $(12)-(5)$

3. $(-3)-(-2)$

4. $(-4)-(-3)$

5. $(-6)-(2)$

6. $(-8)-(5)$

7. $(9)-(-5)$

8. $(7)-(-3)$

9. $(2)-(-7)$

10. $(3)-(-5)$

11. $(-6)-(-8)$

12. $(-3)-(-5)$

13. $(-9)-(-2)$

14. $(-7)-(-4)$

15. $(3)-(7)$

16. $(1)-(9)$

17. $(-6)-(-9)$

18. $(-5)-(-1)$

19. $(-7)-(5)$

20. $(-8)-(2)$

21. $(-8) - (-1)$ **22.** $(-4) - (-2)$ **23.** $(-9) - (-3)$ **24.** $(-5) - (-2)$

25. $(6) - (-8)$ **26.** $(-3) - (7)$ **27.** $(-2) - (-9)$ **28.** $(4) - (6)$

29. $(-5) - (7)$ **30.** $(-9) - (-4)$ **31.** $(5) - (8)$ **32.** $(7) - (-7)$

33. $(-10) - (-6)$ **34.** $(9) - (15)$ **35.** $(-15) - (-20)$ **36.** $(8) - (-17)$

37. $(-16) - (9)$ **38.** $(-26) - (8)$ **39.** $(34) - (89)$ **40.** $(47) - (53)$

41. $(-156) - (-97)$ **42.** $(-203) - (-168)$ **43.** $(384) - (-279)$ **44.** $(136) - (-275)$

45. $\left(-4\frac{2}{3}\right) - \left(2\frac{1}{6}\right)$ **46.** $\left(5\frac{7}{8}\right) - \left(1\frac{1}{4}\right)$ **47.** $\left(-3\frac{3}{4}\right) - \left(-2\frac{1}{6}\right)$ **48.** $\left(9\frac{5}{6}\right) - \left(-6\frac{4}{9}\right)$

49. Subtract (-7) from (-10).

50. Subtract (-6) from (8).

51. $(-7.3) - (2.06)$

52. $(9.48) - (-26.4)$

53. $(-56.2) - (-8.53)$

54. $(0.375) - (0.972)$

55. Subtract $\left(3\frac{1}{4}\right)$ from $\left(-5\frac{3}{10}\right)$.

56. Subtract $\left(1\frac{1}{2}\right)$ from $\left(4\frac{5}{6}\right)$.

57. At 5 A.M. the temperature at Mammoth Mountain, California was -7°F. At noon the temperature was 42°F. What was the rise in temperature?

58. At 4 A.M. the temperature at Massena, New York, was -5.6°F. At 1 P.M. the temperature was 37.5°F. What was the rise in temperature?

59. A jeep starting from the shore of the Dead Sea ($-1{,}299$ ft) is driven to the top of a nearby hill having an elevation of 723 ft. What was the change in the jeep's altitude?

60. A dune buggy starting from the floor of Death Valley (-282 ft) is driven to the top of a nearby mountain having an elevation of 5,782 ft. What was the change in the dune buggy's altitude?

61. A scuba diver descends to a depth of 141 ft below sea level. His buddy dives 68 ft deeper. What is his buddy's altitude at the deepest point of his dive?

62. Mt. Everest (the highest known point on earth) has an altitude of 29,028 ft. The Mariana Trench in the Pacific Ocean (the lowest known point on earth) has an altitude of $-36{,}198$ ft. Find the difference in altitude of these two places.

 In Exercise 63–66, use a calculator to find the differences.

63. $(8) - (-2)$ **64.** $(-4) - (3)$ **65.** $(-17) - (-6)$ **66.** $(-12) - (-20)$

6.4 Multiplying Signed Numbers

Terms Used in Multiplication

Factors
Product

$6 \times 2 = 12$

factors (6, 2) — product (12)

6 and 2 are **factors** of 12; 12 is the **product** of 6 and 2. 3 and 4 are also factors of 12 because $3 \times 4 = 12$.

Multiplicative Identity

Because multiplying any real number by 1 gives the identical real number, 1 is called the **multiplicative identity**.

MULTIPLICATIVE IDENTITY

If a represents any real number

$$1 \cdot a = a \cdot 1 = a$$

Symbols Used in Multiplication Multiplication can be shown in several different ways.

1. $3 \times 2 = 6$
2. $3 \cdot 2 = 6$ The multiplication dot " · " is written a little higher than the decimal point.
3. $ab = a \cdot b$ When two expressions are written next to each other with no symbol of operation, it is understood that they are to be multiplied. *Exception*: When two *numbers* are written next to each other, they are *not* to be multiplied. For example: 23 does *not* mean $2 \cdot 3 = 6$.
4. $3x = 3 \cdot x$
5. $3(2) = 6$ The symbols ( ) are called *parentheses*.
6. $(3)(2) = 6$ In this kind of multiplication, the double parentheses are not necessary.

Multiplying Two Signed Numbers

Multiplication (by a positive integer) is a short method for doing repeated addition of the same number.

Example 1 $3 \cdot 5 =$ three 5's $= 5 + 5 + 5 = 15$ ■

-2 + -2 = -4

Example 2 $6 \cdot 2 =$ six 2's $= 2 + 2 + 2 + 2 + 2 + 2 = 12$ ■

Therefore, carrying this same idea over into multiplying signed numbers, we have:

Example 3 $3 \cdot (-2) =$ three -2's $= (-2) + (-2) + (-2) = -6$ ■

Example 4 $2 \cdot (-6) =$ two -6's $= (-6) + (-6) = -12$ ■

From Examples 3 and 4 we see that *when two numbers having opposite signs are multiplied, their product is negative.*

opposite signs = negative

Consider the products:

Example 5 $(-1)(5) = -5$ the negative of 5 ■

Example 6 $(-1)(8) = -8$ the negative of 8 ■

From Examples 5 and 6 we see that multiplying a number by -1 gives the negative of that number.

$$(-1)\cdot a = a\cdot(-1) = -a$$

Example 7 $(-5)(-4) = (-1)(5)(-4)$*

$(-5)(-4) = (-1)(5)(-4)$*	Because $-5 = (-1)(5)$
$= (-1)(-20)$	Because $(5)(-4) = -20$
$=$ negative of -20	Because -1 times a number gives the negative of that number
$= 20$	Because the negative of a number is found by changing its sign (Section 6.3) ■

Example 8 $(-10)(-15) = (-1)(10)(-15)^* = (-1)(-150) = 150$ ■

From Examples 7 and 8 we see that *when two negative numbers are multiplied, their product is positive.*

Example 9 The fact that the product of two negative numbers is positive can also be seen from the following pattern:

$4\,(-2) = -8$
$3\,(-2) = -6$
$2\,(-2) = -4$
$1\,(-2) = -2$
$0\,(-2) = 0$ — Zero as a factor is discussed in Section 6.7
$-1\,(-2) = 2$
$-2\,(-2) = 4$
$-3\,(-2) = 6$

(left column: ← decreasing by 1; right column: increasing by 2 →)

The last three lines: The product of two negative numbers is positive ■

The rules for multiplying two signed numbers are summarized as follows:

TO MULTIPLY TWO SIGNED NUMBERS

1. Multiply their absolute values.
2. The product is *positive* when the signed numbers have the same sign.

   The product is *negative* when the signed numbers have different signs.

$$+\cdot+=+ \qquad +\cdot-=-$$
$$-\cdot-=+ \qquad -\cdot+=-$$

* These arguments depend upon the commutative and associative properties, which are discussed in Section 6.6

Example 10 Multiply $(-7)(4)$.

Solution $(-7)(4) = \ominus\ (28)$

Product of their absolute values: $7 \cdot 4 = 28$

Product negative because the numbers have different signs ■

Example 11 Multiply $(-14)(-10)$.

Solution $(-14)(-10) = \oplus\ (140)$

$14 \cdot 10 = 140$

Product positive because the numbers have the same sign ■

Example 12 Multiply $\left(4\frac{1}{2}\right)\left(-1\frac{1}{3}\right)$.

Solution $\left(4\frac{1}{2}\right)\left(-1\frac{1}{3}\right) = -\left(\frac{\overset{3}{\cancel{9}}}{\underset{1}{\cancel{2}}} \cdot \frac{\overset{2}{\cancel{4}}}{\underset{1}{\cancel{3}}}\right) = -\frac{6}{1} = -6$ ■

Example 13 Multiply $(-2.7)(-4.6)$.

Solution $(-2.7)(-4.6) = +(2.7 \times 4.6) = 12.42$ ■

Example 14 Multiply $(-2)(-5)(-3)$.

Solution $(-2)(-5)(-3) =$

$(+10)(-3) = -30$ ■

NOTE With an *odd* number of negative signs, the product is *negative*. ☑

Example 15 Multiply $(-2)(-5)(-3)(-4)$.

Solution $(-2)(-5)(-3)(-4) =$

$(+10)(-3)(-4) =$

$(-30)(-4) = +120$ ■

NOTE With an *even* number of negative signs, the product is *positive*. ☑

EXERCISES 6.4

In Exercises 1–44, find the products.

1. $3(-2)$ **2.** $4(-6)$ **3.** $(-5)(2)$ **4.** $(-7)(5)$

5. $(-8)(-2)$ **6.** $(-6)(-7)$ **7.** $8(-4)$ **8.** $9(-5)$

9. $(-6)(3)$ **10.** $(-8)(-5)$ **11.** $(9)(-6)$ **12.** $(-8)(8)$

13. $(-7)(-4)$ **14.** $9(-3)$ **15.** $(-5)(6)$ **16.** $(-3)(-9)$

17. $(-7)(9)$ **18.** $(-6)(8)$ **19.** $(-10)(-10)$ **20.** $(-9)(-9)$

21. $(8)(-7)$ **22.** $(12)(-6)$ **23.** $(-26)(10)$ **24.** $(-11)(12)$

25. $(-20)(-10)$ **26.** $(-30)(-20)$ **27.** $(75)(-15)$ **28.** $(86)(-13)$

29. $(-30)(5)$ **30.** $(-50)(6)$ **31.** $(-7)(-20)$ **32.** $(-9)(-40)$

33. $(-5)(-4)(-2)$ **34.** $(-3)(2)(-8)$ **35.** $(-4)(-2)(-1)(-7)$ **36.** $(3)(-2)(-6)(-5)$

37. $\left(2\frac{1}{4}\right)\left(-1\frac{1}{3}\right)$ **38.** $\left(-3\frac{3}{4}\right)\left(-2\frac{2}{5}\right)$ **39.** $\left(-1\frac{7}{8}\right)\left(-2\frac{4}{5}\right)$ **40.** $\left(-3\frac{1}{3}\right)\left(2\frac{1}{4}\right)$

41. $(2.74)(-100)$ **42.** $(-3.04)(-1000)$ **43.** $(-4.6)(-8.3)$ **44.** $(-9.7)(0.52)$

In Exercises 45–48, use a calculator to find the products.

45. $(-9)(8)$ **46.** $(7)(-6)$ **47.** $(-15)(-25)$ **48.** $(-5)(-9)(-7)$

6.5 Dividing Signed Numbers

Division may be shown in several ways.

$$12 \div 4 = \frac{12}{4} = 4\overline{)12}$$

The quotient in this case is 3.

Terms Used in Division

Quotient
Divisor
Dividend

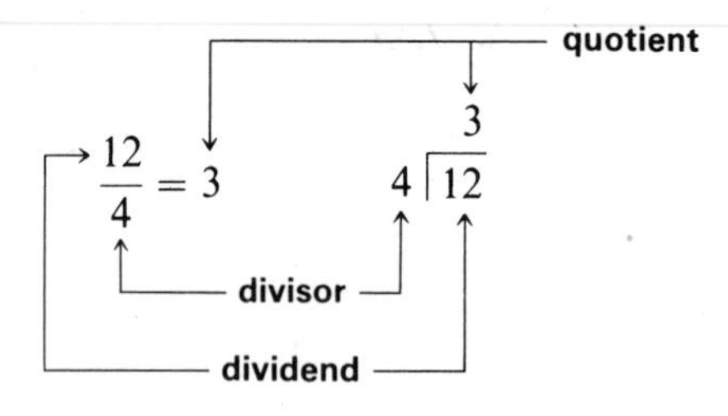

Division is the *inverse* operation of multiplication. Every division problem is related to a multiplication problem.

Because	$4 \cdot 3 = 12,$	it follows that	$12 \div 3 = 4$
Because	$(4) \cdot (-3) = -12,$	it follows that	$(-12) \div (-3) = 4$
Because	$(-4) \cdot (-3) = 12,$	it follows that	$(12) \div (-3) = -4$
Because	$(-4) \cdot (3) = -12,$	it follows that	$(-12) \div (3) = -4$

Dividing Two Signed Numbers

From these examples we see that we need the same rules for *dividing* signed numbers that we used for *multiplying* signed numbers.

TO DIVIDE ONE SIGNED NUMBER BY ANOTHER

1. Divide their absolute values.
2. The quotient is *positive* when the signed numbers have the same sign.

   The quotient is *negative* when the signed numbers have different signs.

$$+ \div + = + \qquad + \div - = -$$
$$- \div - = + \qquad - \div + = -$$

Example 1 $(-30) \div (5)$

Solution $(-30) \div (5) = \ominus 6$

- 6: Quotient of their absolute values: $30 \div 5 = 6$
- $\ominus$: Quotient negative because the numbers have different signs ■

Example 2 $(-64) \div (-8)$

Solution $(-64) \div (-8) = \oplus 8$

- 8: $64 \div 8 = 8$
- $\oplus$: Quotient positive because the numbers have the same sign ■

Example 3 $\dfrac{2\frac{2}{3}}{-1\frac{1}{9}}$

Solution $\dfrac{2\frac{2}{3}}{-1\frac{1}{9}} = \dfrac{\frac{8}{3}}{-\frac{10}{9}} = \dfrac{8}{3} \div \left(-\dfrac{10}{9}\right) = -\left(\dfrac{\cancel{8}^{4}}{\cancel{3}_{1}} \cdot \dfrac{\cancel{9}^{3}}{\cancel{10}_{5}}\right) = -\dfrac{12}{5} = -2\dfrac{2}{5}$ ■

EXERCISES 6.5

In Exercises 1–36, find the following quotients.

1. $(-10) \div (-5)$ **2.** $(-12) \div (-4)$ **3.** $(-8) \div (2)$ **4.** $(-6) \div (3)$

5. $\dfrac{+10}{-2}$ **6.** $\dfrac{+8}{-4}$ **7.** $\dfrac{-6}{-3}$ **8.** $\dfrac{-10}{-2}$

9. $(-40) \div (8)$ **10.** $(-60) \div (10)$ **11.** $16 \div (-4)$ **12.** $25 \div (-5)$

13. $(-15) \div (-5)$ **14.** $(-27) \div (-9)$ **15.** $12 \div (-4)$ **16.** $24 \div (-6)$

17. $\dfrac{-18}{-2}$ **18.** $\dfrac{-49}{-7}$ **19.** $\dfrac{-150}{10}$ **20.** $\dfrac{-250}{100}$

21. $36 \div (-12)$ **22.** $56 \div (-8)$ **23.** $(-45) \div 15$ **24.** $(-39) \div 13$

25. $\dfrac{-15}{6}$ **26.** $\dfrac{-27}{12}$ **27.** $\dfrac{7.5}{-0.5}$ **28.** $\dfrac{1.25}{-0.25}$

29. $\dfrac{-6.3}{-0.9}$ **30.** $\dfrac{-4.8}{-0.6}$ **31.** $\dfrac{-367}{100}$ **32.** $\dfrac{-4{,}860}{1{,}000}$

33. $\dfrac{2\frac{1}{2}}{-5}$ **34.** $\dfrac{3\frac{2}{5}}{-17}$ **35.** $\dfrac{-4\frac{1}{2}}{-1\frac{7}{8}}$ **36.** $\dfrac{-3\frac{3}{4}}{-2\frac{1}{10}}$

In Exercises 37–40, use a calculator to find the quotients.

37. $(-84) \div (7)$ **38.** $(-72) \div (-4)$ **39.** $(-132) \div (-12)$ **40.** $(75) \div (-1.5)$

6.6 Properties

Commutative Properties

Addition

If you change the order of the two numbers in an addition problem, you get the same sum. This is called the **commutative property of addition.**

Example 1 Addition *is* commutative.

$$\left.\begin{aligned}(-6) + (2) &= -4\\ (2) + (-6) &= -4\end{aligned}\right\} \quad \text{Therefore, } (-6) + (2) = (2) + (-6)$$ ■

COMMUTATIVE PROPERTY OF ADDITION

If a and b represent any real numbers, then

$$a + b = b + a$$

Subtraction

If you change the order of the two numbers in a subtraction problem, you *do not* get the same difference (except when the two numbers are equal). Therefore, *subtraction is not commutative.*

Example 2 Subtraction is *not* commutative.

$$\left.\begin{aligned}3 - 2 &= 1\\ 2 - 3 &= -1\end{aligned}\right\} \quad \text{Therefore, } 3 - 2 \neq 2 - 3$$ ■

Multiplication

If you change the order of the two numbers in a multiplication problem, you get the same product. This is called the **commutative property of multiplication.**

Example 3 Multiplication *is* commutative.

$$\left.\begin{aligned}(-9)(3) &= -27\\ (3)(-9) &= -27\end{aligned}\right\} \quad \text{Therefore, } (-9)(3) = (3)(-9)$$ ■

COMMUTATIVE PROPERTY OF MULTIPLICATION

If a and b represent any real numbers, then

$$a \cdot b = b \cdot a$$

Division

If you change the order of the two numbers in a division problem, you *do not* get the same quotient (except when the two numbers are equal). Therefore, *division is not commutative.*

Example 4 Division is *not* commutative.

$$\left.\begin{aligned}10 \div 5 &= 2\\ 5 \div 10 &= \frac{1}{2}\end{aligned}\right\} \quad \text{Therefore, } 10 \div 5 \neq 5 \div 10$$ ■

Associative Properties

Addition

In an addition problem, the sum of three numbers is unchanged no matter how we group the numbers. This is called the **associative property of addition**. Parentheses are used to show which two numbers are to be added first.

Example 5 Addition *is* associative.

$$\left.\begin{aligned}(2+3)+4 &= 5+4 = 9\\ 2+(3+4) &= 2+7 = 9\end{aligned}\right\} \quad \text{Therefore, } (2+3)+4 = 2+(3+4)$$ ■

> ASSOCIATIVE PROPERTY OF ADDITION
>
> If a, b, and c represent any real numbers, then
>
> $$(a+b)+c = a+(b+c)$$

Subtraction

A single example will show that *subtraction is not associative.*

Example 6 Subtraction is *not* associative.

$$\left.\begin{aligned}(7-4)-2 &= 3-2 = 1\\ 7-(4-2) &= 7-2 = 5\end{aligned}\right\} \quad \text{Therefore, } (7-4)-2 \neq 7-(4-2)$$ ■

Multiplication

In a multiplication problem, the product of three numbers is unchanged no matter how we group the numbers. This is called the **associative property of multiplication.**

Example 7 Multiplication *is* associative.

$$\left.\begin{aligned}(3\cdot 4)\cdot 2 &= 12\cdot 2 = 24\\ 3\cdot(4\cdot 2) &= 3\cdot 8 \ \ = 24\end{aligned}\right\} \quad \begin{aligned}&\text{Therefore,}\\ &(3\cdot 4)\cdot 2 = 3\cdot(4\cdot 2)\end{aligned}$$ ■

> ASSOCIATIVE PROPERTY OF MULTIPLICATION
>
> If a, b, and c represent any real numbers, then
>
> $$(a\cdot b)\cdot c = a\cdot(b\cdot c)$$

Division

A single example will show that *division is not associative.*

Example 8 Division is *not* associative.

$$\left.\begin{aligned}(16\div 4)\div 2 &= 4\div 2 \ \ = 2\\ 16\div(4\div 2) &= 16\div 2 = 8\end{aligned}\right\} \quad \text{Therefore, } (16\div 4)\div 2 \neq 16\div(4\div 2)$$ ■

SUMMARY

1. The *commutative property* says that *changing the order* of the numbers in an addition or multiplication problem gives the same answer.
2. The *associative property* says that *changing the grouping* of the numbers in an addition or multiplication problem gives the same answer.

Example 9 State whether each of the following is true or false, and give the reason.

a. $(-7) + 5 = 5 + (-7)$ — *True* because of the commutative property of addition. (*Order* of numbers changed.)

b. $(+6)(-8) = (-8)(+6)$ — *True* because of the commutative property of multiplication. (*Order* of numbers changed.)

c. $(9 + 4) + 5 = 9 + (4 + 5)$ — *True* because of the associative property of addition. (*Grouping* changed.)

d. $(5 \cdot 2) \cdot 3 = 5 \cdot (2 \cdot 3)$ — *True* because of the associative property of multiplication. (*Grouping* changed.)

e. $(+8) - (-7) = (-7) - (+8)$ — *False.* Subtraction is *not* commutative.

f. $a + (b + c) = (a + b) + c$ — *True* because of the associative property of addition. (*Grouping* changed.)

g. $y \div z = z \div y$ — *False.* Division is *not* commutative.

h. $p \cdot (s \cdot r) = p \cdot (r \cdot s)$ — *True* because of the commutative property of multiplication. (*Order* changed.) ■

EXERCISES 6.6

State whether each of the following is true or false, and give the reason.

1. $7 + 5 = 5 + 7$
2. $9 + 4 = 4 + 9$
3. $(2 + 6) + 3 = 2 + (6 + 3)$
4. $(1 + 8) + 7 = 1 + (8 + 7)$
5. $6 - 2 = 2 - 6$
6. $4 - 7 = 7 - 4$
7. $(a \cdot b) \cdot c = a \cdot (b \cdot c)$
8. $(p \cdot q) \cdot r = p \cdot (q \cdot r)$
9. $8 \div 4 = 4 \div 8$
10. $3 \div 6 = 6 \div 3$
11. $(p)(t) = (t)(p)$
12. $(m)(n) = (n)(m)$
13. $(4)(-5) = (-5) + (4)$
14. $(-7)(2) = (2) + (-7)$
15. $5 + (3 + 4) = 5 + (4 + 3)$
16. $6 + (8 + 2) = 6 + (2 + 8)$
17. $e + f = f + e$
18. $j + k = k + j$
19. $9 + (5 + 6) = (9 + 6) + 5$
20. $3 \cdot (8 \cdot 4) = (3 \cdot 4) \cdot 8$
21. $x - 4 = 4 - x$
22. $5 - y = y - 5$
23. $4 \cdot (6a) = (4 \cdot 6)a$
24. $(3 \cdot 5)x = 3 \cdot (5x)$
25. $H + 8 = 8 + H$
26. $4 + P = P + 4$
27. $(3)(-2) = (-2)(3)$
28. $a - 2 = 2 - a$
29. $a \cdot (b \cdot c) = a \cdot (c \cdot b)$
30. $3 \div x = x \div 3$

6.7 Operations with Zero

Additive Identity

Addition Involving Zero Because adding zero to a number gives the identical number for the sum, *zero* is called the **additive identity**.

Example 1

a. $9 + 0 = 9$

b. $0 + (-5) = -5$ ■

> ADDITIVE IDENTITY
>
> If a represents any real number, then
>
> $$a + 0 = 0 + a = a$$

Subtraction Involving Zero Because the subtraction $a - b$ has been defined as $a + (-b)$, the rules for subtractions involving zero are derived from the rules for addition.

> SUBTRACTION INVOLVING ZERO
>
> If a represents any real number, then
>
> **1.** $a - 0 = a$
>
> **2.** $0 - a = 0 + (-a) = -a$

Multiplication Involving Zero Because multiplication is a method for doing repeated addition of the same number, multiplying a number by zero gives a product of zero.

Example 2 $4 \cdot 0 = 0 + 0 + 0 + 0 = 0$

Because of the commutative property of multiplication, it follows that

$$4 \cdot 0 = 0 \cdot 4 \quad = 0$$

and

$$(-2) \cdot 0 = 0 \cdot (-2) = 0 \quad ■$$

> MULTIPLICATION BY ZERO
>
> If a represents any real number, then
>
> $$a \cdot 0 = 0 \cdot a = 0$$

Division Involving Zero Every division problem is related to a multiplication problem.

$$\frac{6}{2} = 3 \qquad \text{because } 3 \cdot 2 = 6$$

Zero divided by a nonzero number is possible, and the quotient is zero.

Example 3 $\frac{0}{2} = 0$ because $0 \cdot 2 = 0$ ■

A nonzero number divided by zero is impossible.

Example 4 $\frac{6}{0} = ? \longleftarrow$ Suppose the quotient is some number x

Then $\frac{6}{0} = x$ means $x \cdot 0 = 6$, which is certainly false

$x \cdot 0 \neq 6$, because any number multiplied by zero $= 0$. Therefore, dividing any nonzero number by zero is undefined. ■

Zero divided by zero cannot be determined.

Example 5 $\frac{0}{0} = 0$ means $0 \cdot 0 = 0$, which is true.

$\frac{0}{0} = 1$ means $1 \cdot 0 = 0$, which is true.

$\frac{0}{0} = 5$ means $5 \cdot 0 = 0$, which is true.

In other words, $\frac{0}{0} = 0$, 1, and also 5. In fact, it can be any number. Because there is *no unique* answer, we say that $\frac{0}{0}$ is undefined. ■

DIVISION INVOLVING ZERO

If a represents any real number, except 0, then

1. $\frac{0}{a} = 0$

2. $\frac{a}{0}$ is undefined } cannot divide by 0

3. $\frac{0}{0}$ is undefined }

EXERCISES 6.7

In Exercises 1–20, find the value of each of the following (if it has one). If an expression does not have a value, give a reason.

1. $5 \cdot 0$	**2.** $0 \cdot 7$	**3.** $4 + 0$	**4.** $0 + 9$	**5.** $0 - 6$
6. $0 - 10$	**7.** $0 \div 12$	**8.** $0 \div (-15)$	**9.** $5 + (0 + 6)$	**10.** $(3 + 0) + 7$
11. $\frac{4}{0}$	**12.** $\frac{8}{0}$	**13.** $(0)(-15)$	**14.** $(-13)(0)$	**15.** $\frac{0}{0}$

16. $-\left(\frac{0}{0}\right)$ **17.** $0 + (-789)$ **18.** $(-546) + 0$ **19.** $\frac{-1}{0}$ **20.** $\frac{-156}{0}$

In Exercises 21–25, use a calculator to find the following:

21. $\frac{9}{0}$ **22.** $\frac{0}{-8}$ **23.** $\frac{0}{0}$ **24.** $0 - 74$ **25.** $(-6.5)(0)$

6.8 Powers of Signed Numbers

Now that we have learned to multiply signed numbers, it is possible to consider products in which the same number is repeated as a factor. For example:

$$3 \cdot 3 \cdot 3 \cdot 3 = 3^4 = 81$$

Base
Exponent
Power

In the symbol 3^4, the 3 is called the **base**. The 4 is called the **exponent** and is written above and to the right of the base 3. The entire symbol 3^4 is called the *fourth* **power** *of three* and is commonly read "three to the fourth power" (Fig. 6.8.1).

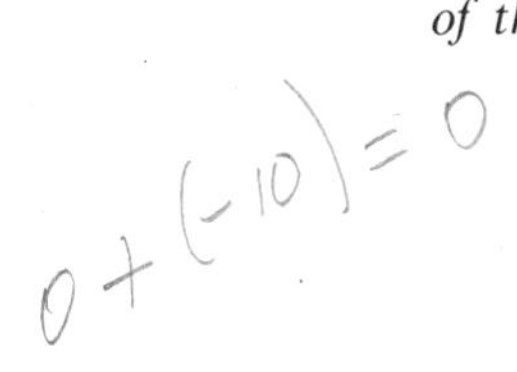

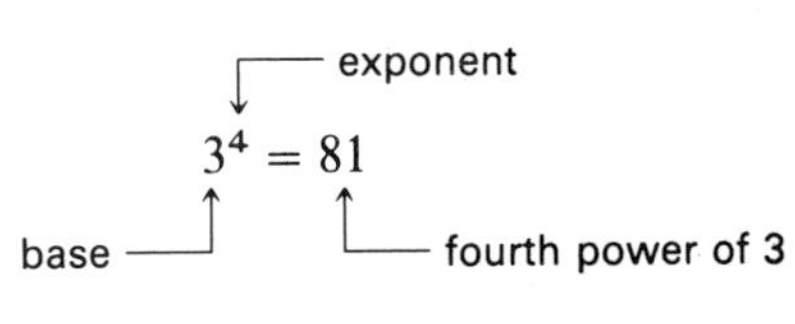

FIGURE 6.8.1

Even Power

If a base has an exponent that is exactly divisible by two, we say that it is an **even power** of the base. For example: 3^2, 5^4, $(-2)^6$ are even powers.

Odd Power

If a base has an exponent that is *not* exactly divisible by two, we say that it is an **odd power** of the base. For example: 3^1, 10^3, $(-4)^5$ are odd powers.

Example 1 Powers of signed numbers

a. $2^3 = 2 \cdot 2 \cdot 2 = 8$

b. $4^2 = 4 \cdot 4 = 16$

c. $(-3)^2 = (-3)(-3) = 9$

d. $(-1)^4 = (-1)(-1)(-1)(-1) = 1$

NOTE An *even* power of a negative number is positive. ☑

e. $(-2)^3 = (-2)(-2)(-2) = -8$

f. $(-1)^5 = (-1)(-1)(-1)(-1)(-1) = -1$

NOTE An *odd* power of a negative number is negative. ☑

g. $\left(\frac{2}{3}\right)^2 = \frac{2}{3} \cdot \frac{2}{3} = \frac{4}{9}$ ■

A WORD OF CAUTION Students often think that expressions such as $(-6)^2$ and -6^2 are the same. They are *not* the same. *The exponent applies only to the symbol immediately preceding it.*

$(-6)^2 = (-6)(-6) = 36$ The exponent applies to the ( )

$-6^2 = -(6 \cdot 6) = -36$ The exponent applies to the 6

Therefore, $-6^2 \neq (-6)^2$. ☑

Powers of Zero

> If a is any positive real number,
>
> $$0^a = 0$$

Example 2 $0^5 = 0 \cdot 0 \cdot 0 \cdot 0 \cdot 0 = 0$ ■

Zero as an Exponent

When any real number (other than 0) is raised to the 0 power, we get 1 (Example 3). The reason for this definition will be covered in Section 15.1.

Example 3

a. $2^0 = 1$ b. $10^0 = 1$ c. $(-5)^0 = 1$ ■

In general,

> If a represents any real number, except 0,
>
> $$a^0 = 1$$

NOTE The symbol 0^0 is not defined or used in this book.

Example 4 Use a calculator to find the following powers. (See Section 1.7 for a discussion on the power key $\boxed{y^x}$.)

a. 2^9
Key in: 2 $\boxed{y^x}$ 9 $\boxed{=}$
Answer: 512

b. $(-8)^5$
Key in: 8 $\boxed{+/-}$ $\boxed{y^x}$ 5 $\boxed{=}$
Answer: $-32{,}768$ ■

EXERCISES 6.8

In Exercises 1–36, find the value of each expression.

1. 3^3 **2.** 2^4 **3.** $(-5)^2$ **4.** $(-6)^3$ **5.** 7^2 **6.** 3^4

7. 0^3 **8.** 0^4 **9.** $(-10)^1$ **10.** $(-10)^2$ **11.** 10^3 **12.** 10^4

13. $(-10)^5$ **14.** $(-10)^6$ **15.** $(-2)^2$ **16.** $(-2)^3$ **17.** $-(-3)^2$ **18.** $-(-3)^3$

19. 5^4 **20.** 25^2 **21.** 40^3 **22.** 400^2 **23.** $(-12)^3$ **24.** $(-15)^2$

25. $(-1)^5$ **26.** $(-1)^7$ **27.** -2^2 **28.** -3^2 **29.** $(-1)^{99}$ **30.** $(-1)^{98}$

31. 3^0 **32.** $(-4)^0$ **33.** $\left(\frac{3}{4}\right)^2$ **34.** $\left(\frac{2}{3}\right)^3$ **35.** $\left(-\frac{1}{10}\right)^3$ **36.** $\left(-\frac{1}{2}\right)^4$

In Exercises 37–42, use a calculator to find the value of each expression.

37. 3^{10} **38.** 12^5 **39.** $(-4)^7$ **40.** $(-6)^9$ **41.** $(-1.5)^4$ **42.** $(-2.4)^0$

6.9 Square Roots of Signed Numbers

Finding the square root of a number is the *inverse* operation of squaring a number. Every square root problem has a related problem involving squaring a number.

Principal Square Root

Every positive number has both a positive and a negative square root. The positive square root is called the **principal square root**.

Example 1 The number 9 has two square roots: $+3$ and -3.

The square root of 9 is $+3$ because $(+3)^2 = 9$.
The square root of 9 is -3 because $(-3)^2 = 9$.

3 is the principal square root of 9 because it is positive. ■

The radical sign, $\sqrt{\ }$, indicates the square root of the number under it, and it always represents the principal square root. Therefore, $\sqrt{9} = 3$.

Example 2 Find the square roots.

a. $\sqrt{16} = 4$ because $4^2 = 16$

b. $\sqrt{36} = 6$ because $6^2 = 36$

c. $\sqrt{1} = 1$ because $1^2 = 1$

d. $\sqrt{0} = 0$ because $0^2 = 0$

e. $\sqrt{\frac{4}{9}} = \frac{2}{3}$ because $\left(\frac{2}{3}\right)^2 = \frac{4}{9}$

f. $-\sqrt{4} = -2$ We know $\sqrt{4} = 2$

therefore $-\sqrt{4} = -2$ ■

NOTE Square roots of negative numbers, such as $\sqrt{-4}$, are imaginary numbers. Imaginary numbers are not real and cannot be graphed on the number line. ☑

Rational Numbers

A **rational number** is a number that can be written in the form $\frac{a}{b}$, where a and b are integers ($b \neq 0$).

Example 3 Examples of rational numbers.

a. $\frac{2}{3}$ All fractions are rational numbers.

b. $-3 = \frac{-3}{1}$ All integers are rational numbers.

c. $2\frac{1}{2} = \frac{5}{2}$ All mixed numbers are rational numbers.

d. $0.25 = \frac{25}{100}$ All terminating decimals are rational numbers.

e. $0.333\ldots = \frac{1}{3}$ All repeating decimals are rational numbers.

f. $\sqrt{25} = 5 = \frac{5}{1}$ $\sqrt{25}$ simplifies to the integer 5 which is a rational number. ■

The decimal representation of any rational number is either a terminating decimal or a repeating decimal. (See Section 3.9.)

Irrational Numbers

An **irrational number** is a number whose decimal representation is a nonterminating nonrepeating decimal.

Example 4 Examples of irrational numbers

a. $\sqrt{2} = 1.414213562\ldots$

b. $\sqrt{3} = 1.732050807\ldots$

c. $\pi = 3.1415926535\ldots$

d. $-\sqrt{5} = -2.236067977\ldots$ ■

Finding Square Roots by Table or Calculator Square roots can be approximated by using Table I (inside back cover) or by using a calculator with a square root key $\boxed{\sqrt{}}$.

Example 5 Find $\sqrt{3}$: a. using Table I and b. using a calculator.

Solution

a. Locate 3 in the column headed N. Read the value of $\sqrt{3}$ in the column headed $\sqrt{N}$. Table I gives the square roots rounded off to three decimal places. Therefore, $\sqrt{3} \approx 1.732$

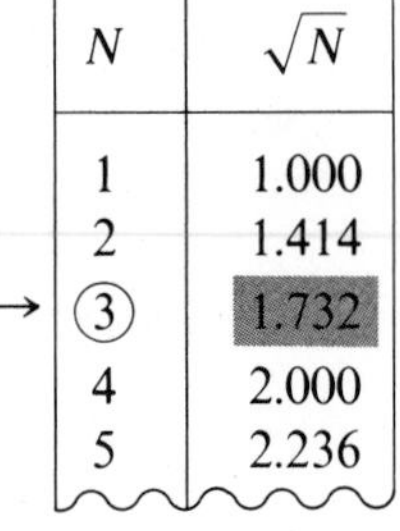

N	$\sqrt{N}$
1	1.000
2	1.414
(3)	1.732
4	2.000
5	2.236

b. Key in: 3 $\boxed{\sqrt{}}$

Answer: $1.7320508 \approx 1.732$

When rounded off to 3 decimal places, we get the same number obtained from the table ■

Real Numbers

Together, the rational numbers and the irrational numbers form the set of **real numbers**. There is a one-to-one correspondence between the real numbers and the points on the number line. (Figure 6.9.1)

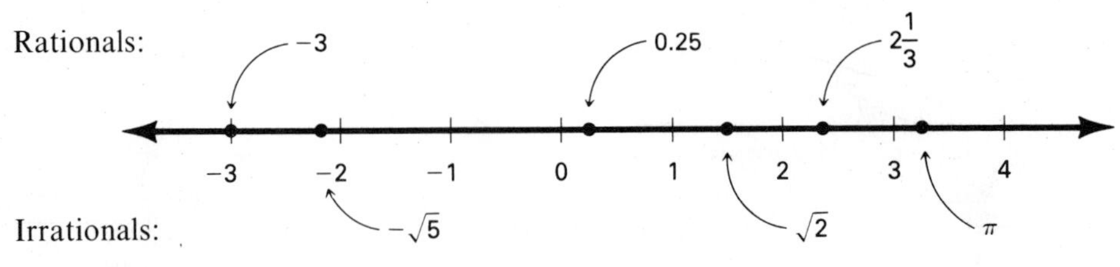

FIGURE 6.9.1

EXERCISES 6.9

In Exercises 1–16, find the square roots.

1. $\sqrt{16}$ **2.** $\sqrt{25}$ **3.** $-\sqrt{4}$ **4.** $-\sqrt{9}$

5. $\sqrt{81}$ **6.** $\sqrt{36}$ **7.** $\sqrt{100}$ **8.** $\sqrt{144}$

9. $-\sqrt{49}$ **10.** $-\sqrt{121}$ **11.** $\sqrt{64}$ **12.** $\sqrt{169}$

13. $-\sqrt{100}$ **14.** $-\sqrt{144}$ **15.** $\sqrt{\frac{16}{25}}$ **16.** $\sqrt{\frac{49}{100}}$

In Exercises 17–24, find the square roots using Table I (inside back cover) or a calculator. Round off answers to 3 decimal places.

17. $\sqrt{13}$ **18.** $\sqrt{18}$ **19.** $\sqrt{7}$ **20.** $\sqrt{6}$

21. $\sqrt{50}$ **22.** $\sqrt{75}$ **23.** $\sqrt{184}$ **24.** $\sqrt{155}$

6.10 Chapter Summary

Natural Numbers 6.1 $\{1, 2, 3, \ldots\}$

Whole Numbers 6.1 $\{0, 1, 2, \ldots\}$

Integers 6.1 $\{\ldots, -3, -2, -1, 0, 1, 2, 3, \ldots\}$

Digits 6.1 $\{0, 1, 2, 3, 4, 5, 6, 7, 8, 9\}$

Rational Numbers 6.9 **Rational numbers** are numbers that can be written in the form $\frac{a}{b}$ where a and b are integers and $b \neq 0$. The decimal representation of a rational number is either a terminating or a repeating decimal.

Irrational Numbers 6.9 **Irrational numbers** are numbers the decimal representation of which is a nonterminating nonrepeating decimal.

Real Numbers 6.9 **Real numbers** are all the rational numbers and all the irrational numbers. There is a one-to-one correspondence between the real numbers and the points on the number line.

Operations Involving Zero 6.7 6.8 Zero is a real number that is neither positive nor negative.

If a is any real number:

a. $a + 0 = 0 + a = a$

b. $a - 0 = a$

c. $0 - a = 0 + (-a) = -a$

d. $a \cdot 0 = 0 \cdot a = 0$

If a is any real number $\neq 0$:

e. $\frac{0}{a} = 0$

f. $\frac{a}{0}$ is undefined

g. $\frac{0}{0}$ is undefined

h. $0^a = 0$

i. $a^0 = 1$

Additive Identity 6.7 The **additive identity** is 0.

$$0 + a = a + 0 = 0$$

Multiplicative Identity 6.4 The **multiplicative identity** is 1.

$$1 \cdot a = a \cdot 1 = a$$

Commutative Property of Addition
6.6

$$a + b = b + a$$

Commutative Property of Multiplication
6.6

$$a \cdot b = b \cdot a$$

Associative Property of Addition
6.6

$$(a + b) + c = a + (b + c)$$

Associative Property of Multiplication
6.6

$$(a \cdot b) \cdot c = a \cdot (b \cdot c)$$

Absolute Value
6.2

The **absolute value** of a number is the distance between that number and 0 on the number line with no regard for direction.
The absolute value of a number can never be negative.
The absolute value of a real number X is written $|X|$.

To Add Signed Numbers
6.2

1. When the numbers have the *same sign*
   - First: Add their absolute values.
   - Second: The sum has the same sign as both numbers.
2. When the numbers have *different signs*
   - First: Subtract the smaller absolute value from the larger.
   - Second: The sum has the sign of the number with the larger absolute value.

To Subtract Signed Numbers
6.3

$$a - b = a + (-b)$$

1. Change the subtraction symbol to an addition symbol, and change the sign of the number being subtracted.
2. Add the resulting signed numbers as shown in Section 6.2.

To Multiply Two Signed Numbers
6.4

1. Multiply their absolute values.
2. The product is *positive* when the signed numbers have the same sign.

   The product is *negative* when the signed numbers have different signs.

To Divide Signed Numbers
6.5

1. Divide their absolute values.
2. The quotient is *positive* when the signed numbers have the same sign.

   The quotient is *negative* when the signed numbers have different signs.

Powers of Signed Numbers
6.8

Raising a number to a power is repeated multiplication.

$$3^4 = 3 \cdot 3 \cdot 3 \cdot 3 = 81$$

exponent (4); base (3); fourth power of 3 (81)

Square Roots of Signed Numbers
6.9

Finding the square root of a number is the inverse of squaring a number.

$$\sqrt{9} = 3 \quad \text{because} \quad 3^2 = 9$$

Review Exercises 6.10

1. Write all the digits greater than 7.

2. Write all the whole numbers less than 3.

3. Write the largest digit.

4. Write the smallest natural number.

5. Write the smallest integer.

6. Write the largest integer less than zero.

7. Write the correct symbol, $<$ or $>$, to make each statement true.

a. $2 \,?\, 6$ b. $-3 \,?\, 0$ c. $-3 \,?\, -5$

d. $2 \,?\, -1$ e. $-1\frac{1}{2} \,?\, -1$

8. State whether each of the following is true or false, and give the reason.

a. $(6 \cdot 3) \cdot 4 = (6)(3 \cdot 4)$ b. $5 + (-2) = (-2) + 5$

c. $5 - (-2) = (-2) - 5$ d. $a + (b + c) = (a + b) + c$

e. $(c \cdot d) \cdot e = e \cdot (c \cdot d)$ f. $5 + (x + 7) = (x + 7) + 5$

In Exercises 9–66, perform the indicated operations. If the indicated operation cannot be done, give a reason.

9. $(-6) \div (-2)$ **10.** $(-5) - (-3)$ **11.** $(5) + (-2)$

12. $(-3)(-4)$ **13.** $(-7) + (-4)$ **14.** $(8)(-9)$

15. $(-4) - (-8)$ **16.** $18 \div (-3)$ **17.** $(6) - (15)$

18. $(-9) + (3)$ **19.** $(-7)(6)$ **20.** $(-2) - (8)$

21. -3^2 **22.** $(-10) + (6)$ **23.** $(-5) - (-8)$

24. $(-6) \div 2$ **25.** 0^3 **26.** $(-5)^2$

27. $(-10) + (-2)$ **28.** $(-4)(6)$ **29.** $(-25) \div (-5)$

30. $0 \div (-4)$ **31.** $\sqrt{36}$ **32.** -2^4

33. $(4) - (-9)$ **34.** $(1) + (-6)$ **35.** $(-8)(-6)$

36. $(-5) + (14)$ **37.** $(-56) \div 8$ **38.** $(0)(-5)$

39. $(-9) + (-5)$ **40.** $\sqrt{25}$ **41.** $(-4) - (7)$

42. $(-3) + (-6)$ **43.** 10^0 **44.** $(2) - (9)$

45. $(-9)(-7)(-2)$ **46.** $(-3)^3$ **47.** $(7) - (-5)$

48. $(-8) + (3)$ **49.** $-\sqrt{9}$ **50.** $(-1) - (-6)$

51. 4^1 **52.** $(-5)(-8)(-2)(-1)$ **53.** $(-9) + (-8)$

54. 10^4 **55.** $(-3) - (8)$ **56.** $(27) + (-43)$

57. $\frac{0}{0}$ **58.** $\frac{-24}{18}$ **59.** $\left(4\frac{5}{6}\right) + \left(-2\frac{2}{3}\right)$

60. $\left(-\frac{9}{10}\right) - \left(\frac{3}{4}\right)$ **61.** $\left(-3\frac{3}{4}\right)\left(3\frac{1}{3}\right)$ **62.** $\frac{-3}{0}$

63. $\left(-2\frac{1}{4}\right) \div \left(-1\frac{5}{6}\right)$

64. $\left(-3\frac{1}{6}\right) + \left(-2\frac{3}{4}\right)$

65. Subtract 9 from -7.

66. Subtract -2 from -10.

In Exercises 67–70, use Table I or a calculator. Round off answers to 3 decimal places.

67. $\sqrt{53}$

68. $\sqrt{92}$

69. $\sqrt{153}$

70. $\sqrt{185}$

Chapter 6 Diagnostic Test

Allow yourself about 40 minutes to do these problems. Complete solutions for every problem, together with section references, are given in the answer section at the end of the book.

In Problems 1–3, write the correct symbol, $<$ or $>$, that should be used to make the statement true.

1. $3 \; ? \; 5$

2. $0 \; ? \; -4$

3. $-5 \; ? \; -4$

4. Write all the digits less than 3.

5. Write the smallest positive integer.

6. At 5 A.M. the temperature at Denver, Colorado, was $-20°$F. At noon the temperature was $38°$F. What was the rise in temperature?

7. Write -35 in words.

In Problems 8–11, state whether each is true or false, and give the reason.

8. $(-9) + (3) = (3) + (-9)$

9. $(3 + 5) + 6 = 3 + (5 + 6)$

10. $a \cdot (b \cdot c) = a \cdot (c \cdot b)$

11. $7 - 4 = 4 - 7$

In Problems 12–47, perform the indicated operation. If the indicated operation cannot be done, give a reason.

12. Add -6 and 2.

13. Subtract -10 from 20.

14. $(-10)(-5)$

15. $(42) \div (-14)$

16. 6^2

17. $(8) + (-5)$

18. Multiply (-12) times 3.

19. Subtract 5 from -10.

20. $\dfrac{2}{0}$

21. $(-5)(8)$

22. $\dfrac{15}{-18}$

23. $(-4)^3$

24. $(12) + (-5)$

25. $(-9) - (-7)$

26. $(7)(-6)$

27. $(5) - (12)$

28. $(-2) - (-6)$

29. $(-3) + (-8)$

30. 2^0

31. $(-9)(-3)(-2)$

32. $(-24) \div (-6)$

33. $(6) - (10)$

34. $(-7) + (-2)$

35. $(6)(-4)$

36. $(-4) + (9)$

37. 0^2

38. $(-4) - (8)$

39. $(-11) + (4)$

40. $\left(-1\frac{7}{8}\right)\left(2\frac{2}{5}\right)$

41. $\left(-2\frac{3}{4}\right) + \left(-1\frac{5}{8}\right)$

42. $\dfrac{-33}{11}$

43. $(-2)^4$

44. $0 \div (-12)$

45. $(-5)(0)$

46. $(9) - (-5)$

47. $(3) + (-8)$

In Problems 48–50, find each of the indicated roots.

48. $\sqrt{49}$

49. $\sqrt{81}$

50. $-\sqrt{64}$

7 Evaluating Expressions

One of the first places that a student uses algebra is in the evaluation of formulas in courses taken in science, business, and so on. In this chapter we apply the operations with signed numbers discussed in Chapter 6 to evaluate algebraic expressions and formulas.

In Chapter 6 we learned how to perform all six arithmetic operations with signed numbers: addition, subtraction, multiplication, division, taking powers, and finding square roots; but each problem dealt with only one kind of operation.

Example 1

a. $2 + (-4) = -2$ Addition only.

b. $(-3) \cdot (4 \cdot 5) = (-3)(20) = -60$ Multiplication only.

c. $(-7) - 4 = -11$ Subtraction only. ■

In this chapter we will evaluate expressions in which more than one kind of operation is used.

Example 2 What does $5 + 4 \cdot 6 = ?$ Multiplication *and* addition.

Solution

If the addition is done first,
we get $5 + 4 \cdot 6 =$
$9 \cdot 6 = 54$

If the multiplication is done first,
we get $5 + 4 \cdot 6 =$
$5 + 24 = 29$

Which is correct?

Obviously, both answers cannot be correct. In Chapter 1 we showed that multiplication is done before addition. Therefore, $5 + 4 \cdot 6 = 5 + 24 = 29$. ■

In the next section we will cover the correct order of operations with signed numbers.

7.1 Order of Operations

In evaluating expressions with more than one operation, we use the following order of operations.

ORDER OF OPERATIONS

1. Any expressions in parentheses are evaluated first.

2. Evaluations are done in this order.
First: Powers and roots
Second: Multiplication and division in order from *left to right*
Third: Addition and subtraction in order from *left to right*

A WORD OF CAUTION It is important for students to realize that an expression such as

$$8 - 6 - 4 + 7$$

is evaluated by doing the additions and subtractions in order from left to right (because subtraction is not commutative or associative).

This same expression may also be considered as a sum

$$(8) + (-6) + (-4) + (7)$$

If the expression is considered as a sum, then the terms can be added in any order (because addition is commutative and associative).

Only evaluated left to right	*Added in any order*
$8 - 6 - 4 + 7 =$	$(8) + (-6) + (-4) + (7) =$
$2 \quad - 4 + 7 =$	$(8) + (7) + (-6) + (-4) =$
$-2 \quad + 7 = 5$	$15 \quad + \quad (-10) \quad = 5$

☑

Example 1 Showing the correct order of operations

a. $(7 + 3) \cdot 5 =$
$10 \quad \cdot 5 = 50$

We do the part in parentheses first

b. $7 - 3 \cdot 5 \quad =$
$7 - \quad 15 \quad =$
$7 + (-15) = -8$

Multiplication is done before subtraction

c. $4^2 + \sqrt{25} - 6 =$
$16 + \quad 5 \quad - 6 =$
$21 \quad - 6 = 15$

Powers and roots are done first

d. $-16 \div 2 \cdot 4 =$
$-8 \quad \cdot 4 = -32$

Multiplication and division are done *left to right*

e. $-4 + (-8) \div 2 =$
$-4 + \quad (-4) \quad = -8$

Division is done before addition

f. $4(-3)^2 - 3(-2)^3 =$
$4(9) \quad - 3(-8) \quad =$
$36 \quad - \quad (-24) \quad =$
$36 \quad + \quad 24 \quad = 60$

Powers are done first
Multiplication is done before subtraction

g. $-5^2 + (4 + 2 \cdot 3) =$
$-5^2 + (4 + \quad 6 \quad) =$
$-5^2 + \quad 10 \quad =$
$-25 + \quad 10 \quad = -15$

Multiply inside ( )
Add inside ( )
In -5^2, the exponent applies only to the 5 ■

Example 2 Find the value of each expression using a calculator.

a. $3(-4) - \sqrt{225}$
Key in: 3 [×] 4 [+/−] [−] 225 [√] [=]
Answer: -27

b. $(-2)^4 - (9 + 8)$
Key in: 2 [+/−] [y^x] 4 [−] [(] 9 [+] 8 [)] [=]
Answer: -1 ■

EXERCISES 7.1

In Exercises 1–42, be sure to perform the operations in the correct order.

1. $12 - 8 - 6$

2. $15 - 9 - 4$

3. $-7 - 11 + 13 - 9$

4. $-2 - 8 + 14 - 6$

5. $7 + 2 \cdot 4$

6. $10 + 3 \cdot 6$

7. $9 - 3(-2)$
8. $-4 - 8(-3)$
9. $10 \div 2 \cdot 5$
10. $20 \cdot 15 \div 5$
11. $12 \div 6 \div (-2)$
12. $-24 \div 12 \div (-2)$
13. $(-12) \div 2 \cdot (-3)$
14. $(-18) \div (-3) \cdot (-6)$
15. $(8 - 2) \cdot 6$
16. $(10 - 6) \cdot (-5)$
17. $(-40)^2 \cdot 0 \cdot (-5)^2$
18. $(-500)^2 \cdot 0 \cdot (-3)^2$
19. $12 \cdot 4 + 16 \div 8$
20. $4(-3) + 15 \div 5$
21. $28 \div 4 \cdot 2(6)$
22. $48 \div 16 \cdot 2(-8)$
23. $-3^2 - 4^2$
24. $-6^2 + (-5)^2$
25. $(-2)^2 + 5(-2)$
26. $(-4)3 - 3^2$
27. $2\sqrt{9} - 5(-6)$
28. $-3(4) - 4\sqrt{25}$
29. $2^3 + (3 - 2 \cdot 8)$
30. $4 - (7 \cdot 2 - 3^3)$
31. $-2(5 - 3) - 5(7 - 3)$
32. $3(2 - 8) + 2(4 - 8)$
33. $-6 - 8 \div 2 \cdot 4$
34. $-6 - 8 \div (2 \cdot 4)$
35. $(-6 - 8) \div 2 \cdot 4$
36. $(-6 - 8 \div 2) \cdot 4$
37. $2 \cdot \frac{7}{16} + \frac{9}{20} \div \frac{3}{5}$
38. $\frac{15}{16} \div 3 - \frac{2}{3} \cdot \frac{3}{8}$
39. $\left(\frac{2}{3}\right)^2 + 3\frac{1}{3} \cdot \frac{1}{4}$
40. $\left(\frac{3}{4}\right)^2 - \frac{5}{8} \cdot 1\frac{1}{5}$
41. $2.3 + 5(3.7) \div 100$
42. $7.3 - 9(4.6) \div 10$

In Exercises 43–51, use a calculator to evaluate each expression.

43. $-18 - 16 \cdot 5$
44. $3^4 + 9(-12)$
45. $-2\sqrt{196} - 6$
46. $-50 + 2 \cdot 5^3$
47. $(-2)^6 - 7 \cdot 3$
48. $15(-6) - 18(-3)$
49. $(-23 + 17) \cdot 12$
50. $-6(34 - 4 \cdot 16)$
51. $(-19 + 8) \cdot (15 - 27)$

7.2 Grouping Symbols

Other symbols along with parentheses can be used to indicate grouping.

Grouping Symbols

Symbol	Name
( )	**Parentheses**
[ ]	**Brackets**
{ }	**Braces**
—	**Bar**

Fraction line → $\frac{8 + 7}{9 - 4}$

Bar in a square root ↓ $\sqrt{6(4) - 8}$

All grouping symbols have the same meaning.

$$(8 + 4) - (9 - 7) =$$
$$[8 + 4] - [9 - 7] =$$
$$\{8 + 4\} - \{9 - 7\} =$$
$$\overline{8 + 4} - \overline{9 - 7} =$$
$$12 - 2 = 10$$

Different grouping symbols can be used in the same expression.

$$(8 + 4) - \{9 - 7\} =$$
$$[8 + 4] - (9 - 7) =$$
$$\{8 + 4\} - [9 - 7] =$$
$$12 - 2 = 10$$

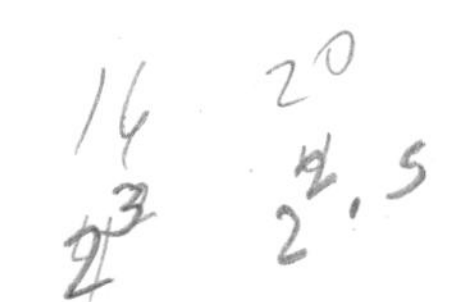

When grouping symbols appear within other grouping symbols, *evaluate the inner grouping first.*

Example 1

$$10 - [3 - (2 - 7)] =$$ Evaluate the inner grouping first
$$10 - [3 - (-5)] =$$
$$10 - [3 + 5] =$$
$$10 - 8 = 2 \quad ■$$

Example 2

$$\frac{(-4) + (-2)}{8 - 5} = \leftarrow$$ This bar is a grouping symbol for both $\overline{(-4) + (-2)}$ and for $\overline{8 - 5}$. Notice that the bar can be used either above or below the numbers being grouped ■
$$\frac{-6}{3} = -2$$

Example 3

$$5 - 2\{9 - [3 - (-2)]\} =$$
$$5 - 2\{9 - [3 + 2]\} =$$
$$5 - 2\{9 - 5\} =$$
$$5 - 2\{4\} =$$
$$5 - 8 = -3 \quad ■$$

Example 4

$$\sqrt{13^2 - 12^2} = \leftarrow$$ This bar is a grouping symbol The expression under it is evaluated first, then the square root is taken
$$\sqrt{169 - 144} =$$
$$\sqrt{25} = 5 \quad ■$$

A WORD OF CAUTION

$$\sqrt{13^2 - 12^2} \neq \sqrt{13^2} - \sqrt{12^2}$$
$$\sqrt{169 - 144} \neq \sqrt{169} - \sqrt{144}$$
$$\sqrt{25} \neq 13 - 12$$
$$5 \neq 1$$

✓

EXERCISES 7.2

Evaluate each expression.

1. $-3[4 - (-5)]$
2. $-2[-8 - (-3)]$
3. $24 - [(-6) + 18]$
4. $17 - [(-9) + 15]$
5. $[12 - (-19)] - 16$
6. $[21 - (-14)] - 29$
7. $[11 - (5 + 8)] - 24$
8. $[16 - (7 + 12)] - 22$
9. $20 - [5 - (7 - 10)]$
10. $16 - [8 - (2 - 7)]$
11. $\sqrt{3^2 + 4^2}$
12. $\sqrt{10^2 - 8^2}$
13. $\dfrac{7 + (-12)}{8 - 3}$
14. $\dfrac{(-14) + (-2)}{9 - 5}$
15. $\dfrac{10 + (-4)}{2 - 5} + 3$
16. $\dfrac{8 - 2}{5 - 7} - 1$
17. $\dfrac{3^2 + 5}{2} - \dfrac{(-4)^2}{8}$
18. $\dfrac{2^3 - 2}{-3} + \dfrac{5^2 - 10}{5}$
19. $6 - 3\{8 - [4 - (-1)]\}$
20. $4 - \{7 - 2[3 - (-4)]\}$
21. $\sqrt{7^2 - 4(2)(3)}$
22. $\sqrt{5^2 - 4(4)(1)}$
23. $-4 + \sqrt{4^2 - 4(1)(3)}$
24. $-2 + \sqrt{2^2 - 4(1)(-8)}$
25. $15 - \{4 - [2 - 3(6 - 4)]\}$
26. $17 - \{6 - [9 - 2(2 - 7)]\}$

7.3 Finding the Value of Expressions Having Variables and Numbers

Variables

In algebra we use letters to represent numbers. Such letters are called **variables.** The value of the variable may change in a particular problem or discussion.

Constants

A **constant** is an object or symbol that does not change its value in a particular problem or discussion. It is usually represented by a number symbol. In the expression $4x - 3y$, the constants are 4 and -3, and the variables are x and y.

Algebraic Expression

An **algebraic expression** consists of numbers, variables, signs of operation, and signs of grouping (not *all* of those need be present).

In this section we make use of what we have already learned about signed numbers to help us find the value of algebraic expressions when the values of the variables are given.

16
33
49

> TO FIND THE VALUE OF AN EXPRESSION HAVING VARIABLES AND NUMBERS
>
> **1.** Replace each variable by its number value in parentheses.
>
> **2.** Carry out all arithmetic operations using the correct order of operations (Section 7.1).

Example 1 Find the value of $3x - 5y$ if $x = 10$ and $y = 4$.

Solution

$$3x - 5y =$$
$3x = 3 \cdot x$; $5y = 5 \cdot y$

$$3(10) - 5(4) =$$

$$30 - 20 = 10 \quad \blacksquare$$

Notice that we simply replace each letter by its number value in parentheses, then carry out the arithmetic operations as we have done before.

A WORD OF CAUTION When replacing a variable by a number, we enclose the number in *parentheses* to avoid the following common errors.

Evaluate $3x$ when $x = -2$.

Correct	*Common error*
$3x = 3(-2) = -6$	$3x \neq 3 - 2 = 1$

Evaluate $4x^2$ when $x = -3$.

Correct	*Common errors*
$4x^2 = 4(-3)^2 = 4 \cdot 9 = 36$	$4x^2 \neq 4 - 3^2 = 4 - 9 = -5$ or $4x^2 \neq 4 - 3^2 = 4 + 9 = 13$

☑

Example 2 Find the value of $\frac{2a-b}{10c}$ if $a = -1$, $b = 3$, and $c = -2$.

Remember this bar is a grouping symbol

Solution $\frac{2a-b}{10c} = \frac{2(-1)-(3)}{10(-2)} = \frac{-2-3}{-20} = \frac{-5}{-20} = \frac{1}{4}$ or 0.25 ■

Example 3 Find the value of $2a - [b - (3x - 4y)]$ for $a = -3$, $b = 4$, $x = -5$, and $y = 2$.

Solution

$$2a - [b - (3x - 4y)] =$$
$$2(-3) - [4 - \{3(-5) - 4(2)\}] =$$ ← Notice that { } were used in place of ( ) to clarify the grouping
$$2(-3) - [4 - \{-15 - 8\}] =$$
$$2(-3) - [4 - \{-23\}] =$$
$$2(-3) - [4 + 23] =$$
$$-6 - [27] = -33$$ ■

Example 4 Evaluate $b - \sqrt{b^2 - 4ac}$ when $a = 3$, $b = -7$, and $c = 2$.

Solution

$$b - \sqrt{b^2 - 4ac} =$$ ← This bar is a grouping symbol for $b^2 - 4ac$
$$(-7) - \sqrt{(-7)^2 - 4(3)(2)} =$$
$$(-7) - \sqrt{49 - 24} =$$
$$(-7) - \sqrt{25} =$$
$$(-7) - 5 = -12$$ ■

A WORD OF CAUTION A common error often made in evaluation problems is to mistake $(-7)^2$ for -7^2.

$(-7)^2 = (-7)(-7) = 49$ The exponent 2 applies to -7

$-7^2 = -(7)(7) = -49$ The exponent 2 applies to 7 ✓

EXERCISES 7.3

In Exercises 1–28, evaluate the expression when $a = 3$, $b = -5$, $c = -1$, $x = 4$, and $y = -7$.

1. $4b$
2. $5c$
3. $2a - 10$
4. $3b - 4$
5. b^2
6. $-y^2$
7. $2a - 3b$
8. $3x - 2y$
9. $x - y - 2b$
10. $a - b - 3y$
11. $3b - ab + xy$
12. $4c + ax - by$
13. $x^2 - y^2$
14. $b^2 - c^2$
15. $4 + 3(x + y)$
16. $5 - 2(a + c)$
17. $2(a - b) - 3c$
18. $3(a - x) - 4b$
19. $3x^2 - 10x + 5$
20. $2y^2 - 7y + 9$
21. $a^2 - 2ab + b^2$
22. $x^2 - 2xy + y^2$
23. $\frac{3x}{y+b}$
24. $\frac{4a}{c-b}$
25. $2(a - 6) + 3(b + 7)$
26. $4(x - 1) + 3(y - 2)$
27. $a - [b - (c + x)]$
28. $y - [c - (a - b)]$

In Exercises 29–36, find the value of the expression when $E = -1$, $F = 3$, $G = -5$, $H = -4$, and $K = 0$.

29. $\dfrac{E + F}{EF}$ **30.** $\dfrac{G + H}{GH}$ **31.** $\dfrac{(1 + G)^2 - 1}{H}$ **32.** $\dfrac{1 - (1 + E)^2}{F}$

33. $2E - [F - (3K - H)]$ **34.** $3H - [K - (4F - E)]$

35. $G - \sqrt{G^2 - 4EH}$ **36.** $F - \sqrt{F^2 + 4HE}$

In Exercises 37–40, find the value of the expression when $x = \dfrac{1}{2}$, $y = -\dfrac{3}{4}$, $z = \dfrac{3}{8}$.

37. $\dfrac{3x^2}{z}$ **38.** $\dfrac{y^2}{xz}$ **39.** $(x + y + z)^2$ **40.** $x^2 + y^2 + z^2$

7.4 Evaluating Formulas

One reason for studying algebra is to prepare us to use formulas. You will encounter formulas in many courses you take, as well as in real-life situations. In the examples and exercises we have listed the subject areas where the formulas are used.

Formulas are evaluated in the same way any expression having numbers and variables is evaluated.

Example 1 Given the formula $A = P(1 + rt)$, find A when $P = 1{,}000$, $r = 0.08$, and $t = 1.5$. (Business)

Solution

$$A = P(1 + r \cdot t)$$
$$A = 1000[1 + (0.08)(1.5)]$$

Notice [ ] were used in place of ( ) to clarify the grouping

$$A = 1000[1 + 0.12]$$
$$A = 1000[1.12]$$
$$A = 1{,}120 \quad \blacksquare$$

Example 2 Given the formula $C = \dfrac{5}{9}(F - 32)$, find C when $F = -13$. (Science)

Solution

$$C = \frac{5}{9}(F - 32)$$
$$C = \frac{5}{9}(-13 - 32)$$
$$C = \frac{5}{\cancel{9}_{1}}(\cancel{-45}^{-5})$$
$$C = -25 \quad \blacksquare$$

Example 3 Given the formula $T = \pi\sqrt{\dfrac{L}{g}}$, find T when $\pi \approx 3.14$, $L = 128$, and $g = 32$. (Physics)

Solution

$$T = \pi\sqrt{\frac{L}{g}}$$

$$T \approx (3.14)\sqrt{\frac{128}{32}}$$

$$T \approx (3.14)\sqrt{4}$$

$$T \approx (3.14)(2)$$

$$T \approx 6.28 \quad \blacksquare$$

Example 4 Given the formula $S = \dfrac{a(1 - r^n)}{1 - r}$, find S when $a = -4$, $r = \dfrac{1}{2}$, and $n = 3$. (Mathematics)

Solution

$$S = \frac{a(1 - r^n)}{1 - r}$$

$$S = \frac{(-4)\left[1 - \left(\frac{1}{2}\right)^3\right]}{1 - \frac{1}{2}} \qquad \left(\frac{1}{2}\right)^3 = \left(\frac{1}{2}\right)\left(\frac{1}{2}\right)\left(\frac{1}{2}\right) = \frac{1}{8}$$

$$S = \frac{(-4)\left[1 - \frac{1}{8}\right]}{1 - \frac{1}{2}}$$

$$S = \frac{(-4)\left[\frac{8}{8} - \frac{1}{8}\right]}{\frac{2}{2} - \frac{1}{2}}$$

$$S = \frac{\frac{\overset{-1}{\cancel{-4}}}{1}\left[\frac{7}{\underset{2}{\cancel{8}}}\right]}{\frac{1}{2}}$$

$$S = \frac{-7}{2} \div \frac{1}{2}$$

$$S = \frac{-7}{\underset{1}{\cancel{2}}} \cdot \frac{\overset{1}{\cancel{2}}}{1}$$

$$S = -7 \quad \blacksquare$$

EXERCISES 7.4

Evaluate each formula using the values of the letters given with the formula.

(Geometry) The area of a triangle is given by the formula

$$A = \frac{1}{2}bh \qquad \text{where } A \text{ is the area, } b \text{ is the base, and } h \text{ is the height.}$$

1. Find A when $b = 15$ and $h = 14$.

2. Find A when $b = 27$ and $h = 36$.

(Electricity) Ohm's law states that

$$I = \frac{E}{R}$$ where I is the current, E is the electromotive force, and R is the resistance.

3. Find I when $E = 110$ and $R = 22$.

4. Find I when $E = 220$ and $R = 33$.

(Business) The formula for simple interest is

$$I = prt$$ where I is the interest, p is the principal, r is the rate, and t is the time.

5. Find I when $p = 600$, $r = 0.09$, $t = 4.5$.

6. Find I when $p = 700$, $r = 0.08$, $t = 2.5$.

(Chemistry) The formula to change Celsius to Fahrenheit is

$$F = \frac{9}{5}C + 32$$ where F is degrees Fahrenheit and C is degrees Celsius.

7. Find F when $C = 25$.

8. Find F when $C = -25$.

(Geometry) The area of a circle is given by the formula

$$A = \pi R^2$$ where A is the area and R is the radius.

9. Find A when $\pi \approx 3.14$ and $R = 10$.

10. Find A when $\pi \approx 3.14$ and $R = 20$.

(Physics) The distance a free-falling object travels is given by the formula

$$s = \frac{1}{2}gt^2$$ where s is the distance, g is the force of gravity, and t is the time.

11. Find s when $g = 32$ and $t = 3$.

12. Find s when $g = 32$ and $t = 5$.

(Business) The value of an item that is depreciated each year is given by the formula

$$V = C - Crt$$ where V is the present value, C is the original cost, r is the rate of depreciation, and t is the time.

13. Find V when $C = 500$, $r = 0.1$, $t = 2$.

14. Find V when $C = 1{,}000$, $r = 0.08$, $t = 5$.

(Statistics) The standard deviation of a binomial distribution is given by the formula

$$\sigma = \sqrt{npq}$$ where σ is the standard deviation, n is the number of trials, p is the probability of a success, and q is the probability of a failure.

15. Find σ when $n = 100$, $p = 0.9$, $q = 0.1$.

16. Find σ when $n = 100$, $p = 0.8$, $q = 0.2$.

(Chemistry) The formula to change Fahrenheit to Celsius is

$$C = \frac{5}{9}(F - 32)$$ where C is degrees Celsius and F is degrees Fahrenheit.

17. Find C when $F = -13$.

18. Find C when $F = 5$.

(Nursing) The formula to determine the dosage for a child is

$$C = \frac{a}{a + 12} \cdot A$$ where C is the child's dosage, a is the age of the child, and A is the adult dosage.

19. Find C when $a = 6$ and $A = 30$.

20. Find C when $a = 4$ and $A = 48$.

(Business) The formula for compound interest is

$$A = P(1 + r)^t$$ where A is the amount, P is the principal, r is the rate, and t is the time. Round off answers to 2 decimal places.

21. Find A when $P = 1000$, $r = 0.05$, $t = 10$.

22. Find A when $P = 5{,}000$, $r = 0.12$, $t = 5$.

(Geometry) The Pythagorean theorem for right triangles is

$$c = \sqrt{a^2 + b^2}$$ where c is the hypotenuse (longest side), and a and b are the other sides. Round off answers to 1 decimal place.

23. Find c when $a = 5$ and $b = 8$.

24. Find c when $a = 12$ and $b = 15$.

7.5 Chapter Summary

Grouping Symbols 7.2

( ) Parentheses
[ ] Brackets
{ } Braces
— Bar

Fraction line $\longrightarrow \dfrac{3x-2}{5+7y}$

Bar in a square root $\longrightarrow \sqrt{5x-12}$

Order of Operations 7.1

1. Any expressions in parentheses are evaluated first.
2. Evaluations are done in this order.
   First: Powers and roots
   Second: Multiplication and division in order from left to right
   Third: Addition and subtraction in order from left to right

To Evaluate an Expression or Formula 7.3 7.4

1. Replace each variable by its number value in parentheses.
2. Carry out all arithmetic operations using the correct order of operations.

Review Exercises 7.5

In Exercises 1–14, evaluate each expression.

1. $26 - 14 + 8 - 11$

2. $11 - 7 \cdot 3$

3. $15 \div 5 \cdot 3$

4. $[16 - (-8)] - 22$

5. $6 - [8 - (3 - 4)]$

6. $7 - 3(5 - 4 \cdot 2)$

7. $\dfrac{6 + (-14)}{3 - 7}$

8. $\sqrt{6^2 + 8^2}$

9. $2 \cdot 4 - 5^2 + 6 \cdot 2$

10. $4\sqrt{36} - 7(-3)$

11. $(-2)^3 + (-3)^2 - 4(-5)$

12. $\dfrac{-5 - 3}{8} + 2(-8)$

13. $9 - \{6 - [4 - (7 - 8)]\}$

14. $\left(\dfrac{3}{4} - \dfrac{1}{3}\right) \div \left(\dfrac{4}{9} + \dfrac{2}{3}\right)$

In Exercises 15–22, find the value of each expression when $x = -2$, $y = 3$, and $z = -4$.

15. $3x - y + z$

16. $6 - x(y - z)$

17. $5x^2 - 3x + 10$

18. $y^2 - 2yz + z^2$

19. $x - 2(y - xz)$

20. $\dfrac{(x + y)^2 - z^2}{x - 2y}$

21. $(x^2 + y^2)(x^2 - y^2)$

22. $(x - y)(x^2 + xy + y^2)$

In Exercises 23–30, evaluate each formula using the values of the variables given with the formula.

23. $C = \dfrac{a}{a + 12} \cdot A$ $\quad a = 8,\ A = 35$

24. $I = Prt$ $\quad P = 100,\ r = 0.07,\ t = 4.5$

25. $C = \dfrac{5}{9}(F - 32)$ $\quad F = 15\dfrac{1}{2}$

26. $\sigma = \sqrt{npq}$ $\quad n = 100,\ p = 0.5,\ q = 0.5$

27. $A = P(1 + rt)$ $\quad P = 1000,\ r = 0.06,\ t = 5$

28. $S = \dfrac{1}{2}gt^2$ $\quad g = 32,\ t = 1\dfrac{1}{2}$

29. $F = \dfrac{9}{5}C + 32$ $\quad C = -15$

30. $V = \pi r^2 h$ $\quad \pi \approx 3.14,\ r = 5,\ h = 8$

Chapter 7 Diagnostic Test

Allow yourself about 50 minutes to do these problems. Complete solutions for every problem, together with section references, are given in the answer section at the end of the book.

In Problems 1–12, evaluate each expression.

1. $17 - 9 - 6 + 11$

2. $5 + 2 \cdot 3$

3. $-12 \div 2 \cdot 3$

4. $-5^2 + (-4)^2$

5. $2 \cdot 3^2 - 4$

6. $3\sqrt{25} - 5(-4)$

7. $\dfrac{8 - 12}{-6 + 2}$

8. $(2^3 - 7)(5^2 + 4^2)$

9. $10 - [6 - (5 - 7)]$

10. $2 + (6 - 3 \cdot 4)$

11. $\sqrt{10^2 - 6^2}$

12. $5 - 2[4 - (6 - 9)]$

In Problems 13–15, find the value of each expression when $a = -2$, $b = 4$, $c = -3$, $x = 5$, and $y = -6$.

13. $3a + bx - cy$

14. $4x - [a - (3c - b)]$

15. $x^2 + 2xy - y^2$

In Problems 16–20, evaluate each formula using the values of the variables given with the formula.

16. $C = \dfrac{5}{9}(F - 32)$ $\quad F = -4$

17. $A = \pi R^2$ $\quad \pi \approx 3.14, R = 20$

18. $C = \dfrac{a}{a + 12} \cdot A$ $\quad a = 7, A = 38$

19. $A = P(1 + rt)$ $\quad P = 600, r = 0.10, t = 2.5$

20. $V = C - Crt$ $\quad C = 600, r = 0.04, t = 10$

8 Polynomials

In this chapter we look in detail at a particular type of algebraic expression called a *polynomial.* Polynomials have the same importance in algebra that whole numbers have in arithmetic. Most of the work in arithmetic involves operations with whole numbers. In the same way, most of the work in algebra involves operations with polynomials.

8.1 Basic Definitions

Terms

The + and − signs in an algebraic expression break it into smaller pieces called **terms**. Each + and − sign is part of the term that follows it. *Exception*: an expression within grouping symbols is considered as a single piece even though it may contain + and − signs. See Examples 1 (b), 1 (c), and Example 2.

Example 1 The terms of algebraic expressions

a. In the algebraic expression $3x^2y - 5xy^3 + 7xy$:

The − and + signs separate the algebraic expression into three terms

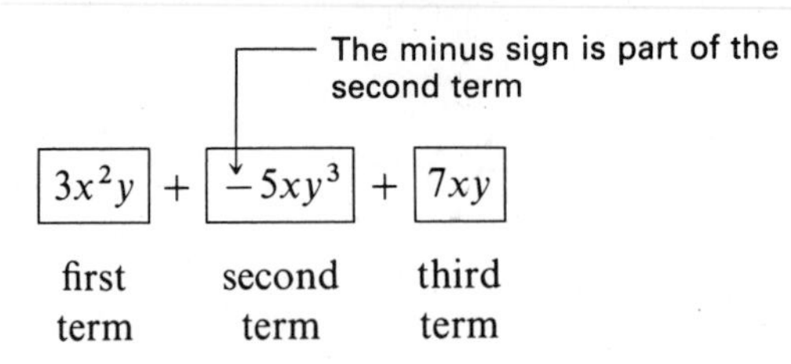

b. In the algebraic expression $3x^2 - 9x(2y + 5z)$,

$$\boxed{3x^2} + \boxed{-9x(2y + 5z)}$$

first term   second term

c. In the algebraic expression $\dfrac{2 - x}{xy} + 5(2x^2 - y)$,

$$\boxed{\frac{2 - x}{xy}} + \boxed{5(2x^2 - y)}$$

first term   second term ■

Example 2 The number of terms in an algebraic expression

a. $3 + 2x - 1$ has three terms.

b. $3 + (2x - 1)$ has two terms.

c. $(3 + 2x - 1)$ has one term. ■

Factors

Recall from Section 1.6 that numbers that are multiplied to give a product are called the **factors** of that product.

Example 3 Factors of a product

a. $(3)(5) = 15$ — 3 and 5 are factors of 15.

factors of 15

b. $(2)(x) = 2x$

factors of $2x$

c. $(7)(a)(b)(c) = 7abc$

product of factors

factors of $7abc$ ■

Coefficients

In a term having *two* factors, the **coefficient** of one factor is the other factor. In a term having *more than two* factors, the coefficient of each factor is the product of all the other factors in that term.

Example 4 Coefficients

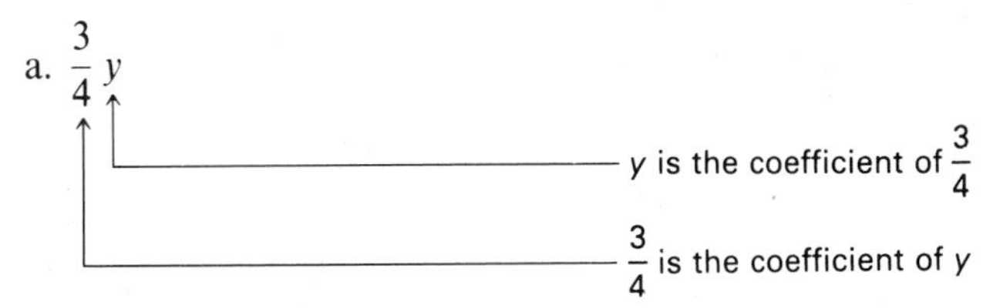

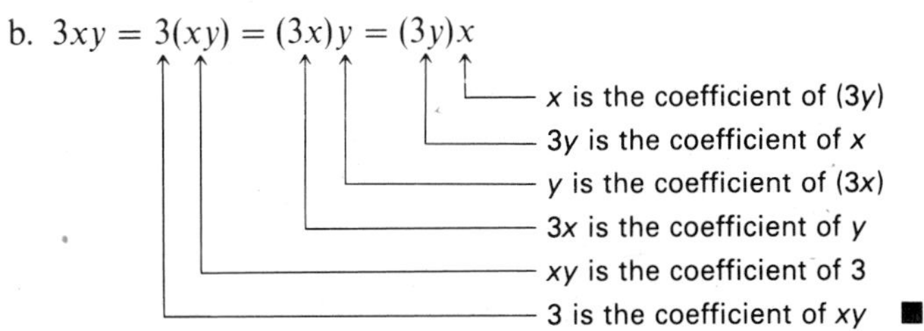

Numerical Coefficients

A **numerical coefficient** is a coefficient that is a number. If we say "*the* coefficient" of a term, it is understood to mean the *numerical* coefficient of that term.

Example 5 Numerical coefficient

Term	*Numerical coefficient*
a. $6w$	6
b. $-12xy^2$	-12
c. $\frac{3xy}{4} = \frac{3}{4}xy$	$\frac{3}{4}$
d. xy	1
e. $-a^2$	-1
f. $\frac{c}{5} = \frac{1}{5}c$	$\frac{1}{5}$ ■

Note to d: Even though no number is written, the numerical coefficient can be considered to be 1 because $xy = 1(xy)$. Also see part (e)

Note to e: -1 is the numerical coefficient of a^2 because $-a^2 = -1 \cdot a^2$

Polynomials A *polynomial in x* is an algebraic expression having only terms of the form ax^n, where a is any real number and n is a whole number.

Example 6 Examples of polynomials

a. $3x$ — **Monomial**: A polynomial of one term is called a **monomial**.

b. $4x^4 - 2x^2$ — **Binomial**: A polynomial of two unlike terms is called a **binomial**.

c. $7x^2 - 5x + 2$ — **Trinomial**: A polynomial of three unlike terms is called a **trinomial**.

d. $x^3 + 3x^2 + 3x + 1$ — We will not use special names for polynomials of more than three terms.

e. 5 — This is a polynomial of one term (monomial) because its only term has the form $5x^0 = 5 \cdot 1 = 5$.

f. $6z^3 - 3z + 1$ — Polynomials can be in any letter. This is a polynomial in z. ■

Example 7 Algebraic expressions that are *not* polynomials

a. $4x^{-2}$ — This expression is *not* a polynomial because the exponent -2 is not a whole number. Negative exponents are discussed in Section 15.1.

b. $\dfrac{2}{x-5}$ — This is *not* a polynomial because the variable x is in the denominator. The term does not have the form ax^n.

c. $\sqrt{2x+3}$ — This is not a polynomial because the variable x is under the radical sign. ■

Polynomials often have terms containing more than one variable.

A *polynomial in x and y* is an algebraic expression having only terms of the form ax^ny^m, where a is any real number and n, m are whole numbers.

Example 8 Polynomials in more than one variable

a. $5x^2y$ — Monomial

b. $-3xy^3 + 2x^2y$ — Binomial

c. $4x^2y^2 - 7xy + 6$ — Trinomial

d. $x^3y^2 - 2x + 3y^2 - 1$

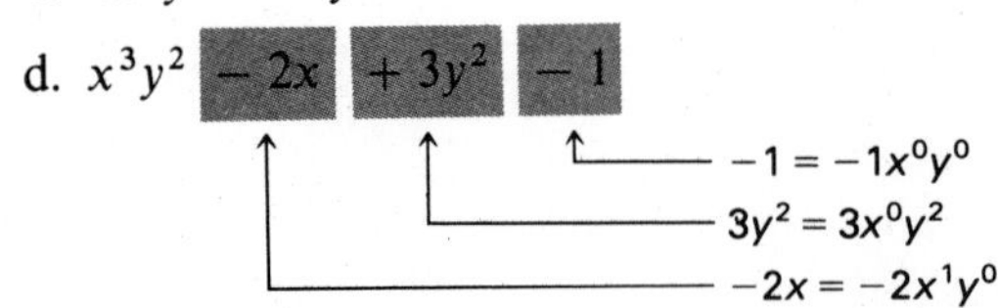

e. $7uv^4 - 5u^2v + 2u$ — Polynomials can be in any variables. This is a polynomial in u and v. ■

Degree of a Term

The **degree of a term** in a polynomial is the sum of the exponents of its variables.

Example 9 Degree of a term.

a. $5x^3$ — 3rd degree

b. $6x^2y$ — 3rd degree because $6x^2y = 6x^2y^1$ ($2 + 1 = 3$)

c. 14 — 0 degree because $14 = 14x^0$

d. $-2u^3vw^2$ — 6th degree because $-2u^3vw^2 = -2u^3v^1w^2$ ($3 + 1 + 2 = 6$) ■

Degree of a Polynomial

The **degree of a polynomial** is the same as that of its highest-degree term (provided like terms have been combined).

Example 10 Degree of a polynomial

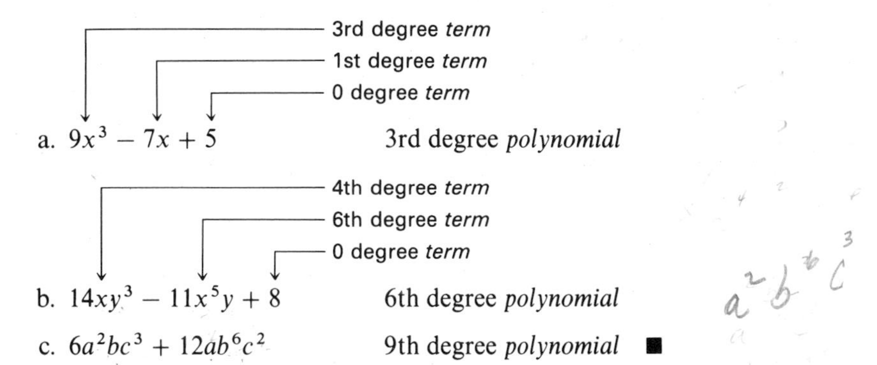

a. $9x^3 - 7x + 5$ 3rd degree *polynomial*

b. $14xy^3 - 11x^5y + 8$ 6th degree *polynomial*

c. $6a^2bc^3 + 12ab^6c^2$ 9th degree *polynomial* ■

Descending Powers

Polynomials are usually written in **descending powers** of one of the variables. For example:

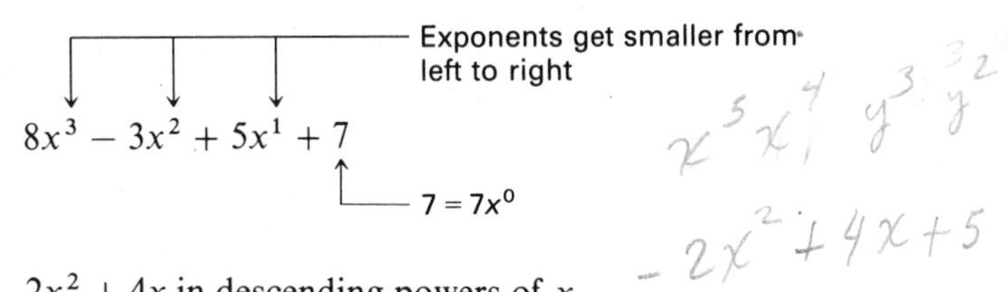

$$8x^3 - 3x^2 + 5x^1 + 7$$

Example 11 Arrange $5 - 2x^2 + 4x$ in descending powers of x.
Solution $-2x^2 + 4x + 5$ ■

When a polynomial has more than one variable, it can be arranged in descending powers of any one of its variables.

Example 12 Arrange $3x^3y - 5xy + 2x^2y^2 - 10$: (a) in descending powers of x, then (b) in descending powers of y.

a. $3x^3y + 2x^2y^2 - 5xy - 10$ Arranged in descending powers of x.

b. $2x^2y^2 + 3x^3y - 5xy - 10$ Arranged in descending powers of y.

Because y is the same power in both terms, the higher-degree term is written first ■

EXERCISES 8.1

In Exercises 1–6, (a) determine the number of terms; (b) write the second term if there is one.

1. $7xy$

2. $[x^2 - (x + y)]$

3. $E - 5F - 3$

4. $5 - (x + y)$

5. $3x^2y + \frac{2x + y}{3xy} + 4(3x^2 - y)$

6. $3E^3 - 2E(8E + F^2)$

In Exercises 7–14, (a) write the degree of the first term; (b) write the numerical coefficient of the second term.

7. $x^4 - 3x^2$

8. $4x^3 + 2x$

9. $x^2 + 2xy + y^2$

10. $a^3 - 5a^2 + a$

11. $x^2y + xy - xy^2$ **12.** $x^3y^2 - x^2y^2$ **13.** $\frac{r}{2} - \frac{s}{3} + \frac{t}{4}$ **14.** $\frac{4a}{5} + \frac{2b}{3}$

In Exercises 15–22, if the expression is a polynomial, find the degree of the polynomial.

15. $2x^2 + 3x$ **16.** $5y^3 + 4y$ **17.** $3x^4 - x^5 + 5$

18. $5x^3 - x^6 + 2$ **19.** $8m^3n - 12m^2n + 6mn^2 - n^3$ **20.** $x^3y^3 - 3x^2y + 3xy^2 - y^3$

21. $x^{-2} + 5x^{-1} + 4$ **22.** $\frac{1}{2x^2 - 5x}$

In Exercises 23–26, write each polynomial in descending powers of the indicated letter.

23. $7x^3 - 4x - 5 + 8x^5$ Powers of x. **24.** $10 - 3y^5 + 4y^2 - 2y^3$ Powers of y.

25. $8xy^2 + xy^3 - 4x^2y$ Powers of y. **26.** $3x^3y + x^4y^2 - 3xy^3$ Powers of x.

8.2 Positive Exponents

In Section 6.8 we discussed bases, exponents, and powers of signed numbers.

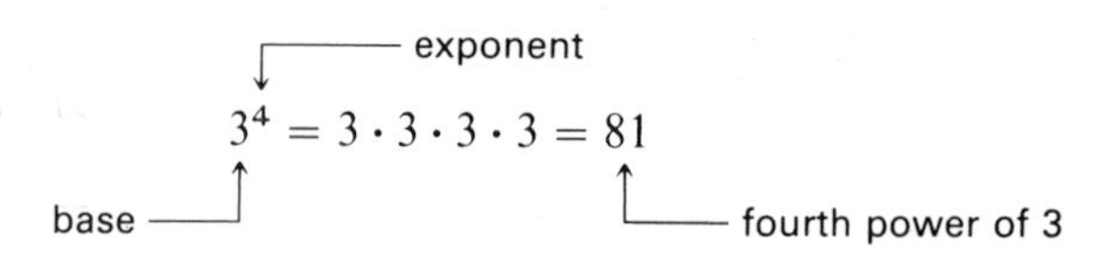

The same definitions are carried over to expressions with variables.

exponent

$$x^4 = x \cdot x \cdot x \cdot x = x^4$$

base — fourth power of x, also read "x to the fourth power"

Consider the product $x^3 \cdot x^2$

$$x^3 \cdot x^2 = (xxx)(xx) = xxxxx = x^5$$

3 factors
+2 factors
5 factors

Therefore, $x^3 \cdot x^2 = x^{3+2} = x^5$

This leads to the following rule of exponents:

Rule 1

MULTIPLYING POWERS HAVING THE SAME BASE

$$x^a \cdot x^b = x^{a+b}$$

When like bases are multiplied, the exponents are added.

Rule 1 of exponents can be extended to include more factors.

$$x^a x^b x^c \cdots = x^{a+b+c+\cdots}$$

Example 1 The use of Rule 1

a. $x^5 \cdot x^2 = x^{5+2} = x^7$

b. $x \cdot x^2 = x^{1+2} = x^3$

$x = x^1$ — NOTE When no exponent is written, the exponent is understood to be 1. ☑

c. $x^3 \cdot x^7 \cdot x^4 = x^{3+7+4} = x^{14}$

d. $10^7 \cdot 10^5 = 10^{7+5} = 10^{12} = \underbrace{1,000,000,000,000}_{\text{12 zeros}}$

e. $2 \cdot 2^3 \cdot 2^2 = 2^{1+3+2} = 2^6 = 64$

f. $2^a \cdot 2^b = 2^{a+b}$

g. $x^3 \cdot y^2$ — NOTE *The rule does not apply* because the bases are different. ☑ ■

Consider the expression $(x^4)^2$.

$$(x^4)^2 = (x^4)(x^4) = x^4 \cdot x^4 = x^{4+4} = x^{2 \cdot 4} = x^{4 \cdot 2} = x^8$$

Note that $(x^4)^2 = x^{4 \cdot 2}$. In this case the exponents are multiplied. This leads to the rule:

Rule 2

> POWER OF A POWER
>
> $$(x^a)^b = x^{ab}$$
>
> When a power is raised to a power, the exponents are multiplied.

Example 2 The use of Rule 2

a. $(x^5)^4 = x^{5 \cdot 4} = x^{20}$

b. $(x)^4 = (x^1)^4 = x^{1 \cdot 4} = x^4$

c. $(y^4)^3 = y^{4 \cdot 3} = y^{12}$

d. $(10^3)^2 = 10^{3 \cdot 2} = 10^6 = \underbrace{1,000,000}_{\text{6 zeros}}$

e. $(2^a)^b = 2^{a \cdot b} = 2^{ab}$ ■

Consider the expression $\frac{x^5}{x^3}$.

$$\frac{x^5}{x^3} = \frac{xxxxx}{xxx} = \frac{xxx \cdot xx}{xxx \cdot 1} = \frac{xxx}{xxx} \cdot \frac{xx}{1} = 1 \cdot \frac{xx}{1} = xx = x^2$$

The value of this fraction $\left(\frac{xxx}{xxx}\right)$ is 1 (for $x \neq 0$).

Note that $\frac{x^5}{x^3} = x^{5-3} = x^2$. In this case, three of the factors of the denominator canceled with three of the five factors of the numerator, leaving $5 - 3 = 2$ factors of x.

This leads to the rule:

Rule 3

DIVIDING POWERS HAVING THE SAME BASE

$$\frac{x^a}{x^b} = x^{a-b} \qquad (x \neq 0)$$

When like bases are divided, the exponents are subtracted.

Example 3 The use of Rule 3

a. $\dfrac{x^6}{x^2} = x^{6-2} = x^4$ — The exponent in the denominator is *subtracted from* the exponent in the numerator

b. $\dfrac{y^3}{y} = \dfrac{y^3}{y^1} = y^{3-1} = y^2$

c. $\dfrac{10^7}{10^3} = 10^{7-3} = 10^4 = 10{,}000$

d. $\dfrac{x^5}{y^2}$ — Rule 3 does not apply when the bases are different

e. $\dfrac{8x^3}{2x} = \dfrac{\overset{4}{\cancel{8}}}{\underset{1}{\cancel{2}}} \cdot \dfrac{x^3}{x} = 4 \cdot x^2 = 4x^2$

f. $\dfrac{6a^3b^4}{9ab^2} = \dfrac{\overset{2}{\cancel{6}}}{\underset{3}{\cancel{9}}} \cdot \dfrac{a^3}{a} \cdot \dfrac{b^4}{b^2} = \dfrac{2}{3} \cdot \dfrac{a^2}{1} \cdot \dfrac{b^2}{1} = \dfrac{2}{3}a^2b^2$ or $\dfrac{2a^2b^2}{3}$

g. $\dfrac{2^a}{2^b} = 2^{a-b}$ ■

A WORD OF CAUTION An expression like $\dfrac{x^5 - y^3}{x^2}$ is usually left unchanged. Rules 1 and 3 cannot be applied here because the numerator is not a product. This expression could be changed as follows:

$$\frac{x^5 - y^3}{x^2} = \frac{x^5}{x^2} - \frac{y^3}{x^2} = x^3 - \frac{y^3}{x^2}$$

A common mistake students make is to divide only x^5 by x^2, instead of dividing both x^5 and y^3 by x^2. Expressions of this form are discussed in Section 8.8. ☑

Why $x \neq 0$ in Rule 3 When writing Rule 3, we added the restriction $x \neq 0$. Consider the example $\dfrac{0^5}{0^2}$. By Rule 3, $\dfrac{0^5}{0^2} = 0^{5-2} = 0^3 = 0$. However, $\dfrac{0^5}{0^2} = \dfrac{0 \cdot 0 \cdot 0 \cdot 0 \cdot 0}{0 \cdot 0} = \dfrac{0}{0}$ is undefined (Section 6.7). For this reason, x cannot be zero in Rule 3.

In this book, unless otherwise noted, none of the variables has a value that makes a denominator zero.

When Rule 3 was used, we were careful to choose the exponent in the numerator larger than the exponent in the denominator, so that the quotient always had a positive exponent. When the exponent in the denominator is larger than the exponent in the numerator, the quotient has a negative exponent. Negative exponents are discussed in Section 15.1.

EXERCISES 8.2

Use the rules of exponents to simplify each expression.

1. $x^5 \cdot x^8$ **2.** $H^3 \cdot H^4$ **3.** $(y^2)^5$ **4.** $(N^3)^4$ **5.** $\frac{x^7}{x^2}$

6. $\frac{y^8}{y^6}$ **7.** $a \cdot a^4$ **8.** $B^7 \cdot B$ **9.** $(x^4)^7$ **10.** $(v^3)^8$

11. $\frac{a^5}{a}$ **12.** $\frac{b^7}{b}$ **13.** $\frac{z^5}{z^4}$ **14.** $\frac{x^8}{x^7}$ **15.** $10^2 \cdot 10^5$

16. $2^3 \cdot 2^2$ **17.** $(10^2)^3$ **18.** $(10^7)^2$ **19.** $\frac{10^{11}}{10}$ **20.** $\frac{2^6}{2}$

21. x^2y^3 **22.** a^4b **23.** $\frac{6x^2}{2x}$ **24.** $\frac{9y^3}{3y}$ **25.** $\frac{a^3}{b^2}$

26. $\frac{x^5}{y^3}$ **27.** $\frac{10x^4}{5x^3}$ **28.** $\frac{15y^5}{9y^2}$ **29.** $\frac{12h^4k^3}{8h^2k}$ **30.** $\frac{16a^5b^3}{12ab^2}$

31. $5^u \cdot 5^v$ **32.** $\frac{x^c}{x^d}$ **33.** $\frac{a^4 - b^3}{a^2}$ **34.** $\frac{x^6 + y^4}{y^2}$ **35.** $y^2 \cdot y^4$

36. $(x^3)^2$ **37.** $\frac{a^6}{a^4}$ **38.** $x^5 \cdot x$ **39.** $(z^2)^4$ **40.** $\frac{b^4}{b}$

41. $\frac{y^5}{y^3}$ **42.** $10^3 \cdot 10^2$ **43.** $(3^3)^2$ **44.** $\frac{5^3}{5^2}$ **45.** hk^2

46. $\frac{12x^3}{10x}$ **47.** $\frac{m^4}{n^2}$ **48.** $\frac{18z^6}{10z^4}$ **49.** $\frac{25a^3b^4}{15ab^3}$ **50.** $3^2 \cdot 5^3$

51. $2 \cdot 2^3 \cdot 2^2$ **52.** $3 \cdot 3^2 \cdot 3^3$ **53.** $x \cdot x^3 \cdot x^4$ **54.** $y \cdot y^5 \cdot y^3$ **55.** $x^2y^3x^5$

56. $z^3z^4w^2$ **57.** ab^3a^5 **58.** x^8yx^4 **59.** $(3^a)^b$ **60.** $2^a \cdot 2^b \cdot 2^c$

8.3 The Distributive Rule

Product of Factors

In working with polynomials, you must be able to simplify a **product of factors**.

TO SIMPLIFY A PRODUCT OF FACTORS

1. *Write the sign.* The sign of the product is negative if there are an odd number of negative factors. The sign is positive if there are an even number of negative factors (or no negative factors).
2. *Write the numerical coefficient.* (Its sign was found in Step 1.) The coefficient is the product of all the absolute values of the numbers in the factors.
3. *Write the variables.* (Usually in alphabetical order.) Use Rule 1 of exponents to determine the exponent of each variable.

Example 1 Simplify the following products.

a. $(-2a^2b)(5a^3b^3)$

$= -(2 \cdot 5)(a^2a^3)(bb^3)$

$= -(10)(a^5)(b^4) = -10a^5b^4$

Rule 1 of exponents

10 is the product of the absolute values of the numbers in the factors

The sign is negative because there are an odd number of negative factors

b. $(4xy^2)(-3x^2y^3)(-2x^3y)$

$= +(4 \cdot 3 \cdot 2)(xx^2x^3)(y^2y^3y)$

$= +(24)(x^6)(y^6) = 24x^6y^6$

Positive, because there are an even number of negative factors

c. $(-2xy^2)(5xyz)(-y^2z^3)(-10x^2z^2)$

$= -(2 \cdot 5 \cdot 10)(xxx^2)(y^2yy^2)(zz^3z^2)$

$= -(100)(x^4)(y^5)(z^6) = -100x^4y^5z^6$

Negative, because there are an odd number of negative factors ■

This method of simplifying products of factors makes use of the associative and commutative properties of multiplication.

The Distributive Rule

One property of real numbers combines both addition and multiplication. Consider $3(10 + 1)$.

$$3(10 + 1) = 3(11) = 33$$

If we distribute the 3 and multiply, we get

$$3 \cdot 10 + 3 \cdot 1 = 30 + 3 = 33$$

Therefore, $3(10 + 1) = 3 \cdot 10 + 3 \cdot 1$

This leads to the fundamental property of real numbers called the **distributive rule.**

THE DISTRIBUTIVE RULE

$$a(b + c) = ab + ac$$

The distributive rule can be extended to have any number of terms within the parentheses.

Meaning of the Distributive Rule When a factor is multiplied by an expression enclosed within grouping symbols, the factor must be multiplied by *each term* within the grouping symbols; then the products are added.

$$5(2-7) = (5)(2 + -7)$$

first product + second product

$$= (5)(2) + (5)(-7)$$
$$= 10 + (-35)$$
$$= -25$$

Each term inside the parentheses is multiplied by the factor outside the parentheses; then these products are added

Example 2 Using the distributive rule

a. $2(x+y) = (2)(x) + (2)(y)$
$\quad = 2x + 2y$

b. $a(x-y) = (a)(x) + (a)(-y)$
$\quad = ax + (-ay)$
$\quad = ax - ay$

c. $x(x^2+y) = (x)(x^2) + (x)(y)$
$\quad = x^3 + xy$

d. $3x(4x + x^3y) = (3x)(4x) + (3x)(x^3y)$
$\quad = 12x^2 + 3x^4y$ ■

Example 3 Using the distributive rule when more than two terms appear within the parentheses

$$-5a(4a^3 - 2a^2b + b^2) = (-5a)(4a^3 + -2a^2b + b^2)$$
$$= (-5a)(4a^3) + (-5a)(-2a^2b) + (-5a)(b^2)$$
$$= -20a^4 + 10a^3b - 5ab^2 \quad ■$$

Further Extensions of the Distributive Rule Because of the distributive rule and the commutative property of multiplication, it follows that

$$(b+c)a = a(b+c) = ab + ac = ba + ca$$

Therefore, $(b+c)a = ba + ca$.

This can be extended to include more terms.

$$(b + c + d + \cdots)a = ba + ca + da + \cdots$$

Example 4 Using the distributive rule when the factor appears on the right

a. $(2x-5)(-4x) = (2x + -5)(-4x)$
$\quad = (2x)(-4x) + (-5)(-4x)$
$\quad = -8x^2 + 20x$

b. $(-2x^2 + xy - 5y^2)(-3xy)$

$$= (-2x^2 + xy + -5y^2)(-3xy)$$
$$= (-2x^2)(-3xy) + (xy)(-3xy) + (-5y^2)(-3xy)$$
$$= 6x^3y - 3x^2y^2 + 15xy^3$$ ■

Multiplying a number by -1 gives the negative of that number (Section 6.4).

$$-1 \cdot a = -a$$

To find the negative of a polynomial, multiply the polynomial by -1.

Example 5 Negative of a polynomial

a. $-(2x + 3) = -1(2x + 3)$
$$= (-1)(2x) + (-1)(3)$$
$$= -2x - 3$$

b. $-(a - b) = -1(a - b)$
$$= (-1)(a) + (-1)(-b)$$
$$= -a + b$$

NOTE The *negative* of a polynomial changes the sign of each term. ☑

c. $-(3x^2 - 5x + 4) = -3x^2 + 5x - 4$ ■

A WORD OF CAUTION A common mistake students make is to think that the distributive rule applies to expressions like $2(3 \cdot 4)$.

The distributive rule only applies when this is an addition

$$2(3 \cdot 4) \neq (2 \cdot 3)(2 \cdot 4)$$
$$2(12) \neq 6 \cdot 8$$
$$24 \neq 48$$ ☑

EXERCISES 8.3

In Exercises 1–12, simplify each product of factors.

1. $(-2a)(4a^2)$ **2.** $(-3x)(5x^3)$ **3.** $(5x^2)(-7y)$
4. $(-3x^3)(4y)$ **5.** $(4x^2)^3$ **6.** $(3x^2)^2$
7. $(-5x^3)(2x^2)(-3x)$ **8.** $(4x^2)(-7x^2)(2x)$ **9.** $(2mn^2)(-4m^2n^2)(2m^3n)$
10. $(5x^2y)(-2xy^2)(-3xy)$ **11.** $(-5x^2y)(2yz^3)(-xz)$ **12.** $(3xyz)(-2x^2y)(-yz^2)$

In Exercises 13–50, find each product using the distributive rule.

13. $5(a + 6)$ **14.** $4(x + 10)$ **15.** $3(m - 4)$
16. $3(a - 5)$ **17.** $-2(x - 3)$ **18.** $-3(x - 5)$
19. $-3(2x^2 - 4x + 5)$ **20.** $-5(3x^2 - 2x - 7)$ **21.** $4x(3x^2 - 6)$
22. $3x(5x^2 - 10)$ **23.** $-2x(5x^2 + 3x - 4)$ **24.** $-4x(2x^2 - 5x + 3)$
25. $-1(x^2 - y^2)$ **26.** $-1(5x + 2y)$ **27.** $-(x + 3)$

28. $-(y-7)$
29. $-(2x^2+4x-7)$
30. $-(z^2-5z+6)$
31. $(x-4)6$
32. $(3-2x)(-5)$
33. $(y^2-4y+3)7$
34. $(-9+z-2z^2)(-7)$
35. $(2x^2-3x+5)4x$
36. $(3w^2+2w-8)5w$
37. $x(xy-3)$
38. $a(ab-4)$
39. $3a(ab-2a^2)$
40. $4x(3x-2y^2)$
41. $(-2x+4x^2y)(-3y)$
42. $(-3a+2a^2z)(-2z)$
43. $-2xy(x^2y-y^2x-y-5)$
44. $-3ab(8-a^2-b^2+ab)$
45. $(3x^3-2x^2y+y^3)(-2xy)$
46. $(4z^3-z^2y-y^3)(-2yz)$
47. $4ab^2(3a^4b^2-2ab-5b)$
48. $6x^2y(2x^5y-4x^3y^2+xy^3)$
49. $(-2xy^2z^2)(6x^2y-3yz-4xz^2)$
50. $(-5a^2bc)(2b^2c-5ac^3-3a^2b^3)$

8.4 Combining Like Terms

Like Terms

Like terms have the same variables with the same exponents. Only the numerical coefficients may differ.

Example 1 Examples of like terms

a. $3x, 4x, x$ are like terms. They are called "x-terms."

b. $2x^2, 10x^2, \frac{3}{4}x^2$ are like terms. They are called "x^2-terms."

c. $5xy, 2xy, 5.6xy$ are like terms. They are called "xy-terms."

d. $4x^2y, 8x^2y, x^2y$ are like terms. They are called "x^2y-terms."

NOTE All constants are like terms. ☑

e. $5, -3, \frac{1}{4}, 2.6$ are like terms. They are called "constant terms." ■

Example 2 Examples of unlike terms

a. $2x, 3y$ are unlike terms. The variables are different.

b. $5x^2, 7x$ are unlike terms. The variable x has different exponents.

c. $4x^2y, 10xy^2$ are unlike terms. ■

Combining Like Terms

3 dollars + 5 dollars = (3 + 5) dollars = 8 dollars

3 cars + 5 cars = (3 + 5) cars = 8 cars

$3x + 5x = (3+5)x = 8x$

This is an application of the distributive rule

TO COMBINE LIKE TERMS

1. Identify the like terms.
2. Find the sum of each group of like terms by adding their numerical coefficients, then multiply that sum by the common variables. (When no coefficient is written, it is understood to be 1.)

Example 3 Combining like terms

a. $2x + 4x = (2 + 4)x = 6x$

b. $e - 9e = (1 - 9)e = -8e$

When no coefficient is written, it is understood to be 1

c. $5x^2y - 7x^2y = (5 - 7)x^2y = -2x^2y$

d. $4ab - 3ab + 6ab = (4 - 3 + 6)ab = 7ab$

e. $\underbrace{9x - 3x} + 5y = (9 - 3)x + 5y = 6x + 5y$

These are the only like terms in the expression. Unlike terms cannot be combined

When combining like terms, you do not need to show the step that uses the distributive rule. Simply add the coefficients mentally and multiply this sum by the common variables. For example, in the expression $7x - 4x + 2x$,

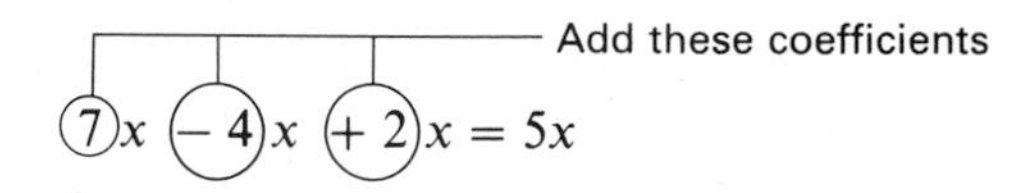

When we combine like terms, we are usually changing the grouping and the order in which the terms appear. The commutative and associative properties of addition guarantee that when we do this, the sum remains unchanged.

f. $$\begin{aligned} 12a - 7b - 9a + 4b &= (12a - 9a) + (-7b + 4b) \\ &= 3a + (-3b) \\ &= 3a - 3b \end{aligned}$$

g. $$\begin{aligned} 3x^2 - 2x + 5 - 4x + 6 &= 3x^2 + (-2x - 4x) + (5 + 6) \\ &= 3x^2 + (-6x) + 11 \\ &= 3x^2 - 6x + 11 \end{aligned}$$ ■

EXERCISES 8.4

Combine like terms.

1. $15x - 3x$

2. $12a - 9a$

3. $5a - 12a$

4. $10x - 24x$

5. $-2x^2 + 6x^2$

6. $-4a^2 + 8a^2$

7. $x^3 + 7x^3$

8. $9y^4 + y^4$

9. $3xy - 10xy$

10. $5ab - 8ab$

11. $4x^2 + 3x$

12. $8a^3 - 2a^2$

13. $2a - 5a + 6a$

14. $3y - 4y + 5y$

15. $5x - 8x + x + 3x$

16. $3a - 5a + a - 2a$

17. $3x + 2y - 3x$

18. $4a - 2b + 2b$

19. $3mn - 5mn + 2mn$

20. $5cd - 8cd + 3cd$

21. $2xy - 5yx + xy$

22. $8mn - 7nm + 3nm$

23. $a^2b - 3a^2b$

24. $x^2y^2 - 5x^2y^2$

25. $5x^2 - 3x + 7 - 2x^2 + 8x - 9$

26. $7y^2 + 4y - 6 - 9y^2 - 2y + 7$

27. $x - 3x^3 + 2x^2 - 5x - 4x^2 + x^3$

28. $y - 2y^2 - 5y^3 - y + 3y^3 - y^2$

29. $x^3 - 2x^2 + 6x - 3x^2 - 7x + 3$

30. $a^2 - 4a - 6 + 7a^3 - 5a^2 - 3$

31. $5x^2 - xy + y^2 + 6xy - 7y^2$

32. $3a^2 + 6ab + b^2 - 2a^2 - 9ab$

33. $7x^2y - 2xy^2 - 4x^2y - xy^2$

34. $4xy^2 - 5x^2y - 2xy^2 + 3x^2y$

35. $\frac{2}{3}y^2 - \frac{1}{2}y^2 + \frac{5}{6}y^2$

36. $\frac{1}{2}x - \frac{3}{4}x + \frac{5}{8}x - \frac{3}{16}x$

37. 12.67 sec + 9.08 sec − 6.73 sec

38. 18.7 ft + 69.5 ft − 42.1 ft − 26.3 ft

8.5 Addition and Subtraction of Polynomials

Addition of Polynomials

The commutative and associative properties allow us to add polynomials by combining like terms.

> TO ADD POLYNOMIALS
>
> **1.** Remove grouping symbols.
>
> **2.** Combine like terms.
>
> It is helpful to underline like terms with the same kind of line before adding.

Example 1 Adding polynomials

a. $(3x^2 + 5x - 4) + (2x + 5) + (x^3 - 4x^2 + x)$
$= \underline{3x^2} + \underline{5x} - \underline{4} + \underline{2x} + \underline{5} + \underline{x^3} - \underline{4x^2} + \underline{x}$
$= x^3 - x^2 + 8x + 1$

b. $(5x^3y^2 - 3x^2y^2 + 4xy^3) + (4x^2y^2 - 2xy^2) + (-7x^3y^2 + 6xy^2 - 3xy^3)$
$= \underline{5x^3y^2} - \underline{3x^2y^2} + \underline{4xy^3} + \underline{4x^2y^2} - \underline{2xy^2} - \underline{7x^3y^2} + \underline{6xy^2} - \underline{3xy^3}$
$= -2x^3y^2 + x^2y^2 + xy^3 + 4xy^2$ ■

Most addition of polynomials will be done horizontally as already shown. However, in a few cases it is convenient to use *vertical addition.* See Section 8.6, Multiplication of Polynomials.

To Add Polynomials Vertically

1. Arrange them under one another so that like terms are in the same vertical line.
2. Find the sum of the terms in each vertical line by adding their numerical coefficients.

Example 2 Add $(3x^2 + 2x - 1)$, $(2x + 5)$, and $(4x^3 + 7x^2 - 6)$ vertically.

$$\begin{array}{r} 3x^2 + 2x - 1 \\ 2x + 5 \\ \underline{4x^3 + 7x^2 \qquad - 6} \\ 4x^3 + 10x^2 + 4x - 2 \end{array}$$ ■

Subtraction of Polynomials

We subtract polynomials the same way we subtract signed numbers. This means we change the signs in the polynomial being subtracted, then add the resulting polynomials.

> TO SUBTRACT ONE POLYNOMIAL FROM ANOTHER POLYNOMIAL
>
> **1.** Change the sign of *each* term in the polynomial being subtracted. Then:
>
> **2.** Add the resulting polynomials.

Example 3 Subtracting polynomials

a. $(-4x^3 + 8x^2 - 2x - 3) - (4x^3 - 7x^2 + 6x + 5)$ ← This is the polynomial being subtracted

$= -4x^3 + 8x^2 - 2x - 3 - 4x^3 + 7x^2 - 6x - 5$

$= -8x^3 + 15x^2 - 8x - 8$

b. Subtract $(-4x^2y + 10xy^2 + 9xy - 7)$ from $(11x^2y - 8xy^2 + 7xy + 2)$. ← This is the polynomial being subtracted

Solution $(11x^2y - 8xy^2 + 7xy + 2) - (-4x^2y + 10xy^2 + 9xy - 7)$

$= 11x^2y - 8xy^2 + 7xy + 2 + 4x^2y - 10xy^2 - 9xy + 7$

$= 15x^2y - 18xy^2 - 2xy + 9$

c. Subtract $(2x^2 - 5x + 3)$ from the sum of $(8x^2 - 6x - 1)$ and $(4x^2 + 7x - 9)$.

Solution $(8x^2 - 6x - 1) + (4x^2 + 7x - 9) - (2x^2 - 5x + 3)$

$= 8x^2 - 6x - 1 + 4x^2 + 7x - 9 - 2x^2 + 5x - 3$

$= 10x^2 + 6x - 13$

d. $(x^2 + 5) - [(x^2 - 3) + (2x^2 - 1)]$

$= (x^2 + 5) - [x^2 - 3 + 2x^2 - 1]$ Removing innermost ( )

$= (x^2 + 5) - [3x^2 - 4]$ Combining like terms inside [ ]

$= x^2 + 5 - 3x^2 + 4$ Removing grouping symbols

$= -2x^2 + 9$ Combining like terms ■

Most subtraction of polynomials will be done horizontally as already shown, but in a few cases it is convenient to use *vertical subtraction.* See Section 8.8, Division of Polynomials.

To Subtract Polynomials Vertically

1. Write the polynomial being subtracted *under* the polynomial it is being subtracted from. Write like terms in the same vertical line.
2. Change the sign of *each* term in the polynomial being subtracted.
3. Find the sum of the resulting terms in each vertical line by adding their numerical coefficients.

Example 4 Subtract
$$\begin{array}{r} 5x^2 + 3x - 6 \\ \underline{3x^2 - 5x + 2} \end{array}$$

Solution

$$\begin{array}{r} 5x^2 + 3x - 6 \\ \underline{-(3x^2 - 5x + 2)} \end{array} \Rightarrow \begin{array}{r} 5x^2 + 3x - 6 \\ \underline{-3x^2 + 5x - 2} \\ 2x^2 + 8x - 8 \end{array}$$

Change the sign of *each* term in the polynomial being subtracted, then *add* the resulting terms ■

Example 5 Subtract $(3x^2 - 2x - 7)$ from $(x^3 - 2x - 5)$ vertically.

Solution

$$\begin{array}{rrr} x^3 \quad\quad - 2x - 5 & & x^3 \quad\quad - 2x - 5 \\ -(\quad + 3x^2 - 2x - 7) & \Rightarrow & \underline{\quad - 3x^2 + 2x + 7} \\ & & x^3 - 3x^2 \quad\quad + 2 \end{array}$$

$\longleftarrow$ *Change* signs and *add* ■

EXERCISES 8.5

Work the following addition and subtraction problems.

1. $(2m^2 - m + 4) + (3m^2 + m - 5)$
2. $(5n^2 + 8n - 7) + (6n^2 - 6n + 10)$
3. $(3x^2 + 4x - 10) - (5x^2 - 3x + 7)$
4. $(2a^2 - 3a + 9) - (3a^2 + 4a - 5)$
5. Subtract $(-5b^2 + 4b + 8)$ from $(8b^2 + 2b - 14)$.
6. Subtract $(-8c^2 - 9c + 6)$ from $(11c^2 - 4c + 7)$.
7. $(2x^3 - 4) + (4x^2 + 8x) + (-9x + 7)$
8. $(5 + 8z^2) + (4 - 7z) + (z^2 + 7z)$
9. $(6a - 5a^2 + 6) + (4a^2 + 6 - 3a)$
10. $(2b + 7b^2 - 5) + (4b^2 - 2b + 8)$
11. $(4a^2 + 6 - 3a) - (5a + 3a^2 - 4)$
12. $(8 + 3b^2 - 7b) - (2b + b^2 - 9)$
13. $(3x^2 - 4x + 8) - (9 - x^2 + 6x)$
14. $(8x^2 - 6 - x) - (7x + 3x^2 + 5)$
15. $(3x - 7) + (5 - x) - (4x - 2)$
16. $(7a + 2) - (3a - 8) + (a - 7)$
17. $(y^2 - 3y) - (4y - 6) - (2 - y^2)$
18. $(x^3 - 4x^2) - (5x^2 - x) - (7x - 9)$

In Exercises 19–22, add the polynomials vertically.

19. $\begin{array}{r} 17a^3 \quad\quad + 4a - 9 \\ \underline{8a^2 - 6a + 9} \end{array}$

20. $\begin{array}{r} - b^3 + 5b^2 - 8 \\ \underline{-20b^4 + 2b^3 \quad\quad + 7} \end{array}$

21. $\begin{array}{r} 14x^2y^3 - 11xy^2 + 8xy \\ -9x^2y^3 + 6xy^2 - 3xy \\ \underline{7x^2y^3 - 4xy^2 - 5xy} \end{array}$

22. $\begin{array}{r} 12a^2b - 8ab^2 + 6ab \\ -7a^2b + 11ab^2 - 3ab \\ \underline{4a^2b - ab^2 - 13ab} \end{array}$

In Exercises 23–26, subtract the lower polynomial from the upper polynomial vertically.

23. $\begin{array}{r} 15x^3 - 4x^2 \quad\quad + 12 \\ \underline{8x^3 \quad\quad + 9x - 5} \end{array}$

24. $\begin{array}{r} - 14y^2 + 6y - 24 \\ \underline{7y^3 + 14y^2 - 13y} \end{array}$

25. $\begin{array}{r} 8a^3 - 3a^2 + 2a + 9 \\ \underline{5a^3 - 7a^2 - a + 6} \end{array}$

26. $\begin{array}{r} 4x^2y^2 + x^2y - 6xy^2 - 2xy \\ \underline{3x^2y^2 + 5x^2y - 2xy^2 + 7xy} \end{array}$

27. $(7 - 8v^3 + 9v^2 + 4v) + (9v^3 - 8v^2 + 4v + 6)$
28. $(15 - 10w + w^2 - 3w^3) + (18 + 4w^3 + 7w^2 + 10w)$
29. $(7x^2y^2 - 3x^2y + xy + 7) - (3x^2y^2 + 7x^2y - 5xy + 4)$
30. $(4x^2y^2 + x^2y - 5xy - 4) - (5x^2y^2 - 3x^2y + 9)$
31. $(5x - 3) - [(x + 2) + (2x - 6)]$
32. $(8y + 1) - [(2y - 3) + (y + 7)]$
33. $(x^2 + 4) - [(x^2 - 5) - (3x^2 + 1)]$
34. $(3x^2 - 2) - [(4 - x^2) - (2x^2 - 1)]$

35. $(4x^2 - 5) - [(2 - x) - (x^2 + 3x)]$

36. $(9a^2 + 2) - [(7 + 3a) - (4a^2 - 5a)]$

37. Subtract $(2x^2 - 4x + 3)$ from the sum of $(5x^2 - 2x + 1)$ and $(-4x^2 + 6x - 8)$.

38. Subtract $(6y^2 + 3y - 4)$ from the sum of $(-2y^2 + y - 9)$ and $(8y^2 - 2y + 5)$.

39. Subtract $(10a^3 - 8a + 12)$ from the sum of $(11a^2 + 9a - 14)$ and $(-6a^3 + 17a)$.

40. Subtract $(15b^3 - 17b^2 + 6)$ from the sum of $(14b^2 - 8b - 11)$ and $(-9b^3 + 12b)$.

41. Subtract the sum of $(x^3y + 3xy^2 - 4)$ and $(2x^3y - xy^2 + 5)$ from the sum of $(5 + xy^2 + x^3y)$ and $(-6 - 3xy^2 + 4x^3y)$.

42. Subtract the sum of $(2m^2n - 4mn^2 + 6)$ and $(-3m^2n + 5mn^2 - 4)$ from the sum of $(5 + m^2n - mn^2)$ and $(3 + 4m^2n + 2mn^2)$.

43. Given the polynomials $(2x^2 - 5x - 7)$, $(-4x^2 + 8x - 3)$, and $(6x^2 - 2x + 1)$, subtract the sum of the first two from the sum of the last two.

44. Given the polynomials $(8y^2 + 10y - 9)$, $(13y^2 - 11y + 6)$, and $(-2y^2 - 16y + 4)$, subtract the sum of the first two from the sum of the last two.

45. $(7.2x^2 - 4.8x + 6.5) + (-2.8x^2 + 8.9x + 5.3)$

46. $(9.6x^2 + 5.7x - 9.8) + (7.9x^2 - 9.6x - 3.2)$

47. $(2.62x^2 - 8.95x - 6.08) - (7.4x^2 + 11.9x - 8.54)$

48. $(3.56x^2 + 17.4x - 5.84) - (14.7x^2 - 3.09x + 12.6)$

8.6 Multiplication of Polynomials

Multiplication of a Polynomial by a Monomial

The basis of multiplication of polynomials is the distributive rule and its extensions.

$$a(b + c + \cdots) = ab + ac + \cdots$$

$$(b + c + \cdots)a = ba + ca + \cdots$$

> TO MULTIPLY A POLYNOMIAL BY A MONOMIAL
>
> Multiply *each* term in the polynomial by the monomial, then add the results.

Example 1 Multiplying a polynomial by a monomial

a. $3x^2(2x + 7) = (3x^2)(2x) + (3x^2)(7) = 6x^3 + 21x^2$

b. $-5x(3x^2 - 2x + 6) = (-5x)(3x^2) + (-5x)(-2x) + (-5x)(6)$
$= -15x^3 + 10x^2 - 30x$

For practice on this type of multiplication, rework Exercises 8.3 (numbers 13–50). ■

Multiplication of a Polynomial by a Polynomial

Consider the product $(x + 2)(x + 5)$. To find this product we will use the distributive rule twice.

First consider $(x + 2)$ as a single quantity and distribute it to each term in $(x + 5)$.

$$(x + 2)(x + 5) = (x + 2)(x) + (x + 2)(5)$$

Now use the distributive rule again.

$$= (x)(x) + (2)(x) + (x)(5) + (2)(5)$$
$$= x^2 + 2x + 5x + 10$$
$$= x^2 + 7x + 10$$

The result was that *each term* of the first polynomial was multiplied by *each term* of the second polynomial, and then these products were added.

A convenient arrangement for multiplying two polynomials is the one used in arithmetic for multiplying two whole numbers.

$$\begin{array}{r} 56 \\ \times\ 23 \\ \hline 168 \\ 112 \\ \hline 1288 \end{array}$$

Product of 56 and 3

Product of 56 and 2

Notice that the second line in this arithmetic example is moved one place to the left. Why?

We use this arrangement in multiplying the following binomials (Examples 2–5).

Example 2 $(2x^2 - 5x)(3x + 7)$

$$\begin{array}{rrrr} & 2x^2 & -\ 5x & \\ \times & 3x & +\ 7 & \\ \hline & 14x^2 & -\ 35x & \Leftarrow (2x^2 - 5x)7 \\ 6x^3 & -\ 15x^2 & & \Leftarrow (2x^2 - 5x)3x \\ \hline 6x^3 & -\ x^2 & -\ 35x & \end{array}$$

Notice that the second line is moved one place to the left so we have like terms in the same vertical line ■

Example 3 $(3a^2b - 6ab^2)(2ab - 5)$

$$\begin{array}{rrrr} & & 3a^2b & -\ 6ab^2 \\ & \times & 2ab & -\ 5 \\ \hline & & -\ 15a^2b & +\ 30ab^2 \\ 6a^3b^2 & -\ 12a^2b^3 & & \\ \hline 6a^3b^2 & -\ 12a^2b^3 & -\ 15a^2b & +\ 30ab^2 \end{array}$$

Notice that the second line is moved over far enough so only like terms are in the same vertical line ■

Example 4 $(3x^2 + 2x - 4)(5x + 2)$

$$\begin{array}{rrrr} & 3x^2 & +\ 2x & -\ 4 \\ \times & & 5x & +\ 2 \\ \hline & 6x^2 & +\ 4x & -\ 8 \\ 15x^3 & +\ 10x^2 & -\ 20x & \\ \hline 15x^3 & +\ 16x^2 & -\ 16x & -\ 8 \end{array}$$ ■

The higher degree polynomial is placed above the other, as in arithmetic

Example 5 $(2m - 5 - 3m^2)(2 + m^2 - 3m)$

$$\begin{array}{rrrrr} & & -3m^2 & +\ 2m & -\ 5 \\ & \times & m^2 & -\ 3m & +\ 2 \\ \hline & & -\ 6m^2 & +\ 4m & -\ 10 \\ & 9m^3 & -\ 6m^2 & +\ 15m & \\ -\ 3m^4 & +\ 2m^3 & -\ 5m^2 & & \\ \hline -\ 3m^4 & +\ 11m^3 & -\ 17m^2 & +\ 19m & -\ 10 \end{array}$$ ■

Multiplication is simplified by first arranging the polynomials in descending powers of *m*

Powers of Polynomials Because raising to a power is repeated multiplication, we now have a method for finding powers of polynomials.

Example 6

$$(a-b)^3 = \underbrace{(a-b)(a-b)}_{\text{First find } (a-b)^2,}\underbrace{(a-b)}_{\text{then multiply by } (a-b)}$$

$$\begin{array}{rrrr} & & a & -b \\ \times & & a & -b \\ \hline & & -ab & +b^2 \\ & a^2 & -ab & \\ \hline & a^2 & -2ab & +b^2 \end{array} \qquad \begin{array}{rrrr} a^2 & -2ab & +b^2 & \\ \times & a & -b & \\ \hline & -a^2b & +2ab^2 & -b^3 \\ a^3 & -2a^2b & +ab^2 & \\ \hline a^3 & -3a^2b & +3ab^2 & -b^3 \end{array}$$

■

EXERCISES 8.6

Find the following products.

1. $-3z^2(5z^3 - 4z^2 + 2z - 8)$
2. $-4m^3(2m^3 - 3m^2 + m - 5)$
3. $(-3x^2y + xy^2 - 4y^3)(-2xy)$
4. $(-4xy^2 - x^2y + 3x^3)(-3xy)$
5. $5ab^2(3a^2 - 6ab + 5b^2)$
6. $7x^2y(9x^2 + 3xy - 2y^2)$
7. $3xy^2z^3(x^4y^2z + 4x^3y^2z^3 - 2xyz^5)$
8. $4rs^4t^2(5r^6s^3t - r^4s^2t^2 + 8rst^5)$
9. $(x + 3)(x - 2)$
10. $(a - 4)(a + 3)$
11. $(2y + 5)(y - 4)$
12. $(3z - 2)(z + 5)$
13. $(x + 4)^2$
14. $(x - 5)^2$
15. $(2x^2 - 5)(x + 2)$
16. $(3y^2 - 2)(y - 2)$
17. $(x^2 - 4x + 2)(x + 3)$
18. $(x^2 + 5x - 8)(x - 2)$
19. $(z - 4)(z^2 - 4z + 16)$
20. $(a + 5)(a^2 + 5a + 25)$
21. $(2x + 5)(3x^2 + x - 4)$
22. $(3x - 1)(5x^2 - 6x - 2)$
23. $(7x + x^2 - 6)(4x - 3)$
24. $(2 - x^2 - 6x)(5x + 2)$
25. $(9 - x + 2x^2)(5 - x)$
26. $(4x - 1 + 3x^2)(3 - x)$
27. $(7 + 2x^3 + 3x)(x - 4)$
28. $(4 + a^3 - 2a^2)(a - 6)$
29. $(y^3 - 7y^2 + y - 4)(2y - 3)$
30. $(3x^3 + x^2 - 5x - 3)(3x - 1)$
31. $(x^2 - x + 6)(x^2 + 3x - 4)$
32. $(4y^2 + 3y - 5)(y^2 - 8y - 2)$
33.
34. $(z^2 - 3z - 4)^2$
35. $(x + 2)^3$
36. $(x + 3)^3$
37. $(x + y)^2(x - y)^2$
38. $(x - 2)^2(x + 2)^2$

8.7 Product of Two Binomials

The product of two binomials occurs so often that it is useful to learn a method that will allow you to obtain their product quickly.

When the first terms of each binomial are like terms and the last terms of each binomial are like terms, their product is a trinomial (Example 1).

Example 1

like terms (first terms x and x); like terms (last terms 2 and 3)

$$\begin{aligned}(x+2)(x+3) &= x(x+3) + 2(x+3) \quad \text{distributive rule}\\ &= x^2 + 3x + 2x + 6\\ &= \underbrace{x^2 + 5x + 6}_{\text{trinomial}}\end{aligned}$$

■

Notice that each term in the first binomial is multiplied by each term in the second binomial. The product $(x + 2)(x + 3)$ can also be found by the following method.

Product of *First* terms

$(x + 2)(x + 3) = x^2$

Product of *Outer* terms

$(x + 2)(x + 3) = x^2 + 3x$

Product of *Inner* terms

$(x + 2)(x + 3) = x^2 + 3x + 2x$

Product of *Last* terms

$(x + 2)(x + 3) = x^2 + 3x + 2x + 6 = x^2 + 5x + 6$

The method above is often referred to as *FOIL* because we are finding the sum of the products of the *F*irst, *O*uter, *I*nner, and *L*ast terms.

Example 2 Multiply $(x - 3)(x - 4)$.

Solution

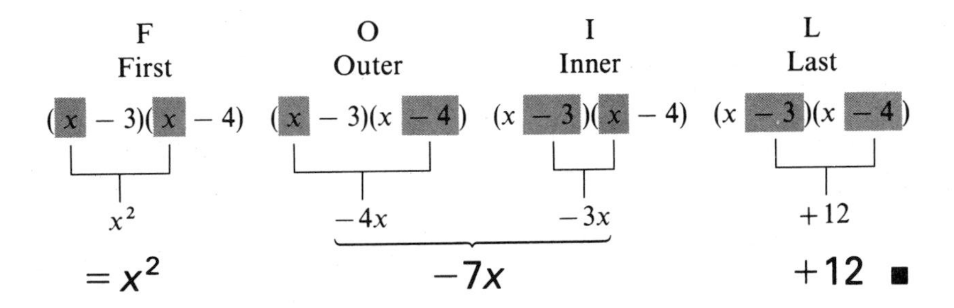

This procedure can be shortened by adding the inner and outer terms mentally.

Example 3 Multiply $(3x + 2)(4x - 5)$.

Solution

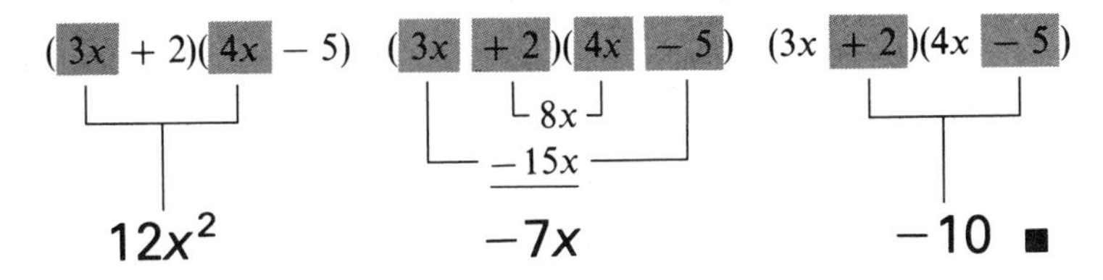

TO MULTIPLY TWO BINOMIALS

1. The first term of the product is the *product* of the first terms of the binomials.
2. The middle term of the product is the *sum* of the inner and outer products.
3. The last term of the product is the *product* of the last terms of the binomials.

Example 4 Multiply $(5x - 4y)(6x + 7y)$.

Solution

$$(5x - 4y)(6x + 7y) \quad (5x - 4y)(6x + 7y) \quad (5x - 4y)(6x + 7y)$$

$$-24xy$$
$$+35xy$$

$$30x^2 \qquad +11xy \qquad -28y^2$$ ∎

Example 5 Multiply $(3x - 8y)(4x - 5y)$.

Solution

$$(3x - 8y) \qquad (4x - 5y).$$

$$-32xy$$
$$-15xy$$

$$12x^2 \qquad -47xy \qquad +40y^2$$ ∎

Practice this procedure until you can find the three terms of the product without having to write anything down.

The Square of a Binomial A binomial can be squared by multiplying it by itself.

Example 6

a. $(a + b)^2 = (a + b) \quad (a + b) = a^2 + 2ab + b^2$

$$ab$$
$$ab$$

$$a^2 + 2ab + b^2$$

b. $(a - b)^2 = (a - b) \quad (a - b) = a^2 - 2ab + b^2$

$$-ab$$
$$-ab$$

$$a^2 - 2ab + b^2$$ ∎

The two special products shown in Example 6 occur so often that they are worth remembering.

TO SQUARE A BINOMIAL

1. The first term of the product is the *square of the first term* of the binomial.
2. The middle term of the product is *twice* the product of the two terms of the binomial.
3. The last term of the product is the *square of the last term* of the binomial.

$$(a + b)^2 = a^2 + 2ab + b^2$$
$$(a - b)^2 = a^2 - 2ab + b^2$$

Example 7 Finding the square of a binomial by using the formulas given in the preceding box

a. $(m + n)^2 = (m)^2 + 2(m)(n) + (n)^2 = m^2 + 2mn + n^2$

b. $(a - 3)^2 = (a)^2 + 2(a)(-3) + (-3)^2 = a^2 - 6a + 9$

c. $(2x - 5)^2 = (2x)^2 + 2(2x)(-5) + (-5)^2 = 4x^2 - 20x + 25$ ■

A WORD OF CAUTION Students remember that

$(ab)^2 = a^2b^2$ from using the rules for exponents.

↑ Here a and b are *factors*

They try to apply this rule of exponents to the expression $(a + b)^2$. But

$(a + b)^2$ cannot be found simply by squaring a and b.

↑ Here a and b are *terms*

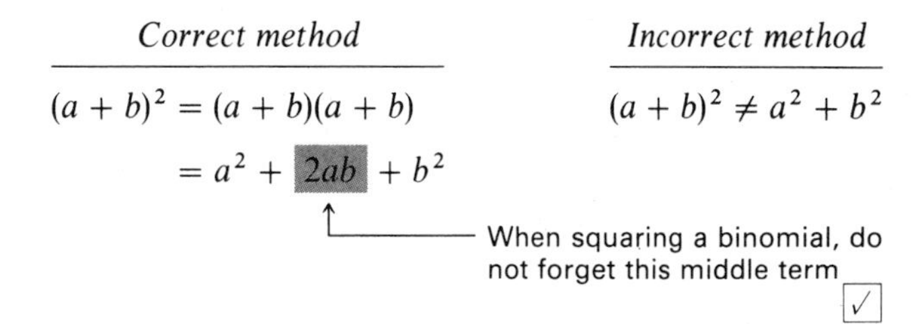

The Product of the Sum and Difference of Two Terms

Example 8

sum difference F O I L

$$(a + b)(a - b) = a^2 - ab + ab - b^2$$

$$= a^2 - b^2$$ ■

Notice that the inner and outer terms add to zero. The product is the difference of two squares.

> THE PRODUCT OF THE SUM AND DIFFERENCE OF TWO TERMS
>
> Is equal to the square of the first term minus the square of the second term:
>
> $$(a + b)(a - b) = a^2 - b^2$$

Example 9 The product of the sum and difference of two terms

a. $(x + 4)(x - 4) = (x)^2 - (4)^2 = x^2 - 16$

b. $(2x + 3y)(2x - 3y) = (2x)^2 - (3y)^2 = 4x^2 - 9y^2$

c. $(m^2 - 5n^2)(m^2 + 5n^2) = (m^2)^2 - (5n^2)^2 = m^4 - 25n^4$ ■

EXERCISES 8.7

Find the following products by inspection.

1. $(x + 1)(x + 4)$
2. $(x + 3)(x + 1)$
3. $(a + 5)(a + 2)$
4. $(a + 7)(a + 1)$
5. $(m - 4)(m + 2)$
6. $(n - 3)(n + 7)$
7. $(8 + y)(9 - y)$
8. $(3 - z)(10 + z)$
9. $(x + 2)(x - 2)$
10. $(x + 3)(x - 3)$
11. $(h - 6)(h + 6)$
12. $(w - 7)(w + 7)$
13. $(a + 4)^2$
14. $(y - 5)^2$
15. $(x + 3)^2$
16. $(x + 5)^2$
17. $(4 - b)^2$
18. $(6 - b)^2$
19. $(3x + 4)(2x - 5)$
20. $(2y + 5)(4y - 3)$
21. $(2x - 3y)(2x + 3y)$
22. $(3x + 4y)(3x - 4y)$
23. $(2a + 5b)(a + b)$
24. $(3c + 2d)(c + d)$
25. $(4x - y)(2x + 7y)$
26. $(3x - 2y)(4x + 5y)$
27. $(7x - 10y)(7x - 10y)$
28. $(4u - 9v)(4u - 9v)$
29. $(3x + 4)^2$
30. $(2x + 5)^2$
31. $(5m - 2n)^2$
32. $(7a - 3b)^2$
33. $(x^2 + y^2)(2x^2 + 3y^2)$
34. $(4a^2 + b^2)(a^2 + 5b^2)$
35. $(m^2 - 3n^2)(4m^2 + 2n^2)$
36. $(8x^2 + y^2)(2x^2 - 3y^2)$
37. $(3x^2 + y^2)(3x^2 - y^2)$
38. $(a^2 - 4b^2)(a^2 + 4b^2)$
39. $(x^2 + y^2)^2$
40. $(2x^2 - 3y^2)^2$
41. $\left(\frac{2x}{3} + 1\right)\left(\frac{2x}{3} - 1\right)$
42. $\left(1 + \frac{3x}{5}\right)^2$
43. $\left(2 - \frac{x}{3}\right)^2$
44. $\left(\frac{3x}{4} - 3\right)\left(\frac{3x}{4} + 3\right)$

8.8 Division of Polynomials

8.8A Division of a Polynomial by a Monomial

Consider the example:

$$\frac{4x^3 - 6x^2}{2x} = \frac{1}{2x} \cdot \frac{4x^3 - 6x^2}{1}$$

$$= \frac{1}{2x}(4x^3 - 6x^2)$$

$$= \left(\frac{1}{2x}\right)(4x^3) + \left(\frac{1}{2x}\right)(-6x^2) \quad \text{By the distributive rule}$$

$$= \frac{4x^3}{2x} + \frac{-6x^2}{2x} \quad \text{Dividing each term of the polynomial by the monomial}$$

$$= 2x^2 - 3x$$

> TO DIVIDE A POLYNOMIAL BY A MONOMIAL
>
> Divide *each* term in the polynomial by the monomial, then add the results.

Example 1 Dividing a polynomial by a monomial

a. $\dfrac{4x+2}{2} = \dfrac{4x}{2} + \dfrac{2}{2} = 2x + 1$

b. $\dfrac{9x^3 - 6x^2 + 12x}{3x} = \dfrac{9x^3}{3x} + \dfrac{-6x^2}{3x} + \dfrac{12x}{3x} = 3x^2 - 2x + 4$

c. $\dfrac{4x^2 - 8x + 16}{-4x} = \dfrac{4x^2}{-4x} + \dfrac{-8x}{-4x} + \dfrac{16}{-4x} = -x + 2 - \dfrac{4}{x}$

d. $\dfrac{4a^2b^2c^2 - 6ab^2c^2 + 12abc^2}{-6abc} = \dfrac{4a^2b^2c^2}{-6abc} + \dfrac{-6ab^2c^2}{-6abc} + \dfrac{12abc^2}{-6abc}$

$= -\dfrac{2}{3}abc + bc - 2c$ ■

EXERCISES 8.8A

Perform the indicated divisions.

1. $\dfrac{3x+6}{3}$ **2.** $\dfrac{10x+15}{5}$ **3.** $\dfrac{4+8x}{4}$ **4.** $\dfrac{5-10x}{5}$

5. $\dfrac{6x-8y}{-2}$ **6.** $\dfrac{5x-10y}{-5}$ **7.** $\dfrac{2x^2+3x}{x}$ **8.** $\dfrac{4y^2-3y}{y}$

9. $\dfrac{3a^2b - ab}{ab}$ **10.** $\dfrac{5mn^2 - mn}{mn}$ **11.** $\dfrac{12z^3 - 16z^2 + 8z}{4z}$

12. $\dfrac{-15a^4 + 12a^3 - 18a^2}{3a^2}$ **13.** $\dfrac{5x^3 - 4x^2 + 10}{5x^2}$ **14.** $\dfrac{7y^3 - 5y^2 + 14}{7y^2}$

15. $\dfrac{-15x^2y^2z^2 - 30xyz}{-5xyz}$ **16.** $\dfrac{-24a^2b^2c^2 - 16abc}{-8abc}$ **17.** $\dfrac{13x^2y^2 - 26x^2y^3 + 39x^2y^2}{13x^2y^2}$

18. $\dfrac{21m^2n^3 - 35m^3n^2 - 14m^2n^2}{7m^2n^2}$ **19.** $\dfrac{10x^7y^4 + 18x^4y^2 + 2x^3y}{2x^3y}$ **20.** $\dfrac{24a^6b^6 - 8a^3b^2 + 4ab^2}{4ab^2}$

8.8B Division of a Polynomial by a Polynomial

The method used to divide a polynomial by a polynomial is like long division of whole numbers in arithmetic.

Example 2 $966 \div 23$

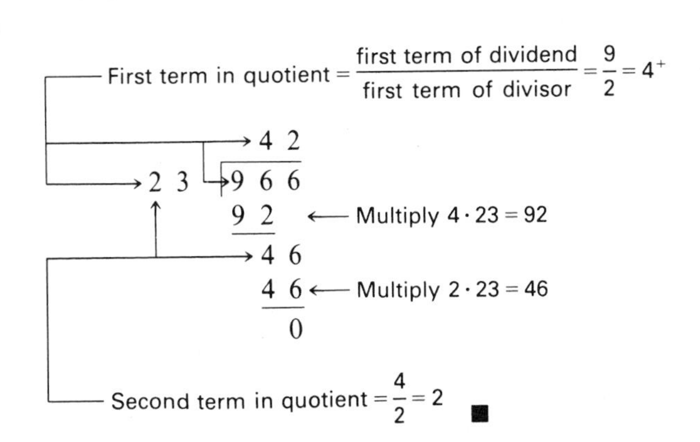

■

Example 3 $(x^2 - 3x - 10) \div (x + 2)$

First term in quotient $= \dfrac{x^2}{x} = x$

Step 1

$$\begin{array}{r|l} & x \\ \hline x + 2 & x^2 - 3x - 10 \end{array}$$

Step 2 Multiply

$$\begin{array}{r|l} & x \\ \hline x + 2 & x^2 - 3x - 10 \\ & x^2 + 2x \end{array}$$

$x^2 + 2x \longleftarrow$ Multiply $x(x + 2) = x^2 + 2x$

Step 3

$$\begin{array}{r|l} & x \\ \hline x + 2 & x^2 - 3x - 10 \\ & \underline{-x^2 \mp 2x} \\ & \quad -5x - 10 \end{array}$$

To subtract *change* all signs and *add*

Bring down next term

Step 4

Second term in quotient $= \dfrac{-5x}{x} = -5$

$$\begin{array}{r|l} & x - 5 \\ \hline x + 2 & x^2 - 3x - 10 \\ & \underline{-x^2 \mp 2x} \\ & \quad -5x - 10 \end{array}$$

Step 5 Multiply

$$\begin{array}{r|l} & x - 5 \\ \hline x + 2 & x^2 - 3x - 10 \\ & \underline{-x^2 \mp 2x} \\ & \quad -5x - 10 \\ & \quad \underline{-5x - 10} \end{array}$$

$-5x - 10 \longleftarrow$ Multiply $-5(x + 2) = -5x - 10$

Step 6

$$\begin{array}{r|l} & x - 5 \\ \hline x + 2 & x^2 - 3x - 10 \\ & \underline{-x^2 \mp 2x} \\ & \quad -5x - 10 \\ & \quad \underline{\pm 5x \pm 10} \\ & \qquad\quad 0 \end{array}$$

Change signs and *add* ■

Example 4 $(6x^2 + x - 10) \div (2x + 3)$

$$\begin{array}{r|l} & 3x - 4 \quad R\ 2 \\ \hline 2x + 3 & 6x^2 + x - 10 \\ & \underline{-6x^2 \mp 9x} \\ & \quad -8x - 10 \\ & \quad \underline{\pm 8x \pm 12} \\ & \qquad\quad 2 \end{array}$$

Remainder ■

or $\quad 3x - 4 + \dfrac{2}{2x + 3}$

Example 5 $(27x - 19x^2 + 6x^3 + 10) \div (5 - 3x)$

The terms of the dividend and divisor should be arranged in *descending powers* of the variable *before* beginning the division.

$$
\begin{array}{r|l}
 & -2x^2 + 3x - 4 \quad \text{R } 30 \\ \hline
-3x+5 & 6x^3 - 19x^2 + 27x + 10 \\
 & \boxed{-}\ 6x^3 \overset{\boxed{+}}{-} 10x^2 \\ \cline{2-2}
 & \qquad -9x^2 + 27x \\
 & \qquad \overset{\boxed{+}}{-} 9x^2 \overset{\boxed{-}}{+} 15x \\ \cline{2-2}
 & \qquad\qquad 12x + 10 \\
 & \qquad\qquad \boxed{-}\ 12x \overset{\boxed{+}}{-} 20 \\ \cline{2-2}
 & \qquad\qquad\qquad 30
\end{array}
$$

Therefore, the answer is $-2x^2 + 3x - 4 \quad \text{R } 30$

or $-2x^2 + 3x - 4 + \dfrac{30}{-3x+5}$ ■

Example 6 $(17ab^2 + 12a^3 - 10b^3 - 11a^2b) \div (3a - 2b)$

Arrange the terms of the dividend and divisor in descending powers of a before beginning the division.

$$
\begin{array}{r|l}
 & 4a^2 - ab + 5b^2 \\ \hline
3a-2b & 12a^3 - 11a^2b + 17ab^2 - 10b^3 \\
 & \boxed{-}\ 12a^3 \overset{\boxed{+}}{-} 8a^2b \\ \cline{2-2}
 & \qquad -3a^2b + 17ab^2 \\
 & \qquad \overset{\boxed{+}}{-} 3a^2b \overset{\boxed{-}}{+} 2ab^2 \\ \cline{2-2}
 & \qquad\qquad 15ab^2 - 10b^3 \\
 & \qquad\qquad \boxed{-}\ 15ab^2 \overset{\boxed{+}}{-} 10b^3 \\ \cline{2-2}
 & \qquad\qquad\qquad 0
\end{array}
$$
■

Example 7 $(x^3 - 1) \div (x - 1)$

$$
\begin{array}{r|l}
 & x^2 + x + 1 \\ \hline
x-1 & x^3 + \boxed{0x^2} + \boxed{0x} - 1 \\
 & \boxed{-}\ x^3 \overset{\boxed{+}}{-} x^2 \\ \cline{2-2}
 & \qquad x^2 + 0x \\
 & \qquad \boxed{-}\ x^2 \overset{\boxed{+}}{-} x \\ \cline{2-2}
 & \qquad\qquad x - 1 \\
 & \qquad\qquad \boxed{-}\ x \overset{\boxed{+}}{-} 1 \\ \cline{2-2}
 & \qquad\qquad\qquad 0
\end{array}
$$
■

It is helpful to leave space for missing powers by using zeros in this way

Example 8 $(2x^4 + x^3 - 8x^2 - 5x - 2) \div (x^2 - x - 2)$

When the divisor is a polynomial of more than two terms, exactly the same procedure is used.

$$
\begin{array}{r|l}
 & 2x^2 + 3x - 1 \quad \text{R } -4 \\ \hline
x^2-x-2 & 2x^4 + x^3 - 8x^2 - 5x - 2 \\
 & \boxed{-}\ 2x^4 \overset{\boxed{+}}{-} 2x^3 \overset{\boxed{+}}{-} 4x^2 \\ \cline{2-2}
 & \qquad 3x^3 - 4x^2 - 5x \\
 & \qquad \boxed{-}\ 3x^3 \overset{\boxed{+}}{-} 3x^2 \overset{\boxed{+}}{-} 6x \\ \cline{2-2}
 & \qquad\qquad -x^2 + x - 2 \\
 & \qquad\qquad \overset{\boxed{+}}{-} x^2 \overset{\boxed{-}}{+} x \overset{\boxed{-}}{+} 2 \\ \cline{2-2}
 & \qquad\qquad\qquad -4
\end{array}
$$
■

EXERCISES 8.8B

Perform the indicated divisions.

1. $(x^2 + 5x + 6) \div (x + 2)$
2. $(x^2 + 5x + 6) \div (x + 3)$
3. $(x^2 - 9x + 20) \div (x - 4)$
4. $(x^2 - x - 12) \div (x + 3)$
5. $(6x^2 + 5x - 6) \div (3x - 2)$
6. $(20x^2 + 13x - 15) \div (5x - 3)$
7. $(15v^2 + 19v + 10) \div (5v - 7)$
8. $(15v^2 + 19v - 4) \div (3v + 8)$
9. $(8a^2 - 2ab - b^2) \div (2a - b)$
10. $(8a^2 - 2ab - b^2) \div (4a + b)$
11. $(6a^2 + 5ab + b^2) \div (2a + 3b)$
12. $(6a^2 + 5ab - b^2) \div (3a - 2b)$
13. $(x^3 - 10x + 24) \div (x + 4)$
14. $(x^3 - 2x - 4) \div (x - 2)$
15. $(8x - 4x^3 + 10) \div (2 - x)$
16. $(12x - 15 - x^3) \div (3 - x)$
17. $(a^3 - 8) \div (a - 2)$
18. $(c^3 - 27) \div (c - 3)$
19. $(4x^3 - 4x^2 + 5x + 4) \div (2x + 1)$
20. $(3x^3 - 5x^2 + 11x - 6) \div (3x - 2)$
21. $(5 + 3x^3 + 9x - 17x^2) \div (x - 5)$
22. $(12 - 8x + 2x^3 - 11x^2) \div (x - 6)$
23. $(x^4 + 2x^3 - x^2 - 2x + 4) \div (x^2 + x - 1)$
24. $(x^4 - 2x^3 + 3x^2 - 2x + 7) \div (x^2 - x + 1)$

8.9 Chapter Summary

Terms 8.1 The + and − signs in an algebraic expression break it up into smaller pieces called **terms**. A minus sign is part of the term that follows it. *Exception*: An expression within grouping symbols is considered as a single piece even though it may contain + and − signs.

Like Terms 8.4 Terms having the same variables with the same exponents are called **like terms.**

Factors 8.1 The numbers that are multiplied together to give a product are called the **factors** of that product.

Coefficients 8.1 The **coefficient** of one factor in a term is the product of all the other factors in that term.

Numerical Coefficients 8.1 A **numerical coefficient** is a coefficient that is a number. If we say *the coefficient* of a term, it is understood to mean the *numerical* coefficient of that term.

Distributive Rule 8.3

$$a(b + c) = ab + ac$$

When a factor is multiplied by an expression enclosed within grouping symbols, the factor must be multiplied by each term within the grouping symbols, then those products are added.

To Combine Like Terms 8.4

1. Identify the like terms.
2. Find the sum of each group of like terms by adding their numerical coefficients, then multiply that sum by the common variables.

Rules for Positive Exponents 8.2

1. $x^a x^b = x^{a+b}$
2. $(x^a)^b = x^{ab}$
3. $\dfrac{x^a}{x^b} = x^{a-b}, \quad (x \neq 0)$

Polynomials 8.1 A **polynomial** *in* x is an algebraic expression having only terms of the form ax^n, where a is any real number and n is a whole number. A *polynomial in* x *and* y is an algebraic expression having only terms of the form ax^ny^m, where a is any real number and n, m are whole numbers.

A *monomial* is a polynomial of one term.

A *binomial* is a polynomial of two unlike terms.

A *trinomial* is a polynomial of three unlike terms.

The Degree of a Term 8.1 *The degree of a term* in a polynomial is the sum of the exponents of its variables. *The degree of a polynomial* is the same as that of its highest-degree term (provided like terms have been combined).

To Add Polynomials 8.5 To add polynomials, remove grouping symbols, then combine like terms.

To Subtract Polynomials 8.5 To subtract polynomials, change the sign of each term in the polynomial being subtracted, then add the resulting polynomials.

To Multiply a Polynomial by a Monomial 8.6 To multiply a polynomial by a monomial, multiply each term in the polynomial by the monomial, then add the results.

To Multiply a Polynomial by a Polynomial 8.6 To multiply a polynomial by a polynomial, multiply each term of the first polynomial by each term of the second polynomial, then add the results.

To Multiply Two Binomials 8.7

1. The first term of the product is the product of the first terms of the binomials.
2. The middle term of the product is the sum of the inner and outer products.
3. The last term of the product is the product of the last terms of the binomials.

To Square a Binomial 8.7

$$(a+b)^2 = a^2 + 2ab + b^2$$

$$(a-b)^2 = a^2 - 2ab + b^2$$

Product of Sum and Difference of Two Terms 8.7

$$(a+b)(a-b) = a^2 - b^2$$

To Divide a Polynomial by a Monomial 8.8A To divide a polynomial by a monomial, divide each term in the polynomial by the monomial, then add the results.

To Divide a Polynomial by a Polynomial 8.8B See Section 8.8B for the method.

Review Exercises 8.9

In Exercises 1–3, (a) determine the number of terms; (b) write the second term if there is one.

1. $a^2 + 2ab + b^2$ **2.** $6x - 2(x^2 + y^2)$ **3.** $\dfrac{2x+y}{3} - 4x$

4. Write the polynomial $3x^4 - 6 + 7x^2 - x$ in descending powers of x.

5. In the polynomial $2xy + 3x^2y$, find:
a. the degree of the first term
b. the degree of the polynomial

In Exercises 6–11, use the rules of exponents to simplify each expression.

6. $x \cdot x^3$ **7.** $m^6 \cdot m^3$ **8.** $(m^6)^3$

9. $\dfrac{m^6}{m^3}$ **10.** $2x^2 \cdot 3x^3$ **11.** $\dfrac{8x^8}{2x^2}$

In Exercises 12–20, simplify the following addition and subtraction problems.

12. $(5x^2y + 3xy^2 - 4y^3) + (2xy^2 + 4y^3 + 3x^2y)$

13. $(7ab^2 + 3a^3 - 5) + (6a^3 + 10 - 10ab^2)$

14. $(5x^2y + 3xy^2 - 4 + y^2) - (8 - 4x^2y + 2xy^2 - y^2)$

15. $(8a + 5a^2b - 3 + 6ab^2) - (5 + 4a - 3ab^2 + a^2b)$

16. $(4a^2 - 5a) - (6a + 3) + (2a^2 + 5)$

17. $(2x + 9) - [(3x - 5) - (2 - x)]$

18. Subtract $(3y^2 - 5y + 1)$ from $(5y^2 - 6y - 4)$

19. Add:
$$\begin{array}{rrrr} 3a^3 & +4a^2 & & -3 \\ -6a^3 & & -7a & +5 \\ 5a^3 & -9a^2 & -4a & -6 \\ \hline \end{array}$$

20. Subtract:
$$\begin{array}{rrr} 5x^2y^2 & -\ x^2y & -3xy \\ 7x^2y^2 & +4x^2y & -8xy \\ \hline \end{array}$$

In Exercises 21–35, find the products.

21. $(9x^8y^3)(-7x^4y^5)$

22. $(-5ab^2)(-a^7b^3)(4a^2b)$

23. $2x(3x^2 - 4x)$

24. $5x^2y(3xy^2 + 4x - 2z)$

25. $(2a - 4)(3a + 5)$

26. $(3x + 4)^2$

27. $(x - 5y)(x + 5y)$

28. $(m - 3)(4m - 6)$

29. $(3u + 1)(5u + 2)$

30. $(7x + 4)(7x - 4)$

31. $(x^2 - 6)^2$

32. $(5x - 4y)(7x + 2y)$

33. $(x^2 + y^2)(2x^2 - 3y^2)$

34. $\left(\dfrac{2x}{3} + 4\right)^2$

35. $(x - 2)(x^2 + 2x + 4)$

In Exercises 36–41, perform the divisions.

36. $\dfrac{3x^2y - 6xy^2}{3xy}$

37. $\dfrac{-15a^2b^3 + 4ab^2 - 10ab}{-5ab}$

38. $(4x^2 - 1) \div (2x + 1)$

39. $(6x^2 - 9x + 10) \div (2x - 5)$

40. $(10a^2 + 23ab - 5b^2) \div (5a - b)$

41. $(2a^4 - a^3 + a^2 + a - 3) \div (2a^2 - a + 3)$

Chapter 8 Diagnostic Test

Allow yourself about 50 minutes to do these problems. Complete solutions for every problem, together with section references, are given in the answer section at the end of the book.

If a problem has an answer that is a polynomial in one letter, write that answer in descending powers.

1. In the polynomial $x^2 - 4xy^2 + 5$, find:
   a. the degree of the first term
   b. the degree of the polynomial
   c. the numerical coefficient of the second term

2. Subtract $(10 - z + 2z^2)$ from $(-4z^2 - 5z + 10)$

3. Add:
$$\begin{array}{rrrr} -5x^3 & +2x^2 & & -5 \\ 7x^3 & & +5x & -8 \\ & 3x^2 & -6x & +10 \\ \hline \end{array}$$

4. Subtract:
$$\begin{array}{r} 9a^2b - 2ab^2 - 3ab \\ 4a^2b + 6ab^2 - 7ab \\ \hline \end{array}$$

In Problems 5–8, simplify the addition and subtraction problems.

5. $(6x^2 - 5 - 11x) + (15x - 4x^2 + 7)$

6. $(3xy^2 - 4xy) - (6xy - 3xy^2) + (4xy + x^3)$

7. $(2m^2 - 5) - [(7 - m^2) - (4m^2 - 3)]$

8. Subtract $(5x^2 - 9x + 6)$ from the sum of $(2x - 8 - 7x^2)$ and $(12 - 2x^2 + 11x)$.

In Problems 9–13, use the rules of exponents to simplify.

9. $10^2 \cdot 10^3$

10. $(a^3)^5$

11. $\dfrac{12x^6}{2x^2}$

12. $x^3 \cdot x^5$

13. $\dfrac{2^8}{2^5}$

In Problems 14–22, find the products.

14. $(-3xy)(5x^3y)(-2xy^4)$

15. $4x(3x^2 - 2)$

16. $-2ab(5a^2 - 3ab^2 + 4b)$

17. $(2x - 5)(3x + 4)$

18. $(2y - 3)^2$

19. $(m + 3)(2m + 6)$

20. $(5a + 2)(5a - 2)$

21. $(4x - 2y)(3x - 5y)$

22. $(w - 3)(w^2 + 3w + 8)$

In Problems 23–25, perform the divisions.

23. $\dfrac{6x^3 - 4x^2 + 8x}{2x}$

24. $(10x^2 + x - 5) \div (2x - 1)$

25. $(2x^3 - 20x - 6) \div (x + 3)$

9 Equations and Inequalities

The main reason for studying algebra is to equip oneself with the tools necessary for solving problems. Most problems are solved by the use of equations. Some problems are solved by using inequalities.

In this chapter we show how to solve simple equations and inequalities. Methods for solving more difficult equations and inequalities will be given in later chapters.

9.1 Solving Equations Having Only One Number on the Same Side as the Variable

Equation

In algebra, an **equation** is a statement that two algebraic expressions are equal.

The Parts of an Equation

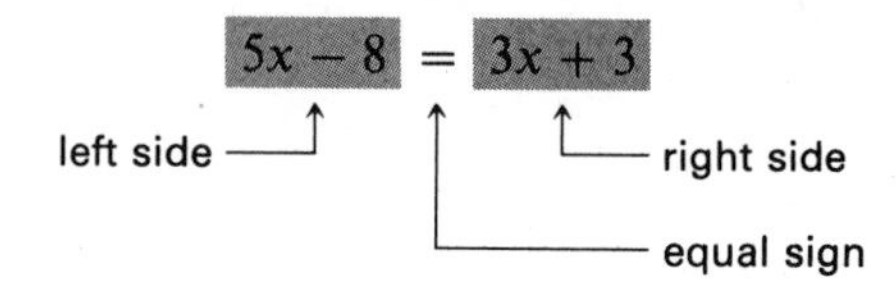

An equation is made up of three parts:

1. The equal sign (=).
2. The expression to the left of the = sign, called the left side (or left member) of the equation.
3. The expression to the right of the = sign, called the right side (or right member) of the equation.

The equal sign (=) in an equation means that the number represented by the left side *is the same* as the number represented by the right side.

Solution

A **solution** of an equation is a number that, when put in place of the variable, makes a true statement.

If x is replaced with 3 in the equation

$$x + 2 = 5$$

we have $3 + 2 = 5$, which is a true statement.

Therefore, 3 is a solution of the equation.

TO CHECK A SOLUTION OF AN EQUATION

1. Replace the variable in the given equation by the solution in parentheses.
2. Perform the indicated operations on both sides of the = sign.
3. If the resulting number on each side of the = sign is the same, the solution is correct.

Example 1 Is -3 a solution of the equation $5 - x = 2$?

Solution

$$5 - x = 2$$

$$5 - (-3) \stackrel{?}{=} 2 \qquad x \text{ was replaced with } (-3)$$

$$5 + 3 \stackrel{?}{=} 2$$

$$8 \neq 2 \qquad \text{False}$$

Therefore, -3 is *not* a solution of $5 - x = 2$. ■

Example 2 Is -4 a solution of the equation $2x + 3 = -5$?

Solution

$$2x + 3 = -5$$

$$2(-4) + 3 \stackrel{?}{=} -5 \qquad x \text{ was replaced with } (-4)$$

$$-8 + 3 \stackrel{?}{=} -5$$

$$-5 = -5 \qquad \text{True}$$

Therefore, -4 *is* a solution of $2x + 3 = -5$. ■

Solving Equations

To *solve an equation* means to find the value of the variable that makes the statement true. An equation is solved when we succeed in getting the variable by itself on one side of the equal sign and only a single number on the other side.

If we add the same signed number to both sides of an equation, the resulting sums must be equal.

Example 3

$$\begin{array}{rcr} 7 & = & 7 \\ +5 & & +5 \\ \hline 12 & = & 12 \end{array}$$

We are adding $+5$ to both sides

The resulting sums are equal ■

If we subtract the same signed number from both sides of an equation, the resulting differences must be equal.

Example 4

$$\begin{array}{rcr} 7 & = & 7 \\ -5 & & -5 \\ \hline 2 & = & 2 \end{array}$$

We are subtracting 5 from both sides

The resulting differences are equal

We could also work Example 4 by *adding* -5 to both sides. ■

If we multiply both sides of an equation by the same signed number, the resulting products must be equal.

Example 5

$$4 = 4$$

$$2(4) = 2(4) \qquad \text{Multiplying both sides by 2}$$

$$8 = 8 \qquad \text{The resulting products are equal}$$ ■

If we divide both sides of an equation by the same signed number (not zero), the resulting quotients must be equal.

Example 6

$$6 = 6$$

$$\frac{6}{-2} = \frac{6}{-2}$$ We are dividing both sides by −2

$$-3 = -3$$ The resulting quotients are equal ■

We summarize these results in the following box.

FOUR BASIC RULES USED IN SOLVING EQUATIONS

1. *Addition rule:* The same number may be added to both sides.
2. *Subtraction rule:* The same number may be subtracted from both sides.
3. *Multiplication rule:* Both sides may be multiplied by the same nonzero number.
4. *Division rule:* Both sides may be divided by the same nonzero number.

To know which rule to use in solving a particular equation, we make use of inverse operations.

Addition and subtraction are inverse operations.

Multiplication and division are inverse operations.

If you have $x - 3$, this is a subtraction of 3. The -3 can be removed by its inverse: the addition of 3. (Example 7)

If you have $x + 5$, this is an addition of 5. The $+5$ can be removed by its inverse: the subtraction of 5. (Example 8)

If you have $7 \cdot x$, this is a multiplication by 7. The $7 \cdot$ can be removed by its inverse: a division by 7. (Example 9)

If you have $\frac{x}{3}$, this is a division by 3. The division by 3 can be removed by its inverse: a multiplication by 3. (Example 10)

Example 7 Solve $x - 3 = 6$.

Solution

Remove by an addition of 3

$$\begin{aligned} x - 3 &= 6 \\ +3 &\quad +3 \\ x &= 9 \end{aligned}$$

Adding 3 to both sides gets x by itself on the left side

Check for x = 9

$$x - 3 = 6$$

$$(9) - 3 \stackrel{?}{=} 6$$

$$6 = 6$$ True, therefore, the solution is correct ■

Example 8 Solve $x + 5 = 3$.

Remove by a subtraction of 5

Solution

$$\begin{array}{rcr} x + 5 & = & 3 \\ -5 & & -5 \\ \hline x & = & -2 \end{array}$$

Check for x = −2

$$x + 5 = 3$$

$$(-2) + 5 \stackrel{?}{=} 3$$

$$3 = 3$$ The solution is correct ■

Example 9 Solve $7x = 56$.

Remove by a division by 7

Solution

$$7x = 56$$

$$\frac{\cancel{7}x}{\cancel{7}} = \frac{56}{7}$$ Dividing both sides by 7

$$x = 8$$ Solution

Check for x = 8

$$7x = 56$$

$$7(8) \stackrel{?}{=} 56$$

$$56 = 56$$ ■

Example 10 Solve $\frac{x}{3} = -5$.

Remove by a multiplication of 3

Solution

$$\frac{x}{3} = -5$$

$$3\left(\frac{x}{3}\right) = 3(-5)$$ Multiplying both sides by 3

$$\cancel{3}\left(\frac{x}{\cancel{3}}\right) = 3(-5)$$

$$x = -15$$ Solution

Check for x = −15

$$\frac{x}{3} = -5$$

$$\frac{-15}{3} \stackrel{?}{=} -5$$

$$-5 = -5$$ ■

In Examples 11 and 12, the variable x is on the right side of the equation.

Example 11 Solve $-6 = 2 + x$.

Solution Because 2 is being added to x, we must subtract 2 from both sides to get x by itself on the right side.

$$\begin{array}{rr} -6 = & 2 + x \\ \underline{-2} & \underline{-2} \\ -8 = & x \end{array} \quad \text{Solution}$$

Check for x = −8

$$-6 = 2 + x$$
$$-6 \stackrel{?}{=} 2 + (-8)$$
$$-6 = -6 \quad \blacksquare$$

Example 12 Solve $-6 = 2x$.

Solution Because 2 is being multiplied by x, we must divide both sides by 2 to get x by itself.

$$-6 = 2x$$
$$\frac{-6}{2} = \frac{2x}{2}$$
$$-3 = x \quad \text{Solution}$$

Check for x = −3

$$-6 = 2x$$
$$-6 \stackrel{?}{=} 2(-3)$$
$$-6 = -6 \quad \blacksquare$$

EXERCISES 9.1

1. Is -3 a solution of the equation $x + 8 = 5$?

2. Is 2 a solution of the equation $x - 6 = -4$?

3. Is -4 a solution of the equation $3x = 12$?

4. Is -6 a solution of the equation $\frac{x}{2} = -3$?

5. Is 2 a solution of the equation $3x - 8 = -2$?

6. Is -5 a solution of the equation $2x + 4 = 6$?

In Exercises 7–50, solve and check the following equations.

7. $x + 5 = 8$
8. $x + 4 = 9$
9. $x - 3 = 4$
10. $x - 7 = 2$
11. $3 + x = -4$
12. $2 + x = -5$
13. $2x = 8$
14. $3x = 15$
15. $21 = 7x$
16. $42 = 6x$
17. $11x = -33$
18. $12x = -48$
19. $x - 35 = 7$
20. $x - 42 = 9$
21. $9 = x + 5$
22. $11 = x + 8$
23. $12 = x - 11$
24. $14 = x - 15$
25. $\frac{x}{3} = 4$
26. $\frac{x}{5} = 3$
27. $\frac{x}{5} = -2$
28. $\frac{x}{6} = -4$
29. $4 = \frac{x}{7}$
30. $3 = \frac{x}{8}$
31. $-28 = -15 + x$
32. $-47 = -18 + x$
33. $-17 + x = 28$
34. $-14 + x = 33$
35. $5.6 + x = 2.8$
36. $3.04 + x = 2.96$
37. $5x = 35$
38. $-24 = 6x$
39. $12x = -36$
40. $54 = 18x$
41. $-13 = \frac{x}{9}$
42. $-15 = \frac{x}{8}$
43. $\frac{x}{10} = 3.14$
44. $\frac{x}{5} = 7.8$
45. $0.25x = 2$
46. $0.4x = 16$
47. $x - \frac{1}{2} = \frac{2}{5}$
48. $x - \frac{3}{5} = \frac{1}{3}$
49. $x + \frac{1}{3} = \frac{3}{4}$
50. $x + \frac{2}{9} = \frac{5}{6}$

9.2 Solving Equations Having Two Numbers on the Same Side as the Variable

TO SOLVE AN EQUATION WHEN TWO NUMBERS APPEAR ON THE SAME SIDE AS THE VARIABLE

All numbers that appear on the same side as the variable must be removed.

1. Remove those numbers being added or subtracted by using the inverse operation.
2. Remove the number multiplied by the variable by dividing both sides by that signed number.

Example 1 Solve the equation $2x - 3 = 11$.

Solution The numbers 2 and -3 must be removed from the side containing the x.

1. Because the 3 is subtracted, it is removed first.

$$\begin{aligned} 2x - 3 &= 11 \\ +3 &\quad +3 \qquad \text{Adding 3 to both sides} \\ \hline 2x &= 14 \end{aligned}$$

2. Because the 2 is multiplied by the x, it is removed by dividing both sides by 2.

$$2x = 14$$

$$\frac{\cancel{2}x}{\cancel{2}} = \frac{14}{2} \qquad \text{Dividing both sides by 2}$$

$$x = 7$$

Check for x = 7

$$2x - 3 = 11$$

$$2(7) - 3 \stackrel{?}{=} 11$$

$$14 - 3 \stackrel{?}{=} 11$$

$$11 = 11 \quad \blacksquare$$

Example 2 Solve the equation $-12 = 3x + 15$.

Solution 15 and 3 must be removed from the side with the x.

1. $$\begin{aligned} -12 &= 3x + 15 \\ -15 &\quad\;\; -15 \qquad \text{Subtracting 15 from both sides} \\ \hline -27 &= 3x \end{aligned}$$

2. $$\frac{-27}{3} = \frac{\cancel{3}x}{\cancel{3}} \qquad \text{Dividing both sides by 3}$$

$$-9 = x$$

Check for x = −9

$$-12 = 3x + 15$$

$$-12 \stackrel{?}{=} 3(-9) + 15$$

$$-12 \stackrel{?}{=} -27 + 15$$

$$-12 = -12 \quad \blacksquare$$

EXERCISES 9.2

Solve and check the following equations.

1. $4x + 1 = 9$	**2.** $5x + 2 = 12$	**3.** $6x - 2 = 10$	**4.** $7x - 3 = 4$
5. $2x - 15 = 11$	**6.** $3x - 4 = 14$	**7.** $4x + 2 = -14$	**8.** $5x + 5 = -10$
9. $14 = 9x - 13$	**10.** $25 = 8x - 15$	**11.** $12x + 17 = 65$	**12.** $11x + 19 = 41$
13. $8x - 23 = 31$	**14.** $6x - 33 = 29$	**15.** $-14 + 4x = 28$	**16.** $-18 + 6x = 44$
17. $-8 = 3x - 25$	**18.** $-10 = 2x - 27$	**19.** $-73 = 24x + 31$	**20.** $-48 = 36x + 42$
21. $18x - 4.8 = 6$	**22.** $15x - 7.5 = 8$	**23.** $-19 + 10x = 26$	**24.** $12x - 17 = 13$
25. $-67 = 18x + 29$	**26.** $-27 = 15x - 32$	**27.** $-2x + 7 = -13$	**28.** $-3x - 8 = 16$
29. $0.3x - 1.8 = -5.04$	**30.** $0.5x + 2.05 = -1.6$		

9.3 Solving Equations in Which the Variable Appears on Both Sides

All the equations previously discussed in this chapter have had the variable on only *one side* of the equation.

TO SOLVE AN EQUATION IN WHICH THE VARIABLE APPEARS ON BOTH SIDES

1. Combine like terms on each side of the equation (if there are any).
2. Remove from one side the term containing the variable by using the inverse operation.
3. Solve the resulting equation by the method given in the last section (Section 9.2).

Example 1 Solve $6x - 15 = -23 + 2x$.

Solution

1. We first remove *the entire term* $2x$ from the right side.

$$\begin{array}{rcl} 6x - 15 &=& -23 + 2x \\ \underline{-2x } && \underline{ - 2x} \\ 4x - 15 &=& -23 \end{array}$$

Subtracting $2x$ from both sides

2. The numbers 4 and -15 must be removed from the side containing the x (-15 first, 4 second).

$$\begin{array}{rcl} 4x - 15 &=& -23 \\ \underline{ + 15} && \underline{+15} \\ 4x &=& -\ 8 \end{array}$$

Adding 15 to both sides

$$\frac{\cancel{4}x}{\cancel{4}} = \frac{-8}{4}$$

Dividing both sides by 4

$$x = -2$$

Check for x = −2

$$6x - 15 = -23 + 2x$$
$$6(-2) - 15 \stackrel{?}{=} -23 + 2(-2)$$
$$-12 - 15 \stackrel{?}{=} -23 - 4$$
$$-27 = -27$$

The answer is obtained whether the x-term is removed from the left side or the right side. We now solve Example 1 by removing the x-term from the *left* side.

$$\begin{array}{rcl} 6x - 15 &=& -23 + 2x \\ \underline{-6x \quad\quad} & & \underline{\quad\quad - 6x} \\ -15 &=& -23 - 4x \\ \underline{+23} & & \underline{+23 \quad\quad} \\ 8 &=& -4x \end{array}$$

$$\frac{8}{-4} = \frac{-4x}{-4} \quad \text{Dividing both sides by } -4$$

$$-2 = x \quad \text{Same answer} \quad ■$$

Example 2 Solve $4x - 5 - x = 13 - 2x - 3$.

Solution

$$4x - 5 - x = 13 - 2x - 3 \quad \text{First combine like terms on each side}$$

$$\begin{array}{rcl} 3x - 5 &=& 10 - 2x \\ \underline{+ 2x \quad\quad} & & \underline{\quad\quad + 2x} \\ 5x - 5 &=& 10 \\ \underline{+5} & & \underline{+5} \\ 5x &=& 15 \end{array}$$

Add $2x$ to both sides so that an x-term remains on only one side

Divide both sides by 5

$$x = 3$$

Check for x = 3

$$4x - 5 - x = 13 - 2x - 3$$
$$4(3) - 5 - (3) \stackrel{?}{=} 13 - 2(3) - 3$$
$$12 - 5 - 3 \stackrel{?}{=} 13 - 6 - 3$$
$$4 = 4 \quad ■$$

EXERCISES 9.3

Solve the following equations.

1. $9x = 5x + 12$
2. $4x = x + 15$
3. $x = 9 - 2x$
4. $3x = 8 - 5x$
5. $5x = 3x - 4$
6. $7x = 4x - 9$
7. $2x - 7 = x$
8. $5x - 8 = x$
9. $2x + 3 = x + 8$
10. $6x + 5 = 4x + 9$
11. $3x - 4 = 2x + 5$
12. $5x - 6 = 3x + 6$
13. $6x + 7 = 3 + 8x$
14. $4x + 28 = 7 + x$
15. $7x - 8 = 8 - 9x$
16. $5x - 7 = 7 - 9x$
17. $3x - 7 - x = 15 - 2x - 6$
18. $5x - 2 - x = 4 - 3x - 27$
19. $8x - 13 + 3x = 12 + 5x - 7$
20. $9x - 16 + 6x = 11 + 4x - 5$
21. $7 - 9x - 12 = 3x + 5 - 8x$
22. $13 - 11x - 17 = 5x + 4 - 10x$
23. $6x - 2 - x = 21 - 3x - 7$
24. $4x + 14 + 2x = 12 - 3x - 8$
25. $16 - 7x - 4 = 5x + 6 - 4x$
26. $3x + 15 - 6x - 10 = 9x - 6 - 3x + 5$

9.4 Solving Equations Containing Grouping Symbols

In many equations it is necessary to simplify the members before solving the equation.

Everything we have already discussed about solving equations is summarized in the following box.

> TO SOLVE AN EQUATION
>
> **1.** Remove grouping symbols.
>
> **2.** Combine like terms on each side of the equation.
>
> **3.** If the variable appears on both sides of the equation, remove the term containing the variable from one side by using the inverse operation.
>
> **4.** Remove all numbers that appear on the same side as the variable.
> *First:* Remove those numbers that are added or subtracted (by using the inverse operation).
> *Second:* Remove the number that is multiplied by the variable (by dividing both sides by that signed number).
>
> Check the solution in the *original* equation.

Example 1 Solve $10x - 2(3 + 4x) = 7 - (x - 2)$.
Solution

$$\begin{array}{rcll}
10x - 2(3 + 4x) &=& 7 - (x - 2) & \\
10x - 6 - 8x &=& 7 - x + 2 & \text{Removing grouping symbols} \\
2x - 6 &=& 9 - x & \text{Combining like terms on each side} \\
+\ x & & +\ x & \text{To get the } x\text{-term on only one side} \\
\hline
3x - 6 &=& 9 & \\
+\ 6 & & +6 & \\
\hline
3x &=& 15 & \text{Divide both sides by 3} \\
x &=& 5 &
\end{array}$$

Check for x = 5

$$\begin{aligned}
10x - 2(3 + 4x) &= 7 - (x - 2) \\
10(5) - 2(3 + 4 \cdot 5) &\stackrel{?}{=} 7 - (5 - 2) \\
10(5) - 2(3 + 20) &\stackrel{?}{=} 7 - (3) \\
10(5) - 2(23) &\stackrel{?}{=} 7 - 3 \\
50 - 46 &\stackrel{?}{=} 4 \\
4 &= 4 \quad ■
\end{aligned}$$

Example 2 Solve $5(2 - 3x) - 4 = 5x + [-(2x - 10) + 8]$.

Solution

$$5(2-3x)-4 = 5x + [-(2x-10)+8]$$

$$10-15x-4 = 5x + [-2x+10+8]$$
$$10-15x-4 = 5x + [-2x+18]$$
Removing grouping symbols

$$10-15x-4 = 5x-2x+18$$

$$-15x+6 = 3x+18$$
Combining like terms on both sides

$$\begin{array}{rl} +15x & +15x \\ \hline 6 = & 18x+18 \\ -18 & -18 \\ \hline -12 = & 18x \end{array}$$
To get the x-term on only one side

$$\frac{-12}{18} = \frac{18x}{18}$$

$$-\frac{2}{3} = x$$

You should check to see that $x = -\frac{2}{3}$ makes the two sides of the equation equal. ■

EXERCISES 9.4

Solve the following equation.

1. $5x - 3(2+3x) = 6$
2. $7x - 2(5+4x) = 8$
3. $6x + 2(3-8x) = -14$
4. $4x + 5(4-5x) = -22$
5. $7x + 5 = 3(3x+5)$
6. $8x + 6 = 2(7x+9)$
7. $9 - 4x = 5(9-8x)$
8. $10 - 7x = 4(11-6x)$
9. $3y - 2(2y-7) = 2(3+y) - 4$
10. $4a - 3(5a-14) = 5(7+a) - 9$
11. $6(3-4x) + 12 = 10x - 2(5-3x)$
12. $7(2-5x) + 27 = 18x - 3(8-4x)$
13. $2(3x-6) - 3(5x+4) = 5(7x-8)$
14. $4(7z-9) - 7(4z+3) = 6(9z-10)$
15. $6(5-4h) = 3(4h-2) - 7(6+8h)$
16. $5(3-2k) = 8(3k-4) - 4(1+7k)$
17. $2y - 3(4y-8) = 2(5+y) - 10$
18. $5(6-3a) + 18 = -9a - 3(4-2a)$
19. $3(2z-6) - 2(6z+4) = 5(z+8)$
20. $7(3-5h) = 4(3h-2) - 6(7+9h)$
21. $2[3-5(x-4)] = 10 - 5x$
22. $3[2-4(x-7)] = 26 - 8x$
23. $3[2h-6] = 2\{2(3-h)-5\}$
24. $6(3h-5) = 3\{4(1-h)-7\}$
25. $5(3-2x) - 10 = 4x + [-(2x-5)+15]$
26. $4(2-6x) - 6 = 8x + [-(3x-11)+20]$
27. $9 - 3(2x-7) - 9x = 5x - 2[6x-(4-x)-20]$
28. $14 - 2(7-4x) - 4x = 8x - 3[2x-(5-x)-30]$
29. $10 - 6\left(\frac{1}{2}x - \frac{2}{3}\right) = 8\left(\frac{3}{4}x + \frac{5}{8}\right)$
30. $3\left[2 - 4\left(\frac{1}{4}x - \frac{1}{2}\right)\right] = 12\left(\frac{2}{3}x - \frac{5}{6}\right)$

9.5 Conditional Equations, Identities, and Equations with No Solution

There are different kinds of equations. In this section, we discuss three types of equations: conditional equations, identical equations, and equations with no solution.

Conditional Equations

Consider the equation $x + 1 = 3$.

If $x = 5$, then	$x + 1 = 3$
becomes	$5 + 1 = 3$???
But we know that	$6 \neq 3$
If $x = -1$, then	$x + 1 = 3$
becomes	$-1 + 1 = 3$???
But we know that	$0 \neq 3$
However, if $x = 2$, then	$x + 1 = 3$
becomes	$2 + 1 = 3$
and	$3 = 3$

The two sides of $x + 1 = 3$ are not always equal. The *condition* $x = 2$ *must exist* for the two sides of $x + 1 = 3$ to be equal. A **conditional equation** is an equation whose two sides are equal only when certain numbers (called *solutions*) are substituted for the variable. All equations given in this chapter so far are conditional equations.

Identities

An **identity** (or *identical equation*) is an equation whose two sides are equal no matter what permissible number is substituted for the variable. Therefore, an identity has an endless number of solutions because any real number will make its two sides equal.

Example 1 The equation $2(5x - 7) = 10x - 14$ is an identity because

$$10x - 14 = 10x - 14$$ Both sides identical ■

Example 2 The equation $4(3x - 5) - 2x = 2(5x + 1) - 22$ is an identity because

$$12x - 20 - 2x = 10x + 2 - 22$$

$$10x - 20 = 10x - 20$$ Both sides identical ■

No Solution

A solution of an equation is a number that, when put in place of the variable makes the two sides of the equation equal. For some equations no such number can be found. In these cases we say that *the equation has* **no solution**.

Example 3 Solve $x + 1 = x + 2$.

Solution

$$\begin{array}{rcr} x + 1 & = & x + 2 \\ -x & & -x \\ \hline 1 & \neq & 2 \end{array}$$

Trying to get the x-term on only one side

The two sides are unequal ■

In this case we say that *the equation has no solution* because no number substituted for x will make the two sides of the equation equal.

> WHEN SOLVING AN EQUATION THAT HAS ONLY ONE VARIABLE
>
> **1.** Attempt to solve the equation (by the method given in Section 9.4).
>
> **2.** There are three possible results.
>
> a. *Conditional equation:* If a *single solution* is obtained, the equation is a conditional equation.
>
> b. *Identity:* If the two sides of the equation simplify *to the same expression*, the equation is an identity.
>
> c. *No solution:* If the two sides of the equation simplify *to unequal expressions*, the equation has no solution.

Example 4 Solve $4x - 2(3 - x) = 12$.

Solution
$$4x - 6 + 2x = 12$$
$$6x - 6 = 12$$
$$6x = 18$$
$$x = 3$$
Conditional equation (Single solution) ■

Example 5 Solve $4x - 2(3 + 2x) = -6$.

Solution
$$4x - 6 - 4x = -6$$
$$-6 = -6$$
Identity (Both sides the same) ■

Example 6 Solve $4x - 2(3 + 2x) = 8$.

Solution
$$4x - 6 - 4x = 8$$
$$-6 \neq 8$$
No solution (Sides unequal) ■

EXERCISES 9.5

Find the solution of each conditional equation. If the equation is an identity, write *identity*. If the equation has no solution, write *no solution.*

1. $x + 3 = 8$
2. $4 - x = 6$
3. $2x + 5 = 7 + 2x$
4. $10 - 5y = 8 - 5y$
5. $6 + 4x = 4x + 6$
6. $7x + 12 = 12 + 7x$
7. $5x - 2(4 - x) = 6$
8. $8x - 3(5 - x) = 7$
9. $6x - 3(5 + 2x) = -15$
10. $4x - 2(6 + 2x) = -12$
11. $4x - 2(6 + 2x) = -15$
12. $6x - 3(5 + 2x) = -12$
13. $7(2 - 5x) - 32 = 10x - 3(6 + 15x)$
14. $6(3 - 4x) + 10 = 8x - 3(2 - 3x)$
15. $2(2x - 5) - 3(4 - x) = 7x - 20$
16. $3(x - 4) - 5(6 - x) = 2(4x - 21)$
17. $2[3 - 4(5 - x)] = 2(3x - 11)$
18. $3[5 - 2(7 - x)] = 6(x - 7)$
19. $5x - 3(2x - 1) = 7 - (x + 4)$
20. $7x - 4(x - 3) = 6x - (8 - x)$

9.6 Inequalities

An *equation* is a statement that two expressions are *equal.* An *inequality* is a statement that two expressions are *not equal.*

The Parts of an Inequality

An inequality has three parts:

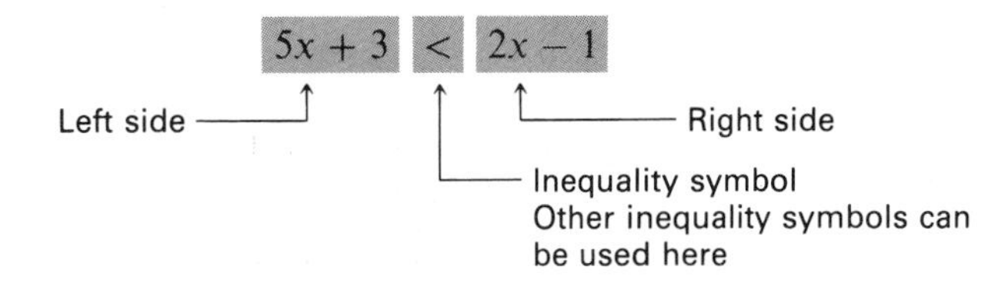

Inequality Symbols

In this text we will discuss only inequalities that have symbols $>$, $<$, $\leq$, or $\geq$.

Types of Inequalities

$>$

Greater-than symbol ($>$) $a > b$ is read "a is greater than b."

Example 1 Greater-than symbol

a. Write $11 > 2$
Read "11 is greater than 2."

b. Write $3x - 4 > 7$
Read "$3x$ minus 4 is greater than 7." ■

$<$

Less-than symbol ($<$) $a < b$ is read "a is less than b."

Example 2 Less-than symbol

a. Write $3 < 6$
Read "3 is less than 6."

b. Write $2x < 5 - x$
Read "$2x$ is less than $5 - x$." ■

$\leq$

Less-than-or-equal-to symbol ($\leq$) $a \leq b$ is read "a is less than *or* equal to b."

This means if $\left\{\begin{matrix} \text{either } a < b \\ \text{or } \quad a = b \end{matrix}\right\}$ is true, then $a \leq b$ is true.

Example 3 Less-than-or-equal-to symbol

a. $2 \leq 3$ is true, because $2 < 3$ is true.

b. $4 \leq 4$ is true, because $4 = 4$ is true.

c. Write $7 \leq 5x - 2$
Read "7 is less than or equal to $5x - 2$." ■

$\geq$

Greater-than-or-equal-to symbol ($\geq$) $a \geq b$ is read "a is greater than *or* equal to b."

This means if $\left\{\begin{matrix} \text{either } a > b \\ \text{or } \quad a = b \end{matrix}\right\}$ is true, then $a \geq b$ is true.

Example 4 Greater-than-or-equal-to symbol

a. $5 \geq 1$ is true, because $5 > 1$ is true.

b. $7 \geq 7$ is true, because $7 = 7$ is true.

c. Write $x + 6 \geq 10$
Read "x plus 6 is greater than or equal to 10." ■

Sense

The **sense** of an inequality symbol refers to the direction the symbol points.

$\left.\begin{matrix} a > b \\ c > d \end{matrix}\right\}$ same sense $\qquad$ $\left.\begin{matrix} a < b \\ c > d \end{matrix}\right\}$ opposite sense

$\left.\begin{matrix} e < f \\ g < h \end{matrix}\right\}$ same sense $\qquad$ $\left.\begin{matrix} e > f \\ g < h \end{matrix}\right\}$ opposite sense

Facts About Inequalities

The basic rules used to solve inequalities are the same as those used to solve equations, with the exception that *the sense must be changed when multiplying or dividing both sides by a negative number.*

Example 5 Illustrating the rules used with inequalities

a. $$\begin{array}{rr} 10 > & 5 \\ +\ 6 & +\ 6 \\ \hline 16 > & 11 \end{array}$$

Adding same number (6) to both sides

Sense is *not* changed

b. $$\begin{array}{rr} 7 < & 12 \\ -\ 2 & -\ 2 \\ \hline 5 < & 10 \end{array}$$

Subtracting the same number (2) from both sides

Sense is *not* changed

c. $3 < 4$

$2(3) < 2(4)$

$6 < 8$

Multiplying both sides by the same *positive* number (2)

Sense is *not* changed

d. $3 < 4$

$(-2)(3)\ ?\ (-2)(4)$

$-6 > -8$

Multiplying both sides by the same *negative* number (−2)

Sense is changed

e. $9 > 6$

$\frac{9}{3} > \frac{6}{3}$

$3 > 2$

Dividing both sides by the same *positive* number (3)

Sense is *not* changed

f. $9 > 6$

$\frac{9}{-3}\ ?\ \frac{6}{-3}$

$-3 < -2$

Dividing both sides by the same *negative* number (−3)

Sense is changed ■

Solving Inequalities

An inequality is solved when we have nothing but the variable on one side of the inequality symbol and everything else on the other side.

The method of solving inequalities is very much like the method used for solving equations. We show how the methods are alike or different in the following summary.

IN SOLVING EQUATIONS	IN SOLVING INEQUALITIES
Addition rule: The same number may be added to both sides.	**Addition rule:** The same number may be added to both sides.
Subtraction rule: The same number may be subtracted from both sides.	**Subtraction rule:** The same number may be subtracted from both sides.
Multiplication rule: Both sides may be multiplied the same nonzero number.	**Multiplication rule:** **1.** Both sides may be multiplied by the same *positive* number. **2.** When both sides are multiplied by the same *negative* number, *the sense must be changed.*
Division rule: Both sides may be divided by the same nonzero number.	**Division rule:** **1.** Both sides may be divided by the same *positive* number. **2.** When both sides are divided by the same *negative* number, *the sense must be changed.*

Example 6 Solve $x + 3 < 7$, and graph the solution on the number line.

Solution

$$\begin{array}{rcll} x + 3 & < & 7 & \\ -3 & & -3 & \text{Adding } -3 \text{ to both sides} \\ \hline x & < & 4 & \text{Solution} \end{array}$$

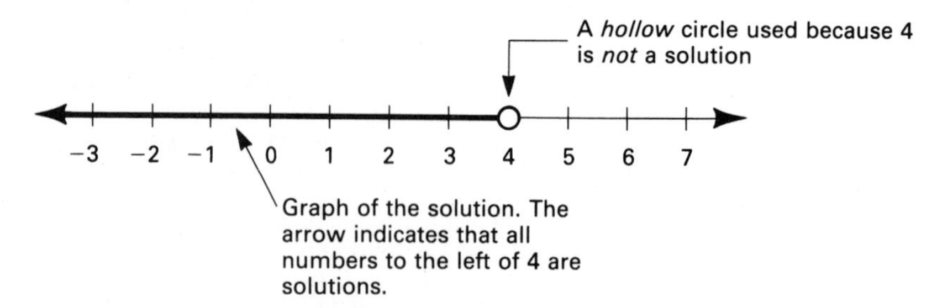

■

Solution of an Inequality

The solution $x < 4$ means that if we replace x by any number less than 4 in the given inequality, we get a true statement. For example, if we replace x by 3 (which is less than 4):

$$x + 3 < 7$$

$$3 + 3 < 7$$

$$6 < 7 \qquad \textit{True}$$

Example 7 Solve $2x - 1 \geq -7$, and graph the solution on the number line.

Solution

$$\begin{array}{rcll} 2x - 1 & \geq & -7 & \\ +1 & & +1 & \text{Adding 1 to both sides} \\ \hline 2x & \geq & -6 & \end{array}$$

$$\frac{2x}{2} \geq \frac{-6}{2} \qquad \text{Dividing both sides by 2. Sense is } \textit{not} \text{ changed}$$

$$x \geq -3 \qquad \text{Solution}$$

A *solid* circle is used because −3 *is* a solution

−6 −5 −4 −3 −2 −1 0 1 2 3 4

The arrow together with the solid circle indicates that −3 and all numbers to the right of −3 are solutions

■

Example 8 Solve $x + 8 > 6x - 2$, and graph the solution on the number line.

Solution

$$\begin{array}{rcll} x + 8 & > & 6x - 2 & \\ -6x & & -6x & \text{To get } x\text{-terms on left} \\ \hline -5x + 8 & > & -2 & \\ -8 & & -8 & \\ \hline -5x & > & -10 & \end{array}$$

$$\frac{-5x}{-5} < \frac{-10}{-5} \qquad \text{Dividing both sides by } -5. \ \textit{Sense is changed}$$

$$x < 2 \qquad \text{Solution}$$

Alternate Solution

$$\begin{array}{rcl} x + 8 & > & 6x - 2 \\ -x & & -x \\ \hline 8 & > & 5x - 2 \\ +2 & & +2 \\ \hline 10 & > & 5x \end{array} \qquad \text{To get } x\text{-terms on right}$$

$$\frac{10}{5} > \frac{5x}{5}$$ Dividing both sides by 5
Sense is *not* changed

$$2 > x$$ Solution

Note that the two inequalities $x < 2$ and $2 > x$ have the same meaning.
Graphing the solution.

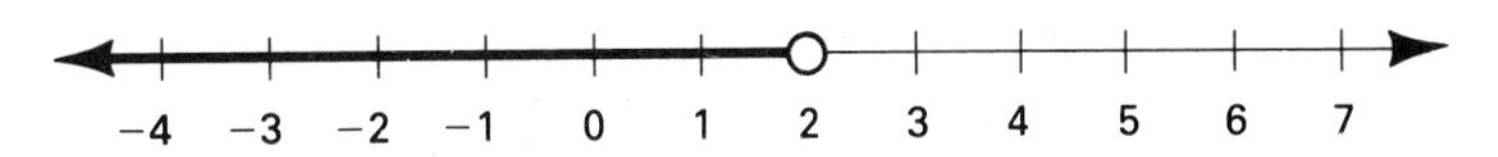

■

Example 9 Solve $3x - 2(2x - 7) \leq 2(3 + x) - 4$.

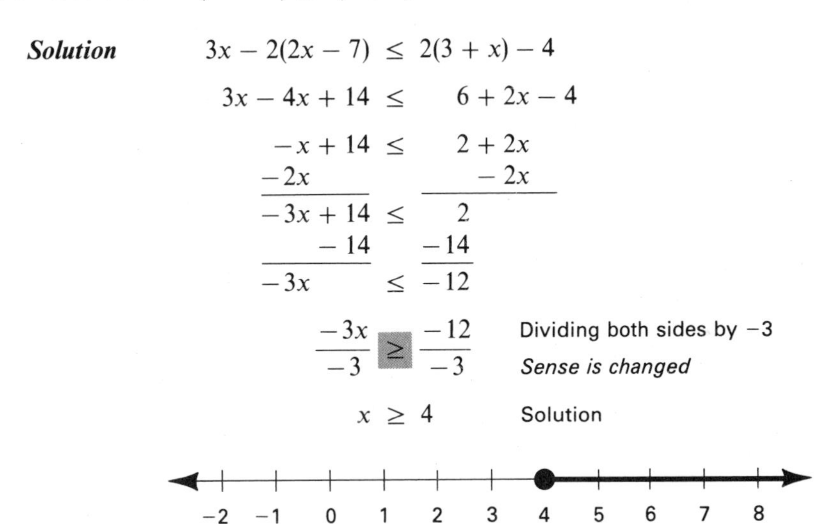

Solution

$$\begin{array}{rcl} 3x - 2(2x - 7) & \leq & 2(3 + x) - 4 \\ 3x - 4x + 14 & \leq & 6 + 2x - 4 \\ -x + 14 & \leq & 2 + 2x \\ -2x & & -2x \\ \hline -3x + 14 & \leq & 2 \\ -14 & & -14 \\ \hline -3x & \leq & -12 \end{array}$$

$$\frac{-3x}{-3} \geq \frac{-12}{-3}$$ Dividing both sides by -3
Sense is changed

$$x \geq 4$$ Solution

■

The method used to solve inequalities may be summarized as follows.

> TO SOLVE AN INEQUALITY
>
> Proceed in the same way used to solve equations, with the *exception* that the sense must be changed when multiplying or dividing both sides by a negative number.

EXERCISES 9.6

Solve the following inequalities and graph the solution on the number line.

1. $x - 5 < 2$ **2.** $x + 3 \geq -4$ **3.** $-3x < 12$ **4.** $-2x \geq -6$

5. $-x \geq 5$ **6.** $-x \geq -8$ **7.** $5x + 4 \leq 19$ **8.** $7x - 3 < 18$

9. $6 - x > 2$

10. $2 - x \geq -3$

11. $6x + 2 \geq x - 8$

12. $4x + 28 > 7 + x$

13. $5x + 7 > 13 + 11x$

14. $6x + 7 \geq 3 + 8x$

15. $2x - 9 > 3(x - 2)$

16. $3x - 11 > 5(x - 1)$

17. $4(6 - 2x) \leq 5x - 2$

18. $3x - 6 \leq 3(2 - x)$

19. $2x - 3(x + 4) \leq -7$

20. $4x - 2(3 - x) \leq 12$

21. $6(3 - 4x) + 12 \geq 10x - 2(5 - 3x)$

22. $6(2x - 5) + 29 > 3x - 7(11 - 4x)$

23. $5(3 - 2x) + 25 > 4x - 6(10 - 3x)$

24. $7(2 - 5x) + 27 > 18x - 3(8 - 4x)$

25. $2[3 - 5(x - 4)] < 10 - 5x$

26. $3[2 - 4(x - 7)] < 26 - 8x$

27. $4x - \{6 - [2x - (x + 3)]\} \leq 6$

28. $2x - \{4 - [5x - (8 - x)]\} \geq -20$

9.7 Chapter Summary

Equation 9.1
An **equation** is a statement that two expressions are *equal.*

Solution of an Equation 9.1
A **solution** of an equation is a number that, when put in place of the variable, makes a true statement.

Conditional Equations 9.5
A **conditional equation** is an equation whose two sides are equal only when certain numbers (called *solutions*) are substituted for the variable.

Identities 9.5
An **identity** (or *identical equation*) is an equation whose two sides are equal no matter what permissible number is substituted for the variable.

Equations With No Solution 9.5
An equation has **no solution** if there is no number which when put in place of the variable makes the two sides of the equation equal.

To Solve an Equation 9.1–9.4

1. Remove grouping symbols.
2. Combine like terms on each side of the equation.
3. Attempt to solve the resulting equation.
   There are three possible results.
   a. **Conditional equation:** If a *single solution* is obtained, the equation is a conditional equation.
   b. **Identity:** If the two sides of the equation simplify *to the same expression*, the equation is an identity.
   c. **No solution:** If the two sides of the equation simplify to *unequal expressions*, the equation has no solution.

To Check the Solution of an Equation 9.1

1. Replace the variable in the original equation by the solution in parentheses.
2. Perform the indicated operations on both sides of the equal sign.
3. If the resulting number on each side of the equal sign is the same, the solution is correct.

Inequality 9.6
An **inequality** is a statement that two expressions are *not equal.*

Solution of an Inequality 9.6
A solution of an inequality is a number that, when put in place of the variable, makes the inequality a true statement.

To Solve an Inequality 9.6 To solve an inequality, proceed in the same way used to solve equations, with the *exception* that the sense must be changed when multiplying or dividing both sides by a *negative* number.

To Graph an Inequality 9.6 To graph a first-degree inequality on the number line

To graph the solution $x > c$:

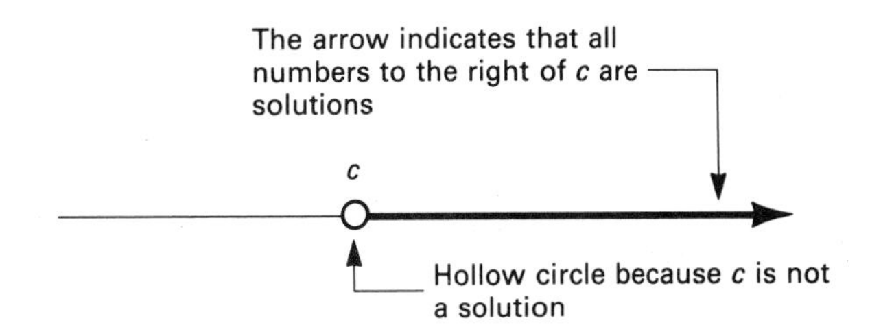

To graph the solution $x \leq b$:

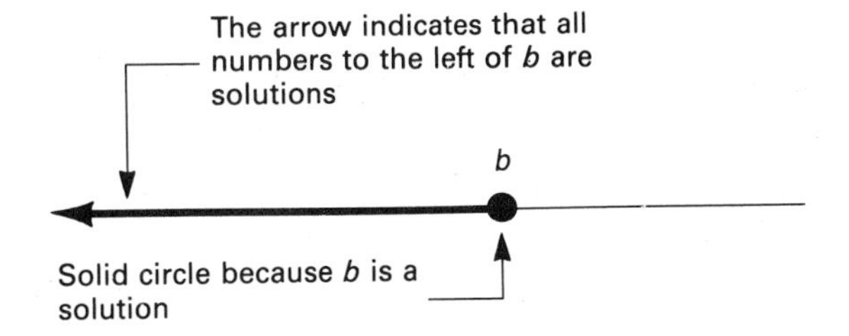

Review Exercises 9.7

1. Is -4 a solution of the equation $2x - 3 = -11$?

2. Is -5 a solution of the equation $6 - 2x = -4$?

In Exercises 3–20, find the solution of each conditional equation, and check your answers. Identify the equations that are identities and those that have no solution.

3. $3x - 5 = 4$

4. $17 - 5x = 2$

5. $2 = 20 - 9x$

6. $2x + 1 = 2x + 7$

7. $7.5 = \dfrac{A}{10}$

8. $\dfrac{C}{7} - 15 = 13$

9. $7 - 2(M - 4) = 5$

10. $10 - 4(2 - 3x) = 2 + 12x$

11. $6R - 8 = 6(2 - 3R)$

12. $7x - (2x - 4) = 3(x + 1)$

13. $4(2x - 5) - (3 - x) = -17$

14. $6 - \{2x - [3 + x]\} = -5$

15. $15(4 - 5V) = 16(4 - 6V) + 10$

16. $56T - 18 = 7(8T - 4)$

17. $5x - 7(4 - 2x) + 8 = 10 - 9(11 - x)$

18. $9 - 3(x - 2) = 3(5 - x)$

19. $4[-24 - 6(3x - 5) + 22x] = 0$

20. $2[-7y - 3(5 - 4y) + 10] = 10y - 12$

In Exercises 21–26, solve each inequality and graph the solution on the number line.

21. $3x - 5 < 4$

22. $x + 4 \geq -2$

23. $9x - 4(2 + 2x) \geq -15$

24. $2x + 7 > 11 + 4x$

25. $3[2 - 5(x - 1)] > 3 - 6x$

26. $2(x + 4) - 13 \leq 7x - 5(2x - 3)$

Chapter 9 Diagnostic Test

Allow yourself about an hour to do these problems. Complete solutions for every problem, together with section references, are given in the answer section at the end of the book.

1. Is -3 a solution of the equation $4x - 5 = -7$?

In Problems 2–16, solve each equation. If the equation is an identity, write *identity*. If the equation has no solution, write *no solution*.

2. $x - 3 = 7$

3. $10 = x + 4$

4. $3x + 2 = 14$

5. $15 - 2x = 7$

6. $-5 = \frac{x}{7}$

7. $4x + 5 = 17 - 2x$

8. $6x - 3 = 5x + 2$

9. $9 + 7x = 5x - 1$

10. $2x - 5 = 8x - 3$

11. $3z - 21 + 5z = 4 - 6z + 17$

12. $6k - 3(4 - 5k) = 9$

13. $6x - 2(3x - 5) = 10$

14. $2m - 4(3m - 2) = 5(6 + m) - 7$

15. $2(3x + 5) = 14 + 3(2x - 1)$

16. $3[7 - 6(y - 2)] = -3 + 2y$

In Problems 17–20, solve each inequality and graph the solution on the number line.

17. $5x - 2 > 10 - x$

18. $4x + 5 < x - 4$

19. $2(x - 4) - 5 \geq 6 + 3(2x - 1)$

20. $7x - 2(5 + 4x) \leq -7$

10 Word Problems

The main reason for studying algebra is to equip oneself with the tools necessary to solve problems. Most problems are expressed in words. In this chapter we show how to change the words of a written problem into an equation. That equation can then be solved by the methods learned in Chapter 9.

10.1 Changing Word Expressions into Algebraic Expressions

It is helpful in solving word problems to break them up into smaller expressions. In this section we show how you can change these small *word* expressions into *algebraic* expressions.

10.1A Representing Word Expressions

A list of key word expressions and their corresponding algebraic operations follows.

+	−	×	÷
add	subtract	multiply	divided by
sum	difference	times	quotient
plus	minus	product	per
increased by	decreased by	of	
more than	diminished by		
	less than		
	subtracted from		

When the expressions *less than* and *subtracted from* are used, the terms are written in reverse order. [See Examples 1(d) and 1(f).]

Example 1 Change each word expression into an algebraic expression.

a. The sum of A and B

Solution $A + B$

b. The product of L and W

Solution LW

c. Two decreased by C

Solution $2 - C$

d. Two less than C

Solution $C - 2$

e. Three times the square of x, plus ten

Solution $3x^2 + 10$

f. Five subtracted from the quotient of S by T

Solution $\frac{S}{T} - 5$

g. Twice the sum of x and 3

Solution $2(x + 3)$ ■

EXERCISES 10.1A

Change each word expression into an algebraic expression.

1. The sum of x and ten
2. A added to B
3. Five less than A
4. B diminished by C
5. The product of six and z
6. A multiplied by B
7. C decreased by D
8. 10 less than A
9. Ten added to x
10. x diminished by 4
11. Subtract the product of U and V from x
12. Subtract x from the product of P and Q.
13. The product of five and the square of x
14. The product of ten and the cube of x
15. The square of the sum of A and B
16. The square of the quotient of A divided by B
17. The sum of x and seven, divided by y
18. T divided by the sum of x and nine
19. The quotient of A divided by the sum of C and ten
20. The quotient of the sum of A and C divided by B
21. Fifteen more than twice F
22. Twice the sum of x and y
23. Five times y subtracted from three times x
24. Ten times x decreased by ten times y
25. The product of x and the difference, six less than y
26. The product of seven and the difference, x less than y
27. Three times the sum of A and five
28. Five times the difference of two and C
29. Twice the difference of six and B
30. The sum of x and four subtracted from ten

10.1B Representing Word Expressions Containing Unknown Numbers

Word expressions may contain unknown numbers. To change such a word expression into an algebraic expression, each unknown number must first be represented in terms of the same variable.

> TO CHANGE A WORD EXPRESSION INTO AN ALGEBRAIC EXPRESSION
>
> 1. Identify which number or numbers are unknown.
> 2. Represent one of the unknown numbers by a variable. Express any other unknown number in terms of the same variable.
> 3. Change the word expression into an algebraic expression using the variable representations in place of the unknown numbers.

Example 2 Change the word expression *the cost of five stamps* into an algebraic expression.

Solution

1. The cost of one stamp is the unknown number.
2. Let c represent the cost of one stamp.
3. Then $5c$ is the algebraic expression for *the cost of five stamps*. Because one stamp costs c cents, then 5 stamps will cost 5 times c cents $= 5c$. ■

Example 3 In the word expression *Mary is ten years older than Nancy*, represent both unknown numbers in terms of the same variable.
First Solution

1. There are two unknown numbers in this expression:
   Mary's age and Nancy's age.
2. Let N represent Nancy's age.
   Then $N + 10$ represents Mary's age, because Mary is 10 years older than Nancy.

Second Solution

1. There are two unknown numbers in this expression:
   Mary's age and Nancy's age.
2. Let M represent Mary's age.
   Then $M - 10$ represents Nancy's age, because Nancy is 10 years younger than Mary. ■

Example 4 In the word expression *The sum of two numbers is ten*, represent both unknown numbers in terms of the same variable.
Solution

1. There are two unknown numbers.
2. Let x = one of the unknown numbers
   Then $10 - x$ = the other unknown number
   Notice that the sum of the two numbers is 10; $x + (10 - x) = 10$. ■

Example 5 In the word expression *The length of a rectangle is three feet more than twice the width*, represent the length and the width in terms of the same variable.
Solution

1. There are two unknown numbers: the length and the width.
2. Let w = width
   $2w + 3$ = length ■

Consecutive Integers

Consecutive integers follow one another in order, such as 5, 6, and 7. **Consecutive integers** are integers that differ by 1.

If x = first consecutive integer,
then $x + 1$ = second consecutive integer,
and $x + 2$ = third consecutive integer.

Consecutive Even Integers

Even integers are divisible by 2. **Consecutive even integers** are even integers that differ by 2, such as 4, 6, and 8.

If x = first consecutive even,
then $x + 2$ = second consecutive even,
and $x + 4$ = third consecutive even.

Consecutive Odd Integers

Odd integers are not divisible by 2. **Consecutive odd integers** are odd integers that differ by 2, such as 5, 7, and 9.

If x = first consecutive odd,
then $x + 2$ = second consecutive odd,
and $x + 4$ = third consecutive odd.

Example 6 Change the word expression *the sum of three consecutive integers* into an algebraic expression.

Solution

1. There are three unknown numbers.
2. Let $x =$ first consecutive integer
   $x + 1 =$ second consecutive integer
   $x + 2 =$ third consecutive integer
3. Then the sum is $x + (x + 1) + (x + 2)$. ■

Example 7 Change the word expression *the product of two consecutive odd integers* into an algebraic expression.
Solution

1. There are two unknown numbers.
2. Let $x =$ first consecutive odd
   $x + 2 =$ second consecutive odd
3. Then their product is $x(x + 2)$. ■

EXERCISES 10.1B

In the following exercises: (a) *If there is only one unknown number*, represent it by a variable and then change the word expression into an algebraic expression. (b) *If there is more than one unknown number*, represent each number in terms of the same variable.

1. Fred's salary plus seventy-five dollars
2. Jaime's salary less forty-two dollars
3. Two less than the number of children in Mr. Moore's family
4. Two more players were added to Jerry's team.
5. Four times Joyce's age
6. One-fourth of Rene's age
7. Twenty times the cost of a record increased by eighty-nine cents
8. Seventeen cents less than five times the cost of a ballpoint pen
9. One-fifth the cost of a hamburger
10. The length of the building divided by eight
11. Five times the speed of the car, plus one hundred miles per hour
12. Twice the speed of a car, diminished by forty miles per hour
13. Ten less than five times the square of an unknown number
14. Eight more than four times the cube of an uknown number
15. The sum of eight and an unknown number is divided by the square of that unknown number.
16. The sum of the square of an unknown number and eleven is divided by the unknown number.
17. The sum of thirty-two and nine-fifths the Celsius temperature
18. Five-ninths times the result of subtracting thirty-two from the Fahrenheit temperature
19. Twice the result of subtracting an unknown number from five
20. The square of the result of subtracting twelve from an unknown number
21. The length of a rectangle is twelve centimeters more than its width.
22. The altitude of a triangle is seven centimeters less than its base.
23. The sum of two numbers is sixty.
24. The sum of two numbers is -22.
25. Henry is five years younger than his brother.
26. Mrs. Lopez is twenty-one years older than her daughter.
27. Tom has \$139 more than Linda.
28. Pete has \$53 less than Ann.
29. The product of two consecutive integers
30. The product of two consecutive even integers

31. The sum of three consecutive odd integers

32. The sum of four consecutive integers

33. The length of a rectangle is four inches less than twice the width.

34. The width of a rectangle is ten inches less than three times the length.

10.2 Solving Word Problems

In this section we show how to change the words of a written problem into an equation. Then we solve the equation using the method discussed in Chapter 9.

> TO SOLVE A WORD PROBLEM
>
> **1.** Represent the unknown number by a variable. Express any other unknown numbers in terms of the same variable.
>
> **2.** Break the word problem into small pieces.
>
> **3.** Represent each piece by an algebraic expression.
>
> **4.** Arrange the algebraic expressions into an equation.
>
> **5.** Solve the equation.
>
> **6.** Check the solution in the word statement.

Number Problems

Example 1 Seven increased by three times an unknown number is thirteen. What is the unknown number?

Solution First we break the word problem into small pieces. Then we represent each piece by an algebraic expression.

Let x stand for the unknown number.

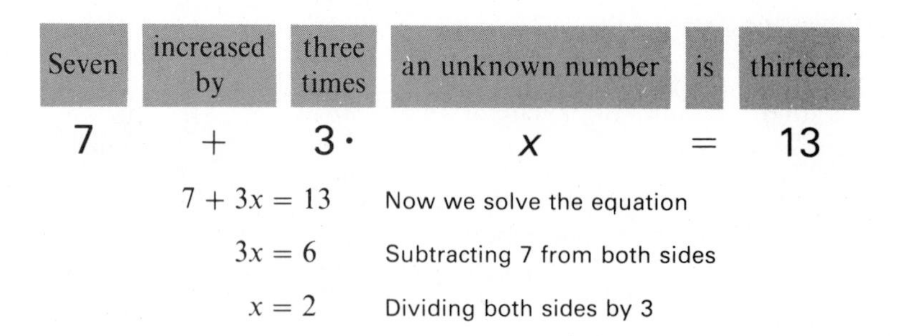

$7 + 3x = 13$ Now we solve the equation

$3x = 6$ Subtracting 7 from both sides

$x = 2$ Dividing both sides by 3

Therefore, the unknown number is 2.
Last, we check $x = 2$ in the word statement.

Seven	increased by	three times	an unknown number	is	thirteen.
7	+	3·	(2)	=	13

$7 + 3(2) \stackrel{?}{=} 13$ The unknown number was replaced by 2

$7 + 6 \stackrel{?}{=} 13$

$13 = 13$ ■

NOTE To check a word problem, the solution must be checked in the word statement. The reason for checking in the word statement is that an error may have been made in writing the equation that would not be discovered if you substitute the solution in the equation. ☑

Example 2 Four times an unknown number is equal to twice the sum of five and that unknown number. Find the unknown number.
Solution Let x = the unknown number.

Four times	an unknown number	is equal to	twice the sum of five and that unknown number.
$4\cdot$	x	$=$	$2\cdot(5+x)$

$4x = 2(5 + x)$

$4x = 10 + 2x$ Using the distributive rule

$2x = 10$ Subtracting $2x$ from both sides

$x = 5$ Dividing both sides by 2

Therefore, the unknown is 5.
Check Checking the solution in the word statement is left to the student. ■

Example 3 When seven is subtracted from six times an unknown number, the result is eleven. What is the unknown number?
Solution Let x = the unknown number.

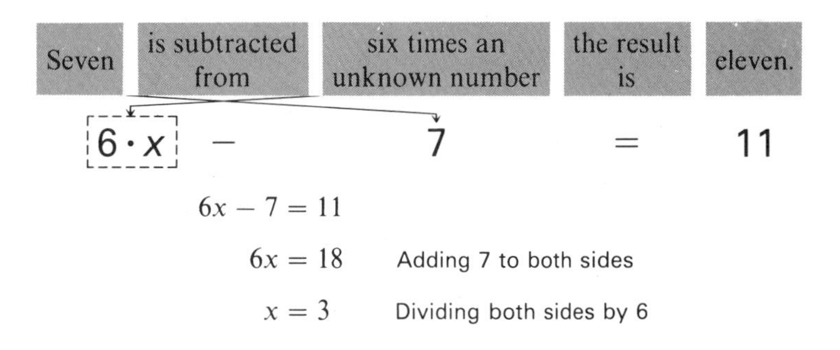

$6x - 7 = 11$

$6x = 18$ Adding 7 to both sides

$x = 3$ Dividing both sides by 6

Therefore, the unknown number is 3. ■

Consecutive Integer Problems

Example 4 The sum of three consecutive even integers is 42. Find the integers.
Solution Let x = first consecutive even
$x + 2$ = second consecutive even
$x + 4$ = third consecutive even

sum of three consecutive even integers	is	42.
$x + (x + 2) + (x + 4)$	$=$	42

$$x + (x + 2) + (x + 4) = 42$$

$3x + 6 = 42$ Combining like terms

$3x = 36$ Subtracting 6 from both sides

$x = 12$ Dividing both sides by 3

Therefore, $x = 12$ First integer

$x + 2 = 14$ Second integer

$x + 4 = 16$ Third integer ■

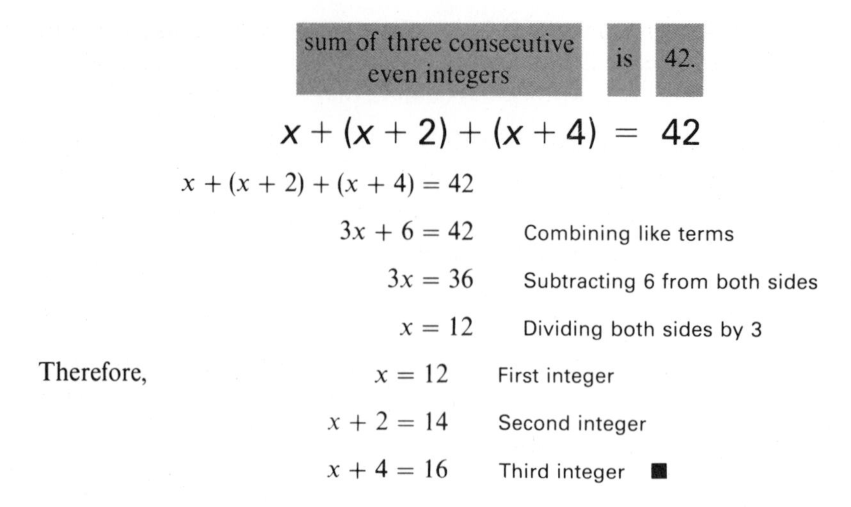

Geometry Problems

A list of geometric formulas is given on the inside front cover of your book. For a complete discussion of geometric formulas see Chapter 16.

Example 5 The length of a rectangle is four inches more than the width. If the perimeter is forty inches, find the length and the width.

Solution There are two unknown numbers, the width and the length.

Let $w = \text{width}$, $w + 4 = \text{length}$ } because the length is 4 more than the width

$P =$	2ℓ	$+ 2w$

$$40 = 2(w + 4) + 2w$$

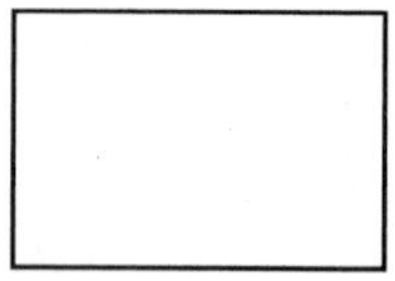

$40 = 2(w + 4) + 2w$

$40 = 2w + 8 + 2w$ Using the distributive rule

$40 = 4w + 8$ Combining like terms

$32 = 4w$ Subtracting 8 from both sides

$8 = w$ Dividing both sides by 4

Therefore, width $= w = 8$ in.,
and the length $= w + 4 = 12$ in. ■

EXERCISES 10.2

In the following word problems: (a) Represent the unknown numbers by a variable. (b) Set up an equation and solve. (c) Answer the question.

1. When twice an unknown number is added to thirteen, the sum is twenty-five. Find the unknown number.
2. When twenty-five is added to three times an unknown number, the sum is thirty-four. Find the unknown number.
3. Five times an unknown number, decreased by eight, is twenty-two. What is the unknown number?
4. Four times an unknown number, decreased by five, is fifteen. What is the unknown number?
5. An unknown number divided by twelve equals six. What is the unknown number?
6. The quotient of an unknown number and five is ten. What is the unknown number?

7. Seven minus an unknown number is equal to the unknown number plus one. Find the unknown number.

8. Six plus an unknown number is equal to twelve decreased by the unknown number. Find the unknown number.

9. When seven is added to an unknown number, the result is twice that unknown number. Find the unknown number.

10. When three times an unknown number is subtracted from twenty, the result is the unknown number. Find the unknown number.

11. Find the unknown number if twice the sum of five and an unknown number is equal to twenty-six.

12. Find the unknown number if four times the sum of nine and an unknown number is equal to twenty.

13. Twice the sum of four and an unknown number is equal to ten less than the unknown number. What is the unknown number?

14. Six subtracted from an unknown number is equal to three times the difference of the unknown number and eight. Find the unknown number.

15. Find three consecutive integers whose sum is 63.

16. Find three consecutive odd integers whose sum is 21.

17. Find three consecutive even integers such that the sum of the first two integers minus the third integer is four.

18. Find two consecutive integers such that twice the first integer minus the second integer is eight.

19. Find three consecutive odd integers such that the sum of the first two integers is equal to four times the third.

20. Find three consecutive even integers such that the sum of the first two integers is equal to three times the third.

21. The length of a rectangle is 12 inches. Find the width, if the perimeter is 36 inches.

22. The perimeter of a square is 64 centimeters. Find the length of a side of the square.

23. The length of a rectangle is three feet more than the width. If the perimeter is 50 feet, find the length and the width.

24. The width of a rectangle is six inches less than the length. Find the length and the width if the perimeter is 40 inches.

25. The area of a triangle is 24 square meters, and the base is 8 meters. What is the height of the triangle?

26. The length of a rectangle is three inches less than twice the width. If the perimeter is 24 inches, what are the length and the width?

10.3 Mixture Problems

Mixture problems involve mixing two or more dry ingredients.

TWO IMPORTANT FACTS NECESSARY TO SOLVE MIXTURE PROBLEMS

1. $\begin{pmatrix}\textit{Amount} \text{ of}\\ \text{ingredient A}\end{pmatrix} + \begin{pmatrix}\textit{Amount} \text{ of}\\ \text{ingredient B}\end{pmatrix} = \begin{pmatrix}\textit{Amount} \text{ of}\\ \text{mixture}\end{pmatrix}$

2. $\begin{pmatrix}\textit{Cost} \text{ of}\\ \text{ingredient A}\end{pmatrix} + \begin{pmatrix}\textit{Cost} \text{ of}\\ \text{ingredient B}\end{pmatrix} = \begin{pmatrix}\textit{Cost} \text{ of}\\ \text{mixture}\end{pmatrix}$

Example 1 A wholesaler makes up a 60-lb mixture of two kinds of coffee. Brand A costs \$3 per pound, and Brand B costs \$6 per pound. How many pounds of each kind must be used if the mixture is to cost \$5 per pound?

Solution Let x = pounds of Brand A

$60 - x$ = pounds of Brand B

Some students find it helpful to use a chart to keep track of the useful information.

	Cost per lb ·	*Number of lbs* =	*Total Cost*
Brand A	3	x	$3x$
Brand B	6	$60 - x$	$6(60 - x)$
Mixture	5	60	$5(60)$

cost of Brand A + cost of Brand B = cost of mixture

$$3x + 6(60 - x) = 5(60)$$

$3x + 360 - 6x = 300$ Using the distributive rule

$360 - 3x = 300$ Combining like terms

$-3x = -60$ Subtracting 360 from both sides

$x = 20$ Dividing both sides by -3

Therefore, $x = 20$ lb of Brand A
$60 - x = 40$ lb of Brand B

Check Cost of Brand A $= 3(20) =$ \$ 60
Cost of Brand B $= 6(40) =$ + 240
Cost of mixture $= 5(60) =$ \$300 ■

Example 2 A farmer wants to mix 10 bushels of soybeans costing \$7.50 per bushel with corn costing \$6.00 per bushel. How many bushels of corn must he use to make a mixture costing \$7.25 per bushel?

Solution Let $x =$ bushels of corn

	Cost per Bushel ·	*Number of Bushels* =	*Total Cost*
Soybeans	7.50	10	$7.50(10)$
Corn	6.00	x	$6.00x$
Mixture	7.25	$x + 10$	$7.25(x + 10)$

cost of soybeans + cost of corn = cost of mixture

$$7.50(10) + 6.00x = 7.25(x + 10)$$

$750(10) + 600x = 725(x + 10)$ Multiply both sides by 100

$7500 + 600x = 725x + 7250$

$$7500 = 125x + 7250$$
$$250 = 125x$$
$$2 = x$$

Therefore, $x = 2$ bushels of corn. ■

The next example involves mixing coins.

Example 3 Dianne has \$3.20 in nickels, dimes, and quarters. If there are 7 more dimes than quarters and 3 times as many nickels as quarters, how many of each kind of coin does she have?

Solution Let Q = number of quarters

then, $Q + 7$ = number of dimes — Because there are 7 more dimes than quarters

and $3Q$ = number of nickels — Because there are 3 times as many nickels as quarters

	Value of One Coin ·	*Number of Coins* =	*Total Value*
Quarters	25	Q	$25Q$
Dimes	10	$Q + 7$	$10(Q + 7)$
Nickels	5	$3Q$	$5(3Q)$
Mixture			320

← This time we are given the *total* value of the *mixture*

value of quarters + value of dimes + value of nickels = 320¢ ← \$3.20 = 320¢

$$25Q \quad + 10(Q + 7) + \quad 5(3Q) \quad = 320$$

$$25Q + 10(Q + 7) + 5(3Q) = 320$$
$$25Q + 10Q + 70 + 15Q = 320$$
$$50Q = 250$$
$$Q = 5$$

Therefore, she has

$Q = 5$ quarters

$Q + 7 = 12$ dimes

$3Q = 15$ nickels

Check
5 quarters = 5(25) = \$1.25
12 dimes = 12(10) = 1.20
15 nickels = 15(5) = + .75
\$3.20 ■

EXERCISES 10.3

In the following mixture problems: (a) Represent the unknown number(s) by a variable. (b) Set up an equation and solve. (c) Answer the question.

1. Rene wants to make a 12-lb mixture of peanuts and cashews. Peanuts cost \$4 per pound and cashews cost \$8 per pound. How many pounds of each kind of nut must be used if the mixture is to cost \$5 per pound?
2. Randy wants to mix dried figs with dried apricots to make an 8-pound mixture costing \$2.70 per pound. If dried figs cost \$1.80 per pound and dried apricots cost \$4.20 per pound, how many pounds of each are used?
3. Mrs. Martinez mixes 15 lb of English toffee candy costing \$1.25 per pound with caramels costing \$1.50 per pound. How many pounds of caramels must she use to make a mixture costing \$1.35 per pound?
4. 20 lb of kidney beans costing 45¢ per pound are mixed with green beans costing 70¢ per pound. How many pounds of green beans should be used to make a mixture costing 60¢ per pound?
5. A 10-pound mixture of nuts and raisins costs \$25. If raisins cost \$1.90 per pound and nuts \$3.40 per pound, how many pounds of each are used?
6. A 50-lb mixture of Delicious and Jonathan apples costs \$14.50. If the Delicious apples cost 30¢ per pound and the Jonathan apples cost 20¢ per pound, how many pounds of each kind are there?
7. Mr. Wong wants to mix 30 bushels of soybeans with corn to make a 100-bushel mixture costing \$4.85 per bushel. How much can he afford to pay for each bushel of corn if soybeans cost \$8.00 per bushel?
8. Mrs. Lavalle wants to mix 6 pounds of Brand A with Brand B to make a 10-pound mixture costing \$11.50. How much can she afford to pay per pound for Brand B if Brand A costs \$1.23 per pound?
9. Bill has 13 coins in his pocket that have a total value of 95¢. If these coins consist of nickels and dimes, how many of each kind are there?
10. Miko has 11 coins that have a total value of 85¢. If the coins are only nickels and dimes, how many of each kind are there?
11. Jennifer has 12 coins that have a total value of \$2.20. The coins are nickels and quarters. How many of each kind of coin are there?
12. Brian has 18 coins consisting of nickels and quarters. If the total value of the coins is \$2.50, how many of each kind of coin does he have?
13. Derek has \$4.00 in nickels, dimes, and quarters. If there are 4 more quarters than nickels and 3 times as many dimes as nickels, how many of each kind of coin does he have?
14. Staci has \$5.50 in nickels, dimes, and quarters. If there are 7 more dimes than nickels and twice as many quarters as dimes, how many of each kind of coin does she have?

10.4 Solution Problems

Solution problems involve mixing two or more liquid ingredients. They can be solved by the same method we used for mixture problems.

Example 1 How many liters of a 20% alcohol solution must be added to 3 liters of a 90% alcohol solution to make an 80% solution?
Solution Let x = liters of 20% solution

	Percent of Alcohol ·	*Number of Liters* =	*Total Amount of Alcohol*
20% Solution	0.20	x	$0.20x$
90% Solution	0.90	3	0.90(3)
Mixture	0.80	$x + 3$	$0.80(x + 3)$

amount of alcohol in 20% solution	+	amount of alcohol in 90% solution	=	amount of alcohol in 80% solution
$0.20x$	+	$0.90(3)$	=	$0.80(x+3)$

$$2x + 9(3) = 8(x + 3) \quad \text{Multiply by 10}$$
$$2x + 27 = 8x + 24$$
$$27 = 6x + 24$$
$$3 = 6x$$
$$\frac{1}{2} = x$$

Therefore, we need $\frac{1}{2}$ liter of the 20% solution.

Check

$\frac{1}{2}$ liter of 20% solution $= \frac{1}{2}(0.20) = 0.10$

3 liters of 90% solution $= 3(0.90) = +\,2.70$

$3\frac{1}{2}$ liters of 80% solution $= 3.5(0.80) = 2.80$ ■

Example 2 How many milliliters (mℓ) of water must be added to a 25% solution of glycerin to make 10 milliliters of a 5% solution? (Note: Water is 0% glycerin.)

Solution Let x = milliliters of water

Then $10 - x$ = milliliters of 25% solution

	Percent of Glycerin ·	*Number of Milliliters* =	*Total Amount of Glycerin*
Water	0	x	0
25% Solution	0.25	$10 - x$	$0.25(10 - x)$
Mixture	0.05	10	$0.05(10)$

amount of glycerin in water	+	amount of glycerin in 25% solution	=	amount of glycerin in 5% solution
0	+	$0.25(10-x)$	=	$0.05(10)$

$$0 + 25(10 - x) = 5(10) \quad \text{Multiply by 100}$$
$$250 - 25x = 50$$
$$-25x = -200$$
$$x = 8$$

Therefore, we need 8 mℓ of water.

Check

8 mℓ of water (0%) $= 8(0) = 0$

2 mℓ of 25% solution $= 2(0.25) = +\,0.50$

10 mℓ of 5% solution $= 10(0.05) = 0.50$ ■

EXERCISES 10.4

In the following solution problems: (a) Represent the unknown number(s) by a variable. (b) Set up an equation and solve. (c) Answer the question.

1. How many cubic centimeters (cc) of a 20% solution of sulfuric acid must be mixed with 100 cc of a 50% solution to make a 25% solution of sulfuric acid?
2. How many pints of a 2% solution of disinfectant must be mixed with 5 pints of a 12% solution to make a 4% solution of disinfectant?
3. If 100 gal of 75% glycerin solution is made up by combining a 30% glycerin solution with a 90% glycerin solution, how much of each solution must be used?
4. If 1600 cc of 10% dextrose solution is made up by combining a 20% dextrose solution with a 4% dextrose solution, how much of each solution must be used?
5. How many milliliters of water must be added to 500 ml of a 40% solution of sodium bromide to reduce it to a 25% solution? (Hint: Water is a 0% solution.)
6. How many liters of pure alcohol must be added to 10 liters of a 20% solution of alcohol to make a 50% solution? (Hint: Pure alcohol is a 100% solution.)
7. How many ounces of a 10% salt solution should be mixed with 20 ounces of a 25% solution to make a 16% solution?
8. How many milliliters of a 4% acid solution should be mixed with 100 milliliters of a 20% solution to make an 8% solution?
9. A chemist needs 50 milliliters of a 30% iodine solution. If she has only a 50% solution and a 25% solution in stock, how much of each will be needed?
10. How many quarts of pure antifreeze must be added to a 20% antifreeze solution to make 8 quarts of a 50% solution? (Hint: Pure antifreeze is a 100% solution.)

10.5 Distance Problems

A physical law relating *distance* traveled *d*, *rate* of travel *r*, and *time* of travel *t* is

Distance Formula

$$r \cdot t = d$$

For example, you know that if you are driving your car at an average speed of 50 mph, then $\quad r \cdot t = d$
you travel a distance of 100 miles in 2 hr: $50(2) = 100$
you travel a distance of 150 miles in 3 hr: $50(3) = 150$
and so on.

METHOD FOR SOLVING DISTANCE-RATE-TIME PROBLEMS

1. Draw the blank chart:

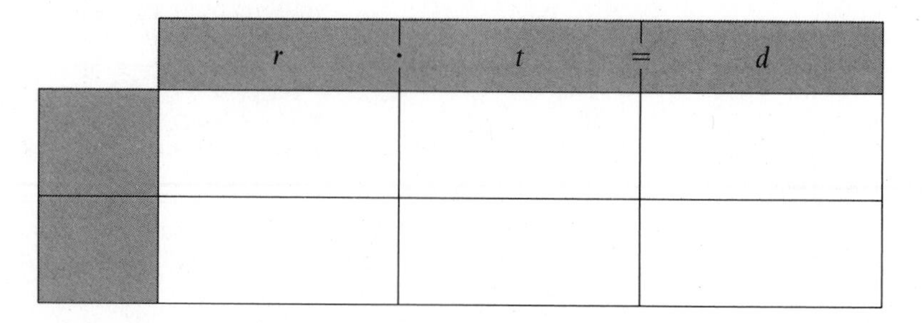

	r ·	t =	d

2. Fill in two columns of the chart using a single variable and the given information.
3. Use the formula $r \cdot t = d$ to fill in the remaining column.
4. Write the equation by using information in the chart with an unused fact given in the problem.
5. Solve the resulting equation.

Example 1 Mr. Maxwell takes 2 hours to drive to the airport in the morning, but he takes 3 hours to return home over the same route during the evening rush hour. If his average morning speed is 20 mph faster than his average evening speed, how far is it from his home to the airport?

Solution Let x = speed returning home

$x + 20$ = speed to airport

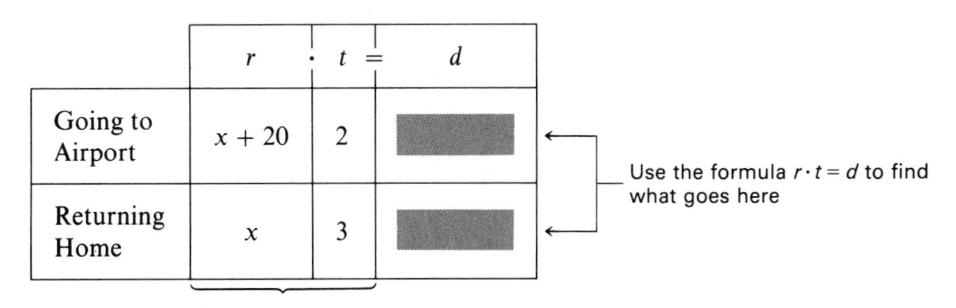

	r	$\cdot$ t $=$	d
Going to Airport	$x + 20$	2	$2(x + 20)$
Returning Home	x	3	$3x$

We also know that the two distances are equal.

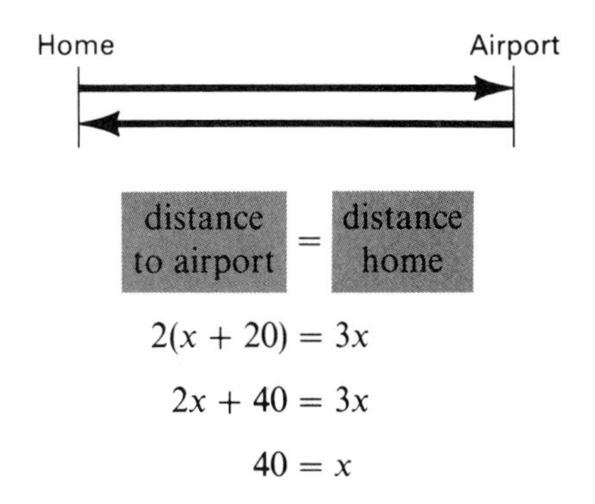

$$2(x + 20) = 3x$$

$$2x + 40 = 3x$$

$$40 = x$$

The problem asks for the *distance* from his home to the airport.

$$\text{distance to airport} = 2(x + 20) = 2(40 + 20) = 2(60) = 120 \text{ mi}$$

$$\text{distance home} = 3x = 3(40) = 120 \text{ mi} \longleftarrow \text{distances are equal} \quad \blacksquare$$

Example 2 Two trains are 600 miles apart and traveling toward each other. Train A's speed is 50 mph and train B's speed is 70 mph. How long will it take for the trains to meet?
Solution Let x = time for each train

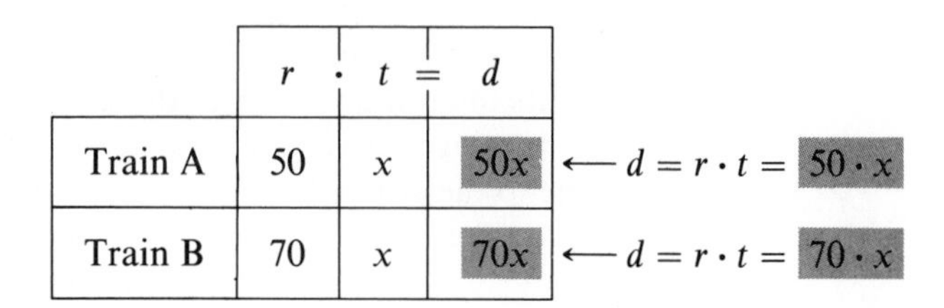

	r ·	t =	d	
Train A	50	x	$50x$	$\longleftarrow d = r \cdot t = 50 \cdot x$
Train B	70	x	$70x$	$\longleftarrow d = r \cdot t = 70 \cdot x$

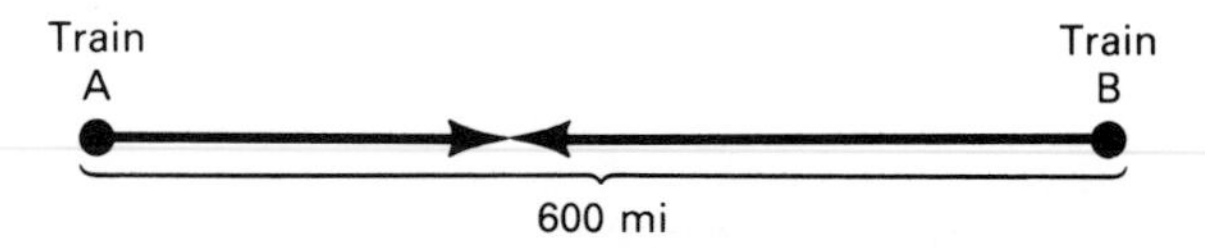

$$\text{distance for train A} + \text{distance for train B} = 600$$

$$50x + 70x = 600$$

$$120x = 600$$

$$x = 5$$

Therefore, they will meet in 5 hr.

Check Distance for train A = 50(5) = 250 mi
Distance for train B = 70(5) = + 350 mi
600 mi ■

Example 3 A boat cruises downstream for 4 hours. Due to the stream's current, it takes the boat 5 hours to go back upstream. If the speed of the stream is 3 mph, find the speed of the boat in still water.
Solution Let x = speed of boat in still water
then $x + 3$ = speed of boat downstream
and $x - 3$ = speed of boat upstream

	r ·	t =	d
Downstream	$x + 3$	4	$4(x + 3)$
Upstream	$x - 3$	5	$5(x - 3)$

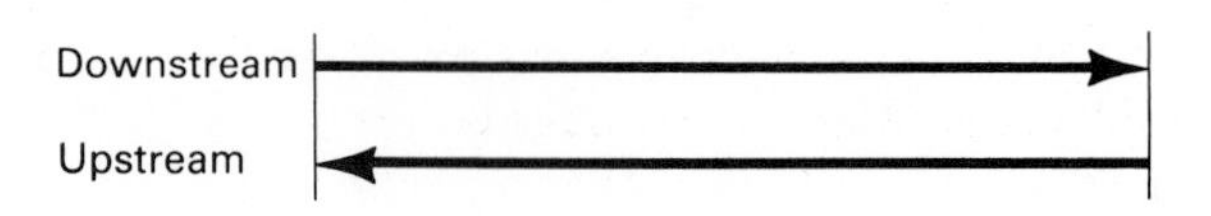

distance downstream	=	distance upstream
$4(x + 3)$	=	$5(x - 3)$
$4x + 12$	=	$5x - 15$
12	=	$x - 15$
27	=	x

Therefore, the speed of the boat is 27 mph.

Check Distance downstream $= 4(x + 3) = 4(27 + 3) = 4(30) = 120$ mi
Distance upstream $= 5(x - 3) = 5(27 - 3) = 5(24) = 120$ mi ■

EXERCISES 10.5

In the following word problems: (a) Represent the unknown number(s) using a variable. (b) Set up an equation and solve. (c) Answer the question.

1. The Malone family left San Diego by car at 7 A.M., bound for San Francisco. Their neighbors the King family left in their car at 8 A.M., also bound for San Francisco. By traveling 9 mph faster, the Kings overtook the Malones at 1 P.M.
a. Find the average speed of each car.
b. Find the total distance traveled by each car before they met.

2. The Duran family left Ames, Iowa, by car at 6 A.M., bound for Yellowstone National Park. Their neighbors the Silva family left in their car at 8 A.M., also bound for Yellowstone. By traveling 10 mph faster, the Silvas overtook the Durans at 4 P.M.
a. Find the average speed of each car.
b. Find the total distance traveled before they met.

3. Eric hiked from his camp to a lake in the mountains and returned to camp later in the day. He walked at a rate of 2 mph going to the lake and 5 mph coming back. If the trip to the lake took 3 hr longer than the trip back:
a. How long did it take him to hike to the lake?
b. How far is it from his camp to the lake?

4. Lee hiked from her camp up to an observation tower in the mountains and returned to camp later in the day. She walked up at the rate of 2 mph and jogged back at the rate of 6 mph. The trip to the tower took 2 hr longer than the return trip.
a. How long did it take her to hike to the tower?
b. How far is it from her camp to the tower?

5. Fran and Ron live 54 miles apart. Both leave their homes at 7 A.M. by bicycle, riding toward one another. They meet at 10 A.M. If Ron's average speed is 2 mph faster than Fran's speed, how fast does each cycle?

6. Danny and Cathy live 60 miles apart. Both leave their homes at 10 A.M. by bicycle, riding toward one another. They meet at 2 P.M. If Cathy's average speed is 3 mph slower than Danny's speed, how fast does each cycle?

7. Colin paddles a kayak downstream for 3 hr. After having lunch, he takes 5 hr to paddle back upstream. If the speed of the stream is 2 mph, how fast does Colin row in still water? How far downstream did he travel?

8. The Wright family sails their houseboat upstream for 4 hr, but it takes them only 2 hr to sail back downstream. If the speed of the houseboat in still water is 15 mph, what is the speed of the stream? How far upstream did the Wrights travel?

9. Mr. Zaleva flew his private plane from his office to his company's storage facility bucking a 20-mph head wind all the way. He flew home the same day with the same wind at his back. The *round trip* took 10 hr of flying time. If the plane makes 100 mph in still air, how far is the storage facility from his office?

10. A motor boat cruised from the marina to an island and then back to the marina. The speed of the boat in still water was 30 mph, and the speed of the current was 6 mph. If the *total* trip took 5 hr, how far is it from the marina to the island?

10.6 Variation Problems

A *variation* is an equation that relates one variable to one or more other variables by means of multiplication, division, or both.

Direct Variation Direct variation is a type of variation relating one variable to another by the formula

$$y = kx$$

(k — constant of proportionality)

Example 1 The circumference C of a circle varies directly with the diameter d according to the formula

$$C = \pi d$$

(π — constant of proportionality)

If the diameter is 2 in., the circumference $C = 3.14(2) = 6.28$ in. If the diameter is 3 in., the circumference $C = 3.14(3) = 9.42$ in. ■

In a direct variation problem where k is positive, when one variable increases, the other variable will also increase.

Example 2 y varies directly with x. If $y = 10$ when $x = 2$, find y when $x = 6$.

Solution

Step 1 Find k.

$$y = kx$$

$$10 = k(2) \quad \text{Substitute } y = 10 \text{ and } x = 2$$

$$\frac{10}{2} = \frac{2k}{2}$$

$$5 = k$$

Step 2 Find y when $x = 6$.

$$y = kx$$

$$y = 5(6) \quad \text{Substitute } k = 5 \text{ and } x = 6$$

$$y = 30$$ ■

Example 3 The distance s a spring stretches varies directly with the force F applied. If a 4-pound force stretches a spring 2 inches, how far will a 10-pound force stretch it?

Solution

Step 1 Find k.

$$s = kF$$

$$2 = k(4) \quad \text{Substitute } s = 2 \text{ and } F = 4$$

$$\frac{2}{4} = \frac{4k}{4}$$

$$\frac{1}{2} = k$$

Step 2 Find s when $F = 10$.

$$s = kF$$

$$s = \frac{1}{2}(10) \quad \text{Substitute } k = \frac{1}{2} \text{ and } F = 10$$

$$s = 5 \text{ in.} \quad \blacksquare$$

If the formula of the variation is $y = kx$, we say that "y varies directly with x." If the formula of the variation is $y = kx^2$, we say that "y varies directly with x^2."

Example 4 y varies directly with x^2. If $y = 20$ when $x = 2$, find y when $x = 6$.
Solution
Step 1 Find k.

$$y = kx^2$$

$$20 = k(2)^2 \quad \text{Substitute } y = 20 \text{ and } x = 2$$

$$\frac{20}{4} = \frac{4k}{4}$$

$$5 = k$$

Step 2 Find y when $x = 6$.

$$y = kx^2$$

$$y = 5(6)^2 \quad \text{Substitute } k = 5 \text{ and } x = 6$$

$$y = 5(36)$$

$$y = 180 \quad \blacksquare$$

Inverse Variation Inverse variation is a type of variation relating one variable to another by the formula

$$y = \frac{k}{x} \longleftarrow \text{constant of proportionality}$$

Example 5 For a 300-mile trip the formula relating the car's rate r and the time t is

$$r \cdot t = 300$$

or

$$t = \frac{300}{r} \longleftarrow \text{constant of proportionality}$$

If the car averages 50 miles per hour, the trip will take $t = \frac{300}{50} = 6$ hours. If the car averages 60 miles per hour, the trip will take $t = \frac{300}{60} = 5$ hours. $\blacksquare$

In an inverse variation problem where k is positive, when one variable increases, the other variable will decrease.

Example 6 y varies inversely with x. If $y = 6$ when $x = 2$, find y when $x = 3$.

Solution

Step 1 Find k.

$$y = \frac{k}{x}$$

$$6 = \frac{k}{2} \quad \text{Substitute } y = 6 \text{ and } x = 2$$

$$2 \cdot 6 = \frac{k}{\cancel{2}_1} \cdot \cancel{2}^1$$

$$12 = k$$

Step 2 Find y when $x = 3$.

$$y = \frac{k}{x}$$

$$y = \frac{12}{3} \quad \text{Substitute } k = 12 \text{ and } x = 3$$

$$y = 4 \quad \blacksquare$$

Example 7 The pressure P varies inversely with the volume V. If $P = 30$ when $V = 500$, find P when $V = 200$.

Solution

Step 1 Find k.

$$P = \frac{k}{V}$$

$$30 = \frac{k}{500} \quad \text{Substitute } P = 30 \text{ and } V = 500$$

$$500 \cdot 30 = \frac{k}{\cancel{500}_1} \cdot \cancel{500}^1$$

$$15{,}000 = k$$

Step 2 Find P when $V = 200$.

$$P = \frac{k}{V}$$

$$P = \frac{15{,}0\cancel{0}\cancel{0}}{2\cancel{0}\cancel{0}} \quad \text{Substitute } k = 15{,}000 \text{ and } V = 200$$

$$P = 75 \quad \blacksquare$$

If the formula of the variation is $y = \frac{k}{x^2}$, we say "y varies inversely with x^2."

Example 8 y varies inversely with x^2. If $y = -3$ when $x = 4$, find y when $x = -6$.

Solution

Step 1 Find k.

$$y = \frac{k}{x^2}$$

$$-3 = \frac{k}{4^2} \quad \text{Substitute } y = -3 \text{ and } x = 4$$

$$16(-3) = \frac{k}{\cancel{16}_1} \cdot \overset{1}{\cancel{16}}$$

$$-48 = k$$

Step 2 Find y when $x = -6$.

$$y = \frac{k}{x^2}$$

$$y = \frac{-48}{(-6)^2} \quad \text{Substitute } k = -48 \text{ and } x = -6$$

$$y = \frac{-48}{36} = -\frac{4}{3} \quad \blacksquare$$

The formulas for variation can be summarized as follows:

DIRECT VARIATION

	Formula
y varies directly with x	$y = kx$
y varies directly with x^2	$y = kx^2$

INVERSE VARIATION

	Formula
y varies inversely with x	$y = \frac{k}{x}$
y varies inversely with x^2	$y = \frac{k}{x^2}$

EXERCISES 10.6

1. y varies directly with x. If $y = 14$ when $x = 2$, find y when $x = 4$.
2. y varies directly with x. If $y = 6$ when $x = -2$, find y when $x = 10$.
3. y varies directly with x. If $y = -9$ when $x = -3$, find y when $x = 4$.
4. y varies directly with x. If $y = -6$ when $x = 3$, find y when $x = 5$.

5. y varies inversely with x. If $y = 6$ when $x = 5$, find y when $x = 10$.

6. y varies inversely with x. If $y = 2$ when $x = -3$, find y when $x = \frac{1}{2}$.

7. y varies inversely with x. If $y = 3$ when $x = -2$, find y when $x = 3$.

8. y varies inversely with x. If $y = -\frac{1}{2}$ when $x = 16$, find y when $x = -2$.

9. y varies directly with x^2. If $y = -20$ when $x = -2$, find y when $x = 5$.

10. y varies directly with x^2. If $y = -12$ when $x = -2$, find y when $x = 5$.

11. y varies directly with x^2. If $y = 90$ when $x = -3$, find y when $x = 10$.

12. M varies directly with P^2. If $M = 2$ when $P = 2$, find M when $P = 8$.

13. F varies inversely with d^2. If $F = 3$ when $d = -4$, find F when $d = 8$.

14. L varies inversely with r^2. If $L = 16$ when $r = -3$, find L when $r = 4$.

15. C varies inversely with v^2. If $C = 6$ when $v = -3$, find C when $v = 6$.

16. y varies inversely with x^2. If $y = \frac{1}{2}$ when $x = 4$, find y when $x = 8$.

17. The distance s an object falls (in a vacuum) varies directly with the square of the time t it takes to fall. If an object falls 64 feet in 2 seconds, how far will it fall in 3 seconds?

18. The air resistance R on a car varies directly with the square of the car's velocity v. If the air resistance is 400 pounds at 60 miles per hour, find the air resistance at 90 miles per hour.

19. The intensity I of light received from a light source varies inversely with the square of the distance d from the source. If the light intensity is 15 candelas at a distance of 10 feet from the light source, what is the light intensity at a distance of 15 feet?

20. The pressure P of a gas (at constant temperature) varies inversely with its volume V. If the pressure is 15 pounds per square inch when the volume is 350 cubic inches, find the pressure when the volume is 70 cubic inches.

21. The amount s a spring is stretched varies directly with the force F applied. If a 5-pound force stretches a spring 3 inches, how far will a 2-pound force stretch it?

22. The resistance R of a boat moving through water varies directly with the square of its speed S. If its resistance is 50 pounds at a speed of 10 knots, what is the resistance at 20 knots?

23. The volume V of gas (at constant temperature) varies inversely with its pressure P (Boyle's law). If the volume is 1600 cc at a pressure of 250 mm of mercury, find the volume at a pressure of 400 mm of mercury.

24. The sound intensity I (loudness) varies inversely with the square of the distance d from the source. If a pneumatic drill has a sound intensity of 75 decibels at 50 feet, find its sound intensity at 150 feet.

25. The weight W of an object varies inversely with the square of the distance d separating that object from the center of the earth. If a man at the surface of the earth weighs 160 pounds, find his weight at an altitude of 8000 miles. (Hint: At the surface of the earth the man is 4000 miles from the earth's center. Therefore, at an altitude of 8000 miles, he is 12,000 miles from the earth's center.)

26. The pressure p in water varies directly with the depth d. If the pressure at a depth of 100 feet is 43.3 pounds per square inch (neglecting atmospheric pressure), find the pressure at a depth of 60 feet.

10.7 Chapter Summary

To Solve Word Problems 10.2–10.5

1. Read the problem and determine what is unknown.
2. Represent each unknown in terms of the same variable.
3. Break up the word statement into small pieces and represent each piece by an algebraic expression.

4. After each of the pieces has been written as an algebraic expression, fit them together into an equation.
5. Solve the equation for the unknown letter.
6. Check the solution in the *original word* statement.

To Solve Mixture Problems 10.3

1. amount of ingredient A + amount of ingredient B = amount of mixture
2. cost of ingredient A + cost of ingredient B = cost of mixture

To Solve Distance Problems 10.5

1. Draw the blank chart:

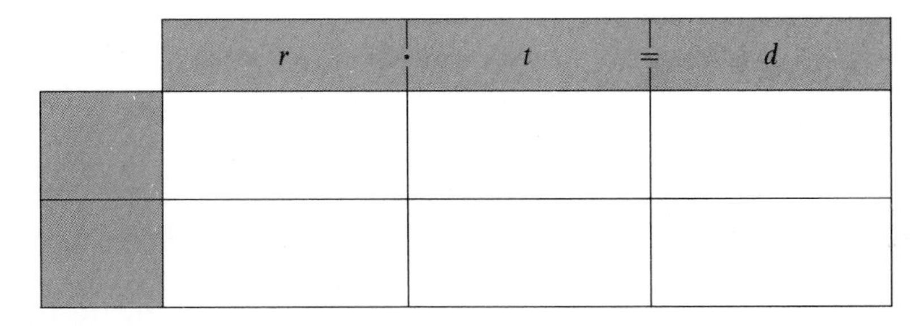

	r ·	t =	d

2. Fill in two columns using a single variable and the given information.
3. Use the formula $r \cdot t = d$ to fill in the remaining column.
4. Write the equation by using information in the chart along with an unused fact given in the problem.
5. Solve the resulting equation.

To Solve a Variation Problem 10.6

1. Choose the correct variation formula for the given problem.

Direct Variation	Formula
y varies directly with x	$y = kx$
y varies directly with x^2	$y = kx^2$

Inverse Variation	Formula
y varies inversely with x	$y = \frac{k}{x}$
y varies inversely with x^2	$y = \frac{k}{x^2}$

2. Substitute the first set of given numbers into the formula and solve for k.
3. Substitute the value of k and the third given number into the formula and solve for the unknown variable.

Review Exercises 10.7

1. Eight minus an unknown number is equal to the unknown number plus four. Find the unknown number.

2. Three times an unknown number is added to fifteen, the sum is forty-two. Find the unknown number.

3. Find three consecutive integers whose sum is -21.

4. Find two consecutive odd integers such that twice the first minus the second is 9.

5. The length of a rectangle is five feet more than the width. If the perimeter is seventy feet, find the length and the width.

6. The area of a triangle is 90 square centimeters, and the base is 18 centimeters. What is the height of the triangle?

7. Doris has 27 coins that have a total value of \$2.15. If all the coins are nickels or dimes, how many of each does she have?

8. Staci has \$2.40 in nickels, dimes, and quarters. If there are 4 more quarters than nickels and 4 times as many dimes as nickels, how many of each kind of coin does she have?

9. A 10-pound mixture of almonds and walnuts costs \$7.20. If walnuts cost 69¢ per pound and almonds 79¢ per pound, how many pounds of each kind are there?

10. A dealer makes up a 100-pound mixture of Colombian coffee costing \$1.70 per pound and Brazilian coffee costing \$1.50 per pound. How many pounds of each kind must he use in order for the mixture to cost \$1.56 per pound?

11. How many cubic centimeters of a 50% phenol solution must be added to 400 cc of a 5% solution to make it a 10% solution?

12. How many milliliters of water must be added to a 40% solution of sodium bromide to make 40 milliliters of a 25% solution?

13. A carload of campers leaves Los Angeles for Lake Havasu at 8:00 A.M. A second carload of campers leaves Los Angeles at 9:00 A.M. and drives 10 mph faster over the same road. If the second car overtakes the first at 1:00 P.M., what is the average speed of each car?

14. Two airplanes take off from the same airport traveling in opposite directions. Plane A is traveling 50 mph faster than plane B, and at the end of 4 hours they are 1400 miles apart. Find the speed of each plane.

15. y varies inversely with x. If $y = -6$ when $x = 2$, find x when $y = -3$.

16. y varies directly with x. If $y = -6$ when $x = 2$, find y when $x = 4$.

17. y varies directly with x^2. If $y = 16$ when $x = 2$, find y when $x = 3$.

18. y varies inversely with x^2. If $y = 2$ when $x = -3$, find y when $x = 2$.

19. The electrical resistance R of a wire varies inversely with the square of its diameter d. If the resistance of a wire having a diameter of 0.02 of an inch is 9 ohms, find the resistance of a wire of the same length and material having a diameter of 0.03 of an inch.

20. In a business, the revenue R varies directly with the number of items sold n, if the price is fixed. If the revenue is \$12,000 when 800 items are sold, how many items must be sold for the revenue to be \$15,000?

21. Don has 10 coins consisting of dimes and quarters. After paying for a 35-cent phone call, he had \$1.10 left. How many dimes did he have after the phone call?

22. It takes a boat 6 hours to go upstream to the marina but only 4 hours to go back to the dock downstream. If the speed of the boat in still water is 20 mph, what is the speed of the current?

Chapter 10 Diagnostic Test

Allow yourself about an hour to do these problems. Complete solutions for every problem, together with section references, are given in the answer section at the end of the book.

1. When 16 is added to three times an unknown number, the sum is 37. Find the unknown number.

2. Find three consecutive even integers such that their sum is 54.

3. Mario has 15 coins consisting of dimes and quarters. If the total value is \$3.00, how many of each kind of coin are there?

4. The length of a rectangle is ten inches more than the width. If the perimeter is sixty inches, find the length and the width.

5. When 4 is subtracted from five times an unknown number, the result is the same as when 6 is added to three times the unknown number. Find the unknown number.

6. A wholesaler wants to make a 10-lb mixture of two kinds of tea. Brand A costs \$2.50 per pound, and Brand B costs \$3.75 per pound. How many pounds of each kind of tea must be used if the mixture is to cost \$3.00 per pound?

7. y varies inversely with x. If $y = 3$ when $x = -4$, find y when $x = \frac{4}{5}$.

8. How many pints of a 2% solution of disinfectant must be mixed with 4 pints of a 12% solution to make a 4% solution of disinfectant?

9. It took John 4 hours to drive to the mountains but 5 hours to drive home. If his average speed driving to the mountains is 12 mph faster than his average speed driving home, how far is it from his home to the mountains?

10. The pressure p in water varies directly with the depth d. If the pressure at a depth of 40 feet is 17.32 pounds per square inch (neglecting atmospheric pressure), find the pressure at a depth of 70 feet.

Cumulative Review Exercises: Chapters 1–10

In Exercises 1–10, perform the indicated operations. Reduce all fractions to lowest terms, and change any improper fractions to mixed numbers.

1. $7\frac{4}{15} + 9\frac{5}{6}$

2. $3\frac{3}{4} \div 1\frac{1}{8}$

3. $4\frac{1}{8} + 12.6$

4. $51 \div 0.06$

5. $-4^2 - 6 \cdot (-2)$

6. $9 - 2[6 - (3 - 5)]$

7. $(3x^2 - 6x - 2) - (x^2 - 4x + 5)$

8. $(2x - 5)^2$

9. $(y^2 - 3y + 4)(y - 2)$

10. $(2x^2 - 10x + 12) \div (x - 3)$

In Exercises 11–14, solve each equation or inequality.

11. $3x + 6 + 2x = 9 - x - 15$

12. $8 + 4(x - 3) = 2(8 + 3x)$

13. $3[2m - 2(6 - m)] = -12$

14. $2x - 3(4 + x) \geq -7$

15. Change $8\frac{1}{3}\%$ to a fraction in lowest terms.

16. 150 is what percent of 120?

17. Find the cost of $4\frac{3}{4}$ yards of molding at $0.69 per yard. Round off answer to the nearest cent.

18. A car burns $2\frac{1}{2}$ quarts of oil on an 800-mile trip. How many quarts of oil can the owner expect to use on a 4,800-mile trip?

19. Seven subtracted from twice an unknown number is equal to nine. Find the number.

20. y varies directly with x^2. If $y = 12$ when $x = 2$, find y when $x = 5$.

21. How many ounces of a 10% acid solution should be mixed with 12 ounces of a 30% solution to make a 25% acid solution?

22. Two boats are 400 miles apart and traveling toward each other. Boat A is traveling 20 mph faster than Boat B. If they meet in 5 hours, find the speed of each boat.

23. Evaluate $xy^2 - 4xy$ if $x = 2$, and $y = -3$.

In Exercises 24 and 25, use the following heights: {62 in., 56 in., 59 in., 68 in., 72 in., 79 in.}

24. Find the mean height.

25. Find the median height.

11 Factoring

In this chapter we discuss factoring. Factoring is essential for working with fractions and solving certain kinds of equations.

11.1 Prime Factorization and Greatest Common Factor (GCF)

Products and Positive Factors We know that $2 \cdot 3 = 6$. There are two ways of looking at this fact. We can think of starting with the $2 \cdot 3$ and *finding the product* 6. Or we can start with the 6 and ask ourselves what positive integers multiplied together will give us 6, then think of $2 \cdot 3$. When we do this we are *finding the factors of* 6, or more simply, *factoring* 6.

Finding the product of 2 and 3:

Product

$$\overrightarrow{2 \cdot 3 = 6}$$

product — Starting with $2 \cdot 3$ and *finding the* **product** 6

Factoring 6:

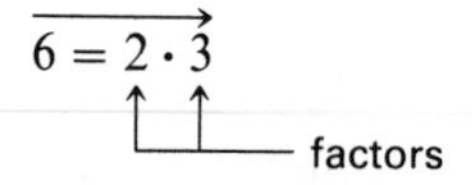

Starting with 6 and *finding the* **factors** 2 and 3

Factors
Divisors

In the latter case, 2 and 3 are called *factors* (or **divisors**) of 6. But because $6 = 1 \cdot 6$, 1 and 6 are also factors of 6. Therefore, 6 has four positive factors: 1, 2, 3, and 6.

Prime Factorization of Positive Integers

Just as prime factoring was useful when working with arithmetic fractions in Chapter 2, prime factoring will be helpful when working with algebraic fractions in Chapter 12.

Prime Numbers

A **prime number** is a positive integer greater than 1 that can be divided evenly only by itself and 1. A prime number has no factors other than itself and 1.

Composite Numbers

A **composite number** is a positive integer that can be divided evenly by some integer other than itself and 1. A composite number has factors other than itself and 1.

Note that 1 is neither prime nor composite.

Example 1 Prime and composite numbers

a. 9 is a composite number because $3 \cdot 3 = 9$, so that 9 has a factor other than itself or 1.

b. 17 is a prime number because 1 and 17 are the only integral factors of 17.

c. 45 is a composite number because it has factors 3, 5, 9, and 15 other than 1 and 45.

d. 31 is a prime number because 1 and 31 are the only integral factors of 31. ■

A partial list of prime numbers is: 2, 3, 5, 7, 11, 13, 17, 19, 23, 29, ...

Prime Factorization

The **prime factorization** of a positive integer is the indicated product of all its factors that are themselves prime numbers. For example,

$$\left.\begin{aligned} 18 &= 2 \cdot 9 \\ 18 &= 3 \cdot 6 \end{aligned}\right\}$$

These *are not prime* factorizations because 9 and 6 are not prime numbers

$$\left.\begin{aligned} 18 &= 2 \cdot 9 = 2 \cdot 3 \cdot 3 = 2 \cdot 3^2 \\ 18 &= 3 \cdot 6 = 3 \cdot 2 \cdot 3 = 2 \cdot 3^2 \end{aligned}\right\}$$ These *are prime* factorizations because all the factors are prime numbers

Note that the two ways we factored 18 led to the *same* prime factorization ($2 \cdot 3^2$). The prime factorization of *any* positive integer (greater than 1) is unique.

Method for Finding the Prime Factorization The smallest prime is 2; the next smallest is 3; the next smallest is 5; and so forth.

NOTE The only even prime number is 2. All other even numbers are composite numbers with 2 among their factors. ☑

The work of finding the prime factorization of a number can be conveniently arranged as shown in Example 2.

Example 2 Finding the prime factorization

a. Find the prime factorization of 30.

Solution

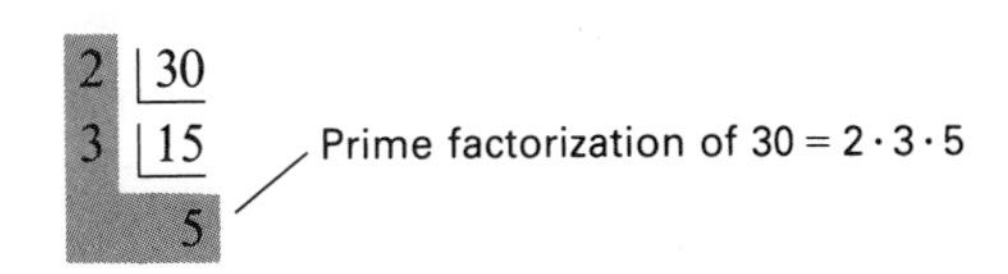

b. Find the prime factorization of 36.

Solution

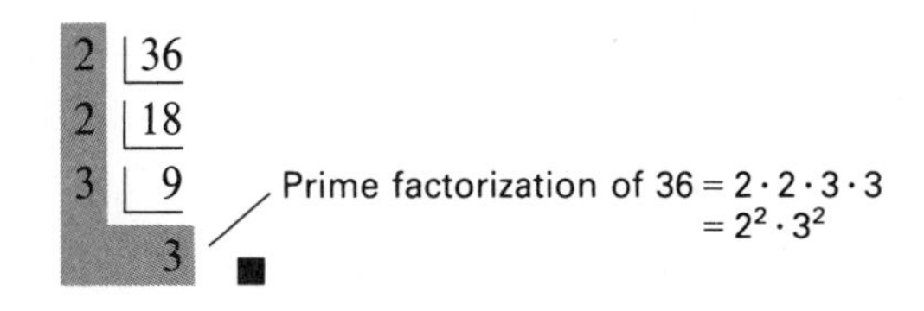

■

When trying to find a prime factor of a number, no prime whose square is greater than that number need be tried (Example 3).

Example 3 Find the prime factorization of 97.
Solution

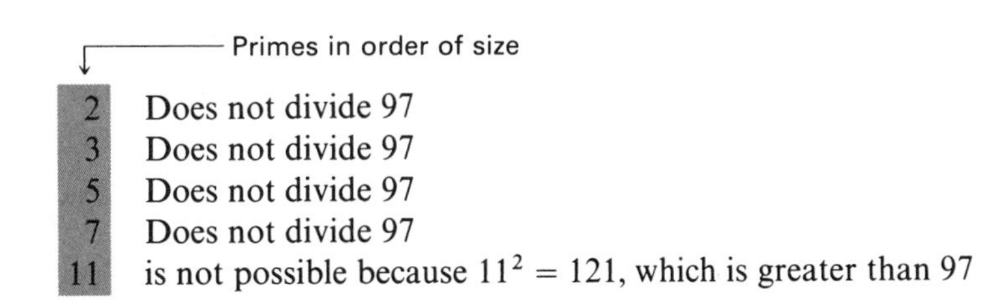

Therefore, 97 is prime. ■

Greatest Common Factor (GCF)

GCF

The **greatest common factor** (*GCF*) of two integers is the greatest integer that is a factor of both integers.

Example 4 Find the GCF of 12 and 16.

Solution

$$\begin{array}{cc} 12 & 16 \\ 2\cdot 2\cdot 3 & 2\cdot 2\cdot 2\cdot 2 \end{array}$$

The greatest factor common to both 12 and 16 is $2 \cdot 2 = 4$

Therefore, the GCF of 12 and 16 is 4. ■

We can also find the GCF of terms in an algebraic expression.

Example 5 Find the GCF for the terms of $6y^3 - 21y$.
Solution

$$6y^3 - 21y = 2 \cdot 3 \cdot y \cdot y \cdot y - 7 \cdot 3 \cdot y$$ Prime factorization of terms

GCF = $3y$ ■

GCF and Polynomial Factors

Example 6 GCF and Polynomial Factors

a. $2(x - 5) = 2x - 10$ By the distributive rule

Therefore, 2 and $x - 5$ are factors of $2x - 10$.

GCF ↑ (2) ↑ Polynomial factor ($x - 5$)

b. $5z(3z^2 + 2z - 1) = 15z^3 + 10z^2 - 5z$ By the distributive rule

Therefore, $5z$ and $3z^2 + 2z - 1$ are factors of $15z^3 + 10z^2 - 5z$.

GCF ↑ ($5z$) ↑ Polynomial factor ($3z^2 + 2z - 1$) ■

TO FIND THE GREATEST COMMON FACTOR (GCF)

1. Write the *prime* factors of each term. Repeated factors should be expressed as powers.
2. Write each different prime factor that is common to all terms.
3. Raise each prime factor (selected in Step 2) to the *lowest* power it occurs anywhere in the expression.
4. The greatest common factor is the product of all the powers found in Step 3.

TO FIND THE POLYNOMIAL FACTOR

5. Divide the expression being factored by the greatest common factor found in Step 4.

Check. Find the product of the greatest common factor and the polynomial factor by using the distributive rule.

Example 7 Factor $15x^3 + 9x$.
Solution Finding the GCF

Step 1: $15x^3 + 9x = 3 \cdot 5x^3 + 3^2x$ Each term in prime factored form

Steps 2 and 3:
- 3 is common to both terms
- 3^1 is the *lowest* power of 3 that occurs in any term
- x is common to both terms
- x^1 is the *lowest* power of x that occurs in any term

Step 4: GCF $= 3^1x^1 = 3x$.

<u>Finding the Polynomial Factor</u>

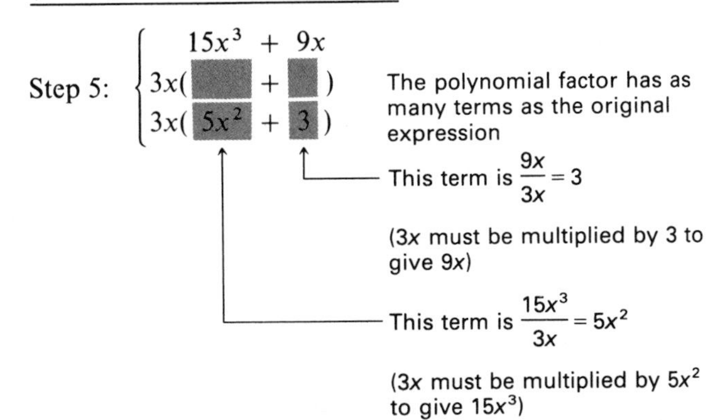

Therefore, the factors of $15x^3 + 9x$ are $3x$ and $(5x^2 + 3)$. Thus

$$15x^3 + 9x = \underbrace{3x}_{\text{GCF}}(\underbrace{5x^2 + 3}_{\text{Polynomial factor}})$$

Check $3x(5x^2 + 3) = (3x)(5x^2) + (3x)(3) = 15x^3 + 9x$ ■

Example 8 Factor $2x^4 + 4x^3 - 8x^2$.

Solution

$$2x^4 + 4x^3 - 8x^2$$
$$= 2^1x^4 + 2^2x^3 - 2^3x^2 \quad \text{Each term in prime factored form}$$

2^1 is the *lowest* power of 2 that occurs in any term.
x^2 is the *lowest* power of x that occurs in any term.
GCF $= 2^1x^2 = 2x^2$.

The polynomial factor $= \dfrac{2x^4}{2x^2} + \dfrac{4x^3}{2x^2} - \dfrac{8x^2}{2x^2} = x^2 + 2x - 4$.

Therefore, $2x^4 + 4x^3 - 8x^2 = \underbrace{2x^2}_{\text{GCF}}(\underbrace{x^2 + 2x - 4}_{\text{Polynomial factor}})$. ■

Example 9 Factor $3xy - 6y^2 - 3y$.

Solution

$$3xy - 6y^2 - 3y$$
$$= 3xy - 2 \cdot 3y^2 - 3y$$

GCF $= 3 \cdot y = 3y$.

The polynomial factor $= \dfrac{3xy}{3y} - \dfrac{6y^2}{3y} - \dfrac{3y}{3y}$
$= x - 2y - 1$.

Therefore, $3xy - 6y^2 - 3y = 3y(x - 2y - 1)$. ■

Sometimes a common factor is not a monomial.

Example 10 Factor $a(x + y) + b(x + y)$. This is a binomial.

Solution

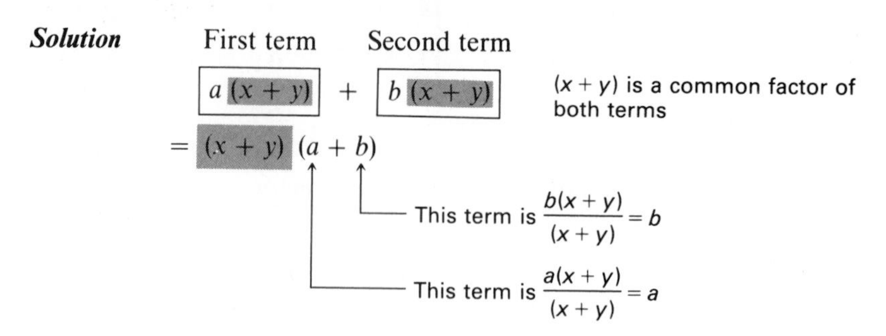

Therefore, $a(x + y) + b(x + y) = (x + y)(a + b)$. ■

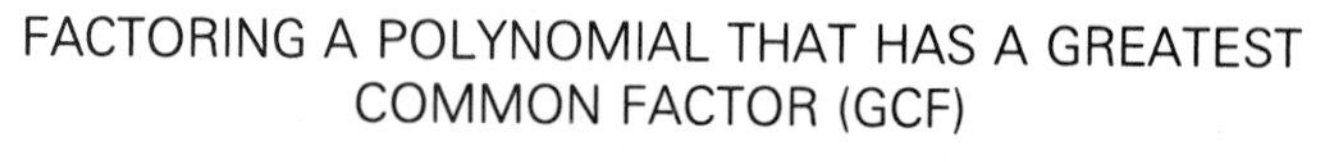

FACTORING A POLYNOMIAL THAT HAS A GREATEST COMMON FACTOR (GCF)

1. Find the GCF.

	1st term		2nd term	
2. Write the polynomial to be factored here.	▭	+	▭	
GCF (	▭	+	▭	)

3. Write the term that the GCF must be multiplied by to give the 1st term of the polynomial to be factored. (arrow to 1st box)

Write the term that the GCF must be multiplied by to give the 2nd term of the polynomial to be factored. (arrow to 2nd box)

4. Check by multiplying the GCF and the polynomial factor by using the distributive rule.

EXERCISES 11.1

In Exercises 1–16, state whether each of the numbers is prime or composite. To justify your answer, give the set of all positive integral factors for each number.

1. 5 **2.** 8 **3.** 13 **4.** 15 **5.** 12 **6.** 11 **7.** 21 **8.** 23

9. 55 **10.** 41 **11.** 49 **12.** 31 **13.** 51 **14.** 42 **15.** 111 **16.** 101

In Exercises 17–32, find the prime factorization of each number.

17. 14 **18.** 15 **19.** 21 **20.** 22 **21.** 26 **22.** 27 **23.** 29 **24.** 31

25. 32 **26.** 33 **27.** 34 **28.** 35 **29.** 84 **30.** 75 **31.** 144 **32.** 180

In Exercises 33–66, factor each expression if possible.

33. $2x - 8$ **34.** $3x - 9$ **35.** $5a + 10$ **36.** $7b + 14$

37. $6y - 3$ **38.** $15z - 5$ **39.** $9x^2 + 3x$ **40.** $8y^2 - 4y$

41. $10a^3 - 25a^2$ **42.** $27b^2 - 18b^4$ **43.** $21w^2 - 20z^2$ **44.** $15x^3 - 16y^3$

45. $2a^2b + 4ab^2$ **46.** $3mn^2 + 6m^2n^2$ **47.** $12c^3d^2 - 18c^2d^3$ **48.** $15ab^3 - 45a^2b^4$

49. $4x^3 - 12x - 24x^2$ **50.** $18y - 6y^2 - 30y^3$ **51.** $24a^4 + 8a^2 - 40$ **52.** $45b^3 - 15b^4 - 30$

53. $-14x^8y^9 + 42x^5y^4 - 28xy^3$ **54.** $-21u^7v^8 - 63uv^5 + 35u^2v^5$

55. $15h^2k - 8hk^2 + 9st$ **56.** $10uv^3 + 5u^2v - 4wz$

57. $18x^{10}y^5 + 24x^7y^4 - 12x^5y^2$ **58.** $30a^3b^4 - 15a^6b^3 + 45a^8b^2$

59. $32m^5n^7 - 24m^8n^9 - 40m^3n^6$ **60.** $18x^3y^4 - 12x^2y^3 - 48x^4y^3$

61. $m(a + b) + n(a + b)$ **62.** $x(m - n) - y(m - n)$

63. $x(y + 1) - 1(y + 1)$ **64.** $a(c - d) + b(c - d)$

65. $3a(a - 2b) + 2(a - 2b)$ **66.** $2e(3e - f) - 3(3e - f)$

11.2 Factoring by Grouping

Sometimes it is possible to factor an expression by rearranging its terms into smaller groups and then finding the GCF of *each group*. The GCF does not have to be a monomial.

> TO FACTOR AN EXPRESSION OF FOUR TERMS BY GROUPING
>
> 1. Arrange the four terms into two groups of two terms each. Each group of two terms must have a GCF.
> 2. Factor each group by using its GCF.
> 3. Factor the two-term expression resulting from Step 2 if the two terms have a GCF.

Example 1 Factor $ax + ay + bx + by$.

Solution

GCF $= a$ GCF $= b$

$$ax + ay \quad bx + by$$

$$= a(x + y) + b(x + y) \qquad (x + y) \text{ is the GCF of these two terms}$$

$$= (x + y)(a + b)$$

This term is $\dfrac{b(x+y)}{(x+y)} = b$

This term is $\dfrac{a(x+y)}{(x+y)} = a$

Therefore, $ax + ay + bx + by = (x + y)(a + b)$. ■

It is sometimes possible to group terms differently and still be able to factor the expression. The same factors are obtained no matter what grouping is used.

Example 2 Factor $ab - b + a - 1$.

Solution

Different Ways of Grouping

One grouping	*A different grouping*
$ab - b + a - 1$	$ab + a - b - 1$
$= b(a-1) + 1(a-1)$	$= a(b+1) - 1(b+1)$
$= (a-1)(b+1)$	$= (b+1)(a-1)$

Same factors

Therefore, $ab - b + a - 1 = (a-1)(b+1) = (b+1)(a-1)$. ■

Example 3 Factor $2x^2 - 6xy + 3x - 9y$.

Solution GCF = $2x$ GCF = 3

$$2x^2 - 6xy + 3x - 9y$$
$$= 2x(x - 3y) + 3(x - 3y) \quad (x - 3y) \text{ is the GCF}$$
$$= (x - 3y)(2x + 3)$$

Therefore, $2x^2 - 6xy + 3x - 9y = (x - 3y)(2x + 3)$. ■

A WORD OF CAUTION An expression is *not* factored until it has been written as a single term that is a product of factors. To illustrate this, consider Example 1 again:

$$ax + ay + bx + by$$
$$= \boxed{a(x+y)} + \boxed{b(x+y)}$$

First term Second term

The expression is *not* in factored form because it has *two* terms

$$= \boxed{(x+y)(a+b)}$$

Single term

Factored form of $ax + ay + bx + by$ ✓

Expressions with more than four terms may also be factored by grouping. However, in this book we only consider factoring expressions of four terms by grouping.

EXERCISES 11.2

Factor each expression if possible.

1. $am + bm + an + bn$
2. $cu + cv + du + dv$
3. $mx - nx - my + ny$
4. $ah - ak - bh + bk$
5. $xy + x - y - 1$
6. $ad - d + a - 1$
7. $3a^2 - 6ab + 2a - 4b$
8. $2h^2 - 6hk + 5h - 15k$
9. $6e^2 - 2ef - 9e + 3f$
10. $8m^2 - 4mn - 6m - 3n$
11. $ax + ay + bx + by$
12. $hw - kw - hz + kz$
13. $ef + f - e + 1$
14. $2s^2 - 6st + 5s - 15t$
15. $10xy - 15y + 8x - 12$
16. $35 - 42m - 18mn + 15n$
17. $x^3 + 3x^2 - 2x - 6$
18. $2a^3 - 8a^2 + 3a - 12$

11.3 Factoring the Difference of Two Squares

Each type of factoring depends upon a particular *special product*. GCF factoring is based on products found by using the distributive rule. The kind of factoring discussed in this section depends upon the special product $(a + b)(a - b)$. (Section 8.7)

$$\begin{aligned}(a + b)(a - b) &= a^2 - ab + ab - b^2 \\ &= a^2 - b^2\end{aligned}$$

Thus, the product of the sum and difference of two terms equals the difference of two squares. To factor the difference of two squares, we need to reverse this process.

Finding the Product →

$$(a + b)(a - b) = a^2 - b^2$$

← Finding the Factors

Therefore, $a^2 - b^2$ *factors into* $(a + b)(a - b)$.

TO FACTOR THE DIFFERENCE OF TWO SQUARES $(a^2 - b^2)$

The factors of the difference of two squares are the sum and difference of the two terms that were squared.

$$a^2 - b^2 = (a + b)(a - b)$$

One factor has +, the other has −

Example 1 Factor $x^2 - 4$.

Solution

$$\begin{aligned}x^2 - 4 &= x^2 - 2^2 \\ &= (x + 2)(x - 2) \quad \blacksquare\end{aligned}$$

Example 2 Factor $25x^2 - 9y^2$.

Solution

$$\begin{aligned}25x^2 - 9y^2 &= (5x)^2 - (3y)^2 \\ &= (5x + 3y)(5x - 3y) \quad \blacksquare\end{aligned}$$

Example 3 Factor $49a^6 - 81b^4$

Solution

$$\begin{aligned}49a^6 - 81b^4 &= (7a^3)^2 - (9b^2)^2 \\ &= (7a^3 + 9b^2)(7a^3 - 9b^2) \quad \blacksquare\end{aligned}$$

Example 4 Factor $50x - 2x^3$.

Solution

$$\begin{aligned}50x - 2x^3 &= 2x(25 - x^2) \quad \text{Remove common factor } \textit{first} \\ &= 2x(5^2 - x^2) \\ &= 2x(5 + x)(5 - x) \quad \blacksquare\end{aligned}$$

EXERCISES 11.3

Factor each expression if possible.

1. $x^2 - 9$
2. $a^2 - 36$
3. $b^2 - 1$
4. $1 - x^2$
5. $m^2 - n^2$
6. $u^2 - v^2$
7. $4c^2 - 25$
8. $16d^2 - 1$
9. $25a^2 - 4b^2$
10. $9x^2 - 100y^2$
11. $64a^2 - 49b^2$
12. $36x^2 - 121y^2$
13. $2x^2 - 8$
14. $3a^2 - 3$
15. $5xy^2 - 5xa^2$
16. $4ab^2 - 4ac^2$
17. $3x^3 - 75x$
18. $2y^3 - 18y$
19. $9h^2 - 10k^2$
20. $16e^2 - 15f^2$
21. $x^6 - a^4$
22. $b^2 - y^6$
23. $49u^4 - 36v^4$
24. $81m^6 - 100n^4$
25. $a^2b^2 - c^2d^2$
26. $m^2n^2 - r^2s^2$
27. $49 - 25w^2z^2$
28. $36 - 25u^2v^2$
29. $3x^4 - 27x^2$
30. $6x^2 - 6x^4$
31. $4a^3b - 9ab^3$
32. $16x^4y - 25x^2y^3$

11.4 Factoring a Trinomial Whose Leading Coefficient Is 1

Leading Coefficient

The **leading coefficient** of a polynomial is the numerical coefficient of its *highest*-degree term.

Example 1 Finding the leading coefficient

a. The leading coefficient of $x^2 - 2x + 8$ is 1.

b. The leading coefficient of $2x^2 - 3x + 5$ is 2.

c. The leading coefficient of $2y - 5y^2 - 7$ is -5. ■

The easiest type of trinomial to factor is one having a leading coefficient of 1. Consider the product $(x + 2)(x + 5)$.

$$(x + 2)(x + 5) = x^2 + 7x + 10$$

Therefore, $x^2 + 7x + 10$ factors into $(x + 2)(x + 5)$.

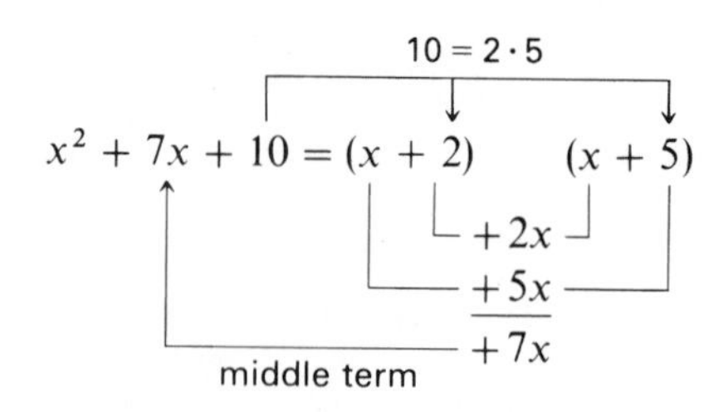

This shows that the last terms of the binomials must be factors of the last term of the trinomial. Also notice that the sum of the inner and outer products must equal the middle term of the trinomial.

Example 2 Factor $x^2 + 5x + 6$.

Solution Each first term of the binomial factors is x in order to give the x^2 in the trinomial.

$$(x \quad)(x \quad)$$

The product of the last terms of the binomials must be 6. Because $6 = 1 \cdot 6$ or $6 = 2 \cdot 3$, we could have either

$$(x + 1)(x + 6) \quad \text{or} \quad (x + 2)(x + 3)$$

The sum of the inner and outer products must equal the middle term of the trinomial, $5x$.

$(x + 1)\ (x + 6)$: inner $+1x$, outer $+6x$; sum $+7x \leftarrow$ incorrect middle term

$(x + 2)\ (x + 3)$: inner $+2x$, outer $+3x$; sum $+5x \leftarrow$ correct middle term

Therefore, $x^2 + 5x + 6$ factors into $(x + 2)(x + 3)$. ■

Example 3 Factor $m^2 - 9m + 8$.

Solution The first terms of the binomial factors must be m in order to give m^2 in the trinomial.

$$(m \quad)(m \quad)$$

The product of the last terms of the binomial must be 8, and their sum must be -9. Therefore, we have either

$(m - 1)\ (m - 8)$: inner $-1m$, outer $-8m$; sum $-9m \leftarrow$ correct middle term

or

$(m - 2)\ (m - 4)$: inner $-2m$, outer $-4m$; sum $-6m \leftarrow$ incorrect middle term

$8 = 1 \cdot 8$ (highlighted)
$= 2 \cdot 4$

Therefore, $m^2 - 9m + 8 = (m - 1)(m - 8)$. ■

Example 4 Factor $a^2 + 4a - 12$.

Solution The first terms of the binomial factors must be a in order to give a^2 in the trinomial.

$$(a \quad)(a \quad)$$

The product of the last term of the binomial must be -12, and their sum must be $+4$. We find the correct pair by trial.

$(a + 6)\ (a - 2)$: inner $+6a$, outer $-2a$; sum $+4a$

$12 = 1 \cdot 12$
$= 2 \cdot 6$ (highlighted)
$= 3 \cdot 4$

Therefore, $a^2 + 4a - 12 = (a + 6)(a - 2)$. ■

Rule of Signs Look at the signs in Examples 2, 3, and 4.

a. If the last term of the trinomial is $+$, the signs of the binomials are the same. And if the middle term of the trinomial is $+$, then both signs are $+$.

$$x^2 + 5x + 6 = (x + 2)(x + 3)$$

(last term $+$: signs the *same*; middle term $+$: both $+$)

b. If the last term of the trinomial is +, the signs of the binomials are the same. And if the middle term of the trinomial is −, then both signs are −.

$$m^2 - 9m + 8 = (m - 1)(m - 8)$$

signs the *same*

both −

c. If the last term of the trinomial is −, then the signs of the binomial are different.

$$a^2 + 4a - 12 = (a + 6)(a - 2)$$

signs *different*

The method of factoring a trinomial having a leading coefficient of 1 is summarized as follows.

TO FACTOR A TRINOMIAL WHOSE LEADING COEFFICIENT IS 1

Arrange the trinomial in descending powers of one of the variables.

1. The product of the first term of each binomial factor equals the first term of the trinomial.
2. List all pairs of factors of the coefficient of the last term of the trinomial.
3. Select the pair of factors so that the sum of the inner and outer products equals the middle term of the trinomial.

Example 5 Factor $x^2 + 9x + 20$.

Solution When the last term of the trinomial is +, the signs of the binomial are the same. Because the middle term is +, both signs are +.

$$(x + \quad)(x + \quad)$$

List all pairs of factors for the coefficient of the last term of the trinomial, +20; then select the pair that gives the correct middle term, $9x$.

$$(x + 4)\quad(x + 5)$$

$$+4x$$
$$+5x$$
$$+9x$$

$$20 = 1 \cdot 20$$
$$= 2 \cdot 10$$
$$= 4 \cdot 5$$

Therefore, $x^2 + 9x + 20 = (x + 4)(x + 5)$. ■

Example 6 Factor $x^2 - 8x - 20$.

Solution Because the last term of the trinomial is −, the signs of the binomial are different.

$$(x + \quad)(x - \quad)$$

From the list of factors for 20, pick the pair that gives the correct middle term, $-8x$.

$$(x+2)\quad(x-10)$$
$$+2x$$
$$-10x$$
$$-8x$$

$$20 = 1 \cdot 20$$
$$= 2 \cdot 10$$
$$= 4 \cdot 5$$

Therefore, $x^2 - 8x - 20 = (x+2)(x-10)$. ■

Example 7 Factor $30 - 13y + y^2$.

Solution

$$y^2 - 13y + 30 \qquad \text{Arrange in descending powers}$$

When the last term of the trinomial is $+$, the signs of the binomial are the same. Since the middle term is $-$, both are $-$

$$(y - \quad)(y - \quad)$$

List all pairs of factors of 30, then select the correct pair that gives the middle term, $-13y$.

$$(y-3)\quad(y-10)$$
$$-3y$$
$$-10y$$
$$-13y$$

$$30 = 1 \cdot 30$$
$$= 2 \cdot 15$$
$$= 3 \cdot 10$$
$$= 5 \cdot 6$$

Therefore, $y^2 - 13y + 30 = (y-3)(y-10)$. ■

EXERCISES 11.4

Factor each expression.

1. $x^2 + 6x + 8$
2. $x^2 + 9x + 8$
3. $x^2 + 5x + 4$
4. $x^2 + 4x + 4$
5. $k^2 + 7k + 6$
6. $k^2 + 5k + 6$
7. $7u + u^2 + 10$
8. $11u + u^2 + 10$
9. $y^2 - 2y + 8$
10. $y^2 - 7y + 8$
11. $b^2 - 9b + 14$
12. $b^2 - 15b + 14$
13. $z^2 - 9z + 20$
14. $z^2 - 12z + 20$
15. $x^2 - 11x + 18$
16. $x^2 - 9x + 18$
17. $x^2 + 9x - 10$
18. $y^2 - 3y - 10$
19. $z^2 - z - 6$
20. $m^2 + 5m - 6$
21. $t^2 + 11t - 30$
22. $m^2 - 17m - 30$
23. $u^4 - 16u^2 + 64$
24. $v^4 - 30v^2 - 64$
25. $16 - 8v + v^2$
26. $16 - 10v + v^2$
27. $b^2 - 11bd - 60d^2$
28. $c^2 + 17cx - 60x^2$
29. $r^2 - 13rs - 48s^2$
30. $s^2 + 22st - 48t^2$
31. $w^2 + 2w - 24$
32. $r^2 - 9r - 20$
33. $n^2 - 10n - 24$
34. $36 - 15y + y^2$
35. $w^2 - 8wz - 48z^2$
36. $p^2 - 9pt - 52t^2$

11.5 Factoring a Trinomial Whose Leading Coefficient Is Greater Than 1

When the leading coefficient equals 1, only the factors of the last term need to be considered. When the leading coefficient is not equal to 1, the factors of the first term as well as the last term must be considered.

Trial Method

TO FACTOR A TRINOMIAL WHOSE LEADING COEFFICIENT IS GREATER THAN 1

Arrange the trinomial in descending powers of one of the variables.

1. Make a blank outline and fill in all obvious information.
2. List all pairs of factors for the coefficient of the first term *and* the last term of the trinomial.
3. Select the correct pairs of factors so that the sum of the inner and outer products equals the middle term of the trinomial.

Check your factoring by multiplying the binomial factors to see if their product is the given trinomial.

Example 1 Factor $2x^2 + 7x + 5$.

Solution The signs for both binomial factors are +.

$$(\quad + \quad)(\quad + \quad)$$

Each first term of the binomial factors must contain an x in order to give the x^2 in the first term of the trinomial $2x^2$.

$$(x + \quad)(x + \quad)$$

The product of the first terms of the binomial factors must be $2x^2$. Because the coefficient 2 has only two factors, 1 and 2, we will use $1x$ and $2x$.

$$(1x + \quad)(2x + \quad)$$

The product of the last terms of the binomial factors must be 5, which has only two factors, 1 and 5. Therefore, we could have either

$(1x + 5)\quad(2x + 1)$: inner $+10x$, outer $+1x$, sum $+11x$ ← Incorrect middle term

or

$(1x + 1)\quad(2x + 5)$: inner $+2x$, outer $+5x$, sum $+7x$ ← Correct middle term

The sum of the inner and outer products must equal the middle term of the trinomial. We found the correct pair by trial. Therefore, $2x^2 + 7x + 5$ factors into $(x + 1)(2x + 5)$. ■

Example 2 Factor $5x^2 + 13x + 6$.

Solution Make a blank outline, and fill in all the obvious information.

$$(\ x + \quad)(\ x + \quad)$$

The letter in each binomial must be x so that the first term of the trinomial contains x^2

The sign in each binomial is +

Next, list all pairs of factors of the first coefficient and the last term of the trinomial.

Factors of the first coefficient	Factors of the last term
$5 = 1 \cdot 5$	$6 = 1 \cdot 6$
	$= 2 \cdot 3$

$(1x + \quad)(5x + \quad)$ Because 1 and 5 are the only factors of 5, we can fill them in next

Now we must select the correct pair of factors of the last term, 6, so that the sum of the inner and outer products is $13x$. We find the correct pair by trial.

$$(1x + 2)(5x + 3) \qquad \begin{array}{r} +10x \\ +\ 3x \\ \hline +13x \end{array} \qquad \begin{aligned} 6 &= 1 \cdot 6 \\ &= 2 \cdot 3 \end{aligned}$$

Therefore, $5x^2 + 13x + 6$ factors into $(x + 2)(5x + 3)$. ■

Example 3 Factor $6x^2 - 19x + 15$.
Solution Make a blank outline.

$(\quad x - \quad)(\quad x - \quad)$ Because the last term is + and the middle term is −, the signs in both binomials must be −

Select correct pairs of factors by trial.

$$\begin{aligned} 6 &= 1 \cdot 6 \\ &\ \ \ 2 \cdot 3 \end{aligned} \qquad (2x - 3)(3x - 5) \qquad \begin{array}{r} -\ 9x \\ -10x \\ \hline -19x \end{array} \qquad \begin{aligned} 15 &= 1 \cdot 15 \\ &\ \ \ 3 \cdot 5 \end{aligned}$$

Therefore, $6x^2 - 19x + 15 = (2x - 3)(3x - 5)$. ■

Example 4 Factor $12a^2 + 7ab - 10b^2$.
Solution

$(\quad a + \quad b)(\quad a - \quad b)$ Because the last term is −, the signs in the binomials must be different

$$\begin{aligned} 12 &= 1 \cdot 12 \\ &= 2 \cdot 6 \\ &= 3 \cdot 4 \end{aligned} \qquad (4a + 5b)(3a - 2b) \qquad \begin{array}{r} +15ab \\ -\ 8ab \\ \hline 7ab \end{array} \qquad \begin{aligned} 10 &= 1 \cdot 10 \\ &= 2 \cdot 5 \end{aligned}$$

Therefore, $12a^2 + 7ab - 10b^2 = (4a + 5b)(3a - 2b)$. ■

Master Product Method (Optional)

The master product method for factoring trinomials makes use of factoring by grouping.

TO FACTOR A TRINOMIAL BY THE MASTER PRODUCT METHOD

Arrange the trinomial in descending powers of one of the variables.

$$ax^2 + bx + c$$

1. Find the master product (MP) by multiplying the first and last coefficients of the trinomial being factored ($MP = a \cdot c$).
2. Write the pairs of factors of the master product (MP).
3. Choose the pair of factors whose sum is the coefficient of the middle term (b).
4. Rewrite the given trinomial, replacing the middle term by the sum of two terms whose coefficients are the pair of factors found in Step 3.
5. Factor the Step 4 expression by grouping.

Check your factoring by multiplying the binomial factors to see if their product is the given trinomial.

Example 5 Factor $5x + 2x^2 + 3$.

Solution $2x^2 + 5x + 3$ Arrange in descending powers

Master product $= (2)(+3) = +6$.

List factors of 6.

$1 \cdot 6$

$2 \cdot 3$ $(+2) + (+3) = 5$ Choose factors whose sum is the middle coefficient, 5

Rewrite the middle term $5x$ as the sum $2x + 3x$.

Therefore, $2x^2 + 5x + 3 = 2x^2 + 2x + 3x + 3$

$= 2x(x + 1) + 3(x + 1)$

$= (x + 1)(2x + 3)$ Factor by grouping

■

Example 6 Factor $3m^2 - 2m - 8$.

Solution

Master product $= (3)(-8) = -24$.

Factors of 24:

$1 \cdot 24$

$2 \cdot 12$

$3 \cdot 8$

$4 \cdot 6$ $(+4) + (-6) = -2$ The middle coefficient

Therefore, $3m^2 - 2m - 8 = 3m^2 + 4m - 6m - 8$

$= m(3m + 4) - 2(3m + 4)$

$= (3m + 4)(m - 2)$ Factor by grouping

■

Example 7 Factor $12a^2 + 7ab - 10b^2$.

Solution

Master product = $(12)(-10) = -120$.

$1 \cdot 120$
$2 \cdot 60$
$3 \cdot 40$
$4 \cdot 30$
$5 \cdot 24$
$6 \cdot 20$
$8 \cdot 15$

Factors of 120

$(-8) + (+15) = +7$ The middle coefficient

Therefore, $12a^2 + 7ab - 10b^2 = 12a^2 - 8ab + 15ab - 10b^2$

$= 4a(3a - 2b) + 5b(3a - 2b)$

$= (3a - 2b)(4a + 5b)$ ■

Factor by grouping

NOTE The master product method of factoring trinomials can also be used with trinomials whose leading coefficient is 1. However, we think the method presented in Section 11.4 is shorter and simpler for trinomials of that type. ☑

EXERCISES 11.5

Factor each expression by trial method or master product method.

1. $3x^2 + 7x + 2$ **2.** $3x^2 + 5x + 2$ **3.** $5x^2 + 7x + 2$ **4.** $5x^2 + 11x + 2$

5. $7x + 4x^2 + 3$ **6.** $13x + 4x^2 + 3$ **7.** $5x^2 + 20x + 4$ **8.** $5x^2 + 11x + 4$

9. $5a^2 - 16a + 3$ **10.** $5m^2 - 8m + 3$ **11.** $3b^2 - 22b + 7$ **12.** $3u^2 - 10u + 7$

13. $5z^2 - 36z + 7$ **14.** $5z^2 - 12z + 7$ **15.** $3n^2 + 14n - 5$ **16.** $3n^2 - 2n - 5$

17. $5k^2 - 34k - 7$ **18.** $5k^2 + 2k - 7$ **19.** $7x^2 + 23xy + 6y^2$ **20.** $7a^2 + 43ab + 6b^2$

21. $7h^2 - 11hk + 4k^2$ **22.** $7h^2 - 16hk + 4k^2$ **23.** $3t^2 + 19tz - 6z^2$ **24.** $3w^2 - 11wx - 6x^2$

25. $6 - 17v + 5v^2$ **26.** $6 - 11v + 5v^2$ **27.** $6e^4 - 7e^2 - 20$ **28.** $10f^4 - 29f^2 - 21$

29. $35k^2 - 12k + 1$ **30.** $3e^2 - 20e - 7$ **31.** $7x^2 + 2x - 5$ **32.** $3a^2 + 8ab + 5b^2$

33. $4m^2 - 16mn + 7n^2$ **34.** $6s^2 - 17st - 5t^2$ **35.** $17u - 12 + 5u^2$ **36.** $8v^4 - 14v^2 - 15$

11.6 Factoring Completely

Consider the following example.

Example 1 Factor $27x^2 - 12y^2$.

Solution $27x^2 - 12y^2$ 3 is the GCF

This factor can be factored again

$= 3(9x^2 - 4y^2)$ $(9x^2 - 4y^2) = (3x + 2y)(3x - 2y)$

$= 3(3x + 2y)(3x - 2y)$ ■

We say that $27x^2 - 12y^2 = 3(3x + 2y)(3x - 2y)$ has been *completely factored.* We will consider an expression to be completely factored if no more factoring can be done (by *any* method we have discussed).

Example 1 illustrates the importance of the statement:
Always remove a greatest common factor first when possible.

TO FACTOR AN EXPRESSION COMPLETELY

1. Factor out any common factors *first*, if possible.
2. If the expression has two terms, look for the difference of squares.
3. If the expression has three terms, look for a factorable trinomial.
4. If the expression has four terms, see if it can be factored by grouping.
5. *Check to see if any factor already obtained can be factored again.*

Example 2 Factor $3x^3 - 27x + 5x^2 - 45$.

Solution

$$3x^3 - 27x + 5x^2 - 45$$ Factor by grouping

$$= 3x(x^2 - 9) + 5(x^2 - 9)$$

This factor can be factored again

$$= (x^2 - 9)(3x + 5)$$ $x^2 - 9 = (x + 3)(x - 3)$

$$= (x + 3)(x - 3)(3x + 5)$$ ■

Example 3 Factor $5x^2 + 10x - 40$.

Solution $5x^2 + 10x - 40$ 5 is the GCF

This factor can be factored again

$$= 5(x^2 + 2x - 8)$$ $(x^2 + 2x - 8) = (x - 2)(x + 4)$

$$= 5(x - 2)(x + 4)$$ ■

Example 4 Factor $a^4 - b^4$.

Solution $a^4 - b^4$

This factor can be factored again

$$= (a^2 + b^2)(a^2 - b^2)$$ $(a^2 - b^2) = (a + b)(a - b)$

$$= (a^2 + b^2)(a + b)(a - b)$$ ■

EXERCISES 11.6

Factor each expression *completely.*

1. $2x^2 - 8y^2$ **2.** $3x^2 - 27y^2$ **3.** $5a^4 - 20b^2$ **4.** $6m^2 - 54n^4$

5. $x^4 - y^4$ **6.** $a^4 - 16$ **7.** $4v^2 + 14v - 8$ **8.** $6v^2 - 27v - 15$

9. $8z^2 - 12z - 8$ **10.** $18z^2 - 21z - 9$ **11.** $12x^2 + 10x - 8$ **12.** $45x^2 - 6x - 24$

13. $ab^2 - 2ab + a$
14. $au^2 - 2au + a$
15. $3a^2 - 75b^2$
16. $4h^4 - 36b^2$
17. $m^4 - 1$
18. $10x^2 + 25x - 15$
19. $10y^2 + 14y - 5$
20. $30w^2 + 27w - 21$
21. $h^2k - 4hk + 4k$
22. $81c^4 + 16$
23. $4m^3n^3 - mn^5$
24. $2t^2r^4 - 18t^4$
25. $5wz^2 + 5w^2z - 10w^3$
26. $12x^2y - 42xy^2 + 36y^3$
27. $x^4 - 81$
28. $16y^8 - z^4$
29. $a^5b^2 - 4a^3b^4$
30. $x^2y^4 - 100x^4y^2$
31. $2ax^2 - 8a^3y^2$
32. $3b^2x^4 - 12b^2y^2$
33. $2u^3 + 2u^2v - 12uv^2$
34. $3m^3 - 3m^2n - 36mn^2$
35. $8h^3 - 20h^2k + 12hk^2$
36. $15h^2k - 35hk^2 + 10k^3$
37. $12 + 4x - 3x^2 - x^3$
38. $45 - 9z - 5z^2 + z^3$
39. $6my - 4nz + 15mz - 5zn$
40. $10xy + 5mn - 6xy - nm$
41. $6ac - 6bd + 6bc - 6ad$
42. $10cy - 6cz + 5dy - 3dz$
43. $24x^2 + 30x - 9$
44. $18x^2 - 24x - 10$
45. $45a^3 - 65a^2 + 20a$
46. $24y^4 - 28y^3 - 40y^2$
47. $x^3 - 4x^2 - 4x + 16$
48. $x^3 + x^2 - 9x - 9$
49. $27a^2 + 36ab + 12b^2$
50. $2x^4 - 16x^3 + 32x^2$
51. $x^4 - 2x^2 + 1$
52. $y^4 - 8y^2 + 16$
53. $8x^3y^2 + 4x^2y^3 - 12xy^4$
54. $3a^4b^2 - 9a^3b^3 - 12a^2b^4$

11.7 Solving Equations by Factoring

Factoring has many applications. In this section we use factoring to solve equations.

Polynomial Equations

A **polynomial equation** is a polynomial set equal to zero. The degree of the equation is the degree of the polynomial.

First Degree
Second Degree

Polynomial equations with a first-degree term as the highest-degree term are called **first-degree** *equations.* Polynomial equations with a second-degree term as the highest-degree term are called **second-degree** or *quadratic equations.*

Example 1 Polynomial equations

a. $5x - 3 = 0$ First-degree equation in one variable

b. $2x^2 - 4x + 7 = 0$ Second-degree equation in one variable (also called *quadratic* equation) ■

We know from arithmetic that if the product of two numbers is zero, one or both of the numbers must be zero.

In algebra:

> If the product of two factors is zero, then one or both of the factors must be zero.
>
> $$\text{If } a \cdot b = 0, \text{ then } \begin{cases} a = 0 \\ \text{or } b = 0 \\ \text{or both } a \text{ and } b = 0 \end{cases}$$

We make use of this fact in solving some polynomial equations.

Example 2 Solve $(x-1)(x-2)=0$.

Solution

Because $(x-1)(x-2)=0$,

then $(x-1)=0$ or $(x-2)=0$.

If
$$\begin{array}{rl} x-1 = & 0 \\ +1 & +1 \\ \hline x \quad = & 1 \end{array}$$
then $x=1$.

If
$$\begin{array}{rl} x-2 = & 0 \\ +2 = & +2 \\ \hline x \quad = & 2 \end{array}$$
then $x=2$.

Therefore, 1 and 2 are solutions for the equation $(x-1)(x-2)=0$.

Check for x = 1	***Check for x = 2***
$(x-1)(x-2)=0$	$(x-1)(x-2)=0$
$(1-1)(1-2)\stackrel{?}{=}0$	$(2-1)(2-2)\stackrel{?}{=}0$
$(0)(-1)\stackrel{?}{=}0$	$(1)(0)\stackrel{?}{=}0$
$0=0$	$0=0$ ■

The same method can be used when a product of more than two factors is equal to zero.

Example 3 Solve $2x(x-3)(x+4)=0$.

Solution

$$2x(x-3)(x+4)=0$$

$$\begin{array}{c|c|c} 2x=0 & x-3 = 0 & x+4 = 0 \\ \dfrac{2x}{2}=\dfrac{0}{2} & \quad +3 \quad +3 & \quad -4 \quad -4 \\ x=0 & x \quad = \quad 3 & x \quad = -4 \end{array}$$ ■

TO SOLVE AN EQUATION BY FACTORING

1. Write all nonzero terms on one side of the equation by adding the same expression to both sides. *Only zero* must remain on the other side. Then arrange the polynomial in descending powers.
2. Factor the polynomial.
3. Set each factor equal to zero, and solve for the unknown variable.
4. Check apparent solutions in the *original* equation.

Example 4 Solve $4x^2-16x=0$.

Solution The polynomial must be factored first.

$$4x^2-16x=0 \qquad \text{A quadratic equation}$$

$$4x(x-4)=0 \qquad \text{Factor}$$

$$\begin{array}{c|c} 4x=0 & x-4 = 0 \\ \dfrac{4x}{4}=\dfrac{0}{4} & \quad +4 \quad +4 \\ x=0 & x \quad = \quad 4 \end{array} \qquad \text{Set each factor equal to 0}$$ ■

Example 5 Solve $x^2-x-6=0$.

Solution

$$x^2 - x - 6 = 0$$

$$(x + 2)(x - 3) = 0 \qquad \text{Factor}$$

$$\begin{array}{rr|rr} x + 2 = & 0 & x - 3 = & 0 \\ -2 & -2 & +3 & +3 \\ \hline x \quad = & -2 & x \quad = & 3 \end{array} \qquad \text{Set each factor equal to 0}$$

■

Sometimes it is more convenient to get the polynomial on the *right* side of the equal sign and the zero on the *left* side.

Example 6 Solve $4 - x = 3x^2$.

Solution

$$0 = 3x^2 + x - 4$$

$$0 = (x - 1)(3x + 4)$$

In this case it is convenient to arrange all nonzero terms on the *right* side in descending order

$$\begin{array}{r|r} x - 1 = 0 & 3x + 4 = 0 \\ x = 1 & 3x = -4 \\ & x = -\dfrac{4}{3} \end{array}$$

■

Example 7 Solve $\dfrac{x-1}{x-3} = \dfrac{12}{x+1}$.

Solution

$$\frac{x-1}{x-3} = \frac{12}{x+1} \qquad \text{This is a proportion}$$

$$(x - 1)(x + 1) = 12(x - 3) \qquad \text{Product of means = product of extremes}$$

$$x^2 - 1 = 12x - 36$$

$$x^2 - 12x + 35 = 0$$

$$(x - 7)(x - 5) = 0$$

$$\begin{array}{rr|rr} x - 7 = & 0 & x - 5 = & 0 \\ +7 & +7 & +5 & +5 \\ \hline x \quad = & 7 & x \quad = & 5 \end{array}$$

■

A WORD OF CAUTION The product must equal *zero*, or no conclusions can be drawn about the factors.

Suppose $\qquad (x - 1)(x - 3) = \boxed{8}$

↑ No conclusion can be drawn because the product $\neq 0$

Some common mistakes Students sometimes think that:

If $\qquad (x - 1)(x - 3) = 8$

then

$$\begin{array}{r|r} x - 1 = 8 & x - 3 = 8 \\ x = 9 & x = 11 \end{array}$$

Both of these assumptions are incorrect.

Check x = 9	*Check x = 11*
$(x-1)(x-3)=8$	$(x-1)(x-3)=8$
$(9-1)(9-3) \stackrel{?}{=} 8$	$(11-1)(11-3) \stackrel{?}{=} 8$
$8 \cdot 6 \stackrel{?}{=} 8$	$10 \cdot 8 \stackrel{?}{=} 8$
$48 \neq 8$	$80 \neq 8$

The correct solution is

$$(x-1)(x-3)=8$$

$$x^2-4x+3=8 \quad \text{Removing parentheses}$$

$$x^2-4x-5=0 \quad \text{Subtracting 8 from both sides}$$

$$(x-5)(x+1)=0 \quad \text{Factor}$$

$$x-5=0 \quad \bigg| \quad x+1=0 \quad \text{Set each factor equal to 0}$$

$$x=5 \quad \bigg| \quad x=-1$$

✓

EXERCISES 11.7

Solve each equation.

1. $(x-5)(x+4)=0$
2. $(x+7)(x-2)=0$
3. $3x(x-4)=0$
4. $5x(x+6)=0$
5. $(x+10)(2x-3)=0$
6. $(x-8)(3x+2)=0$
7. $3(x+2)(x-1)=0$
8. $2x(3x-2)(5x+9)=0$
9. $x^2+9x+8=0$
10. $x^2+6x+8=0$
11. $x^2-x-12=0$
12. $x^2+x-12=0$
13. $x^2-18=7x$
14. $x^2-20=8x$
15. $6x^2-10x=0$
16. $6y^2-21y=0$
17. $24w=4w^2$
18. $20m=5m^2$
19. $5a^2=16a-3$
20. $3z^2=22z-7$
21. $3u^2=2u+5$
22. $5k^2=34k+7$
23. $m^2+3m-18=0$
24. $a^2+8a+12=0$
25. $5h^2-20h=0$
26. $w^2-24=5w$
27. $3n^2=7n+6$
28. $12t=6t^2$
29. $21x^2+60x=18x^3$
30. $13x+3=-4x^2$
31. $(y-3)(y-6)=-2$
32. $u(u-9)=-14$
33. $(x-2)(x-3)=2$
34. $(x-3)(x-5)=3$
35. $x(x-4)=12$
36. $x(x-2)=15$
37. $4x(2x-1)(3x+7)=0$
38. $5x(4x-3)(7x-6)=0$
39. $2x^3+x^2=3x$
40. $4x^3=10x-18x^2$
41. $2a^3-10a^2=0$
42. $4b^3-24b^2=0$
43. $\dfrac{x-1}{x+4}=\dfrac{3}{x}$
44. $\dfrac{2x+5}{3x}=\dfrac{x}{2x-5}$
45. $\dfrac{x+2}{x+1}=\dfrac{10}{x+5}$

11.8 Word Problems Solved by Factoring

Number Problems

Example 1 The difference of two numbers is 3. Their product is 10. What are the two numbers?

Solution

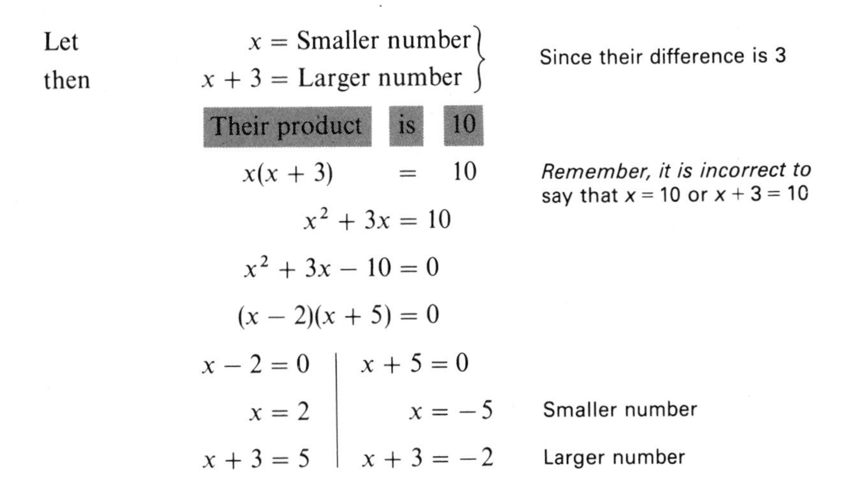

Let $x = \text{Smaller number}$
then $x + 3 = \text{Larger number}$ — Since their difference is 3

Their product | is | 10

$x(x + 3) = 10$ — *Remember, it is incorrect to say that* $x = 10$ *or* $x + 3 = 10$

$x^2 + 3x = 10$

$x^2 + 3x - 10 = 0$

$(x - 2)(x + 5) = 0$

$x - 2 = 0$	$x + 5 = 0$	
$x = 2$	$x = -5$	Smaller number
$x + 3 = 5$	$x + 3 = -2$	Larger number

Therefore the numbers 2 and 5 are a solution and the numbers -5 and -2 are another solution. ■

Consecutive Integer Problems

Example 2 Find two consecutive odd integers such that the sum of their squares is 34.

Solution Let $x = \text{first consecutive odd}$

$x + 2 = \text{second consecutive odd}$

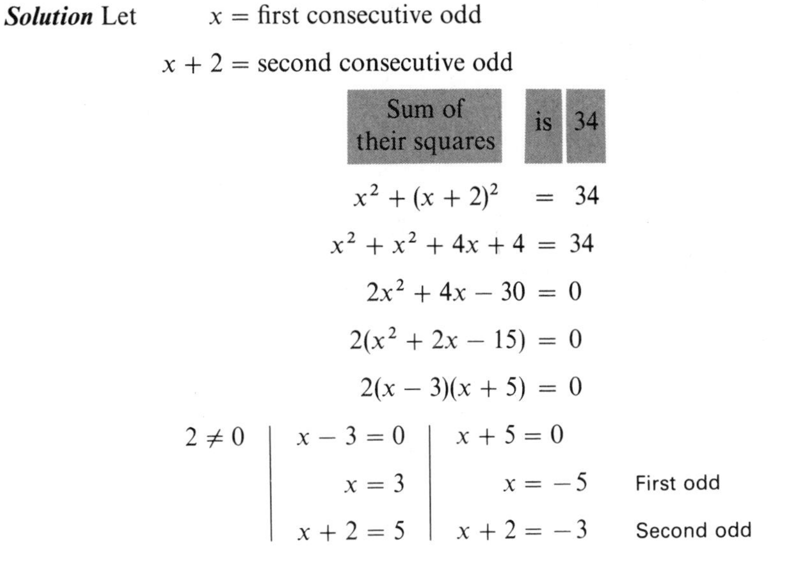

Sum of their squares | is | 34

$x^2 + (x + 2)^2 = 34$

$x^2 + x^2 + 4x + 4 = 34$

$2x^2 + 4x - 30 = 0$

$2(x^2 + 2x - 15) = 0$

$2(x - 3)(x + 5) = 0$

$2 \neq 0$	$x - 3 = 0$	$x + 5 = 0$	
	$x = 3$	$x = -5$	First odd
	$x + 2 = 5$	$x + 2 = -3$	Second odd

There are two solutions: 3 and 5, or -5 and -3. ■

Geometry Problems

Example 3 The length of a rectangle is 2 inches more than its width. The area of the rectangle is 24 square inches. Find the length and the width.

Solution There are two unknown numbers, the width and the length.

Let $w = \text{width}$

then $w + 2 = \text{length}$

Area = 24 — w — $w + 2$

$$\text{Area} = \text{width} \times \text{length}$$
$$24 = w(w + 2)$$
$$24 = w^2 + 2w$$
$$0 = w^2 + 2w - 24$$
$$0 = (w - 4)(w + 6)$$

	$w - 4 = 0$	$w + 6 = 0$
width	$w = 4$	$w = -6$ ← -6 is not a solution because length cannot be negative
length	$w + 2 = 6$	

Therefore, the rectangle has a width of 4 inches and a length of 6 inches. ■

EXERCISES 11.8

In the following exercises: (a) Represent the unknown numbers by variables. (b) Set up an equation and solve. (c) Answer the question.

1. The difference of two numbers is 5. Their product is 14. Find the numbers.

2. The difference of two numbers is 6. Their product is 27. Find the numbers.

3. The sum of two numbers is 12. Their product is 35. Find the numbers.

4. The sum of two numbers is -4. Their product is -12. Find the numbers.

5. Find two consecutive odd integers whose product is 63.

6. Find two consecutive even integers whose product is 48.

7. Find two consecutive integers whose product is 11 more than their sum.

8. Find two consecutive integers whose product is 5 more than their sum.

9. Find three consecutive integers such that the product of the first two is 7 more than the third integer.

10. Find three consecutive integers such that the product of the first two is 23 more than the third integer.

11. Find two consecutive integers such that the sum of their squares is 25.

12. Find two consecutive even integers such that the sum of their squares is 20.

13. One number is three times another number. The sum of their squares is 40. Find the numbers.

14. One number is three more than another number. The sum of their squares is 89. Find the numbers.

15. The length of a rectangle is 5 feet more than its width. Its area is 84 square feet. Find the length and the width.

16. The width of a rectangle is 3 feet less than its length. Its area is 28 square feet. Find the length and the width.

17. The base of a triangle is 3 inches more than its altitude. Its area is 20 square inches. Find the base and altitude.

18. The sum of the base and altitude of a triangle is 15 centimeters. The area of the triangle is 27 square centimeters. Find the base and the altitude.

19. The width of a rectangle is 4 yards less than its length. The area is 17 more (numerically) than its perimeter. Find the length and the width.

20. The area of a square is twice its perimeter (numerically). What is the length of its side?

21. One square has a side 3 centimeters shorter than the side of a second square. The area of the larger square is 4 times as great as the area of the smaller square. Find the length of the side of each square.

22. One square has a side 2 meters longer than the side of a second square. The area of the larger square is 16 times as great as the area of the smaller square. Find the length of the side of each square.

23. The fourth term of a proportion is 7 more than its first term. Find the proportion if its second term is 3 and its third term is 6.

24. The third term of a proportion is 2 more than its second term. Find the proportion if its first term is 3 and its fourth term is 8.

11.9 Chapter Summary

Prime Number 11.1 A **prime number** is a positive integer greater than 1 that can be divided evenly only by itself and 1. A prime number has no factors other than itself and 1.

Composite Number 11.1 A **composite number** is a positive integer that can be exactly divided by some integer other than itself and 1. A composite number has factors other than itself and 1.

Prime Factorization 11.1 The **prime factorization** *of a positive integer* is the indicated product of all its factors that are themselves prime numbers.

Methods of Factoring

11.1 1. Greatest common factor (GCF)

11.3 2. Difference of two squares, $a^2 - b^2 = (a + b)(a - b)$

11.4, 11.5 3. Trinomial $\begin{cases}\text{leading coefficient equal to 1} \\ \text{leading coefficient greater than 1}\end{cases}$

11.2 4. Grouping

Factoring Completely 11.6

1. Factor out any common factors first, if possible.
2. If the expression has 2 terms, look for the difference of squares.
3. If the expression has 3 terms, look for a factorable trinomial.
4. If the expression has 4 terms, see if it can be factored by grouping.
5. Check to see if any factor already obtained can be factored again.

To Solve an Equation by Factoring 11.7

1. Write *all* nonzero terms on one side of the equation by adding the same expression to both sides. *Only zero must remain on the other side.* Then arrange the polynomial in descending powers.
2. Factor the polynomial.
3. Set each factor equal to zero, and solve for the variable.
4. Check apparent solutions in the original equation.

Review Exercises 11.9

In Exercises 1–5, find the prime factorization of each number.

1. 12 **2.** 36 **3.** 31 **4.** 42 **5.** 210

In Exercises 6–23, factor each expression completely.

6. $8x - 4$

7. $m^2 - 4$

8. $x^2 + 10x + 21$

9. $2u^2 + 4u$

10. $z^2 - 7z - 18$

11. $4x^2 - 25x + 6$

12. $9k^2 - 144$

13. $8 - 2a^2$

14. $ab + 2b - a - 2$

15. $xy - 4x + 3y - 12$

16. $3x^2 + 4x - 4$

17. $15u^2v - 3uv$

18. $a^4 - b^4$

19. $x^8 - 1$

20. $x^4 - 5x^2 + 4$

21. $4x^2y - 8xy^2 + 4xy$

22. $15a^2 + 15ab - 30b^2$

23. $6u^3v^2 - 9uv^3 - 12uv$

In Exercises 24–32, solve each of the equations.

24. $(x - 5)(x + 3) = 0$

25. $x^2 - 5x - 14 = 0$

26. $m^2 = 18 + 3m$

27. $x^2 - 9 = 0$

28. $3z^2 = 12z$

29. $6e^2 = 13e + 5$

30. $2u(u + 6)(u - 2) = 0$

31. $6x^3 = 9x^2 + 6x$

32. $\frac{x - 2}{9} = \frac{4}{x + 7}$

33. The difference of two numbers is 3. Their product is 28. Find the numbers.

34. Find three consecutive integers such that the product of the first two minus the third is 7.

35. The width of a rectangle is 3 less than its length. Its area is 40. Find the length and width of the rectangle.

36. One side of a square is 6 feet longer than the side of a second square. The area of the larger square is 16 times as great as the area of the smaller square. Find the length of the side of each square.

37. The area of a square is 3 times its perimeter. What is the length of its side?

38. The length of a rectangle is 6 yards more than its width. The area is 12 more (numerically) than its perimeter. Find the length and the width.

Chapter 11 Diagnostic Test

Allow yourself about an hour to do these problems. Complete solutions for every problem, together with section references, are given in the answer section at the end of the book.

1. State whether each of the following numbers is prime or composite.
a. 18 b. 21 c. 31

In Problems 2 and 3, find the prime factorization of each number.

2. 45

3. 160

In Problems 4–15, factor each expression completely.

4. $5x + 10$

5. $3x^2 - 6x$

6. $16x^2 - 49y^2$

7. $2z^2 - 8$

8. $z^2 + 6z + 8$

9. $m^2 + m - 6$

10. $5w^2 - 12w + 7$

11. $3v^2 + 14v - 5$

12. $5n - mn - 5 + m$

13. $6h^2k - 8hk^2 + 2k^3$

14. $x^4 - 8x^2 - 9$

15. $24a^2 + 2a - 12$

In Problems 16–18, solve each equation by factoring.

16. $x^2 - 12x + 20 = 0$

17. $3x^2 = 12x$

18. $36y = 18y^2 + 18$

19. The length of a rectangle is 3 feet more than its width. Its area is 28 square feet. Find the length and the width.

20. Find two consecutive odd integers such that their product is 7 more than their sum.

12 Algebraic Fractions

In this chapter we define algebraic fractions, how to perform necessary operations with them, and how to solve equations and word problems involving them. A knowledge of the different methods of factoring discussed in Chapter 11 is essential in your work with *algebraic* fractions.

12.1 Simplifying Algebraic Fractions

Rational Expression

A *simple algebraic fraction* (also called a **rational expression**) is an algebraic expression of the form

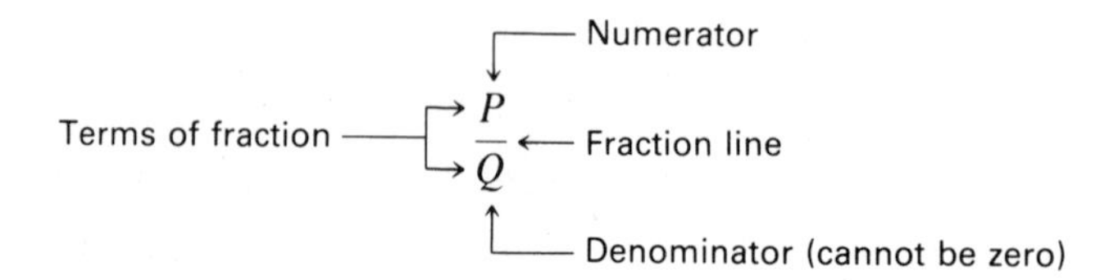

where P and Q are polynomials.

Excluded Values Any value of the variable (or variables) that makes the denominator Q equal to zero must be excluded.

Example 1 Examples of algebraic fractions [some show excluded value(s) of the variable].

a. $\dfrac{x}{3}$ — *No* value of x is excluded because no value of x makes the denominator zero

b. $\dfrac{5}{x}$ — x cannot be 0

c. $\dfrac{2x-5}{x-1}$ — x cannot be 1, because that would make the denominator 0

d. $\dfrac{x^2+2}{x^2-3x-4} = \dfrac{x^2+2}{(x-4)(x+1)}$ — x cannot be 4 or -1, because either value makes the denominator 0

e. $\dfrac{2}{3}$ — Arithmetic fractions are also algebraic fractions; here, 2 and 3 are polynomials of degree 0. ■

NOTE After this section, whenever a fraction is written, it will be understood that the value(s) of the variable(s) that make the denominator zero are excluded. ☑

The Three Signs of a Fraction Every fraction has three signs associated with it.

Sign of numerator

Sign of fraction $\longrightarrow + \dfrac{-2}{+3}$

Sign of denominator

> If any two of the three signs of a fraction are changed, the value of the fraction is unchanged.

We illustrate the rule for signs of a fraction by applying it to the fraction $\frac{8}{4}$.

$$+\frac{+8}{+4} = \begin{cases} +\frac{-8}{-4} = +\left(\frac{-8}{-4}\right) = +(+2) = 2 \\ \text{Here, we changed the signs of numerator and denominator} \\ -\frac{-8}{+4} = -\left(\frac{-8}{+4}\right) = -(-2) = 2 \\ \text{Here, we changed the signs of fraction and numerator} \\ -\frac{+8}{-4} = -\left(\frac{+8}{-4}\right) = -(-2) = 2 \\ \text{Here, we changed the signs of fraction and denominator} \end{cases}$$

This rule of signs is helpful in simplifying some expressions with fractions.

Example 2 Changing two signs of a fraction to obtain an equivalent fraction

a. $-\frac{-5}{xy} = +\frac{+5}{xy} = \frac{5}{xy}$

Here, we changed the signs of fraction and numerator

b. $-\frac{1}{2-x} = +\frac{1}{-(2-x)} = \frac{1}{-2+x} = \frac{1}{x-2}$

Here, we changed the signs of fraction and denominator ■

The simplification in Example 2(b) can be explained as follows:

$$-(2-x) = (-1)(2-x) = -2 + x = x - 2 \qquad \text{Distributive rule}$$

This fact is summarized as follows.

> The negative of a binomial difference interchanges the terms.
>
> $$-(b-a) = a - b$$
>
> negative; terms interchanged

Example 3 Negative of a binomial difference

a. $-(x-y) = y - x$

b. $-(d-5) = 5 - d$ ■

Reducing Fractions to Lowest Terms We reduce fractions to lowest terms in algebra for the same reason we do in arithmetic: It makes them simpler and easier to work with.

After this section, it is understood that *all fractions are to be reduced to lowest terms* (unless otherwise indicated).

In Arithmetic

Reducing fractions is explained by the following example.

$$\frac{6}{8} = \frac{2 \cdot 3}{2 \cdot 4} = \frac{2}{2} \cdot \frac{3}{4} = 1 \cdot \frac{3}{4} = \frac{3}{4}$$

This work is usually shortened by dividing numerator and denominator by the same number.

$$\frac{\overset{3}{\cancel{6}}}{\underset{4}{\cancel{8}}} = \frac{3}{4}$$

This is only possible when the numerator and denominator have a common factor. For example,

$$\frac{6}{8} = \frac{2 \cdot 3}{2 \cdot 4} = \frac{\cancel{2} \cdot 3}{\cancel{2} \cdot 4} = \frac{3}{4}$$

2 is a factor of the numerator

2 is a factor of the denominator

Both numerator and denominator were divided by 2

In Algebra

The procedure for reducing fractions in algebra is the same as the method used in arithmetic.

TO REDUCE A FRACTION TO LOWEST TERMS

1. Factor the numerator and denominator completely.
2. Divide the numerator and denominator by all factors common to both.

Example 4 Reduce to lowest terms.

a. $$\frac{4x^2y}{2xy} = \frac{\overset{2}{\cancel{4}}x^{\cancel{2}}\cancel{y}}{\underset{1}{\cancel{2}\cancel{x}\cancel{y}}} = 2x$$

Here, the numerator and denominator are already factored

b. $$\frac{x-3}{x^2-9} = \frac{\overset{1}{\cancel{(x-3)}}}{(x+3)\underset{1}{\cancel{(x-3)}}} = \frac{1}{x+3}$$

NOTE A factor of 1 will always remain in the numerator. ☑

c. $$\frac{x^2-4x-5}{x^2+5x+4} = \frac{\cancel{(x+1)}(x-5)}{\cancel{(x+1)}(x+4)} = \frac{x-5}{x+4}$$

d. $$\frac{3x^2-5xy-2y^2}{6x^3y+2x^2y^2} = \frac{(x-2y)\cancel{(3x+y)}}{2x^2y\cancel{(3x+y)}} = \frac{x-2y}{2x^2y}$$

e. $$\frac{y-x}{x-y} = \frac{-\overset{1}{\cancel{(x-y)}}}{\underset{1}{\cancel{x-y}}} = \frac{-1}{1} = -1$$

NOTE Since $y - x$ is the negative of $x - y$, they may be canceled directly if a factor of -1 is used. ☑

$$\frac{y-x}{x-y} = \frac{\overset{-1}{\cancel{y-x}}}{\underset{1}{\cancel{x-y}}} = \frac{-1}{1} = -1$$

f. $\dfrac{a^2 - 3ab + 2b^2}{b^2 - a^2} = \dfrac{\overset{-1}{\cancel{(a-b)}}(a-2b)}{\underset{1}{\cancel{(b-a)}}(b+a)} = \dfrac{-1(a-2b)}{b+a} = \dfrac{2b-a}{b+a}$

g. $\dfrac{x+3}{x+6}$ Cannot be reduced {Neither x nor 3 is a *factor* of numerator or denominator

h. $\dfrac{x+y}{x}$ Cannot be reduced (See the following word of caution) ■

(x is *not* a *factor* of the numerator)

A WORD OF CAUTION A common error made in reducing fractions is to forget that the number the numerator and denominator are divided by *must* be a *factor* of *both* [see Examples 4(g) and 4(h)].

Error:

3 is *not* a factor of the numerator

$$\frac{3+2}{3} \neq 2 \quad \textit{incorrect reduction}$$

The above reduction is incorrect because

$$\frac{3+2}{3} = \frac{5}{3} \neq 2$$ ☑

EXERCISES 12.1

In Exercises 1–12, what value(s) of the variable (if any) must be excluded?

1. $\dfrac{3x+4}{x-2}$ **2.** $\dfrac{5-4x}{x+3}$ **3.** $\dfrac{x}{10}$ **4.** $\dfrac{y}{20}$

5. $\dfrac{3+x}{(x-1)(x+2)}$ **6.** $\dfrac{x-4}{3x(x-2)}$ **7.** $\dfrac{z}{4}$ **8.** $\dfrac{x-1}{3}$

9. $\dfrac{x^2+4}{x^2-x-2}$ **10.** $\dfrac{x^2-2}{x^2-4}$ **11.** $\dfrac{x-6}{2x^2+6x}$ **12.** $\dfrac{x+2}{5x^2-5x}$

In Exercises 13–20, use the rule about the three signs of a fraction to find the missing term.

13. $-\dfrac{5}{6} = \dfrac{?}{6}$ **14.** $\dfrac{2}{-x} = \dfrac{?}{x}$ **15.** $\dfrac{5}{-y} = \dfrac{-5}{?}$ **16.** $\dfrac{2-x}{-9} = \dfrac{?}{9}$

17. $\dfrac{7}{5-x} = \dfrac{?}{x-5}$ **18.** $\dfrac{1-x}{-8} = \dfrac{?}{8}$ **19.** $\dfrac{6-y}{5} = \dfrac{y-6}{?}$ **20.** $-\dfrac{3}{x-4} = \dfrac{3}{?}$

In Exercises 21–56, reduce each fraction to lowest terms.

21. $\dfrac{9}{12}$ **22.** $\dfrac{8}{14}$ **23.** $\dfrac{6ab^2}{3ab}$ **24.** $\dfrac{10m^2n}{5mn}$

25. $\dfrac{4x^2y}{2xy}$ **26.** $\dfrac{12x^3y}{4xy}$ **27.** $\dfrac{5x-10}{x-2}$ **28.** $\dfrac{3x+12}{x+4}$

29. $\dfrac{5x-6}{6-5x}$ **30.** $\dfrac{4-3z}{3z-4}$ **31.** $\dfrac{5x^2+30x}{10x^2-40x}$ **32.** $\dfrac{4x^3-4x^2}{12x^2-12x}$

33. $\dfrac{2+4}{4}$ **34.** $\dfrac{3+9}{3}$ **35.** $\dfrac{5+x}{5}$ **36.** $\dfrac{x+8}{8}$

37. $\dfrac{3xy}{6x^2+3xy}$ **38.** $\dfrac{5ab}{5ab+5a^2}$ **39.** $\dfrac{x^2-1}{x+1}$ **40.** $\dfrac{x^2-4}{x-2}$

41. $\dfrac{3x-12}{4-x}$ **42.** $\dfrac{2x-4}{2-x}$ **43.** $\dfrac{x^2-11x+30}{x^2-9x+20}$ **44.** $\dfrac{x^2-9}{x^2+5x+6}$

45. $\dfrac{x^2-16}{x^2-x-12}$ **46.** $\dfrac{x^2+x-20}{x^2+2x-15}$ **47.** $\dfrac{6x^2-x-2}{10x^2+3x-1}$ **48.** $\dfrac{8x^2-10x-3}{12x^2+11x+2}$

49. $\dfrac{x^2-y^2}{(x+y)^2}$ **50.** $\dfrac{a^2-9b^2}{(a-3b)^2}$ **51.** $\dfrac{(a-2b)(b-a)}{(2a+b)(a-b)}$ **52.** $\dfrac{8(n-2m)}{(3n+m)(2m-n)}$

53. $\dfrac{2y^2+xy-6x^2}{3x^2+xy-2y^2}$ **54.** $\dfrac{10y^2+11xy-6x^2}{4x^2-4xy-15y^2}$

55. $\dfrac{8x^2-2y^2}{2ax-ay+2bx-by}$ **56.** $\dfrac{3x^2-12y^2}{ax+2by+2ay+bx}$

12.2 Multiplying and Dividing Fractions

Multiplying Fractions

In Arithmetic

$$\frac{4}{9}\cdot\frac{3}{8}=\frac{\overset{1}{\cancel{4}}\cdot\overset{1}{\cancel{3}}}{\underset{3}{\cancel{9}}\cdot\underset{2}{\cancel{8}}}=\frac{1}{6}$$

In Algebra

We multiply algebraic fractions the same way we multiply fractions in arithmetic.

> TO MULTIPLY FRACTIONS
>
> **1.** Factor the numerator and denominator of the fractions.
>
> **2.** Divide the numerator and denominator by all factors common to both.
>
> **3.** The answer is the product of the factors remaining in the numerator divided by the product of the factors remaining in the denominator. A factor of 1 will always remain in both numerator and denominator [see Example 1(a)].

Example 1 Multiply the fractions.

a. $\dfrac{1}{m^2} \cdot \dfrac{m}{5} = \dfrac{1 \cdot \cancel{m}^{1}}{\cancel{m^2}_{m} \cdot 5} = \dfrac{1}{5m}$

b. $\dfrac{2y^3}{3x^2} \cdot \dfrac{12x}{5y^2} = \dfrac{2y^3 \cdot \cancel{12}^{4}x}{\cancel{3}_{1}x^2 \cdot 5y^2} = \dfrac{8y}{5x}$

c. $\dfrac{x}{2x-6} \cdot \dfrac{4x-12}{x^2} = \dfrac{x}{2(x-3)} \cdot \dfrac{4(x-3)}{x^2} = \dfrac{x \cdot \cancel{4}^{2}\cancel{(x-3)}}{\cancel{2}_{1}\cancel{(x-3)} \cdot x^2} = \dfrac{2}{x}$

d. $\dfrac{10xy^3}{x^2-y^2} \cdot \dfrac{2x^2+xy-y^2}{15x^2y} = \dfrac{\cancel{10}^{2}xy^3}{\cancel{(x+y)}(x-y)} \cdot \dfrac{\cancel{(x+y)}(2x-y)}{\cancel{15}_{3}x^2y}$

$= \dfrac{2y^2(2x-y)}{3x(x-y)}$ ■

In Arithmetic

Dividing Fractions

$$\frac{3}{5} \div \frac{4}{7} = \frac{3}{5} \cdot \boxed{\frac{7}{4}} = \frac{3 \cdot 7}{5 \cdot 4} = \frac{21}{20}$$

↑ Invert the second fraction and multiply

This method works because

$$\frac{3}{5} \div \frac{4}{7} = \frac{\frac{3}{5}}{\frac{4}{7}} = \frac{\frac{3}{5}}{\frac{4}{7}} \cdot \left(\frac{\frac{7}{4}}{\frac{7}{4}}\right) = \frac{\frac{3}{5} \cdot \frac{7}{4}}{\frac{4}{7} \cdot \frac{7}{4}} = \frac{\frac{3}{5} \cdot \frac{7}{4}}{1} = \frac{3}{5} \cdot \frac{7}{4}$$

↑ The value of this fraction is 1

Therefore, $\dfrac{3}{5} \div \dfrac{4}{7} = \dfrac{3}{5} \cdot \dfrac{7}{4}$.

In Algebra

We divide algebraic fractions the same way we divide fractions in arithmetic.

TO DIVIDE FRACTIONS

Invert the second fraction and multiply.

$$\frac{a}{b} \div \frac{c}{d} = \frac{a}{b} \cdot \frac{d}{c}$$

First fraction ↑ ↑ Second fraction

Example 2 Divide the fractions.

a. $\frac{4}{3x} \div \frac{12}{x^3} = \frac{\overset{1}{\cancel{4}}}{3x} \cdot \frac{x^3}{\underset{3}{\cancel{12}}} = \frac{x^2}{9}$

b. $\frac{4r^3}{9s^2} \div \frac{8r^2s^4}{15rs} = \frac{\overset{1}{\cancel{4}}r^3}{\underset{3}{\cancel{9}}s^2} \cdot \frac{\overset{5}{\cancel{15}}rs}{\underset{2}{\cancel{8}}r^2s^4} = \frac{5r^2}{6s^5}$

c. $\frac{3y^3 - 3y^2}{16y^5 + 8y^4} \div \frac{y^2 + 2y - 3}{4y + 12} = \frac{3y^2\cancel{(y-1)}}{\underset{2}{\cancel{8}}y^4(2y+1)} \cdot \frac{\overset{1}{\cancel{4}}\cancel{(y+3)}}{\cancel{(y-1)}\cancel{(y+3)}}$

$$= \frac{3}{2y^2(2y+1)}$$

d. $\frac{y^2 - x^2}{4xy - 2y^2} \div \frac{2x - 2y}{2x^2 + xy - y^2} = \frac{(y+x)\overset{-1}{\cancel{(y-x)}}}{2y\underset{1}{\cancel{(2x-y)}}} \cdot \frac{(x+y)\overset{1}{\cancel{(2x-y)}}}{2\underset{1}{\cancel{(x-y)}}}$

$$= \frac{-(x+y)^2}{4y} \text{ or } -\frac{(x+y)^2}{4y}$$ ■

EXERCISES 12.2

Perform the indicated operations.

1. $\frac{5}{6} \div \frac{5}{3}$
2. $\frac{3}{8} \div \frac{21}{12}$
3. $\frac{4a^3}{5b^2} \cdot \frac{10b}{8a^2}$
4. $\frac{6d}{8c} \cdot \frac{12c^2}{9d}$
5. $\frac{3x^2}{16} \div \frac{x}{8}$
6. $\frac{4y^3}{7} \div \frac{4y^2}{21}$
7. $\frac{3x^4y^2z}{18xy} \cdot \frac{15z}{x^3yz^2}$
8. $\frac{21a^2b^5}{4b^2} \cdot \frac{6c}{7b^2c^3}$
9. $\frac{5x^3}{4y^2} \cdot \frac{8y}{10x^2}$
10. $\frac{a^3}{2a+6} \div \frac{a^2}{a+3}$
11. $\frac{x}{x+2} \cdot \frac{5x+10}{x^3}$
12. $\frac{y-2}{y} \cdot \frac{6}{3y-6}$
13. $\frac{b^3}{a+3} \div \frac{4b^2}{2a+6}$
14. $\frac{3n-6}{15n} \div \frac{n-2}{20n^2}$
15. $\frac{a+2}{a+1} \cdot \frac{a^2+a}{a^2-4}$
16. $\frac{2x-1}{y} \div \frac{2x^2+x-1}{y^2}$
17. $\frac{a+4}{a-4} \div \frac{a^2+8a+16}{a^2-16}$
18. $\frac{x^3-5x}{9x} \div \frac{4x^3-20x}{12x^2}$
19. $\frac{5}{z+4} \cdot \frac{z^2-16}{(z-4)^2}$
20. $\frac{u+3}{u^2-9} \cdot \frac{2u-6}{6}$
21. $\frac{3a-3b}{4c+4d} \cdot \frac{2c+2d}{b-a}$
22. $\frac{2y-2x}{6} \cdot \frac{x+y}{x^2-y^2}$
23. $\frac{4x-8}{4} \cdot \frac{x+2}{x^2-4}$
24. $\frac{2a+2b}{3} \cdot \frac{a-b}{a^2-b^2}$
25. $\frac{4a+4b}{ab^2} \div \frac{3a+3b}{a^2b}$
26. $\frac{x^2-y^2}{x^2-2xy+y^2} \div \frac{x+y}{x-y}$
27. $\frac{2u^2-14u}{5u^2} \cdot \frac{15u^3}{4u-28}$
28. $\frac{-3v}{18v+90} \cdot \frac{6v^3+30v^2}{v^3}$
29. $\frac{3e^2}{28e-42} \div \frac{6e^3}{14e^2-21e}$
30. $\frac{12f+16}{15f} \div \frac{6f^3+8f^2}{20f^4}$
31. $\frac{x^2+x-2}{x-1} \div \frac{x^2+5x+6}{x^2}$
32. $\frac{z^3}{z+4} \div \frac{z-1}{z^2+3z-4}$
33. $\frac{x^2+10x+25}{x^2-25} \cdot \frac{5-x}{x+5}$

34. $\dfrac{3-y}{y+1} \cdot \dfrac{4+5y+y^2}{y^2+y-12}$

35. $\dfrac{x-y}{9x+9y} \div \dfrac{x^2-y^2}{3x^2+6xy+3y^2}$

36. $\dfrac{x^2-y^2}{8x^2-16xy+8y^2} \div \dfrac{x+y}{4x-4y}$

37. $\dfrac{2x^3y+2x^2y^2}{6x} \div \dfrac{x^2y^2-xy^3}{y-x}$

38. $\dfrac{2b^2c-2bc^2}{b+c} \div \dfrac{4bc^2-4b^2c}{4b+4c}$

39. $\dfrac{x^2-8x-9}{x^3-3x^2-x+3} \cdot \dfrac{x^2-9}{x^2-9x}$

12.3 Adding and Subtracting Like Fractions

Like Fractions

Like fractions are fractions that have the same denominator.

Example 1 Examples of like fractions

$$\frac{2}{3}, \frac{5}{3}, \frac{1}{3} \quad \text{are like fractions.}$$

Same denominator ■

Unlike Fractions

Unlike fractions are fractions that have different denominators.

Example 2 Examples of unlike fractions

$$\frac{1}{3}, \frac{2}{7}, \frac{3}{8} \quad \text{are unlike fractions.}$$

Different denominators ■

Adding Like Fractions

We know that

$$1 \text{ car} + 3 \text{ cars} + 7 \text{ cars} = (1 + 3 + 7) \text{ cars} = 11 \text{ cars}.$$

In Arithmetic

In the same way,

$$2 \text{ thirds} + 5 \text{ thirds} + 1 \text{ third} = (2 + 5 + 1) \text{ thirds} = 8 \text{ thirds}$$

so, $$\frac{2}{3} + \frac{5}{3} + \frac{1}{3} = \frac{2+5+1}{3} = \frac{8}{3}$$

TO ADD LIKE FRACTIONS

1. Add the numerators.
2. Write the sum of the numerators over the denominator of the like fractions.

$$\frac{a}{c} + \frac{b}{c} = \frac{a+b}{c}$$

3. Reduce the resulting fraction to lowest terms.

Example 3 Adding like arithmetic fractions

a. Add: $\frac{1}{9}+\frac{4}{9}+\frac{7}{9}=\frac{1+4+7}{9}=\frac{12}{9}=\frac{\overset{4}{\cancel{12}}}{\underset{3}{\cancel{9}}}=\frac{4}{3}$

b. Add: $\frac{11}{23}+\frac{5}{23}=\frac{11+5}{23}=\frac{16}{23}$ Already lowest terms ■

In Algebra

This method is also used for adding like fractions in algebra.

Example 4 Adding like algebraic fractions

a. $\frac{2}{x}+\frac{5}{x}=\frac{2+5}{x}=\frac{7}{x}$

b. $\frac{15}{d-5}+\frac{-3d}{d-5}=\frac{15-3d}{d-5}=\frac{3\overset{-1}{\cancel{(5-d)}}}{\underset{1}{\cancel{d-5}}}=\frac{-3}{1}=-3$ ■

When denominators differ only by a sign, we can make the fractions like fractions by changing signs. (See Example 5.)

Example 5 Add $\frac{9}{x-2}+\frac{5}{2-x}$.

$$\frac{9}{x-2}+\frac{5}{2-x}=\frac{9}{x-2}+\frac{-5}{-(2-x)}=\frac{9}{x-2}+\frac{-5}{x-2}=\frac{9-5}{x-2}=\frac{4}{x-2}$$

Changing sign of numerator and denominator

This denominator is changed to $x-2$ by a sign change ■

Subtracting Like Fractions

TO SUBTRACT ONE LIKE FRACTION FROM ANOTHER

1. Subtract the numerators.
2. Write the difference of the numerators over the denominator of the like fractions.

$$\frac{a}{c}-\frac{b}{c}=\frac{a-b}{c}$$

3. Reduce the resulting fraction to lowest terms.

Example 6 Subtracting like fractions

a. $\frac{3}{4a} - \frac{5}{4a} = \frac{3}{4a} + \frac{-5}{4a} = \frac{3-5}{4a} = \frac{-2}{4a} = -\frac{\overset{1}{\cancel{2}}}{\underset{2}{\cancel{4}}a} = -\frac{1}{2a}$

b. $\frac{4x}{2x-y} - \frac{2y}{2x-y} = \frac{4x-2y}{2x-y} = \frac{2(2x-y)}{(2x-y)} = \frac{2\cancel{(2x-y)}}{\cancel{(2x-y)}} = 2$

c. $\frac{7}{x+2} - \frac{5-x}{x+2} = \frac{7-(5-x)}{x+2} = \frac{7-5+x}{x+2} = \frac{2+x}{x+2} = 1$ ■

A WORD OF CAUTION If the fraction being subtracted has more than one term in the numerator, it is important to put parentheses around that numerator so that the entire numerator will be subtracted.

Incorrect subtraction:

Common error

$$\frac{2x+3}{x+3} - \frac{x-3}{x+3} = \frac{2x+3-x-3}{x+3} = \frac{x}{x+3}$$

Correct subtraction:

$$\frac{2x+3}{x+3} - \frac{x-3}{x+3} = \frac{2x+3-(x-3)}{x+3} = \frac{2x+3-x+3}{x+3} = \frac{x+6}{x+3}$$ ☑

EXERCISES 12.3

Perform the indicated operations.

1. $\frac{7}{a} + \frac{2}{a}$

2. $\frac{5}{b} + \frac{2}{b}$

3. $\frac{2}{3a} + \frac{4}{3a}$

4. $\frac{8}{5z} + \frac{2}{5z}$

5. $\frac{6}{x} - \frac{2}{x}$

6. $\frac{x+1}{a+b} + \frac{x-1}{a+b}$

7. $\frac{6}{x-y} - \frac{2}{x-y}$

8. $\frac{7}{m+n} - \frac{1}{m+n}$

9. $\frac{2y}{y+1} + \frac{2}{y+1}$

10. $\frac{3x}{x-4} - \frac{12}{x-4}$

11. $\frac{3}{x+3} + \frac{x}{x+3}$

12. $\frac{m}{m-4} - \frac{4}{m-4}$

13. $\frac{2a}{2a+3} + \frac{3}{2a+3}$

14. $\frac{2x+1}{x+4} - \frac{x-3}{x+4}$

15. $\frac{9x}{3x-y} - \frac{3y}{3x-y}$

16. $\frac{5a}{a-2b} - \frac{10b}{a-2b}$

17. $\frac{8}{z-4} - \frac{2z}{z-4}$

18. $\frac{15}{m-5} - \frac{3m}{m-5}$

19. $\frac{4x+y}{2x+y} + \frac{2x+2y}{2x+y}$

20. $\frac{7x-1}{3x-1} - \frac{x+1}{3x-1}$

21. $\frac{x-3}{y-2} - \frac{x+5}{y-2}$

22. $\frac{z+4}{a-b} - \frac{z+3}{a-b}$

23. $\frac{a+2}{2a+1} - \frac{1-a}{2a+1}$

24. $\frac{6x-1}{3x-2} - \frac{3x+1}{3x-2}$

25. $\frac{-x}{x-2} - \frac{2}{2-x}$

26. $\frac{-2a}{b-2a} + \frac{-b}{2a-b}$

27. $\frac{-15w}{1-5w} - \frac{3}{5w-1}$

28. $\frac{-35}{6w-7} - \frac{30w}{7-6w}$

29. $\frac{7z}{8z-4} + \frac{6-5z}{4-8z}$

30. $\frac{13-30w}{15-10w} - \frac{10w+17}{10w-15}$

12.4 Lowest Common Denominator (LCD)

We ordinarily use the lowest common denominator (LCD) when adding or subtracting unlike fractions.

In Arithmetic

The *lowest common denominator* (*LCD*) is the smallest number that is exactly divisible by each of the denominators. Consider the expression

$$\frac{1}{2}+\frac{3}{4}$$

In this example it is possible to determine the LCD by inspection: LCD = 4. Because $\frac{1}{2}=\frac{2}{4}$, then

$$\frac{1}{2}+\frac{3}{4}=\frac{2}{4}+\frac{3}{4}=\frac{2+3}{4}=\frac{5}{4}$$

Consider the expression:

$$\frac{7}{12}+\frac{4}{15}$$

Here the LCD is not easily found by inspection. To find the LCD in this case:

1. Find the prime factorization of each denominator.

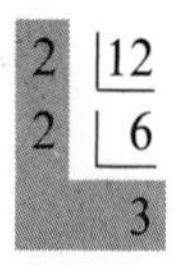

$12 = 2^2 \cdot 3$

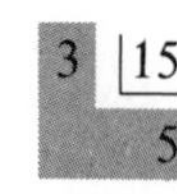

$15 = 3 \cdot 5$

2. Write down each different factor that appears in the prime factorizations.

   2, 3, 5

3. Raise each factor to the highest power it occurs in any denominator.

   2^2, 3, 5

4. LCD $= 2^2 \cdot 3 \cdot 5 = 60$

In Algebra

This same method is used for finding the LCD for algebraic fractions.

> TO FIND THE LCD
>
> **1.** Factor each denominator completely. Repeated factors should be expressed as powers.
>
> **2.** Write each different factor that appears.
>
> **3.** Raise each factor to the *highest* power it occurs in *any* denominator.
>
> **4.** The LCD is the product of all the powers found in Step 3.

Example 1 Find the LCD for $\frac{2}{x} + \frac{5}{x^3}$.

Solution

1. The denominators are already factored.
2. x (x is the only different factor.)
3. x^3 (x^3 is the highest power in any denominator.)
4. LCD $= x^3$ ■

Example 2 Find the LCD for $\frac{7}{18x^2y} + \frac{5}{8xy^4}$.

Solution

1. $2 \cdot 3^2 \cdot x^2 \cdot y, \quad 2^3 \cdot x \cdot y^4$ — Denominators in factored form
2. $2, 3, x, y$ — All the different factors
3. $2^3, 3^2, x^2, y^4$ — Highest powers of factors
4. LCD $= 2^3 \cdot 3^2 \cdot x^2 \cdot y^4 = 72x^2y^4$ ■

Example 3 Find the LCD for $\frac{2}{x} + \frac{x}{x+2}$.

Solution

1. The denominators are already factored.
2. $x, (x + 2)$
3. $x^1, (x + 2)^1$
4. LCD $= x(x + 2)$ ■

Example 4 Find the LCD for $\frac{8}{3x+3} - \frac{5}{x^2 + 2x + 1}$.

Solution

1. $3x + 3 = 3(x + 1); \quad x^2 + 2x + 1 = (x + 1)^2$
2. $3, (x + 1)$
3. $3^1, (x + 1)^2$
4. LCD $= 3(x + 1)^2$ ■

EXERCISES 12.4

Find the LCD in each exercise. Do *not* add the fractions.

1. $\frac{x}{2} - \frac{x}{5}$ **2.** $\frac{3y}{4} + \frac{y}{12}$ **3.** $\frac{3}{2} + \frac{4}{y}$ **4.** $\frac{5}{x^2} - \frac{7}{3}$

5. $\frac{x+1}{4} - \frac{x+3}{2}$ **6.** $\frac{y-2}{3} + \frac{y+5}{9}$ **7.** $\frac{3}{2x} + \frac{5}{3x}$ **8.** $\frac{2}{5z} - \frac{4}{7z}$

9. $\frac{4}{x} + \frac{6}{x^3}$ **10.** $\frac{3}{a^2} + \frac{5}{a^4}$ **11.** $\frac{5}{4y^2} + \frac{9}{6y}$ **12.** $\frac{5}{9x^2y} - \frac{7}{6xy^2}$

13. $\frac{7}{12u^3v^2} - \frac{11}{18uv^3}$ **14.** $\frac{13}{50x^3y^4} - \frac{17}{20x^2y^5}$ **15.** $\frac{4}{a} + \frac{a}{a+3}$ **16.** $\frac{5}{b} + \frac{b}{b-5}$

17. $\frac{x}{2x+4} - \frac{5}{4x}$ **18.** $\frac{4}{3x} + \frac{2x}{3x+6}$ **19.** $\frac{x}{x^2+4x+4} + \frac{1}{x+2}$ **20.** $\frac{2x}{x^2-2x+1} - \frac{5}{x-1}$

21. $\frac{x-4}{x^2+3x+2} + \frac{3x+1}{x^2+2x+1}$ **22.** $\frac{2x+3}{x^2-x-12} + \frac{x-4}{x^2+6x+9}$

23. $\frac{9y}{x^2-y^2} + \frac{6x}{(x+y)^2}$ **24.** $\frac{5}{8z^3} - \frac{8z}{z^2-4} + \frac{5z}{9z+18}$

25. $\frac{x^2+1}{12x^3+24x^2} - \frac{4x+3}{x^2-4x+4} + \frac{1}{x^2-4}$ **26.** $\frac{2y+5}{y^2+6y+9} - \frac{7y}{y^2-9} - \frac{11}{8y^2-24y}$

12.5 Adding and Subtracting Unlike Fractions

Equivalent Fractions

Equivalent fractions are fractions that have the same value. If a fraction is multiplied by 1, its value is unchanged.

$$\frac{2}{2} = 1, \quad \frac{x}{x} = 1, \quad \frac{x+2}{x+2} = 1$$

Multiplying a fraction by expressions like these will produce equivalent fractions

For example:

$$\frac{5}{6} = \frac{5}{6} \cdot \frac{2}{2} = \frac{10}{12}$$ A fraction equivalent to $\frac{5}{6}$

$$\frac{x}{x+2} = \frac{x}{x+2} \cdot \frac{x}{x} = \frac{x^2}{x(x+2)}$$ A fraction equivalent to $\frac{x}{x+2}$

$$\frac{2}{x} = \frac{2}{x} \cdot \frac{x+2}{x+2} = \frac{2x+4}{x(x+2)}$$ A fraction equivalent to $\frac{2}{x}$

TO ADD UNLIKE FRACTIONS

1. Find the LCD.
2. Convert all fractions to equivalent fractions that have the LCD as denominator.
3. Add the resulting like fractions.
4. Reduce the resulting fraction to lowest terms.

Example 1 Add $\frac{5}{4x^2y} + \frac{3}{2xy^3}$.

Solution

Step 1. LCD $= 4x^2y^3$

Step 2. $\dfrac{5}{4x^2y} \qquad + \dfrac{3}{2xy^3} \qquad =$

Multiply numerator and denominator by y^2 to obtain an equivalent fraction whose denominator is the LCD $= 4x^2y^3$

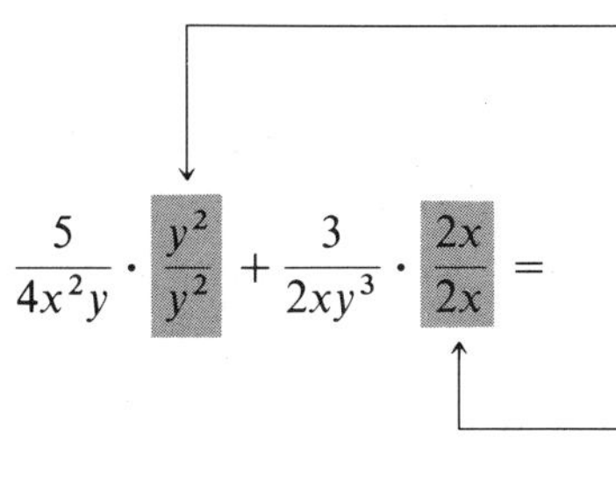

$$\frac{5}{4x^2y} \cdot \frac{y^2}{y^2} + \frac{3}{2xy^3} \cdot \frac{2x}{2x} =$$

Multiply numerator and denominator by $2x$ to obtain an equivalent fraction whose denominator is the LCD $= 4x^2y^3$

Step 3. $\dfrac{5y^2}{4x^2y^3} + \dfrac{6x}{4x^2y^3} = \dfrac{5y^2 + 6x}{4x^2y^3}$ ■

Example 2 Add $\dfrac{2}{x} + \dfrac{x}{x+2}$.

Solution

Step 1. LCD $= x(x + 2)$

Step 2. $\dfrac{2}{x} \qquad + \dfrac{x}{x+2} \qquad =$

Multiply numerator and denominator by $(x + 2)$ to obtain an equivalent fraction whose denominator is the LCD $= x(x + 2)$

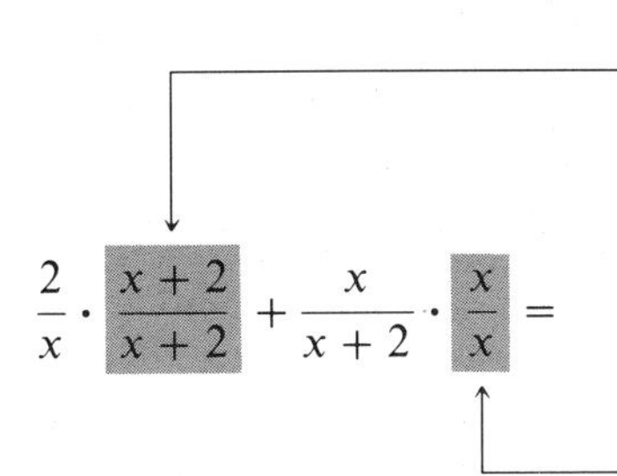

$$\frac{2}{x} \cdot \frac{x+2}{x+2} + \frac{x}{x+2} \cdot \frac{x}{x} =$$

Multiply numerator and denominator by x to obtain an equivalent fraction whose denominator is the LCD $= x(x + 2)$

Step 3. $\dfrac{2x+4}{x(x+2)} + \dfrac{x^2}{x(x+2)} = \dfrac{x^2 + 2x + 4}{x(x+2)}$ ■

Example 3 Add $3 - \dfrac{2a}{a+4}$.

Solution

Step 1. LCD $= a + 4$

Step 2. $3 \qquad - \dfrac{2a}{a+4} =$

$$\frac{3}{1} \cdot \frac{a+4}{a+4} - \frac{2a}{a+4} =$$

$$\frac{3a+12}{a+4} - \frac{2a}{a+4} =$$

Step 3. $\dfrac{3a + 12 - 2a}{a + 4} = \dfrac{a + 12}{a + 4}$ ← NOTE This fraction cannot be reduced because neither a nor 4 is a factor (see Section 12.1, Example 4g). ☑

■

Example 4 Add $\dfrac{20}{x^2 - 25} + \dfrac{2}{x + 5}$.

Solution

Step 1. Factor the denominators to find the LCD.

$$\frac{20}{x^2 - 25} + \frac{2}{x + 5} =$$

$$\frac{20}{(x + 5)(x - 5)} + \frac{2}{x + 5} =$$

The LCD = $(x + 5)(x - 5)$

Step 2. $$\frac{20}{(x + 5)(x - 5)} + \frac{2}{x + 5} \cdot \frac{x - 5}{x - 5} =$$

$$\frac{20}{(x + 5)(x - 5)} + \frac{2x - 10}{(x + 5)(x - 5)} =$$

Step 3. $$\frac{20 + 2x - 10}{(x + 5)(x - 5)} =$$

$$\frac{2x + 10}{(x + 5)(x - 5)} =$$

Step 4. Factor the numerator and reduce.

$$\frac{2\cancel{(x + 5)}}{\cancel{(x + 5)}(x - 5)} = \frac{2}{x - 5}$$ ■

Example 5 Add $\dfrac{6}{x^2 - 9} - \dfrac{2}{x^2 - 4x + 3}$.

Solution

Step 1. Factor the denominators to find the LCD.

$$\frac{6}{x^2 - 9} - \frac{2}{x^2 - 4x + 3} =$$

$$\frac{6}{(x + 3)(x - 3)} - \frac{2}{(x - 3)(x - 1)} =$$

The LCD = $(x + 3)(x - 3)(x - 1)$

Step 2. $$\frac{6}{(x + 3)(x - 3)} \cdot \frac{x - 1}{x - 1} - \frac{2}{(x - 3)(x - 1)} \cdot \frac{x + 3}{x + 3} =$$

$$\frac{6x - 6}{(x + 3)(x - 3)(x - 1)} - \frac{2x + 6}{(x - 3)(x - 1)(x + 3)} =$$

Step 3. $\dfrac{6x - 6 - (2x + 6)}{(x + 3)(x - 3)(x - 1)} =$

$\dfrac{6x - 6 - 2x - 6}{(x + 3)(x - 3)(x - 1)} =$

$\dfrac{4x - 12}{(x + 3)(x - 3)(x - 1)} =$

NOTE When subtracting fractions, be sure to subtract the entire numerator of the second fraction. Thus, $2x + 6$ becomes $-2x - 6$. ☑

Step 4. Factor the numerator and reduce.

$$\frac{4\cancel{(x - 3)}}{(x + 3)\cancel{(x - 3)}(x - 1)} = \frac{4}{(x + 3)(x - 1)}$$ ■

EXERCISES 12.5

In Exercises 1–48, perform the indicated additions and subtractions.

1. $\dfrac{3}{a^2} + \dfrac{2}{a^3}$ **2.** $\dfrac{5}{u} + \dfrac{4}{u^3}$ **3.** $\dfrac{3}{x} + \dfrac{4}{x^2}$ **4.** $\dfrac{1}{3} - \dfrac{1}{a} + \dfrac{2}{a^2}$ **5.** $\dfrac{1}{2} + \dfrac{3}{x} - \dfrac{5}{x^2}$

6. $\dfrac{2}{3} - \dfrac{1}{y} + \dfrac{4}{y^2}$ **7.** $\dfrac{2}{xy} - \dfrac{3}{y}$ **8.** $\dfrac{5}{ab} - \dfrac{4}{a}$ **9.** $\dfrac{5}{xy} - \dfrac{3}{x}$ **10.** $5 + \dfrac{2}{x}$

11. $3 + \dfrac{4}{y}$ **12.** $2 + \dfrac{4}{z}$ **13.** $\dfrac{5}{6xy^2} + \dfrac{7}{8x^2y}$ **14.** $4m - \dfrac{5}{m}$ **15.** $\dfrac{5}{4y^2} + \dfrac{9}{6y}$

16. $\dfrac{10}{5x} + \dfrac{3}{4x^2}$ **17.** $3x - \dfrac{3}{x}$ **18.** $4y - \dfrac{5}{y}$ **19.** $\dfrac{3}{a} + \dfrac{a}{a + 3}$ **20.** $\dfrac{5}{b} + \dfrac{b}{b - 5}$

21. $\dfrac{x}{2x + 4} + \dfrac{-5}{4x}$ **22.** $\dfrac{4}{3x} + \dfrac{-2x}{3x + 6}$ **23.** $x + \dfrac{2}{x} - \dfrac{3}{x - 2}$ **24.** $m - \dfrac{3}{m} + \dfrac{2}{m + 4}$

25. $\dfrac{a + b}{b} + \dfrac{b}{a - b}$ **26.** $\dfrac{x - y}{x} - \dfrac{x}{x + y}$ **27.** $\dfrac{3}{2e - 2} - \dfrac{2}{3e - 3}$ **28.** $\dfrac{2}{3f + 6} - \dfrac{1}{5f + 10}$

29. $\dfrac{3}{m - 2} - \dfrac{5}{2 - m}$ **30.** $\dfrac{7}{n - 5} - \dfrac{2}{5 - n}$ **31.** $\dfrac{x + 1}{x - 1} - \dfrac{x - 1}{x + 1}$ **32.** $\dfrac{x - 5}{x + 5} - \dfrac{x + 5}{x - 5}$

33. $\dfrac{2}{x} + \dfrac{x}{x + 2}$ **34.** $\dfrac{2}{x + 5} - \dfrac{3}{x - 5}$ **35.** $a + \dfrac{3}{a} - \dfrac{2}{a - 2}$ **36.** $\dfrac{2}{a + 3} - \dfrac{4}{a - 1}$

37. $\dfrac{a - b}{b} + \dfrac{b}{a + b}$ **38.** $\dfrac{x + 2}{x - 3} - \dfrac{x + 3}{x - 2}$ **39.** $\dfrac{x - 1}{x + 2} - \dfrac{x - 2}{x + 1}$ **40.** $\dfrac{5}{x - 3} - \dfrac{4}{3 - x}$

41. $\dfrac{y}{x^2 - xy} + \dfrac{x}{y^2 - xy}$ **42.** $\dfrac{b}{ab - a^2} - \dfrac{a}{b^2 - ab}$ **43.** $\dfrac{2x}{x - 3} - \dfrac{2x}{x + 3} + \dfrac{36}{x^2 - 9}$

44. $\dfrac{x}{x + 4} - \dfrac{x}{x - 4} - \dfrac{32}{x^2 - 16}$ **45.** $\dfrac{x}{x - 1} - \dfrac{x}{x + 1} + \dfrac{2}{x^2 - 1}$ **46.** $\dfrac{x + 2}{x^2 + x - 2} + \dfrac{3}{x^2 - 1}$

47. $\dfrac{x}{x^2 + 4x + 4} + \dfrac{1}{x + 2}$ **48.** $\dfrac{2x}{x^2 - 2x + 1} - \dfrac{5}{x - 1}$

In Exercises 49–64, perform the indicated operations.

49. $\dfrac{2x}{x^2 + 3x - 4} \cdot \dfrac{x^2 + 5x - 6}{2x + 12}$ **50.** $\dfrac{x^2 + 5x + 6}{x^2 - 2x - 15} \cdot \dfrac{3x - 15}{15}$ **51.** $\dfrac{2}{x^2 - 1} + \dfrac{1}{x^2 + 3x + 2}$

52. $\dfrac{1}{x^2 - 7x + 12} + \dfrac{6}{x^2 - 9}$

53. $\dfrac{a^2}{5a + 10} \div \dfrac{a^2 - 5a}{a^2 - 3a - 10}$

54. $\dfrac{2b^3}{b^2 + 8b} \div \dfrac{8b + 8}{b^2 + 9b + 8}$

55. $\dfrac{3x}{x^2 - x - 20} - \dfrac{5}{x^2 - 7x + 10}$

56. $\dfrac{x}{x^2 + x - 6} - \dfrac{3}{x^2 + 11x + 24}$

57. $\dfrac{x - 4}{x^2 + 3x + 2} + \dfrac{3x + 1}{x^2 + 2x + 1}$

58. $\dfrac{2x + 3}{x^2 - x - 12} + \dfrac{x - 4}{x^2 + 6x + 9}$

59. $\dfrac{4x^2 - y^2}{4x^2 + y^2} \cdot \dfrac{4x^2 + 2xy}{4x^2 + 4xy + y^2}$

60. $\dfrac{a^2 - 6ab + 9b^2}{3ab - 9b^2} \cdot \dfrac{a^2 + 9b^2}{a^2 - 9b^2}$

61. $\dfrac{2x}{x + 2} + \dfrac{x}{x - 2} - \dfrac{3x + 2}{x^2 - 4}$

62. $\dfrac{3x}{x + 5} - \dfrac{x}{x - 5} + \dfrac{7x + 15}{x^2 - 25}$

63. $\dfrac{2x}{x^2 - 9x + 14} \cdot \dfrac{x^2 - 6x - 7}{x^2 + 4x} \div \dfrac{x^2 - 1}{x^2 + 2x - 8}$

64. $\dfrac{x + 1}{x^2 - 9} \div \dfrac{3x}{x^2 + 8x + 15} \cdot \dfrac{x^2 - 3x}{x^2 + 6x + 5}$

12.6 Complex Fractions

Simple Fractions

A **simple fraction** is a fraction that has only one fraction line.

Examples: $\dfrac{2}{x}$, $\dfrac{3 + y}{12}$, $\dfrac{7a - 7b}{ab^2}$, $\dfrac{5}{x + y}$

Complex Fractions

A **complex fraction** is a fraction that has more than one fraction line.

Examples: $\dfrac{\frac{2}{x}}{3}$, $\dfrac{a}{\frac{1}{c}}$, $\dfrac{\frac{3}{z}}{\frac{5}{z}}$, $\dfrac{\frac{3}{x} - \frac{2}{y}}{\frac{5}{x} + \frac{3}{y}}$

The Parts of a Complex Fraction

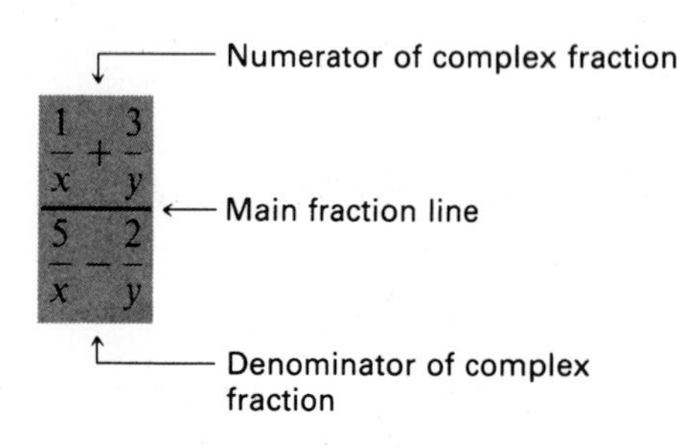

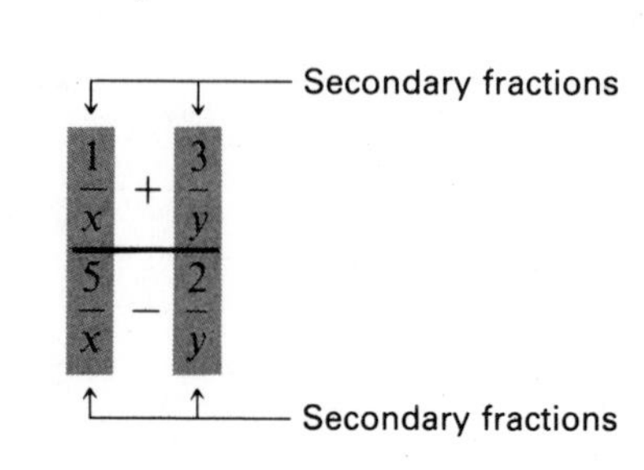

Methods for Simplifying Complex Fractions

> TO SIMPLIFY COMPLEX FRACTIONS
>
> *Method* 1. Multiply both numerator and denominator of the complex fraction by the LCD of the secondary fractions; then simplify the results.
>
> *Method* 2. First, simplify the numerator and denominator of the complex fraction; then divide the simplified numerator by the simplified denominator.

NOTE In some of the following examples, the solution by method 1 is easier than that by method 2. In others, the opposite is true. ☑

Example 1 Simplify the complex fraction.

$$\frac{\frac{1}{2}+\frac{3}{4}}{\frac{5}{6}-\frac{2}{3}}$$

Solution

Method 1. The LCD of the secondary denominators 2, 4, 6, and 3 is 12.

$$\frac{12}{12}\cdot\frac{\frac{1}{2}+\frac{3}{4}}{\frac{5}{6}-\frac{2}{3}}=\frac{12\left(\frac{1}{2}+\frac{3}{4}\right)}{12\left(\frac{5}{6}-\frac{2}{3}\right)}=\frac{\frac{12}{1}\left(\frac{1}{2}\right)+\frac{12}{1}\left(\frac{3}{4}\right)}{\frac{12}{1}\left(\frac{5}{6}\right)-\frac{12}{1}\left(\frac{2}{3}\right)}=\frac{6+9}{10-8}=\frac{15}{2}=7\frac{1}{2}$$

This is 1 (pointing to $\frac{12}{12}$)

Method 2.

$$\frac{\frac{1}{2}+\frac{3}{4}}{\frac{5}{6}-\frac{2}{3}}=\frac{\frac{2}{4}+\frac{3}{4}}{\frac{5}{6}-\frac{4}{6}}=\frac{\frac{5}{4}}{\frac{1}{6}}=\frac{5}{4}\div\frac{1}{6}=\frac{5}{\cancel{4}_2}\cdot\frac{\cancel{6}^3}{1}=\frac{15}{2}=7\frac{1}{2}$$ ■

Example 2 Simplify $\dfrac{\frac{4b^2}{9a^2}}{\frac{8b}{3a^3}}$. ←main fraction line

Solution

Method 1. The LCD of the secondary denominators $9a^2$ and $3a^3$ is $9a^3$.

$$\frac{9a^3}{9a^3}\left(\frac{\frac{4b^2}{9a^2}}{\frac{8b}{3a^3}}\right)=\frac{\frac{9a^3}{1}\left(\frac{4b^2}{9a^2}\right)}{\frac{9a^3}{1}\left(\frac{8b}{3a^3}\right)}=\frac{4ab^2}{24b}=\frac{ab}{6}$$

The value of this fraction is 1 (pointing to $\frac{9a^3}{9a^3}$)

Method 2. $$\frac{\frac{4b^2}{9a^2}}{\frac{8b}{3a^3}}=\frac{4b^2}{9a^2}\div\frac{8b}{3a^3}=\frac{\cancel{4}^1b^2}{\cancel{9}_3a^2}\cdot\frac{\cancel{3}^1a^3}{\cancel{8}_2b}=\frac{ab}{6}$$ ■

Example 3 Simplify $\dfrac{1+\frac{2}{x}}{1-\frac{4}{x^2}}$.

Solution

Method 1. The LCD of the secondary denominators x and x^2 is x^2.

$$\frac{x^2}{x^2}\left[\frac{1+\frac{2}{x}}{1-\frac{4}{x^2}}\right] = \frac{\frac{x^2}{1}\cdot\frac{1}{1}+\frac{x^2}{1}\cdot\frac{2}{x}}{\frac{x^2}{1}\cdot\frac{1}{1}-\frac{x^2}{1}\cdot\frac{4}{x^2}} = \frac{x^2+2x}{x^2-4}$$

$$= \frac{x(\cancel{x+2})}{(\cancel{x+2})(x-2)} = \frac{x}{x-2}$$

Method 2.
$$\frac{1+\frac{2}{x}}{1-\frac{4}{x^2}} = \frac{\frac{x}{x}+\frac{2}{x}}{\frac{x^2}{x^2}-\frac{4}{x^2}} = \frac{\frac{x+2}{x}}{\frac{x^2-4}{x^2}} = \frac{x+2}{x} \div \frac{x^2-4}{x^2}$$

$$= \frac{\overset{1}{\cancel{x+2}}}{x}\cdot\frac{x^2}{\underset{1}{(\cancel{x+2})}(x-2)}$$

$$= \frac{x}{x-2} \quad ■$$

EXERCISES 12.6

Simplify each of the complex fractions.

1. $\dfrac{\frac{3}{4}-\frac{1}{2}}{\frac{5}{8}+\frac{1}{4}}$
2. $\dfrac{\frac{5}{6}-\frac{1}{3}}{\frac{2}{9}+\frac{1}{6}}$
3. $\dfrac{\frac{3}{5}+2}{2-\frac{3}{4}}$
4. $\dfrac{\frac{3}{16}+5}{6-\frac{7}{8}}$
5. $\dfrac{\frac{5x^3}{3y^4}}{\frac{10x}{9y}}$
6. $\dfrac{\frac{8a^4}{5b}}{\frac{4a^3}{15b^2}}$
7. $\dfrac{\frac{18cd^2}{5a^3b}}{\frac{12cd^2}{15ab^2}}$
8. $\dfrac{\frac{8x^2y}{7z^3}}{\frac{12xy^2}{21z^5}}$
9. $\dfrac{\frac{3a^3}{5b^2}}{\frac{6a^2}{10b^3}}$
10. $\dfrac{\frac{x-2}{4}}{\frac{3x-6}{12}}$
11. $\dfrac{\frac{x+3}{5}}{\frac{2x+6}{10}}$
12. $\dfrac{\frac{a-4}{3}}{\frac{2a-8}{9}}$
13. $\dfrac{\frac{a}{b}+1}{\frac{a}{b}-1}$
14. $\dfrac{2+\frac{x}{y}}{2-\frac{x}{y}}$
15. $\dfrac{\frac{c}{d}+2}{\frac{c^2}{d^2}-4}$
16. $\dfrac{\frac{x^2}{y^2}-1}{\frac{x}{y}-1}$
17. $\dfrac{1}{1-\frac{1}{x}}$
18. $\dfrac{1}{1+\frac{1}{y}}$
19. $\dfrac{x+\frac{1}{y}}{x}$
20. $\dfrac{\frac{1}{a}-b}{b}$
21. $\dfrac{\frac{1}{x}+x}{\frac{1}{x}-x}$
22. $\dfrac{a-\frac{4}{a}}{a+\frac{4}{a}}$
23. $\dfrac{1-\frac{1}{b}}{3-\frac{3}{b}}$
24. $\dfrac{x+\frac{x}{y}}{1+\frac{1}{y}}$
25. $\dfrac{1-\frac{1}{a^2}}{\frac{1}{a}-\frac{1}{a^2}}$
26. $\dfrac{\frac{1}{y}+\frac{x}{y^2}}{1-\frac{x^2}{y^2}}$
27. $\dfrac{\frac{x^2}{y^2}-4}{\frac{x}{y}+2}$
28. $\dfrac{\frac{1}{y^2}-9}{\frac{1}{y}+3}$
29. $\dfrac{\frac{1}{a^2}-\frac{1}{b^2}}{\frac{1}{a}+\frac{1}{b}}$
30. $\dfrac{\frac{1}{c^2}-\frac{1}{4}}{\frac{1}{c}-\frac{1}{2}}$

12.7 Solving Equations Having Fractions

Equations Having Fractions That Simplify to First-degree Equations

> TO SOLVE AN EQUATION HAVING FRACTIONS THAT SIMPLIFIES TO A FIRST-DEGREE EQUATION
>
> 1. Remove fractions by multiplying each term by the LCD.
> 2. Remove grouping symbols.
> 3. Combine like terms: all terms with the unknown on one side, all other terms on the other side.
> 4. Divide both sides by the coefficient of the unknown.
>
> *Check* apparent solutions in the original equation. Any value of a variable that makes any denominator in the equation zero is not a solution (Example 4).

Example 1 Solve $\frac{x}{2}+\frac{x}{3}=5$.

Solution The LCD of the fractions is 6. Multiply both sides by the LCD, 6.

$$6\left(\frac{x}{2}\right)+6\left(\frac{x}{3}\right)=6(5)$$

This results in *each term* of the equation being multiplied by the LCD, 6

$$\frac{\overset{3}{\cancel{6}}}{1}\left(\frac{x}{\underset{1}{\cancel{2}}}\right)+\frac{\overset{2}{\cancel{6}}}{1}\left(\frac{x}{\underset{1}{\cancel{3}}}\right)=\frac{6}{1}\left(\frac{5}{1}\right)$$

$$3x+2x=30$$

$$5x=30$$

$$x=6$$

Check

$$\frac{x}{2}+\frac{x}{3}=5$$

$$\frac{6}{2}+\frac{6}{3}\stackrel{?}{=}5$$

$$3+2=5 \quad \blacksquare$$

Example 2 Solve $\frac{x-4}{2}-\frac{x}{5}=\frac{1}{10}$.

Solution LCD = 10

$$\frac{\overset{5}{\cancel{10}}}{1}\left(\frac{x-4}{\underset{1}{\cancel{2}}}\right)-\frac{\overset{2}{\cancel{10}}}{1}\left(\frac{x}{\underset{1}{\cancel{5}}}\right)=\frac{\cancel{10}}{1}\left(\frac{1}{\cancel{10}}\right)$$

Multiplying both sides by the LCD

$$5(x-4)-2x=1$$

$$5x-20-2x=1$$

$$3x=21$$

$$x=7$$

Check

$$\frac{x-4}{2}-\frac{x}{5}=\frac{1}{10}$$

$$\frac{7-4}{2}-\frac{7}{5}\stackrel{?}{=}\frac{1}{10}$$

$$\frac{3}{2}-\frac{7}{5}\stackrel{?}{=}\frac{1}{10}$$

$$\frac{15}{10}-\frac{14}{10}=\frac{1}{10} \quad ■$$

When an equation has only one fraction on each side of the equal sign, it is a proportion. Proportions can be solved by setting the product of the means equal to the product of the extremes. (See Example 3.) Proportions can also be solved by multiplying both sides by the LCD.

Example 3 Solve $\frac{9x}{x-3}=6$.

Solution

$\frac{9x}{x-3}=\frac{6}{1}$	Writing $\frac{6}{1}$ makes the equation a proportion
$9x\cdot 1=6(x-3)$	Product of means = product of extremes
$9x = 6x-18$ $-6x \quad -6x$	Removing ( )
$3x = -18$	Getting x's on one side
$x=-6$	Dividing both sides by 3

Check

$$\frac{9x}{x-3}=6$$

$$\frac{9(-6)}{-6-3}\stackrel{?}{=}6$$

$$\frac{-54}{-9}\stackrel{?}{=}6$$

$$6=6 \quad ■$$

Before beginning to solve an equation, note all excluded values of the variable by inspection. Any excluded value cannot be a solution of the equation.

Example 4 Solve $\frac{x}{x-3}=\frac{3}{x-3}+4$.

NOTE 3 is an excluded value because it makes the denominator $x-3$ zero. Therefore, 3 cannot be a solution of this equation. ☑

Solution LCD $= x-3$

$$\frac{\cancel{(x-3)}}{1}\cdot\frac{x}{\cancel{(x-3)}}=\frac{\cancel{(x-3)}}{1}\cdot\frac{3}{\cancel{(x-3)}}+\frac{(x-3)}{1}\cdot\frac{4}{1}$$

$$x = 3 + 4(x-3)$$

$$x = 3 + 4x - 12$$

$$x = 4x - 9$$

$$9 = 3x$$

$$3 = x$$ Because 3 is an excluded value, this equation has *no solution*

If we try to check the value 3 in the equation, we have

Check

$$\frac{x}{x-3} = \frac{3}{x-3} + 4$$

$$\frac{3}{3-3} \stackrel{?}{=} \frac{3}{3-3} + 4$$

$$\frac{3}{0} \stackrel{?}{=} \frac{3}{0} + 4$$

$\uparrow$ $\uparrow$ not possible ■

A WORD OF CAUTION Because students solve a proportion by cross-multiplication, they sometimes use cross-multiplication incorrectly in a product of fractions.

Correct application of cross-multiplication	*Incorrect application of cross-multiplication*
This is an *equation.*	This is a *product.*
If $\frac{16}{6} = \frac{8}{3}$,	$\frac{16}{6} \cdot \frac{8}{3} \neq \frac{16 \cdot 3}{6 \cdot 8}$
then $16 \cdot 3 = 6 \cdot 8$	*Correct product*
$48 = 48$	$\frac{16}{6} \cdot \frac{8}{3} = \frac{\overset{8}{\cancel{16}} \cdot 8}{\underset{3}{\cancel{6}} \cdot 3} = \frac{64}{9}$

☑

Equations Having Fractions That Simplify to Second-degree Equations

After removing fractions and grouping symbols, there may be second-degree terms. When this is the case, the equation can sometimes be solved by factoring (Section 11.7).

TO SOLVE A QUADRATIC EQUATION BY FACTORING

1. Get *all* nonzero terms to one side by adding the same expression to both sides. *Only zero must remain on the other side.* Then arrange the terms in descending powers.
2. Factor the polynomial.
3. Set each factor equal to zero, and then solve for the unknown.

Check apparent solutions in the original equation. Any value of a variable that makes any denominator in the equation zero is not a solution (Example 4).

Example 5 Solve $\frac{2}{x} + \frac{3}{x^2} = 1$.

Solution LCD $= x^2$

$$\frac{x^2}{1}\left(\frac{2}{x}\right) + \frac{x^2}{1}\left(\frac{3}{x^2}\right) = \frac{x^2}{1}\left(\frac{1}{1}\right)$$

Second-degree term

$$2x + 3 = x^2$$
$$0 = x^2 - 2x - 3$$
$$0 = (x - 3)(x + 1)$$
$$x - 3 = 0 \quad \Big| \quad x + 1 = 0$$
$$x = 3 \quad \Big| \quad x = -1$$

Check for x = 3

$$\frac{2}{x} + \frac{3}{x^2} = 1$$
$$\frac{2}{3} + \frac{3}{3^2} \stackrel{?}{=} 1$$
$$\frac{2}{3} + \frac{1}{3} = 1$$

Check for x = −1

$$\frac{2}{x} + \frac{3}{x^2} = 1$$
$$\frac{2}{-1} + \frac{3}{(-1)^2} \stackrel{?}{=} 1$$
$$-2 + 3 = 1$$

■

Example 6 Solve $\frac{8}{x} = \frac{3}{x+1} + 3$.

Solution LCD $= x(x + 1)$

$$\frac{x(x+1)}{1}\,\frac{8}{x} = \frac{x(x+1)}{1}\,\frac{3}{(x+1)} + \frac{x(x+1)}{1}\,\frac{3}{1}$$
$$8(x + 1) = 3x + 3x(x + 1)$$

Second-degree term

$$8x + 8 = 3x + 3x^2 + 3x$$
$$0 = 3x^2 - 2x - 8$$
$$0 = (3x + 4)(x - 2)$$
$$3x + 4 = 0 \quad \Big| \quad x - 2 = 0$$
$$3x = -4 \quad \Big| \quad x = 2$$
$$x = -\frac{4}{3}$$

You should check to see that both $x = 2$ and $x = -\frac{4}{3}$ make the two sides of the equation equal. ■

EXERCISES 12.7

Solve the equations.

1. $\frac{x}{3} + \frac{x}{4} = 7$

2. $\frac{x}{5} + \frac{x}{3} = 8$

3. $\frac{a}{2} - \frac{a}{5} = 6$

4. $\frac{b}{3} - \frac{b}{7} = 12$

5. $\frac{x}{5} + \frac{x}{2} = 7$

6. $\frac{8}{z} = 4$

7. $\frac{9}{2x} = 3$

8. $\frac{14}{3x} = 7$

9. $z + \frac{1}{z} = \frac{17}{z}$

10. $y + \frac{3}{y} = \frac{12}{y}$

11. $\frac{2}{x} - \frac{2}{x^2} = \frac{1}{2}$

12. $\frac{3}{x} - \frac{4}{x^2} = \frac{1}{2}$

13. $\frac{M-2}{5} + \frac{M}{3} = \frac{1}{5}$

14. $\frac{y+2}{4} + \frac{y}{5} = \frac{1}{4}$

15. $\frac{7}{x+4} = \frac{3}{x}$

16. $\frac{5}{x+6} = \frac{2}{x}$

17. $\frac{3x}{x-2} = 5$

18. $\frac{4}{x+3} = \frac{2}{x}$

19. $\frac{2z-4}{3} + \frac{3z}{2} = \frac{5}{6}$

20. $\frac{y-1}{2} + \frac{y}{5} = \frac{3}{10}$

21. $x + \frac{1}{x} = \frac{10}{x}$

22. $\frac{5}{x} - \frac{1}{x^2} = \frac{9}{4}$

23. $\frac{x}{x+1} = \frac{4x}{3x+2}$

24. $\frac{x}{3x-4} = \frac{3x}{2x+2}$

25. $\frac{2x}{3x+1} = \frac{4x}{5x+1}$

26. $\frac{4y-3}{2} = \frac{5y}{y+2}$

27. $\frac{6}{y-2} = \frac{3}{y}$

28. $\frac{5}{x-3} = \frac{2}{x}$

29. $\frac{x}{x^2+1} = \frac{2}{1+2x}$

30. $\frac{3x}{3x^2+2} = \frac{1}{x+1}$

31. $\frac{2x-1}{3} + \frac{3x}{4} = \frac{5}{6}$

32. $\frac{3z-2}{4} + \frac{3z}{8} = \frac{3}{4}$

33. $\frac{x}{x-2} = \frac{2}{x-2} + 5$

34. $\frac{x}{x+5} = 4 - \frac{5}{x+5}$

35. $\frac{1}{x-2} - \frac{4}{x+2} = \frac{1}{5}$

36. $\frac{4}{x+1} = \frac{3}{x} + \frac{1}{15}$

37. $\frac{1}{x-1} + \frac{2}{x+1} = \frac{5}{3}$

38. $\frac{2}{3x+1} + \frac{1}{x-1} = \frac{7}{10}$

39. $\frac{3}{2x+5} + \frac{x}{4} = \frac{3}{4}$

40. $\frac{5}{2x-1} - \frac{x}{6} = \frac{4}{3}$

12.8 Literal Equations

Literal Equations

Literal equations are equations that have more than one letter.

Example 1 When we solve a literal equation for one of its letters, the solution will contain the other letters as well as numbers.

a. Solve $x + y = 5$ for x

$x + y = 5$ Subtract y from both sides

$x = 5 - y$ Solution for x

b. Solve $x + y = 5$ for y

$x + y = 5$ Subtract x from both sides

$y = 5 - x$ Solution for y ■

TO SOLVE A LITERAL EQUATION

1. *Remove fractions* (if there are any) by multiplying both sides by the LCD.
2. *Remove grouping symbols* (if there are any).
3. *Collect like terms:* all terms containing the letter you are solving for on one side, all other terms on the other side.
4. *Factor out the letter you are solving for* (if it appears in more than one term).
5. *Divide both sides by the coefficient of the letter you are solving for.*

Example 2 Solve $3x + 4y = 12$ for x.

Solution

$3x + 4y = 12$	Subtract $4y$ from both sides
$3x = 12 - 4y$	Divide both sides by 3
$x = \dfrac{12 - 4y}{3}$	Solution for x ■

Example 3 Solve $\dfrac{4ab}{d} = 15$ for a. Sometimes a literal equation can solved as a proportion

Solution

$\dfrac{4ab}{d} = \dfrac{15}{1}$	This is a proportion
$4ab \cdot 1 = 15 \cdot d$	Product of means = product of extremes
$\dfrac{\cancel{4}a\cancel{b}}{\cancel{4b}} = \dfrac{15d}{4b}$	Dividing both sides by $4b$
$a = \dfrac{15d}{4b}$	Solution ■

Example 4 Solve $A = P(1 + rt)$ for t.

Solution

$A = P(1 + rt)$	
$A = P + Prt$	Removed ( ) by using the distributive rule
$A - P = Prt$	Collected terms with the letter being solved for t on one side and all other terms on the other side
$\dfrac{A - P}{Pr} = \dfrac{\cancel{Pr}t}{\cancel{Pr}}$	Divided both sides by Pr
$\dfrac{A - P}{Pr} = t$	Solution ■

Example 5 Solve $I = \dfrac{nE}{R + nr}$ for n.

Solution

$\dfrac{I}{1} = \dfrac{nE}{R + nr}$	This is a proportion
$1(nE) = I(R + nr)$	Product of means = product of extremes
$nE = IR + Inr$	Removed ( ) using distributive rule
$nE - Inr = IR$	Collected terms with the letter being solved for n on one side and all other terms on the other side

$$n(E - Ir) = IR \quad \text{Removed } n \text{ as a common factor}$$

$$\frac{n\cancel{(E - Ir)}}{\cancel{(E - Ir)}} = \frac{IR}{(E - Ir)} \quad \text{Divided both sides by } (E - Ir)$$

$$n = \frac{IR}{E - Ir} \quad \text{Solution} \quad ■$$

EXERCISES 12.8

Solve for the letter listed after each equation.

1. $2x + y = 4; \quad x$

2. $x + 3y = 6; \quad y$

3. $y - z = -8; \quad z$

4. $m - n = -5; \quad n$

5. $2x - y = -4; \quad y$

6. $3y - z = -5; \quad z$

7. $2x - 3y = 6; \quad x$

8. $3x - 2y = 6; \quad x$

9. $x + 2y = 5; \quad x$

10. $x - y = -4; \quad y$

11. $2x - y = -4; \quad x$

12. $3x - 4y = 12; \quad y$

13. $2(x - 3y) = x + 4; \quad x$

14. $3x - 14 = 2(y - 2x); \quad x$

15. $PV = k; \quad V$

16. $I = Prt; \quad P$

17. $\frac{3xy}{z} = 10; \quad x$

18. $\frac{PV}{T} = 100; \quad V$

19. $y = mx + b; \quad x$

20. $P = 2\ell + 2w; \quad w$

21. $A = P(1 + rt); \quad r$

22. $b = c(1 + xy); \quad x$

23. $S = \frac{a}{1 - r}; \quad r$

24. $I = \frac{E}{R + r}; \quad R$

25. $ax + bx = c; \quad x$

26. $az - 2z = 4; \quad z$

27. $z = \frac{Rr}{R + r}; \quad R$

28. $c = \frac{ax}{x + 3}; \quad x$

29. $\frac{1}{F} = \frac{1}{u} + \frac{1}{v}; \quad u$

30. $\frac{1}{a} = \frac{1}{b} - \frac{1}{c}; \quad b$

31. $C = \frac{5}{9}(F - 32); \quad F$

32. $A = \frac{h}{2}(B + b); \quad B$

33. $\frac{m + n}{x} - a = \frac{m - n}{x} + c; \quad x$

34. $\frac{a - b}{x} + c = \frac{a + b}{x} - h; \quad x$

12.9 Word Problems Involving Fractions

Number Problems

Example 1 The sum of a number and its reciprocal is $\frac{13}{6}$. Find the number.

Solution Let $\quad x =$ the number,

then $\quad \frac{1}{x} =$ its reciprocal.

Sum of a number and its reciprocal	is	$\frac{13}{6}$
$x + \frac{1}{x}$	$=$	$\frac{13}{6}$

LCD $= 6x$

$$6x(x) + 6x\left(\frac{1}{x}\right) = \not{6}x\left(\frac{13}{\not{6}}\right)$$

$$6x^2 + 6 = 13x$$

$$6x^2 - 13x + 6 = 0$$

$$(2x - 3)(3x - 2) = 0$$

$$2x - 3 = 0 \quad \Big| \quad 3x - 2 = 0$$

$$2x = 3 \quad \Big| \quad 3x = 2$$

$$x = \frac{3}{2} \quad \Big| \quad x = \frac{2}{3}$$

Therefore, there are two answers: $\frac{3}{2}$ and $\frac{2}{3}$.

Check Number + Reciprocal

$$\frac{3}{2} + \frac{2}{3} = \frac{9}{6} + \frac{4}{6} = \frac{13}{6}$$

Number + Reciprocal

$$\frac{2}{3} + \frac{3}{2} = \frac{4}{6} + \frac{9}{6} = \frac{13}{6} \quad \blacksquare$$

Example 2 The denominator of a fraction exceeds the numerator by 3. If 4 is added to the numerator and 2 is subtracted from the denominator, the resulting fraction is $\frac{3}{2}$. Find the original fraction.

Solution Let $x =$ numerator
and $x + 3 =$ denominator.

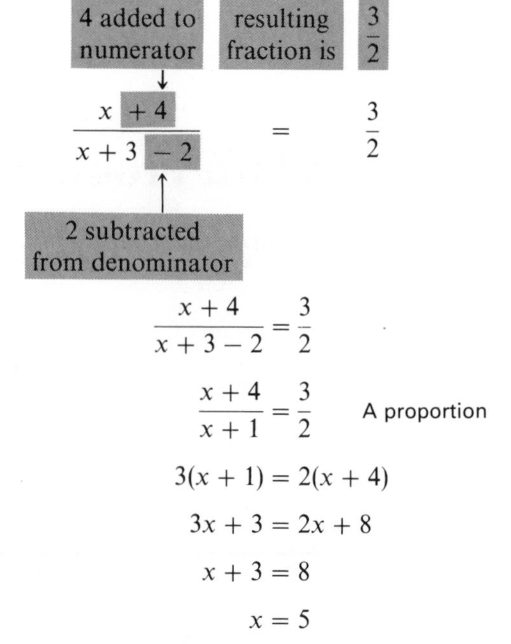

$$\frac{x+4}{x+3-2} = \frac{3}{2}$$

$$\frac{x+4}{x+1} = \frac{3}{2} \qquad \text{A proportion}$$

$$3(x + 1) = 2(x + 4)$$

$$3x + 3 = 2x + 8$$

$$x + 3 = 8$$

$$x = 5$$

Therefore, the original fraction is $\frac{x}{x+3} = \frac{5}{8}$.

Check

$$\frac{5+4}{8-2} \stackrel{?}{=} \frac{3}{2}$$

$$\frac{9}{6} \stackrel{?}{=} \frac{3}{2}$$

$$\frac{3}{2} = \frac{3}{2} \quad \blacksquare$$

Work Problems

The basic idea used to solve work problems is

$$\text{Rate} \times \text{Time} = \text{Amount of work}$$

Suppose John can mow a lawn in 2 hours. Therefore,

$$\text{John's } \textit{rate} = \frac{1 \text{ lawn}}{2 \text{ hour}} = \frac{1}{2} \text{ lawn per hour}$$

If John works for 8 hours,

Rate	×	Time	=	Amount of work
$\frac{1}{2}$	·	8	=	4 lawns

If John works for x hours,

Rate	×	Time	=	Amount of work
$\frac{1}{2}$	·	x	=	$\frac{x}{2}$ lawns

Example 3 John can mow a lawn in 2 hours, and Mike can mow the same lawn in 4 hours. How long will it take them to mow the lawn working together?

Solution Let x = time each works.

$$\text{John's rate} = \frac{1 \text{ lawn}}{2 \text{ hours}} = \frac{1}{2} \text{ lawn per hr}$$

$$\text{Mike's rate} = \frac{1 \text{ lawn}}{4 \text{ hours}} = \frac{1}{4} \text{ lawn per hr}$$

	Rate ·	*Time* =	*Amount of work*
John	$\frac{1}{2}$	x	$\frac{x}{2}$
Mike	$\frac{1}{4}$	x	$\frac{x}{4}$

$$\boxed{\text{amount John mows}} + \boxed{\text{amount Mike mows}} = \boxed{\text{1 lawn}}$$

$$\frac{x}{2} + \frac{x}{4} = 1$$

LCD = 4

$$\overset{2}{\cancel{4}}\left(\frac{x}{\cancel{2}}\right) + \cancel{4}\left(\frac{x}{\cancel{4}}\right) = 4(1)$$

$$2x + x = 4$$

$$3x = 4$$

$$x = \frac{4}{3} = 1\frac{1}{3}$$

Therefore, it will take them $1\frac{1}{3}$ hours working together.

Check

$$\text{John's work} = \frac{1 \text{ lawn}}{2 \text{ hours}} \cdot \frac{4}{3} \text{ hours} = \frac{2}{3} \text{ lawn}$$

$$\text{Mike's work} = \frac{1 \text{ lawn}}{4 \text{ hours}} \cdot \frac{4}{3} \text{ hours} = +\frac{1}{3} \text{ lawn}$$

$$1 \text{ lawn}$$ ■

Example 4 Mary can wash a car in 45 minutes, and Sue can wash a car in 30 minutes. How long will it take them to wash 10 cars working together?

Solution Let x = time each works.

$$\text{Mary's rate} = \frac{1 \text{ car}}{45 \text{ min}} = \frac{1}{45} \text{ car per min.}$$

$$\text{Sue's rate} = \frac{1 \text{ car}}{30 \text{ min}} = \frac{1}{30} \text{ car per min.}$$

	Rate ·	*Time* =	*Amount of work*
Mary	$\frac{1}{45}$	x	$\frac{x}{45}$
Sue	$\frac{1}{30}$	x	$\frac{x}{30}$

$$\boxed{\text{amount Mary washes}} \quad \boxed{\text{amount Sue washes}} \quad \boxed{\text{10 cars}}$$

$$\frac{x}{45} + \frac{x}{30} = 10$$

LCD = 90

$$\overset{2}{\cancel{90}}\left(\frac{x}{\underset{1}{\cancel{45}}}\right) + \overset{3}{\cancel{90}}\left(\frac{x}{\underset{1}{\cancel{30}}}\right) = 90(10)$$

$$2x + 3x = 900$$

$$5x = 900$$

$$x = 180$$

Therefore, it will take them 180 min = 3 hr.

Check

$$\text{Mary's work} = \frac{1 \text{ car}}{45 \text{ min}} \cdot 180 \text{ min} = \quad 4 \text{ cars}$$

$$\text{Sue's work} = \frac{1 \text{ car}}{30 \text{ min}} \cdot 180 \text{ min} = +\ 6 \text{ cars}$$

$$10 \text{ cars}$$ ■

Example 5 The hot water faucet can fill the bath tub in 10 minutes, and the cold water faucet can fill the tub in 5 minutes. The drain can empty the tub in 15 minutes. How long will it take to fill the tub if both faucets are turned on, but the drain is accidentally left open for 3 minutes?

Solution Let x = time both faucets are on.

$$\text{Hot water's rate} = \frac{1 \text{ tub}}{10 \text{ min}} = \frac{1}{10} \text{ tub per min}$$

$$\text{Cold water's rate} = \frac{1 \text{ tub}}{5 \text{ min}} = \frac{1}{5} \text{ tub per min}$$

$$\text{Drain's rate} = \frac{1 \text{ tub}}{15 \text{ min}} = \frac{1}{15} \text{ tub per min}$$

	Rate ·	*Time* =	*Amount of work*
Hot Water	$\frac{1}{10}$	x	$\frac{x}{10}$
Cold Water	$\frac{1}{5}$	x	$\frac{x}{5}$
Drain	$\frac{1}{15}$	3	$\frac{3}{15} = \frac{1}{5}$

amount hot water fills + amount cold water fills − amount drain empties = 1 full tub

(−: Because drain *empties*)

$$\frac{x}{10} + \frac{x}{5} - \frac{1}{5} = 1$$

LCD = 10

$$\overset{1}{\cancel{10}}\left(\frac{x}{\underset{1}{\cancel{10}}}\right) + \overset{2}{\cancel{10}}\left(\frac{x}{\underset{1}{\cancel{5}}}\right) - \overset{2}{\cancel{10}}\left(\frac{1}{\underset{1}{\cancel{5}}}\right) = 10(1)$$

$$x + 2x - 2 = 10$$

$$3x - 2 = 10$$

$$3x = 12$$

$$x = 4$$

Therefore, it will take 4 min to fill the tub.

Check $\quad \text{Hot water} = \dfrac{1 \text{ tub}}{10 \cancel{\text{min}}} \cdot 4 \cancel{\text{min}} = \dfrac{4}{10} = \dfrac{2}{5} \text{ tub}$

$$\text{Cold water} = \frac{1 \text{ tub}}{5 \cancel{\text{min}}} \cdot 4 \cancel{\text{min}} = \frac{4}{5} \text{ tub}$$

$$\text{Drain} = \frac{1 \text{ tub}}{15 \cancel{\text{min}}} \cdot 3 \cancel{\text{min}} = \frac{3}{15} = \frac{1}{5} \text{ tub}$$

And

$$\frac{2}{5} + \frac{4}{5} - \frac{1}{5} = \frac{5}{5} = 1 \quad \blacksquare$$

EXERCISES 12.9

In the following word problems: (a) Represent the unknown numbers using a variable. (b) Set up an equation and solve. (c) Answer the question.

1. The sum of a number and its reciprocal is $\frac{25}{12}$. Find the number.

2. The sum of a number and its reciprocal is $\frac{29}{10}$. Find the number.

3. The difference between a number and its reciprocal is $\frac{8}{3}$. Find the number.

4. The difference between a number and its reciprocal is $\frac{3}{2}$. Find the number.

5. The denominator of a fraction exceeds the numerator by 2. If 1 is subtracted from the numerator, and 5 is added to the denominator, the resulting fraction is $\frac{1}{5}$. Find the original fraction.

6. The denominator of a fraction exceeds the numerator by 4. If 3 is added to the numerator and the denominator, the resulting fraction is $\frac{2}{3}$. Find the original fraction.

7. The denominator of a fraction is twice the numerator. If 2 is added to the numerator, and 8 is added to the denominator, the resulting fraction is $\frac{2}{5}$. Find the original fraction.

8. The denominator of a fraction is three times the numerator. If 1 is subtracted from the numerator, and 3 is added to the denominator, the resulting fraction is $\frac{2}{9}$. Find the original fraction.

9. Lee can do a job in 2 hours, and Mark can do the same job in 6 hours. How long will it take them to do the job working together?

10. Ken can do a job in 4 hours, and Betty can do the same job in 6 hours. How long will it take them to do the job working together?

11. Pipe A can fill a tank in 3 hours, Pipe B can fill the tank in 2 hours, and Pipe C can fill the tank in 6 hours. How long will it take to fill the tank if all three pipes are turned on?

12. Pipe A can fill a tank in 6 hours, Pipe B can fill the tank in 3 hours, and Pipe C can drain the tank in 8 hours. How long will it take to fill the tank if all three pipes are turned on?

13. Ralph can paint a wall in 30 minutes, Marcia can paint a wall in 50 minutes. How long will it take them to paint 4 walls working together?

14. Greg can type a page in 10 minutes, and Debbie can type a page in 15 minutes. How long will it take them to type a 20-page report?

15. The old machine can sort the mail in 4 hours, and the new machine can sort the mail in 3 hours. How long will it take to sort the mail with both machines working if the old machine breaks down after 1 hour?

16. Pipe A can fill a water tank in 6 hours, and Pipe B can fill the tank in 3 hours. Pipe C can drain the tank in 8 hours. How long will it take Pipe A and B to fill the tank if Pipe C is accidentally left open for 1 hour?

17. Joe and Ron live 54 miles apart. Both leave their homes at 7 A.M. by bicycle, riding toward one another. They meet at 10 A.M. If Ron's average speed is four-fifths of Joe's, how fast does each cycle?

18. Tim and Cathy live 60 miles apart. Both leave their homes at 10 A.M. by bicycle, riding toward one another. They meet at 2 P.M. If Cathy's average speed is two-thirds of Tim's, how fast does each cycle?

19. A student scored 70, 85, and 83 on three exams. If he must average at least 80 to get a B in the class, what must he score on the fourth exam to receive a B?

20. A student scored 95, 88, and 91 on three exams. If he must average at least 90 to get an A in the class, what must he score on the final exam to receive an A if the final exam is counted as two tests?

12.10 Chapter Summary

Simple Fraction 12.6

A **simple fraction** is a fraction having only one fraction line.

Complex Fraction 12.6

A **complex fraction** is a fraction having more than one fraction line.

The Three Signs of a Fraction 12.1

Every fraction has three signs associated with it: the sign of the entire fraction, the sign of the numerator, and the sign of the denominator. *If any two of the three signs of a fraction are changed, the value of the fraction is unchanged.*

To Reduce a Fraction to Lowest Terms 12.1

1. Factor the numerator and denominator completely.
2. Divide the numerator and denominator by all factors common to both.

To Multiply Fractions 12.2

1. Factor the numerator and denominator of the fractions.
2. Divide the numerator and denominator by all factors common to both.
3. The answer is the product of factors remaining in the numerator divided by the product of factors remaining in the denominator. A factor of 1 will always remain in both numerator and denominator.

To Divide Fractions 12.2

To divide fractions, invert the second fraction and multiply.

$$\frac{a}{b} \div \frac{c}{d} = \frac{a}{b} \cdot \frac{d}{c}$$

First fraction ($\frac{a}{b}$) — Second fraction ($\frac{c}{d}$)

To Find The LCD 12.4

1. Factor each denominator completely. Repeated factors should be expressed as powers.
2. Write down each different factor that appears.
3. Raise each factor to the highest power it occurs in *any* denominator.
4. The LCD is the product of all the powers found in Step 3.

To Add Like Fractions 12.3

1. Add their numerators.
2. Write the sum of the numerators over the denominator of the like fractions.
3. Reduce the resulting fraction to lowest terms.

To Add Unlike Fractions 12.5

1. Find the LCD.
2. Convert all fractions to equivalent fractions that have the LCD as denominator.
3. Add the resulting like fractions.
4. Reduce the fraction found in Step 3 to lowest terms.

To Simplify Complex Fractions 12.6

Method 1. Multiply both numerator and denominator of the complex fraction by the LCD of the secondary fractions; then simplify the results.

Method 2. First, simplify the numerator and denominator of the complex fraction; then divide the simplified numerator by the simplified denominator.

To Solve an Equation That Has Fractions 12.7

1. Remove fractions by multiplying each term by the LCD.
2. Remove grouping symbols.
3. Combine like terms.

If a *first-degree equation* is obtained in Step 3:

4. Get all terms with the unknown on one side and the remaining terms on the other side.
5. Divide both sides by the coefficient of the unknown.

If a *second-degree equation* (quadratic) is obtained in Step 3:

4. Get all nonzero terms on one side, and arrange them in descending powers. *Only zero must remain on the other side.*
5. Factor the polynomial.
6. Set each factor equal to zero, and then solve for unknown.

Check apparent solutions in the original equation. Any value of the variable that makes any denominator in the equation zero is not a solution.

Literal Equations 12.8

Literal equations are equations that have more than one letter.

To Solve a Literal Equation 12.8

To solve a literal equation, proceed in the same way used to solve an equation with a single letter. The solution will be expressed in terms of the other letters given in the literal equation.

Review Exercises 12.10

In Exercises 1–3, what value(s) of the variable (if any) must be excluded?

1. $\dfrac{2x-1}{x+4}$

2. $7x+\dfrac{2}{3x}$

3. $\dfrac{x-1}{x^2-3x-10}$

In Exercises 4–6, use the rule about the three signs of a fraction to find the missing term.

4. $-\dfrac{2}{5}=\dfrac{-2}{?}$

5. $\dfrac{5}{2-x}=\dfrac{-5}{?}$

6. $\dfrac{2}{a-b}=-\dfrac{?}{b-a}$

In Exercises 7–12, reduce each fraction to lowest terms.

7. $\dfrac{4x^3y}{2xy^2}$

8. $\dfrac{2+4m}{2}$

9. $\dfrac{a^2-4}{a+2}$

10. $\dfrac{x+3}{x^2-x-12}$

11. $\dfrac{x^2-2xy+y^2}{y^2-x^2}$

12. $\dfrac{a-b}{ax+ay-bx-by}$

In Exercises 13–23, perform the indicated operations.

13. $\dfrac{7}{z}-\dfrac{2}{z}$

14. $5-\dfrac{3}{2x}$

15. $\dfrac{3x}{x+1}-\dfrac{2x-1}{x+1}$

16. $\dfrac{-5a^2}{3b}\div\dfrac{10a}{9b^2}$

17. $\dfrac{3x+6}{6}\cdot\dfrac{2x^2}{4x+8}$

18. $\dfrac{x+4}{5}-\dfrac{x-2}{3}$

19. $\dfrac{a-2}{a-1}+\dfrac{a+1}{a+2}$

20. $\dfrac{2x^2-6x}{x+2}\div\dfrac{x}{4x+8}$

21. $\dfrac{3x}{x-3}+\dfrac{9}{3-x}$

22. $\dfrac{x-y}{xy^2}-\dfrac{y-x}{x^2y}$

23. $\dfrac{3}{x^2+5x+4}-\dfrac{2}{x^2+6x+8}$

In Exercises 24–26, simplify the complex fractions.

24. $\dfrac{\dfrac{5k^2}{3m^2}}{\dfrac{10k}{9m}}$

25. $\dfrac{\dfrac{x}{y}+3}{\dfrac{x}{y}-3}$

26. $\dfrac{\dfrac{y}{x}-\dfrac{y^2}{x^2}}{\dfrac{1}{x}-\dfrac{y}{x^2}}$

In Exercises 27–32, solve the equations for the unknown letter.

27. $\dfrac{6}{m}=5$

28. $\dfrac{2m}{3}-m=1$

29. $\dfrac{z}{5}-\dfrac{z}{8}=3$

30. $\dfrac{2x+1}{3}=\dfrac{5x-4}{2}$

31. $\dfrac{4}{2z}+\dfrac{2}{z}=1$

32. $\dfrac{3}{x}-\dfrac{8}{x^2}=\dfrac{1}{4}$

In Exercises 33–38, solve for the letter listed after each equation.

33. $3x-4y=12;\ x$

34. $\dfrac{2m}{n}=P;\ n$

35. $V=LWH;\ H$

36. $E = \frac{mv^2}{gr}; m$ **37.** $\frac{F - 32}{C} = \frac{9}{5}; C$ **38.** $\frac{ax + b}{x} = c; x$

39. The denominator of a fraction exceeds the numerator by twenty. If seven is added to the numerator and subtracted from the denominator, the resulting fraction equals $\frac{6}{7}$. Find the fraction.

40. Twice a number plus three times its reciprocal is 7. Find the number.

41. Ed can wash a window in 10 minutes, and Cindy can wash a window in 8 minutes. How long will it take them to wash the 45 office windows working together?

42. Debbie and Wendy live 21 miles apart. Both leave their homes, riding their bicycles toward one another. They meet in $1\frac{1}{2}$ hours. If Wendy's average speed is 2 mph less than Debbie's, how fast does each cycle?

Chapter 12 Diagnostic Test

Allow yourself about an hour to do these problems. Complete solutions for every problem, together with section references, are given in the answer section at the end of the book.

1. What value(s) of the variable must be excluded, if any?

a. $\dfrac{3x}{x-4}$ b. $\dfrac{5x+4}{x^2+2x}$

2. Use the rule about the three signs of a fraction to find the missing term in each of the following expressions.

a. $-\dfrac{-4}{5} = \dfrac{4}{?}$ b. $\dfrac{-3}{x-y} = \dfrac{?}{y-x}$

In Problems 3–5, reduce each fraction to lowest terms.

3. $\dfrac{6x^3y}{9x^2y^2}$

4. $\dfrac{x^2+8x+16}{x^2-16}$

5. $\dfrac{6a^2+11ab-10b^2}{6a^2b-4ab^2}$

In Problems 6–11, perform the indicated operations. (Be sure to reduce fractions to lowest terms.)

6. $\dfrac{a}{a+2} \cdot \dfrac{4a+8}{6a^2}$

7. $\dfrac{-15}{4y-5} - \dfrac{12y}{5-4y}$

8. $4 + \dfrac{2}{x}$

9. $\dfrac{b+1}{b} - \dfrac{b}{b-1}$

10. $\dfrac{2x}{x^2-9} \div \dfrac{4x^2}{x-3}$

11. $\dfrac{2}{x^2+4x+3} - \dfrac{1}{x^2+5x+6}$

In Problems 12 and 13, simplify each complex fraction.

12. $\dfrac{\dfrac{9x^5}{10y}}{\dfrac{3x^2}{20y^3}}$

13. $\dfrac{1+\dfrac{2}{a}}{1-\dfrac{4}{a^2}}$

In Problems 14–19, solve each equation.

14. $\dfrac{y}{3} - \dfrac{y}{4} = 1$

15. $\dfrac{x-2}{5} = \dfrac{x+1}{2} + \dfrac{3}{5}$

16. $\dfrac{3}{a+4} = \dfrac{5}{a}$

17. $\dfrac{2x-5}{x} = \dfrac{x-2}{3}$

18. Solve for y: $3x - 4y = 9$

19. Solve for P: $PM + Q = PN$

20. Sid can unload a truck full of dirt in 20 minutes, and George can unload the same truck in 30 minutes. How long would it take them to unload the truck working together?

13 Graphing

Many algebraic relationships are easier to understand if a picture called a *graph* is drawn. In this chapter we discusss how to draw such graphs.

13.1 The Rectangular Coordinate System

In Section 6.1 we discussed how any real number can be represented by a point on a *number line* (Figure 13.1.1).

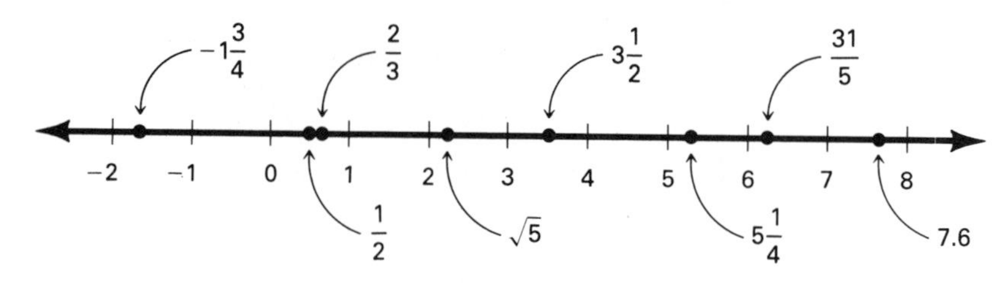

FIGURE 13.1.1
HORIZONTAL NUMBER LINE

We draw another number line vertically with its zero point at the zero point of the horizontal number line. These two lines form *axes* of a *rectangular coordinate system.* The rectangular coordinate system consists of a vertical number line called the *vertical axis* or *y-axis* and a horizontal line called the *horizontal axis* or *x-axis* that meet at a point called the *origin.* The vertical and horizontal axes determine the *plane* of the rectangular coordinate system (Figure 13.1.2).

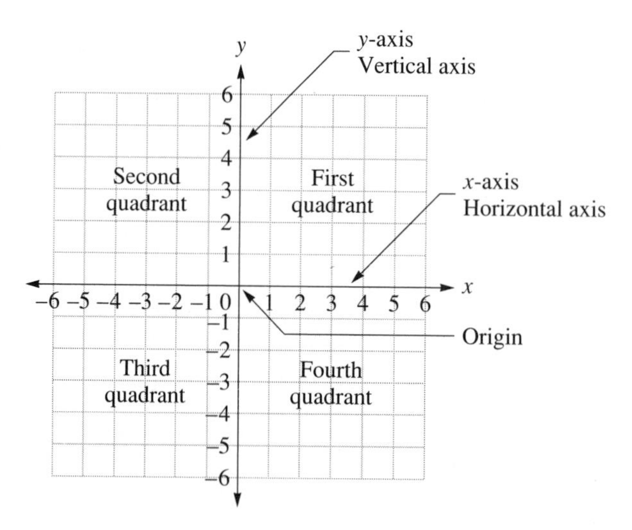

FIGURE 13.1.2
RECTANGULAR COORDINATE SYSTEM

With a single number line, we needed only a *single* real number to represent a point on that line. With two lines forming a rectangular coordinate system, we need a *pair* of real numbers to represent a point in the plane.

Ordered Pair
Coordinates

Graphing Points A point is represented by an **ordered pair** of numbers. The point (3, 2) is shown in Figure 13.1.3. We call 3 and 2 the **coordinates** of the point (3, 2). The first number, 3, is called the *x-coordinate* (or horizontal coordinate, or abscissa). The second number, 2, is called the *y-coordinate* (or vertical coordinate, or ordinate).

To graph the point (3, 2), start at the origin and move 3 unit *right*, then move 2 units *up*.

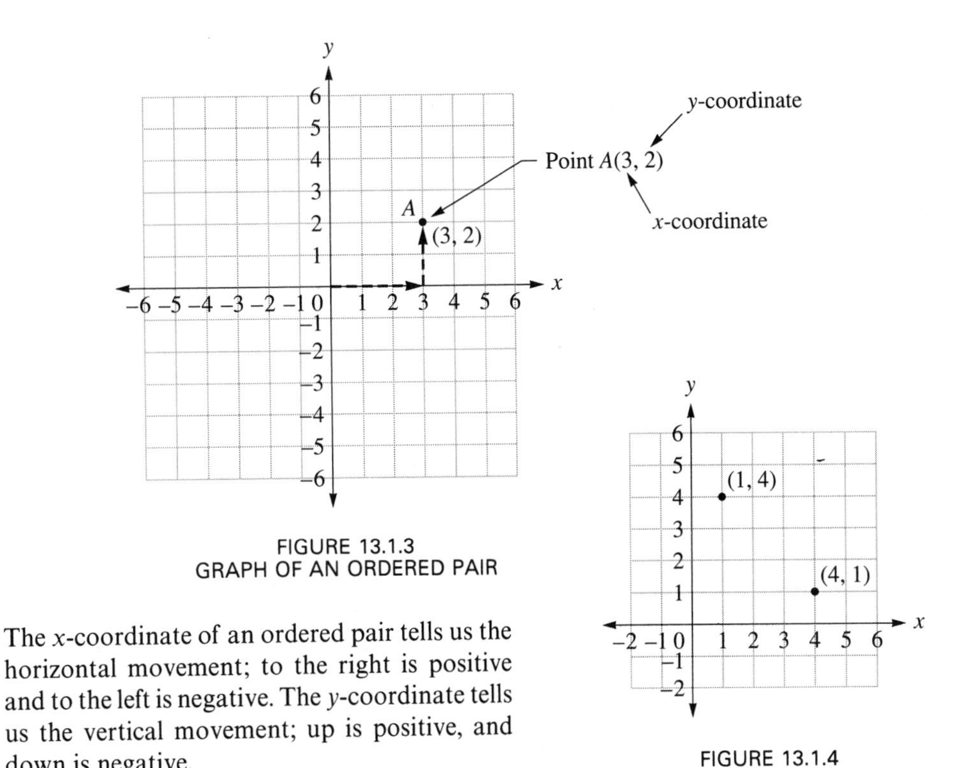

FIGURE 13.1.3
GRAPH OF AN ORDERED PAIR

FIGURE 13.1.4

The x-coordinate of an ordered pair tells us the horizontal movement; to the right is positive and to the left is negative. The y-coordinate tells us the vertical movement; up is positive, and down is negative.

NOTE When the order is changed in an ordered pair, we get a different point. For example, (1, 4) and (4, 1) are two different points (Figure 13.1.4). ☑

Example 1 Graph the following points.

a. (3, 5) Start at the origin and move *right* 3 units, then move *up* 5 units (point A in Figure 13.1.5).

b. $(-5, 2)$ Start at the origin and move *left* 5 units, then move *up* 2 units (point B in Figure 13.1.5).

c. $(-5, -4)$ Start at the origin and move *left* 5 units, then move *down* 4 units (point C in Figure 13.1.5).

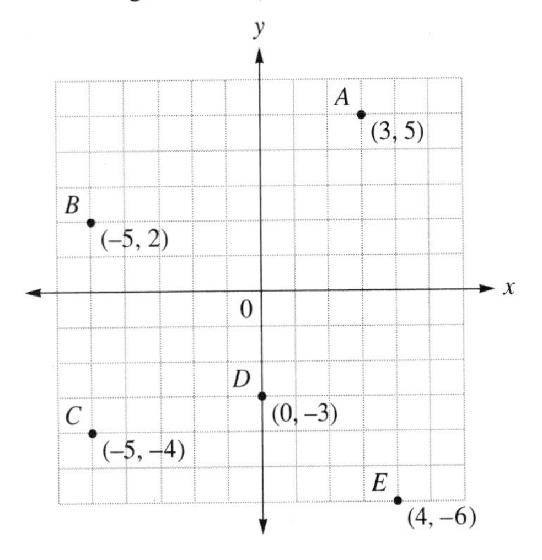

FIGURE 13.1.5

d. (0, −3) Start at the origin, but because the first number is zero, do not move either right or left. Just move *down* 3 units (point *D* in Figure 13.1.5).

e. (4, −6) Start at the origin and move *right* 4 units, then move *down* 6 units (point *E* in Figure 13.1.5). ■

The phrase *plot the points* means the same as *graph the points.*

EXERCISES 13.1

1. Graph each of the following points.
a. (3, 1) b. (−4, −2) c. (0, 3)
d. (5, −4) e. (4, 0) f. (−2, 4)

2. Graph each of the following points.
a. (2, 4) b. (2, −4) c. (3, 0)
d. (−3, −2) e. (0, 0) f. (0, −4)

In Exercises 3 and 4, use Figure 13.1.6.

3. Give the coordinates of each of the following points.
a. *R* b. *N* c. *U* d. *S*

4. Give the coordinates of each of the following points.
a. *M* b. *P* c. *Q* d. *T*

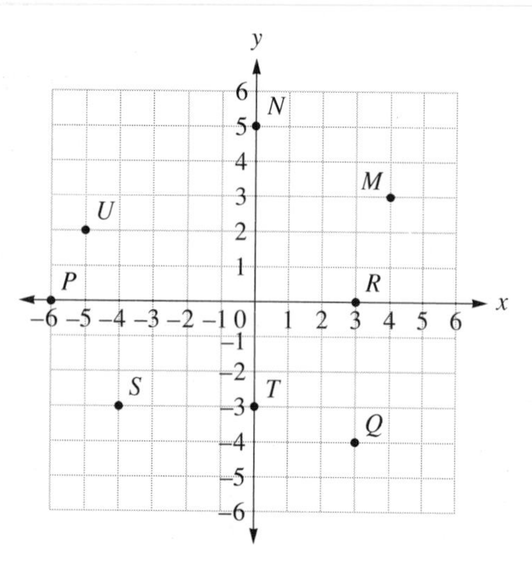

FIGURE 13.1.6

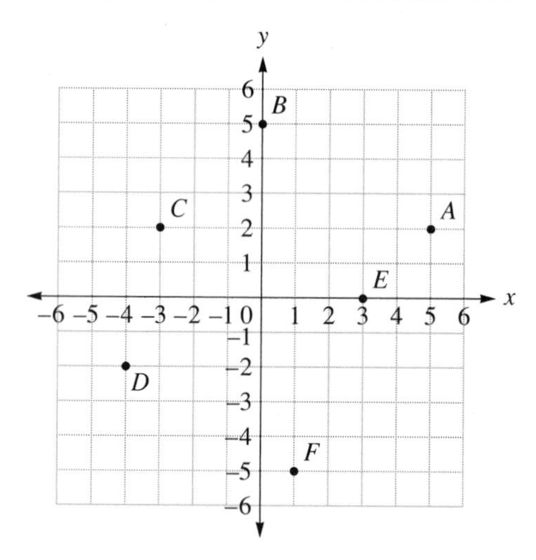

FIGURE 13.1.7

In Exercises 5–8, use Figure 13.1.7.

5. Write the *x*-coordinate of each of the following points.
a. *A* b. *C* c. *E* d. *F*

6. Write the *y*-coordinate of each of the following points.
a. *B* b. *D* c. *E* d. *F*

7. a. What is the abscissa of point *F*?
b. What is the ordinate of point *C*?

8. a. What is the ordinate of point *B*?
b. What is the abscissa of point *D*?

9. What name is given to the point (0, 0)?

10. What is the *x*-coordinate of the origin?

11. List some other names for the *x*-coordinate of a point.

12. List some other names for the *y*-coordinate of a point.

13. Draw the triangle whose vertices have the following coordinates?
A(0, 0) *B*(3, 2) *C*(−4, 5)

14. Draw the triangle whose vertices have the following coordinates:
A(−2, −3) *B*(−2, 4) *C*(3, 5)

13.2 Graphing Lines

In the preceding section we showed how to graph points. In this section we show how to graph straight lines.

Table of Values

Consider the equation $y = x + 1$. To graph this equation, we need to find values for x and y that make the equation true. For example if $x = 0$, then

$$y = x + 1$$
$$y = 0 + 1 = 1$$

These values, $x = 0$ and $y = 1$, form an ordered pair (0, 1) that represents a point on the graph of the equation $y = x + 1$. By choosing several values for x and finding the *corresponding values* for y, we obtain a set of points (ordered pairs) on the graph of $y = x + 1$. These ordered pairs are listed in the following *table of values*.

Equation: $y = x + 1$

When $x = 0$, $y = 0 + 1 = 1$

When $x = 2$, $y = 2 + 1 = 3$

When $x = 5$, $y = 5 + 1 = 6$

When $x = -3$, $y = -3 + 1 = -2$

Table of Values

x	y
0	1
2	3
5	6
−3	−2

Each of these ordered pairs, called "a pair of corresponding values," represents a point on the graph of $y = x + 1$.

We plot the four points contained in the table of values, in Figure 13.2.1. Note that the points appear to lie in a straight line. This suggests that the graph of $y = x + 1$ is a straight line. In fact any first-degree equation (in no more than two variables) has a graph that is a straight line. Such equations are called **linear equations**.

Linear Equations

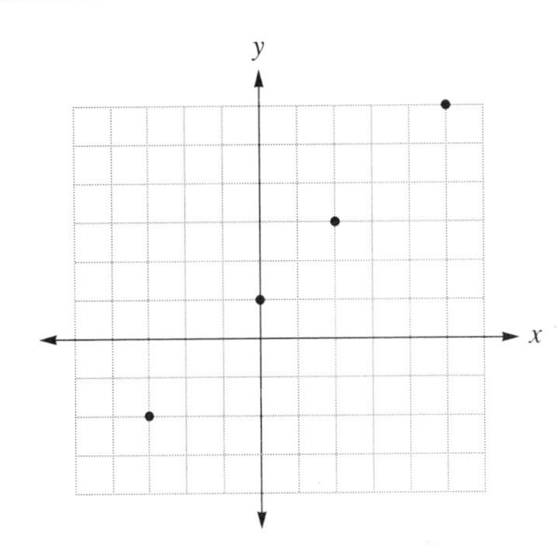

FIGURE 13.2.1

A straight line can be drawn if we know two points that lie on that line.

Check Point

Although two points are all that are *necessary* to draw the line, it is *advisable* to plot a third point as a **check point**. If the three points do not lie in a straight line, a mistake has been made in some calculation of the coordinates of the points. In Figure 13.2.2 we have drawn the line through the points listed in the table of values. This line represents the graph of the equation $y = x + 1$.

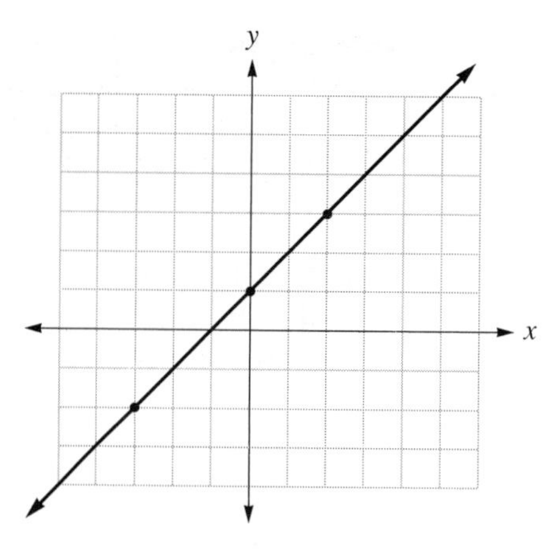

FIGURE 13.2.2

Example 1 Graph the equation $3x - 5y = 15$.

Solution Substitute three values for x, and find the corresponding values for y.

Equation: $3x - 5y = 15$

If $x = 0$, $3(0) - 5y = 15$
$$-5y = 15$$
$$y = -3$$

If $x = 2$, $3(2) - 5y = 15$
$$6 - 5y = 15$$
$$-5y = 9$$
$$y = -\frac{9}{5} = -1\frac{4}{5}$$

If $x = 5$, $3(5) - 5y = 15$
$$15 - 5y = 15$$
$$-5y = 0$$
$$y = 0$$

x	y
0	-3
2	$-1\frac{4}{5}$
5	0

Plot the three points from the table of values, and draw a straight line through them (Figure 13.2.3).

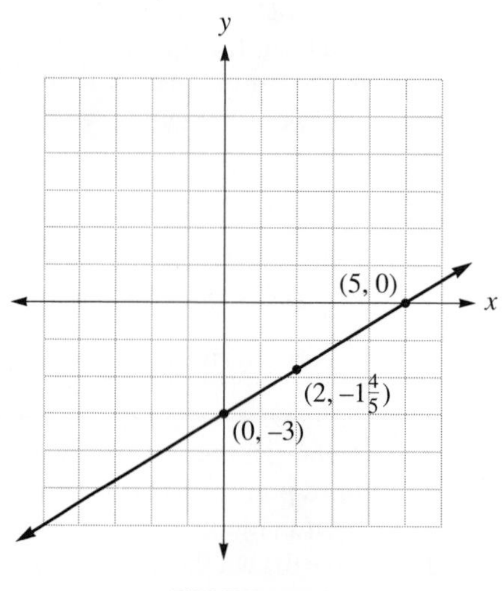

FIGURE 13.2.3

■

Intercepts

Students often ask which points to use when plotting the graph of a straight line. We usually start by choosing zeros.

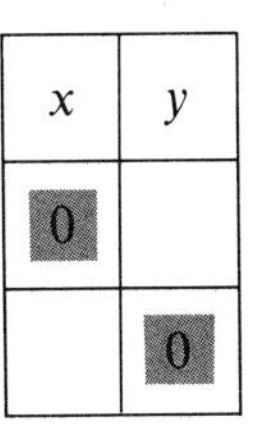

x	y
0	
	0

For example, if the equation of the line is $3x - 2y = -6$

If $x = 0$, $3(0) - 2y = -6$

$$-2y = -6$$

$$y = 3$$

If $y = 0$, $3x - 2(0) = -6$

$$3x = -6$$

$$x = -2$$

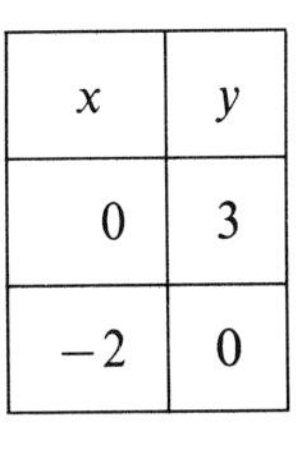

x	y
0	3
−2	0

The points we find by this method are called the *x-intercept* and the *y-intercept*. The x-intercept of an equation is the point where its graph meets the x-axis. The y-value at this point is zero (Figure 13.2.4). The y-intercept of an equation is the point where its graph meets the y-axis. The x-value at this point is zero (Figure 13.2.4).

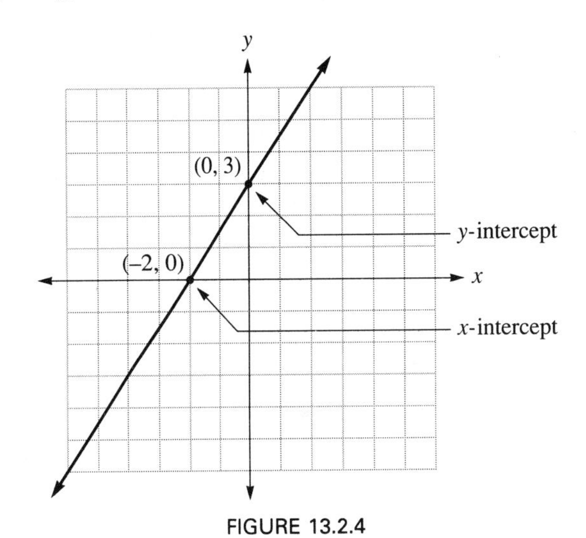

FIGURE 13.2.4

Example 2 Graph the equation $4x + 3y = 12$.

Solution The *intercept method* of graphing a straight line is often easier to use than the method shown in Example 1.

<u>x-intercept:</u> Set $y = 0$. Then $4x + 3y = 12$

becomes $4x + 3(0) = 12$

$$4x = 12$$

$$x = 3$$

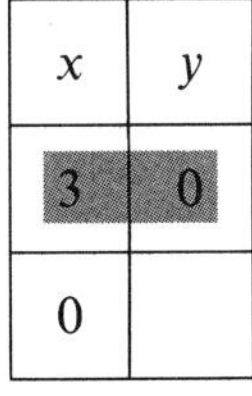

x	y
3	0
0	

Therefore, the x-intercept is (3, 0).

<u>y-intercept:</u> Set $x = 0$. Then $4x + 3y = 12$

becomes $4(0) + 3y = 12$

$$3y = 12$$

$$y = 4$$

x	y
3	0
0	4

Therefore, the y-intercept is (0, 4).

Check point: Set $x = 6$. Then $4x + 3y = 12$

becomes

$$4(6) + 3y = 12$$
$$24 + 3y = 12$$
$$3y = -12$$
$$y = -4$$

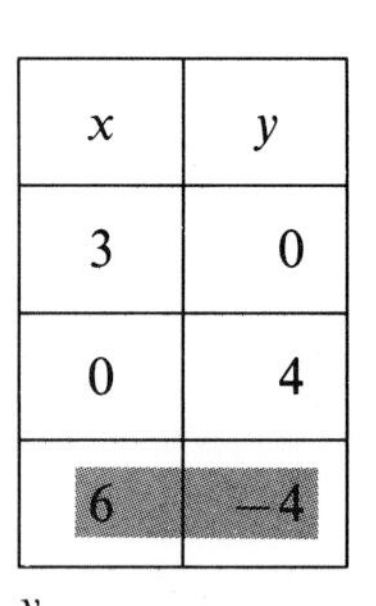

x	y
3	0
0	4
6	−4

Therefore, this check point is (6, −4).

Plot the x-intercept (3, 0), the y-intercept (0, 4), and the checkpoint (6, −4); then draw the straight line through them (Figure 13.2.5).

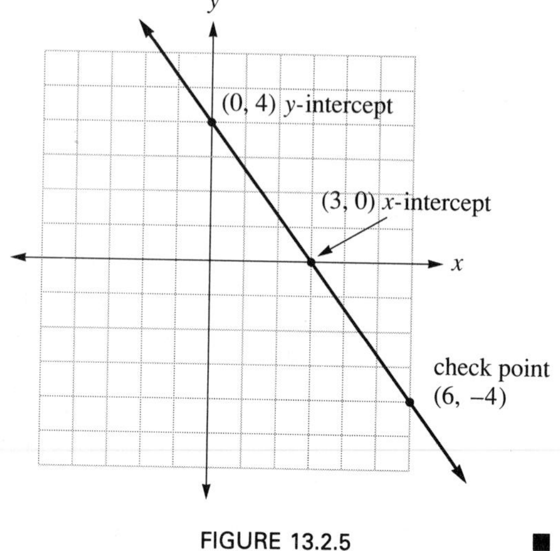

FIGURE 13.2.5

■

Example 3 Graph the equation $3x - 4y = 0$.

Solution

x-intercept: Set $y = 0$. Then $3x - 4y = 0$

becomes

$$3x - 4(0) = 0$$
$$3x = 0$$
$$x = 0$$

Therefore, the x-intercept is (0, 0) (origin). Because the line goes through the origin, the y-intercept is also (0, 0).

We have found only one point on the line: (0, 0). Therefore, we must find another point on the line. To find another point, we must set either variable equal to a number and then solve the equation for the other variable. For example,

Set $y = 3$. Then $3x - 4y = 0$

becomes

$$3x - 4(3) = 0$$
$$3x = 12$$
$$x = 4$$

x	y
0	0
4	3
−4	−3

This gives the point (4, 3) on the line.

Check point: Set $x = -4$, then $3x - 4y = 0$

becomes

$$3(-4) - 4y = 0$$
$$-12 - 4y = 0$$
$$-4y = 12$$
$$y = -3$$

Therefore, the check point is (−4, −3).

Plot the points $(0, 0)$, $(4, 3)$, and $(-4, -3)$; then draw the straight line through them (Figure 13.2.6).

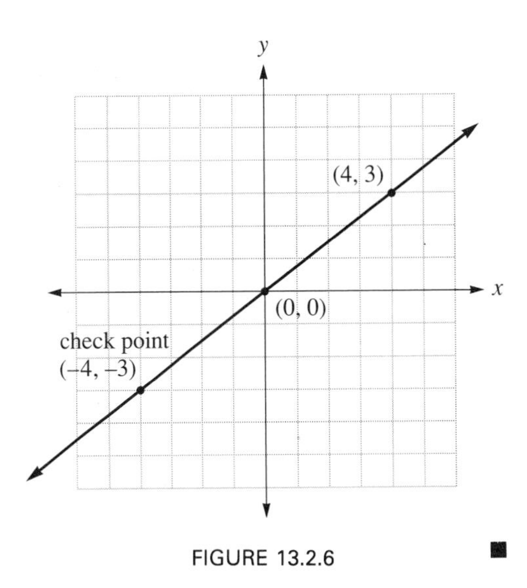

FIGURE 13.2.6 ■

Some equations of a line have only one variable. Such equations have graphs that are either vertical or horizontal lines (Examples 4 and 5).

Example 4 Graph the equation $x = 3$.

Solution The equation $x = 3$ is equivalent to

$$0y + x = 3.$$

If $y = 5$, $\quad 0(5) + x = 3$

$$0 + x = 3$$

$$x = 3$$

If $y = -2$, $\quad 0(-2) + x = 3$

$$0 + x = 3$$

$$x = 3$$

x	y
3	5
3	-2

You can see that no matter what value y has in this equation, x is always 3. Therefore, all the points having an x-value of 3 lie in a vertical line whose x-intercept is $(3, 0)$ (Figure 13.2.7).

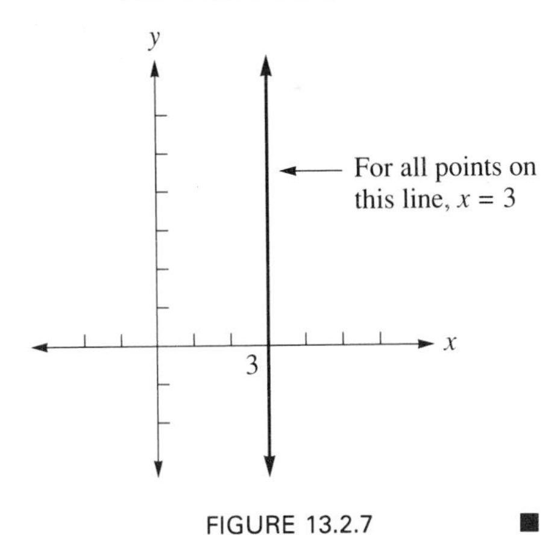

FIGURE 13.2.7 ■

Example 5 Graph the equation $y + 4 = 0$.

Solution $\quad y + 4 = 0$

$$y = -4$$

In the equation $y + 4 = 0$, no matter what value x has, y is always -4. Therefore, all the points having a y-value of -4 lie in a horizontal line whose y-intercept is $(0, -4)$ (Figure 13.2.8).

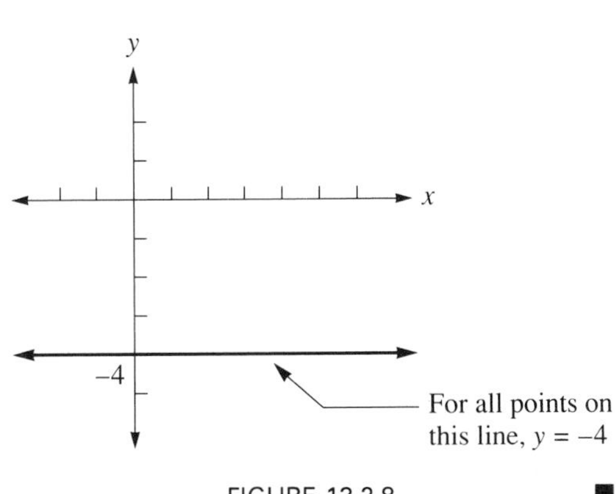

FIGURE 13.2.8 ■

The methods used in these examples are summarized as follows:

TO GRAPH A STRAIGHT LINE

Table of values

1. Substitute a number in the equation for one of the variables and solve for the corresponding value of the other variable. Find three points in this way, and list them in a table of values. (The third point is a check point.)
2. Graph the points that are listed in the table of values.
3. Draw a straight line through the points.

Intercept method

1. Find the x-intercept: Set $y = 0$; then solve for x.
2. Find the y-intercept: Set $x = 0$; then solve for y.
3. Draw a straight line through the x- and y-intercepts.
4. If both intercepts are (0, 0), an additional point must be found before the line can be drawn (Example 3).

Exceptions:
The graph of $x = a$ is a vertical line (at $x = a$).
The graph of $y = b$ is a horizontal line (at $y = b$).

NOTE Sometimes the x- and y-intercepts are very close together. To draw the line through them accurately would be very difficult. In this case, find another point on the line far enough away from the intercepts so that it is easy to draw an accurate line. To find the other point, set either variable equal to a number, and then solve the equation for the other variable. ☑

EXERCISES 13.2

In Exercises 1–28, graph each of the equations.

1. $x + y = 3$ **2.** $x - y = 4$ **3.** $2x - 3y = 6$ **4.** $3x - 4y = 12$

5. $y = 8$ **6.** $x = 9$ **7.** $x + 5 = 0$ **8.** $y + 2 = 0$

9. $3x - 5y = 15$ **10.** $2x - 5y = 10$ **11.** $y = -\frac{1}{2}x$ **12.** $y = -2x$

13. $4 - x = y$ **14.** $6 - x = y$ **15.** $y = x$ **16.** $x + y = 0$

17. $x - y = 0$ **18.** $x = 0$ **19.** $y = \frac{2}{3}x$ **20.** $y = \frac{7}{3}x$

21. $4x - 3y = 12$ **22.** $6x - 3y = 18$ **23.** $3x - 2y = 12$ **24.** $3x - 4y = -4$

25. $3x - 4y = 24$ **26.** $7x + 2y = 14$ **27.** $y = -\frac{3}{2}x + 4$ **28.** $y = -\frac{3}{4}x + 1$

In Exercises 29 and 30: (a) Graph the two equations for each exercise on the same set of axes. (b) What are the coordinates of the point where the two lines cross?

29. $x - y = 5$
$x + y = 1$

30. $3x - 4y = -12$
$3x + y = 18$

13.3 Slope of a Line

Consider the two points $P_1(x_1, y_1)$ and $P_2(x_2, y_2)$ shown in Figure 13.3.1. From this figure it can be seen that:

The change in x (horizontal movement) from P_1 to $P_2 = x_2 - x_1$

The change in y (vertical movement) from P_1 to $P_2 = y_2 - y_1$

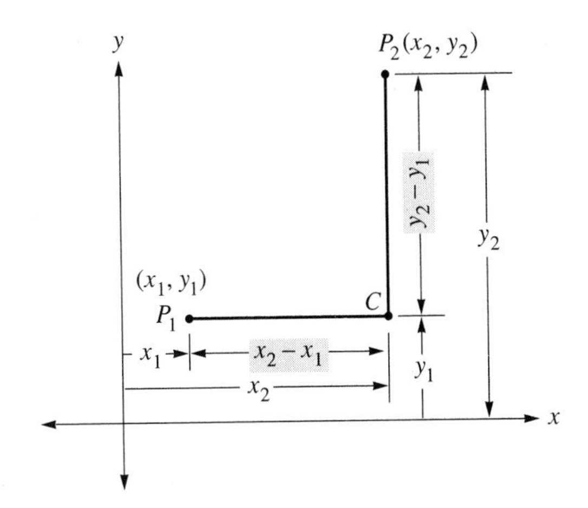

FIGURE 13.3.1

Slope

If we imagine the line as representing a hill, then the **slope** of the line is a measure of the steepness of the hill. To measure the slope of a line, we choose any two points on the line, $P_1(x_1, y_1)$ and $P_2(x_2, y_2)$ (Figure 13.3.2). The change in x and y between the two points is used to find the slope of that line.

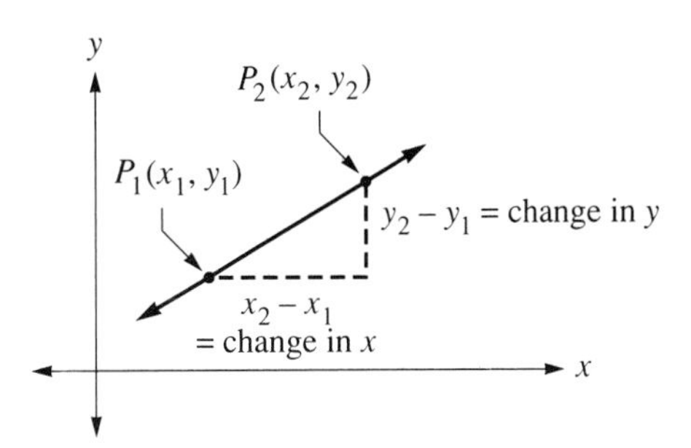

FIGURE 13.3.2

The letter m is used to represent the slope of a line. The slope is defined as follows:

SLOPE OF A LINE

$$\text{Slope} = \frac{\text{The change in } y}{\text{The change in } x}$$

$$m = \frac{y_2 - y_1}{x_2 - x_1}$$

Example 1 Find the slope of the line through the points $(-3, 5)$ and $(6, -1)$ (Figure 13.3.3).

Solution

Let $\quad P_1 = (-3, 5)$

and $\quad P_2 = (6, -1)$

$$m = \frac{y_2 - y_1}{x_2 - x_1} = \frac{(-1) - (5)}{(6) - (-3)} = \frac{-6}{9} = -\frac{2}{3}$$

FIGURE 13.3.3

The slope is not changed if the points P_1 and P_2 are interchanged.

Let $\quad P_1 = (6, -1)$

and $\quad P_2 = (-3, 5)$

then $\quad m = \dfrac{y_2 - y_1}{x_2 - x_1} = \dfrac{(5) - (-1)}{(-3) - (6)} = \dfrac{6}{-9} = -\dfrac{2}{3}$ ■

Example 2 Find the slope of the line through the points $A(-2, -4)$ and $B(5, 1)$ (Figure 13.3.4).

Solution

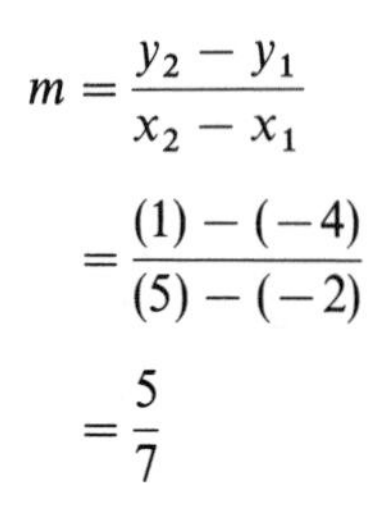

$$m = \frac{y_2 - y_1}{x_2 - x_1} = \frac{(1) - (-4)}{(5) - (-2)} = \frac{5}{7}$$

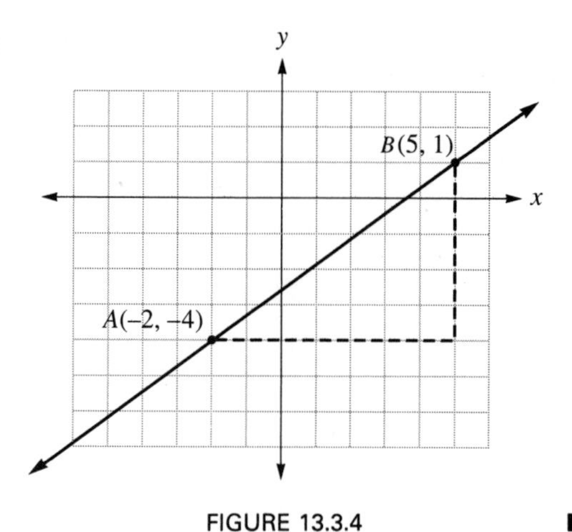

FIGURE 13.3.4 ■

Example 3 Find the slope of the horizontal line through the points $E(-4, -3)$ and $F(2, -3)$ (Figure 13.3.5).

Solution

$$m = \frac{y_2 - y_1}{x_2 - x_1} = \frac{(-3) - (-3)}{(2) - (-4)} = \frac{0}{6} = 0$$

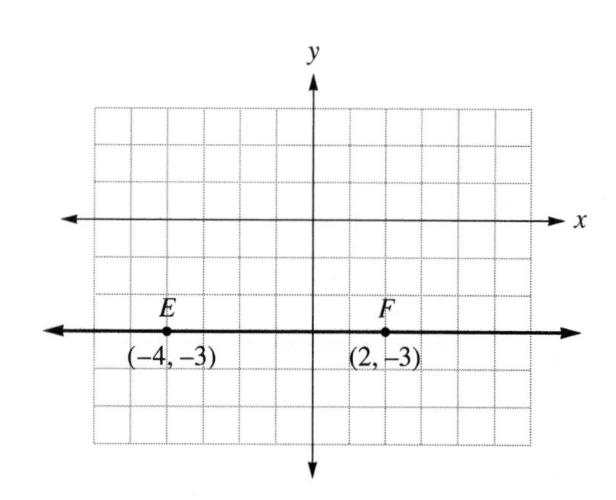

FIGURE 13.3.5 ■

NOTE The slope of a horizontal line is zero. ☑

Example 4 Find the slope of the vertical line through the points $R(4, 5)$ and $S(4, -2)$ (Figure 13.3.6).

Solution

$$m = \frac{y_2 - y_1}{x_2 - x_1}$$

$$= \frac{(-2) - (5)}{(4) - (4)}$$

$$= \frac{-7}{0} \quad \text{undefined}$$

NOTE Because we cannot divide by zero, a vertical line has no slope. ☑

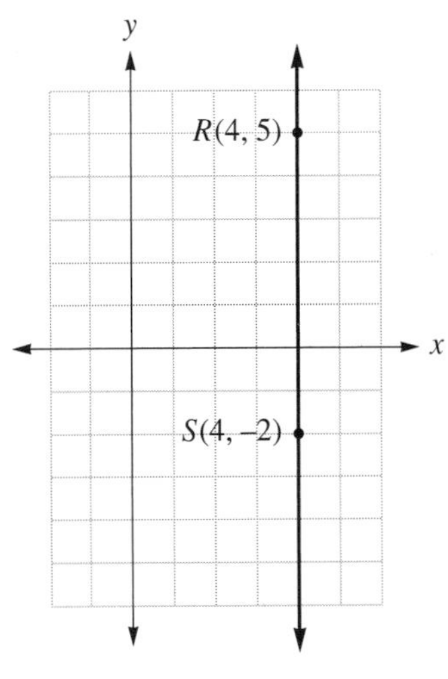

FIGURE 13.3.6 ■

Meaning of the Signs of the Slope

The slope of a line is positive if a point moving along the line in the positive x-direction (to the right) rises (Figure 13.3.4).
The slope of a line is negative if a point moving along the line in the positive x-direction falls (Figure 13.3.3).
The slope is zero if the line is horizontal (Figure 13.3.5).
There is no slope if the line is vertical (Figure 13.3.6).

Slope-Intercept Form

Let $(0, b)$ be the y-intercept of a line whose slope is m. Let $P(x, y)$ represent any other point on the line (Figure 13.3.7).

$$m = \frac{y - b}{x - 0}$$

$$m = \frac{y - b}{x}$$

$$m \cdot x = \frac{y - b}{\cancel{x}} \cdot \cancel{x}$$

$$mx = y - b$$

$$mx + b = y$$

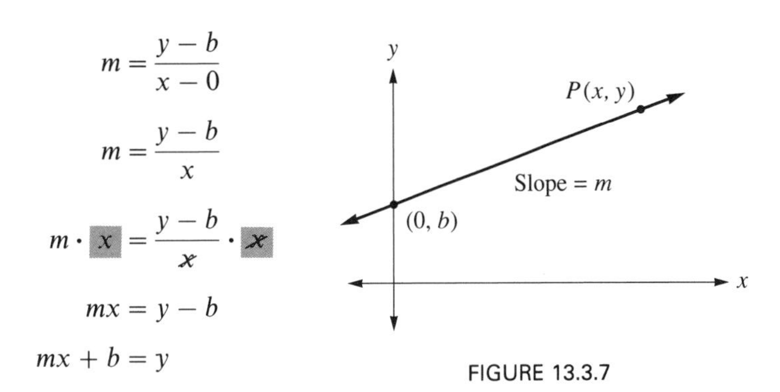

FIGURE 13.3.7

Or

$$y = m\,x + b$$ Slope-intercept form

Slope ↑ (m) ↑ (b) y-intercept

SLOPE-INTERCEPT FORM OF THE EQUATION OF A LINE

$$y = mx + b$$

where m = slope of the line,
and b = y-intercept of the line.

Example 5 Identify the slope and the y-intercept.

Equation of line	*Slope*	*y-intercept*
a. $y = 2x + 3$ (slope ↑ 2, ↑ y-intercept 3)	2	3
b. $y = -\frac{1}{2}x + 5$	$-\frac{1}{2}$	5
c. $y = x - 6$	1	-6
d. $y = 4x \quad (y = 4x + 0)$	4	0
e. $y = 4 \quad (y = 0 \cdot x + 4)$	0	4

■

Graphing Lines Using Slope and *Y*-Intercept

For the equation $y = \frac{2}{3}x + 1$, we know that the slope $= \frac{2}{3}$ and the y-intercept $= 1$. First graph the y-intercept (0, 1) (Figure 13.3.8). Now we can locate a second point on the line by using the slope.

$$\text{Slope} = \frac{\text{Change in } y}{\text{Change in } x} = \frac{2}{3}$$

Since the change in y is 2 and the change in x is 3, starting from the y-intercept, we will move *up* 2 units and then to the *right* 3 units (Figure 13.3.9). Now we have the second point (3, 3), and we can draw the line.

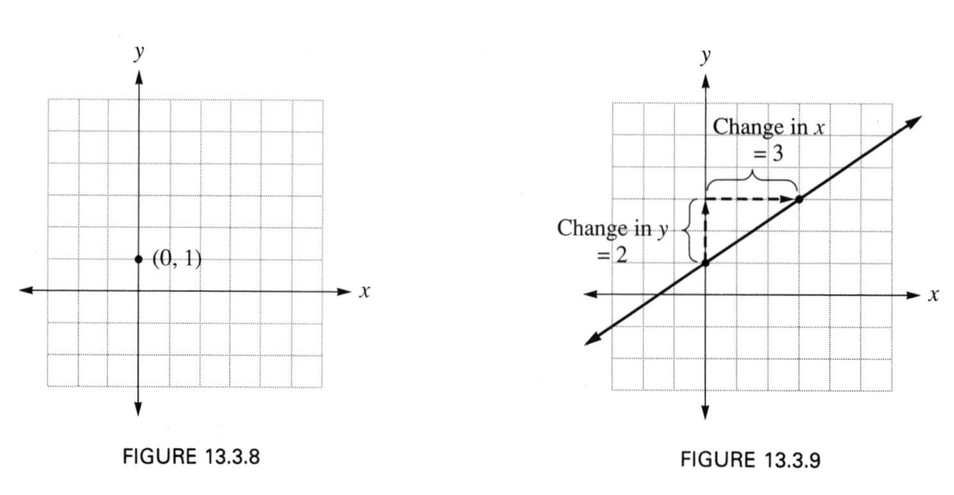

FIGURE 13.3.8

FIGURE 13.3.9

> GRAPHING LINES USING SLOPE-INTERCEPT METHOD
>
> **1.** Graph the y-intercept.
>
> **2.** To locate a second point, start from the y-intercept and move *up* the number of units indicated by the numerator of the slope (move *down* if the numerator is negative), then move *right* the number of units indicated by the denominator of the slope.
>
> **3.** Draw a straight line through the points.

Example 6 Graph $y = -\frac{1}{2}x + 5$ using slope-intercept method.

Solution

Step 1 Graph the y-intercept $(0, 5)$.

Step 2 Slope $= -\frac{1}{2}$

$$\frac{\text{Change in } y}{\text{Change in } x} = \frac{-1}{2}$$

From the y-intercept, move *down* 1 unit, then move *right* 2 units.

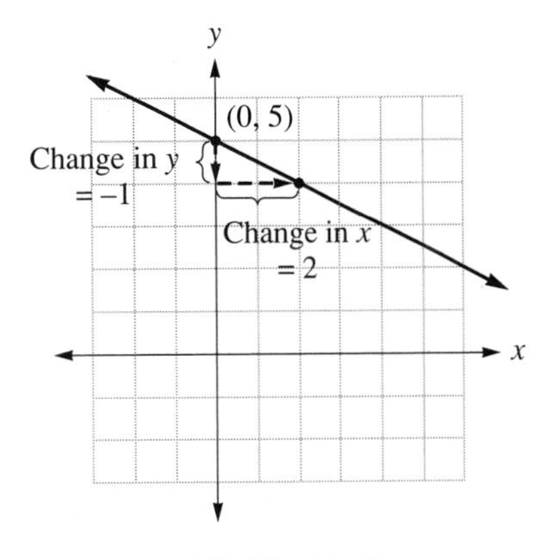

FIGURE 13.3.10 ■

Example 7 Graph $3x - y = 4$ using slope-intercept method.

Solution To identify the slope and the y-intercept we must put the equation in slope-intercept form; that is, solve for y.

$$3x - y = 4$$
$$-y = -3x + 4$$
$$\frac{-y}{-1} = \frac{-3x + 4}{-1}$$
$$y = 3x - 4$$

Step 1 Graph the y-intercept $(0, -4)$.

Step 2 Slope $= 3$

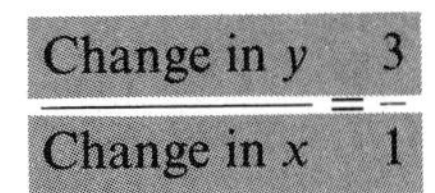

From the y-intercept, move *up* 3 units, then move *right* 1 unit.

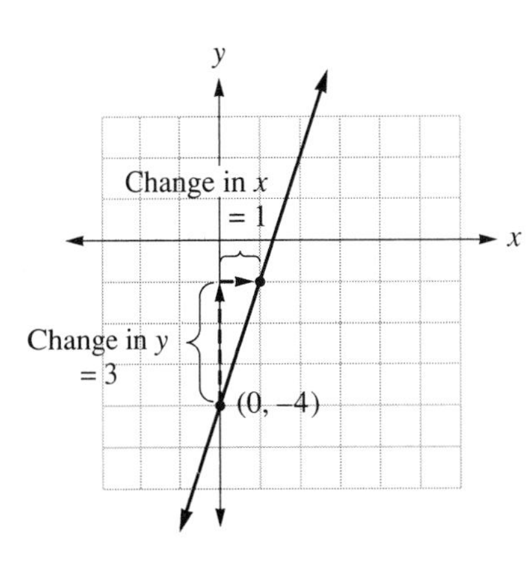

FIGURE 13.3.11 ■

A WORD OF CAUTION When graphing the slope, be sure to *start from the y-intercept*, NOT the origin.

Graph $y = \frac{1}{4}x - 3$

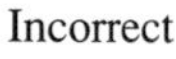

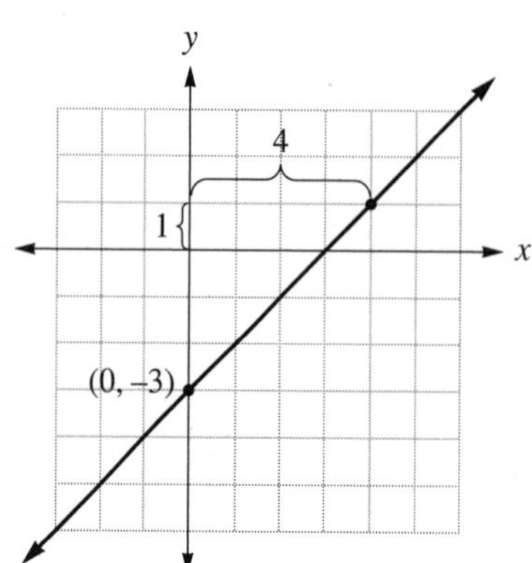

Correct

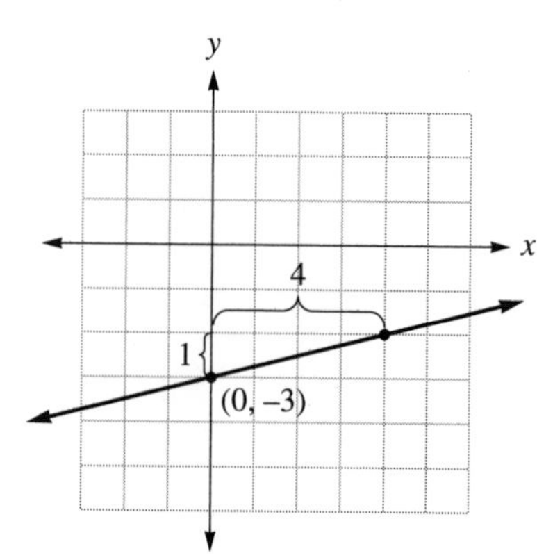

FIGURE 13.3.12 ✓

EXERCISES 13.3

In Exercises 1–18, find the slope of the line through the given pair of points.

1. $(-4, 8)$ and $(7, -9)$
2. $(-1, -3)$ and $(5, 2)$
3. $(-1, -1)$ and $(5, -3)$
4. $(16, -14)$ and $(8, -11)$
5. $(-5, -3)$ and $(4, -3)$
6. $(-2, -4)$ and $(3, -4)$
7. $(-7, 5)$ and $(8, -3)$
8. $(12, -9)$ and $(-5, -4)$
9. $(-6, -15)$ and $(4, -5)$
10. $(-7, 8)$ and $(-4, 5)$
11. $(-2, 5)$ and $(-2, 8)$
12. $(6, -4)$ and $(6, -11)$
13. $(-4, 3)$ and $(-1, 1)$
14. $(6, 3)$ and $(10, -2)$
15. $(-3, 2)$ and $(6, 4)$
16. $(-2, -2)$ and $(4, -1)$
17. $(-10, 2)$ and $(-4, -8)$
18. $(-3, 8)$ and $(9, 12)$

In Exercises 19–22, find the slope of the line by using any two points on the line.

19. $2x + 3y = 6$
20. $x + 4y = 8$
21. $4x + 5y = 20$
22. $3x - 5y = 15$

In Exercises 23–30, identify the slope and the y-intercept.

23. $y = \frac{2}{3}x + 4$
24. $y = -\frac{3}{4}x + 6$
25. $y = -x$
26. $y = -3$
27. $3x + 4y = 12$
28. $x + 4y = 16$
29. $2x - 3y = 6$
30. $5x - 4y = -20$

In Exercises 31–42, graph using the slope-intercept method.

31. $y = \frac{3}{4}x - 2$
32. $y = \frac{2}{5}x - 3$
33. $y = -\frac{1}{2}x + 5$
34. $y = -\frac{2}{3}x + 4$
35. $y = 3x - 6$
36. $y = 2x + 3$
37. $y = -x + 2$
38. $y = -3x$
39. $3x + 2y = 12$
40. $2x + 5y = -10$
41. $3x - 5y = -15$
42. $5x - 4y = 20$

13.4 Equations of Lines

In Sections 13.2 and 13.3 we discussed the *graph* of a straight line. In this section we show how to write the *equation* of a line when certain facts about the line are known.

General Form

> GENERAL FORM OF THE EQUATION OF A LINE
>
> $$Ax + By = C$$
>
> where A, B, and C are real numbers, and A and B are not both 0.

Whenever possible, write the general form having A *positive* and A, B, and C *integers.*

Example 1 Write $-\frac{2}{3}x + \frac{1}{2}y = 1$ in the general form.

Solution

LCD = 6

$$\frac{6}{1}\left(-\frac{2}{3}x\right) + \frac{6}{1}\left(\frac{1}{2}y\right) = \frac{6}{1}\left(\frac{1}{1}\right)$$

$$-4x + 3y = 6$$

$$4x - 3y = -6 \quad \text{General form} \ ■$$

Point-Slope Form

Let $P_1(x_1, y_1)$ be a known point on a line whose slope is m. Let $P(x, y)$ represent any other point on the line (Figure 13.4.1).

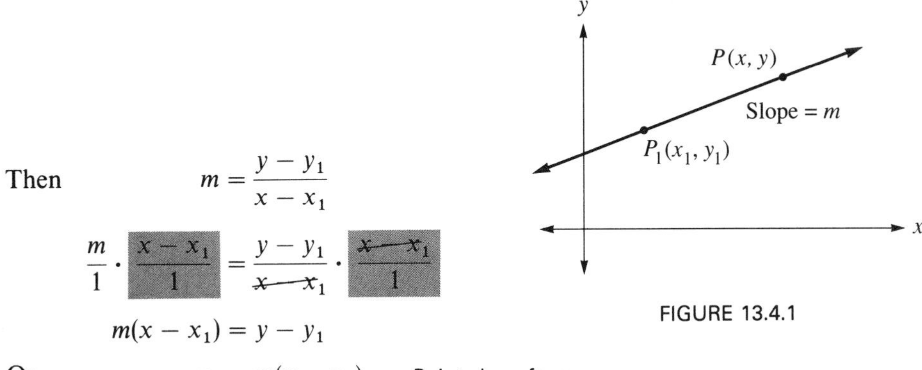

Then $$m = \frac{y - y_1}{x - x_1}$$

$$\frac{m}{1} \cdot \frac{x - x_1}{1} = \frac{y - y_1}{\cancel{x - x_1}} \cdot \frac{\cancel{x - x_1}}{1}$$

$$m(x - x_1) = y - y_1$$

Or $$y - y_1 = m(x - x_1) \quad \text{Point-slope form}$$

FIGURE 13.4.1

> POINT-SLOPE FORM OF THE EQUATION OF A LINE
>
> $$y - y_1 = m(x - x_1)$$
>
> where m = slope of the line, and $P_1(x_1, y_1)$ is a known point on the line.

Example 2 Write the equation of the line (in general form) that passes through $(2, -3)$ and has a slope of 4.

Solution

$$\begin{aligned} y - y_1 &= m(x - x_1) && \text{Point-slope form} \\ y - (-3) &= 4(x - 2) \\ y + 3 &= 4x - 8 \\ -4x + y &= -11 \\ 4x - y &= 11 && \text{General form} \end{aligned}$$

■

Example 3 Write the equation of the line (in general form) that passes through $(-1, 4)$ and has a slope of $-\frac{2}{3}$.

Solution

$$\begin{aligned} y - y_1 &= m(x - x_1) && \text{Point-slope form} \\ y - (4) &= -\frac{2}{3}[x - (-1)] \\ 3y - 12 &= -2(x + 1) && \text{Multiplied by 3} \\ 3y - 12 &= -2x - 2 \\ 2x + 3y &= 10 && \text{General form} \end{aligned}$$

■

In the following examples we choose the particular form of the equation of the line that makes the best use of the given information.

Example 4 Find the equation of the line that passes through the points $(-15, -9)$ and $(-5, 3)$.

Solution

1. Find the slope from the two given points.

$$m = \frac{(-9) - (3)}{(-15) - (-5)} = \frac{-12}{-10} = \frac{6}{5}$$

2. Use this slope with *either* given point to find the equation of the line.

Using the point $(-5, 3)$	*Using the point* $(-15, -9)$
$y - y_1 = m(x - x_1)$	$y - y_1 = m(x - x_1)$
$y - (3) = \frac{6}{5}[x - (-5)]$	$y - (-9) = \frac{6}{5}[x - (-15)]$
$5y - 15 = 6(x + 5)$	$5(y + 9) = 6(x + 15)$
$5y - 15 = 6x + 30$	$5y + 45 = 6x + 90$
$-6x + 5y = 45$	$-6x + 5y = 45$
$6x - 5y = -45$	$6x - 5y = -45$

This shows that the same equation is obtained no matter which of the two given points is used. ■

Example 5 Write the equation of the line (in general form) that has a slope of $-\frac{3}{4}$ and a y-intercept of -2.

Solution

$$y = mx + b \qquad \text{Slope-intercept form}$$

$$y = \left(-\frac{3}{4}\right)x + (-2)$$

$$4y = -3x - 8 \qquad \text{Multiplied by LCD, 4}$$

$$3x + 4y = -8 \qquad \text{General form} \quad \blacksquare$$

Recall from Section 13.2.

> The equation of a horizontal line is $y = b$. The slope of a horizontal line is 0.
>
> The equation of a vertical line is $x = a$. A vertical line has no slope.

Example 6 Find the equation of the horizontal line through $(-5, 3)$.
Solution A horizontal line has an equation in the form $y = b$. Because the y-coordinate of $(-5, 3)$ is 3, the equation of the line is

$$y = 3 \quad \blacksquare$$

Example 7 Find the equation of the vertical line through $(-5, 3)$.
Solution A vertical line has an equation in the form $x = a$. Because the x-coordinate of $(-5, 3)$ is -5, the equation of the line is

$$x = -5 \quad \blacksquare$$

EXERCISES 13.4

In Exercises 1–8, write each equation in general form.

1. $3x = 2y - 4$ **2.** $2x = 3y + 7$ **3.** $y = -\frac{3}{4}x - 2$ **4.** $y = -\frac{3}{5}x - 4$

5. $y = \frac{2}{3}x - \frac{1}{6}$ **6.** $y = \frac{1}{2}x + \frac{3}{4}$ **7.** $y - 2 = \frac{3}{4}(x + 1)$ **8.** $y + 1 = -\frac{1}{3}(x - 2)$

In Exercises 9–14, write the equation of the line through the given point and having the indicated slope. (Write the equation in general form.)

9. $(3, 4)$, $m = \frac{1}{2}$ **10.** $(5, 6)$, $m = \frac{1}{3}$ **11.** $(-1, -2)$, $m = -\frac{2}{3}$

12. $(-2, -3)$, $m = -\frac{5}{4}$ **13.** $(-6, 3)$, $m = -\frac{1}{2}$ **14.** $(5, -7)$, $m = -\frac{3}{4}$

In Exercises 15–20, write the equation of the line having the indicated slope and y-intercept. (Write the equation in general form.)

15. $m = \frac{3}{4}$, y-intercept $= -3$ **16.** $m = -\frac{2}{3}$, y-intercept $= -4$

17. $m = \frac{2}{7}$, y-intercept $= -2$

18. $m = \frac{5}{4}$, y-intercept $= -3$

19. $m = -\frac{2}{5}$, y-intercept $= \frac{1}{2}$

20. $m = -\frac{5}{3}$, y-intercept $= \frac{3}{4}$

In Exercises 21–28, find the equation of the line that passes through the given points. (Write the equation in general form.)

21. $(4, -1)$ and $(2, 4)$

22. $(5, -2)$ and $(3, 1)$

23. $(0, 0)$ and $(3, 4)$

24. $(0, 0)$ and $(-2, -5)$

25. $(4, 3)$ and $(4, -2)$

26. $(-5, 2)$ and $(5, 2)$

27. $(-3, 4)$ and $(5, -2)$

28. $(5, -3)$ and $(-2, -4)$

29. Write the equation of the horizontal line through $(3, 5)$.

30. Write the equation of the vertical line through $(3, 5)$.

31. Write the equation of the vertical line through $(-6, -2)$.

32. Write the equation of the horizontal line through $(-6, -2)$.

13.5 Graphing Curves

In Section 13.2 we showed how to graph straight lines. In this section we show how to graph *curves*.

Two points are all that we need to draw a straight line. To draw a curve line, we must find more than two points.

> TO GRAPH A CURVE
>
> **1.** Use the equation to make a table of values.
>
> **2.** Plot the points from the table of values.
>
> **3.** Draw a smooth curve through the points, joining them in order from left to right.

Example 1 Graph the equation $y = x^2 - x - 2$.

Solution Make a table of values by substituting values of x in the equation and finding the corresponding values for y.

$$y = x^2 - x - 2$$

If $x = -2$, then $y = (-2)^2 - (-2) - 2 = 4 + 2 - 2 = 4$

If $x = -1$, then $y = (-1)^2 - (-1) - 2 = 1 + 1 - 2 = 0$

If $x = 0$, then $y = (0)^2 - (0) - 2 = -2$

If $x = 1$, then $y = (1)^2 - (1) - 2 = 1 - 1 - 2 = -2$

If $x = 2$, then $y = (2)^2 - (2) - 2 = 4 - 2 - 2 = 0$

If $x = 3$, then $y = (3)^2 - (3) - 2 = 9 - 3 - 2 = 4$

x	y
-2	4
-1	0
0	-2
1	-2
2	0
3	4

In Figure 13.5.1, we graph these points and draw a smooth curve through them. In drawing the smooth curve, start with the point in the table of values having the smallest x-value. Draw to the point having the next larger x-value. Continue in this way through all the points. The graph of the equation $y = x^2 - x - 2$ is called a **parabola**.

Parabola

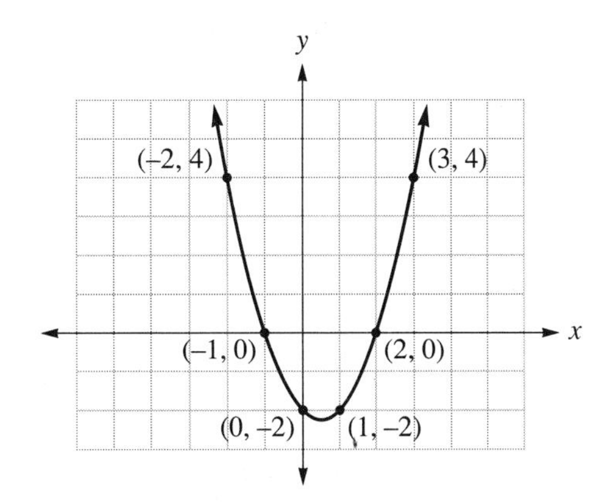

FIGURE 13.5.1 ■

Example 2 Graph the equation $y = x^3 - 4x$.
Solution First, make a table of values.

$$y = x^3 - 4x$$

If $x = -3$, then $y = (-3)^3 - 4(-3) = -27 + 12 = -15$

If $x = -2$, then $y = (-2)^3 - 4(-2) = -8 + 8 = 0$

If $x = -1$, then $y = (-1)^3 - 4(-1) = -1 + 4 = 3$

If $x = \ \ 0$, then $y = (0)^3 - 4(0) = 0$

If $x = \ \ 1$, then $y = (1)^3 - 4(1) = 1 - 4 = -3$

If $x = \ \ 2$, then $y = (2)^3 - 4(2) = 8 - 8 = 0$

If $x = \ \ 3$, then $y = (3)^3 - 4(3) = 27 - 12 = 15$

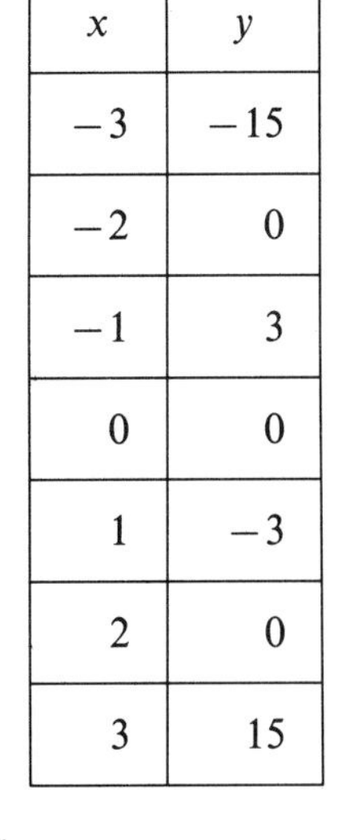

x	y
-3	-15
-2	0
-1	3
0	0
1	-3
2	0
3	15

In Figure 13.5.2, we graph these points and draw a smooth curve through them, joining the points in order from left to right.

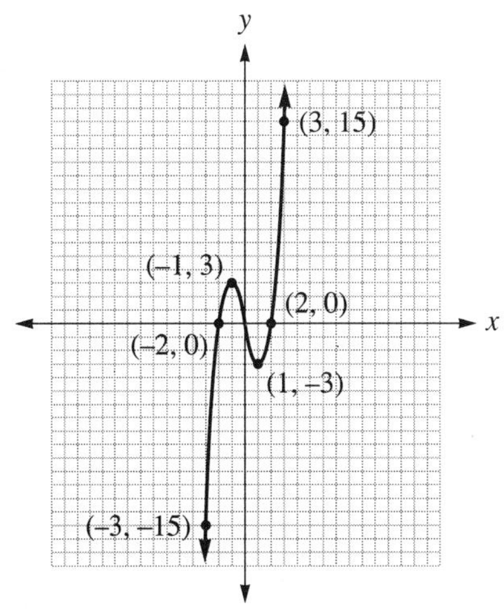

FIGURE 13.5.2 ■

EXERCISES 13.5

In Exercises 1–8, complete the table of values for each equation, and then draw its graph.

1. $y = x^2$

x	y
-3	
-2	
-1	
0	
1	
2	
3	

2. $y = \frac{x^2}{4}$

x	y
-3	
-2	
-1	
0	
1	
2	
3	

3. $y = \frac{x^2}{2}$

x	y
-3	
-2	
-1	
0	
1	
2	
3	

4. $y = x^2 + 4x$

x	y
-5	
-4	
-3	
-2	
-1	
0	
1	

5. $y = x^2 - 2x$

x	y
-2	
-1	
0	
1	
2	
3	
4	

6. $y = 3x - x^2$

x	y
-2	
-1	
0	
1	
2	
3	
4	

7. $y = 2x - x^2$

x	y
-2	
-1	
0	
1	
2	
3	
4	

8. $y = 2x + x^2$

x	y
-4	
-3	
-2	
-1	
0	
1	
2	

9. Use integer values of x from -2 to $+2$ to make a table of values for the equation $y = x^3$. Graph the points and draw a smooth curve through them.

10. Use integer values of x from -2 to $+2$ to make a table of values for the equation $y = -x^3$. Graph the points and draw a smooth curve through them.

11. Use integer values of x from -2 to $+2$ to make a table of values for the equation $y = x^3 - 3x + 4$. Graph the points and draw a smooth curve through them.

12. Use integer values of x from -2 to $+2$ to make a table of values for the equation $y = 1 - 2x - x^3$. Graph the points and draw a smooth curve through them.

13.6 Graphing Inequalities in the Plane

Half-Planes

Any line in a plane divides that plane into two **half-planes**. For example, in Figure 13.6.1, the line AB divides the plane into the two half-planes shown.

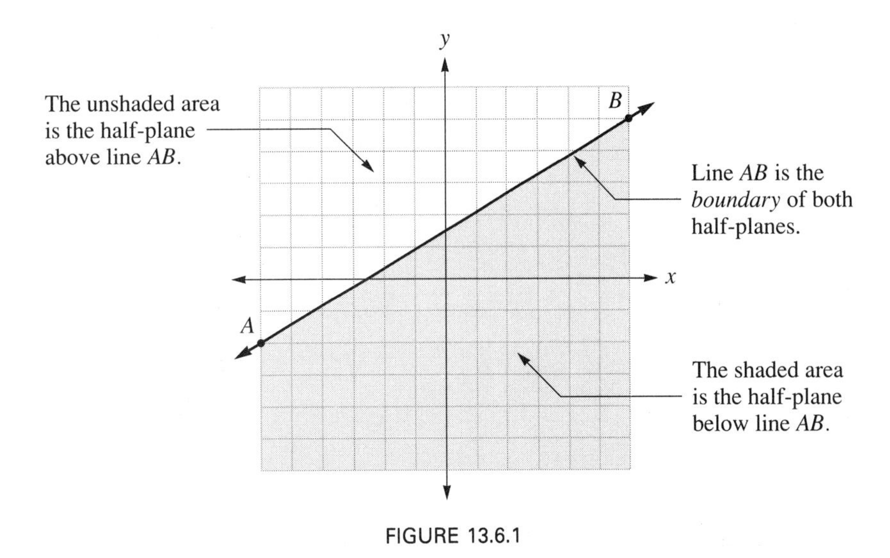

FIGURE 13.6.1

Any first-degree *inequality* (in no more than two variables) has a graph that is a half-plane.

Boundary Line

The equation of the **boundary line** of the half-plane is obtained by replacing the inequality sign by an equal sign.

How to Determine When the Boundary Is a Dashed or a Solid Line

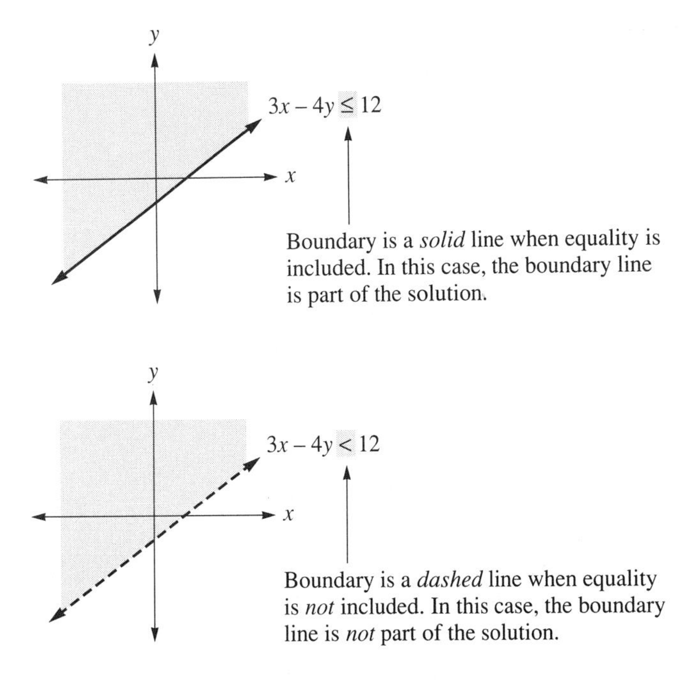

How to Determine the Correct Half-Plane

1. *If the boundary does not go through the origin*, substitute the coordinates of the origin (0, 0) into the inequality.
   If the resulting inequality is *true*, the solution is the half-plane containing (0, 0).
   If the resulting inequality is *false*, the solution is the half-plane *not* containing (0, 0).

2. *If the boundary goes through the origin*, select a point *not* on the boundary. Substitute the coordinate of this point into the inequality.
   If the resulting inequality is *true*, the solution is the half-plane containing the point selected.
   If the resulting inequality is *false*, the solution is the half-plane *not* containing the point selected.

TO GRAPH A FIRST-DEGREE INEQUALITY IN A PLANE

1. *Graph the boundary line.* The equation of the boundary line is obtained by replacing the inequality sign by an equal sign.

The boundary line is *solid* if the equality is included ($\leq, \geq$).

The boundary line is *dashed* if the equality is not included ($<, >$).

2. *Select and shade the correct half-plane.* Choose a test point [usually (0, 0] not on the boundary line and substitute its coordinates into the inequality.

If the resulting inequality is true, the solution is the half-plane containing the test point.

If the resulting inequality is false, the solution is the half-plane not containing the test point.

Example 1 Graph the inequality $2x - 3y < 6$.
Solution

Step 1. To graph the boundary line, replace the inequality sign by an equal sign.

$$2x - 3y < 6$$
$$\downarrow$$
$$2x - 3y = 6 \quad \text{Boundary line}$$

x-intercept: Set $y = 0$. Then $2x - 3y = 6$

becomes $2x - 3(0) = 6$

$$2x = 6$$
$$x = 3$$

y-intercept: Set $x = 0$. Then $2x - 3y = 6$

becomes $2(0) - 3y = 6$

$$-3y = 6$$
$$y = -2$$

x	y
3	0
0	-2

The boundary is a *dashed* line because the equality is *not* included.

$$2x - 3y < 6$$

Step 2. *Select the correct half-plane.* The solution of the inequality is only one of the two half-planes determined by the boundary line. Substitute the coordinates of the origin (0, 0) into the inequality.

$$2x - 3y < 6$$
$$2(0) - 3(0) < 6$$
$$0 < 6 \quad \textit{True}$$

Therefore, the half-plane containing the origin is the solution. The solution is the shaded area in Figure 13.6.2.

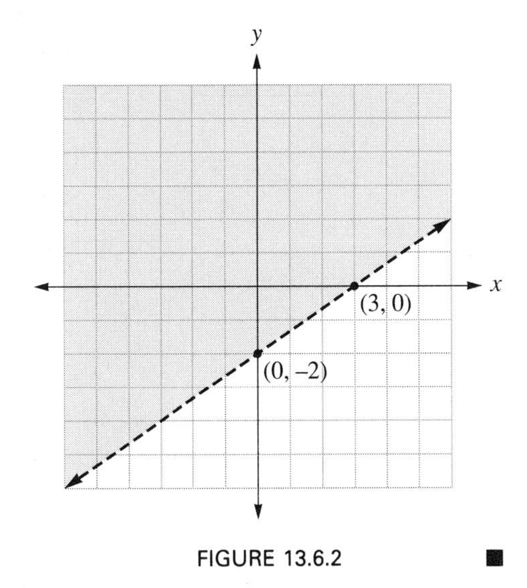

FIGURE 13.6.2 ■

Example 2 Graph the inequality $3x + 4y \leq -12$.
Solution

Step 1. Graph the boundary line $3x + 4y = -12$.

<u>x-intercept</u>: Set $y = 0$. Then $3x + 4y = -12$

becomes $3x + 4(0) = -12$

$$3x = -12$$
$$x = -4$$

<u>y-intercept</u>: Set $x = 0$. $3(0) + 4y = -12$

$$4y = -12$$
$$y = -3$$

x	y
-4	0
0	-3

The boundary is a solid line because the equality is included.

$$3x + 4y \leq -12$$

Step 2. Select the correct half-plane. Substitute the coordinates of the origin (0, 0):

$$3x + 4y \leq -12$$
$$3(0) + 4(0) \leq -12$$
$$0 \leq -12 \quad \textit{False}$$

Therefore, the solution is the half-plane *not* containing (0, 0).

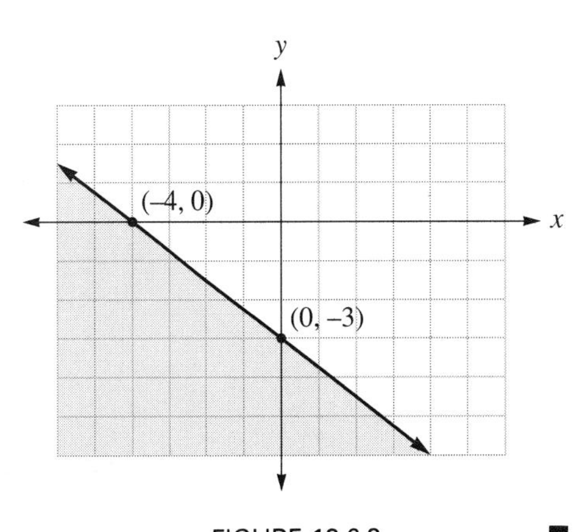

FIGURE 13.6.3 ■

Some inequalities have equations with *only one variable*. Such inequalities have graphs whose *boundaries* are either *vertical* or *horizontal* lines.

Example 3 Graph the inequality $x + 4 < 0$.

Solution

Step 1. Graph the boundary line $x + 4 = 0$

$$x = -4$$

The boundary is a *dashed* line because the equality is *not* included.

$$x + 4 < 0$$

Step 2. Select the correct half-plane. Substitute the coordinates of the origin (0, 0):

$$x + 4 < 0$$

$$(0) + 4 < 0$$

$$4 < 0 \quad \textit{False}$$

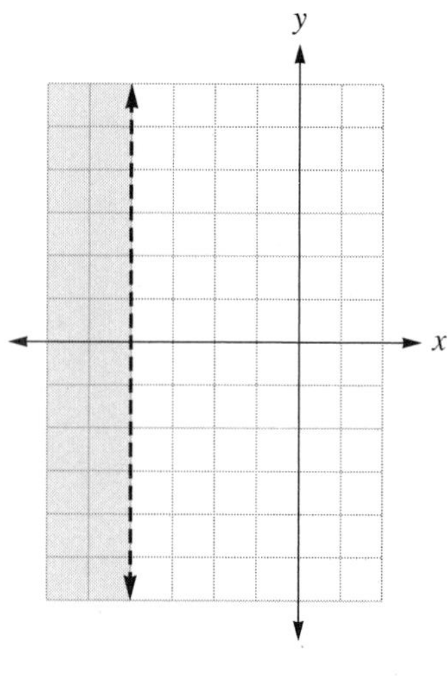

FIGURE 13.6.4

Therefore, the solution is the half-plane *not* containing (0, 0). ■

In Section 9.6, we discussed how to graph an inequality such as $x + 4 < 0$ on a *single number line.*

Because $x + 4 < 0$

$$x < -4$$

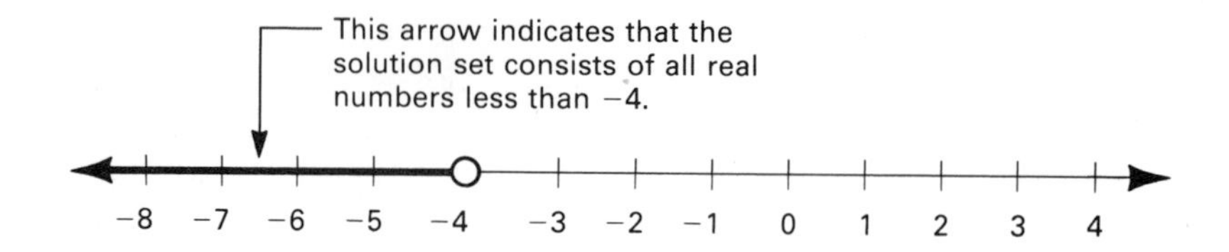

Example 3 of *this* section shows that the solution set of this same inequality, $x + 4 < 0$, represents an entire half-plane when it is plotted in the rectangular coordinate system (Figure 13.6.4).

Example 4 Graph the inequality $2x - 5y \geq 0$.

Solution

Step 1. Graph the boundary line $2y - 5x = 0$

x-intercept: Set $y = 0$. $2(0) - 5x = 0$

$$-5x = 0$$

$$x = 0$$

y-intercept: Set $x = 0$. $2y - 5(0) = 0$

$$2y = 0$$

$$y = 0$$

Therefore, the boundary lines pass through the origin (0, 0).

To find another point on the line

$$2y - 5x = 0$$

Set $x = 2$. $\quad 2y - 5(2) = 0$

$$2y - 10 = 0$$

$$2y = 10$$

$$y = 5$$

x	y
0	0
0	0
2	5

This gives the point (2, 5) on the line.
The boundary is a *solid* line because the equality is included.

$$2y - 5x \geq 0$$

Step 2. Select the correct half-plane. Since the boundary goes through the origin (0, 0), select a point *not* on the boundary, say (1, 0). Substitute the coordinate of (1, 0) into $2y - 5x \geq 0$.

$$2(0) - 5(1) \geq 0$$

$$-5 \geq 0 \qquad \textit{False}$$

Therefore, the solution is the half-plane *not* containing (1, 0) (Figure 13.6.5).

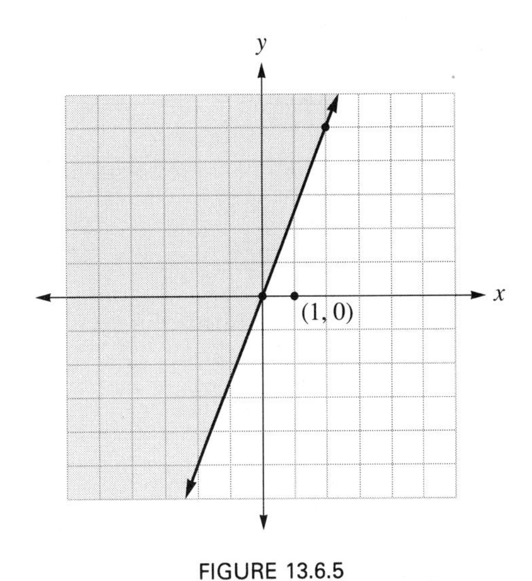

FIGURE 13.6.5

■

EXERCISES 13.6

Graph each of the following inequalities in the plane.

1. $x + 2y < 4$

2. $3x + y < 6$

3. $2x - 3y > 6$

4. $5x - 2y > 10$

5. $x \geq -2$

6. $y \geq -3$

7. $3y - 4x \geq 12$

8. $3x + 2y \geq -6$

9. $x + y > 0$

10. $x - y > 0$

11. $3x - 4y \geq 10$

12. $2x - 3y \leq 11$

13. $4x + 2y < 8$

14. $2x - 7y \geq 14$

15. $5x - 3y \geq 15$

16. $4x - y \leq 0$

17. $y > \frac{2}{3}x + 3$ **18.** $y < \frac{1}{2}x - 4$ **19.** $y \leq -\frac{1}{4}x + 2$ **20.** $y \geq -\frac{4}{3}x - 1$

21. $y - 4 \geq 0$ **22.** $x + 1 \leq 0$ **23.** $\frac{x}{5} - \frac{y}{3} > 1$ **24.** $\frac{y}{4} - \frac{x}{5} \geq 1$

13.7 Chapter Summary

Ordered Pairs 13.1

An **ordered pair** of numbers is used to represent a point in the plane.

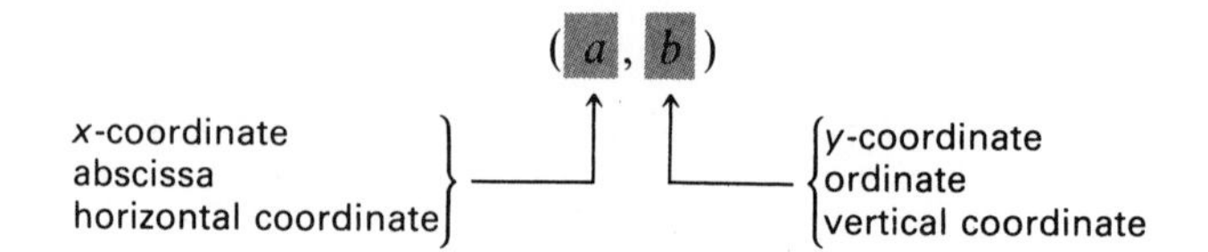

To Graph a Straight Line 13.2,13.3

Method 1 *(Table of Values)*

1. Substitute a number in the equation for one of the variables and solve for the corresponding value of the other variable. Find three points in this way, and list them in a table of values. (The third point is a check point.)
2. Graph the points that are listed in the table of values.
3. Draw a straight line through the points.

Method 2 *(Intercepts Method)*

1. Find the x-intercept: Set $y = 0$, then solve for x.
2. Find the y-intercept: Set $x = 0$, then solve for y.
3. Draw a straight line through the x- and y-intercepts.
4. If both intercepts are (0, 0), an additional point must be found before the line can be drawn.

Method 3 *(Slope-Intercept Method)*

1. Graph the y-intercept.
2. To locate a second point, start from the y-intercept and move up the number of units indicated by the numerator of the slope (move down if the numerator is negative), then move to the right the number of units indicated by the denominator of the slope.
3. Draw a straight line through the points.

To Graph a Curve 13.5

1. Use the equation to make a table of values.
2. Plot the points from the table of values.
3. Draw a smooth curve through the points, joining them in order from left to right.

To Graph a First-Degree Inequality in a Plane 13.6

1. *Graph the boundary line.* The equation of the boundary line is obtained by replacing the inequality sign by an equal sign.

   The boundary line is *solid* if the equality is included ($\leq$, $\geq$).

   The boundary line is *dashed* if the equality is not included ($<$, $>$).

2. *Select and shade the correct half-plane.* Choose a test point [usually (0, 0)] not on the boundary line and substitute its coordinates into the inequality.

   If the resulting inequality is true, the solution is the half-plane containing the test point.

   If the resulting inequality is false, the solution is the half-plane not-containing the test point.

Slope of a Line 13.3

The slope of the line through points $P_1(x_1, y_1)$ and $P_2(x_2, y_2)$

$$m = \frac{y_2 - y_1}{x_2 - x_1}$$

Equations of a Line 13.3,13.4

1. *General form:* $Ax + By = C$, where A and B are not both 0.
2. *Point-slope form:* $y - y_1 = m(x - x_1)$, where (x_1, y_1) is a known point on the line and $m =$ slope.
3. *Slope-intercept form:* $y = mx + b$, where $m =$ slope and b is the y-intercept of the line.
4. $y = b$ is the equation of a *horizontal* line. The slope of a horizontal line is 0.
5. $x = a$ is the equation of a *vertical* line. A vertical line has no slope.

Review Exercises 13.7

1. Draw a rectangle whose vertices have the following coordinates: $A(2, -5)$, $B(2, 1)$, $C(-3, 1)$, and $D(-3, -5)$.

2. Given the point $E(5, -3)$ find:
a. the abscissa of E
b. the ordinate of E
c. the x-coordinate of E

In Exercises 3–10, graph each equation by any convenient method.

3. $x - y = 5$ **4.** $x = y$ **5.** $x = -2$

6. $y + 3 = 0$ **7.** $4y - 5x = 20$ **8.** $x + 2y = 0$

9. $y = \frac{1}{3}x - 4$ **10.** $y = -\frac{3}{2}x + 5$

In Exercises 11 and 12, complete the table of values and then draw the graph.

11. $y = \dfrac{x^2}{2}$

x	y
-4	
-2	
-1	
0	
1	
2	
4	

12. $y = x^3 - 3x$

x	y
-3	
-2	
-1	
0	
1	
2	
3	

13. Find the slope of the line through $(2, -6)$ and $(-3, 5)$.

14. Find the slope and y-intercept of the line $4x - 15y = 30$.

15. Write the equation of the line having a slope of $-\dfrac{1}{2}$ and a y-intercept of 6. (Write your answer in general form.)

16. Write the equation of the line through $(3, -5)$ and having a slope of $-\dfrac{2}{3}$. (Write your answer in general form.)

17. Find the equation of the line through $(-3, 4)$ and $(1, -2)$. (Write your answer in general form.)

18. Find the equation of the vertical line through $(-7, 3)$.

In Exercises 19–22, graph the inequalities in the plane.

19. $2x - 5y > 10$

20. $x + 4y \leq 0$

21. $y > 3$

22. $\dfrac{x}{2} - \dfrac{y}{6} \geq 1$

Chapter 13 Diagnostic Test

Allow yourself about 50 minutes to do these problems. Complete solutions for every problem, together with section references, are given in the answer section at the end of the book.

1. Graph the points: $A(0, -3)$, $B(4, 0)$, $C(-3, 2)$, and $D(-1, -4)$.

2. Find the slope and y-intercept of the line $5x + 2y = 8$.

In Problems 3–5, graph each equation by any convenient method.

3. $2x - 3y = -12$

4. $y = -\frac{2}{5}x + 4$

5. $x - 3 = 0$

6. Complete the table of values and draw the graph of $y = x^2 + x - 6$.

x	y
-4	
-3	
-2	
-1	
0	
1	
2	
3	

7. Graph the inequality $3x - 2y \leq 6$ in the plane.

8. Find the slope of the line through $(-5, 2)$ and $(1, -1)$.

9. Find the equation of the horizontal line through $(2, 5)$.

10. Find the equation of the line (in general form) with slope $-\frac{2}{3}$ and a y-intercept of -4.

11. Find the equation of the line (in general form) through $(-2, 5)$ with slope $\frac{1}{4}$.

12. Find the equation of the line (in general form) through $(-4, 2)$ and $(1, -3)$.

14 Systems of Equations

In previous chapters we showed how to solve a single equation for a single variable. In this chapter we show how to solve systems of two linear equations in two variables.

14.1 Graphical Method

One Equation in One Variable A *solution* of one equation in one variable is a number that, when put in place of the variable, makes the two sides of the equation equal.

Example 1 3 is a solution for $2x + 1 = 7$
because $2(3) + 1 = 7$ is a true statement. ■

Two Equations in Two Variables

$\begin{Bmatrix} x + y = 6 \\ x - y = 2 \end{Bmatrix}$ is called *a system of two equations in two variables.*

Solution of a System

A **solution of a system** of two equations in two variables is an ordered pair that, when substituted into each equation, makes *both* equations true statements.

Example 2 Is (2, 3) a solution of the following system?

$$\begin{Bmatrix} 2x + y = 7 \\ 3x - y = 3 \end{Bmatrix}$$

Solution In Equation (1):

$$\begin{aligned} 2x + y &= 7 \\ 2(2) + 3 &\stackrel{?}{=} 7 \\ 4 + 3 &= 7 \qquad \textit{True} \end{aligned}$$

In Equation (2):

$$\begin{aligned} 3x - y &= 3 \\ 3(2) - 3 &\stackrel{?}{=} 3 \\ 6 - 3 &= 3 \qquad \textit{True} \end{aligned}$$

Therefore, (2, 3) is a solution of the system. ■

Example 3 Is $(-4, 2)$ a solution of the following system?

$$\begin{Bmatrix} 3x + 2y = -8 \\ x - 3y = -2 \end{Bmatrix}$$

Solution In Equation (1):

$$\begin{aligned} 3x + 2y &= -8 \\ 3(-4) + 2(2) &\stackrel{?}{=} -8 \\ -12 + 4 &\stackrel{?}{=} -8 \\ -8 &= -8 \qquad \textit{True} \end{aligned}$$

In Equation (2):

$$x - 3y = -2$$
$$-4 - 3(2) \stackrel{?}{=} -2$$
$$-4 - 6 \stackrel{?}{=} -2$$
$$-10 = -2 \qquad \textit{False}$$

Therefore, $(-4, 2)$ is *not* a solution of the system. ■

In Section 13.2 we showed how to graph a straight line. The graph of each equation in a system of linear equations in two variables is a straight line.

Example 4 Solve the system $\begin{Bmatrix} x + y = 6 \\ x - y = 2 \end{Bmatrix}$ graphically.

Solution Draw the graph of each equation on the same set of axes (Figure 14.1.1).

Line (1): $x + y = 6$

x-intercept: If $y = 0$, then $x = 6$.

y-intercept: If $x = 0$, then $y = 6$.

x	y
6	0
0	6

Therefore, line (1) goes through (6, 0) and (0, 6).

Line (2): $x - y = 2$

x-intercept: If $y = 0$, then $x = 2$.

y-intercept: If $x = 0$, then $y = -2$.

x	y
2	0
0	-2

Therefore, line (2) goes through (2, 0) and $(0, -2)$.

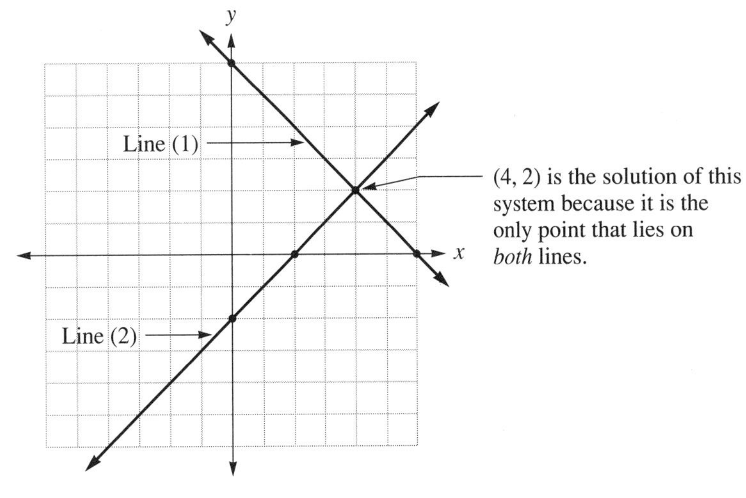

FIGURE 14.1.1
INTERSECTING LINES

The coordinates of any point on line (1) satisfy the equation of line (1). The coordinates of any point on line (2) satisfy the equation of line (2). The only point that lies on *both* lines is (4, 2). Therefore, it is the only point whose coordinates satisfy *both* equations.

Check for (4, 2) (4, 2) must satisfy *both* equations to be a solution.

$$(1):\quad x + y = 6 \qquad\qquad (2):\quad x - y = 2$$
$$4 + 2 = 6 \quad \textit{True} \qquad \text{and} \qquad 4 - 2 = 2 \quad \textit{True}$$ ■

Consistent System

When the system has a solution, it is called a **consistent system**.

Independent Equations

When each equation in the system has a different graph, they are called **independent equations**.

Example 5 Solve the system $\begin{Bmatrix} 2x - 3y = 6 \\ 6x - 9y = 36 \end{Bmatrix}$ graphically.

Solution Draw the graph of each equation on the same set of axes (Figure 14.1.2).

Line (1): $2x - 3y = 6$ has intercepts (3, 0) and (0, −2).

Line (2): $6x - 9y = 36$ has intercepts (6, 0) and (0, −4).

There is no solution because these lines never meet. Lines that never meet, such as these, are called *parallel lines*.

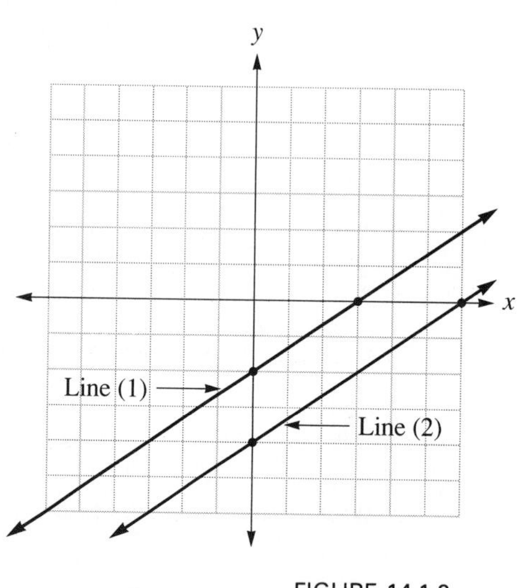

FIGURE 14.1.2
PARALLEL LINES

■

Inconsistent System

When the system has no solution, it is called an **inconsistent system.**

Because each equation in this system has a different graph, they are called *independent equations.*

Example 6 Solve the system $\begin{Bmatrix} 3x + 5y = 15 \\ 6x + 10y = 30 \end{Bmatrix}$ graphically.

Solution Draw the graph of each equation on the same set of axes (Figure 14.1.3).

Line (1): $3x + 5y = 15$ has intercepts (5, 0) and (0, 3).

Line (2): $6x + 10y = 30$ has intercepts (5, 0) and (0, 3).

Because each line goes through the same two points, they must be the same line.

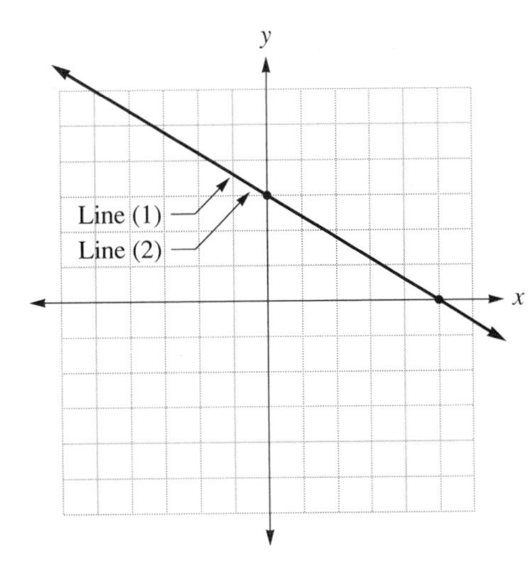

Because every point of each line lies on *both* lines, *every* point on each line is a solution.

FIGURE 14.1.3
BOTH EQUATIONS HAVE SAME GRAPH

To find one of the many solutions, pick a value for one of the variables and substitute it into either equation. For example, let $x = 1$ in Equation (1).

$$3x + 5y = 15$$
$$3(1) + 5y = 15$$
$$3 + 5y = 15$$
$$5y = 12$$
$$y = \frac{12}{5} = 2\frac{2}{5}$$

Therefore, $\left(1, 2\frac{2}{5}\right)$ is a solution for this system. The intercepts (5, 0) and (0, 3) are also solutions. Because this system has solutions, it is a consistent system. ■

Dependent Equations

When each equation in the system has the same graph, they are called **dependent equations**.

The graphical method is summarized in the following box.

Graphical Method

TO SOLVE A SYSTEM OF EQUATIONS BY THE GRAPHICAL METHOD

1. Graph each equation of the system on the same set of axes.
2. There are three possibilities:
   a. *The lines intersect at one point.* The solution is the ordered pair representing the point of intersection. (See Figure 14.1.1.)
   b. *The lines never cross* (they are parallel). There is no solution. (See Figure 14.1.2.)
   c. *Both equations have the same line for their graph.* Any ordered pair that represents a point on the line is a solution. (See Figure 14.1.3.)

EXERCISES 14.1

In Exercises 1–4, determine whether the given ordered pair is a solution of the system.

1. $2x + 3y = -5$
$5x - 2y = 4$
$(2, -3)$

2. $x - 4y = 8$
$2x - 3y = 5$
$(4, -1)$

3. $3x - y = -5$
$4x + y = -2$
$(-1, 2)$

4. $x + 2y = -10$
$6x - 3y = 0$
$(-2, -4)$

In Exercises 5–20, find the solution of each system graphically. Check your solution in both equations. Write *inconsistent* if no solution exists (parallel lines). Write *dependent* if many solutions exist (same line).

5. $2x + y = 6$
$2x - y = -2$

6. $2x - y = -4$
$x + y = 1$

7. $x - 2y = -6$
$4x + 3y = 20$

8. $x - 3y = 6$
$4x + 3y = 9$

9. $x + 2y = 0$
$x - 2y = -2$

10. $2x + y = 0$
$2x - y = -6$

11. $4y - 2x = 8$
$3x - 6y = -18$

12. $4x - 2y = 8$
$-6x + 3y = 6$

13. $10x - 4y = 20$
$6y - 15x = -30$

14. $12x - 9y = 36$
$6y - 8x = -24$

15. $2x + y = -4$
$x - y = -5$

16. $3x + 2y = 0$
$3x - 2y = 12$

17. $2x + y = 4$
$y = \frac{1}{3}x - 3$

18. $3x + 2y = 12$
$y = 2x - 1$

19. $x - 2y = -4$
$y = -\frac{1}{2}x + 6$

20. $x + 2y = -4$
$y = \frac{3}{2}x + 2$

14.2 Addition Method

The graphical method for solving a system of equations has two disadvantages: (1) it is slow, and (2) it is not an exact method of solution. The method we discuss in this section has neither of these disadvantages.

The *addition method* for solving a system is one in which the equations are added to eliminate one variable.

Example 1 Solve the system $\begin{Bmatrix} x + y = 6 \\ x - y = 2 \end{Bmatrix}$.

Solution

$$\begin{array}{lrl} \text{Equation (1):} & x + y = 6 & \\ \text{Equation (2):} & x - y = 2 & \\ \hline & 2x \quad = 8 & \text{By adding the equations vertically} \\ & x = 4 & \end{array}$$

Then, substituting $x = 4$ into Equation (1), we have

$$\begin{array}{lr} \text{Equation (1):} & x + y = 6 \\ & 4 + y = 6 \\ & y = 2 \end{array}$$

Therefore, the solution for the system is $\begin{Bmatrix} x = 4 \\ y = 2 \end{Bmatrix}$, written as the ordered pair (4, 2). ■

The system in Example 1 is simple because one variable can be *eliminated* by adding the equations directly. Sometimes we must multiply one or both equations by a number to make the coefficients of one of the variables add to zero.

Systems That Have Only One Solution (Consistent Systems with Independent Equations)

Example 2 Solve the system $\begin{Bmatrix} x - 2y = -6 \\ 4x + 3y = 20 \end{Bmatrix}$.

Solution

$$\text{Equation (1):} \quad x - 2y = -6$$
$$\text{Equation (2):} \quad 4x + 3y = 20$$

If both sides of Equation (1) are multiplied by -4, the coefficients of x will add to zero

$$\begin{aligned} -4\{x - 2y = -6\} &\Rightarrow^* & -4x + 8y &= 24 \\ 4x + 3y = 20 &\Rightarrow & 4x + 3y &= 20 \\ \hline & & 11y &= 44 \\ & & y &= 4 \end{aligned}$$

Adding the equations

When we have found the value of one variable, that value may be substituted into *either* Equation (1) or Equation (2) to find the value of the other variable. Usually one equation is easier to work with than the other.

Substituting $y = 4$ in Equation (1), we have

$$\begin{aligned} \text{Equation (1):} \quad x - 2y &= -6 \\ x - 2(4) &= -6 \\ x - 8 &= -6 \\ x &= 2 \end{aligned}$$

Therefore, the solution of the system is $\begin{Bmatrix} x = 2 \\ y = 4 \end{Bmatrix}$, written (2, 4).

Check for (2, 4) (2, 4) must satisfy *both* equations to be a solution.

$$\begin{aligned} (1)\text{:} \quad x - 2y &= -6 \\ 2 - 2(4) &\stackrel{?}{=} -6 \\ 2 - 8 &= -6 \quad \textit{True} \end{aligned}$$

$$\begin{aligned} (2)\text{:} \quad 4x + 3y &= 20 \\ 4(2) + 3(4) &\stackrel{?}{=} 20 \\ 8 + 12 &= 20 \quad \textit{True} \end{aligned}$$

■

Example 3 Solve the system $\begin{Bmatrix} 3x + 4y = 6 \\ 2x + 3y = 5 \end{Bmatrix}$.

Solution

$$\text{Equation (1):} \quad 3x + 4y = 6$$
$$\text{Equation (2):} \quad 2x + 3y = 5$$

If Equation (1) is multiplied by 2 and Equation (2) is multiplied by -3, the coefficients of x will add to zero

This symbol means the equation is to be multiplied by 2

$$\begin{aligned} 2\,]\quad 3x + 4y = 6 &\Rightarrow & 6x + 8y &= 12 \\ -3\,]\quad 2x + 3y = 5 &\Rightarrow & -6x - 9y &= -15 \\ \hline & & -y &= -3 \\ & & y &= 3 \end{aligned}$$

Adding the equations

* The symbol $\Rightarrow$, read *implies*, means the second statement is true if the first statement is true. For example:

$$x - 2 = 0 \quad \Rightarrow \quad x = 2$$

is read "$x - 2 = 0$ implies $x = 2$"

and means $x = 2$ is true if $x - 2 = 0$ is true.

The value $y = 3$ may be substituted into either Equation (1) or Equation (2) to find the value of x. It is usually easier to substitute in the equation having the smaller coefficients.

We substitute $y = 3$ into Equation (2).

$$\text{Equation (2):} \quad \begin{aligned} 2x + 3y &= 5 \\ 2x + 3(3) &= 5 \\ 2x + 9 &= 5 \\ 2x &= -4 \\ x &= -2 \end{aligned}$$

Therefore, the solution of the system is $\begin{Bmatrix} x = -2 \\ y = 3 \end{Bmatrix}$, written as the ordered pair $(-2, 3)$.

Check for (−2, 3)

$$\text{(1):} \quad \begin{aligned} 3x + 4y &= 6 \\ 3(-2) + 4(3) &\stackrel{?}{=} 6 \\ -6 + 12 &\stackrel{?}{=} 6 \\ 6 &= 6 \end{aligned} \quad \textit{True}$$

$$\text{(2):} \quad \begin{aligned} 2x + 3y &= 5 \\ 2(-2) + 3(3) &\stackrel{?}{=} 5 \\ -4 + 9 &\stackrel{?}{=} 5 \\ 5 &= 5 \end{aligned} \quad \textit{True} \ \blacksquare$$

How to Choose the Numbers Multiplying Each Equation In Example 3 we multiplied Equation (1) by 2 and Equation (2) by -3. How were these numbers found?

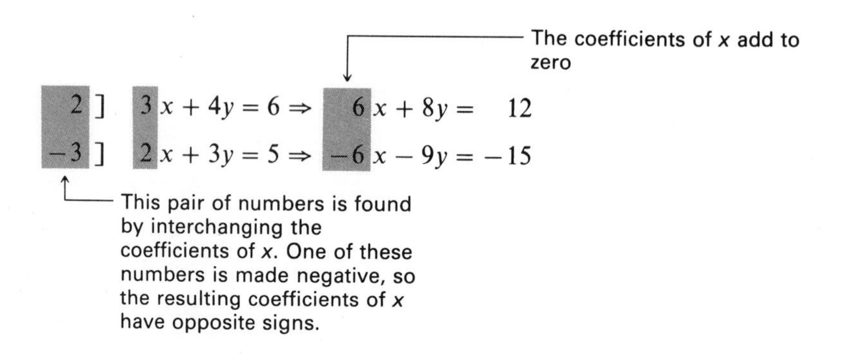

In this same system we show how to make the coefficients of y add to zero.

The coefficients of y add to zero

$$\begin{aligned} 3\,]\quad 3x + 4y = 6 &\Rightarrow 9x + 12y = 18 \\ -4\,]\quad 2x + 3y = 5 &\Rightarrow -8x - 12y = -20 \end{aligned}$$

This pair of numbers is found by interchanging the coefficients of y. One of these numbers is made negative, so the resulting coefficients of y have opposite signs.

Example 4 Consider the system: $\begin{Bmatrix} 10x - 9y = 5 \\ 15x + 6y = 4 \end{Bmatrix}$.

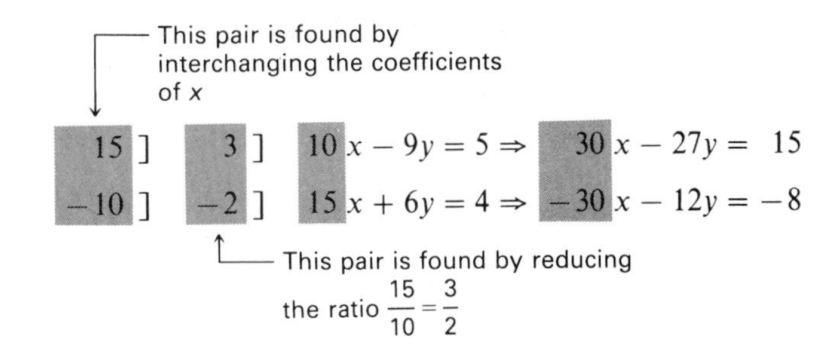

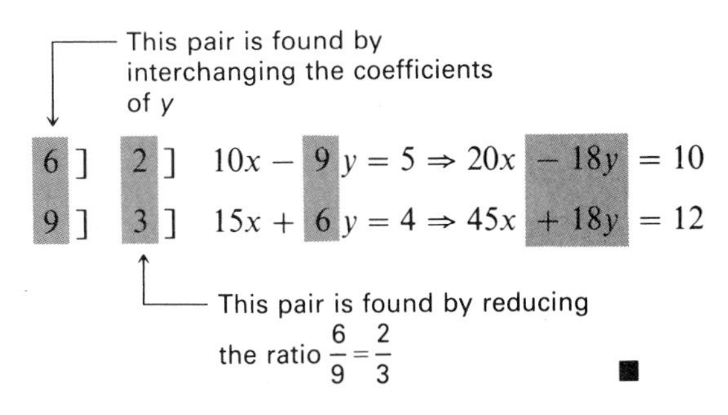

■

Example 5 Solve the system $\begin{Bmatrix} 2x + 3y = 4 \\ 5x + 6y = 11 \end{Bmatrix}$.

Solution

$$\begin{array}{rl} 5\,] & 2x + 3y = 4 \Rightarrow \\ -2\,] & 5x + 6y = 11 \Rightarrow \end{array} \begin{array}{rcr} 10x + 15y &=& 20 \\ -10x - 12y &=& -22 \\ \hline 3y &=& -2 \\ y &=& -\dfrac{2}{3} \end{array}$$

Substitute $y = -\dfrac{2}{3}$ in $2x + 3y = 4$.

$$\begin{aligned} 2x + 3y &= 4 \\ 2x + \frac{\not{3}}{1}\left(\frac{-2}{\not{3}}\right) &= 4 \\ 2x - 2 &= 4 \\ 2x &= 6 \\ x &= 3 \end{aligned}$$

Therefore, the solution of the system is $\left(3, -\dfrac{2}{3}\right)$. The check is left for the student. ■

Systems That Have No Solution (Inconsistent Systems with Independent Equations)

In Section 14.1 (Example 5) we found that a system whose graphs are parallel lines has no solution. Here, we show how to identify systems that have no solutions by using the *addition method.*

Example 6 Solve the system $\begin{Bmatrix} 2x - 3y = 6 \\ 6x - 9y = 36 \end{Bmatrix}$.

Solution

$$\begin{array}{lrrrr} \text{Equation (1):} & 6\,] & 3\,] & 2x - 3y = 6 \Rightarrow & 6x - 9y = 18 \\ \text{Equation (2):} & -2\,] & -1\,] & 6x - 9y = 36 \Rightarrow & -6x + 9y = -36 \\ & & & & \hline 0 = -18 \end{array}$$

False

No values for x and y can make $0 = -18$.

When both variables are eliminated and a *false* statement occurs, there is *no solution* for the system. ■

Systems That Have More Than One Solution (Consistent Systems with Dependent Equations)

In Section 14.1 (Example 6) we found that a system whose equations have the same line for a graph has an infinite number of solutions. Here, we show how to identify such systems using the *addition method*.

Example 7 Solve the system $\begin{Bmatrix} 4x + 6y = 4 \\ 6x + 9y = 6 \end{Bmatrix}$.

Solution

$$\begin{array}{lrrrr} \text{Equation (1):} & 9\,] & 3\,] & 4x + 6y = 4 \Rightarrow & 12x + 18y = 12 \\ \text{Equation (2):} & -6\,] & -2\,] & 6x + 9y = 6 \Rightarrow & -12x - 18y = -12 \\ & & & & \hline 0 = 0 \end{array}$$

True

When both variables are eliminated and a *true* statement occurs, there are *many solutions* for the system. Any ordered pair that satisfies Equation (1) or Equation (2) is a solution of the system. ■

The addition method is summarized in the following box.

Addition Method

TO SOLVE A SYSTEM OF EQUATIONS BY THE ADDITION METHOD

1. Multiply the equations by numbers that make the coefficients of one of the variables add to zero.
2. Add the equations.
3. There are three possibilities:
   a. *The resulting equation can be solved for one of the variables.* Substitute this value into either of the equations to find the value of the other variable.
   b. *Both variables are eliminated and a false statement occurs.* There is no solution (inconsistent system).
   c. *Both variables are eliminated and a true statement occurs.* There are many solutions (dependent equations).

Check your solutions in *both* of the original equations.

EXERCISES 14.2

Find the solution of each system by the addition method. Check your solutions. Write *inconsistent* if no solution exists. Write *dependent* if many solutions exist.

1. $2x - y = -4$, $x + y = -2$

2. $2x + y = 6$, $x - y = 0$

3. $x - 2y = 10$, $x + y = 4$

4. $x + 4y = 4$, $x - 2y = -2$

5. $x - 3y = 6$, $4x + 3y = 9$

6. $2x + 5y = 2$, $3x - 5y = 3$

7. $x + y = 2$, $3x - 2y = -9$

8. $x + 2y = -4$, $2x - y = -3$

9. $x + 2y = 0$, $2x - y = 0$

10. $2x + y = 0$, $x - 3y = 0$

11. $4x + 3y = 2$, $3x + 5y = -4$

12. $5x + 7y = 1$, $3x + 4y = 1$

13. $6x - 10y = 6$, $9x - 15y = -4$

14. $7x - 2y = 7$, $21x - 6y = 6$

15. $3x - 5y = -2$, $10y - 6x = 4$

16. $15x - 9y = -3$, $6y - 10x = 2$

17. $x - y = -5$, $3x + y = -3$

18. $x - 2y = 3$, $3x + 7y = -4$

19. $x - 2y = 6$, $3x - 2y = 12$

20. $4x - 3y = 7$, $2x + 7y = 12$

21. $2x + 6y = 2$, $3x + 9y = 3$

22. $2x - 5y = 4$, $5y - 2x = 6$

23. $2x - 5y = 4$, $3x - 4y = -1$

24. $5x - 2y = -9$, $7x + 3y = -1$

14.3 Substitution Method

Another algebraic method of solving a system of equations is called the substitution method. We use the substitution method when we can easily solve for one of the variables in terms of the other variable.

Example 1 Solve the system $\begin{Bmatrix} y = x + 4 \\ x + 2y = 5 \end{Bmatrix}$.

Solution Equation (1), $y = x + 4$, is solved for y in terms of x. If we substitute this expression for y in the second equation, we will have an equation in one variable, x.

Step 1 Substitute $x + 4$ in place of y in Equation (2).

Equation (2):

$$\begin{aligned} x + 2y &= 5 \\ x + 2(x + 4) &= 5 \\ x + 2x + 8 &= 5 \\ 3x + 8 &= 5 \\ 3x &= -3 \\ x &= -1 \end{aligned}$$

Step 2 Then substitute $x = -1$ in

$$\begin{aligned} y &= x + 4 \\ y &= -1 + 4 \\ y &= 3 \end{aligned}$$

Check (−1, 3)

(1): $y = x + 4$

$3 = -1 + 4$ *True*

(2): $x + 2y = 5$

$-1 + 2(3) \stackrel{?}{=} 5$

$-1 + 6 = 5$ *True*

Therefore, the solution is $(-1, 3)$. ∎

Example 2 Solve the system $\begin{Bmatrix} x - 2y = 11 \\ 3x + 5y = -11 \end{Bmatrix}$.

Solution Look at Equation (1) below. Because the coefficient of x is 1, we can easily solve for x in terms of y.

Equation (1): $\boxed{x} - 2y = 11$

↑ x has a coefficient of 1

Step 1 Solve Equation (1) for x.

$$x - 2y = 11$$
$$x = \boxed{11 + 2y}$$

Step 2 Substitute $\boxed{11 + 2y}$ in place of x in Equation (2).

Equation (2):
$$3x + 5y = -11$$
$$3(\boxed{11 + 2y}) + 5y = -11$$
$$33 + 6y + 5y = -11$$
$$11y = -44$$
$$y = -4$$

Step 3 Substitute $y = -4$ in

$$x = \boxed{11 + 2y}$$
$$x = 11 + 2(-4)$$
$$x = 11 - 8$$
$$x = 3$$

Check

(1): $x - 2y = 11$
$$3 - 2(-4) \stackrel{?}{=} 11$$
$$3 + 8 = 11 \quad \textit{True}$$

(2): $3x + 5y = -11$
$$3(3) + 5(-4) \stackrel{?}{=} -11$$
$$9 - 20 = -11 \quad \textit{True}$$

Therefore, the solution is $(3, -4)$. ■

Example 3 Solve the system $\begin{Bmatrix} \dfrac{x}{2} + \dfrac{y}{3} = 2 \\ \dfrac{x}{4} + \dfrac{y}{2} = 2 \end{Bmatrix}$

Solution First, remove fractions by multiplying each equation by its LCD.

The LCD for Equation (1) is 6.

$$\boxed{6}\left(\frac{x}{2}\right) + \boxed{6}\left(\frac{y}{3}\right) = \boxed{6}(2)$$
$$3x + 2y = 12$$

The LCD for Equation (2) is 4.

$$\boxed{4}\left(\frac{x}{4}\right) + \boxed{4}\left(\frac{y}{2}\right) = \boxed{4}(2)$$
$$x + 2y = 8$$

↑ coefficient is 1

Step 1 Solve Equation (2) for x.

$$x + 2y = 8$$
$$x = \boxed{8 - 2y}$$

Step 2 Substitute $8 - 2y$ in place of x in Equation (1).

Equation (1):
$$3x + 2y = 12$$
$$3(8 - 2y) + 2y = 12$$
$$24 - 6y + 2y = 12$$
$$24 - 4y = 12$$
$$-4y = -12$$
$$y = 3$$

Step 3 Substitute $y = 3$ in
$$x = 8 - 2y$$
$$x = 8 - 2(3)$$
$$x = 8 - 6$$
$$x = 2$$

Check

(1): $\frac{x}{2} + \frac{y}{3} = 2$

$\frac{2}{2} + \frac{3}{3} \stackrel{?}{=} 2$

$1 + 1 = 2$ *True*

(2): $\frac{x}{4} + \frac{y}{2} = 2$

$\frac{2}{4} + \frac{3}{2} \stackrel{?}{=} 2$

$\frac{1}{2} + \frac{3}{2} \stackrel{?}{=} 2$

$\frac{4}{2} = 2$ *True*

Therefore, the solution is (2, 3). ■

Systems That Have No Solution (Inconsistent Systems)

We now show how to identify systems that have no solution by using the *substitution method.*

Example 4 Solve the system $\begin{Bmatrix} 6x - 2y = 3 \\ 3x - 1 \ = y \end{Bmatrix}$.

Solution

Step 1 Equation (2), $3x - 1 = y$, is already solved for y.

Step 2 Substitute $3x - 1$ in place of y in Equation (1).

Equation (1):
$$6x - 2y = 3$$
$$6x - 2(3x - 1) = 3$$
$$6x - 6x + 2 = 3$$
$$2 = 3 \quad \textit{False}$$

No values of x and y can make $2 = 3$.

When both variables are eliminated and a *false* statement occurs, there is *no solution* for the system. ■

Systems That Have More Than One Solution (Dependent Equations)

We now show how to identify systems that have more than one solution by using the *substitution method.*

Example 5 Solve the system $\begin{Bmatrix} 3x - 12y = 6 \\ x - 4y = 2 \end{Bmatrix}$.

Solution (coefficient of x in Equation (2) is 1)

Step 1 Solve Equation (2) for x.

$$x - 4y = 2$$
$$x = 4y + 2$$

Step 2 Substitute $4y + 2$ in place of x in Equation (1).

Equation (1):

$$3x - 12y = 6$$
$$3(4y + 2) - 12y = 6$$
$$12y + 6 - 12y = 6$$
$$6 = 6 \quad \textit{True}$$

When both variables are eliminated and a *true* statement occurs, there are *many solutions* for the system. ■

The substitution method is summarized in the following box.

Substitution Method

TO SOLVE A SYSTEM OF EQUATIONS BY THE SUBSTITUTION METHOD

1. Solve one equation for one of the variables in terms of the other variable.
2. Substitute the expression obtained in Step 1 into the *other* equation (in place of the variable solved for in Step 1).
3. There are three possibilities:
   a. *The resulting equation can be solved for one of the variables.* Substitute this value into either of the system's equations to find the value of the other variable.
   b. *Both variables are eliminated and a false statement occurs.* There is no solution (inconsistent system).
   c. *Both variables are eliminated and a true statement occurs.* There are many solutions (dependent equations).

Check your solutions in *both* of the original equations.

EXERCISES 14.3

Find the solution of each system using the substitution method. Write *inconsistent* if no solution exists. Write *dependent* if many solutions exist.

1. $2x - 3y = 1$
$x = y + 2$

2. $y = 2x + 3$
$3x + 2y = 20$

3. $3x + 4y = 2$
$y = x - 3$

4. $2x + 3y = 11$
$x = y - 2$

5. $4x + y = 2$
$7x + 3y = 1$

6. $5x + 7y = 1$
$x + 4y = -5$

7. $4x + y = 3$
$8x + 2y = 6$

8. $x + 2y = 3$
$4x + 8y = 12$

9. $x - y = 1$
$y = 2x - 3$

10. $x + 2y = -1$
$x = 5 + y$

11. $x - 4y = 9$
$3x + 8y = 7$

12. $4x + y = 11$
$3x - 4y = -6$

13. $x - 3y = 8$
$3y - x = 8$

14. $5x - y = 4$
$y - 5x = 2$

15. $3x = y$
$2x - y = 2$

16. $5y = x$
$3y - x = 2$

17. $2x + y = 2$
$4x - y = 1$

18. $6x + y = 7$
$3x - y = -1$

19. $2x - 3y = -1$
$3x = y + 2$

20. $5x - 3y = -6$
$2y = x + 4$

21. $\frac{x}{2} + \frac{y}{8} = 1$
$\frac{x}{2} + \frac{y}{5} = 4$

22. $\frac{x}{3} - \frac{y}{2} = 6$
$\frac{x}{2} + \frac{y}{8} = 2$

23. $\frac{x}{4} + \frac{y+1}{2} = 2$
$\frac{x}{2} + \frac{y-2}{3} = 1$

24. $\frac{x}{3} + \frac{y-6}{9} = 2$
$\frac{x}{3} + \frac{y-5}{4} = 1$

14.4 Using Systems of Equations to Solve Word Problems

In solving word problems that involve more than one unknown, it is sometimes difficult to represent each unknown in terms of a single variable. In this section we eliminate that difficulty by using a different variable for each unknown.

TO SOLVE A WORD PROBLEM USING A SYSTEM OF EQUATIONS

1. Read the problem completely and determine *how many* unknown numbers there are.
2. Draw a diagram showing the relationships in the problem whenever possible.
3. Represent *each* unknown number by a *different* variable.
4. Use the word statements to write a system of equations. *There must be as many equations as variables.*
5. Solve the system of equations using one of the following:
   a. Addition method (Section 14.2)
   b. Substitution method (Section 14.3)
   c. Graphical method (Section 14.1)

In Example 1 the word problem is solved first by using a single variable and a single equation; then it is solved by using two variables and a system of equations.

Example 1 The sum of two numbers is 20. Their difference is 6. What are the numbers?

Solution

Using One Variable

Let $x =$ Larger number

$20 - x =$ Smaller number

Their difference | is | 6

$x - (20 - x) = 6$

$x - 20 + x = 6$

$2x = 26$

Larger number $x = 13$

Smaller number $20 - x = 7$

The difficulty in using the one-variable method to solve this problem is that some students have difficulty deciding whether to represent the second unknown number by $x - 20$ or by $20 - x$. ($x - 20$ is incorrect.)

Using Two Variables

Let $x =$ Larger number

$y =$ Smaller number

The sum of two numbers | is | 20

(1) $x + y = 20$

Their difference | is | 6

(2) $x - y = 6$

Using addition method:

$$\begin{aligned} x + y &= 20 \\ x - y &= 6 \\ \hline 2x \quad &= 26 \\ x &= 13 \quad \text{Larger number} \end{aligned}$$

Substitute $x = 13$ into (1).

(1) $x + y = 20$

$13 + y = 20$

$y = 7$ Smaller number ■

Example 2 Doris has 17 coins in her purse that have a total value of \$1.15. If she has only nickels and dimes, how many of each are there?

Solution Let $D =$ Number of dimes

$N =$ Number of nickels

number of nickels + number of dimes = 17 coins

(1) $N + D = 17$

The coins in her purse have a total value of \$1.15 (115¢).

amount of money in nickels + amount of money in dimes = total amount of money

(2) $5N + 10D = 115$

Multiply Equation (1) by -5.

$$\begin{aligned} -5]\ \ (N + D = 17) &\Rightarrow -5N - 5D = -85 \\ 5N + 10D = 115 &\Rightarrow \ \ 5N + 10D = 115 \\ &\qquad\qquad\quad 5D = 30 \\ &\qquad\qquad\quad\ D = 6 \text{ dimes} \end{aligned}$$

Substitute $D = 6$ into Equation (1).

$N + D = 17$

$N + 6 = 17$

$N = 11$ nickels

Therefore, she has 11 nickels and 6 dimes.

Check

11 nickels	= 11(5) =	\$ .55
+ 6 dimes	= 6(10) =	+ .60
17 coins	=	\$1.15 ■

Example 3 A 50-pound mixture of two different grades of coffee costs \$40.50. If grade A costs 95¢ a pound and grade B costs 75¢ a pound, how many pounds of each grade were used?

Solution Let a = number of pounds of grade A
and b = number of pounds of grade B.

Because the sum of the weights of the two ingredients must equal the weight of the mixture,

$$\boxed{\text{pounds of A}} + \boxed{\text{pounds of B}} = \boxed{\text{pounds of mixture}}$$

(1) $$a + b = 50 \quad \Rightarrow b = \boxed{50 - a}$$

Because the sum of the costs of the two ingredients must equal the cost of the mixture (all costs are given in cents),

$$\boxed{\text{cost of A}} + \boxed{\text{cost of B}} = \boxed{\text{cost of mixture}}$$

(2) $$95a + 75b = 4050$$

$$95a + 75b = 4050$$

$$95a + 75(\boxed{50 - a}) = 4050$$

$$95a + 3750 - 75a = 4050$$

$$20a = 300$$

$$a = 15 \text{ pounds of grade A}$$

and $$b = \boxed{50 - a} = 50 - 15$$

$$b = 35 \text{ pounds of grade B}$$

Check

15 lb of grade A
+ 35 lb of grade B
50 lb of mixture

Cost of grade A	= \$.95(15) =	\$14.25
Cost of grade B	= \$.75(35) =	+ 26.25
Cost of mixture	=	\$40.50 ■

EXERCISES 14.4

Solve the following word problems by using a system of equations: (a) Represent the unknown numbers using two variables. (b) Set up two equations and solve. (c) Answer the question.

1. The sum of two numbers is 30. Their difference is 12. What are the numbers?

2. Half the sum of two numbers is 15. Half their difference is 8. Find the numbers.

3. The sum of two angles is 90°. Their difference is 40°. Find the angles.

4. The sum of two angles is 180°. Their difference is 70°. Find the angles.

5. Find two numbers such that twice the smaller plus three times the larger is 34, and five times the smaller minus twice the larger is 9.

6. Find two numbers such that five times the larger plus three times the smaller is 47, and four times the larger minus twice the smaller is 20.

7. A 20-pound mixture of almonds and hazel nuts costs $19.75. If almonds cost 85¢ a pound and hazel nuts cost $1.40 a pound, find the number of pounds of each.

8. A 100-pound mixture of two different grades of coffee costs $145.00. If grade A costs $1.80 a pound and grade B costs $1.30 a pound, how many pounds of each grade were used?

9. The length of a rectangle is 1 foot 6 inches longer than its width. Its perimeter is 19 feet. Find its dimensions.

10. The length of a rectangle is 2 feet 6 inches longer than its width. Its perimeter is 25 feet. Find its dimensions.

11. Don spent $3.40 for 22 stamps. If he bought only 10-cent and 25-cent stamps, how many of each kind did he buy?

12. Sue spent $10.25 for 50 stamps. If she bought only 25-cent and 10-cent stamps, how many of each kind did she buy?

13. A fraction has the value $\frac{2}{3}$. If 4 is added to the numerator and the denominator is decreased by 2, the resulting fraction has the value $\frac{6}{7}$. What is the original fraction?

14. A fraction has the value $\frac{3}{4}$. If 4 is added to its numerator and 8 is subtracted from its denominator, the value of the resulting fraction is one. What is the original fraction?

15. Several families went to a movie together. They spent $10.60 for 8 tickets. If adult tickets cost $1.95 and children's tickets cost 95¢, how many of each kind of ticket were bought?

16. A class received $233 for selling 200 tickets to the school play. If student tickets cost $1 each and nonstudent tickets cost $2 each, how many nonstudents attended the play?

17. A mail-order office paid $725 for a total of 40 rolls of stamps in two denominations. If one kind costs $15 per roll and the other kind costs $20 per roll, how many rolls of each kind were bought?

18. An office manager paid $176 for 20 boxes of legal-size and letter-size file folders. If legal-size folders cost $10 a box and letter-size folders cost $8 a box, how many boxes of each kind were bought?

19. Bobby worked at two jobs during the week for a total of 30 hours. For this he received a total of $140.00. If he was paid $4.00 an hour as a tutor and $5.00 an hour as a cashier, how many hours did he work at each job?

20. Linda worked at two jobs during the week for a total of 26 hours. For this she received a total of $90.00. If she was paid $3.00 an hour as a lab assistant and $4.00 an hour as a clerk-typist, how many hours did she work at each job?

21. A tie and a pin cost $1.10. The tie cost $1 more than the pin. What is the cost of each?

22. A number of birds are resting on two limbs of a tree. One limb is above the other. A bird on the lower limb says to the birds on the upper limb, "If one of you will come down here, we will have an equal number on each limb." A bird from above replies, "If one of you will come up here we will have twice as many up here as you will have down there." How many birds were sitting on each limb?

14.5 Chapter Summary

Solution of a System 14.1 A **solution of a system** of two equations in two unknowns is an ordered pair that, when substituted into each equation, makes them both true.

Solving a System of Equations 14.1 14.2 14.3

In solving a system of equations, there are three possibilities.

1. *There is only one solution.*
   a. Graphical method: The lines intersect at one point.
   b. Algebraic method:
      Addition / Substitution — The equations can be solved for a single ordered pair.
2. *There is no solution.*
   a. Graphical method: The lines are parallel.
   b. Algebraic method:
      Addition / Substitution — Both variables are eliminated and a *false* statement occurs.
3. *There are many solutions.*
   a. Graphical method: Both equations have the same line for a graph.
   b. Algebraic method:
      Addition / Substitution — Both variables are eliminated and a *true* statement occurs.

Word Problems Using a System of Equations 14.4

To solve a word problem using a system of equations:

1. Read the problem completely and determine how many unknown numbers there are.
2. Draw a diagram showing the relationships in the problem, whenever possible.
3. Represent each unknown number by a different variable.
4. Use the word statements to write a system of equations. *There must be as many equations as variables.*
5. Solve the system of equations using one of the following:
   a. Addition method (14.2)
   b. Substitution method (14.3)
   c. Graphical method (14.1)

Review Exercises 14.5

In Exercises 1–4, find the solution of each system graphically. Write *inconsistent* if no solution exists. Write *dependent* if many solutions exist.

1. $x + y = 6$
$x - y = 4$

2. $4x + 5y = 22$
$3x - 2y = 5$

3. $2x - 3y = 3$
$3y - 2x = 6$

4. $3x - 4y = 12$
$y = \frac{3}{4}x - 3$

In Exercises 5–8, find the solution of each system using the addition method. Write *inconsistent* if no solution exists. Write *dependent* if many solutions exist.

5. $x + 5y = 11$
$3x + 4y = 11$

6. $4x - 8y = 4$
$3x - 6y = 3$

7. $3x - 5y = 15$
$5y - 3x = 8$

8. $\frac{x}{2} - \frac{y}{4} = 4$
$\frac{x}{3} + \frac{y}{4} = 1$

In Exercises 9–12, solve each system using the substitution method. Write *inconsistent* if no solution exists. Write *dependent* if many solutions exist.

9. $x = y + 2$
$4x - 5y = 3$

10. $8x - 3y = 1$
$y = 2x - 3$

11. $x + 2y = 4$
$4x - y = -2$

12. $\frac{x}{3} + \frac{y}{6} = -1$
$\frac{x}{4} + \frac{y}{12} = -1$

In Exercises 13–16, solve each system by any convenient method. Write *inconsistent* if no solution exists. Write *dependent* if many solutions exist.

13. $4x + 3y = 8$
$8x + 7y = 12$

14. $6x - 2y = -2$
$3x = y + 1$

15. $5x - 4y = -7$
$-6x + 8y = 2$

16. $\frac{x - 1}{6} + \frac{y + 5}{9} = 2$
$\frac{x + 1}{4} - \frac{y + 2}{6} = 1$

17. The sum of two numbers is 84. Their difference is 22. What are the numbers?

18. The sum of two numbers is 13. If three times the larger number minus four times the smaller number is 4, what are the numbers?

19. The length of a rectangle is 4 feet longer than its width. Its perimeter is 36 feet. Find the length and the width.

20. Brian worked at two jobs during the week for a total of 32 hours. For this he received a total of $216. If he was paid $8 an hour as a tutor and $6 an hour as a waiter, how many hours did he work at each job?

21. A mail-order office paid $730 for a total of 80 rolls of stamps in two denominations. If one kind costs $10 per roll and the other kind costs $8 per roll, how many rolls of each kind were bought?

22. An office manager paid $148 for 20 boxes of legal-size and letter-size file folders. If legal-size folders cost $9 a box and letter-size folders cost $7 a box, how many boxes of each kind were bought?

Chapter 14 Diagnostic Test

Allow yourself fifty minutes to do these problems. Complete solutions for every problem, together with section references, are given at the end of the book.

In Problems 1–7, write *inconsistent* if no solution exists. Write *dependent* if many solutions exist.

1. Solve graphically.
$$\begin{Bmatrix} 3x + 2y = 2 \\ 2x - 3y = 10 \end{Bmatrix}$$

2. Solve by addition.
$$\begin{Bmatrix} 3x - 4y = 1 \\ 5x - 3y = 9 \end{Bmatrix}$$

3. Solve by addition.
$$\begin{Bmatrix} 3x - 4y = 1 \\ 5x - 6y = 5 \end{Bmatrix}$$

4. Solve by substitution.
$$\begin{Bmatrix} 3x - 5y = 14 \\ x = y + 2 \end{Bmatrix}$$

5. Solve by substitution.
$$\begin{Bmatrix} 4x + y = 2 \\ 2x - 3y = 8 \end{Bmatrix}$$

6. Solve by any convenient method.
$$\begin{Bmatrix} 4x - 3y = 7 \\ x - 2y = -2 \end{Bmatrix}$$

7. Solve by any convenient method.
$$\begin{Bmatrix} 6x - 9y = 2 \\ 15y - 10x = -5 \end{Bmatrix}$$

8. The sum of two numbers is 18. Their difference is 42. What are the numbers?

9. Linda paid \$19.53 for 15 records at a special sale. If classical records sold for \$2.99 per disc and pop records for 88¢ per disc, how many of each kind did she buy?

10. The length of a rectangle is three centimeters more than its width. Its perimeter is 102 centimeters. Find the dimensions of the rectangle.

15 Exponents and Radicals

In this chapter we complete the discussion of exponents, then we discuss the simplification of and the operations with square roots.

15.1 Negative and Zero Exponents

In Section 8.2 we discussed *positive* exponents. At that time the following three rules were given.

RULE 1 $x^a \cdot x^b = x^{a+b}$

When powers of the same base are *multiplied*, their exponents are *added*. For example:

$$x^2 \cdot x^5 = x^{2+5} = x^7$$

RULE 2 $(x^a)^b = x^{a \cdot b}$

When a power is raised to a power, the exponents are *multiplied*. For example:

$$(x^3)^2 = x^{3 \cdot 2} = x^6$$

RULE 3 $\dfrac{x^a}{x^b} = x^{a-b}$

When powers of the same base are *divided*, the exponent in the denominator is *subtracted from* the exponent in the numerator. For example:

$$\frac{x^5}{x^2} = x^{5-2} = x^3$$

Zero Exponents In the example just shown the exponent of the numerator is larger than the exponent of the denominator. Now we consider the case in which the exponents of the numerator and denominator are the same.

$\dfrac{x^4}{x^4} = \dfrac{xxxx}{xxxx} = 1$ Because a nonzero number divided by itself is 1.

and $\dfrac{x^4}{x^4} = x^{4-4} = x^0$ Using Rule 3

Therefore, we define $x^0 = 1$.

Rule 4

ZERO EXPONENT

$$x^0 = 1 \qquad (x \neq 0)^*$$

Example 1 Showing zero as an exponent

a. $a^0 = 1$ Provided $a \neq 0$

b. $10^0 = 1$

c. $6x^0 = 6 \cdot 1 = 6$ Provided $x \neq 0$

↑ The 0 exponent applies only to x ■

Negative Exponents Now we consider the case in which the exponent of the numerator is smaller than the exponent of the denominator. Consider the expression $\frac{x^3}{x^5}$.

$$\frac{x^3}{x^5} = \frac{xxx}{xxxxx} = \frac{xxx \cdot 1}{xxx \cdot xx} = \frac{xxx}{xxx} \cdot \frac{1}{xx} = 1 \cdot \frac{1}{xx} = \frac{1}{x^2}$$

The value of this fraction is 1

However, if we use Rule 3,

$$\frac{x^3}{x^5} = x^{3-5} = x^{-2}$$

Therefore, we define $x^{-2} = \frac{1}{x^2}$.

This leads to the definition of negative exponents.

Rule 5a

NEGATIVE EXPONENT

$$x^{-n} = \frac{1}{x^n} \qquad (x \neq 0)^*$$

Example 2 Using Rule 5a

a. $x^{-5} = \frac{1}{x^5}$

b. $10^{-4} = \frac{1}{10^4}$ ■

Suppose we have a negative exponent in the denominator. Consider $\frac{1}{x^{-2}}$.

$$\frac{1}{x^{-2}} = \frac{1}{\frac{1}{x^2}} = 1 \div \frac{1}{x^2} = 1 \cdot \frac{x^2}{1} = x^2$$

Rule 5b

RECIPROCAL OF NEGATIVE EXPONENT

$$\frac{1}{x^{-n}} = x^n \qquad (x \neq 0)^*$$

Example 3 Using Rule 5b

a. $\frac{1}{x^{-4}} = x^4$

b. $\frac{1}{y^{-3}} = y^3$ ■

---

* In Rules 4, 5a, and 5b, the base $x \neq 0$. If the base were 0, this would lead to division by 0, which is undefined. (Section 6.7).

For example: $0^{-2} = \frac{1}{0^2} = \frac{1}{0 \cdot 0} = \frac{1}{0}$ undefined

Or: $0^0 = 0^{2-2} = \frac{0^2}{0^2} = \frac{0 \cdot 0}{0 \cdot 0} = \frac{0}{0}$ undefined

In this book, unless otherwise noted, none of the variables has a value that makes the denominator zero.

Rules 5a and 5b lead to the following statement.

> A *factor* can be moved either from the numerator to the denominator or from the denominator to the numerator simply by changing the sign of its exponent.
>
> NOTE This does not change the sign of the *expression*. ☑

Example 4 Writing expressions with only positive exponents

a. $a^{-3}b^4 = \dfrac{a^{-3}}{1} \cdot \dfrac{b^4}{1} = \dfrac{1}{a^3} \cdot \dfrac{b^4}{1} = \dfrac{b^4}{a^3}$

The factor a^{-3} was moved from the numerator to the denominator by changing the sign of its exponent

b. $\dfrac{h^5}{k^{-4}} = \dfrac{h^5}{1} \cdot \dfrac{1}{k^{-4}} = \dfrac{h^5}{1} \cdot \dfrac{k^4}{1} = h^5 k^4$

k^{-4} was moved from the denominator to the numerator by changing the sign of its exponent

These steps need not be written. We use them in these examples to show you why this method works

The exponent -3 applies *only* to x

c. $2x^{-3} = \dfrac{2}{1} \cdot \dfrac{x^{-3}}{1} = \dfrac{2}{1} \cdot \dfrac{1}{x^3} = \dfrac{2}{x^3}$

d. $7^{-1}x^4 = \dfrac{7^{-1}}{1} \cdot \dfrac{x^4}{1} = \dfrac{1}{7} \cdot \dfrac{x^4}{1} = \dfrac{x^4}{7}$

e. $\dfrac{a^{-2}b^4}{c^5d^{-3}} = \dfrac{a^{-2}}{1} \cdot \dfrac{b^4}{1} \cdot \dfrac{1}{c^5} \cdot \dfrac{1}{d^{-3}} = \dfrac{1}{a^2} \cdot \dfrac{b^4}{1} \cdot \dfrac{1}{c^5} \cdot \dfrac{d^3}{1} = \dfrac{b^4d^3}{a^2c^5}$ ■

Example 5 Writing expressions without fractions, using negative exponents if necessary

a. $\dfrac{m^5}{n^2} = \dfrac{m^5}{1} \cdot \dfrac{1}{n^2} = \dfrac{m^5}{1} \cdot \dfrac{n^{-2}}{1} = m^5n^{-2}$

b. $\dfrac{a}{bc^2} = \dfrac{a^1}{1} \cdot \dfrac{1}{b^1} \cdot \dfrac{1}{c^2} = \dfrac{a^1}{1} \cdot \dfrac{b^{-1}}{1} \cdot \dfrac{c^{-2}}{1} = ab^{-1}c^{-2}$ ■

A WORD OF CAUTION An expression that is *not* a factor *cannot* be moved from the numerator to the denominator of a fraction simply by changing the sign of its exponent.

$$\frac{a^{-2} + b^5}{c^4} = \frac{\dfrac{1}{a^2} + b^5}{c^4}$$

a^{-2} *cannot* be moved to the denominator because it is not a *factor* of the numerator. (The + sign indicates that a^{-2} is a *term* rather than a factor of the numerator.)

Expressions of this kind were simplified in Section 12.6

Using the Rules of Exponents with Positive, Negative, and Zero Exponents All the rules for positive exponents can also be used with negative and zero exponents.

Example 6 Applying the rules of exponents

a. $a^4 \cdot a^{-3} = a^{4+(-3)} = a^1 = a$ Rule 1

b. $x^{-5} \cdot x^2 = x^{-5+2} = x^{-3} = \dfrac{1}{x^3}$ Rules 1 and 5

c. $(y^{-2})^{-1} = y^{(-2)(-1)} = y^2$ Rule 2

d. $(x^2)^{-4} = x^{2(-4)} = x^{-8} = \dfrac{1}{x^8}$ Rules 2 and 5

e. $\dfrac{y^{-2}}{y^{-6}} = y^{(-2)-(-6)} = y^{-2+6} = y^4$ Rule 3

f. $\dfrac{z^{-4}}{z^{-2}} = z^{(-4)-(-2)} = z^{-4+2} = z^{-2} = \dfrac{1}{z^2}$ Rules 3 and 5

g. $h^3h^{-1}h^{-2} = h^{3+(-1)+(-2)} = h^0 = 1$ Rules 1 and 4

h. $x^{2n} \cdot x^n = x^{2n+n} = x^{3n}$ Rule 1 ■

Example 7 Simplifying fractions using the rules of exponents
Write results using only positive exponents.

a. $\dfrac{12x^{-2}}{4x^{-3}} = \dfrac{\overset{3}{\cancel{12}}}{\underset{1}{\cancel{4}}} \cdot \dfrac{x^{-2}}{x^{-3}} = \dfrac{3}{1} \cdot \dfrac{x^{-2-(-3)}}{1} = 3x$

b. $\dfrac{5a^4b^{-3}}{10a^{-2}b^{-4}} = \dfrac{\overset{1}{\cancel{5}}}{\underset{2}{\cancel{10}}} \cdot \dfrac{a^4}{a^{-2}} \cdot \dfrac{b^{-3}}{b^{-4}} = \dfrac{1}{2} \cdot a^{4-(-2)}b^{-3-(-4)} = \dfrac{a^6b}{2}$

c. $\dfrac{9xy^{-3}}{15x^3y^{-4}} = \dfrac{\overset{3}{\cancel{9}}}{\underset{5}{\cancel{15}}} \cdot \dfrac{x}{x^3} \cdot \dfrac{y^{-3}}{y^{-4}} = \dfrac{3}{5} \cdot \dfrac{x^{1-3}}{1} \cdot \dfrac{y^{-3-(-4)}}{1}$

$= \dfrac{3}{5} \cdot \dfrac{x^{-2}}{1} \cdot \dfrac{y}{1} = \dfrac{3}{5} \cdot \dfrac{1}{x^2} \cdot \dfrac{y}{1} = \dfrac{3y}{5x^2}$ ■

Evaluating Expressions That Have Numerical Bases

Example 8 Evaluate

a. $10^3 \cdot 10^2 = 10^5 = 10 \cdot 10 \cdot 10 \cdot 10 \cdot 10 = \underbrace{100{,}000}_{\text{5 zeros}}$

b. $10^{-2} = \dfrac{1}{10^2} = \dfrac{1}{10 \cdot 10} = \dfrac{1}{\underbrace{100}_{\text{2 zeros}}}$

c. $(2^3)^{-1} = 2^{-3} = \dfrac{1}{2^3} = \dfrac{1}{2 \cdot 2 \cdot 2} = \dfrac{1}{8}$

d. $\dfrac{3^{-1}}{3^{-4}} = 3^{-1-(-4)} = 3^3 = 3 \cdot 3 \cdot 3 = 27$

This exponent applies only to the 5

e. $\dfrac{-5^2}{(-5)^2} = \dfrac{-(5 \cdot 5)}{(-5)(-5)} = \dfrac{-25}{25} = -1$

This exponent applies to the (-5) ■

A WORD OF CAUTION A common mistake students make is shown by the following examples.

Correct method	*Incorrect method*
a. $2^3 \cdot 2^2 = 2^{3+2} = 2^5 = 32$	$2^3 \cdot 2^2 \neq (2 \cdot 2)^{3+2} = 4^5 = 1024$
b. $10^2 \cdot 10 = 10^{2+1} = 10^3 = 1{,}000$	$10^2 \cdot 10 \neq (10 \cdot 10)^{2+1} = 100^3 = 1{,}000{,}000$

In other words: When multiplying powers of the same base, add the exponents and keep the base the same; do *not* multiply the bases. ☑

Simplifying Exponential Expressions

An expression with exponents is considered **simplified** when each different base appears only once, and its exponent is a single integer.

Example 9 Simplify. Write answer using only positive exponents.

Solution

$$\frac{x^{-4}y^3}{y^2} = \frac{y^3}{x^4y^2} \quad \longleftarrow \text{Not simplified because the base } y \text{ appears twice}$$

$$= \frac{y}{x^4} \quad \longleftarrow \text{Simplified form}$$ ■

EXERCISES 15.1

In Exercises 1–42, simplify each expression. Write answers using only positive exponents.

1. x^{-4} **2.** y^{-7} **3.** $4a^{-3}$ **4.** $\dfrac{1}{x^{-2}}$ **5.** $\dfrac{1}{a^{-4}}$ **6.** $\dfrac{3}{b^{-5}}$

7. $r^{-4}st^{-2}$ **8.** $r^{-5}s^{-3}t$ **9.** $x^{-2}y^3$ **10.** x^3y^{-2} **11.** $r^{-2}st^{-4}$ **12.** $h^{-3}k^5$

13. $\dfrac{x}{y^{-3}}$ **14.** $\dfrac{a^{-2}}{b}$ **15.** $\dfrac{h^2}{k^{-4}}$ **16.** $\dfrac{m^3}{n^{-2}}$ **17.** $\dfrac{3x^{-4}}{y}$ **18.** $\dfrac{5a^{-5}}{b}$

19. $ab^{-2}c^0$ **20.** $x^{-3}y^0z$ **21.** $x^{-3} \cdot x^4$ **22.** $y^6 \cdot y^{-2}$ **23.** $10^3 \cdot 10^{-2}$ **24.** $2^{-3} \cdot 2^2$

25. $(x^2)^{-4}$ **26.** $(z^3)^{-2}$ **27.** $(a^{-2})^{-3}$ **28.** $(b^{-5})^{-2}$ **29.** $\dfrac{y^{-2}}{y^5}$ **30.** $\dfrac{z^{-2}}{z^2}$

31. mn^0p^{-4} **32.** $x^{-2} \cdot x^5$ **33.** $10^{-4} \cdot 10^3$ **34.** $(n^4)^{-1}$ **35.** $(z^{-3})^2$ **36.** $\dfrac{y^{-1}}{y^4}$

37. $\dfrac{10^3}{10^{-2}}$ **38.** $p^2p^0p^{-3}$ **39.** $\dfrac{10^2}{10^{-5}}$ **40.** $\dfrac{2^3}{2^{-2}}$ **41.** $x^4x^{-1}x^{-3}$ **42.** $y^{-2}y^{-3}y^5$

In Exercises 43–48, write each expression without fractions, using negative exponents if necessary.

43. $\frac{1}{x^2}$ **44.** $\frac{1}{y^3}$ **45.** $\frac{h}{k}$ **46.** $\frac{m}{n}$ **47.** $\frac{x^2}{yz^5}$ **48.** $\frac{a^3}{b^2c}$

In Exercises 49–66, evaluate each expression.

49. $10^4 \cdot 10^{-2}$ **50.** $3^{-2} \cdot 3^3$ **51.** 10^{-4} **52.** 2^{-3} **53.** $2^{-3} \cdot 2^5$ **54.** 10^{-6}

55. $3^0 \cdot 5^2$ **56.** $5^0 \cdot 7^2$ **57.** $4^3 \cdot 2^0$ **58.** $\frac{8^0}{8^2}$ **59.** $\frac{5^2}{5^0}$ **60.** $\frac{10^0}{10^{-2}}$

61. $(10^{-4})^2$ **62.** $(2^{-3})^2$ **63.** $\frac{10^{-3} \cdot 10^2}{10^5}$ **64.** $\frac{2^3 \cdot 2^{-4}}{2^2}$ **65.** $\frac{-4^2}{(-4)^2}$ **66.** $\frac{(-3)^4}{-3^4}$

In Exercises 67–96, simplify each expression. Write answers using only positive exponents.

67. $\frac{a^3b^0}{c^{-2}}$ **68.** $\frac{d^0e^2}{f^{-3}}$ **69.** $\frac{p^4r^{-1}}{t^{-2}}$ **70.** $\frac{u^5v^{-2}}{w^{-3}}$ **71.** $\frac{8x^{-3}}{12x}$ **72.** $\frac{15y^{-2}}{10y}$

73. $\frac{20h^{-2}}{35h^{-4}}$ **74.** $\frac{35k^{-1}}{28k^{-4}}$ **75.** $\frac{7x^{-3}y}{14y^{-2}}$ **76.** $\frac{24m^{-4}p}{16m^{-2}}$ **77.** $\frac{15m^0n^{-2}}{5m^{-3}n^4}$ **78.** $\frac{14x^0y^{-3}}{12x^{-2}y^{-4}}$

79. $\frac{x^0y^2}{z^{-5}}$ **80.** $\frac{u^{-1}v^2}{w^{-3}}$ **81.** $\frac{16h^{-2}}{10h}$ **82.** $\frac{24m^{-1}}{18m^{-3}}$ **83.** $\frac{22a^{-1}b}{33b^{-3}}$ **84.** $\frac{18w^0z^{-4}}{16w^{-2}z^2}$

85. $x^{2n} \cdot x^{-n}$ **86.** $(k^{-2c})^2$ **87.** $\frac{y^{4n}}{y^{-3n}}$ **88.** $(x^{5n})^0$ **89.** $x^{3m} \cdot x^{-m}$ **90.** $y^{-2n} \cdot y^{5n}$

91. $(x^{3b})^{-2}$ **92.** $(y^{2a})^{-3}$ **93.** $\frac{x^{2a}}{x^{-5a}}$ **94.** $\frac{a^{3x}}{a^{-5x}}$ **95.** $\frac{x + y^{-1}}{y}$ **96.** $\frac{a^{-1} - b}{b}$

15.2 General Rule of Exponents

Power of a Product Consider the expression $(xy)^3$.

$$(xy)^3 = (xy) \cdot (xy) \cdot (xy) = x \cdot x \cdot x \cdot y \cdot y \cdot y = x^3y^3$$

This leads to the following rule.

Rule 6

POWER OF A PRODUCT

$$(xy)^n = x^ny^n$$

To raise a product to a power, distribute the exponent to each factor.

Example 1 Using Rule 6

a. $(ab)^4 = a^4b^4$

b. $(2x)^3 = 2^3x^3 = 8x^3$

Be sure to distribute the exponent to the numerical base as well as the literal base

c. $(4x)^{-1} = 4^{-1}x^{-1} = \frac{1}{4} \cdot \frac{1}{x} = \frac{1}{4x}$

d. $(x^2y^3)^5 = (x^2)^5(y^3)^5 = x^{2\cdot 5}y^{3\cdot 5} = x^{10}y^{15}$ Rules 6 and 2 ■

Power of a Quotient Consider the expression $\left(\frac{x}{y}\right)^3$.

$$\left(\frac{x}{y}\right)^3 = \left(\frac{x}{y}\right) \cdot \left(\frac{x}{y}\right) \cdot \left(\frac{x}{y}\right) = \frac{x \cdot x \cdot x}{y \cdot y \cdot y} = \frac{x^3}{y^3}$$

This leads to the following rule.

Rule 7

POWER OF A QUOTIENT

$$\left(\frac{x}{y}\right)^n = \frac{x^n}{y^n}$$

To raise a quotient to a power, distribute the exponent to each factor.

Example 2 Using Rule 7

a. $\left(\frac{a}{b}\right)^7 = \frac{a^7}{b^7}$

b. $\left(\frac{3}{x}\right)^{-2} = \frac{3^{-2}}{x^{-2}} = \frac{x^2}{3^2} = \frac{x^2}{9}$

c. $\left(\frac{x^4}{y^2}\right)^3 = \frac{(x^4)^3}{(y^2)^3} = \frac{x^{4\cdot 3}}{y^{2\cdot 3}} = \frac{x^{12}}{y^6}$ Rules 7 and 2 ■

Rules 2, 6, and 7 of exponents can be combined into the following general rule.

Rule 8

GENERAL RULE OF EXPONENTS

$$\left(\frac{x^ay^b}{z^c}\right)^n = \frac{x^{an}y^{bn}}{z^{cn}}$$

When using Rule 8, notice the following:

1. x, y, and z are *factors* of the expression within the parentheses. They are *not* separated by + or − signs. See Example 3(g).
2. The exponent of each factor within the parentheses is multiplied by the exponent outside the parentheses.

Example 3 Simplify each expression. Write answers using only positive exponents.

a. $(x^3y^{-1})^5 = x^{3\cdot 5}y^{(-1)5} = x^{15}y^{-5} = \frac{x^{15}}{y^5}$

b. $\left(\frac{x^{-3}}{y^6}\right)^{-2} = \frac{x^{(-3)(-2)}}{y^{6(-2)}} = \frac{x^6}{y^{-12}} = x^6y^{12}$

The same rules of exponents apply to *numerical* bases as well as literal bases

c. $\left(\frac{2a^{-3}b^2}{c^5}\right)^3 = \frac{2^{1\cdot 3}a^{(-3)3}b^{2\cdot 3}}{c^{5\cdot 3}} = \frac{2^3a^{-9}b^6}{c^{15}} = \frac{8b^6}{a^9c^{15}}$

d. $\left(\frac{3^{-7}x^{10}}{y^{-4}}\right)^0 = 1$ The zero power of any nonzero expression is one

e. $\left(\frac{x^5y^4}{x^3y^7}\right)^2 = (x^2y^{-3})^2 = x^4y^{-6} = \frac{x^4}{y^6}$

Simplify the expression within the parentheses first whenever possible

f. $\left(\frac{10^{-2}\cdot 10^5}{10^4}\right)^3 = \left(\frac{10^3}{10^4}\right)^3 = (10^{-1})^3 = 10^{-3} = \frac{1}{10^3}$

g. $(x^2 + y^3)^4$ Rules 2 and 8 *cannot* be used here because the + sign means that x^2 and y^3 are *not* factors; they are *terms* of the expression being raised to the fourth power. ■

EXERCISES 15.2

Simplify each expression. Write the answer using only positive exponents.

1. $(xy)^5$ **2.** $\left(\frac{a}{b}\right)^4$ **3.** $\left(\frac{4}{x}\right)^2$ **4.** $(3x)^3$ **5.** $(5x)^{-1}$ **6.** $(2x)^{-3}$

7. $\left(\frac{5}{x}\right)^{-2}$ **8.** $\left(\frac{2}{m}\right)^{-4}$ **9.** $(a^2b^3)^2$ **10.** $(x^4y^5)^3$ **11.** $(2z^3)^2$ **12.** $(3w^2)^3$

13. $(m^{-2}n)^4$ **14.** $(p^{-3}r)^5$ **15.** $(x^{-2}y^3)^{-4}$ **16.** $(w^{-3}z^4)^{-2}$ **17.** $(3a^{-2})^{-3}$ **18.** $(9x^{-4})^{-1}$

19. $\left(\frac{rs^3}{t^4}\right)^3$ **20.** $\left(\frac{x^3y}{z^2}\right)^2$ **21.** $\left(\frac{R^4}{S^{-2}}\right)^3$ **22.** $\left(\frac{a^3b}{c^2}\right)^3$

23. $\left(\frac{M^{-2}}{N^3}\right)^4$ **24.** $\left(\frac{R^5}{S^{-4}}\right)^3$ **25.** $\left(\frac{a^2b^{-4}}{b^{-5}}\right)^2$ **26.** $\left(\frac{x^{-2}y^2}{x^{-3}}\right)^3$

27. $\left(\frac{10^2\cdot 10^{-1}}{10^{-2}}\right)^2$ **28.** $\left(\frac{3^{-3}\cdot 3^2}{3^{-2}}\right)^3$ **29.** $\left(\frac{mn^{-1}}{m^3}\right)^{-2}$ **30.** $\left(\frac{ab^{-2}}{a^2}\right)^{-3}$

31. $\left(\frac{x^4}{x^{-1}y^{-2}}\right)^{-1}$ **32.** $\left(\frac{x^3}{x^{-2}y^{-4}}\right)^{-1}$ **33.** $\left(\frac{8s^{-3}}{4st^2}\right)^{-2}$ **34.** $\left(\frac{10u^{-4}}{5uv^3}\right)^{-2}$

35. $(10^0k^{-4})^{-2}$ **36.** $(6^0z^{-5})^{-2}$ **37.** $(x^3 + y^4)^5$ **38.** $(a^5 - b^2)^6$

39. $\left(\frac{r^7s^8}{r^9s^6}\right)^0$ **40.** $\left(\frac{t^5u^6}{t^8u^7}\right)^0$ **41.** $\left(\frac{6m^{-4}p}{m^{-2}}\right)^{-1}$ **42.** $\left(\frac{10x^{-6}y}{x^{-4}}\right)^{-1}$

43. $\left(\dfrac{10 \cdot 10^{-2}}{10^{-3}}\right)^3$ **44.** $\left(\dfrac{m^{-1}n^3}{m}\right)^{-2}$ **45.** $\left(\dfrac{x^2}{x^{-3}y^{-2}}\right)^{-1}$ **46.** $\left(\dfrac{a^{-3}b^2}{a^{-4}}\right)^3$

47. $\left(\dfrac{x^2y^{-3}}{x^4y^2}\right)^{-2}$ **48.** $\left(\dfrac{t^{-1}u^2}{t^4u^{-3}}\right)^{-3}$ **49.** $\left(\dfrac{8yz^{-3}}{z^{-2}}\right)^{-1}$ **50.** $\left(\dfrac{9u^{-3}}{uv^2}\right)^{-2}$

15.3 Scientific Notation

In science we often work with very large and very small numbers. To write and calculate with such numbers we need a different system from the one we have used so far. In this section we study a number system used in science called *scientific notation.*

Before we can write numbers in scientific notation we must understand positive and negative powers of 10. In Section 3.1 we showed positive powers of 10. We now extend this system of powers of 10 to include the negative powers of 10. See Figure 15.3.1.

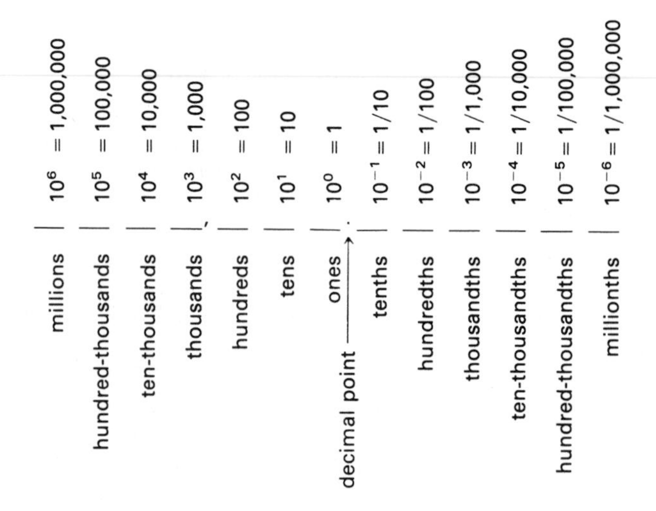

FIGURE 15.3.1 POWERS OF 10

Multiplying by a Positive Power of 10 To multiply a decimal by a *positive* power of 10, move the decimal point to the *right* the same number of places as the power of 10 (Section 3.7).

Example 1

a. $7.5 \times 10^2 = 7.50_\wedge = 750$

b. $0.048 \times 10^3 = 0.048_\wedge = 48$

c. $272 \times 10^5 = 272.00000_\wedge = 27{,}200{,}000$ ■

Multiplying by a Negative Power of 10 Recall from Section 3.7 that when we divided a decimal by a positive power of 10, the decimal point moved to the left. But multiplying by a negative power of 10 is the same as dividing by a positive power of 10. Therefore, to multiply by a *negative* power of 10, move the decimal point to the *left* the same number of places as the power of 10.

Example 2

a. $7.5 \times 10^{-2} = \dfrac{7.5}{10^2} = {}_\wedge 07.5 = 0.075$

b. $6.04 \times 10^{-3} = \dfrac{6.04}{10^3} = {}_\wedge 006.04 = 0.00604$

c. $97 \times 10^{-1} = 9_\wedge 7. = 9.7$

d. $3.2 \times 10^{-4} = {}_\wedge 0003.2 = 0.00032$ ■

Scientific Notation

A number is written in **scientific notation** if it is a number between 1 and 10 multiplied by a power of 10. To write a number in scientific notation, place the decimal point after the first nonzero digit and then multiply by the appropriate power of 10.

Example 3

	Common notation		*Scientific notation*
a.	245 =	$2_\wedge 4\ 5.$ =	2.45×10^2
b.	24.5 =	$2_\wedge 4.5$ =	2.45×10^1
c.	2.45 =	$2.4\ 5$ =	2.45×10^0
d.	0.245 =	$0.2_\wedge 4\ 5$ =	2.45×10^{-1}
e.	0.0245 =	$0.0\ 2_\wedge 4\ 5$ =	2.45×10^{-2}
f.	0.00245 =	$0.0\ 0\ 2_\wedge 4\ 5$ =	2.45×10^{-3} ■

The preceding examples lead us to the following rule:

TO WRITE A NUMBER IN SCIENTIFIC NOTATION

1. Place a caret ($_\wedge$) to the right of the first nonzero digit.
2. Draw an arrow *from* the caret to the actual decimal point.
3. The exponent of 10 is $\begin{cases} (+) \text{ if arrow points right.} \\ (-) \text{ if arrow points left.} \end{cases}$
4. The number part of the exponent of 10 is equal to the number of places between the caret and the actual decimal point.
5. Write the number with decimal point after the first nonzero digit and multiply by the power of 10 found in steps 3 and 4.

NOTE The convention of signs used for the exponent of 10 is the same as that used on the number line.

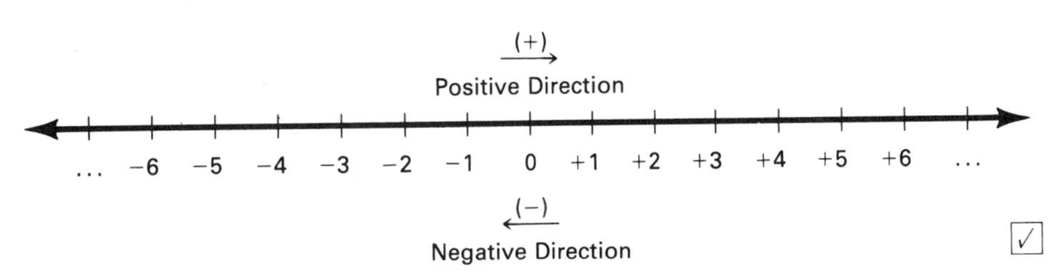

✓

Example 4 Changing numbers from common to scientific notation

a. 92,900,000. = 9‸2 9 0 0 0 0 0. $= 9.29 \times 10^{7}$

b. 0.0056 = 0.0 0 5‸6 $= 5.6 \times 10^{-3}$

c. 0.01745 = 0.0 1‸7 4 5 $= 1.745 \times 10^{-2}$

d. 684.5 = 6‸8 4.5 $= 6.845 \times 10^{2}$ ■

Example 5 Changing numbers from scientific notation to common notation

a. 2.54×10^{-2} = 0.0 2‸5 4 = 0.0254

b. 1.609×10^{3} = 1‸6 0 9. = 1,609

c. 4.67×10^{-5} = 0.0 0 0 0 4‸6 7 = 0.0000467

d. 3.57×10^{5} = 3‸5 7 0 0 0. = 357,000 ■

Calculations Using Scientific Notation

Example 6 Find $2.8 \times 10^{3} \times 1.6 \times 10^{2}$.

Solution

$$2.8 \times 10^{3} \times 1.6 \times 10^{2} = (2.8 \times 1.6) \times (10^{3} \times 10^{2}) = 4.48 \times 10^{5}$$ ■

Example 7 Find $\dfrac{7.5 \times 10^{-2}}{1.5 \times 10^{4}}$.

Solution

$$\frac{7.5 \times 10^{-2}}{1.5 \times 10^{4}} = \frac{7.5}{1.5} \times \frac{10^{-2}}{10^{4}} = 5 \times 10^{-2-4} = 5 \times 10^{-6}$$ ■

Example 8 Find $\dfrac{0.065 \times 24{,}000}{0.0006}$.

Solution Write each number in scientific notation, then perform the indicated operations.

$$\frac{0.065 \times 24{,}000}{0.0006} = \frac{6.5 \times 10^{-2} \times 2.4 \times 10^{4}}{6 \times 10^{-4}} = \frac{15.6 \times 10^{2}}{6 \times 10^{-4}} = 2.6 \times 10^{2-(-4)} = 2.6 \times 10^{6}$$ ■

When the result of a calculation in common notation exceeds the calculator's display, the calculator may automatically convert to scientific notation.

Example 9 Use a calculator to find the following:

a. $6{,}000{,}000 \times 20{,}000$

Key in: 6000000 [×] 20000 [=]

Answer: 1.2 11 or 1.2 ^{11} — Either display means 1.2×10^{11}

b. $\dfrac{0.0000438}{300,000}$

Key in: .0000438 [÷] 300000 [=]

Answer: 1.46 −10 or 1.46 $^{-10}$ $= 1.46 \times 10^{-10}$ ■

EXERCISES 15.3

For Exercises 1–20, complete the following table.

	Common Notation	*Scientific Notation*
1.	748	
2.	25,000	
3.	0.063	
4.		6.7×10^3
5.	0.001732	
6.		2.81×10^{-2}
7.		3.47×10^6
8.	86.48	
9.		1.91×10^{-6}
10.	588,000	
11.	0.0000563	
12.	27,800	
13.	0.000058	
14.	1,761,000	
15.		6.547×10^{-27}
16.	0.00000078	
17.		4.77×10^{-10}
18.	5,780,000,000,000	
19.		8.36×10^7
20.		1.05×10^{-5}

In Exercises 21–30, perform the indicated operations, and write your answers in scientific notation.

21. $3.4 \times 10^2 \times 2.5 \times 10^{-5}$

22. $2.3 \times 10^{-1} \times 4.06 \times 10^5$

23. $\dfrac{8.1 \times 10^3}{2.7 \times 10^{-2}}$

24. $\dfrac{7.8 \times 10^{-1}}{1.3 \times 10^5}$

25. $203{,}000 \times 0.0004$

26. $0.00003 \times 2{,}600$

27. $\dfrac{0.00036}{0.024}$

28. $\dfrac{450{,}000}{0.0018}$

29. $\dfrac{0.0055 \times 1200}{0.00011}$

30. $\dfrac{0.072}{0.12 \times 50{,}000}$

In Exercises 31–36, use a calculator to perform the indicated operations. Write answers in scientific notation.

31. $7{,}200{,}000 \times 5{,}000$

32. 0.0000009×0.00087

33. $\dfrac{0.000024}{6{,}400{,}000}$

34. $\dfrac{66{,}000{,}000}{0.00096}$

35. $\dfrac{380 \times 21{,}000}{0.0000035}$

36. $\dfrac{0.0057 \times 0.0003}{7{,}600{,}000}$

15.4 Square Roots

Finding the square root of a number is the inverse operation of squaring a number (Section 6.9).

Principal Square Root
Radicand

Every positive number has both a positive and a negative square root. The positive square root is called the **principal square root**. The principal square root of N is written $\sqrt{N}$. The negative square root of N is written $-\sqrt{N}$. The number under the radical sign is called the **radicand**.

$\sqrt{N}$ Read "the square root of N."

Radical sign ↑ ↑ Radicand

Example 1 Find the square roots.

a. $\sqrt{25} = 5$ because $5^2 = 25$

b. $\sqrt{100} = 10$ because $10^2 = 100$

c. $\sqrt{\dfrac{4}{9}} = \dfrac{2}{3}$ because $\left(\dfrac{2}{3}\right)^2 = \dfrac{4}{9}$

d. $-\sqrt{64} = -8$ because $\sqrt{64} = 8$

$-\sqrt{64} = -8$ ■

Rational Numbers

A **rational number** is a number that can be written in the form $\dfrac{a}{b}$, where a and b are integers ($b \neq 0$).

The decimal representation of any rational number is either a terminating decimal or a repeating decimal.

Example 2 Examples of rational numbers

a. $\dfrac{2}{3}$ All fractions are rational numbers.

b. $4 = \frac{4}{1}$ All integers are rational numbers.

c. $0.25 = \frac{25}{100}$ All terminating decimals are rational numbers.

d. $0.333\ldots = \frac{1}{3}$ All repeating decimals are rational numbers.

e. $2\frac{1}{2} = \frac{5}{2}$ All mixed numbers are rational numbers.

f. $\sqrt{25} = 5 = \frac{5}{1}$ $\sqrt{25}$ simplifies to the integer 5 which is a rational number. ■

Irrational Numbers

An **irrational number** is a number whose decimal representation is a nonterminating nonrepeating decimal.

Example 3 Examples of irrational numbers

a. $\sqrt{2} = 1.414213562\ldots$

b. $\sqrt{5} = 2.236067977\ldots$

c. $\pi = 3.1415926535\ldots$ ■

Real Numbers

Together, the rational numbers and the irrational numbers form the set of **real numbers** and, therefore, can be represented by points on the number line.

NOTE Square roots of negative numbers, such as $\sqrt{-4}$, are imaginary numbers. Imaginary numbers are not real and cannot be graphed on the number line. ☑

EXERCISES 15.4

1. Which of the following are rational numbers?
$-5;\ \frac{3}{4};\ 2\frac{1}{2};\ \sqrt{2};\ 3.5;\ \sqrt{4};\ 3$

2. Which of the following are irrational numbers?
$\frac{5}{6};\ -4;\ \sqrt{3};\ 3\frac{1}{4};\ \sqrt{9};\ 0.7;\ \pi$

In Exercises 3 and 4, what is the radicand in each expression?

3. a. $\sqrt{17}$ b. $\sqrt{x + 1}$

4. a. $\sqrt{15}$ b. $\sqrt{5x}$

In Exercises 5–14, find the square roots.

5. $\sqrt{36}$ **6.** $\sqrt{64}$ **7.** $\sqrt{1}$ **8.** $\sqrt{0}$ **9.** $\sqrt{81}$

10. $\sqrt{144}$ **11.** $-\sqrt{4}$ **12.** $-\sqrt{25}$ **13.** $\sqrt{\frac{16}{49}}$ **14.** $\sqrt{\frac{9}{100}}$

15.5 Simplifying Square Roots

We consider square roots of two kinds:

1. Square roots *without* fractions in the radicand.
2. Square roots *with* fractions in the radicand.

15.5A Square Roots without Fractions in the Radicand

$$\sqrt{x^2} = x \text{ because } (x)^2 = x^2$$

$$\text{also } \sqrt{x^2} = -x \text{ because } (-x)^2 = x^2$$

Because x can be a positive number or a negative number, we don't know whether x or $-x$ is the principal root. However, $|x|$ must be positive (provided $x \neq 0$); therefore $\sqrt{x^2} = |x|$ is the principal square root of x^2.

In this chapter we assume that all variables represent positive numbers unless otherwise indicated. For this reason we do not use the absolute value symbol to indicate a principal square root.

Example 1 Does $\sqrt{4 \cdot 9} = \sqrt{4} \cdot \sqrt{9}$?

Solution

$$\sqrt{4 \cdot 9} = \sqrt{36} = 6$$

$$\sqrt{4} \cdot \sqrt{9} = 2 \cdot 3 = 6$$

Therefore, $\sqrt{4 \cdot 9} = \sqrt{4} \cdot \sqrt{9}$. ■

This leads to the first rule of radicals.

Rule 1

SQUARE ROOT OF A PRODUCT

$$\sqrt{a \cdot b} = \sqrt{a} \cdot \sqrt{b}$$

We will use the rule just stated to help us find the principal square root of a product.

TO FIND THE PRINCIPAL SQUARE ROOT

1. Find the square root of each factor. (Numerical and literal factors are done the same way.)
   If the exponent of the factor is an even number: Divide the exponent by 2.

   Even exponent

   $$\sqrt{x^6} = x^{6 \div 2} = x^3$$

   If the exponent of the factor is an odd number: Write the factor as the product of two factors—one factor having an even exponent, the other factor having an exponent of 1.

   Odd exponent

   $$\sqrt{x^9} = \sqrt{x^8 x^1} = \sqrt{x^8}\sqrt{x^1}$$
   $$= x^4\sqrt{x}$$

2. Multiply the square roots of all the factors found in step 1.

Example 2 Finding the square root of a factor whose exponent is an *even* number

a. $\sqrt{5^2} = 5^{\frac{2}{2}} = 5^1$ Dividing exponent by 2

b. $\sqrt{x^2} = x^{\frac{2}{2}} = x^1$

c. $\sqrt{y^4} = y^{\frac{4}{2}} = y^2$ ■

Example 3 Finding the square root of a factor whose exponent is an *odd* number

a. $\sqrt{3^5} = \sqrt{3^4 \cdot 3^1} \leftarrow 3^5 = 3^4 \cdot 3^1$

$= \sqrt{3^4}\sqrt{3}$ The factor 3^4 is the highest power of 3 whose exponent (4) is exactly divisible by 2.

$= 3^2\sqrt{3}$

$= 9\sqrt{3}$

b. $\sqrt{x^7} = \sqrt{x^6 \cdot x^1} \leftarrow x^7 = x^6 \cdot x^1$

$= \sqrt{x^6}\sqrt{x}$ The factor x^6 is the highest power of x whose exponent (6) is exactly divisible by 2. ■

$= x^3\sqrt{x}$

When finding the square root of a number, first express it in prime factored form (Section 11.1).

Example 4 Finding the square root of a number

a. $\sqrt{48} = \sqrt{2^4 \cdot 3}$ ← Prime factored form of 48

$= \sqrt{2^4}\sqrt{3}$

$= 2^2\sqrt{3}$

$= 4\sqrt{3}$

$$\begin{array}{r|l} 2 & 48 \\ 2 & 24 \\ 2 & 12 \\ 2 & 6 \\ & 3 \end{array}$$

$48 = 2^4 \cdot 3$

b. $\sqrt{75} = \sqrt{3 \cdot 5^2}$ ← Prime factored form of 75

$= \sqrt{3}\sqrt{5^2}$

$= \sqrt{3}(5)$

$= 5\sqrt{3}$ ■

$$\begin{array}{r|l} 5 & 75 \\ 5 & 15 \\ & 3 \end{array}$$

$75 = 3 \cdot 5^2$

A convenient arrangement of the work for finding the principal square root of a product is shown in Example 5.

Example 5 Finding the square root of a product

a. $\sqrt{360} = \sqrt{2^3 \cdot 3^2 \cdot 5}$

$= \sqrt{2^2 \cdot 2 \cdot 3^2 \cdot 5}$

$= 2 \cdot 3 \cdot \sqrt{2 \cdot 5} = 6\sqrt{10}$

Simplified form

$$\begin{array}{r|l} 2 & 360 \\ 2 & 180 \\ 2 & 90 \\ 3 & 45 \\ 3 & 15 \\ & 5 \end{array}$$

$360 = 2^3 \cdot 3^2 \cdot 5$

b. $\sqrt{12x^4y^3} = \sqrt{2^2 \cdot 3 \cdot x^4 \cdot y^2 \cdot y}$

$= 2 \cdot x^2 \cdot y\sqrt{3y} = 2x^2y\sqrt{3y}$

Simplified form

$$\begin{array}{r|l} 2 & 12 \\ 2 & 6 \\ & 3 \end{array}$$

$12 = 2^2 \cdot 3$

c. $\sqrt{24a^5b^7} = \sqrt{2^3 \cdot 3 \cdot a^5b^7}$
$= \sqrt{2^2 \cdot 2 \cdot 3 \cdot a^4 \cdot a \cdot b^6 \cdot b}$
$= 2 \cdot a^2 \cdot b^3 \cdot \sqrt{2 \cdot 3 \cdot a \cdot b}$
$= 2a^2b^3\sqrt{6ab}$

2	24
2	12
2	6
	3

$24 = 2^3 \cdot 3$ ■

Sometimes finding the square root of a number can be simplified by inspection if you can see that it has a factor that is a perfect square (Example 6).

Example 6

a. $\sqrt{12} = \sqrt{4 \cdot 3}$ 4 is a factor of 12 and is a perfect square
$= \sqrt{4}\sqrt{3}$
$= 2\sqrt{3}$

b. $\sqrt{50} = \sqrt{25 \cdot 2}$ 25 is a factor of 50 and is a perfect square
$= \sqrt{25} \cdot \sqrt{2}$
$= 5\sqrt{2}$

c. $5\sqrt{18} = 5\sqrt{9 \cdot 2}$ 9 is a factor of 18 and is a perfect square
$= 5 \cdot \sqrt{9} \cdot \sqrt{2}$
$= 5 \cdot 3 \cdot \sqrt{2}$
$= 15\sqrt{2}$ ■

It is true for a product that $\sqrt{9 \cdot 16} = \sqrt{9} \cdot \sqrt{16}$ by Rule 1, but what about a sum? (See Example 7.)

Example 7 Does $\sqrt{9 + 16} = \sqrt{9} + \sqrt{16}$?

Solution

$$\sqrt{9 + 16} = \sqrt{25} = 5$$
$$\sqrt{9} + \sqrt{16} = 3 + 4 = 7$$

Because $5 \neq 7$, the square root of a sum does not equal the sum of the square roots. ■

Multiplying a Square Root by Itself

By Rule 1 $\qquad \sqrt{a}\sqrt{b} = \sqrt{ab}$

If $a = b$, then $\qquad \sqrt{a}\sqrt{a} = \sqrt{aa}$

$$\sqrt{a}\sqrt{a} = \sqrt{a^2}$$
$$\sqrt{a}\sqrt{a} = a$$

Rule 2

MULTIPLYING A SQUARE ROOT BY ITSELF

$$\sqrt{a}\sqrt{a} = a$$

or

$$(\sqrt{a})^2 = a$$

Example 8 Examples of multiplying a square root by itself

a. $\sqrt{5}\sqrt{5} = 5$

b. $\sqrt{7}\sqrt{7} = 7$

c. $\sqrt{2x}\sqrt{2x} = 2x$

d. $(\sqrt{x})^2 = x$

e. $(\sqrt{x-1})^2 = x - 1$ ■

Simplifying a Square Root

A square root without fractions is considered **simplified** if no prime factor of the radicand has an exponent equal to or greater than two.

EXERCISES 15.5A

In Exercises 1–40, simplify each square root.

1. $\sqrt{z^2}$ **2.** $\sqrt{b^6}$ **3.** $\sqrt{x^4}$ **4.** $\sqrt{a^8}$ **5.** $\sqrt{16x^6}$

6. $\sqrt{64w^4}$ **7.** $\sqrt{12}$ **8.** $\sqrt{20}$ **9.** $\sqrt{18}$ **10.** $\sqrt{45}$

11. $\sqrt{8}$ **12.** $\sqrt{32}$ **13.** $\sqrt{24}$ **14.** $\sqrt{54}$ **15.** $\sqrt{x^3}$

16. $\sqrt{y^5}$ **17.** $\sqrt{m^7}$ **18.** $\sqrt{n^9}$ **19.** $\sqrt{h^3k^2}$ **20.** $\sqrt{a^2b^5}$

21. $3\sqrt{28}$ **22.** $2\sqrt{27}$ **23.** $4\sqrt{50}$ **24.** $5\sqrt{40}$ **25.** $\sqrt{75x^3}$

26. $\sqrt{63a^5}$ **27.** $\sqrt{20x^4y^6}$ **28.** $\sqrt{8a^8b^2}$ **29.** $\sqrt{36x^{36}}$ **30.** $\sqrt{64y^{64}}$

31. $\sqrt{20x^2y^3}$ **32.** $\sqrt{8a^4b^7}$ **33.** $\sqrt{16x^5y^3}$ **34.** $\sqrt{4m^7n^5}$ **35.** $\sqrt{27a^3b^4}$

36. $\sqrt{18c^5d^2}$ **37.** $\sqrt{250x^4y^3}$ **38.** $\sqrt{72h^6k^5}$ **39.** $\sqrt{25 + 144}$ **40.** $\sqrt{25} + \sqrt{144}$

In Exercises 41–50, find the products.

41. $\sqrt{3}\sqrt{3}$ **42.** $\sqrt{2}\sqrt{2}$ **43.** $\sqrt{x}\sqrt{x}$ **44.** $\sqrt{y}\sqrt{y}$ **45.** $\sqrt{5x}\sqrt{5x}$

46. $\sqrt{7a}\sqrt{7a}$ **47.** $(\sqrt{5})^2$ **48.** $(\sqrt{x})^2$ **49.** $(\sqrt{6x})^2$ **50.** $(\sqrt{x+2})^2$

15.5B Square Roots with Fractions in the Radicand

Example 9 Does $\sqrt{\frac{4}{9}} = \frac{\sqrt{4}}{\sqrt{9}}$?

Solution

$$\sqrt{\frac{4}{9}} = \frac{2}{3} \quad \text{because } \left(\frac{2}{3}\right)^2 = \frac{4}{9}$$

$$\frac{\sqrt{4}}{\sqrt{9}} = \frac{2}{3}$$

Therefore, $\sqrt{\frac{4}{9}} = \frac{\sqrt{4}}{\sqrt{9}}$. ■

This leads to the third rule of radicals.

Rule 3

SQUARE ROOT OF A QUOTIENT

$$\sqrt{\frac{a}{b}} = \frac{\sqrt{a}}{\sqrt{b}} \qquad (b \neq 0)$$

Example 10 Finding the square root of a fraction

a. $\sqrt{\frac{25}{36}} = \frac{\sqrt{25}}{\sqrt{36}} = \frac{5}{6}$

b. $\sqrt{\frac{x^4}{y^6}} = \frac{\sqrt{x^4}}{\sqrt{y^6}} = \frac{x^2}{y^3}$

c. $\sqrt{\frac{50h^3}{2h}} = \sqrt{\frac{\overset{25}{\cancel{50}}h^3}{\cancel{2}h}} = \sqrt{25h^2} = 5h$

Simplify fraction first

d. $\sqrt{\frac{8}{9}} = \frac{\sqrt{4 \cdot 2}}{\sqrt{9}} = \frac{2\sqrt{2}}{3}$

e. $\sqrt{\frac{1}{5}} = \frac{\sqrt{1}}{\sqrt{5}} = \frac{1}{\sqrt{5}}$ ← This denominator is an irrational number ■

An algebraic fraction is not considered simplified if a square root appears in the denominator.

Rationalizing the Denominator

When the denominator of a fraction is not a rational number, the procedure for changing it into a rational number is called **rationalizing the denominator.**

Example 11 Rationalize the denominator of each of the following fractions.

a. $\frac{1}{\sqrt{5}} = \frac{1}{\sqrt{5}} \cdot \frac{\sqrt{5}}{\sqrt{5}} = \frac{\sqrt{5}}{\sqrt{5}\sqrt{5}} = \frac{\sqrt{5}}{5}$

Denominator *is* now a rational number

Multiplying numerator and denominator by $\sqrt{5}$. Because the value of this fraction is 1, multiplying $\frac{1}{\sqrt{5}}$ by 1 does not change its value.

Denominator is *not* a rational number

b. $\sqrt{\frac{1}{x}} = \frac{\sqrt{1}}{\sqrt{x}} \cdot \frac{\sqrt{x}}{\sqrt{x}} = \frac{\sqrt{x}}{x}$

c. $\sqrt{\frac{9}{7}} = \frac{\sqrt{9}}{\sqrt{7}} = \frac{3}{\sqrt{7}} \cdot \frac{\sqrt{7}}{\sqrt{7}} = \frac{3\sqrt{7}}{7}$

d. $\frac{6}{\sqrt{3}} = \frac{6}{\sqrt{3}} \cdot \frac{\sqrt{3}}{\sqrt{3}} = \frac{\overset{2}{\cancel{6}}\sqrt{3}}{\underset{1}{\cancel{3}}} = 2\sqrt{3}$ ■

> THE SIMPLIFIED FORM OF AN EXPRESSION HAVING SQUARE ROOTS
>
> **1.** No prime factor of a radicand has an exponent equal to or greater than 2.
>
> **2.** No radicand contains a fraction.
>
> **3.** No denominator contains a square root.

EXERCISES 15.5B

Simplify each of the following expressions.

1. $\sqrt{\frac{4}{81}}$ **2.** $\sqrt{\frac{16}{25}}$ **3.** $\sqrt{\frac{9}{49}}$ **4.** $\sqrt{\frac{9}{100}}$ **5.** $\sqrt{\frac{a^2}{b^6}}$

6. $\sqrt{\frac{y^4}{x^2}}$ **7.** $\sqrt{\frac{x^6}{v^8}}$ **8.** $\sqrt{\frac{x^4}{y^6}}$ **9.** $\sqrt{\frac{4x^2}{9}}$ **10.** $\sqrt{\frac{36a^4}{49}}$

11. $\sqrt{\frac{2m^2}{18}}$ **12.** $\sqrt{\frac{20}{5x^2}}$ **13.** $\sqrt{\frac{2x^3}{50x}}$ **14.** $\sqrt{\frac{3a}{27a^5}}$ **15.** $\sqrt{\frac{12}{25}}$

16. $\sqrt{\frac{27}{16}}$ **17.** $\sqrt{\frac{50}{81}}$ **18.** $\sqrt{\frac{18}{49}}$ **19.** $\sqrt{\frac{x^3}{y^4}}$ **20.** $\sqrt{\frac{a^5}{b^2}}$

21. $\sqrt{\frac{1}{2}}$ **22.** $\sqrt{\frac{1}{3}}$ **23.** $\sqrt{\frac{1}{a}}$ **24.** $\sqrt{\frac{1}{y}}$ **25.** $\sqrt{\frac{4}{5}}$

26. $\sqrt{\frac{9}{10}}$ **27.** $\frac{8}{\sqrt{2}}$ **28.** $\frac{10}{\sqrt{5}}$ **29.** $\frac{2x}{\sqrt{x}}$ **30.** $\frac{5mn}{\sqrt{m}}$

15.6 Adding Square Roots

Like Square Roots

Like square roots are square roots having the same radicand.

Example 1 Example of like square roots

a. $3\sqrt{5}, 2\sqrt{5}, -7\sqrt{5}$

b. $2\sqrt{x}, -9\sqrt{x}, 11\sqrt{x}$ ■

Unlike Square Roots

Unlike square roots are square roots having different radicands.

Example 2 Examples of unlike square roots

a. $2\sqrt{15}, -6\sqrt{11}, 8\sqrt{24}$

b. $5\sqrt{y}, 3\sqrt{x}, -4\sqrt{13}$ ■

Adding Like Square Roots Like square roots are added in the same way as any other like things: by adding their coefficients and then multiplying that sum by the like square root.

$$\left.\begin{aligned} 3 \text{ cars} + 2 \text{ cars} &= (3+2) \text{ cars} = 5 \text{ cars} \\ 3c + 2c &= (3+2)c = 5c \\ 3\sqrt{7} + 2\sqrt{7} &= (3+2)\sqrt{7} = 5\sqrt{7} \end{aligned}\right\} \text{Applications of the distributive rule}$$

Example 3 Adding like square roots

a. $5\sqrt{2} + 3\sqrt{2} = (5+3)\sqrt{2} = 8\sqrt{2}$

b. $7\sqrt{x} - 3\sqrt{x} = (7-3)\sqrt{x} = 4\sqrt{x}$ ■

Adding Unlike Square Roots Simplifying *unlike* square roots *sometimes* results in *like* square roots, which can then be added.

TO ADD UNLIKE SQUARE ROOTS

1. Simplify each square root.
2. Combine like square roots by adding their coefficients and then multiplying that sum by the like square root.

Example 4

$$\sqrt{8} + \sqrt{18} =$$
$$\sqrt{4 \cdot 2} + \sqrt{9 \cdot 2} =$$
$$2\sqrt{2} + 3\sqrt{2} = 5\sqrt{2} \quad \blacksquare$$

Example 5

$$\sqrt{12} - \sqrt{27} + \sqrt{75} =$$
$$\sqrt{4 \cdot 3} - \sqrt{9 \cdot 3} + \sqrt{25 \cdot 3} =$$
$$2\sqrt{3} - 3\sqrt{3} + 5\sqrt{3} = 4\sqrt{3} \quad \blacksquare$$

Example 6

$$3\sqrt{20} - \sqrt{45} - \sqrt{50} =$$
$$3 \cdot \sqrt{4 \cdot 5} - \sqrt{9 \cdot 5} - \sqrt{25 \cdot 2} =$$
$$3 \cdot 2\sqrt{5} - 3\sqrt{5} - 5\sqrt{2} =$$
$$6\sqrt{5} - 3\sqrt{5} - 5\sqrt{2} = 3\sqrt{5} - 5\sqrt{2} \quad \blacksquare$$

EXERCISES 15.6

Find the sums and simplify.

1. $2\sqrt{3} + 5\sqrt{3}$
2. $4\sqrt{2} + 3\sqrt{2}$
3. $5\sqrt{5} - \sqrt{5}$
4. $8\sqrt{6} - \sqrt{6}$
5. $7\sqrt{x} + 2\sqrt{x}$
6. $3\sqrt{a} - 2\sqrt{a}$
7. $5\sqrt{7} + 2\sqrt{7} - \sqrt{7}$
8. $6\sqrt{3} - \sqrt{3} + 2\sqrt{3}$
9. $6\sqrt{2} + 2\sqrt{2} - 4\sqrt{3}$
10. $4\sqrt{6} + 3\sqrt{5} - 2\sqrt{6}$
11. $5\sqrt{x} - 2\sqrt{y} + 3\sqrt{x}$
12. $\sqrt{a} + 7\sqrt{a} - 6\sqrt{b}$
13. $\sqrt{50} + \sqrt{8}$
14. $\sqrt{20} + \sqrt{125}$
15. $\sqrt{32} - \sqrt{18}$
16. $\sqrt{63} - \sqrt{28}$
17. $5\sqrt{12} + \sqrt{75}$
18. $3\sqrt{24} + \sqrt{54}$
19. $2\sqrt{48} - 4\sqrt{27}$
20. $2\sqrt{250} - 4\sqrt{90}$
21. $\sqrt{16x} + \sqrt{9x}$
22. $\sqrt{64x} + \sqrt{4x}$
23. $\sqrt{25} - \sqrt{20}$
24. $\sqrt{49} - \sqrt{48}$
25. $\sqrt{75} + 2\sqrt{27} - \sqrt{48}$
26. $\sqrt{18} + 5\sqrt{8} - \sqrt{32}$
27. $\sqrt{45} - 3\sqrt{12} + \sqrt{20}$
28. $\sqrt{54} - 5\sqrt{28} + \sqrt{24}$
29. $\sqrt{64} + \sqrt{50} - 2\sqrt{72}$
30. $\sqrt{36} - \sqrt{150} + 3\sqrt{24}$

15.7 Multiplying Square Roots

In Section 15.5A we used Rule 1 in the following direction.

$$\longrightarrow$$
$$\sqrt{ab} = \sqrt{a}\sqrt{b}$$
$$\sqrt{4 \cdot 3} = \sqrt{4}\sqrt{3}$$
$$= 2\sqrt{3}$$

Using Rule 1 to find the square root of a product

In this section we use Rule 1 in the opposite way.

$$\sqrt{a}\sqrt{b} = \sqrt{ab}$$
$$\sqrt{2}\sqrt{8} = \sqrt{2 \cdot 8}$$
$$= \sqrt{16} = 4$$

Using Rule 1 to find the product of square roots

This means:

Product of square roots = Square root of product
$\sqrt{a}\sqrt{b} = \sqrt{ab}$

Example 1 Multiplying square roots

Products of square roots = Square root of product

a. $\sqrt{2}\sqrt{32} = \sqrt{2 \cdot 32} = \sqrt{64} = 8$

b. $\sqrt{4x}\sqrt{x} = \sqrt{4x \cdot x} = \sqrt{4x^2} = 2x$

c. $\sqrt{3y}\sqrt{6y^2} = \sqrt{3y \cdot 6y^2} = \sqrt{18y^3}$ Multiplying square roots
$= \sqrt{9 \cdot 2 \cdot y^2 \cdot y}$
$= 3y\sqrt{2y}$ Simplifying

d. $2\sqrt{3} \cdot 3\sqrt{8} = (2 \cdot 3)\sqrt{3 \cdot 8} = 6\sqrt{24}$ Multiplying square roots
$= 6\sqrt{4 \cdot 6}$
$= 6 \cdot 2\sqrt{6}$
$= 12\sqrt{6}$ Simplifying ■

Example 2 $\sqrt{2}(3\sqrt{2} + 5)$

$\sqrt{2}(3\sqrt{2} + 5) = \sqrt{2} \cdot 3\sqrt{2} + \sqrt{2} \cdot 5$ Distribute $\sqrt{2}$
$= 3\sqrt{4} + 5\sqrt{2}$
$= 3 \cdot 2 + 5\sqrt{2}$
$= 6 + 5\sqrt{2}$ ■

Example 3 $(2\sqrt{3} - 5)(4\sqrt{3} - 6)$

$(2\sqrt{3} - 5)(4\sqrt{3} - 6)$ Using FOIL

$2\sqrt{3} \cdot 4\sqrt{3} = 8 \cdot 3$ $\quad -20\sqrt{3}$ $\quad -12\sqrt{3}$ $\quad (-5)(-6)$

$= 24 - 32\sqrt{3} + 30$
$= 54 - 32\sqrt{3}$ ■

Example 4 $(3+\sqrt{5})^2$

$$\begin{aligned}(3+\sqrt{5})^2 &= (3)^2 + 2(3)(\sqrt{5}) + (\sqrt{5})^2 \\ &= 9 + 6\sqrt{5} + 5 \\ &= 14 + 6\sqrt{5} \quad \blacksquare\end{aligned}$$

Squaring a binomial

EXERCISES 15.7

Find the products and simplify.

1. $\sqrt{3}\sqrt{12}$
2. $\sqrt{2}\sqrt{8}$
3. $\sqrt{9x}\sqrt{x}$
4. $\sqrt{25y}\sqrt{y}$
5. $\sqrt{2a}\sqrt{18a^3}$
6. $\sqrt{5x^5}\sqrt{20x}$
7. $\sqrt{6}\sqrt{2}$
8. $\sqrt{2}\sqrt{10}$
9. $\sqrt{3x}\sqrt{6x}$
10. $\sqrt{8y}\sqrt{3y^3}$
11. $\sqrt{5a}\sqrt{10a^2}$
12. $\sqrt{3x^2}\sqrt{15x^3}$
13. $3\sqrt{5}\cdot\sqrt{5}$
14. $5\sqrt{2}\cdot\sqrt{2}$
15. $2\sqrt{3}\cdot 4\sqrt{3}$
16. $3\sqrt{5}\cdot 2\sqrt{5}$
17. $\sqrt{12}\cdot 5\sqrt{2}$
18. $\sqrt{2}\cdot 3\sqrt{14}$
19. $3\sqrt{8}\cdot 2\sqrt{5}$
20. $2\sqrt{18}\cdot 5\sqrt{3}$
21. $\sqrt{2}(\sqrt{2}+1)$
22. $\sqrt{3}(\sqrt{3}-1)$
23. $\sqrt{5}(2\sqrt{5}+3)$
24. $\sqrt{7}(3\sqrt{7}+2)$
25. $\sqrt{x}(\sqrt{x}-2)$
26. $\sqrt{y}(4-\sqrt{y})$
27. $\sqrt{6}(\sqrt{6}-\sqrt{2})$
28. $\sqrt{2}(\sqrt{10}+\sqrt{2})$
29. $(\sqrt{7}+2)(\sqrt{7}+3)$
30. $(\sqrt{3}+2)(\sqrt{3}+4)$
31. $(3-2\sqrt{5})(4-\sqrt{5})$
32. $(7-3\sqrt{2})(1-\sqrt{2})$
33. $(5\sqrt{3}-2)(2\sqrt{3}+4)$
34. $(2\sqrt{5}+4)(3\sqrt{5}-3)$
35. $(3-\sqrt{2})(3+\sqrt{2})$
36. $(2+\sqrt{7})(2-\sqrt{7})$
37. $(\sqrt{5}+\sqrt{3})(\sqrt{5}-\sqrt{3})$
38. $(\sqrt{6}-\sqrt{2})(\sqrt{6}+\sqrt{2})$
39. $(\sqrt{3}+4)^2$
40. $(2-\sqrt{5})^2$
41. $(3-\sqrt{x})^2$
42. $(\sqrt{y}+5)^2$
43. $(\sqrt{5}+\sqrt{3})^2$
44. $(\sqrt{6}+\sqrt{2})^2$

15.8 Dividing Square Roots

In Section 15.5B we used Rule 3 in the following direction:

$$\longrightarrow$$

$$\sqrt{\frac{a}{b}} = \frac{\sqrt{a}}{\sqrt{b}}$$

Using Rule 3 to find the square root of a quotient

$$\sqrt{\frac{4}{9}} = \frac{\sqrt{4}}{\sqrt{9}} = \frac{2}{3}$$

In this section we use Rule 3 in the opposite direction:

$$\longrightarrow$$

$$\frac{\sqrt{a}}{\sqrt{b}} = \sqrt{\frac{a}{b}}$$

Using Rule 3 to find the quotient of square roots

$$\frac{\sqrt{8}}{\sqrt{2}} = \sqrt{\frac{8}{2}} = \sqrt{4} = 2$$

This means:

$$\text{Quotient of square roots} = \text{Square root of quotient}$$

$$\frac{\sqrt{a}}{\sqrt{b}} = \sqrt{\frac{a}{b}} \qquad (b \neq 0)$$

Example 1 Dividing square roots

Quotient of square roots = Square root of quotient

a. $\dfrac{\sqrt{32}}{\sqrt{2}} = \sqrt{\dfrac{32}{2}} = \sqrt{16} = 4$

b. $\dfrac{\sqrt{x^5}}{\sqrt{x}} = \sqrt{\dfrac{x^5}{x}} = \sqrt{x^4} = x^2$

c. $\dfrac{\sqrt{150}}{\sqrt{3}} = \sqrt{\dfrac{150}{3}} = \sqrt{50} = \sqrt{25 \cdot 2} = 5\sqrt{2}$

d. $\dfrac{\sqrt{10}}{\sqrt{15}} = \sqrt{\dfrac{10}{15}} = \sqrt{\dfrac{2}{3}} = \dfrac{\sqrt{2}}{\sqrt{3}} \cdot \dfrac{\sqrt{3}}{\sqrt{3}} = \dfrac{\sqrt{6}}{3}$

Multiply by $\frac{\sqrt{3}}{\sqrt{3}}$ to rationalize the denominator

e. $\dfrac{2}{\sqrt{20}} = \dfrac{2}{\sqrt{20}} \cdot \dfrac{\sqrt{5}}{\sqrt{5}} = \dfrac{2\sqrt{5}}{\sqrt{100}} = \dfrac{\overset{1}{\cancel{2}}\sqrt{5}}{\underset{5}{\cancel{10}}} = \dfrac{\sqrt{5}}{5}$

Multiply by $\frac{\sqrt{5}}{\sqrt{5}}$ to make the denominator a perfect square ■

NOTE In Example 1(e), we could have simplified the denominator first.

$$\frac{2}{\sqrt{20}} = \frac{2}{\sqrt{4 \cdot 5}} = \frac{\overset{1}{\cancel{2}}}{\underset{1}{\cancel{2}}\sqrt{5}} = \frac{1}{\sqrt{5}} \cdot \frac{\sqrt{5}}{\sqrt{5}} = \frac{\sqrt{5}}{5}$$ ☑

Rationalizing a Binomial Denominator That Contains Square Roots

Conjugate

The *conjugate* of a binomial is a binomial that has the same two terms with the sign of the second term changed. The conjugate of $a + b$ is $a - b$.

Example 2 Examples of conjugates that contain square roots

a. The conjugate of $1 + \sqrt{3}$ is $1 - \sqrt{3}$.

b. The conjugate of $\sqrt{5} - \sqrt{3}$ is $\sqrt{5} + \sqrt{3}$.

c. The conjugate of $\sqrt{x} + 2$ is $\sqrt{x} - 2$. ■

The product of a binomial containing square roots, and its conjugate, is a rational number. For example:

$$(1 - \sqrt{2})(1 + \sqrt{2}) = (1)^2 - (\sqrt{2})^2 = 1 - 2 = -1 \quad \text{A rational number}$$

Because of this fact, the following procedure should be used when a binomial denominator contains a square root.

Rationalizing Binomial Denominators

TO RATIONALIZE A BINOMIAL DENOMINATOR THAT CONTAINS SQUARE ROOTS

Multiply the numerator and the denominator by the conjugate of the denominator:

$$\frac{a}{b + \sqrt{c}} \cdot \frac{b - \sqrt{c}}{b - \sqrt{c}} = \frac{a(b - \sqrt{c})}{b^2 - c}$$

Example 3 Rationalizing binomial denominators containing square roots

a. $$\frac{2}{1 + \sqrt{3}} = \frac{2}{1 + \sqrt{3}} \cdot \frac{1 - \sqrt{3}}{1 - \sqrt{3}} = \frac{2(1 - \sqrt{3})}{1 - 3} = \frac{\overset{1}{\cancel{2}}(1 - \sqrt{3})}{\underset{-1}{\cancel{-2}}} = \sqrt{3} - 1$$

Multiply numerator and denominator by $1 - \sqrt{3}$ (the conjugate of the denominator $1 + \sqrt{3}$)—because the value of this fraction is 1, multiplying $\frac{2}{1 + \sqrt{3}}$ by 1 does not change its value.

b. $$\frac{6}{\sqrt{5} - \sqrt{3}} = \frac{6}{\sqrt{5} - \sqrt{3}} \cdot \frac{\sqrt{5} + \sqrt{3}}{\sqrt{5} + \sqrt{3}} = \frac{6(\sqrt{5} + \sqrt{3})}{5 - 3} = \frac{\overset{3}{\cancel{6}}(\sqrt{5} + \sqrt{3})}{\underset{1}{\cancel{2}}}$$

$$= 3\sqrt{5} + 3\sqrt{3}$$

Multiply numerator and denominator by $\sqrt{5} + \sqrt{3}$ (the conjugate of the denominator $\sqrt{5} - \sqrt{3}$) ■

EXERCISES 15.8

In Exercises 1–24, find the quotients and simplify.

1. $\frac{\sqrt{20}}{\sqrt{5}}$ **2.** $\frac{\sqrt{27}}{\sqrt{3}}$ **3.** $\frac{\sqrt{7}}{\sqrt{28}}$ **4.** $\frac{\sqrt{2}}{\sqrt{50}}$ **5.** $\frac{\sqrt{x^3}}{\sqrt{x}}$ **6.** $\frac{\sqrt{y^7}}{\sqrt{y^3}}$

7. $\frac{\sqrt{a}}{\sqrt{a^5}}$ **8.** $\frac{\sqrt{x^3}}{\sqrt{x^9}}$ **9.** $\frac{\sqrt{18x^5}}{\sqrt{2x}}$ **10.** $\frac{\sqrt{72a^3}}{\sqrt{2a}}$ **11.** $\frac{\sqrt{40}}{\sqrt{5}}$ **12.** $\frac{\sqrt{60}}{\sqrt{3}}$

13. $\frac{\sqrt{24}}{\sqrt{2}}$ **14.** $\frac{\sqrt{48}}{\sqrt{6}}$ **15.** $\frac{\sqrt{2}}{\sqrt{6}}$ **16.** $\frac{\sqrt{3}}{\sqrt{15}}$ **17.** $\frac{\sqrt{6}}{\sqrt{10}}$ **18.** $\frac{\sqrt{20}}{\sqrt{50}}$

19. $\frac{\sqrt{14}}{\sqrt{21}}$ **20.** $\frac{\sqrt{6}}{\sqrt{21}}$ **21.** $\frac{2}{\sqrt{8}}$ **22.** $\frac{5}{\sqrt{20}}$ **23.** $\frac{3}{\sqrt{12}}$ **24.** $\frac{2}{\sqrt{32}}$

In Exercises 25–28, write the conjugate for each expression.

25. $\sqrt{2} - 1$ **26.** $3 - \sqrt{3}$ **27.** $\sqrt{5} + \sqrt{2}$ **28.** $\sqrt{x} + 2$

In Exercises 29–38, rationalize the denominators and simplify.

29. $\frac{3}{\sqrt{2} - 1}$ **30.** $\frac{5}{\sqrt{2} + 1}$ **31.** $\frac{6}{\sqrt{5} + \sqrt{2}}$ **32.** $\frac{12}{\sqrt{6} - \sqrt{2}}$ **33.** $\frac{4}{2 + \sqrt{2}}$

34. $\frac{6}{3 - \sqrt{3}}$ **35.** $\frac{8}{\sqrt{6} - 2}$ **36.** $\frac{9}{3 + \sqrt{5}}$ **37.** $\frac{x - 4}{\sqrt{x} + 2}$ **38.** $\frac{y - 9}{\sqrt{y} - 3}$

15.9 Radical Equations

A *radical equation* is an equation in which the unknown letter appears in a radicand. In this text we will only consider radical equations with square roots.

Example 1 Radical equations

a. $\sqrt{x} = 7$

b. $\sqrt{x + 2} = 3$

c. $\sqrt{2x - 3} = \sqrt{x} + 5$ ■

> If two numbers are equal, *then* their squares are equal.
>
> If $a = b$
>
> then $a^2 = b^2$

Example 2 Using the fact given in the rule just stated to remove the square root symbol

a. If these are equal: $\sqrt{x} = 3$
their squares are equal. $(\sqrt{x})^2 = (3)^2$
$x = 9$

b. If these are equal: $\sqrt{x - 1} = 6$
their squares are equal. $(\sqrt{x - 1})^2 = (6)^2$
$x - 1 = 36$
$x = 37$ ■

> **TO SOLVE A RADICAL EQUATION WITH SQUARE ROOTS**
>
> 1. Arrange the terms so that the term with a radical is by itself on one side of the equation.
> 2. Square both sides of the equation.
> 3. Collect like terms.
> 4. Solve the resulting equation for the variable.
>
> *Check* apparent solutions in the original equation.
>
> (Extra answers may occur due to the squaring process. See Example 5.)

Example 3 Solve $\sqrt{x} = 7$.

Solution

$$\sqrt{x} = 7$$
$$(\sqrt{x})^2 = (7)^2$$
$$x = 49$$

Check

$$\sqrt{x} = 7$$
$$\sqrt{49} \stackrel{?}{=} 7$$
$$7 = 7 \quad \blacksquare$$

Example 4 Solve $\sqrt{x+2} = 3$.

Solution

$$\sqrt{x+2} = 3$$
$$(\sqrt{x+2})^2 = (3)^2$$
$$x + 2 = 9$$
$$x = 7$$

Check

$$\sqrt{x+2} = 3$$
$$\sqrt{7+2} \stackrel{?}{=} 3$$
$$\sqrt{9} \stackrel{?}{=} 3$$
$$3 = 3 \quad \blacksquare$$

A WORD OF CAUTION Any apparent solution *must* be checked by substituting it into the original equation because the squaring process may yield an extra answer, called an *extraneous root*, that does not satisfy the original equation (Example 5). ☑

Example 5 Solve $\sqrt{2x+1} + 1 = x$.

Solution

$$\sqrt{2x+1} = x - 1 \qquad \text{Isolate the radical term}$$
$$(\sqrt{2x+1})^2 = (x-1)^2$$

When squaring $(x - 1)$, do not forget this middle term ($-2x$)

$$2x + 1 = x^2 - 2x + 1$$
$$0 = x^2 - 4x$$
$$0 = x(x - 4)$$
$$x = 0 \quad \Big| \quad x - 4 = 0$$
$$x = 4$$

Check for x = 0:

$$\sqrt{2x+1} + 1 = x$$
$$\sqrt{2(0)+1} + 1 \stackrel{?}{=} 0$$
$$\sqrt{1} + 1 \stackrel{?}{=} 0$$
$$1 + 1 \stackrel{?}{=} 0$$
$$2 \neq 0$$

The symbol $\sqrt{1}$ *always* stands for the *principal* square root of 1, which is 1 (*not* −1)

Therefore, *0 is not a solution* of $\sqrt{2x+1}+1=x$ because it does not satisfy the equation (extraneous root).

Check for x = 4:

$$\sqrt{2x+1}+1=x$$
$$\sqrt{2(4)+1}+1\stackrel{?}{=}4$$
$$\sqrt{9}+1\stackrel{?}{=}4$$
$$3+1\stackrel{?}{=}4$$
$$4=4$$

Therefore, *4 is a solution* because it does satisfy the equation. ■

EXERCISES 15.9

Solve each equation.

1. $\sqrt{x}=5$
2. $\sqrt{x}=10$
3. $\sqrt{x}=8$
4. $\sqrt{5x}=10$
5. $\sqrt{2x}=4$
6. $\sqrt{3x}=6$
7. $\sqrt{x-3}=2$
8. $\sqrt{x+4}=6$
9. $\sqrt{x-6}=3$
10. $\sqrt{6x+1}=5$
11. $\sqrt{2x+1}=9$
12. $\sqrt{5x-4}=4$
13. $\sqrt{3x+1}=5$
14. $\sqrt{7x+8}=6$
15. $\sqrt{9x-5}=7$
16. $\sqrt{9-2x}=\sqrt{5x-12}$
17. $\sqrt{x+1}=\sqrt{2x-7}$
18. $\sqrt{3x-2}=\sqrt{x+4}$
19. $\sqrt{3x-2}=x$
20. $\sqrt{5x-6}=x$
21. $x=\sqrt{3x+10}$
22. $\sqrt{3x+2}=3x$
23. $\sqrt{4x-1}=2x$
24. $\sqrt{6x-1}=3x$
25. $\sqrt{x-3}+5=x$
26. $\sqrt{4x+5}+5=2x$
27. $2x=\sqrt{2x+3}+3$
28. $\sqrt{x-6}+8=x$
29. $x=\sqrt{2x+4}+2$
30. $\sqrt{7-x}+x=1$

15.10 Chapter Summary

THE RULES OF EXPONENTS

Rule	Condition	Section
1. $x^a x^b = x^{a+b}$		(15.1)
2. $(x^a)^b = x^{ab}$		(15.1)
3. $\dfrac{x^a}{x^b} = x^{a-b}$	$x \neq 0$	(15.1)
4. $x^0 = 1$	$x \neq 0$	(15.1)
5a. $x^{-n} = \dfrac{1}{x^n}$	$x \neq 0$	(15.1)
5b. $\dfrac{1}{x^{-n}} = x^n$	$x \neq 0$	(15.1)
6. $(xy)^n = x^n y^n$		(15.2)
7. $\left(\dfrac{x}{y}\right)^n = \dfrac{x^n}{y^n}$	$y \neq 0$	(15.2)
8. $\left(\dfrac{x^a y^b}{z^c}\right)^n = \dfrac{x^{an} y^{bn}}{z^{cn}}$	$z \neq 0$	(15.2)

Simplifying Expressions Having Exponents 15.1

An expression having exponents is considered simplified when each different base appears only once, and its exponent is a single integer.

Scientific Notation 15.3

To write a number in scientific notation, place the decimal point after the first nonzero digit, then multiply by the appropriate power of 10. The exponent in the power of 10 tells how many places (and the direction) to move the decimal point back to its original position.

Square Roots 15.4

The square root of a number N is a number which, when squared, gives N. Every positive number has both a positive and a negative square root. The positive square root is called the *principal square root.* The principal square root of N is written $\sqrt{N}$. The negative square root of N is written $-\sqrt{N}$.

THE RULES FOR SQUARE ROOTS

1. $\sqrt{a \cdot b} = \sqrt{a} \cdot \sqrt{b}$ (15.5A)
2. $\sqrt{a} \cdot \sqrt{a} = a$ (15.5A)
3. $\sqrt{\frac{a}{b}} = \frac{\sqrt{a}}{\sqrt{b}}$ (15.5B)

Simplifying Expressions Having Square Roots 15.5

1. No prime factor of a radicand has an exponent equal to or greater than 2.
2. No radicand contains a fraction.
3. No denominator contains a square root.

To Add Square Roots 15.6

1. Simplify each square root.
2. Combine like square roots by adding their coefficients and then multiplying that sum by the like square root.

To Multiply Square Roots 15.7

Use Rule 1, $\sqrt{a}\sqrt{b} = \sqrt{ab}$. Simplify the results.

To Divide Square Roots 15.8

Use Rule 3, $\frac{\sqrt{a}}{\sqrt{b}} = \sqrt{\frac{a}{b}}$. Simplify the results.

To Rationalize a Denominator 15.5B 15.8

Monomial Denominator $\quad \frac{1}{\sqrt{a}} = \frac{1}{\sqrt{a}} \cdot \frac{\sqrt{a}}{\sqrt{a}} = \frac{\sqrt{a}}{a}$

Binomial Denominator $\quad \frac{a}{b+\sqrt{c}} = \frac{a}{b+\sqrt{c}} \cdot \frac{b-\sqrt{c}}{b-\sqrt{c}} = \frac{a(b-\sqrt{c})}{b^2-c}$

To Solve a Radical Equation 15.9

1. Arrange the terms so that the term with a radical is by itself on one side of the equation.
2. Square both sides of the equation.
3. Collect like terms.
4. Solve the resulting equation for the variable.
5. *Check* apparent solutions in the original equation. (Extra answers may occur due to the squaring process.)

Review Exercises 15.10

In Exercises 1–28, simplify each expression. Write the answers using only positive exponents.

1. $x^4 \cdot x^7$ **2.** $a^5 \cdot a^{-3}$ **3.** $c^{-5} \cdot d^0$ **4.** $\frac{p^5}{p^2}$

5. $\frac{x^{-4}}{x^5}$ **6.** $\frac{m^0}{m^{-3}}$ **7.** $\frac{10^6}{10^2}$ **8.** $\frac{10^4}{10^{-3}}$

9. $(x^2)^5$ **10.** $(p^{-3})^5$ **11.** $(m^{-4})^{-2}$ **12.** $(h^0)^{-4}$

13. $(x^2y^3)^4$ **14.** $(p^{-1}r^3)^{-2}$ **15.** $2a^{-3}$ **16.** $(3x^4)^2$

17. $(4b^3)^{-2}$ **18.** $(10)^{-2}$ **19.** $\left(\frac{x^2y^3}{z^4}\right)^5$ **20.** $\left(\frac{x^{-3}}{y^0z^2}\right)^{-4}$

21. $\left(\frac{u^{-5}}{v^2w^{-4}}\right)^3$ **22.** $(5a^3b^{-4})^{-2}$ **23.** $\left(\frac{4h^2}{ij^{-2}}\right)^{-3}$ **24.** $\left(\frac{x^{10}y^5}{x^5y}\right)^3$

25. $\left(\frac{6x^{-5}y^8}{3x^2y^{-4}}\right)^0$ **26.** $x^{3d} \cdot x^d$ **27.** $(x^{4a})^{-2}$ **28.** $\frac{6^{2x}}{6^{-x}}$

In Exercises 29–31, write each expression without fractions. Use negative exponents if necessary.

29. $\frac{x^2}{y^3}$ **30.** $\frac{m^2}{n^{-3}}$ **31.** $\frac{a^3}{b^2c^5}$

In Exercises 32–37, evaluate each expression.

32. 4^{-2} **33.** $(3^{-1})^3$ **34.** $\frac{2^0}{2^{-3}}$

35. $8^0 \cdot 10^{-2}$ **36.** $\frac{(-8)^2}{-8^2}$ **37.** $\left(\frac{10^{-4} \cdot 10}{10^{-2}}\right)^5$

In Exercises 38–40, write the following numbers in scientific notation.

38. 0.000225 **39.** 960,000 **40.** $\frac{1}{200}$

In Exercises 41–43, write the following numbers in common notation.

41. 7.8×10^3 **42.** 4.06×10^{-5} **43.** 1.207×10^{-2}

In Exercises 44–46, perform the indicated operations and write your answers in scientific notation.

44. $1{,}600 \times 0.00006$ **45.** $\frac{78{,}000}{0.026}$ **46.** $\frac{0.0035 \times 540}{0.00027}$

47. Which of the numbers $\sqrt{3}$, $2\frac{1}{2}$, $2\sqrt{5}$, 3.6, $\sqrt{5}$, $\frac{5}{2}$

a. are irrational numbers? b. have like square roots?

48. What is the radicand in each expression?

a. $\sqrt{9x}$ b. $\sqrt{\frac{1}{2}}$ c. $\sqrt{x-5}$

In Exercises 49–68, simplify each expression.

49. $\sqrt{81}$ **50.** $\sqrt{48}$ **51.** $\sqrt{2}\sqrt{32}$ **52.** $\sqrt{a^6}$

53. $\sqrt{x^3}$ **54.** $\sqrt{16x^2y^4}$ **55.** $\sqrt{a^3b^5}$ **56.** $3\sqrt{5}\cdot 2\sqrt{10}$

57. $3\sqrt{5}+\sqrt{5}$ **58.** $\sqrt{\frac{18x}{2x^3}}$ **59.** $\sqrt{18}-\sqrt{8}$ **60.** $\frac{\sqrt{6}}{\sqrt{15}}$

61. $\frac{6}{\sqrt{3}}$ **62.** $3\sqrt{12}-\sqrt{75}$ **63.** $\sqrt{2}(5\sqrt{2}+3)$

64. $(\sqrt{5}+3)(\sqrt{5}-3)$ **65.** $(\sqrt{3}+2)^2$ **66.** $(3\sqrt{2}+1)(2\sqrt{2}-1)$

67. $\frac{8}{\sqrt{3}-2}$ **68.** $\frac{10}{\sqrt{6}-2}$

In Exercises 69–74, solve each equation.

69. $\sqrt{x}=4$ **70.** $\sqrt{3a}=6$

71. $\sqrt{2x-1}=5$ **72.** $\sqrt{5a-4}=\sqrt{3a+2}$

73. $\sqrt{7x-6}=x$ **74.** $\sqrt{2x-1}+2=x$

Chapter 15 Diagnostic Test

Allow yourself about one hour to do these problems. Complete solutions for every problem, together with section references, are given in the answer section at the end of the book.

In Problems 1–6, simplify each expression. Write your answers using only positive exponents.

1. $a^{-6} \cdot a^2$

2. $(a^{-3}b)^2$

3. $x^{3a} \cdot x^{2a}$

4. $\dfrac{x^5}{x^{-2}}$

5. $\left(\dfrac{3x^{-4}}{y^2}\right)^2$

6. $\left(\dfrac{x^5y}{x^2y^3}\right)^3$

In Problems 7–9, evaluate each expression.

7. $(10^{-3})^2$

8. $\dfrac{2^{-2} \cdot 2^3}{2^{-4}}$

9. $\dfrac{-4^2}{(-4)^2}$

10. Write $\dfrac{h^{-2}}{k^{-3}h^{-4}}$ without fractions, using negative exponents if necessary.

11. Write each of the following in scientific notation.
a. 723,000 b. 0.0048

12. Perform the indicated operation and write your answer in scientific notation.

$$\frac{4{,}200}{0.00014}$$

In Problems 13–23, perform the indicated operations and simplify.

13. $\sqrt{2}\sqrt{18x^2}$

14. $\sqrt{3}(2\sqrt{3} - 5)$

15. $\sqrt{72}$

16. $\sqrt{\dfrac{18}{2m^2}}$

17. $\dfrac{\sqrt{16}}{\sqrt{50}}$

18. $\dfrac{4}{\sqrt{3} + 1}$

19. $\sqrt{12x^4y^3}$

20. $2\sqrt{20} + \sqrt{45}$

21. $(3\sqrt{2} + 5)(2\sqrt{2} + 1)$

22. $\sqrt{\dfrac{4}{5}}$

23. $\sqrt{28} + \sqrt{75} - \sqrt{27}$

In Problems 24 and 25, solve each equation.

24. $\sqrt{4x + 5} = 5$

25. $\sqrt{5x - 6} = x$

16 Geometry

In this chapter we consider some of the applications of arithmetic and algebra to the more common geometric figures. We live in *rectangular* rooms, we use *circular* plates, our roofs have *triangular* shapes, much of our food comes in *cylindrical* cans, we play with *spherical* basketballs, volleyballs, and so on.

16.1 Lines, Angles, and Polygons

In geometry, the terms *point*, *line*, and *plane* are undefined. We will give meaning to these terms by means of a description.

Point

A **point** indicates a position or location. A point has no size. It is usually represented by a dot and named by a capital letter.

• B

• A

Points: A and B

Line

A **line** (or straight line) is a set of points that has no width but extends forever in opposite directions. Two points determine a line. A line is named by identifying any two points on the line or by labeling it with a lowercase letter.

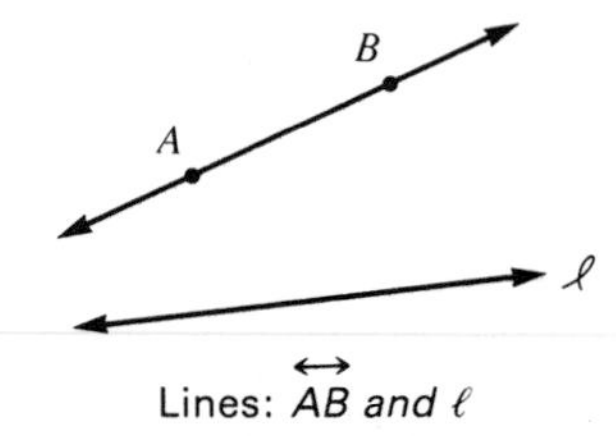

Lines: $\overleftrightarrow{AB}$ and ℓ

Line Segment

A **line segment** is a part of a line that lies between two points. A line segment is named by giving its two endpoints.

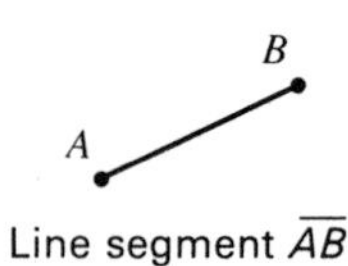

Line segment $\overline{AB}$

Ray

A **ray** is a portion of a line that has one endpoint and continues forever in one direction. A ray is named by giving its endpoint first, then any other point on the ray.

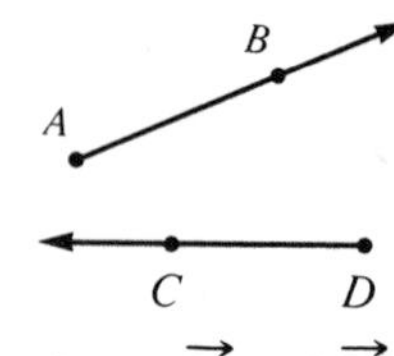

Rays: $\overrightarrow{AB}$ and $\overrightarrow{DC}$

Angle

Vertex

An **angle** is formed by two rays with a common endpoint. The rays are the sides of the angle, and the common point is called the **vertex**. The symbol for angle is $\angle$. Angles may be named by a single letter (the vertex) or by three letters. The middle letter is always the vertex (see figure at right). We may also label the angle with a number.

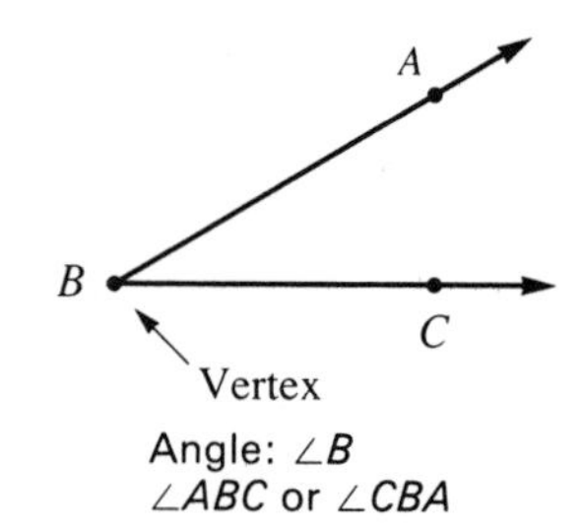

Angle: $\angle B$
$\angle ABC$ or $\angle CBA$

Adjacent Angles

When two angles have the same vertex and a common side between them, they are called **adjacent angles**.

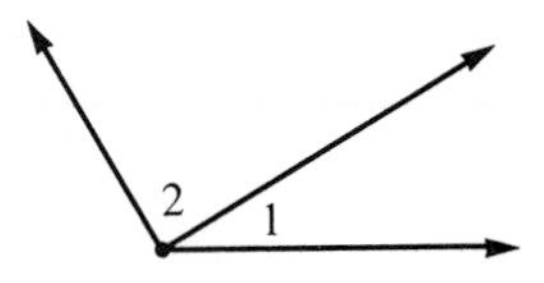

Adjacent angles: $\angle 1$ and $\angle 2$

Measuring Angles Angles may be measured in degrees, °. One revolution measures 360°.

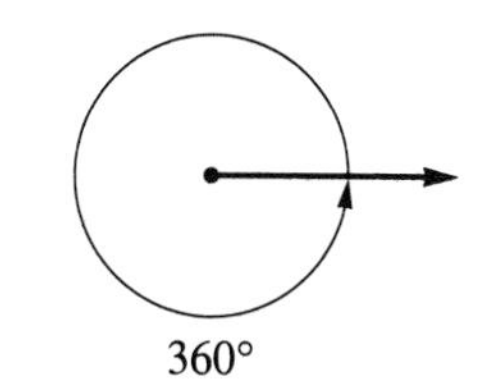

Straight Angle

A **straight angle** $\left(\frac{1}{2} \text{ of a revolution}\right)$ measures 180°.

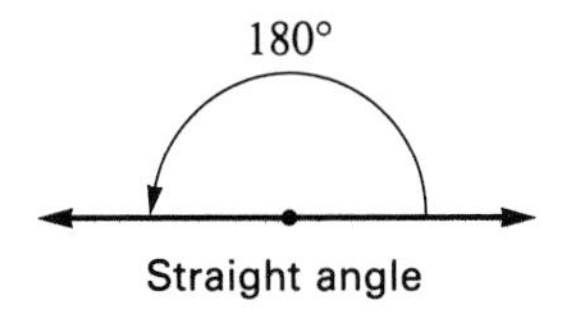

Straight angle

Right Angle

A **right angle** $\left(\frac{1}{4} \text{ of a revolution}\right)$ measures 90°. The square corner denotes a right angle.

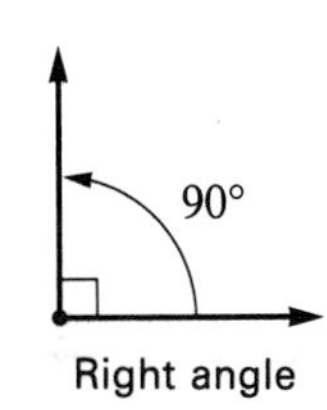

Right angle

Acute Angle

An angle that measures less than 90° is called an **acute angle**.

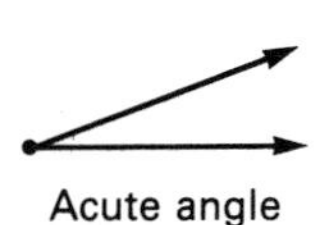
Acute angle

Obtuse Angle

An angle that measures more than 90° and less than 180° is called an **obtuse angle**.

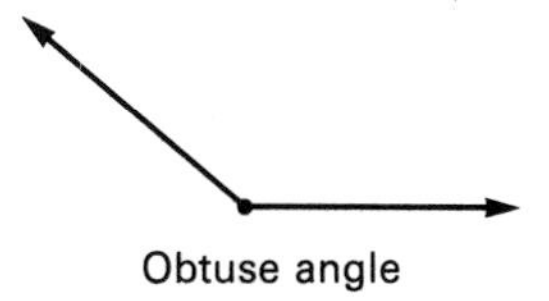
Obtuse angle

Example 1 Classify each angle as acute, right, obtuse, or straight.

a. $\angle A$ acute
b. $\angle DBC$ right
c. $\angle ABC$ obtuse
d. $\angle C$ acute
e. $\angle CDA$ straight
f. $\angle BDA$ obtuse ■

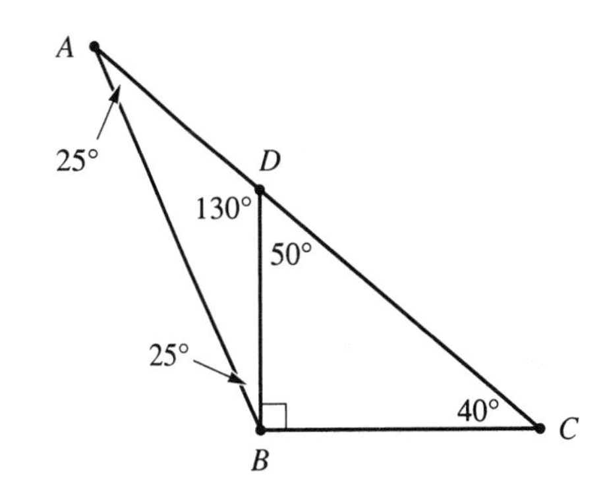

Complementary Angles

Two angles are **complementary** if their sum is 90°.

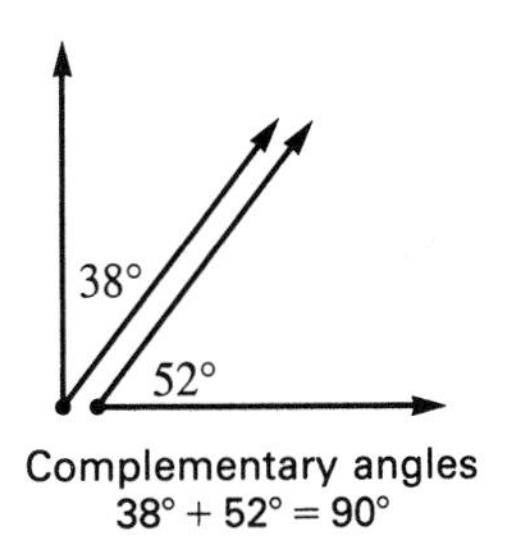

Complementary angles
38° + 52° = 90°

Supplementary Angles

Two angles are **supplementary** if their sum is 180°.

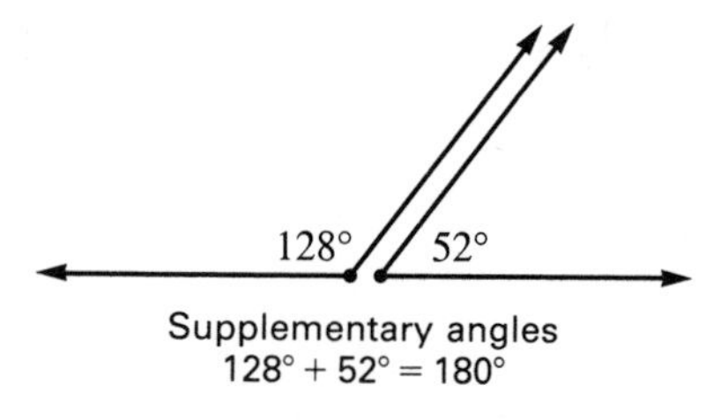

Supplementary angles
$128° + 52° = 180°$

Example 2 Find $\angle 1$ if $\angle 2$ is 67°.

Solution $\angle 1$ and $\angle 2$ form a straight angle. Therefore, their sum is 180° and they are supplementary.

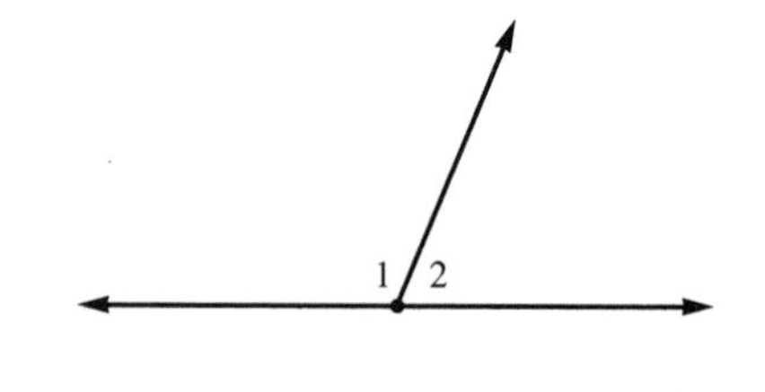

$$\begin{aligned} \angle 1 + \angle 2 &= 180 \\ \angle 1 + 67 &= 180 \\ -67 &= -67 \\ \hline \angle 1 &= 113° \end{aligned}$$ ■

Example 3 Find two complementary angles if one angle is 20° more than the other angle.

Solution Let $x =$ one angle

$x + 20 =$ other angle

Because the two angles are complementary, their sum is 90°.

$$\begin{aligned} x + x + 20 &= 90 \\ 2x + 20 &= 90 \\ 2x &= 70 \\ x &= 35 \\ x + 20 &= 55 \end{aligned}$$

Therefore, the angles are 35° and 55°. ■

Lines

Intersecting Lines

Two lines **intersect** if they have only one point in common. The intersecting lines form four angles.

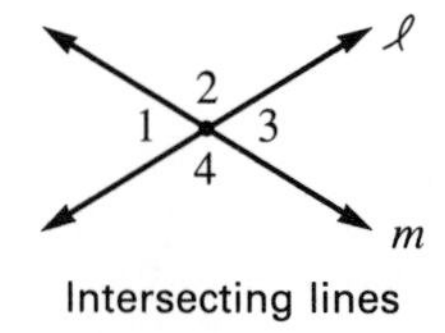

Intersecting lines

Vertical Angles

If two lines intersect, the nonadjacent (or opposite) angles are called **vertical angles.** $\angle 1$ and $\angle 3$ are vertical angles. Also, $\angle 2$ and $\angle 4$ are vertical angles.

An important geometric theorem is stated in the following box.

> When two lines intersect, the vertical angles are equal.
>
> 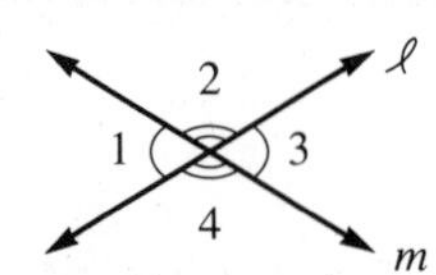
>
>
> $\angle 1 = \angle 3$
>
> $\angle 2 = \angle 4$

Perpendicular Lines

If two intersecting lines form a right angle, the lines are **perpendicular**. The symbol for perpendicular is $\perp$.

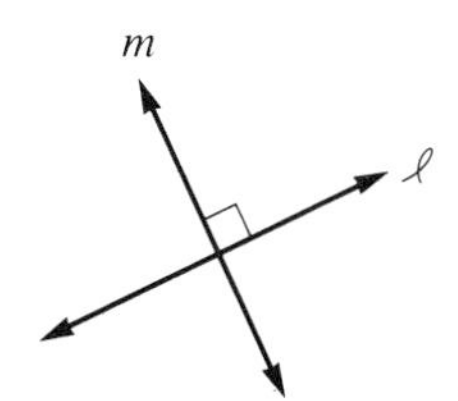

Perpendicular lines
$\ell \perp m$

Parallel Lines

If two lines in a plane do not intersect, even when they are extended, the lines are said to be **parallel**. The symbol for parallel is $\|$.

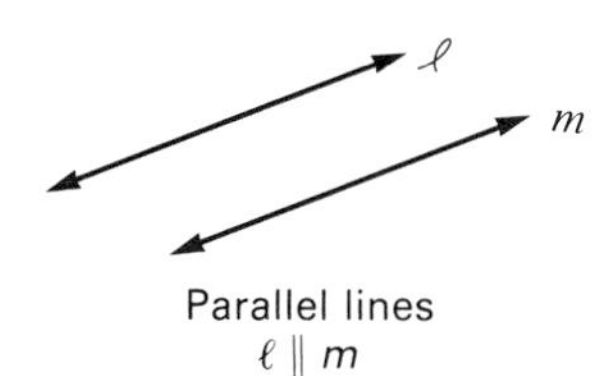

Parallel lines
$\ell \| m$

Transversal

A line that crosses two or more lines is called a **transversal**. The eight angles formed in the figure on the right have special names.

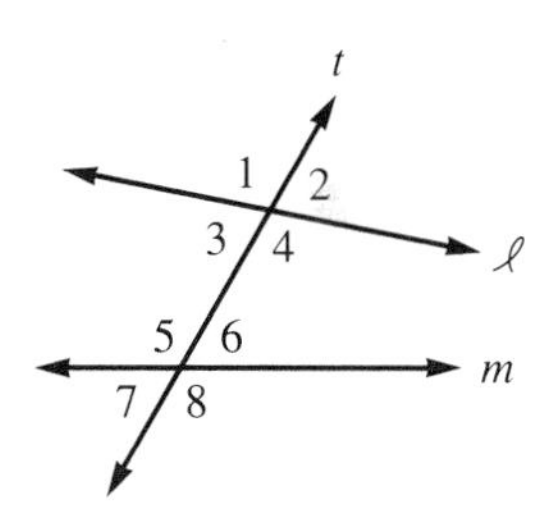

Line t is the transversal of lines ℓ and m

Alternate Interior Angles

Alternate interior angles are nonadjacent angles on different sides of the transversal and between the two lines ℓ and m. $\angle 3$ and $\angle 6$ are alternate interior angles. Another pair of alternate interior angles is $\angle 4$ and $\angle 5$.

Alternate Exterior Angles

Alternate exterior angles are nonadjacent angles on different sides of the transversal and outside of the two lines ℓ and m. $\angle 1$ and $\angle 8$ are alternate exterior angles. Another pair of alternate exterior angles is $\angle 2$ and $\angle 7$.

Corresponding Angles

Corresponding angles are nonadjacent angles on the same side of the transversal. One is an interior angle, and the other is an exterior angle. $\angle 2$ and $\angle 6$ are corresponding angles. Other pairs of corresponding angles are: $\angle 1$ and $\angle 5$, $\angle 4$ and $\angle 8$, $\angle 3$ and $\angle 7$.

The following box lists three properties of parallel lines.

> If two parallel lines are cut by a transversal, then:
>
> a. the alternate interior angles are equal
>
> b. the alternate exterior angles are equal
>
> c. the corresponding angles are equal

NOTE The converse of the above properties is also true. This means: If the alternate interior angles are equal, then the lines are parallel. ✓

Example 4 Given $\ell_1 \parallel \ell_2$, find

a. $\angle 1$,

b. $\angle 2$

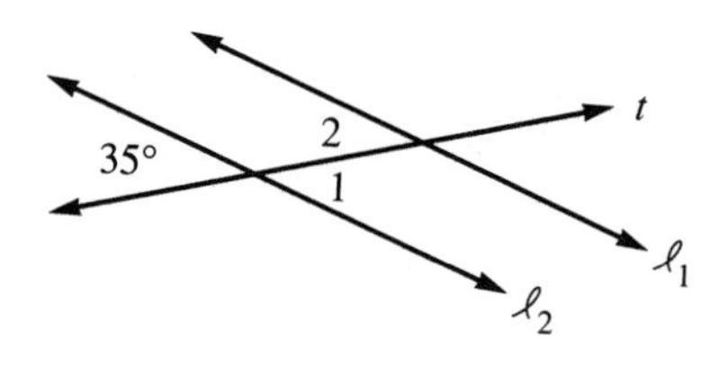

Solution

a. Vertical angles are equal.
$\angle 1 = 35°$

b. Alternate interior angles are equal.
$\angle 2 = \angle 1 = 35°$ ■

Example 5 Given $\overleftrightarrow{AB} \parallel \overleftrightarrow{CD}$ and $\angle ABD = 140°$, find

a. $\angle CDE$

b. $\angle CDB$

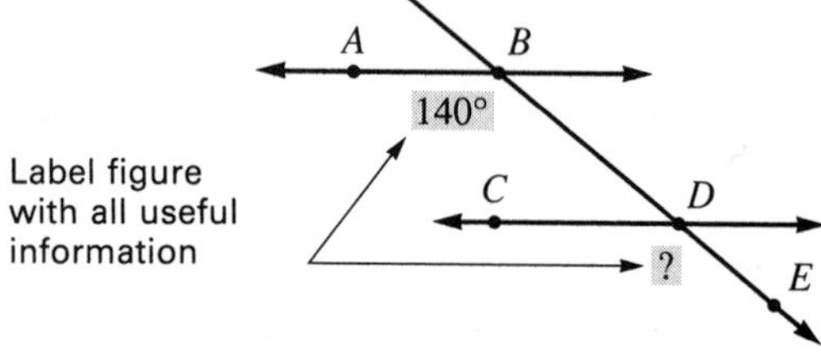

Solution

a. Corresponding angles are equal.

$\angle CDE = \angle ABD = 140°$

b. $\angle CDE$ and $\angle CDB$ form a straight angle, so their sum is 180°.

$$\angle CDE + \angle CDB = 180$$

$$140 + \angle CDB = 180$$

$$\angle CDB = 40° \quad ■$$

NOTE $\angle ABD + \angle CDB = 140° + 40° = 180°$. Example 5 illustrates the property: If two parallel lines are cut by a transversal, the interior angles on the same side of the transversal are supplementary. ☑

Polygons

Polygon

A **polygon** is a simple closed figure bounded by line segments called sides. Polygons are named according to the number of sides.

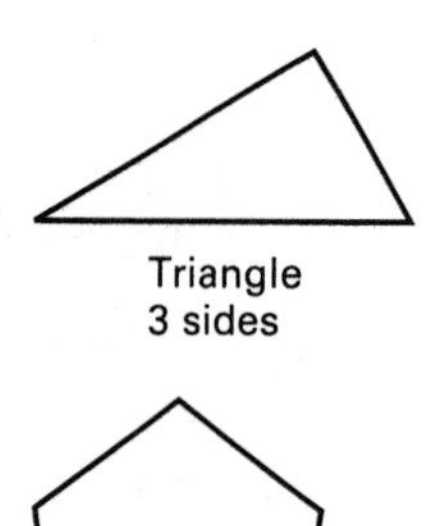

Triangle
3 sides

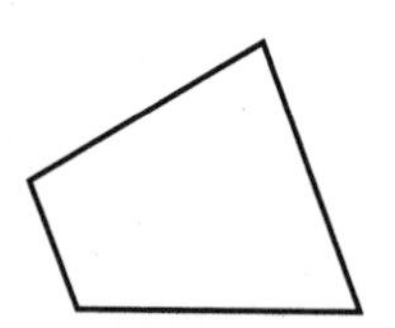

Quadrilateral
4 sides

Pentagon
5 sides

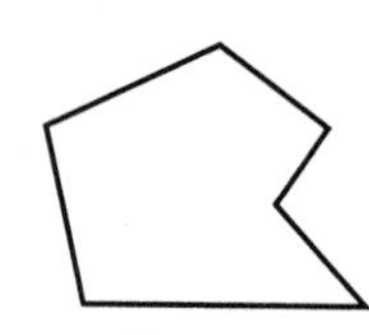

Hexagon
6 sides

Heptagon
7 sides

Octagon
8 sides

Nonagon
9 sides

Decagon
10 sides

Regular Polygon

A **regular polygon** is equilateral (all sides equal) and equiangular (all angles equal).

Diagonal

A **diagonal** of a polygon is a line segment joining two nonadjacent vertices.

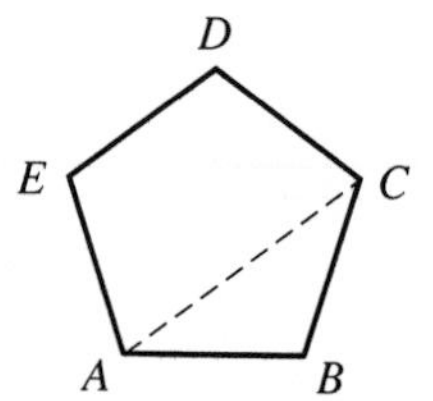

Regular pentagon with diagonal $\overline{AC}$

EXERCISES 16.1

1. Classify each angle as acute, right, obtuse, or straight.

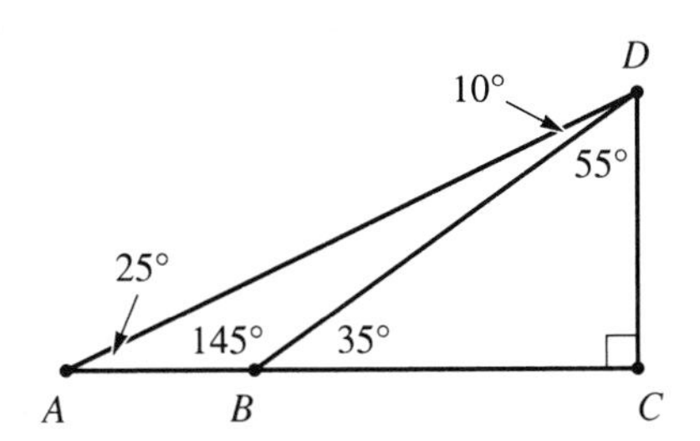

a. $\angle ABD$
b. $\angle ABC$
c. $\angle ACD$
d. $\angle ADC$

2. List pairs of angles that are the following:

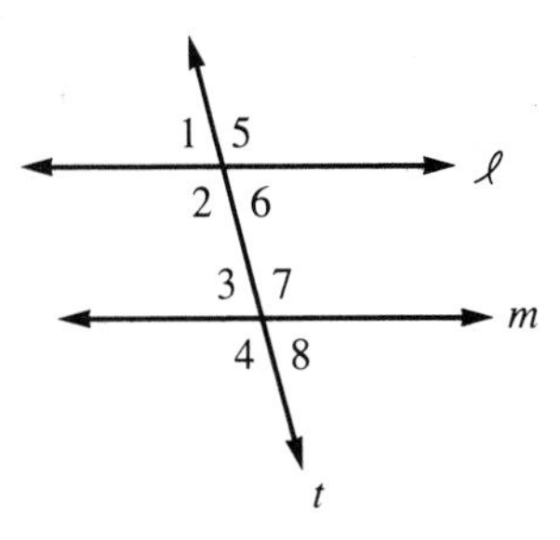

a. Vertical angles
b. Alternate interior angles
c. Alternate exterior angles
d. Corresponding angles

3. Find the complement of 25°.

4. Find the supplement of 25°.

5. Find the supplement of 73°.

6. Find the complement of 73°.

7. Find two supplementary angles if one angle is three times larger than the other angle.

8. Find two complementary angles if one angle is four times larger than the other angle.

9. Find two complementary angles if one angle is 30° more than the other angle.

10. Find two supplementary angles if one angle is 50° less than the other angle.

11. Find $\angle 1$.

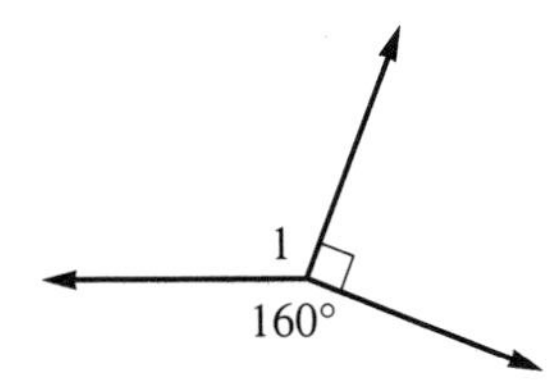

12. Find $\angle 1$.

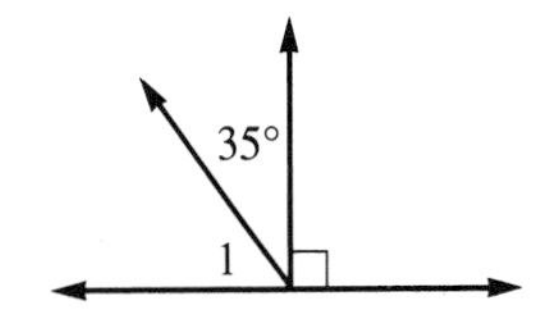

13. Given $\ell \perp m$, find $\angle 1$ and $\angle 2$.

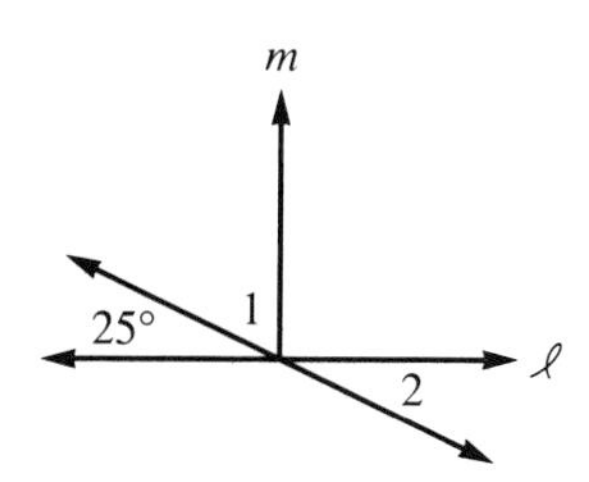

14. Given $\ell \perp m$, find $\angle 1$ and $\angle 2$.

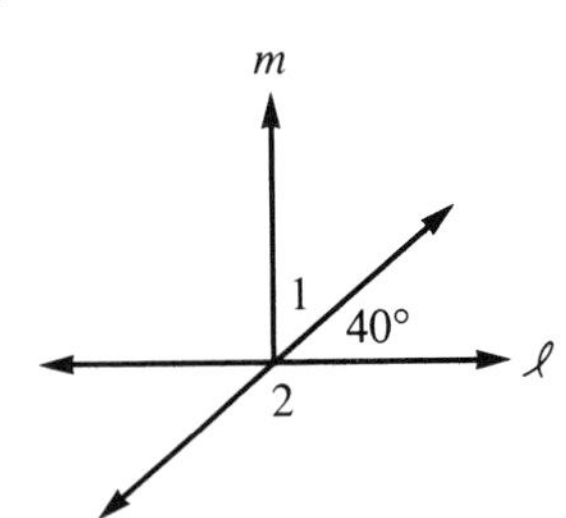

15. Given $\overleftrightarrow{DE} \parallel \overleftrightarrow{BC}$ and $\angle ABC = 145°$, find $\angle CBE$, $\angle BED$, and $\angle DEF$.

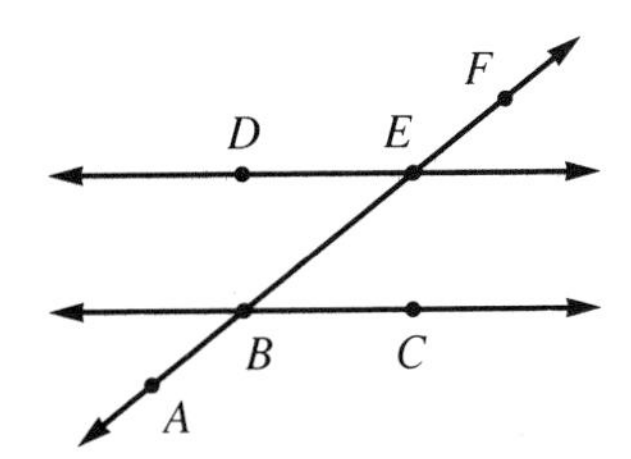

16. Given $\overleftrightarrow{AC} \parallel \overleftrightarrow{DE}$ and $\angle DBC = 60°$, find $\angle FDE$, $\angle EDB$, and $\angle ABD$.

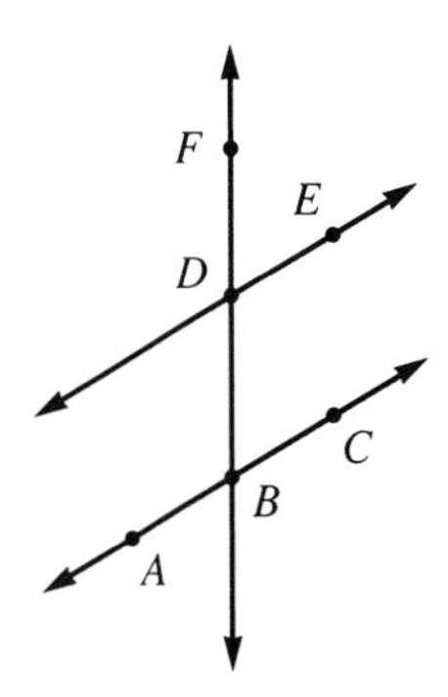

17. Given $\ell_1 \parallel \ell_2$, $\angle 2 = 75°$ and $\angle 6 = 45°$, find the remaining angles.

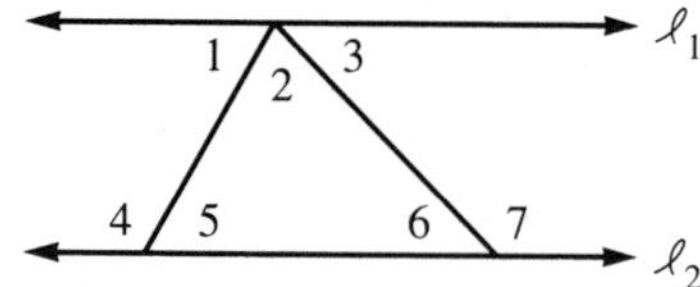

18. Given $\ell_1 \parallel \ell_2$, $m \perp \ell_1$, $\angle 3 = 50°$, find the remaining angles.

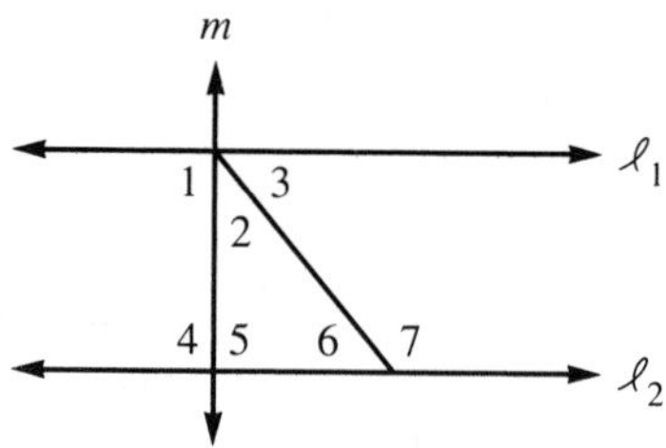

19. Name the polygons.

a.

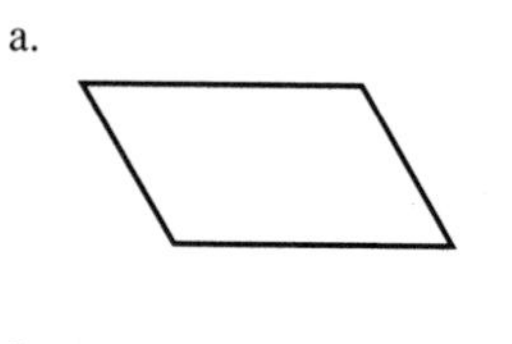

b.

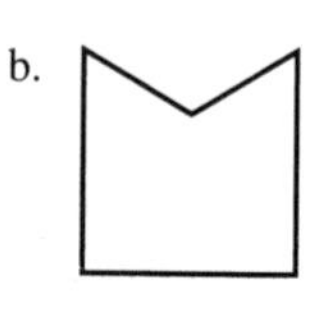

c.

d.

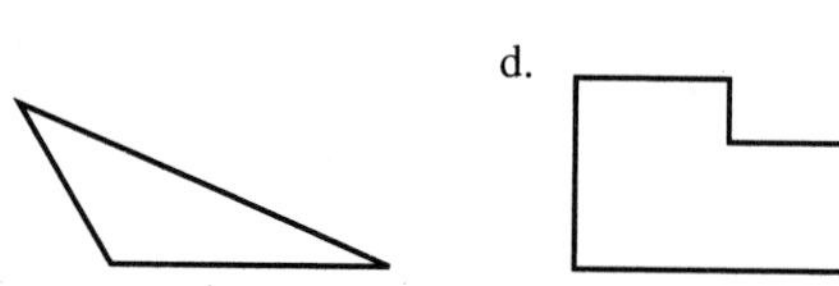

20. Draw the following polygons.
a. A regular hexagon
b. A triangle with one right angle
c. A quadrilateral with opposite sides parallel
d. A quadrilateral with all angles equal

16.2 Triangles

Triangle

A **triangle** has three sides and three angles. The symbol for triangle is $\triangle$.

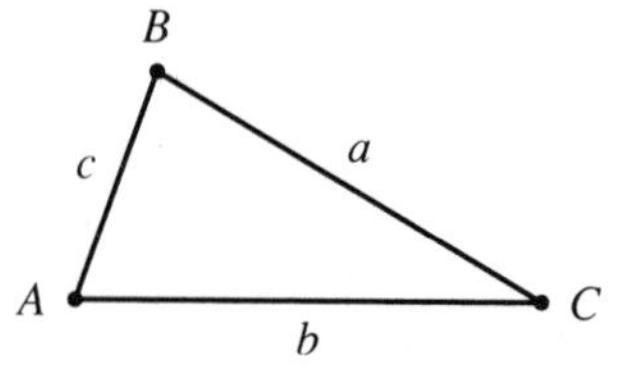

Triangle: $\triangle ABC$

Equilateral Triangle

An **equilateral triangle** is a triangle with all sides equal. All three angles are also equal.

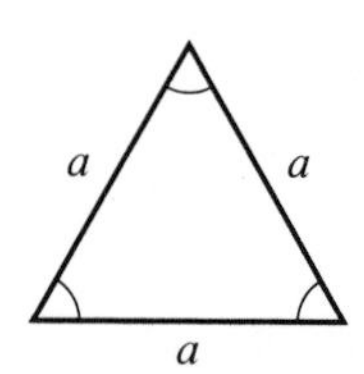

Equilateral triangle

Isosceles Triangle

An **isosceles triangle** has two sides equal. The angles opposite those sides are called base angles, and they are equal.

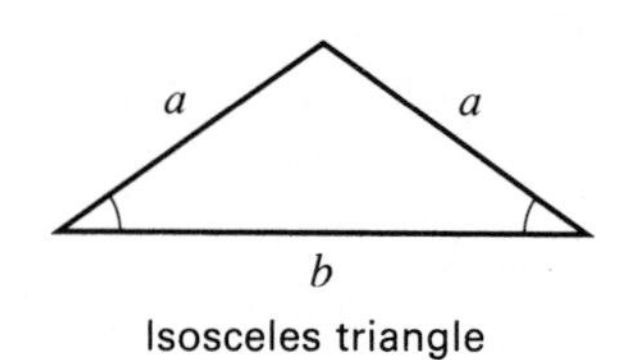

Isosceles triangle

Scalene Triangle

A **scalene triangle** has no sides equal.

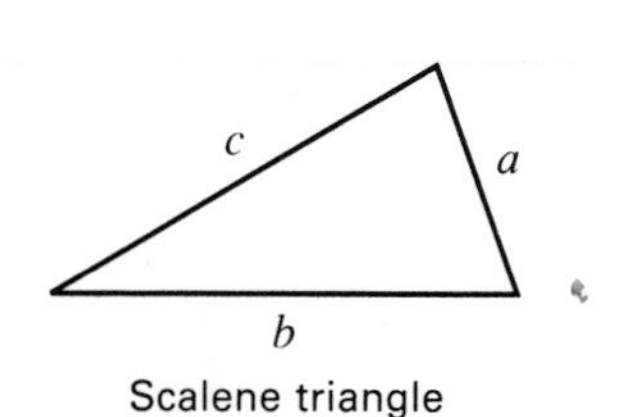

Scalene triangle

Right Triangle

A **right triangle** has one right angle.

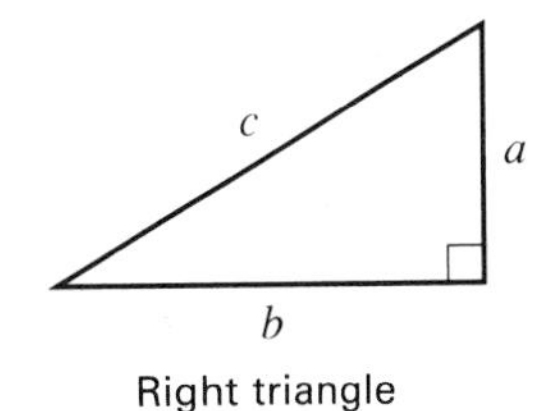

Right triangle

Sum of Angles of a Triangle

Draw line ℓ through C and parallel to $\overline{AB}$. Notice that

$$\left.\begin{array}{l}\angle 1 = \angle 4\\ \angle 3 = \angle 5\end{array}\right\}$$ If two lines are parallel, the alternate interior angles are equal.

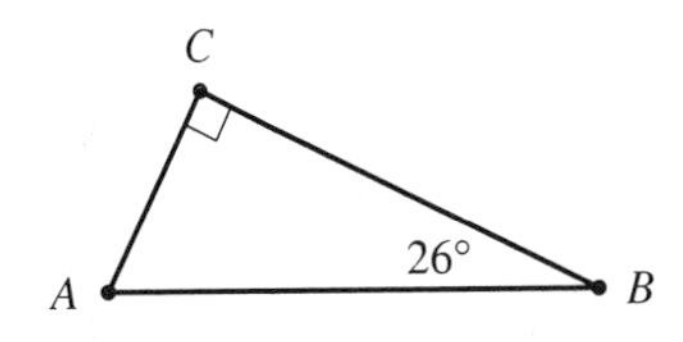

And $\angle 4 + \angle 2 + \angle 5 = 180°$, a straight angle.

Therefore, the sum of angles of a triangle $= \angle 1 + \angle 2 + \angle 3$
$= \angle 4 + \angle 2 + \angle 5$
$= 180°$

> The sum of the angles of a triangle equals 180°.

Example 1 In right $\triangle ABC$, $\angle B = 26°$. Find $\angle A$.

Solution Sum of angles $= 180°$

$$\angle A + \angle B + \angle C = 180$$
$$\angle A + 26 + 90 = 180$$
$$\angle A + 116 = 180$$
$$\angle A = 64°$$ ■

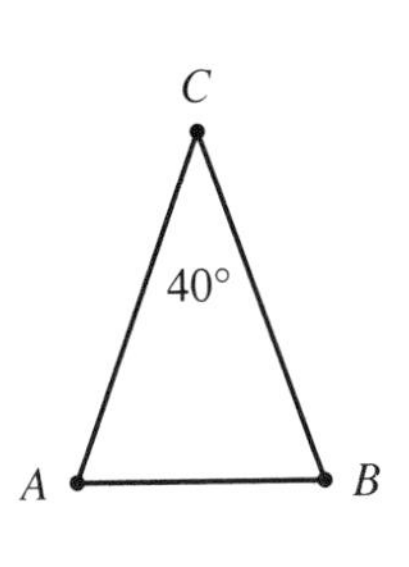

Example 2 In $\triangle ABC$, $AC = BC$, and $\angle C = 40°$. Find $\angle A$ and $\angle B$.

Solution Because $AC = BC$, $\triangle ABC$ is an isosceles triangle, where the base angles are equal ($\angle A = \angle B$).

Let $\angle A = \angle B = x°$

$$\angle A + \angle B + \angle C = 180$$
$$x + x + 40 = 180$$
$$2x + 40 = 180$$
$$2x = 140$$
$$x = 70$$

Therefore, $\angle A = \angle B = 70°$. ■

Sum of Angles of a Quadrilateral

Diagonal $\overline{AC}$ divides quadrilateral $ABCD$ into two triangles.

$$\left.\begin{array}{l}\angle 1 + \angle 2 + \angle 3 = 180\\ \angle 4 + \angle 5 + \angle 6 = 180\end{array}\right\}$$ Sum of angles of a triangle equals 180°.

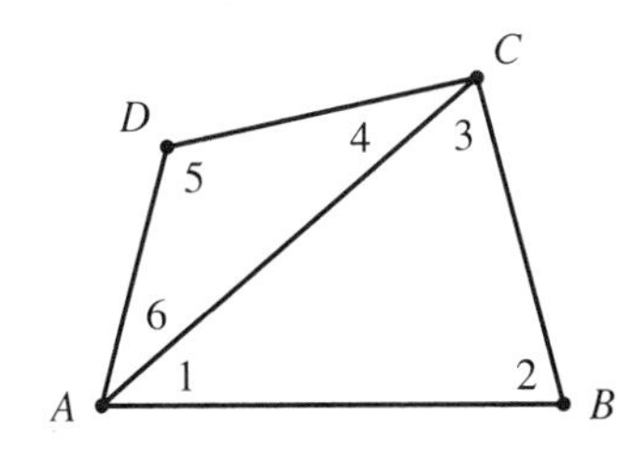

and $\angle A = \angle 6 + \angle 1$
$\angle C = \angle 3 + \angle 4$

Therefore, the sum of the angles of a quadrilateral =

$$\angle A + \angle B + \angle C + \angle D =$$
$$\overbrace{\angle 6 + \angle 1} + \angle 2 + \overbrace{\angle 3 + \angle 4} + \angle 5 = 360^\circ$$

> The sum of the angles of a quadrilateral equals 360°.

Example 3 In quadrilateral $ABCD$, $\overline{AB} \perp AD$, $\angle B = 130^\circ$, and $\angle C = 75^\circ$. Find $\angle D$.

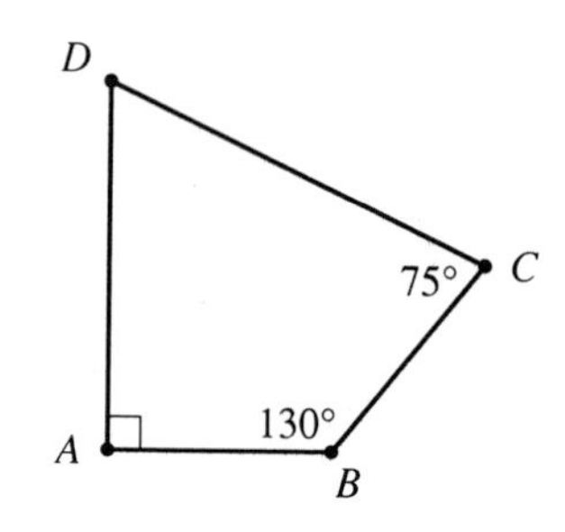

Solution $\angle A + \angle B + \angle C + \angle D = 360$

$$90 + 130 + 75 + \angle D = 360$$
$$295 + \angle D = 360$$
$$\angle D = 65^\circ \quad ■$$

The Pythagorean Theorem

Hypotenuse
Legs

A triangle that has a right angle is called a **right triangle**. The side opposite the right angle is the **hypotenuse**, and the other two sides are called **legs**.

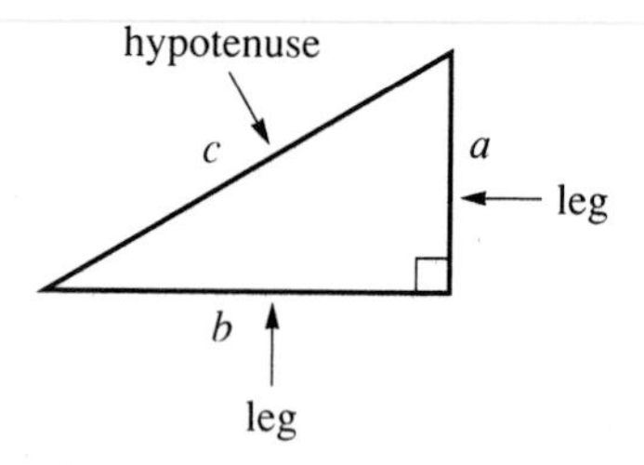

An important relationship between the sides of a right triangle was known to the Greeks about 500 BC and is named after the famous Greek mathematician Pythagoras.

> THE PYTHAGOREAN THEOREM
>
> The square of the hypotenuse of a right triangle is equal to the sum of the squares of the other sides.
>
> $$a^2 + b^2 = c^2$$
>
> c a b

NOTE The Pythagorean theorem applies only to *right triangles*. ☑

Example 4 Using the Pythagorean theorem to show whether or not a given triangle is a right triangle

a. $3^2 + 4^2 \stackrel{?}{=} 5^2$
$9 + 16 \stackrel{?}{=} 25$
$25 = 25$

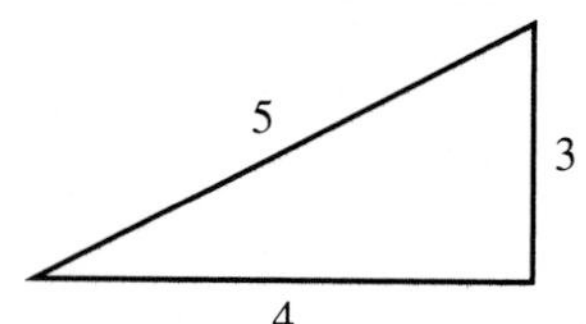

Therefore, the given triangle *is* a right triangle.

b. $2^2 + 3^2 \stackrel{?}{=} (\sqrt{11})^2$
$4 + 9 \stackrel{?}{=} 11$
$13 \neq 11$

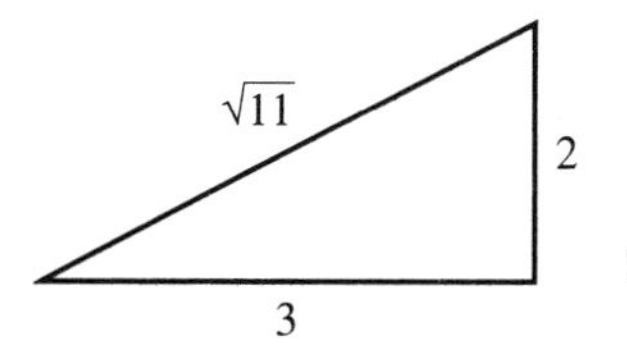

Therefore, the given triangle *is not* a right triangle. ■

Example 5 Find the hypotenuse of a right triangle with legs 8 and 6.

Solution

$a^2 + b^2 = c^2$

$6^2 + 8^2 = c^2$

$36 + 64 = c^2$

$100 = c^2$

$\sqrt{100} = \sqrt{c^2}$ Take the square root of both sides

$10 = c$

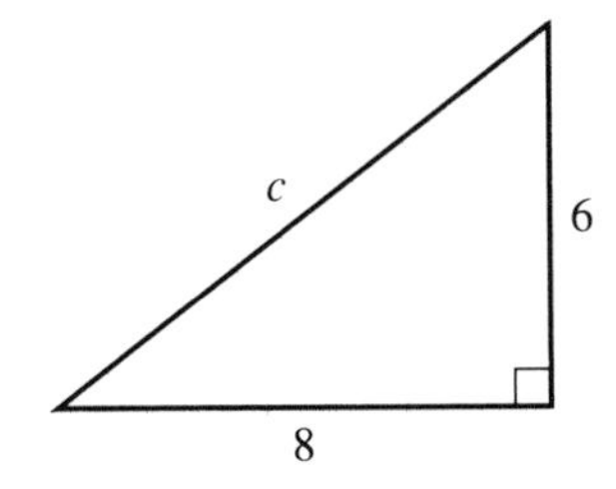

NOTE $100 = c^2$

$0 = c^2 - 100$

$0 = (c + 10)(c - 10)$

$c + 10 = 0$	$c - 10 = 0$
$c = -10$	$c = 10$

However, $c = -10$ cannot be a solution of this geometric problem because we usually consider lengths as positive numbers. For this reason we will take only the positive (principal) square root. ✓ ■

Example 6 Find x using the Pythagorean theorem.

Solution

$a^2 + b^2 = c^2$

$x^2 + 2^2 = 4^2$

$x^2 + 4 = 16$

$x^2 = 12$

$\sqrt{x^2} = \sqrt{12} = \sqrt{4 \cdot 3}$

$x = 2\sqrt{3}$ ■

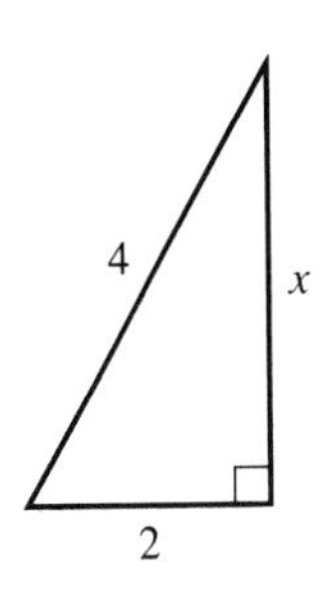

Example 7 In right $\triangle ABC$, $\angle A = \angle B = 45°$, $AB = 8$. Find AC and BC.

Solution Because $\angle A = \angle B$, $\triangle ABC$ is an isosceles triangle, and $AC = BC$.

Let $x = AC = BC$. Then

$a^2 + b^2 = c^2$

$x^2 + x^2 = 8^2$

$2x^2 = 64$

$x^2 = 32$

$\sqrt{x^2} = \sqrt{32} = \sqrt{16 \cdot 2}$

$x = 4\sqrt{2}$

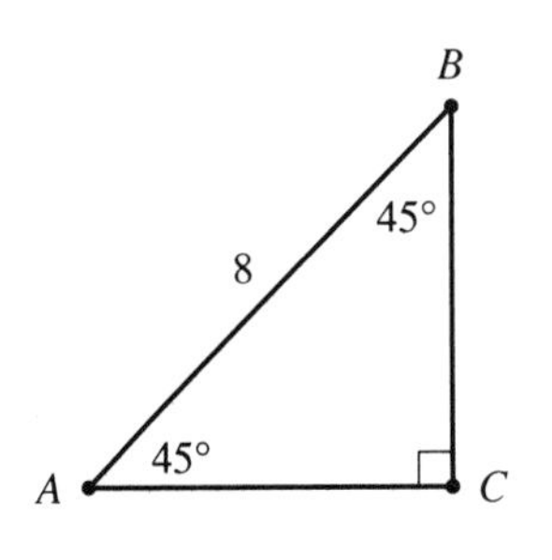

Therefore, $AC = BC = 4\sqrt{2}$. ■

EXERCISES 16.2

1. Classify each triangle as equilateral, isosceles, scalene, or right.

a. 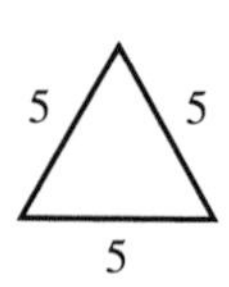

b. 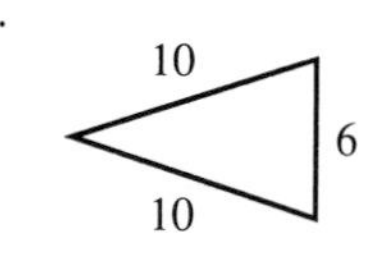

c. 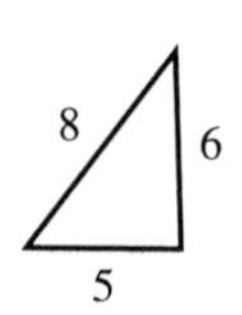

d. 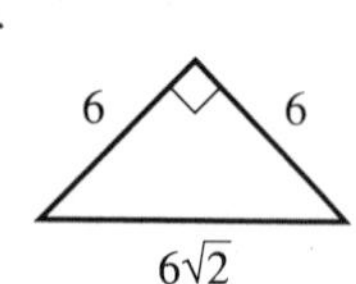

2. For right $\triangle ABC$,
a. name the hypotenuse
b. name the legs

For right $\triangle BCD$,
c. name the hypotenuse
d. name the legs

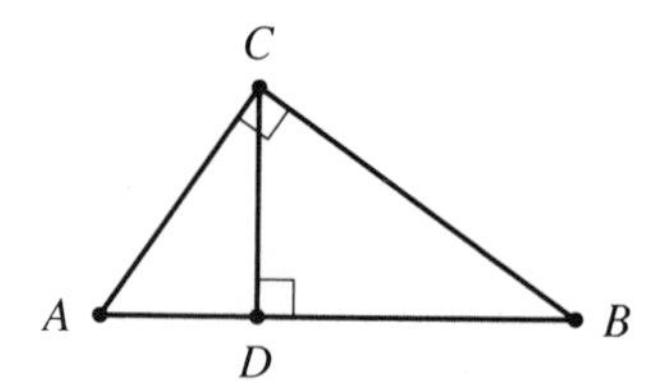

3. Find $\angle A$.

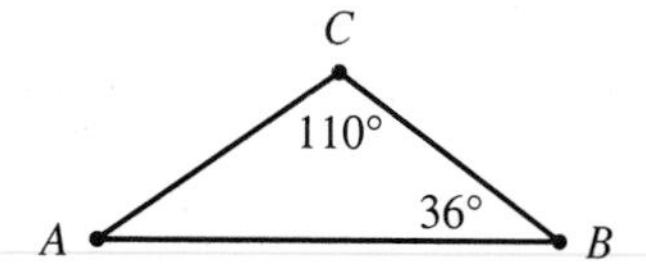

4. Find $\angle X$.

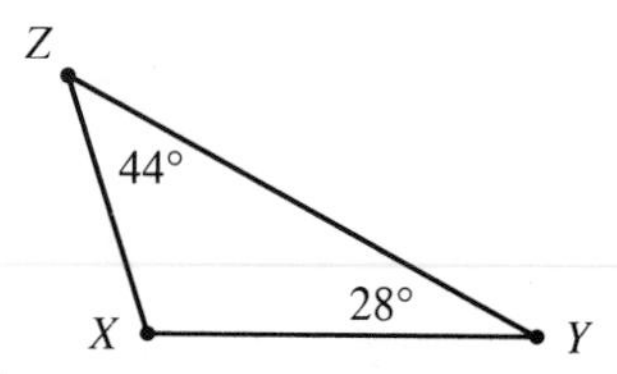

5. Find $\angle C$.

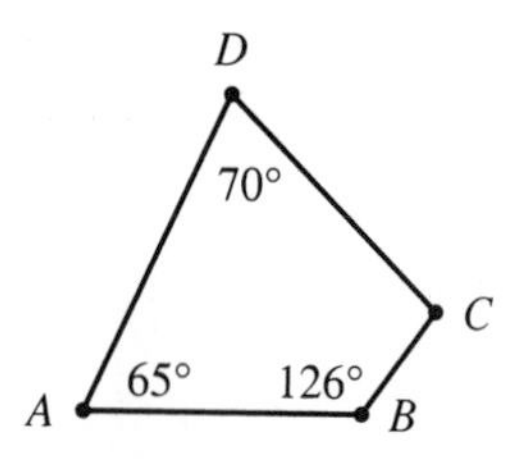

6. Find $\angle D$.

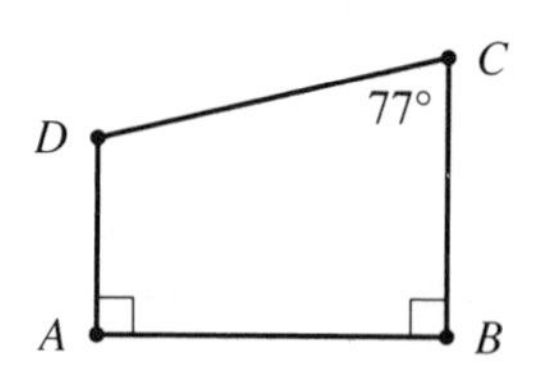

7. $\triangle ABC$ is an equilateral triangle. Find each angle.

8. $\triangle EFG$ is an isosceles triangle. Find the base angles if the other angle is 45°.

9. If one acute angle of a right triangle is 38°, find the other acute angle.

10. If one acute angle of a right triangle is 63°, find the other acute angle.

11. Find both acute angles of a right triangle if one angle is twice the other acute angle.

12. Find both acute angles of a right triangle if one acute angle is 40° more than the other acute angle.

13. In $\triangle ABC$, $\angle A$ is three times larger than $\angle B$, and $\angle C$ is 10° less than $\angle B$. Find all three angles.

14. In $\triangle ABC$, $\angle B$ equals $\angle A$ plus 20°, and $\angle C$ equals the sum of $\angle A$ and $\angle B$. Find all three angles.

15. Given $\overline{AC} \perp \overline{BC}$, $\overline{AB} \perp \overline{CD}$, and $\angle B = 32°$, find $\angle A$, $\angle ACD$, and $\angle BCD$.

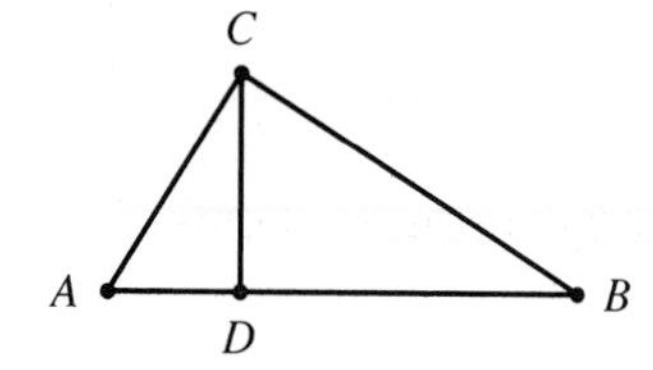

16. Given $\overline{DC} \perp \overline{BC}$, $\overline{EB} \perp \overline{BC}$, and $\angle D = 53°$. Find $\angle A$, $\angle BEA$, and $\angle BED$.

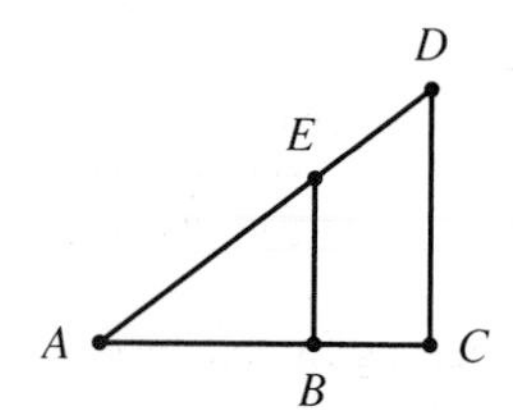

17. Given $\ell \parallel m$, $\angle 1 = 30°$, and $\angle 5 = 80°$. Find the remaining angles.

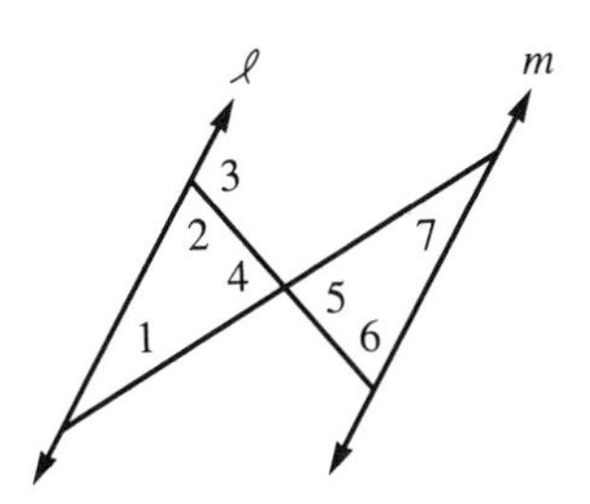

18. Given $\ell \parallel m$, $\angle 1 = 45°$, and $\angle 5 = 35°$. Find the remaining angles.

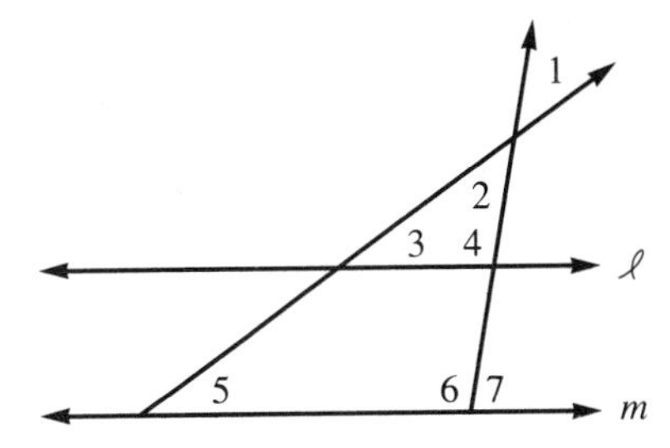

In Exercises 19–22, use the Pythagorean theorem to determine whether or not the given triangle is a right triangle.

19.

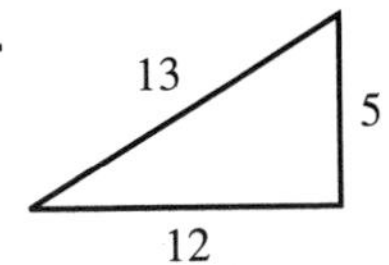

20.

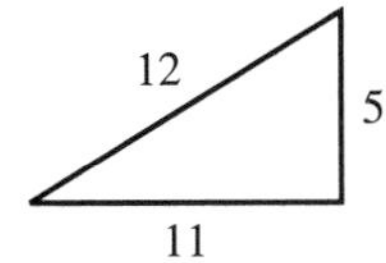

21.

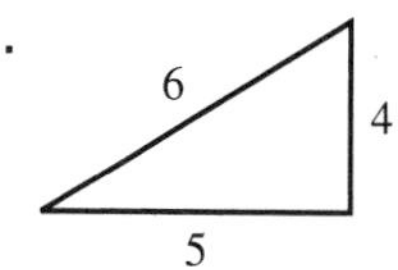

22.

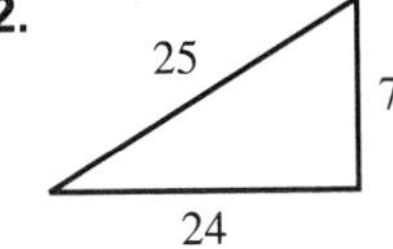

In Exercises 23–30, use the Pythagorean theorem to find x in each figure.

23.

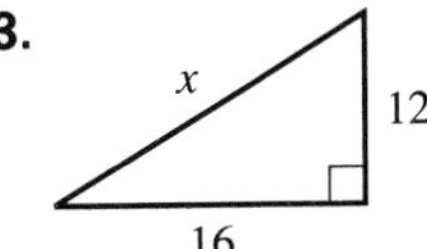

24.

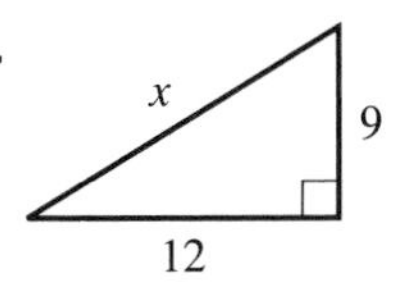

25.

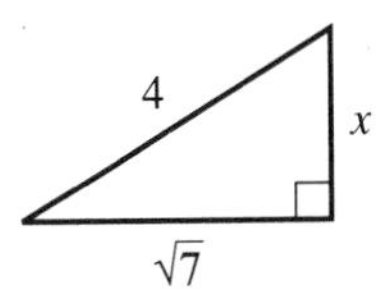

26.

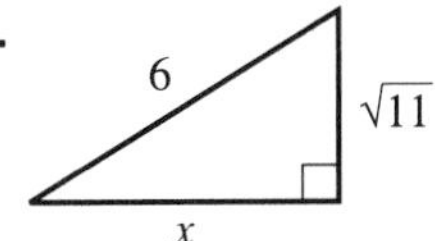

27.

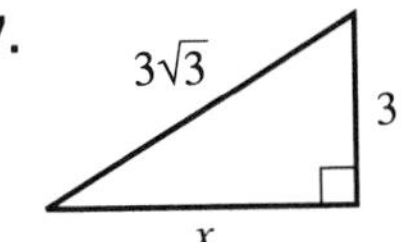

28.

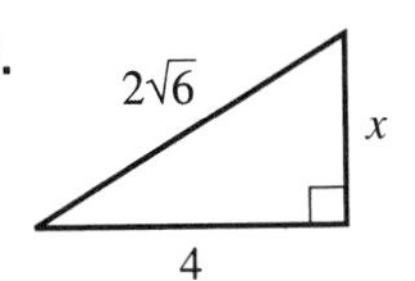

29.

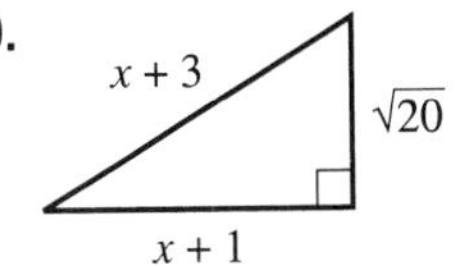

30.

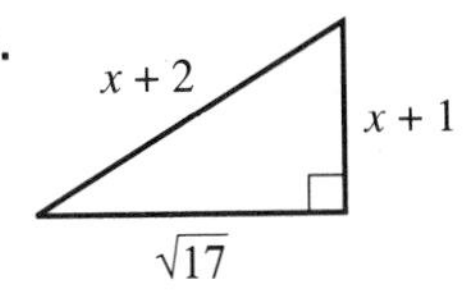

31. One leg of a right triangle is 4 less than twice the other leg. If its hypotenuse is 10, how long are the two legs?

32. One leg of a right triangle is 2 less than twice the other leg. If the hypotenuse is 5, how long are the two legs?

33. In a 30°-60°-90° triangle, the hypotenuse is twice the length of the short leg (opposite the 30° angle).
Let $x =$ short leg
$2x =$ hypotenuse

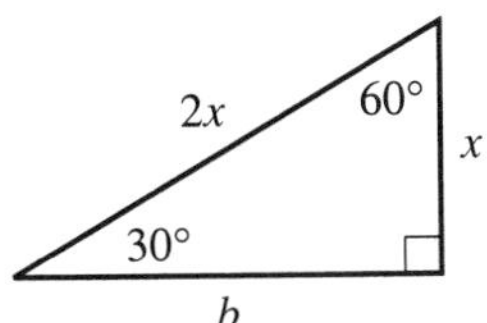

Find the other leg in terms of x.

34. In a 45°-45°-90° triangle the two legs are equal.
Let $x =$ each leg

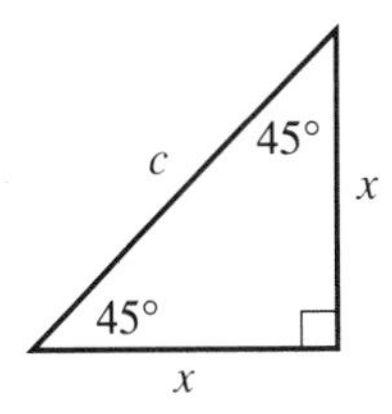

Find the hypotenuse in terms of x.

35. Use the results in Exercise 33 to find the missing sides of the 30°-60°-90° triangles.

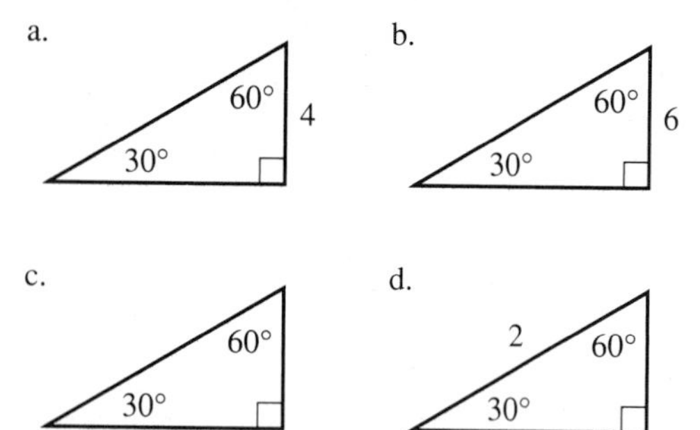

36. Use the results in Exercise 34 to find the missing sides of the 45°-45°-90° triangles.

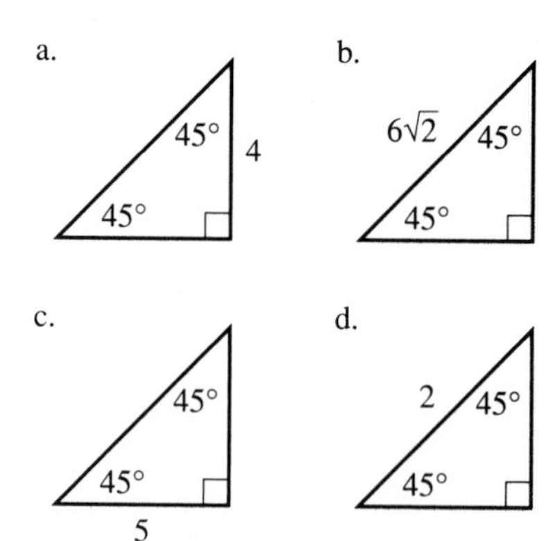

16.3 Congruent Triangles

Congruent Triangles

If two triangles have the same shape and the same size, the triangles are said to be **congruent**. In congruent triangles, the corresponding angles are equal, and the corresponding sides are equal. The symbol for congruent is $\cong$.

> If two triangles are congruent, their corresponding angles are equal, and their corresponding sides are equal.
>
> 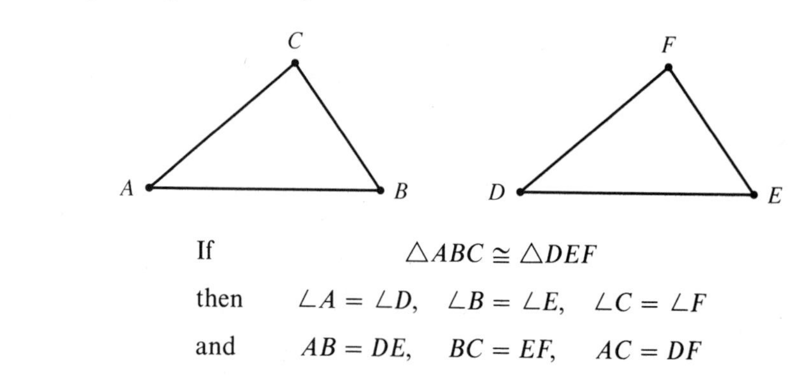
>
>
> If $\triangle ABC \cong \triangle DEF$
>
> then $\angle A = \angle D, \quad \angle B = \angle E, \quad \angle C = \angle F$
>
> and $AB = DE, \quad BC = EF, \quad AC = DF$

Example 1 For the given congruent triangles, name the corresponding parts that are equal.

a. Given $\triangle ABE \cong \triangle DCE$

Solution $\angle A = \angle D$

$\angle ABE = \angle DCE$

$\angle AEB = \angle DEC$

$AB = DC$

$BE = CE$

$AE = DE$

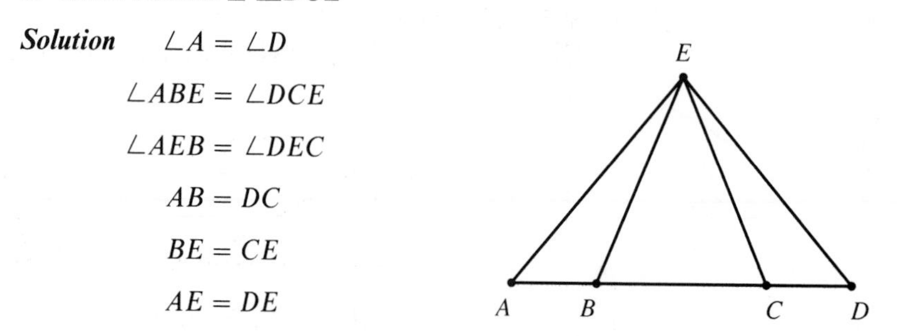

NOTE Corresponding sides lie opposite corresponding angles. ☑

b. Given $\triangle ADB \cong \triangle BEA$

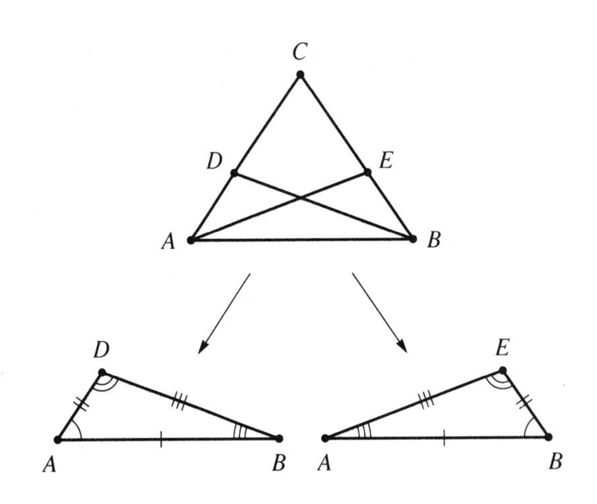

Solution It is helpful to label the corresponding parts that are equal.

$$\angle DAB = \angle EBA$$
$$\angle ADB = \angle BEA$$
$$\angle ABD = \angle BAE$$
$$AB = AB$$
$$AD = BE$$
$$DB = EA \quad \blacksquare$$

Example 2 Given $\triangle ADC \cong \triangle BDC$, find BC and $\angle BCD$.

Solution Corresponding parts of congruent triangles are equal. Therefore, $BC = AC = 6$.

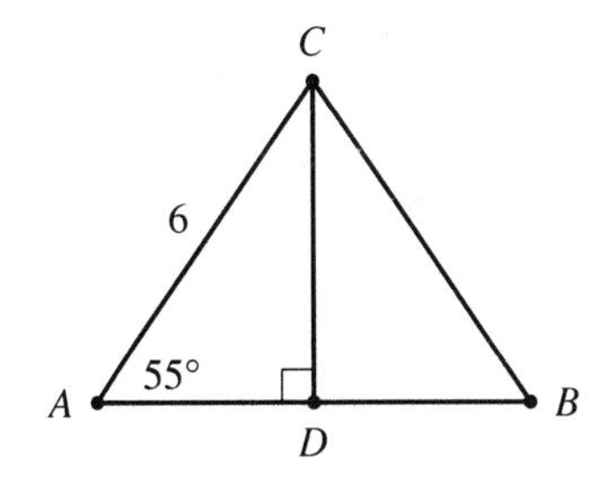

The sum of the angles of a triangle is 180°.

$$\angle A + \angle ADC + \angle ACD = 180$$
$$55 + 90 + \angle ACD = 180$$
$$145 + \angle ACD = 180$$
$$\angle ACD = 35$$

Therefore, $\angle BCD = \angle ACD = 35°$. $\blacksquare$

Example 3 Given $\triangle ABE \cong \triangle CDE$, find x and y.

Solution Corresponding parts of congruent triangles are equal.

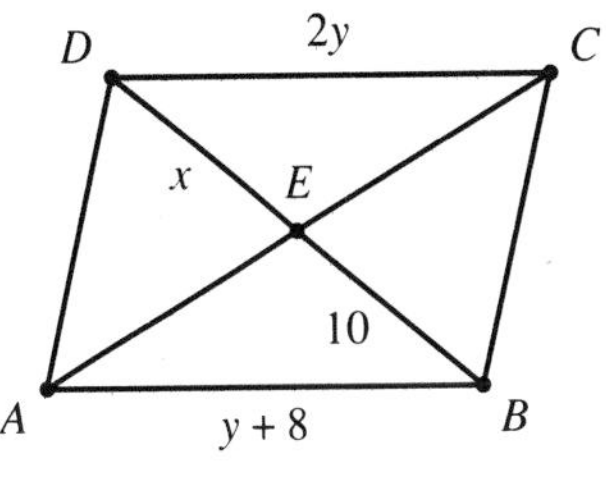

$$DE = BE$$
$$x = 10$$
$$DC = AB$$
$$2y = y + 8$$
$$y = 8 \quad \blacksquare$$

We can prove that two triangles are congruent by using one of the following properties (see below and the top of page 488).

SSS (side-side-side)
If three sides of one triangle are equal to three sides of another triangle, the triangles are congruent.

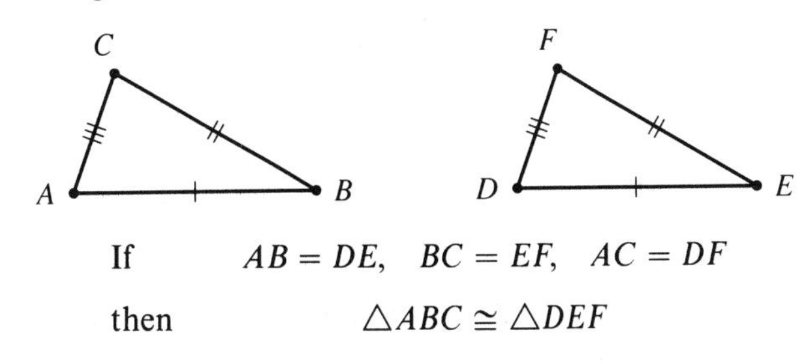

If $AB = DE$, $BC = EF$, $AC = DF$

then $\triangle ABC \cong \triangle DEF$

SAS (side-angle-side)
If two sides and the included angle of one triangle are equal to two sides and the included angle of another triangle, the triangles are congruent.

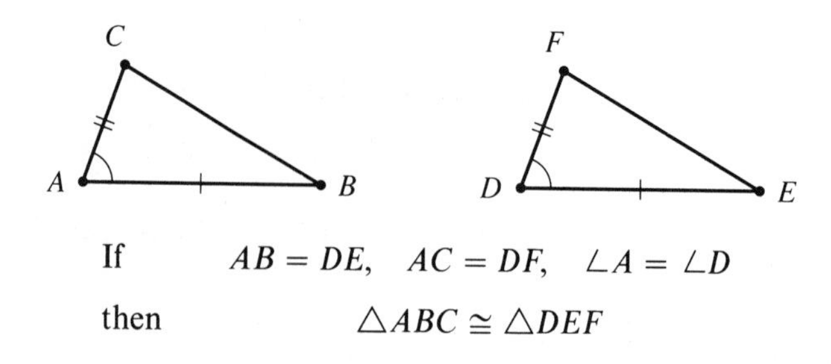

If $AB = DE$, $AC = DF$, $\angle A = \angle D$

then $\triangle ABC \cong \triangle DEF$

ASA (angle-side-angle)
If two angles and the included side of one triangle are equal to two angles and the included side of another triangle, the triangles are congruent.

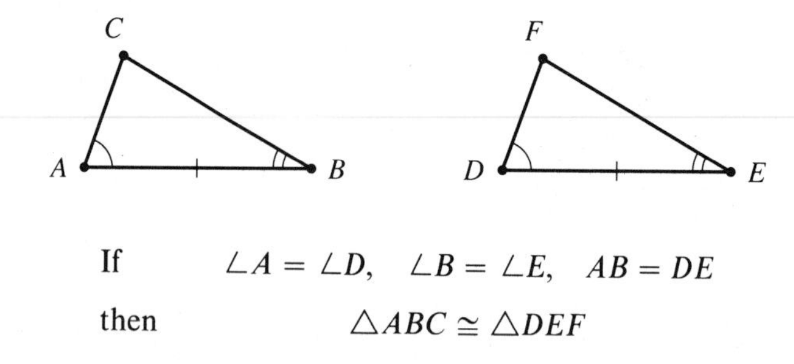

If $\angle A = \angle D$, $\angle B = \angle E$, $AB = DE$

then $\triangle ABC \cong \triangle DEF$

Example 4 Identify the congruent triangles and name the property used.

a. Given $AC = BC$ and $AD = BD$

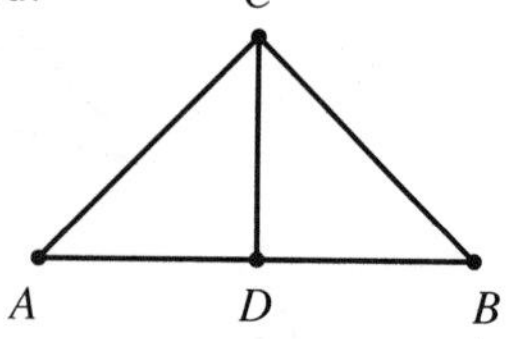

Solution It is helpful to label the corresponding parts that we know are equal.

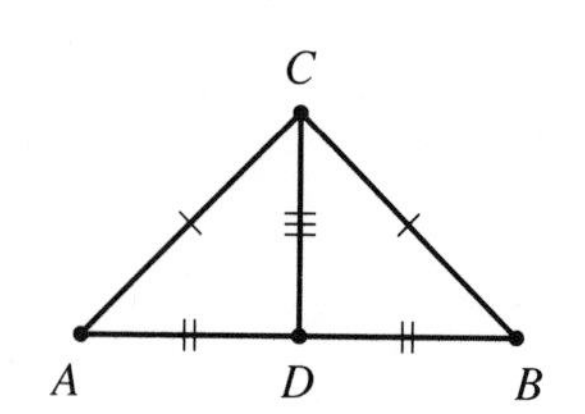

1. $AC = BC$ } is given
2. $AD = BD$ }
3. $DC = DC$ is a common side.

Therefore, $\triangle ADC \cong \triangle BDC$ by SSS

b. Given $AE = BE$ and $CE = DE$

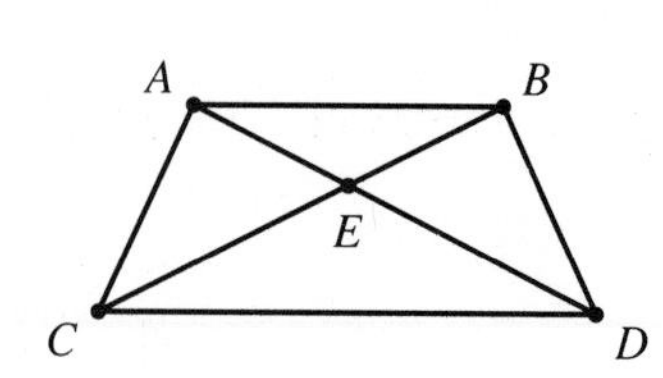

Solution

1. $AE = BE$ } is given
2. $CE = DE$ }
3. $\angle AEC = \angle BED$ Vertical angles are equal.

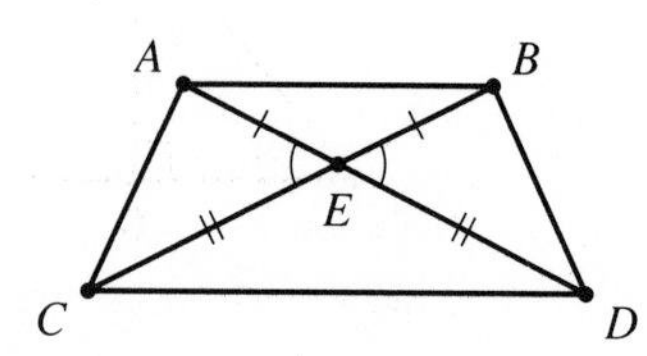

Therefore, $\triangle AEC \cong \triangle BED$ by SAS

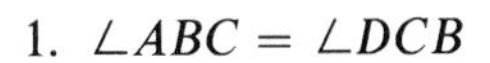
c. Given $\overline{AB} \parallel \overline{CD}$ and $\overline{AC} \parallel \overline{BD}$

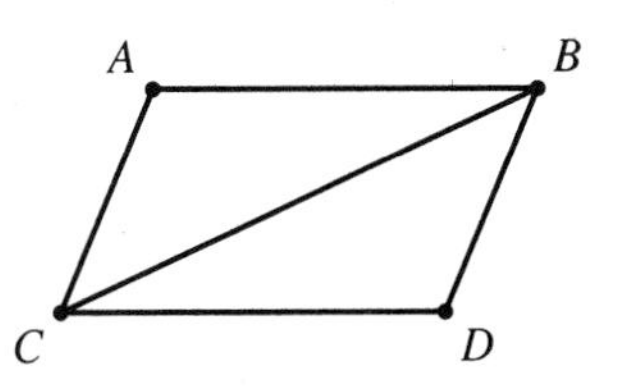

Solution If two lines are parallel, the alternate interior angles are equal.

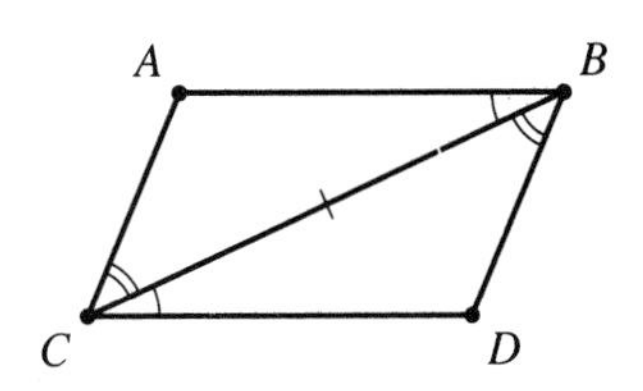

1. $\angle ABC = \angle DCB$
2. $\angle ACB = \angle DBC$
3. $CB = CB$ is a common side.

Therefore, $\triangle ABC \cong \triangle DCB$ by ASA. ■

EXERCISES 16.3

In Exercises 1–6, for the given congruent triangles, name the corresponding parts.

1. $\triangle ABD \cong \triangle CDB$

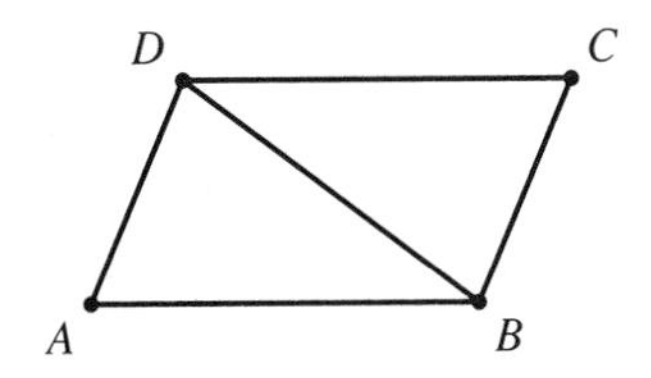

2. $\triangle AFD \cong \triangle BFE$

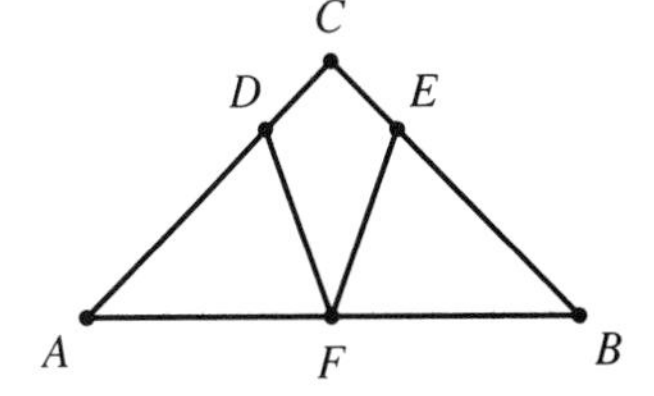

3. $\triangle AFD \cong \triangle BFE$

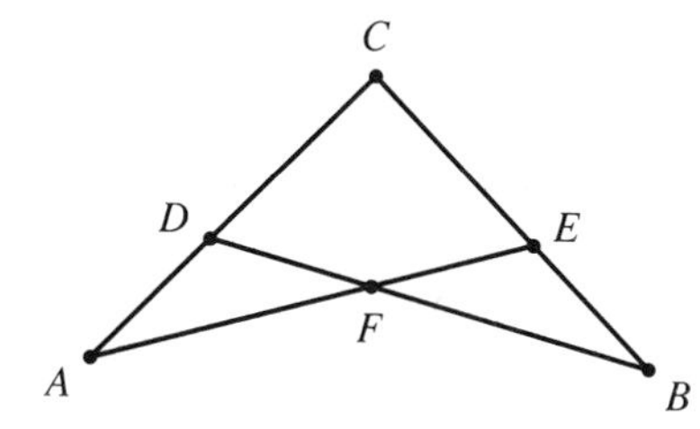

4. $\triangle AEC \cong \triangle BED$

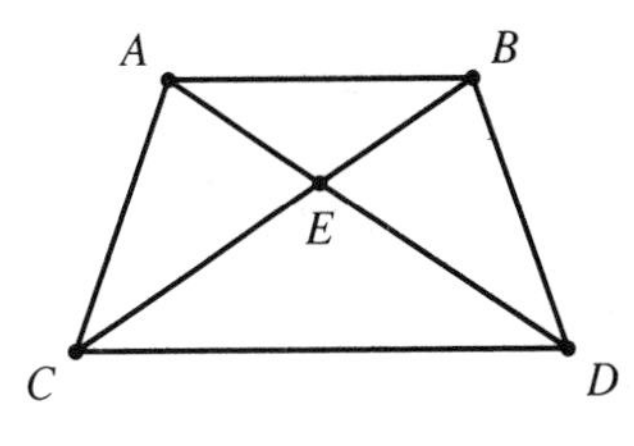

5. $\triangle ACD \cong \triangle CAB$

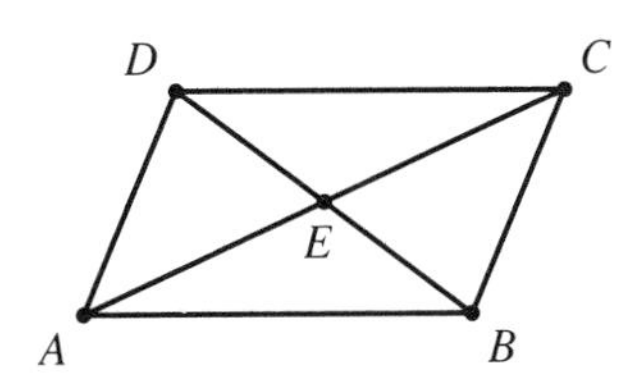

6. $\triangle ABE \cong \triangle CDE$

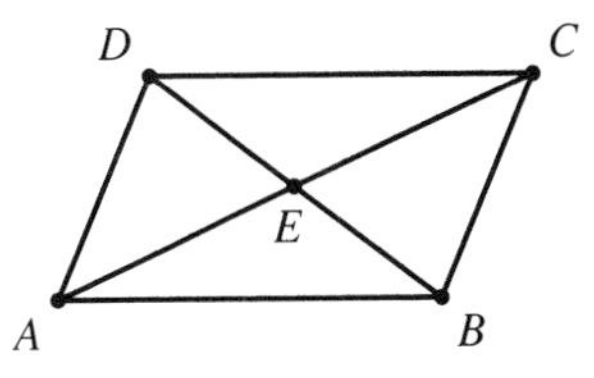

7. Given $\triangle ADC \cong \triangle BEF$, find $\angle F$ and $\angle FBE$.

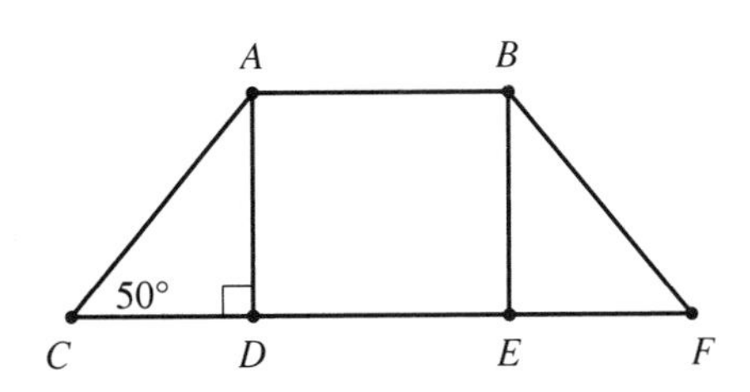

8. Given $\triangle ABC \cong \triangle CDA$, find $\angle DCA$ and $\angle DCB$.

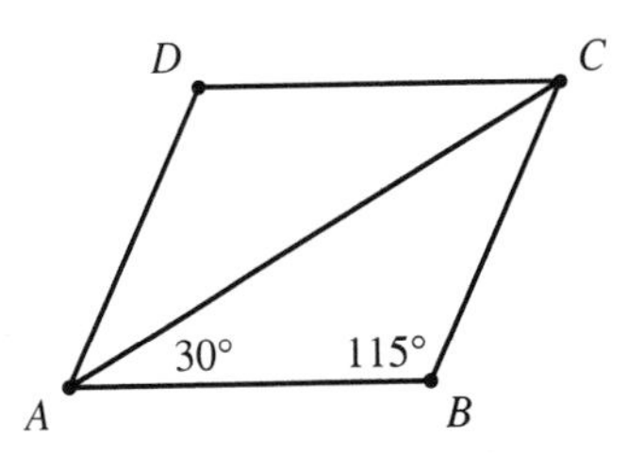

9. Given $\triangle ABE \cong \triangle DCE$, find $\angle D$ and $\angle ECB$.

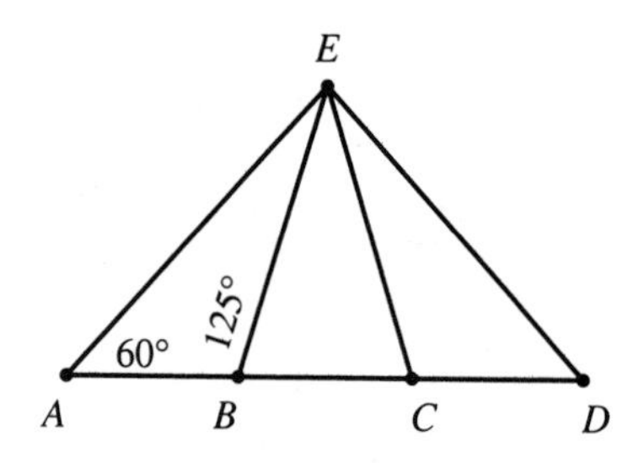

10. Given $\triangle AFD \cong \triangle BFE$, find $\angle B$ and $\angle DFE$.

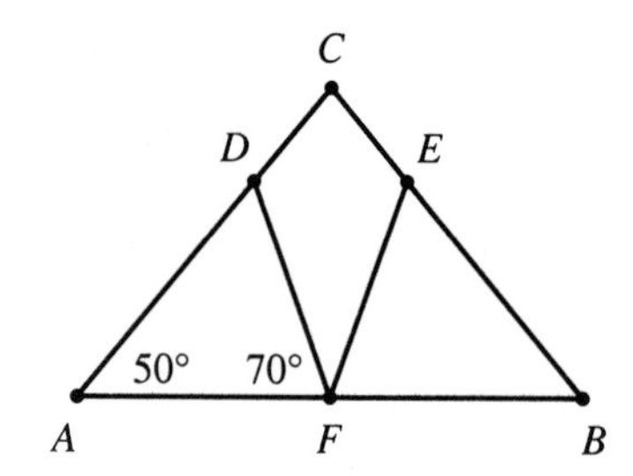

11. Given $\triangle ABE \cong \triangle CBD$, find x and y.

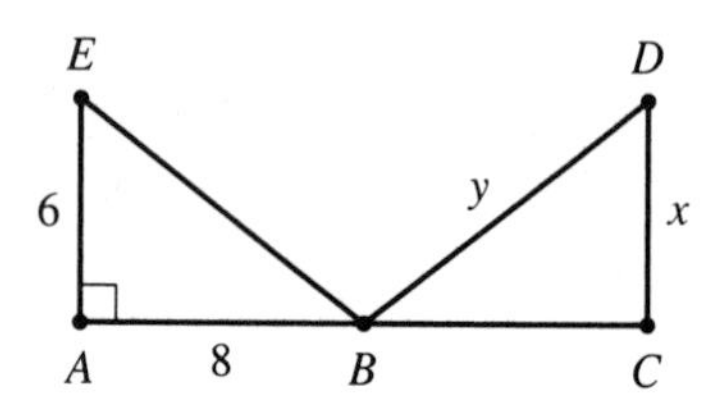

12. Given $\triangle ABD \cong \triangle FEC$, find x and y.

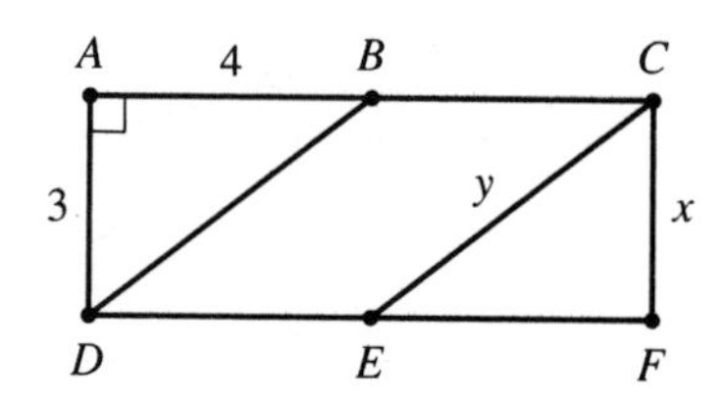

13. Given $\triangle AFD \cong \triangle BFE$, find x and y.

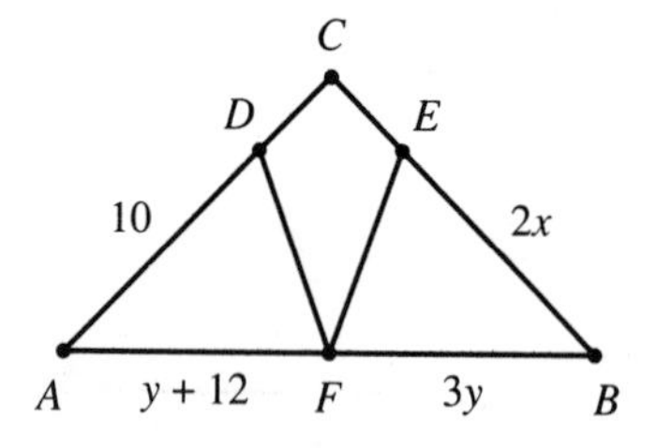

14. Given $\triangle ABE \cong \triangle CBD$, find x and y.

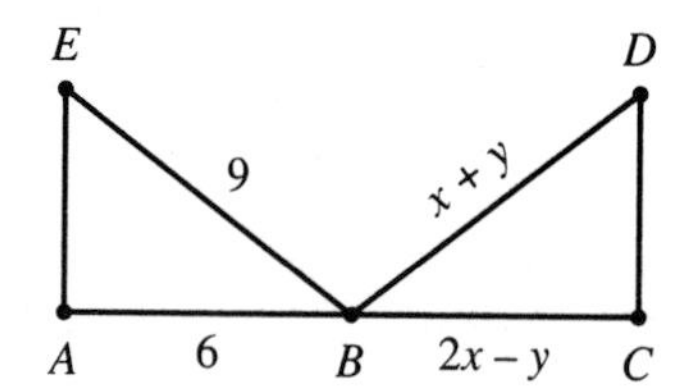

In Exercises 15–22, identify the congruent triangles and name the property used.

15. Given $AB = AD$ and $BC = DC$

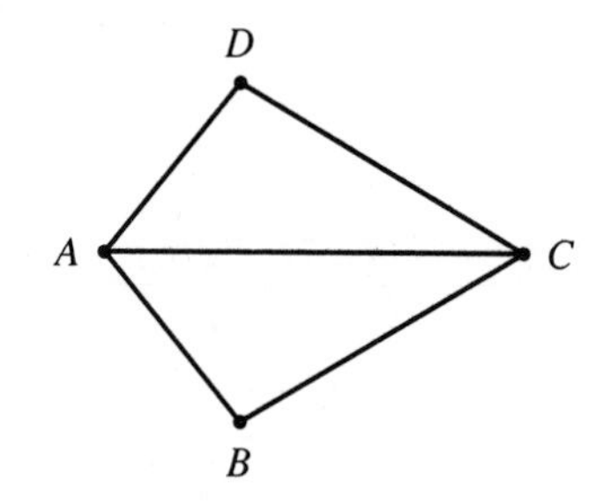

16. Given $AE = DE$ and $BE = CE$

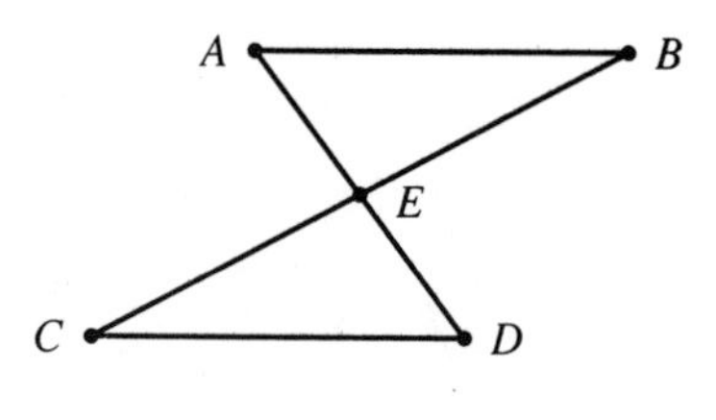

17. Given $\overline{AC} \parallel \overline{BD}$ and $\angle ABC = \angle DCB$

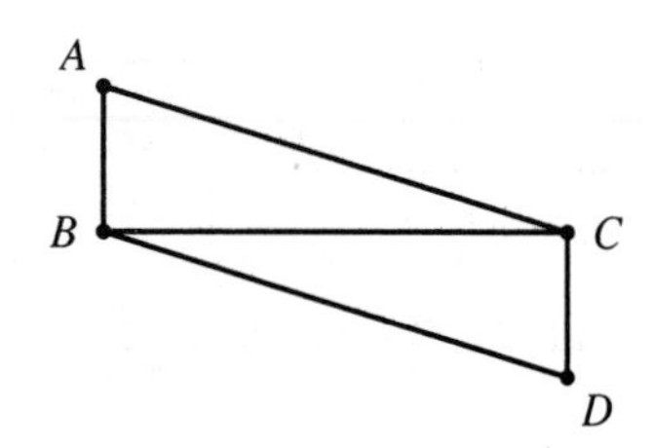

18. Given $\overline{AB} \perp \overline{BC}$, $\overline{DC} \perp \overline{BC}$, and $AB = DC$

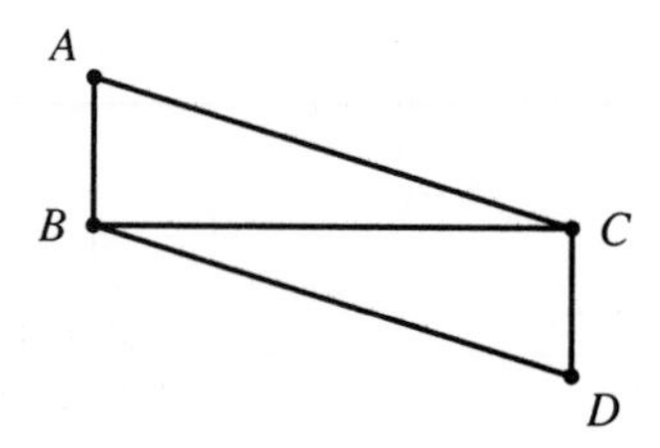

19. Given $\overline{AB} \perp \overline{CD}$, and $AD = BD$

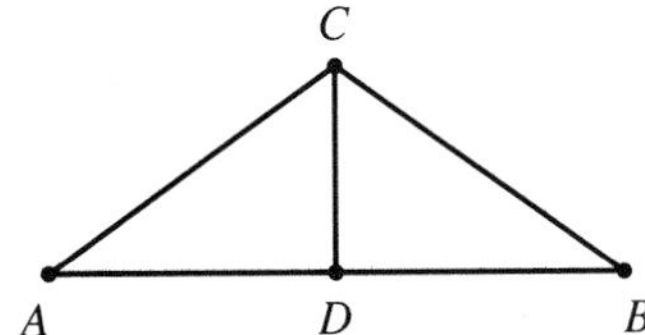

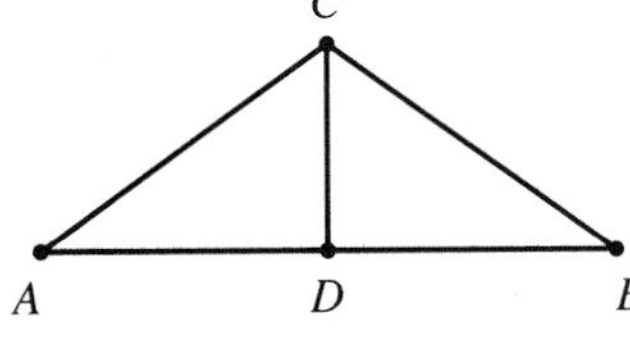

20. Given $\triangle ABC$ is an isosceles triangle and $AD = BD$

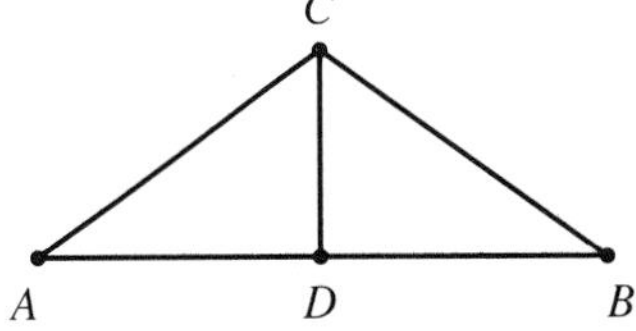

21. Given $\angle A = \angle B$ and $AF = BF$

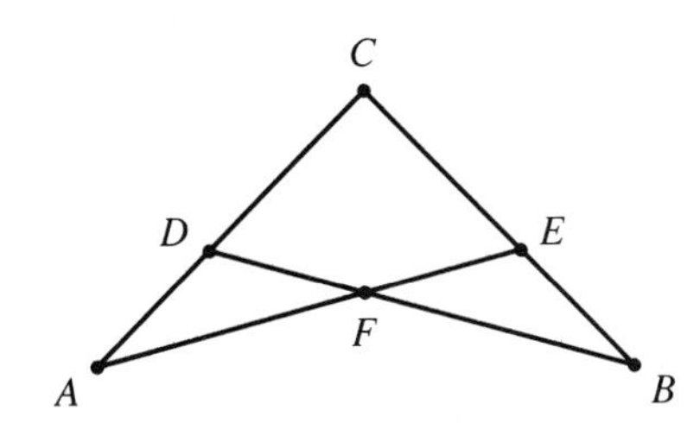

22. Given $\angle A = \angle B$ and $AC = BC$

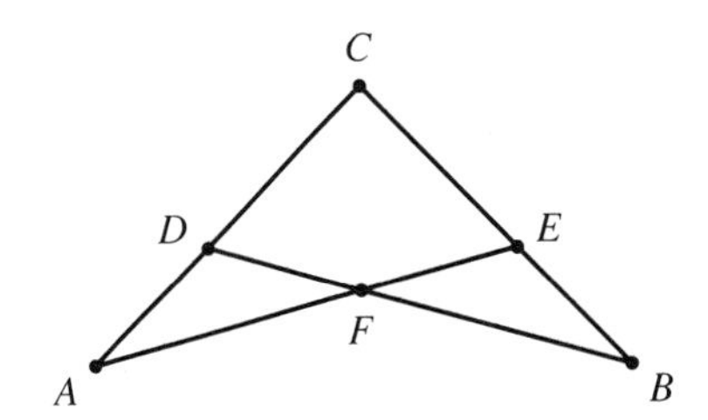

16.4 Similar Triangles

Similar Triangles

If two triangles have the same shape but not necessarily the same size, the triangles are said to be **similar**. In similar triangles, the corresponding angles are equal. The symbol for similar is $\sim$.

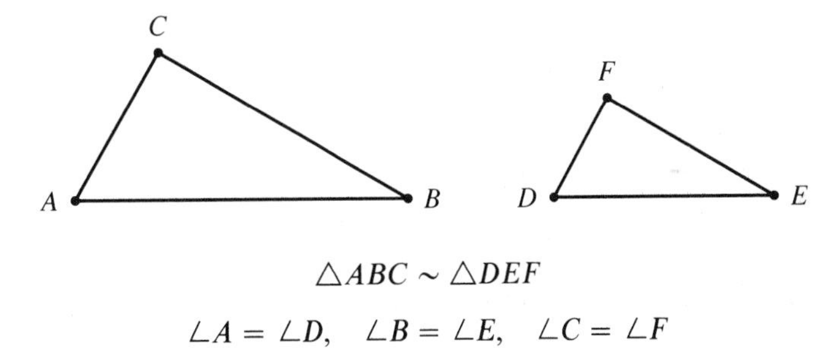

$$\triangle ABC \sim \triangle DEF$$

$$\angle A = \angle D, \quad \angle B = \angle E, \quad \angle C = \angle F$$

Because the sum of the angles of a triangle equals 180° (Section 16.2), when two pairs of corresponding angles are equal, the third pair of corresponding angles must also be equal. Thus,

> If two angles of one triangle are equal to two angles of another triangle, then the triangles are similar.

Example 1 Similar triangles

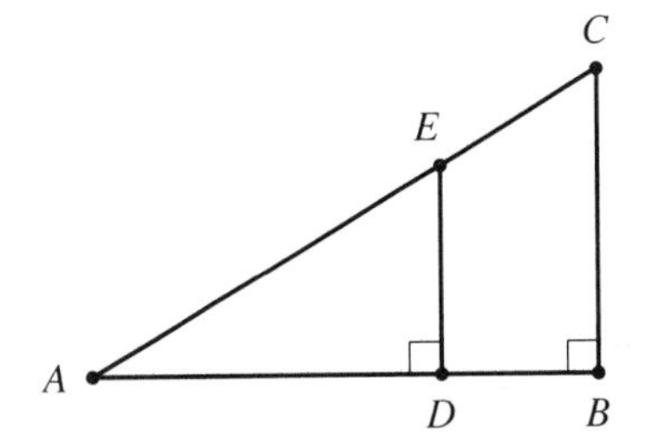

$\angle A$ is a common angle to both $\triangle ADE$ and $\triangle ABC$. Also, $\angle ADE = \angle ABC = 90°$. Therefore, $\triangle ADE \sim \triangle ABC$. ■

An important property of similar triangles is given below.

If two triangles are similar, their corresponding sides are proportional.

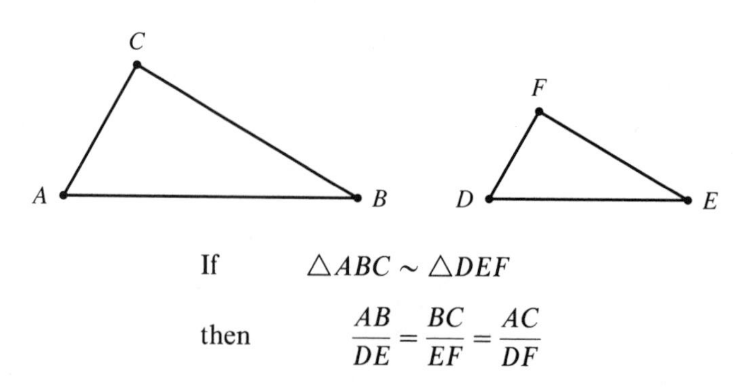

If $\triangle ABC \sim \triangle DEF$

then $\dfrac{AB}{DE} = \dfrac{BC}{EF} = \dfrac{AC}{DF}$

Example 2 Given $\triangle ABC \sim \triangle DEF$, find DF.

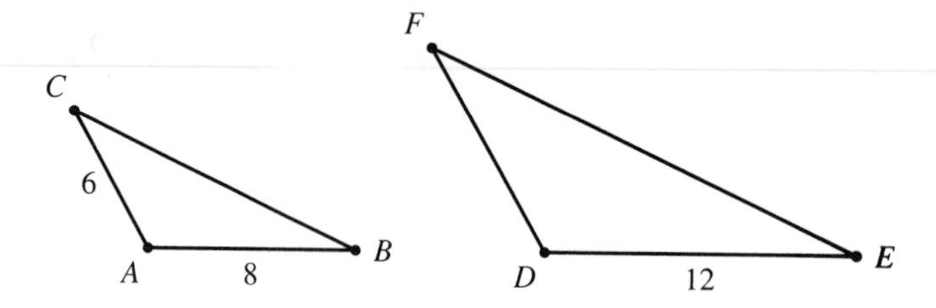

Solution Because the triangles are similar, the corresponding sides are proportional.

$$\frac{AB}{DE} = \frac{AC}{DF}$$

$$\frac{8}{12} = \frac{6}{DF}$$

$$8DF = 72 \quad \text{Product of means = product of extremes}$$

$$DF = 9$$ ■

Example 3 Given $\overline{AB} \parallel \overline{DE}$, find DE.

Solution $\angle A = \angle E$ and $\angle B = \angle D$. (If two lines are parallel, the alternate interior angles are equal.) Therefore, $\triangle ABC \sim \triangle EDC$.

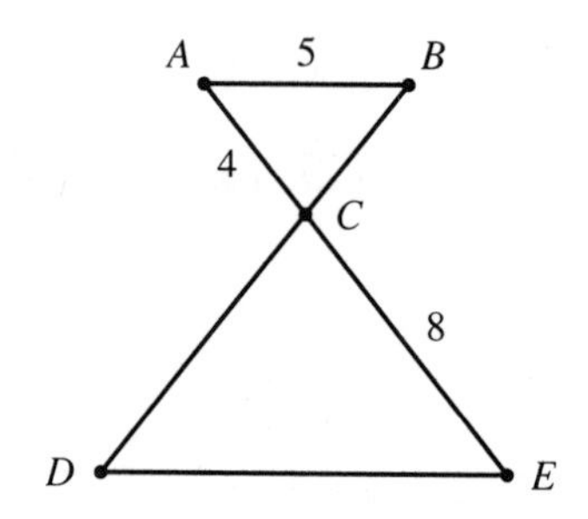

$$\frac{AC}{EC} = \frac{AB}{DE}$$

$$\frac{4}{8} = \frac{5}{DE}$$

$$4DE = 40$$

$$DE = 10$$ ■

Example 4 Find AD.

Solution $\triangle ADE \sim \triangle ABC$ (Example 1)

Let $AD = x$

then $AB = x + 8$

$$\frac{AD}{AB} = \frac{DE}{BC}$$

$$\frac{x}{x+8} = \frac{6}{10}$$

$$10x = 6x + 48$$

$$4x = 48$$

$$x = 12 \quad \blacksquare$$

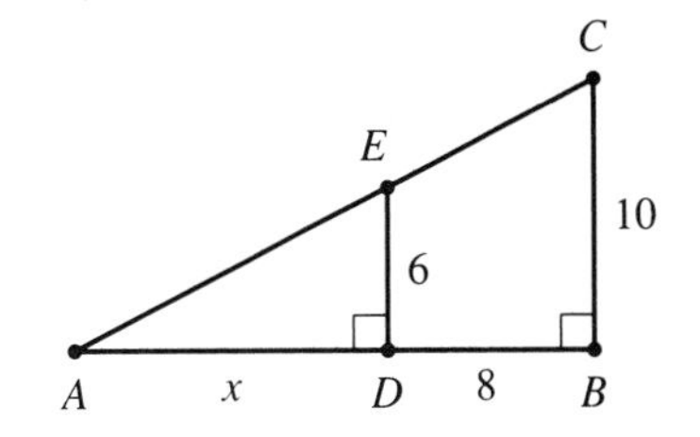

EXERCISES 16.4

In Exercises 1–4, find the missing sides.

1. $\triangle ABC \sim \triangle DEF$

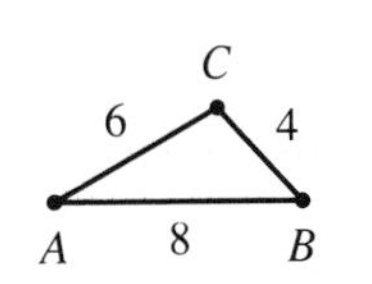

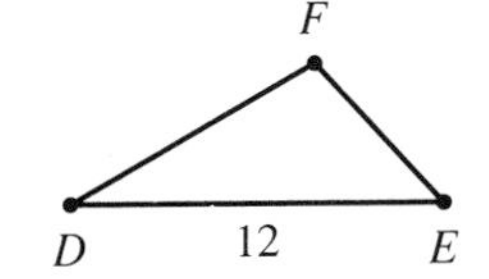

2. $\triangle ABC \sim \triangle DEF$

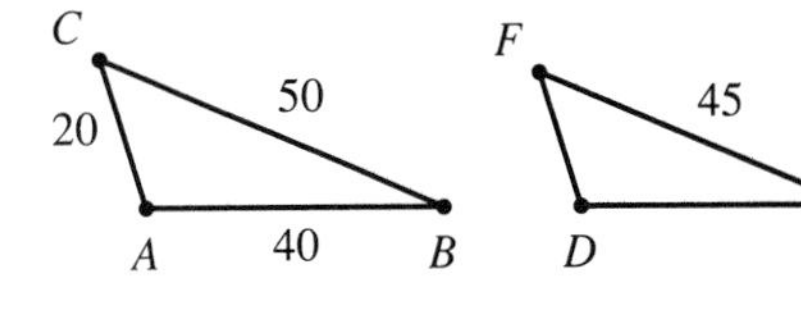

3. $\triangle ABC \sim \triangle DEF$

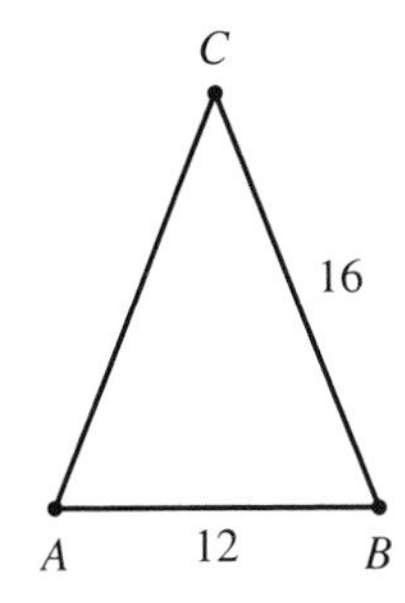

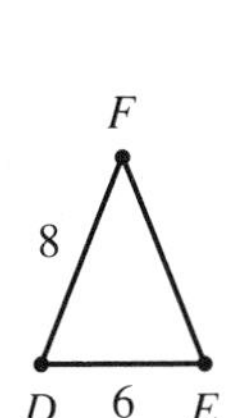

4. $\triangle ABC \sim \triangle XYZ$

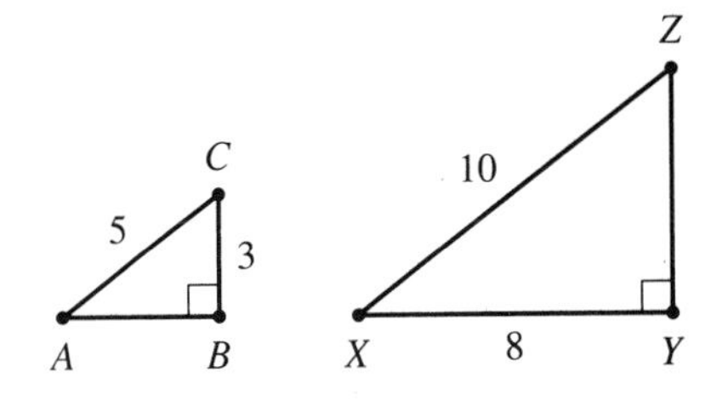

5. Given $\overline{AB} \parallel \overline{CD}$, find AE.

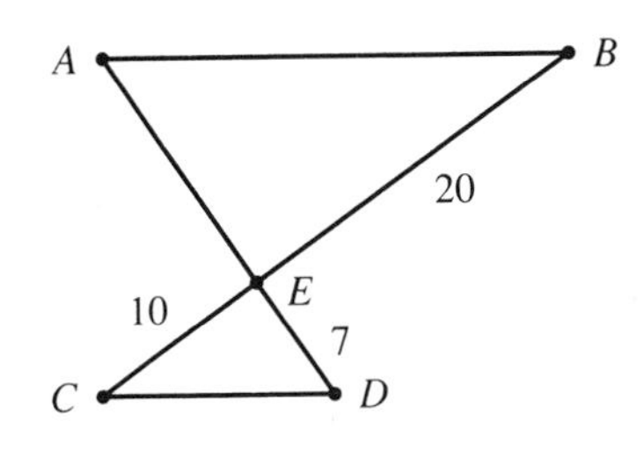

6. Given $\overline{AB} \parallel \overline{CD}$, find DE.

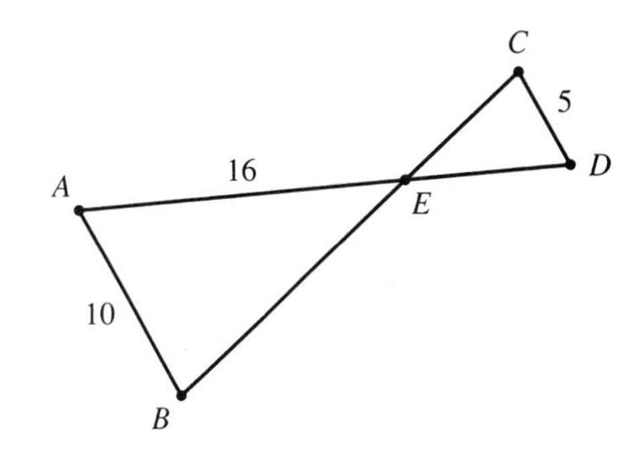

7. Given $\overline{BC} \parallel \overline{DE}$, find BC.

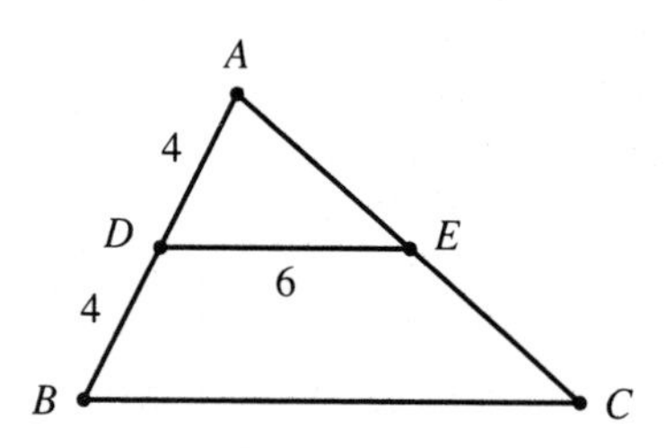

8. Given $\overline{BC} \parallel \overline{DE}$, find BD.

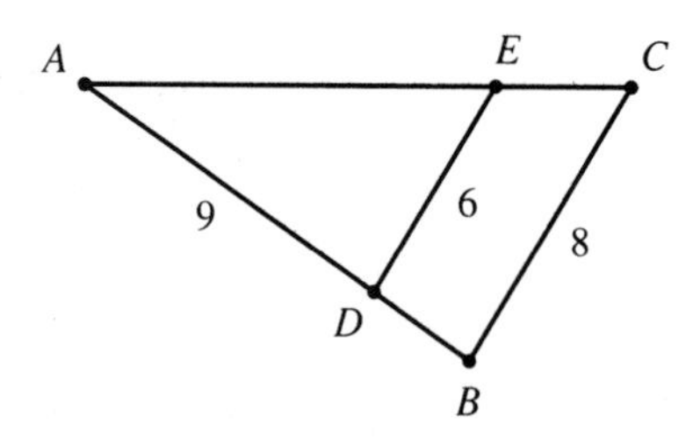

9. Given $\overline{AB} \perp \overline{DE}$, $\overline{AB} \perp \overline{BC}$, find BD.

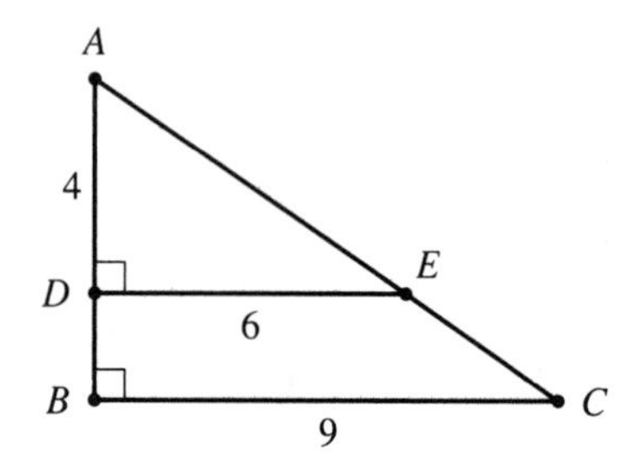

10. Given $\overline{AB} \perp \overline{BC}$, $\overline{DE} \perp \overline{BC}$, find DC.

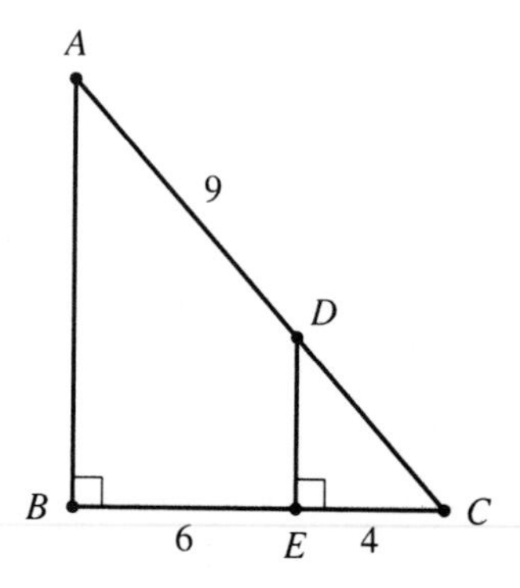

16.5 Perimeter

Perimeter

The word **perimeter** means *the distance around a figure.* To find the perimeter of a geometric figure, we sum the lengths of all its sides. The perimeter of a triangle is the sum of its three sides. Perimeter $= a + b + c$.

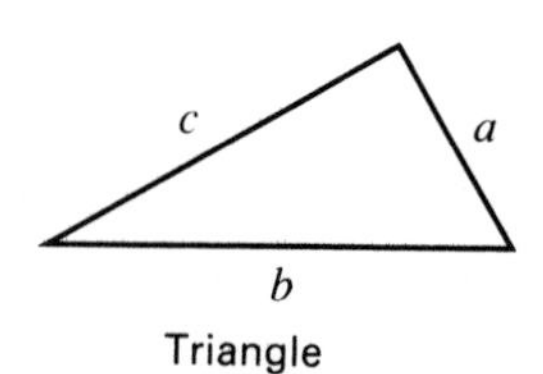

Triangle
$P = a + b + c$

Parallelogram

A **parallelogram** is a quadrilateral whose opposite sides are parallel. It is also true that the opposite sides are equal. The perimeter of a parallelogram $=$ $a + b + a + b = 2a + 2b$.

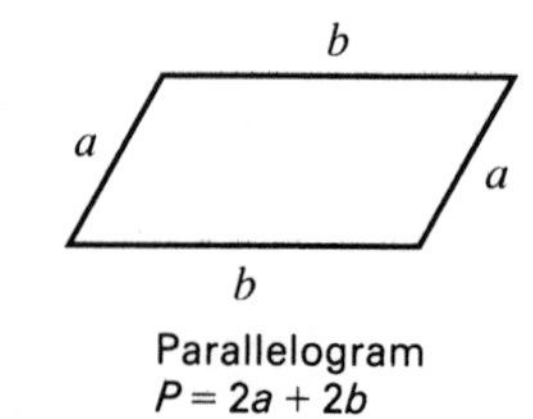

Parallelogram
$P = 2a + 2b$

Rectangle

A **rectangle** is a parallelogram with four right angles. If we label the sides length ℓ and the width w, then the perimeter $= 2\ell + 2w$.

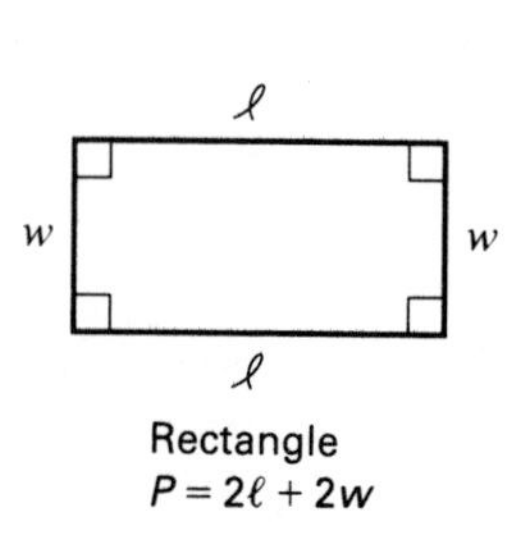

Rectangle
$P = 2\ell + 2w$

Square

A **square** is a rectangle with all sides equal. If we let s represent the length of one side, then the perimeter $=$ $s + s + s + s = 4s$.

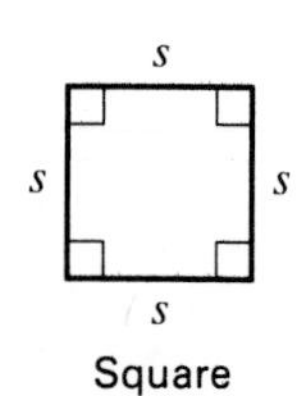

Square
$P = 4s$

Circle

Radius

Diameter

Circumference

An important geometric figure that is not a polygon is the **circle**. All points of a circle are the same distance from a point within it called the *center O*. The **radius** r of the circle is the distance from the center to any point on the circle. The **diameter** d of the circle is the greatest distance across the circle; it is twice the radius. $d = 2r$. The distance around a circle is called its **circumference** C. If we divide the circumference of any circle by its diameter, we always get the same number. That number has been named *pi* (a Greek letter) and is written π.

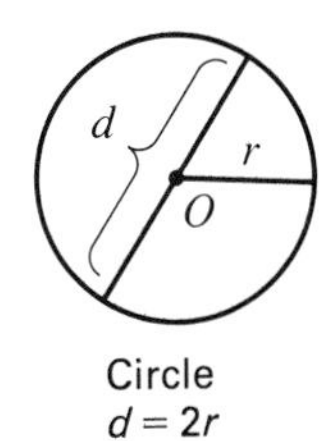

Circle
$d = 2r$
$C = \pi d$
$C = 2\pi r$

$$\frac{\text{circumference}}{\text{diameter}} = \frac{C}{d} = \pi$$

Multiplying both sides by d $\qquad C = \pi d$

And since $d = 2r$ $\qquad C = 2\pi r$

Like $\sqrt{2}$ and $\sqrt{3}$, π is an irrational number (Section 15.4). This means π cannot be written exactly using decimals or fractions. When we want an approximate answer, we will use $\pi \approx 3.14$.

Example 1 A circle has a radius of 5 in. Find the circumference.

Solution $\qquad C = 2\pi r$

$C = 2\pi(5) = 10\pi$ in.

or using $\pi \approx 3.14$, $\qquad C = 10(3.14) = 31.4$ in. ■

Example 2 A rectangle has a length of 15 ft and a perimeter of 50 ft. Find the width.

Solution The formula for the perimeter of a rectangle is $P = 2\ell + 2w$. Substitute the given quantities into the formula, and solve for the unknown.

$$P = 2\ell + 2w$$
$$50 = 2(15) + 2w$$
$$50 = 30 + 2w$$
$$20 = 2w$$
$$10 \text{ ft} = w \quad ■$$

Example 3 Find the distance around each figure.

a.

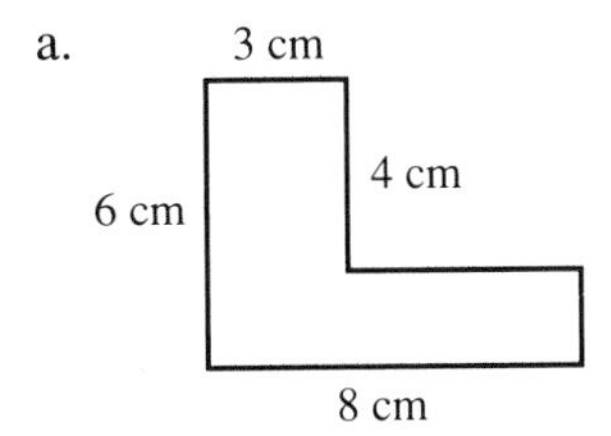

b.

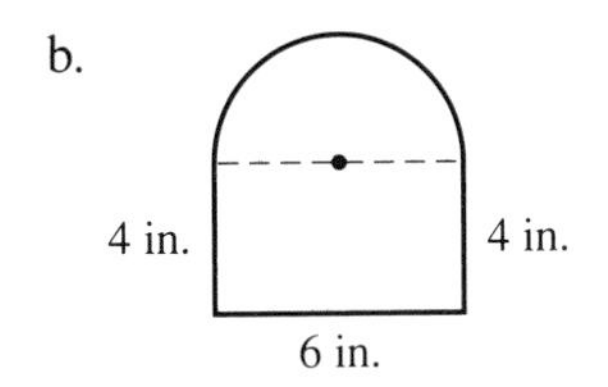

Solution

a. Divide the figure into rectangles to determine the lengths of the missing sides.

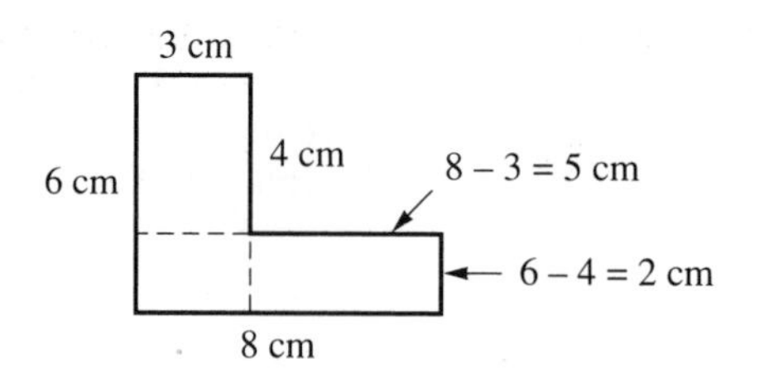

The perimeter is the sum of all the sides.
$P = 6 + 3 + 4 + 5 + 2 + 8$
$P = 28$ cm

b. The diameter of the semicircle (half circle) is 6. The distance along the curved portion of the figure is half the circumference of a circle.

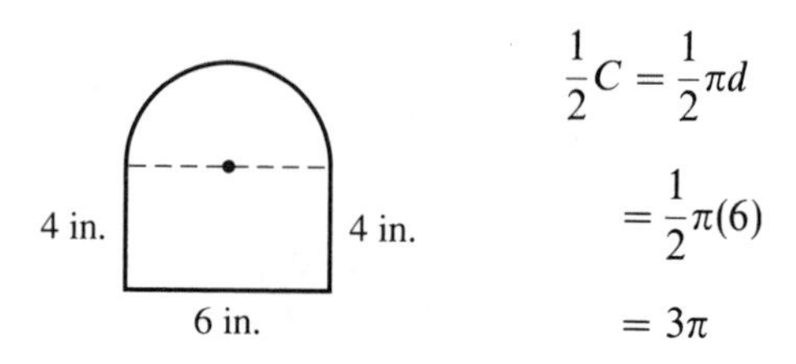

$$\frac{1}{2}C = \frac{1}{2}\pi d$$
$$= \frac{1}{2}\pi(6)$$
$$= 3\pi$$

The total distance around the figure is

$$P = 3\pi + 4 + 6 + 4$$
$$= (3\pi + 14) \text{ in.} \quad ■$$

Example 4 The perimeter of a square is 20 in. Find the length of its diagonal.
Solution First, find the length of a side.

$$P = 4s$$
$$20 = 4s$$
$$5 = s$$

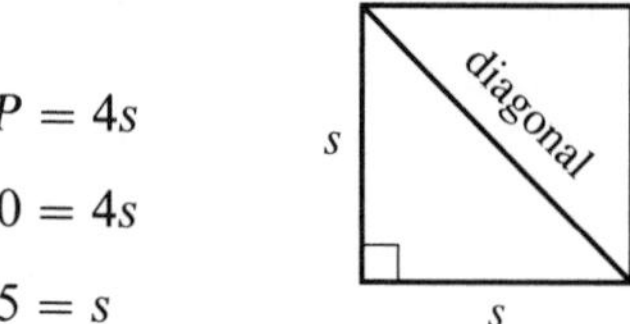

To find the diagonal use the Pythagorean theorem for right triangles.

$$a^2 + b^2 = c^2$$
$$5^2 + 5^2 = c^2$$
$$25 + 25 = c^2$$
$$50 = c^2$$
$$\sqrt{50} = \sqrt{c^2}$$
$$5\sqrt{2} = c$$

Therefore, the diagonal equals $5\sqrt{2}$ in. ■

EXERCISES 16.5

1. Find the perimeter of a triangle with sides 6 ft, 7 ft, and 9 ft.
2. Find the perimeter of a parallelogram with sides 8 m and 12 m.
3. Find the perimeter of a rectangle if the length is 16 in. and the width is 9 in.
4. Find the perimeter of a square if the length of a side is 6 cm.

5. Find the circumference of a circle with a diameter of 4 in.

6. Find the circumference of a circle with a radius of 4 in.

7. If the perimeter of a square is 64 in., find the length of a side.

8. If the perimeter of a rectangle is 32 ft and the width is 7 ft, find the length.

9. The circumference of a circle is 12π ft. Find the radius.

10. The circumference of a circle is 15π cm. Find the diameter.

In Exercises 11–16, find the distance around each figure.

11. 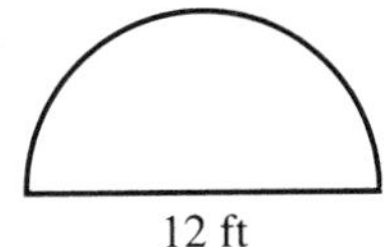

12. 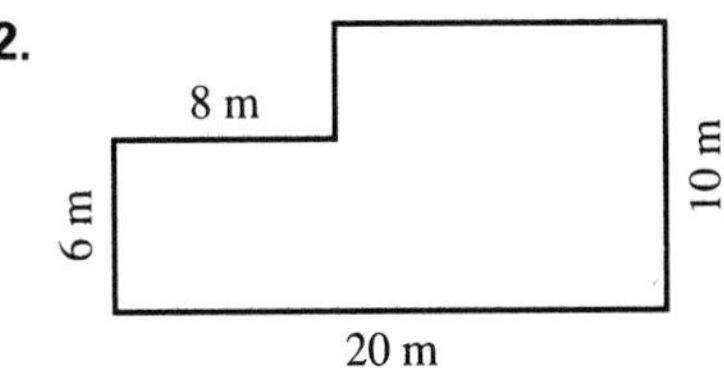

13. 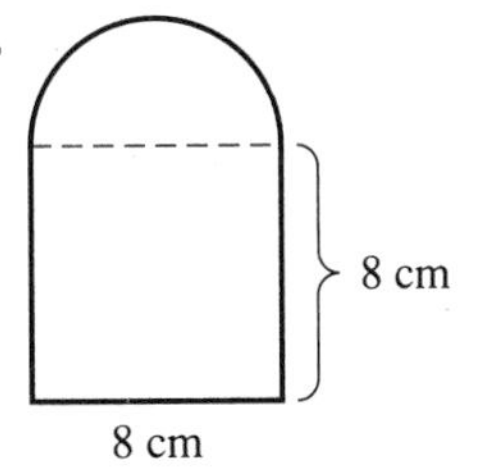

14. 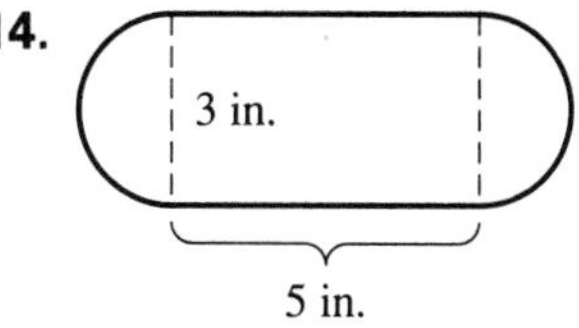

15. 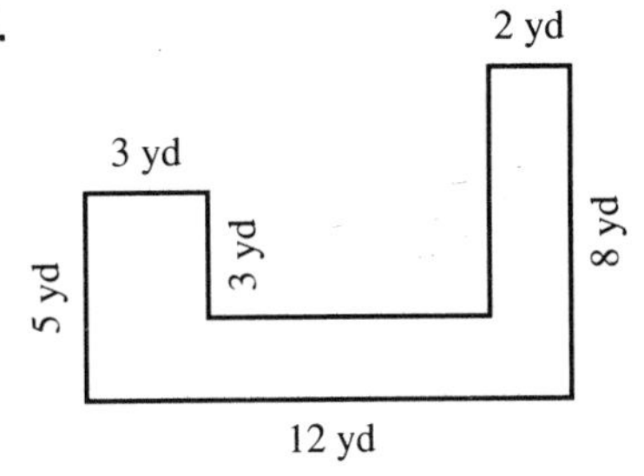

16. 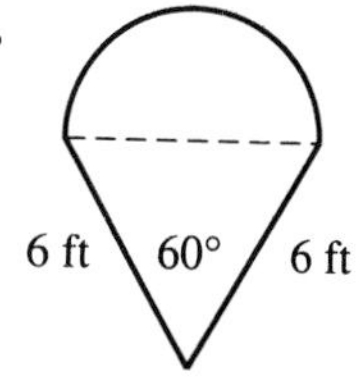

17. Find the perimeter of a right triangle with legs 5 cm and 12 cm.

18. Find the diagonal of a square if its perimeter is 32 ft.

19. In a rectangle, the length is 15 in. and the diagonal is 17 in. Find the perimeter.

20. In a rectangle, the width is 6 m and the diagonal is 10 m. Find the perimeter.

21. The perimeter of a rectangle is 38 yd. If the length is 5 yd less than twice the width, find the length and width.

22. The perimeter of a parallelogram is 80 cm. If one side is 8 cm more than the other side, find the sides of the parallelogram.

16.6 Area

Area

If the side of a square has a length of 1 foot, the square is called a *square foot* (sq. ft). If the side of a square is 1 inch, the square is called a *square inch* (sq. in.), and if the side of a square is 1 meter, the square is called a *square meter* (sq. m). The **area** of a geometric figure is the space inside the lines. Area is measured in square units: square feet, square inches, square meters, and so on.

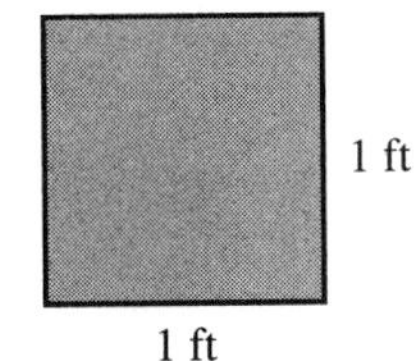

Square foot

To measure the area of the rectangle, we see how many times a unit of area fits into it.

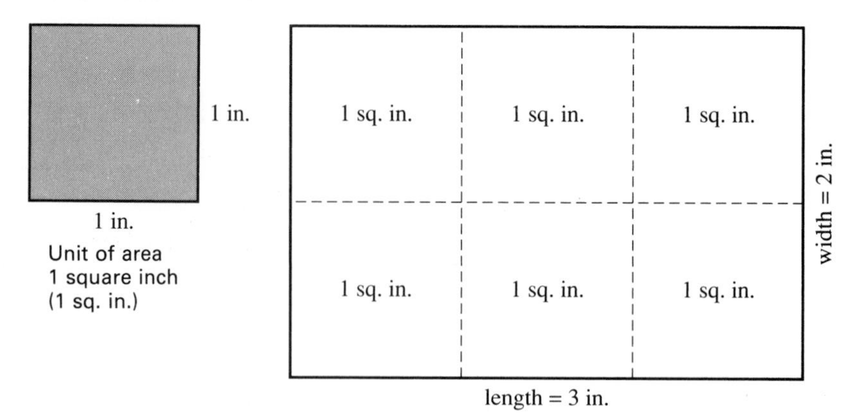

Because the unit of area (1 sq. in.) fits into the space six times, the area of the rectangle is 6 sq. in. We can find this area by multiplying the length times the width.

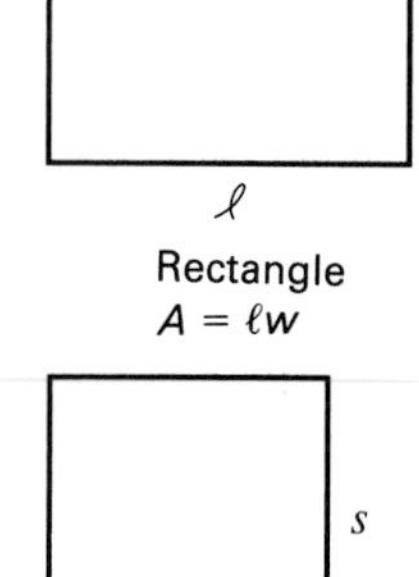

Rectangle
$A = \ell w$

$3 \times 2 = 6$

- Area is 6 sq. in.
- Width is 2 in.
- Length is 3 in.

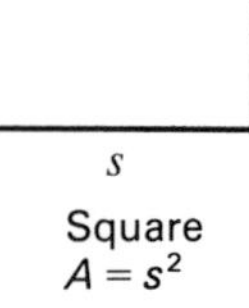

Square
$A = s^2$

A square is a rectangle in which the length and width are equal. If we let s represent the length of a side, then the area of a square $= s \cdot s = s^2$.

Height of a Parallelogram

The **height h of a parallelogram** is the perpendicular distance between a pair of parallel sides. To find the formula for the area of a parallelogram, slide the triangle on the left of the height to the other side of the parallelogram.

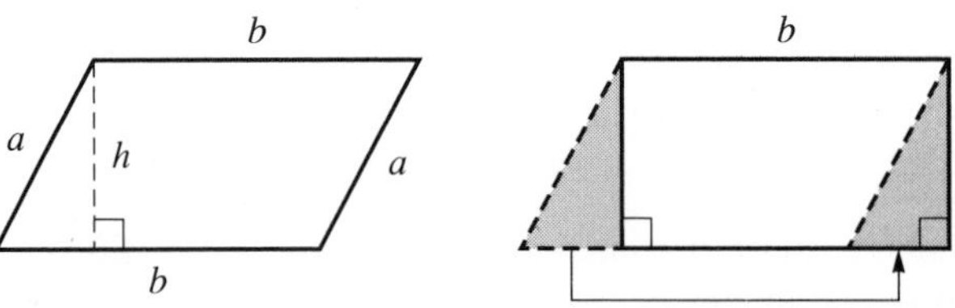

Now we have a rectangle whose area is length times width $= b \cdot h$. Because the areas of the parallelogram and the rectangle are equal, the area of a parallelogram is bh.

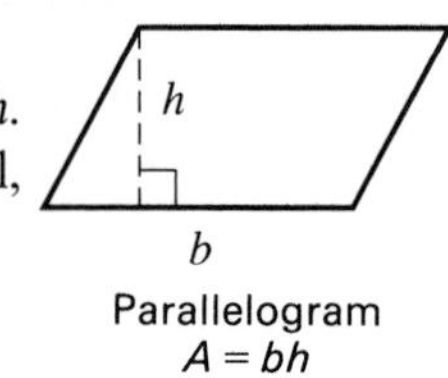

Parallelogram
$A = bh$

Height of a Triangle

The **height h of a triangle** is the perpendicular distance from a vertex to the opposite side.

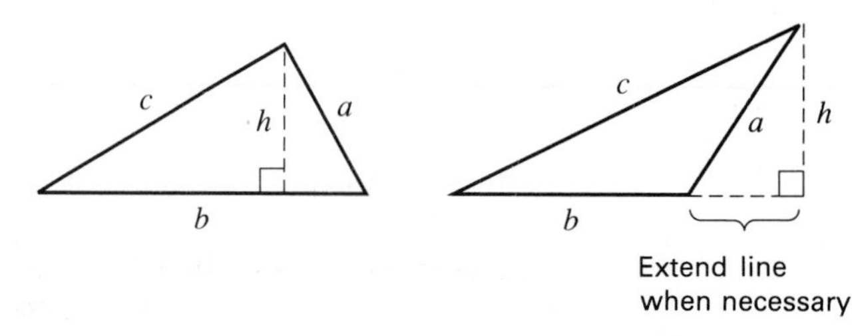

To find the formula for the area of a triangle, draw parallelogram $ABCD$ with diagonal $\overline{BD}$.

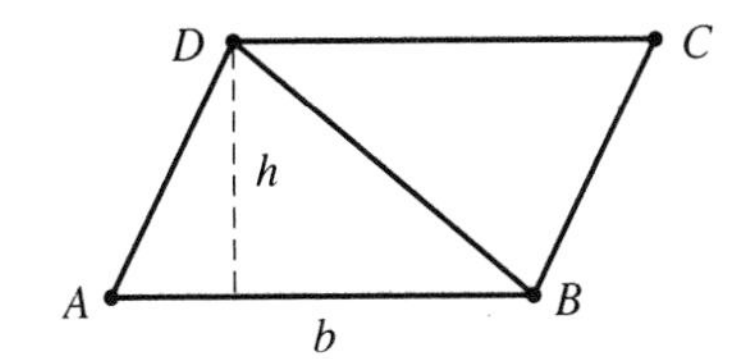

The diagonal $\overline{BD}$ divides the parallelogram into two triangles of equal area. Since the area of the parallelogram is bh, the area of each triangle must be $\frac{1}{2}bh$.

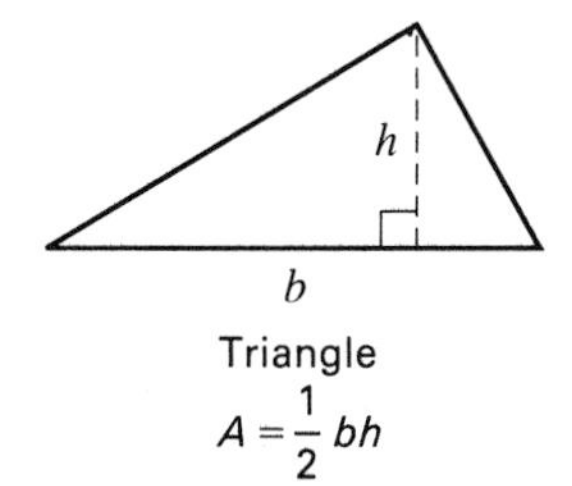

Triangle
$A = \frac{1}{2}bh$

Area of a Circle

A circle with radius r is shown on the right. The formula for the **area of a circle** is πr^2.

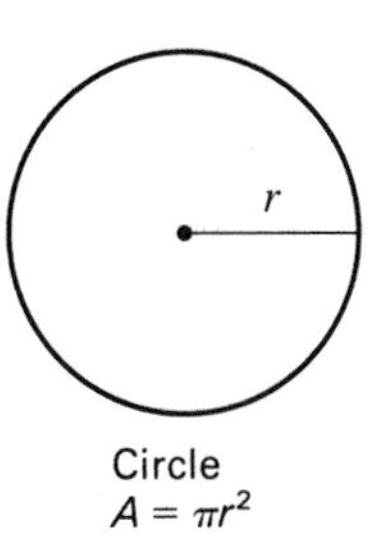

Circle
$A = \pi r^2$

NOTE A complete list of all formulas for this chapter is given in the Chapter Summary, Section 16.8. ☑

Example 1 Find the area of each figure.

a.

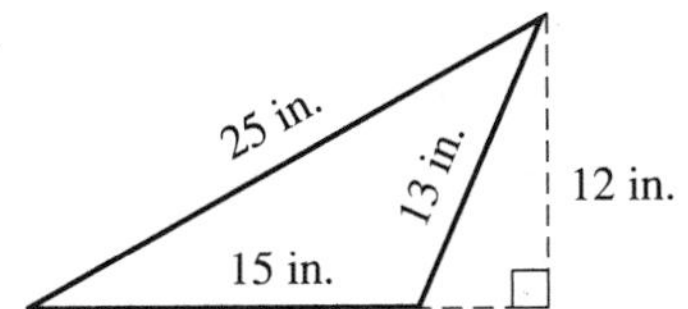

b.

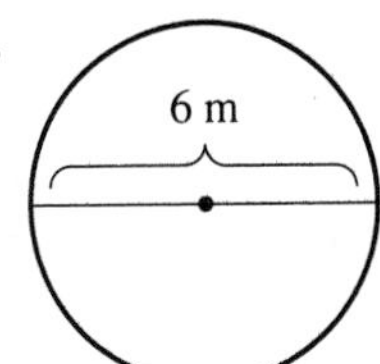

c.

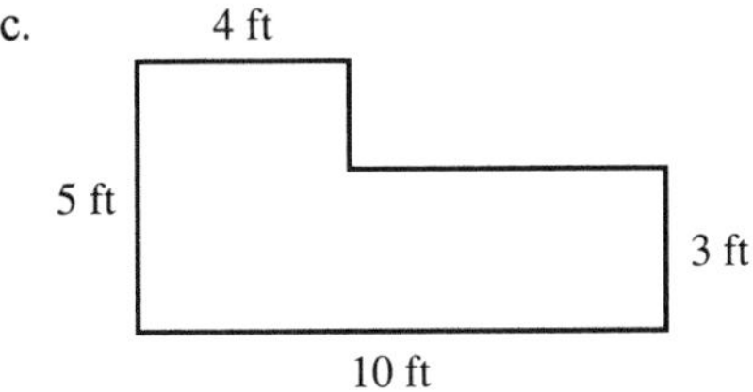

d.

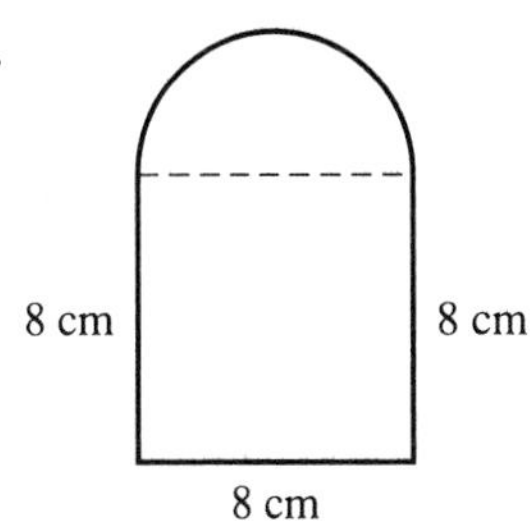

Solution

a. Notice that the height is 12 in. and not 13 in., because the height is measured perpendicularly to the base.

$$A = \frac{1}{2}bh$$

$$= \frac{1}{2} \cdot 15 \cdot 12$$

$$= 90 \text{ sq. in.} \quad \text{or} \quad 90 \text{ in.}^2$$

Area is measured in square units

b. The radius of a circle is half the diameter. Since the diameter = 6 m, the radius = 3 m.

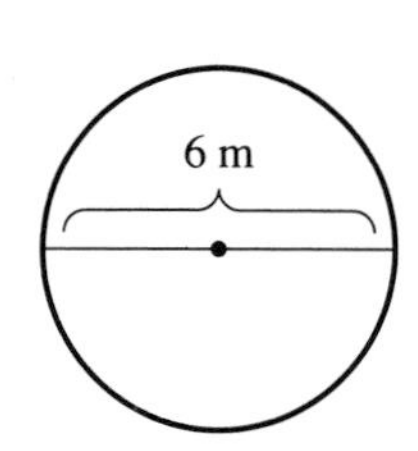

$$A = \pi r^2$$
$$= \pi(3)^2$$
$$= 9\pi \text{ sq. m} \quad \text{or} \quad 9\pi \text{ m}^2$$

c. Divide the figure into two rectangles, then add the areas of the rectangles to find the total area of the figure.

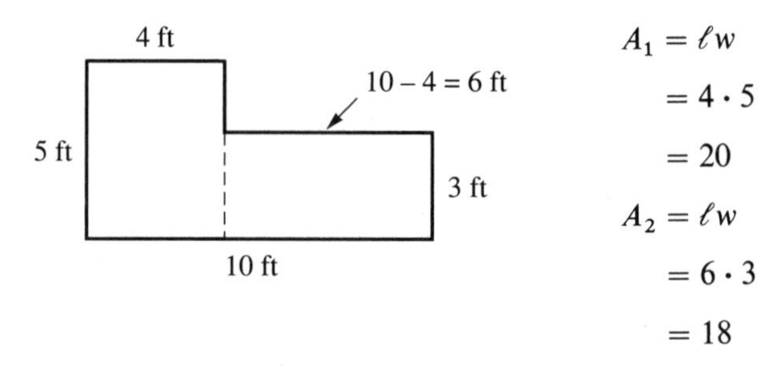

$$A_1 = \ell w$$
$$= 4 \cdot 5$$
$$= 20$$
$$A_2 = \ell w$$
$$= 6 \cdot 3$$
$$= 18$$

Total area $= A_1 + A_2 = 20 + 18 = 38$ sq. ft.

d. Because $d = 8$ cm, $r = 4$ cm. The area of the semicircle is half the area of a circle.

Half area of a circle $= A_1 = \frac{1}{2}\pi r^2$

$$= \frac{1}{2}\pi(4)^2$$
$$= \frac{1}{2}\pi 16$$
$$= 8\pi \text{ cm}^2$$

8 cm 8 cm 8 cm

Area of a square $= A_2 = s^2$

$$= (8)^2$$
$$= 64 \text{ cm}^2$$

Total area $= A_1 + A_2 = (8\pi + 64) \text{ cm}^2$ ■

Example 2 The perimeter of a rectangle is 28 in. and the length is 8 in. Find its area.

Solution Use the formula for perimeter of a rectangle to find the width.

$$P = 2\ell + 2w$$
$$28 = 2(8) + 2w$$
$$28 = 16 + 2w$$
$$12 = 2w$$
$$6 = w$$
$$A = \ell w$$
$$= (8)(6)$$
$$= 48 \text{ sq. in.}$$ ■

w

8 in.

Example 3 Find the area of a right triangle with hypotenuse 5 m and leg 3 m.

Solution When finding the area of a triangle, the height and base must be perpendicular. Therefore, we must find AC first.

Because $\triangle ABC$ is a right triangle, use the Pythagorean theorem to find AC.

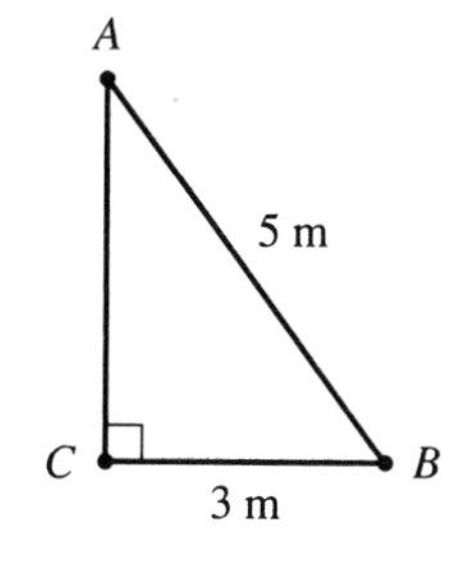

$$\begin{aligned} a^2 + b^2 &= c^2 \\ 3^2 + b^2 &= 5^2 \\ 9 + b^2 &= 25 \\ b^2 &= 16 \\ b &= 4 \\ AC &= 4 \end{aligned}$$

Then, $A = \frac{1}{2}bh$

$= \frac{1}{2}(3)(4) = 6 \text{ m}^2$ ■

EXERCISES 16.6

1. Find the area of a rectangle with length 12 ft and width 8 ft.
2. Find the area of a square with side 6 m.
3. Find the area of a triangle with height 3 in. and base 5 in.
4. Find the area of a parallelogram with height 10 cm and base 18 cm.
5. Find the area of a circle with radius 8 ft.
6. Find the area of a circle with diameter 8 in.

In Exercises 7–15, find the area of each figure.

7.

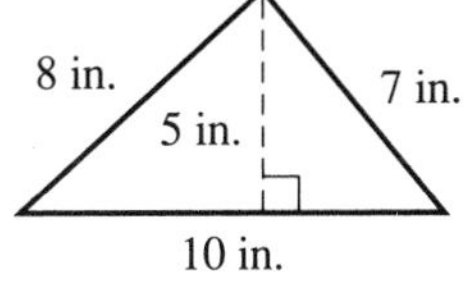

8.

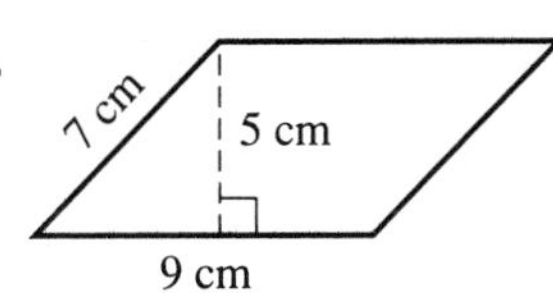

9.

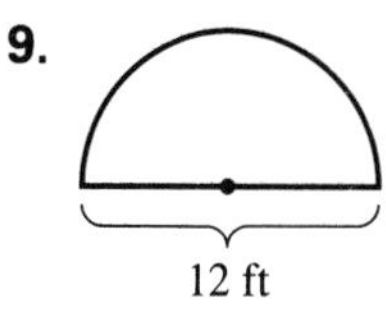

10.

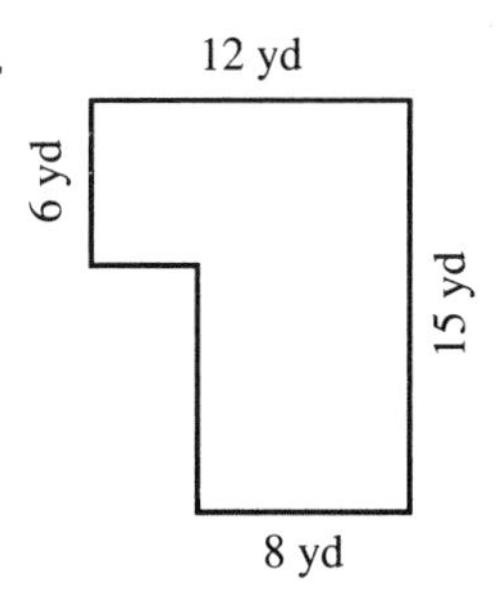

11.

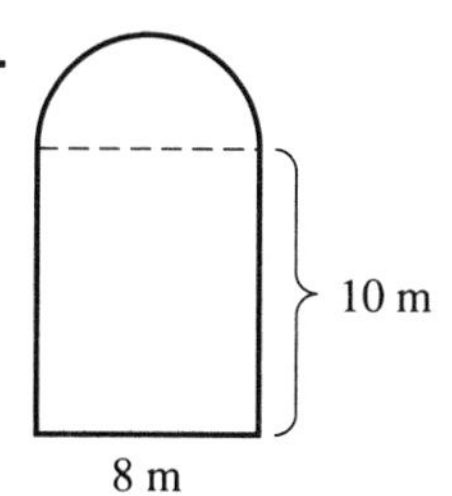

12.

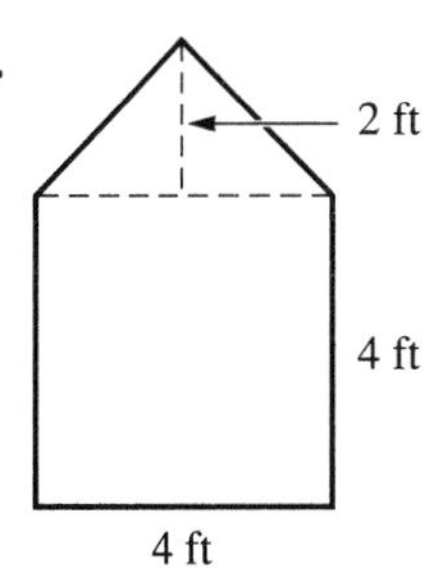

13.

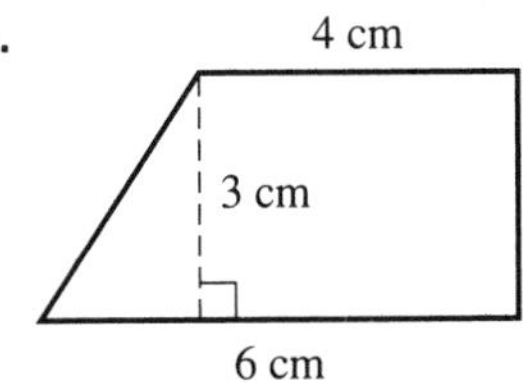

14.

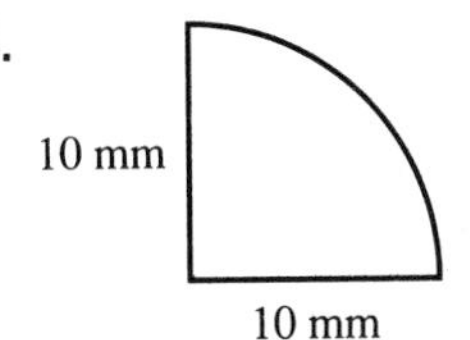

15.

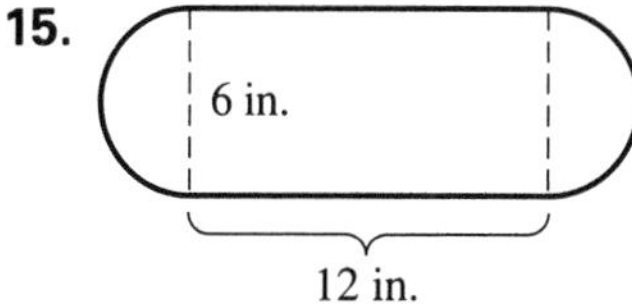

16. If the area of a rectangle is 60 cm^2 and the width is 5 cm, find the length.

17. If the area of a right triangle is 30 sq. in. and one leg is 12 in., find the other leg and the hypotenuse.

18. If the area of a circle is 36π sq. ft, find the radius.

19. If the area of a square is 64 m^2, find the length of a side.

20. Find the number of square inches in a square foot.

21. Find the number of square feet in a square yard.

22. Find the area of the circle if the side of the square is 4 in.

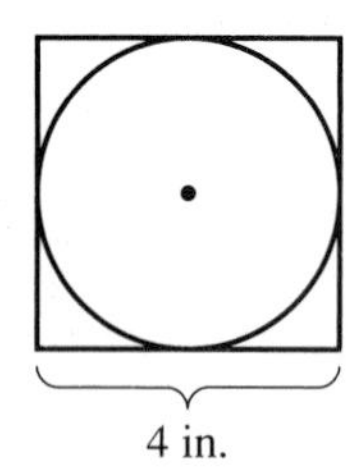

23. Find the area of the square if the radius of the circle is 7 in.

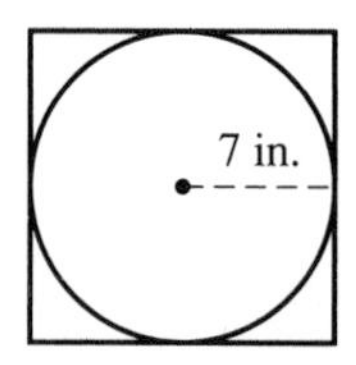

24. If the perimeter of a square is 20 cm, find its area.

25. If the circumference of a circle is 10π ft, find its area.

26. Given isosceles ΔABC with $AC = BC = 10$ m and $CD = 8$ m, find the area.

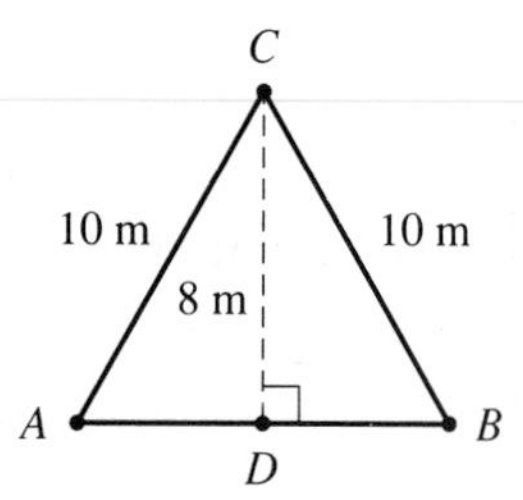

27. Given that the area of parallelogram $ABCD$ is 32 sq. in., $AD = 5$ in., and $DE = 4$ in., find the perimeter.

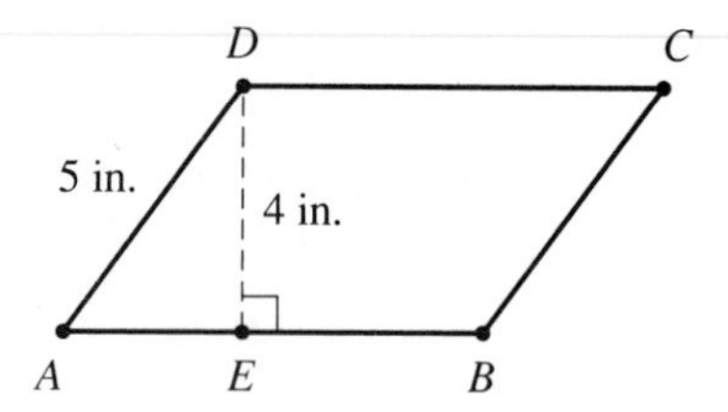

28. Let x = side of a square, then its area $= x^2$. If we double the length of the side, $s = 2x$, find how many times the area will increase.

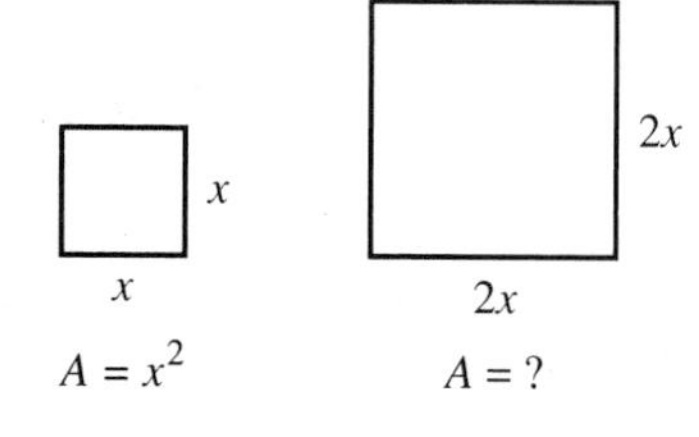

29. Let x = side of a square, then its area $= x^2$. If we triple the length of the side, $s = 3x$, find how many times the area will increase.

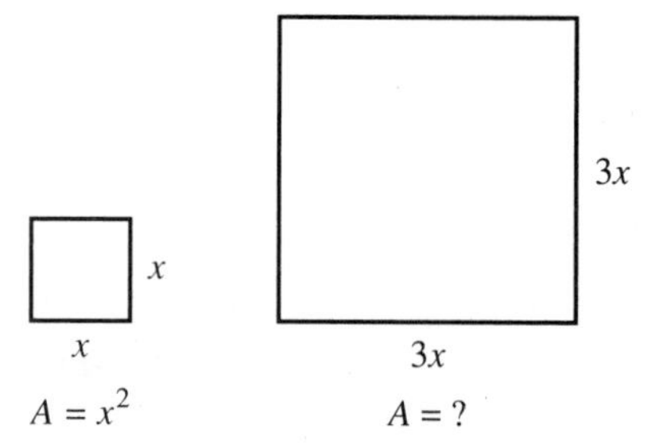

16.7 Volume

Volume

A square box having equal length, width, and height is called a cube. If the length, width, and height are each 1 foot, the cube is called a *cubic foot* (cu. ft). If the length, width, and height are 1 inch, the cube is called a *cubic inch* (cu. in.). If the length, width, and height are each 1 yard, the cube is called a *cubic yard* (cu. yd). The **volume** of any container is a measure of the space inside that container. Volume is often measured in cubic units: cubic feet, cubic inches, cubic meters, and so on.

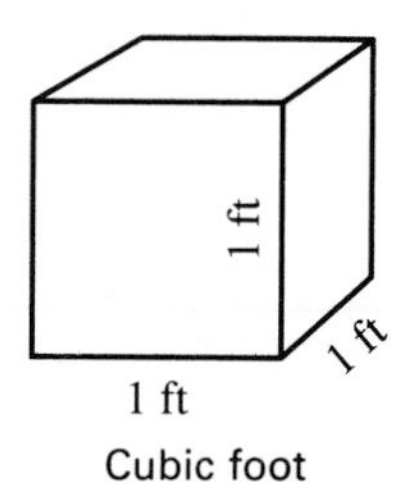

Cubic foot

Rectangular Box

We often need to know the volume of a **rectangular box**. The top, bottom, and all sides of a rectangular box are rectangles. Examples of rectangular boxes are most classrooms, most rooms in houses and apartments, shipping boxes and crates, Kleenex boxes, laundry soap boxes, and so on.

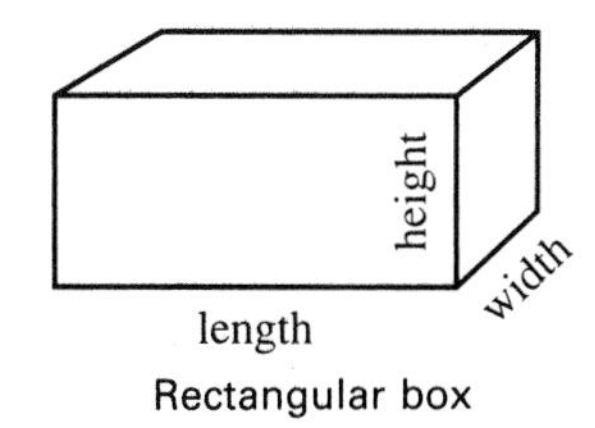

Rectangular box

A rectangular box is shown. It has a length of 4 inches, a width of 3 inches, and a height of 2 inches.

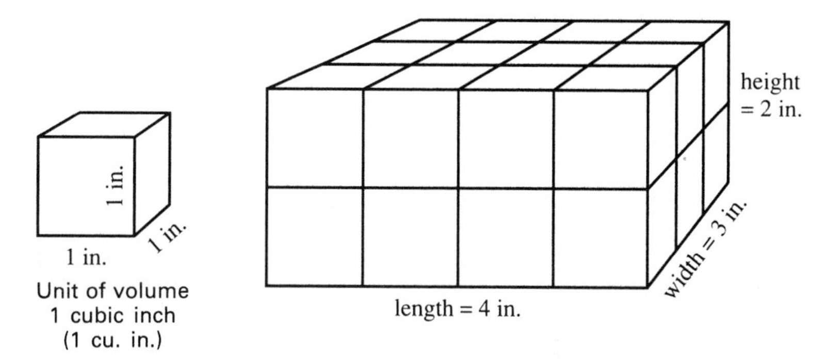

Unit of volume
1 cubic inch
(1 cu. in.)

The unit of volume (1 cu. in.) fits into the top layer of the box 12 times. Since the box is made up of 2 layers each containing 12 cu. in., the volume of the box is $2 \cdot 12 = 24$ cu. in. If the height were 5 inches, there would be 5 layers each containing 12 cu. in., so that the volume would be $5 \cdot 12 = 60$ cu. in.

Thus, we can find the volume of a rectangular box by multiplying the length $\times$ width $\times$ height.

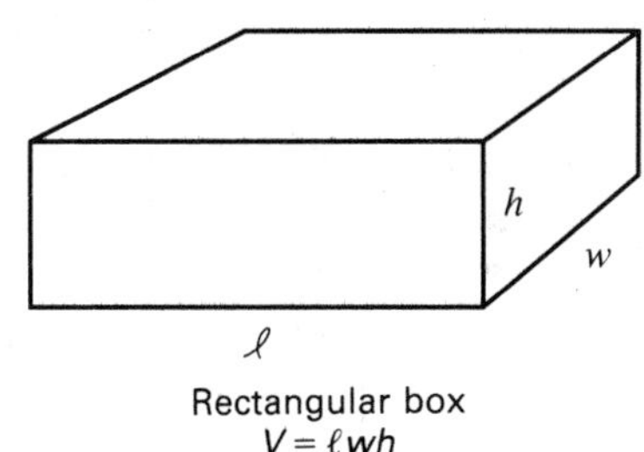

Rectangular box
$V = \ell wh$

Cube

A **cube** is a rectangular box in which the length, width, and height are all equal. If we represent the side by the letter s, then the volume of a cube is $s \cdot s \cdot s = s^3$.

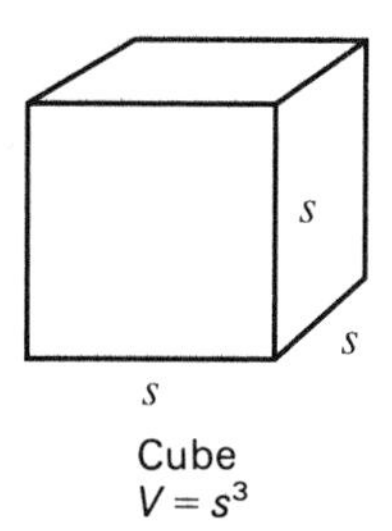

Cube
$V = s^3$

Cylinder

A right circular **cylinder** is shown. This cylinder is called *circular* because the top and bottom are circles. This cylinder is called *right* because the top and bottom form square corners with the sides. We can find the volume of the cylinder by multiplying the area of the bottom (πr^2) times the height h. Thus, $V = \pi r^2 h$.

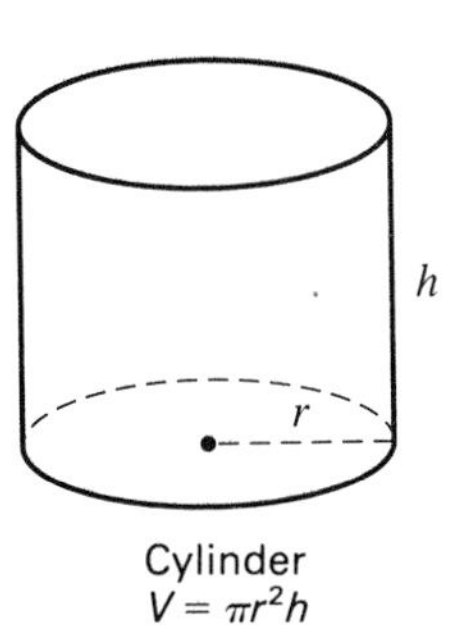

Cylinder
$V = \pi r^2 h$

Sphere

A **sphere** is shown. If the radius (r) is the distance from the center to any point on the sphere, the formula for the volume of a sphere is $V = \frac{4}{3}\pi r^3$.

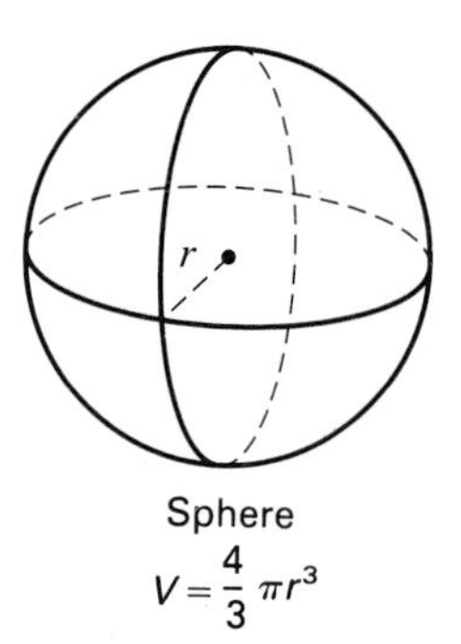

Sphere
$V = \frac{4}{3}\pi r^3$

Example 1 Find the volume of a classroom having a length of 30 feet, a width of 25 feet, and a height of 8 feet.

Solution

$$V = \ell wh$$
$$= 30 \cdot 25 \cdot 8$$
$$= 6{,}000 \text{ cu. ft} \quad \text{or} \quad 6{,}000 \text{ ft}^3$$

Volume is measured in cubic units ■

Example 2 An aquarium in the shape of a cube measures 10 in. on a side.

a. Find the volume of the aquarium.

b. Find the weight of the water in the aquarium if 1 cubic inch of water weighs 0.0361 pounds.

Solution

a.
$$V = s^3$$
$$= (10)^3$$
$$= 1000 \text{ cu. in.} \quad \text{or} \quad 1000 \text{ in.}^3$$

b. Weight of water $= 0.0361 \times 1000$
$$= 36.1 \text{ pounds}$$ ■

Example 3 Find the volume of a sphere with diameter 12 cm.
Solution Because the diameter = 12 cm, the radius = 6 cm.

$$V = \frac{4}{3}\pi r^3$$
$$= \frac{4}{3}\pi(6)^3$$
$$= \frac{4}{\cancel{3}_1} \cdot \pi \cdot \cancel{6}^2 \cdot 6 \cdot 6$$
$$= 288\pi \text{ cm}^3$$ ■

Example 4 The figure shown is a cylinder capped with a hemisphere (half sphere). Find the total volume.
Solution The volume of the hemisphere is half the volume of a sphere.

$$\text{Half volume of a sphere} = V_1 = \frac{1}{2}\cdot\frac{4}{3}\pi r^3$$
$$= \frac{1}{2}\cdot\frac{4}{3}\pi(3)^3$$
$$= \frac{1}{\cancel{2}_1}\cdot\frac{\cancel{4}^2}{\cancel{3}_1}\cdot\pi\cdot\cancel{27}^9$$
$$= 18\pi \text{ m}^3$$

$$\text{Volume of a cylinder} = V_2 = \pi r^2 h$$
$$= \pi(3)^2(5)$$
$$= \pi\cdot 9\cdot 5$$
$$= 45\pi \text{ m}^3$$

$$\text{Total volume} = V_1 + V_2 = 18\pi + 45\pi = 63\pi \text{ m}^3 \quad ■$$

EXERCISES 16.7

1. Find the volume of a rectangular box with length 15 ft, width 10 ft, and height 6 ft.

2. Find the volume of a cube with side 5 m.

3. Find the volume of a cylinder with radius 8 in. and height 10 in.

4. Find the volume of a cylinder with diameter 8 cm and height 12 cm.

5. Find the volume of a sphere with radius 6 in.

6. Find the volume of a sphere with diameter 6 ft.

In Exercises 7–9, find the total volume.

7.

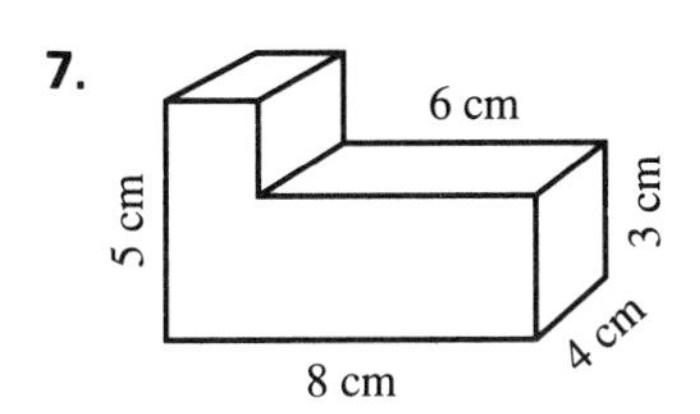

8.

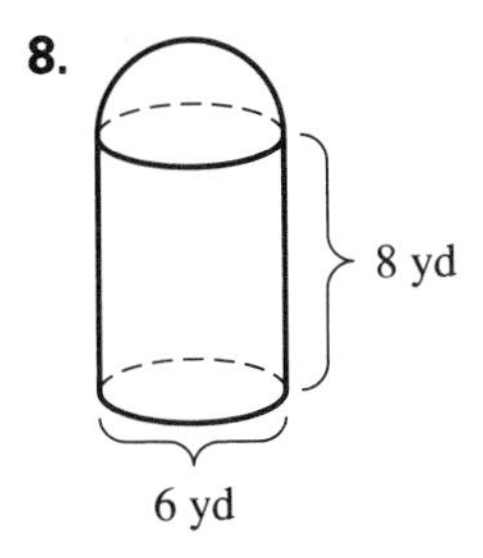

9.

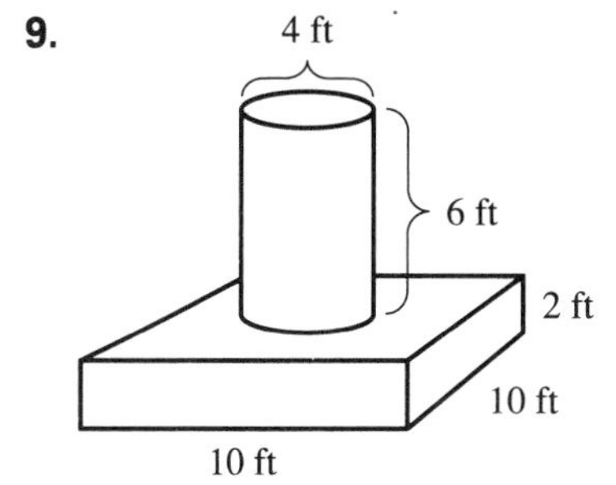

10. Find the number of cubic feet in a cubic yard.

11. Find the number of cubic inches in a cubic foot.

12. The volume of a rectangular box is 90 cm^3. If the length is 6 cm and the width is 5 cm, find the height.

13. The volume of a cylinder is 200π cu. in. If the radius is 5 in., find the height.

14. A classroom measures 10 yards long, 9 yards wide, and 3 yards high.
a. Find the volume.
b. If air weighs about 2 pounds per cubic yard, find the weight of the air in this room.

15. An aquarium measures 24 inches long, 10 inches wide, and is filled to a depth of 15 inches.
a. Find the volume.
b. Find the weight of the water in the tank if 1 cubic inch of water weighs 0.0361 pounds. (Round off to the nearest pound.)

16. A cylindrical cistern is 16 feet deep and 12 feet in diameter.
a. Find its volume.
b. If 1 cubic foot equals 7.48 gallons, how many gallons of water will the cistern hold? (Use $\pi \approx 3.14$ and round off to the nearest gallon.)

17. A cylindrical water tank measures 20 inches in diameter and 50 inches high.
a. Find its volume.
b. If 1 gallon equals 231 cubic inches, how many gallons of water will the tank hold? (Use $\pi \approx 3.14$ and round off to the nearest gallon.)

18. Let x = side of a cube; then its volume $= x^3$. If we double the length of the side, $s = 2x$, find how many times the volume will increase.

19. Let x = side of a cube, then its volume $= x^3$. If we triple the length of the side, $s = 3x$, find how many times the volume will increase.

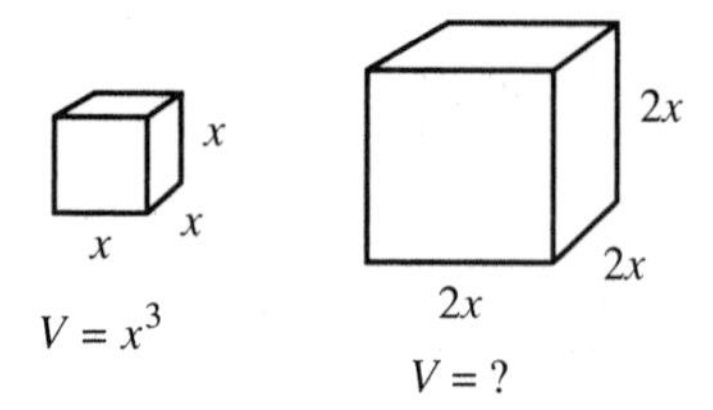

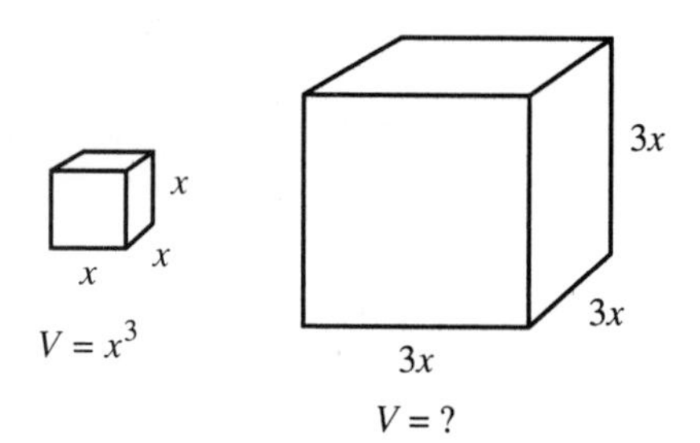

16.8 Chapter Summary

Angles 16.1 Two angles are *complementary* if their sum equals 90°. Two angles are *supplementary* if their sum equals 180°.

Lines 16.1 When two lines intersect, the *vertical angles* are equal.

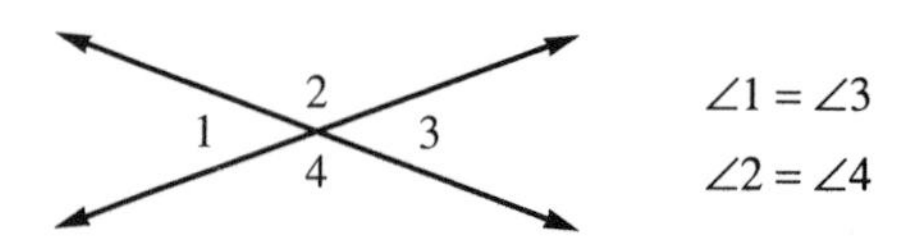

When two lines are *perpendicular*, they form a right angle (90°).

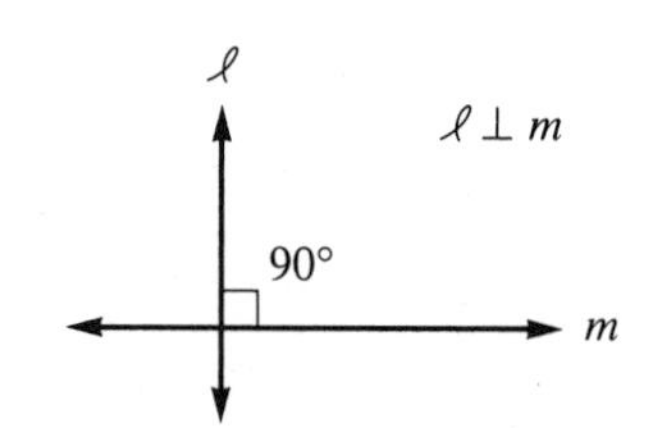

When two lines are *parallel:*

1. The *alternate interior angles* are equal. ($\angle 3 = \angle 6$, and $\angle 4 = \angle 5$)
2. The *alternate exterior angles* are equal. ($\angle 1 = \angle 8$, and $\angle 2 = \angle 7$)
3. The *corresponding angles* are equal. ($\angle 1 = \angle 5$, $\angle 2 = \angle 6$, $\angle 3 = \angle 7$, and $\angle 4 = \angle 8$)

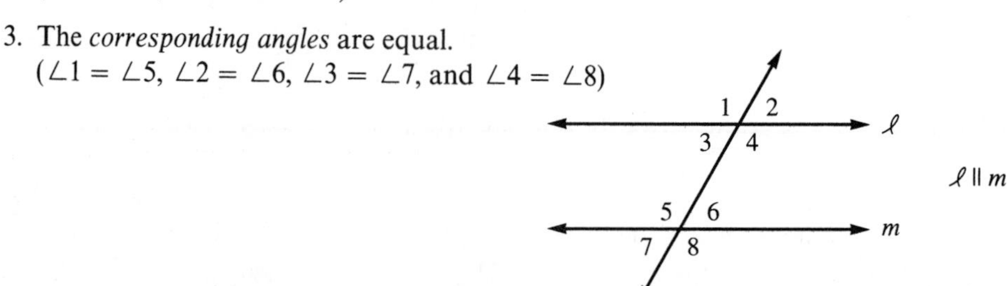

Triangles 16.2 The sum of the angles of a triangle equals 180°.

An *equilateral triangle* is a triangle with three equal sides and three equal angles.

An *isosceles triangle* has two equal sides and two equal angles.

A *scalene triangle* has no sides equal.

A *right triangle* has one right angle.

Pythagorean Theorem 16.2 In a right triangle the square of the hypotenuse equals the sum of the squares of the other two sides.

$$a^2 + b^2 = c^2$$

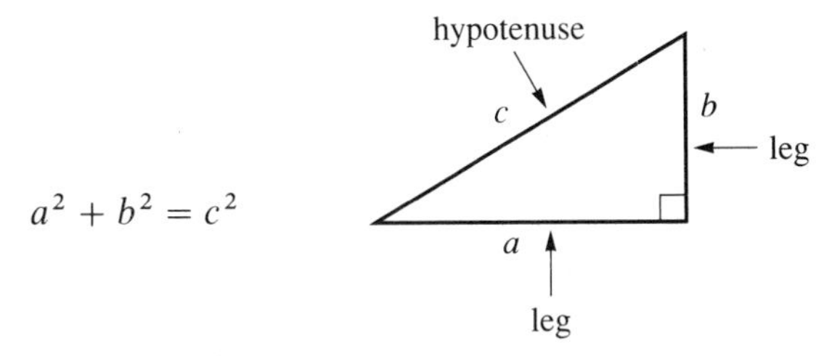

Congruent Triangles 16.3 If two triangles are congruent, their corresponding angles are equal, and their corresponding sides are equal.

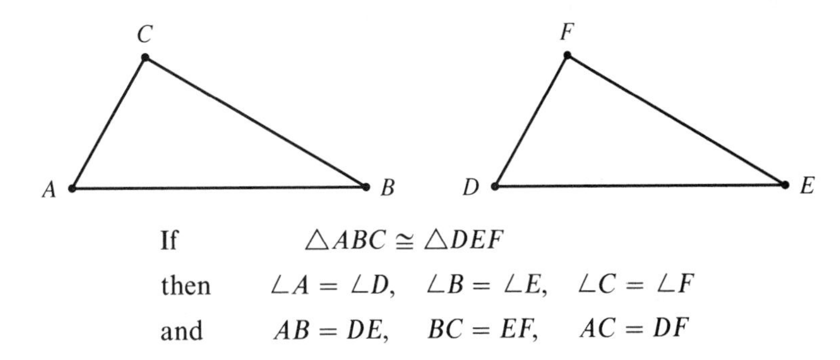

If $\triangle ABC \cong \triangle DEF$

then $\angle A = \angle D, \quad \angle B = \angle E, \quad \angle C = \angle F$

and $AB = DE, \quad BC = EF, \quad AC = DF$

SSS (side-side-side)
If three sides of one triangle are equal to three sides of another triangle, the triangles are congruent.

SAS (side-angle-side)
If two sides and the included angle of one triangle are equal to two sides and the included angle of another triangle, the triangles are congruent.

ASA (angle-side-angle)
If two angles and the included side of one triangle are equal to two angles and the included side of another triangle, the triangles are congruent.

Similar Triangles 16.4 If two angles of one triangle are equal to two angles of another triangle, then the triangles are *similar*.

If two triangles are similar, their corresponding sides are proportional.

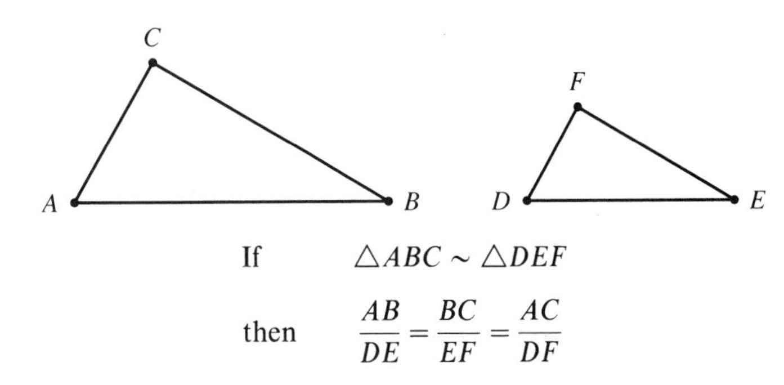

If $\triangle ABC \sim \triangle DEF$

then $$\frac{AB}{DE} = \frac{BC}{EF} = \frac{AC}{DF}$$

Perimeter and Area
16.5,16.6

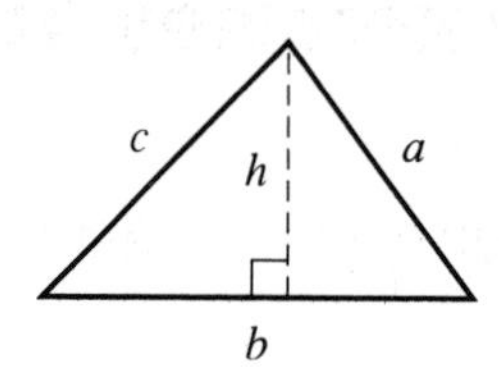

Triangle

$P = a + b + c$

$A = \frac{1}{2}bh$

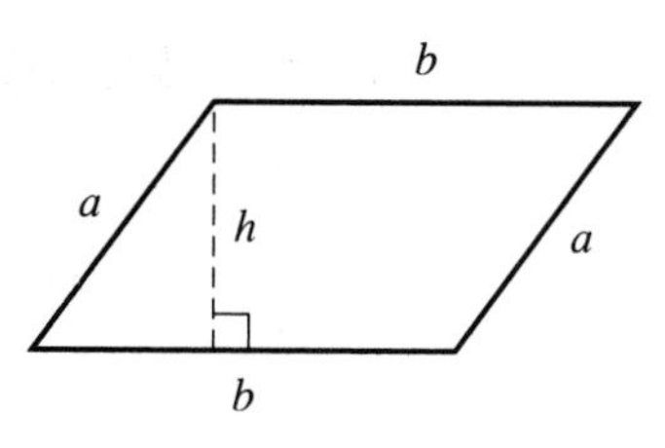

Parallelogram

$P = 2a + 2b$

$A = bh$

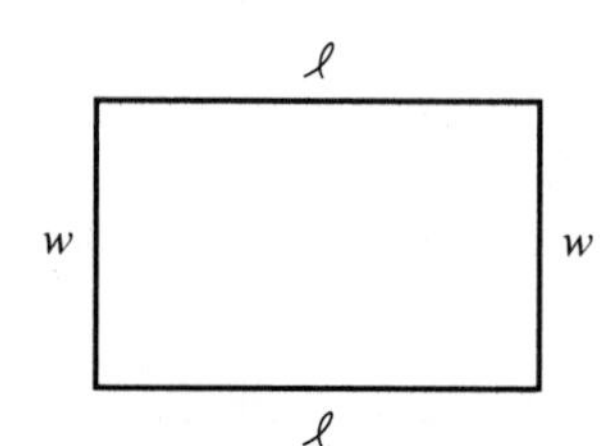

Rectangle

$P = 2\ell + 2w$

$A = \ell w$

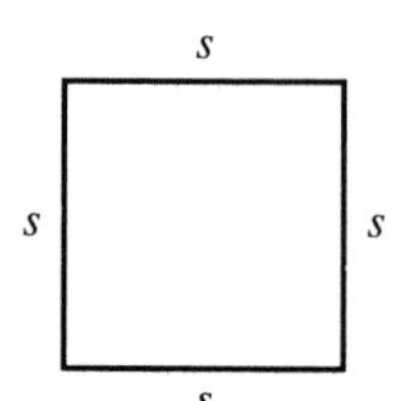

Square

$P = 4s$

$A = s^2$

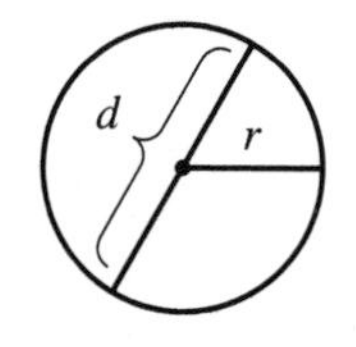

Circle

$d = 2r$

$C = \pi d$

$C = 2\pi r$

$A = \pi r^2$

Volume
16.7

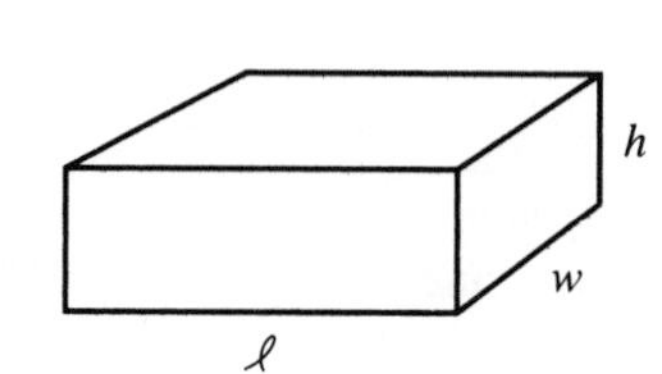

Rectangular box

$V = \ell wh$

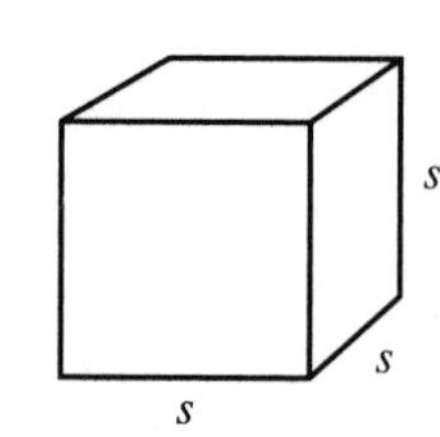

Cube

$V = s^3$

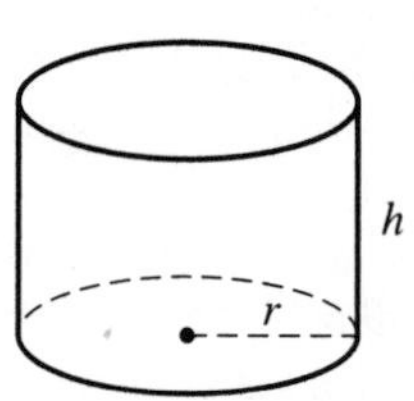

Cylinder

$V = \pi r^2 h$

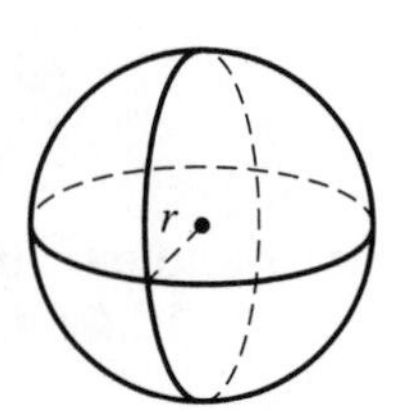

Sphere

$V = \frac{4}{3}\pi r^3$

Review Exercises 16.8

1. Find two complementary angles if one angle is five times larger than the other angle.

2. Find two supplementary angles if one angle is 40° less than the other angle.

3. Given $\overline{AB} \parallel \overline{DE}$, $\angle 6 = 30°$ and $\angle 7 = 65°$. Find the remaining angles.

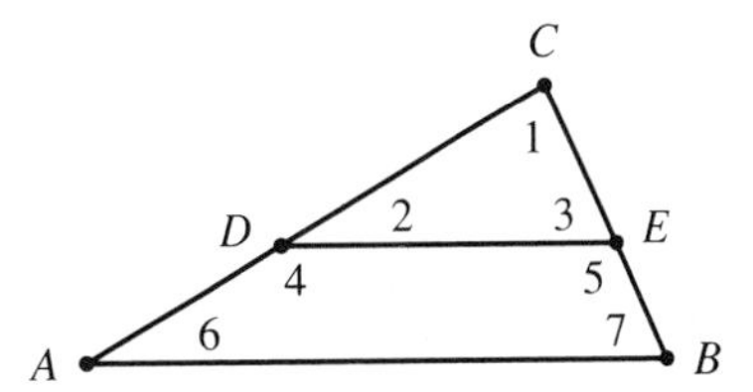

4. Given $\ell \parallel m$, $\angle 1 = 35°$, and $\angle 7 = 75°$. Find the remaining angles.

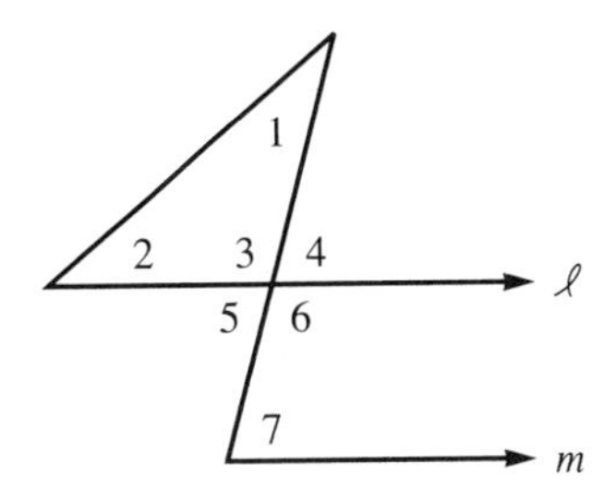

5. In quadrilateral $ABCD$, $\overline{AB} \perp \overline{BC}$, $AB \perp \overline{AD}$, and $\angle C = 52°$. Find $\angle D$.

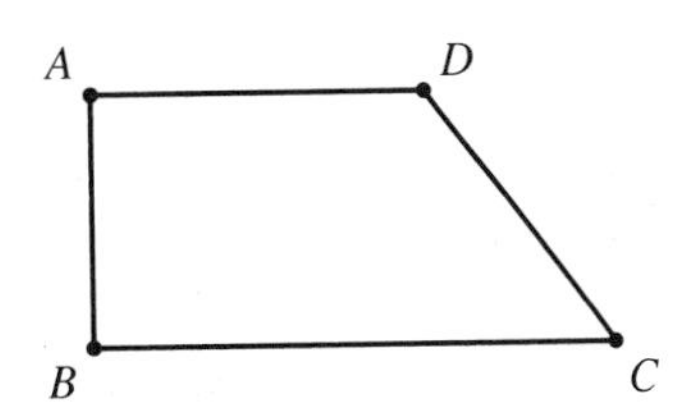

6. In $\triangle ABC$, $AC = AD = CD = DB$. Find $\angle B$.

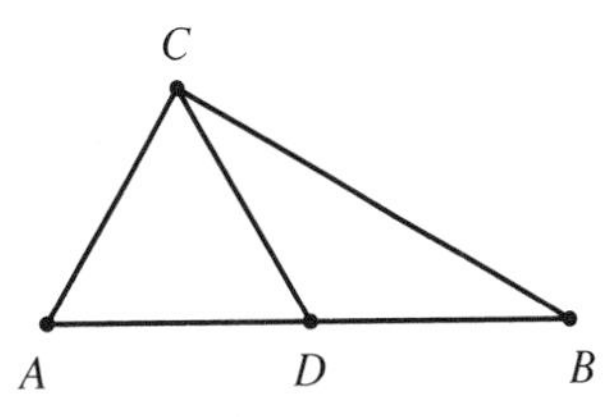

In Exercises 7–9, find the missing side of each right triangle.

7.

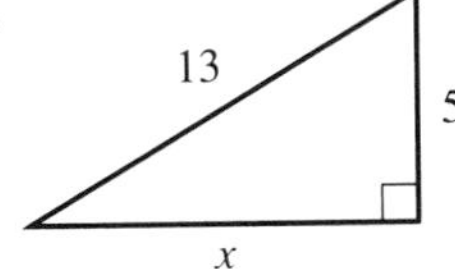

8.

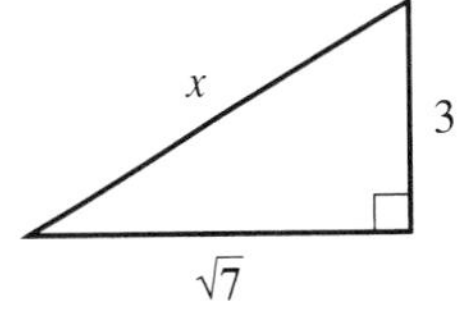

9.

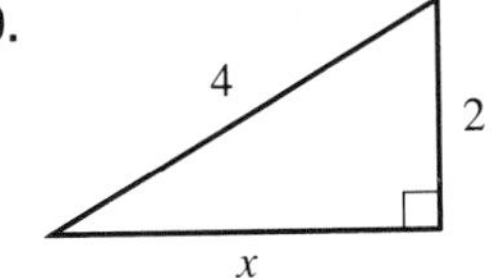

10. Find the diagonal of the rectangle with length 4 cm and width 2 cm.

11. Find the diagonal of a square if its area is 64 sq. in.

12. Given $\triangle ABD \cong \triangle CBE$, find $\angle C$ and $\angle DBE$.

13. Given $\triangle AED \cong \triangle CEB$, find $\angle DAE$ and AE.

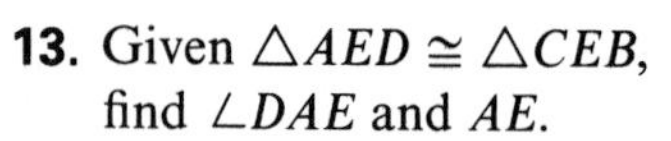

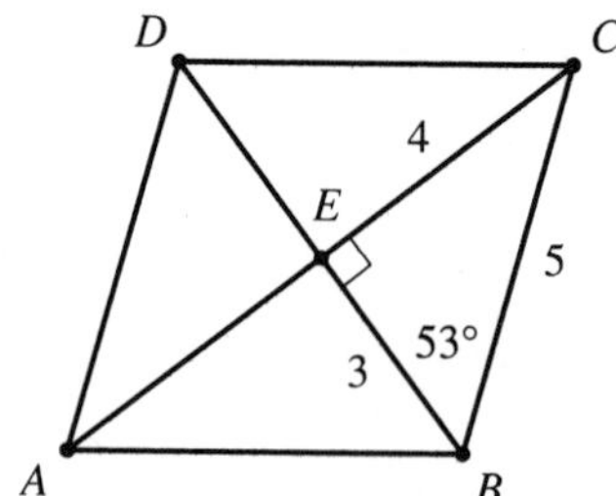

In Exercises 14 and 15, identify the congruent triangles and name the property used.

14. Given $\angle D = \angle C$ and $DE = CE$

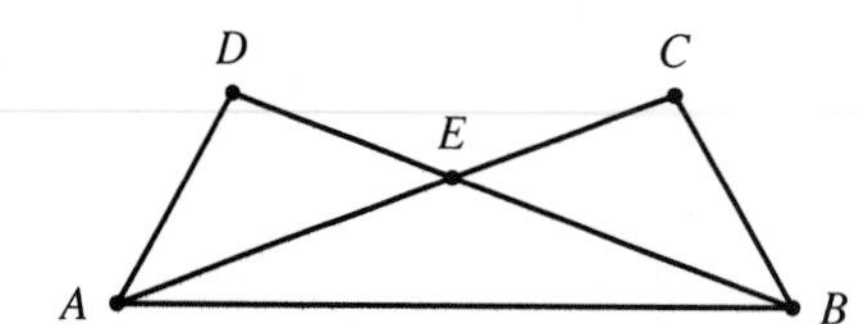

15. Given $\angle DAB = \angle CBA$ and $DA = CB$

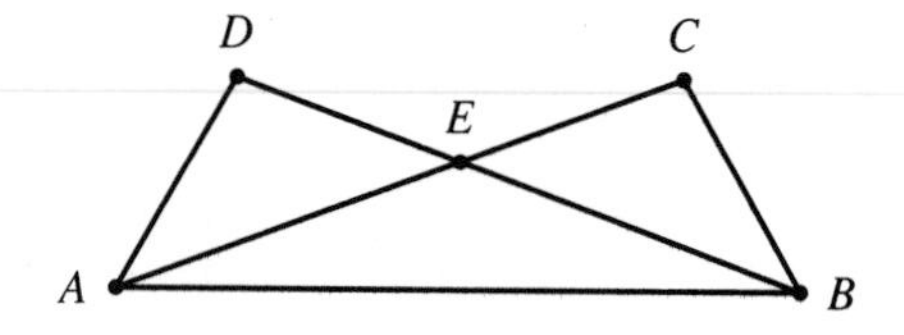

16. Given $\overline{AB} \parallel \overline{CD}$, find DE.

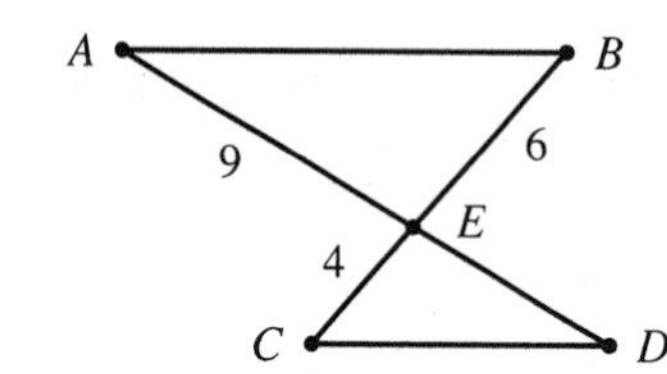

17. Given $\overline{AB} \perp \overline{BC}$, $\overline{DE} \perp \overline{BC}$, find EC.

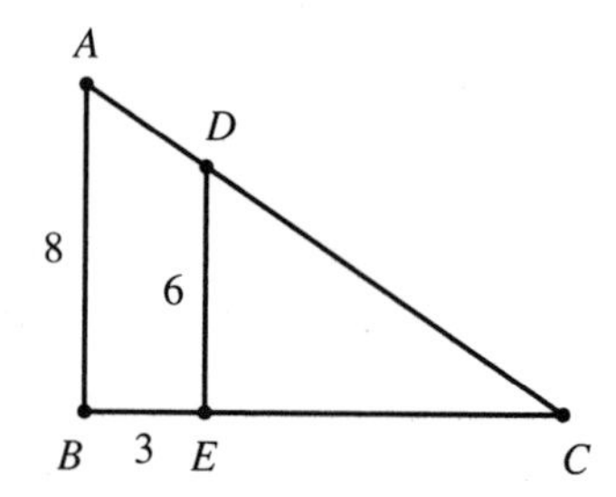

In Exercises 18–23, find the perimeter and the area of each figure.

18.

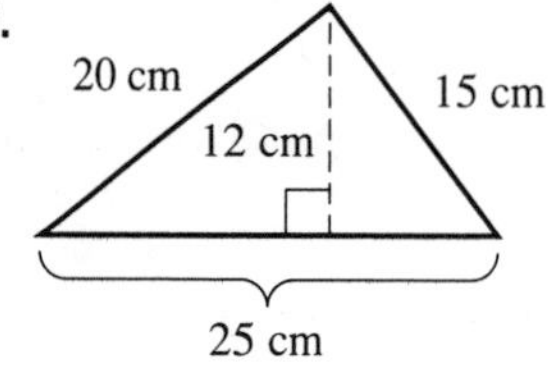

19.

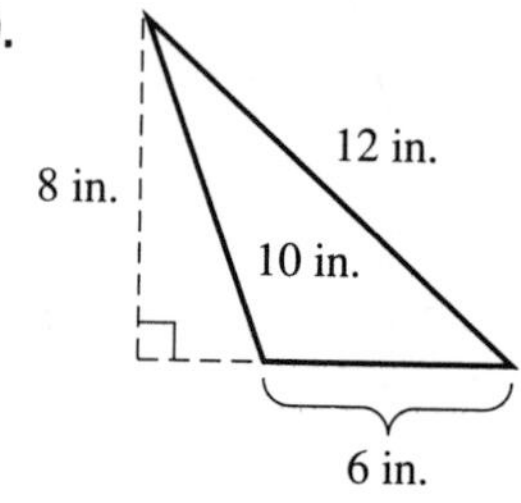

20.

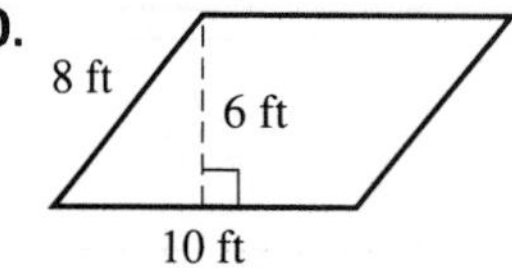

21.

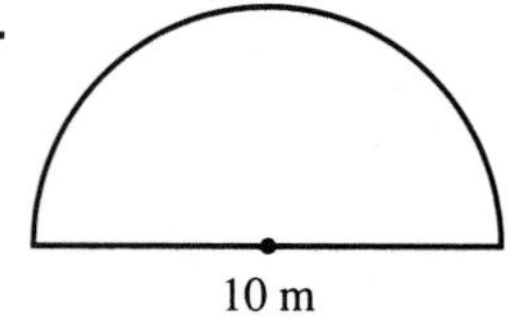

22.

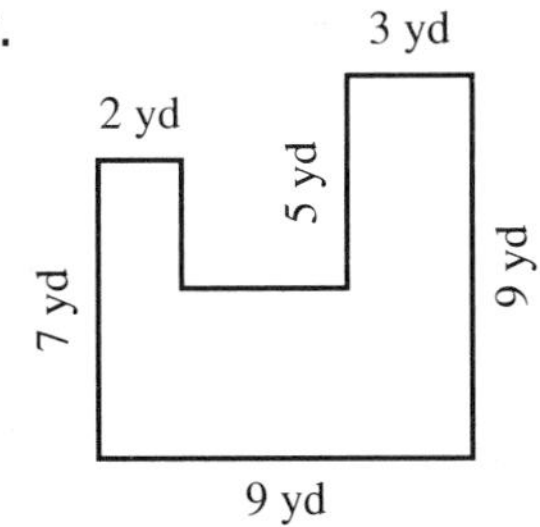

23.

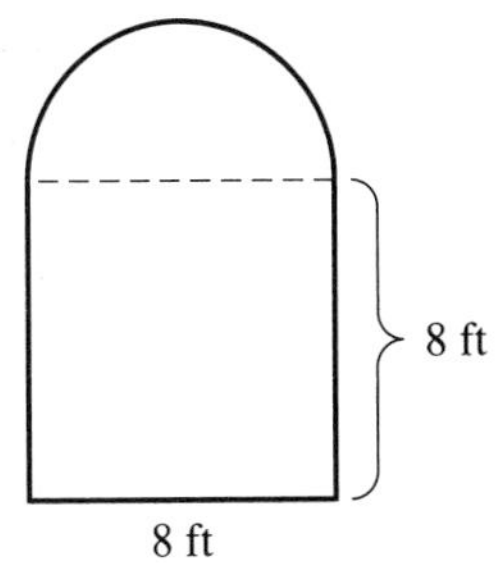

In Exercises 24–26, find the area of the shaded regions.

24.

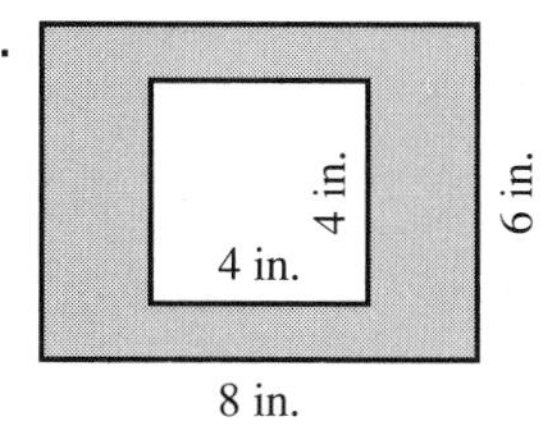

25.

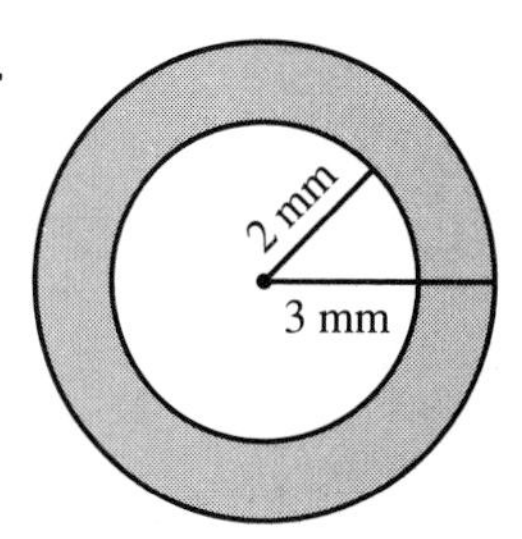

26.

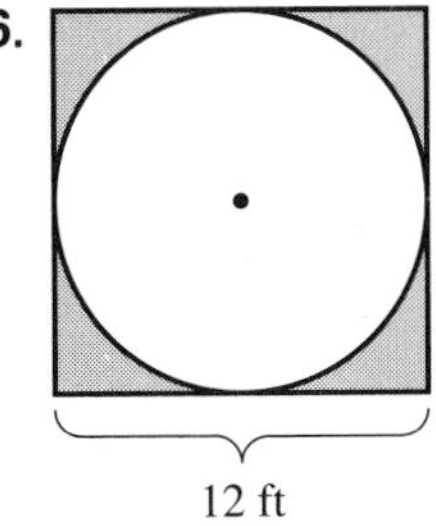

27. The perimeter of a rectangle is 22 m and the length is 7 m. Find the area.

28. The area of a square is 36 cm^2. Find its perimeter.

29. The area of a circle is 64π sq. ft. Find its circumference.

30. The legs of a right triangle are 6 in. and 8 in. Find its perimeter.

31. The area of right $\triangle ABC$ is 10 sq. in. and $BC = 5$ in. Find AC.

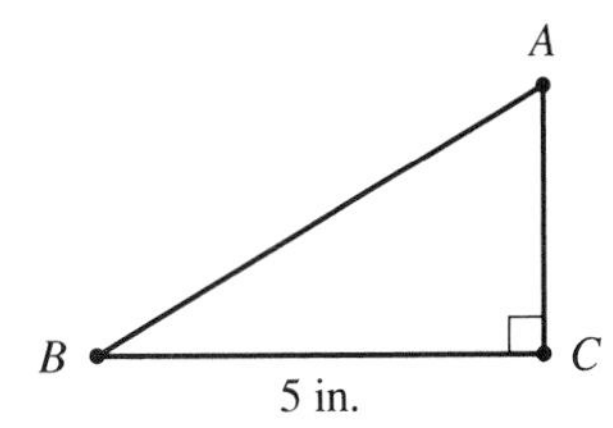

32. The area of parallelogram $ABCD$ is 32 m^2. If $AB = 6$ m and $AE = 4$ m, what is the perimeter?

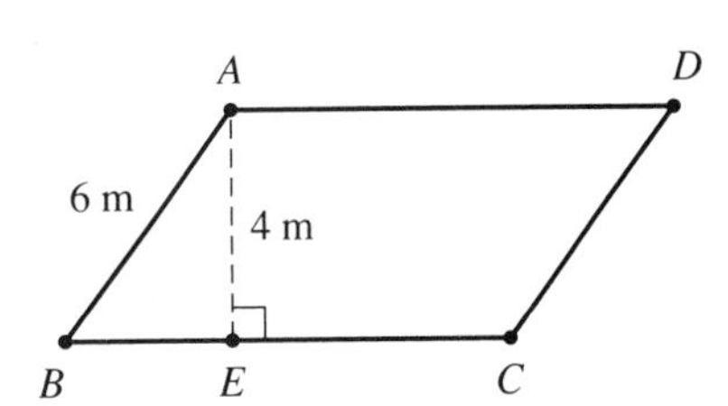

33. Find the volume of a cube with side 4 ft.

34. The volume of a rectangular box is 120 cm^3. If the length is 8 cm and the width is 5 cm, find the height.

35. Find the volume of a hemisphere (half sphere) with diameter 12 m.

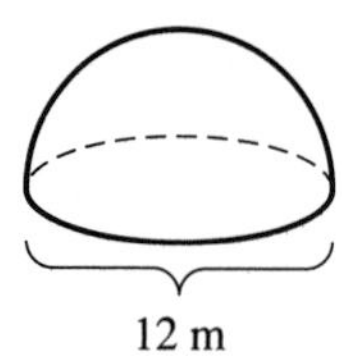

36. Find the volume of water in the cylindrical tank if one fourth of the water has been drained from the tank.

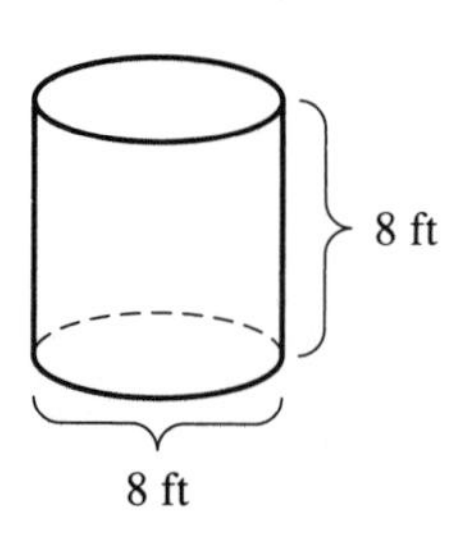

Chapter 16 Diagnostic Test

Allow yourself about 60 minutes to do these problems. Complete solutions for every problem, together with section references, are given in the answer section at the end of the book.

1. Given $\ell \parallel m$, $\angle 1 = 25°$, and $\angle 6 = 80°$. Find the remaining angles.

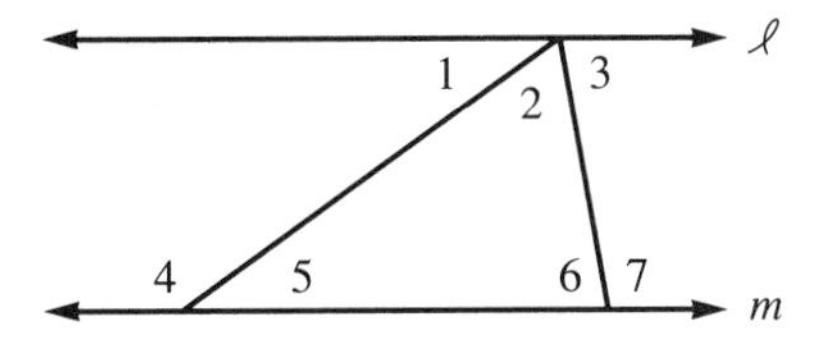

2. Given $\overline{AB} \perp \overline{CD}$, $\angle A = 30°$, and $\angle ACB = 115°$. Find $\angle B$ and $\angle BCD$.

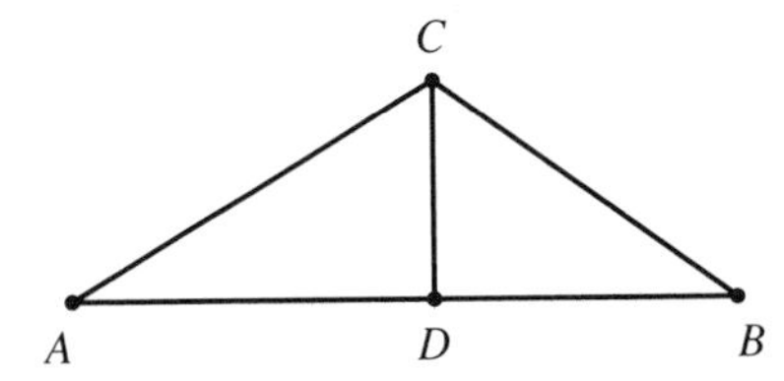

3. In right $\triangle ABC$, find BC.

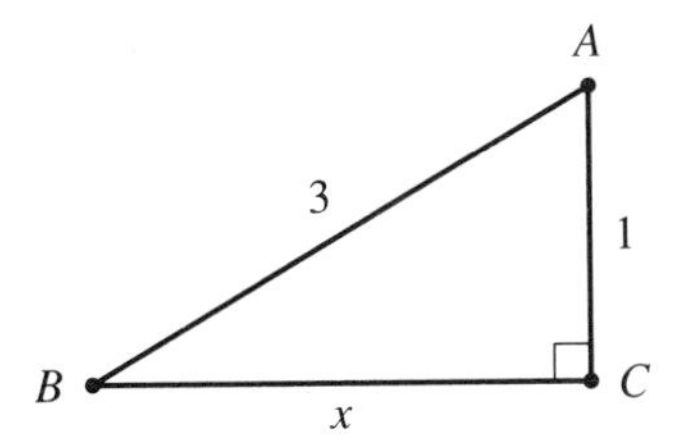

4. In right $\triangle ABC$, find AB.

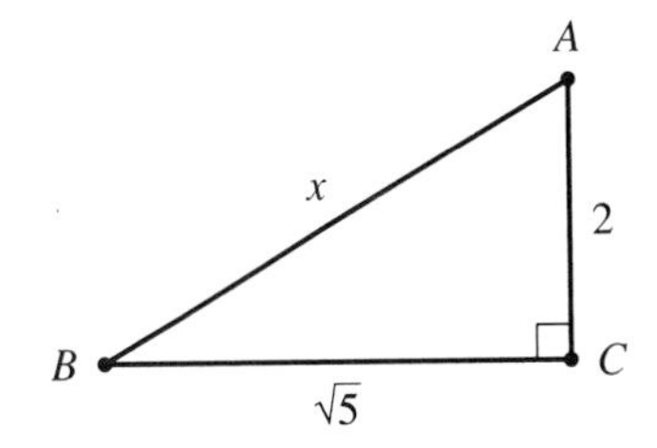

5. Find two supplementary angles if one angle is 30° more than the other angle.

6. Find the perimeter of a right triangle if the legs are 9 in. and 12 in.

7. Find the diagonal of a square if its perimeter is 20 cm.

8. Find the area of parallelogram $ABCD$.

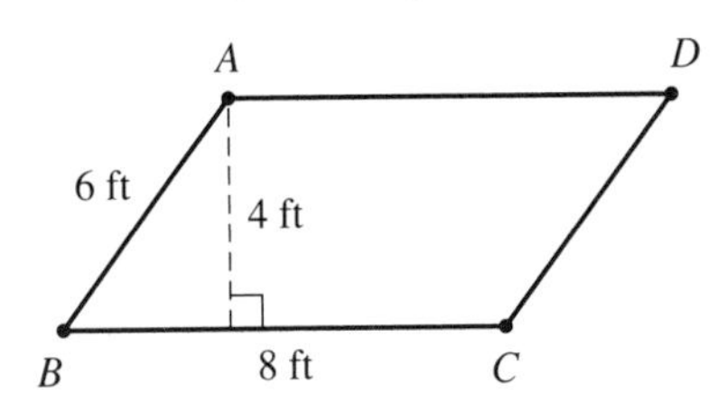

9. Find the perimeter.

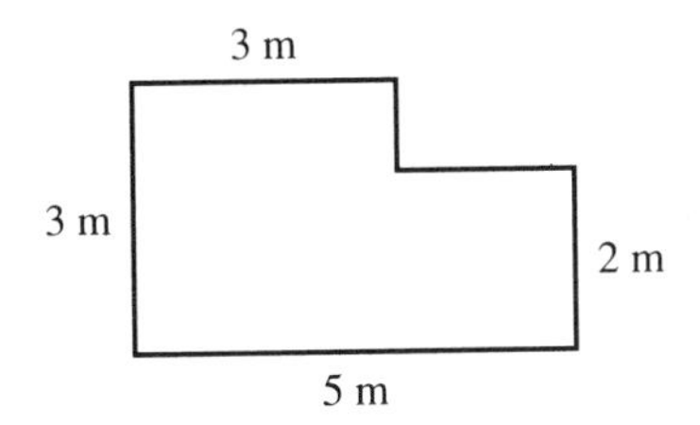

10. Find the area.

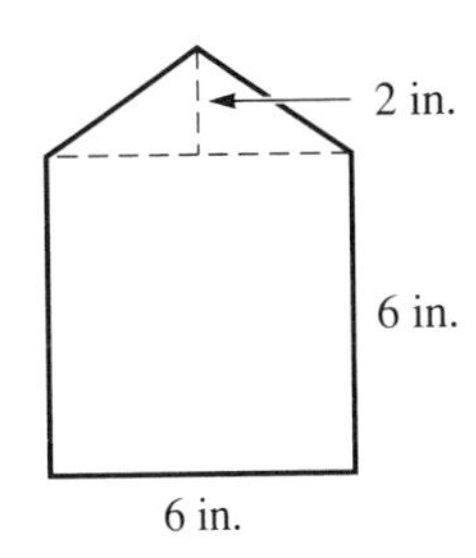

11. The area of a rectangle is 54 sq. ft and the width is 6 ft. Find the perimeter of the rectangle.

12. The circumference of a circle is 12π ft. Find its area.

13. Find the volume of the rectangular box.

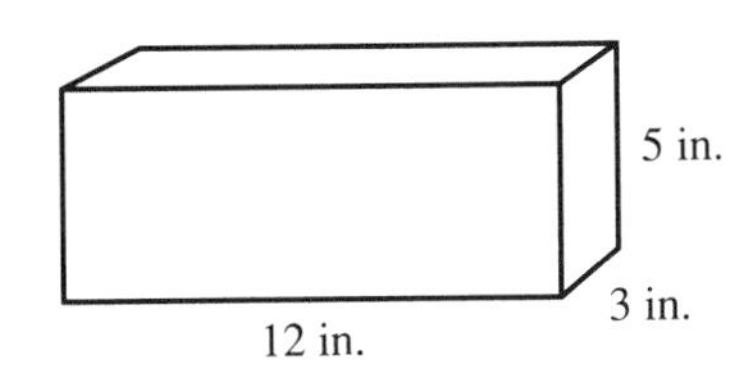

14. Find the volume of a sphere with diameter 6 m.

15. If $\overline{AB} \perp \overline{BC}$, $\overline{DE} \perp \overline{BC}$, find AB.

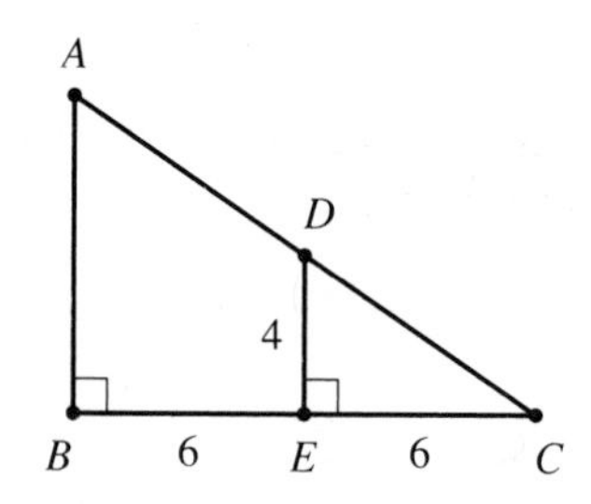

16. If $\overline{AB} \parallel \overline{DE}$, find AD.

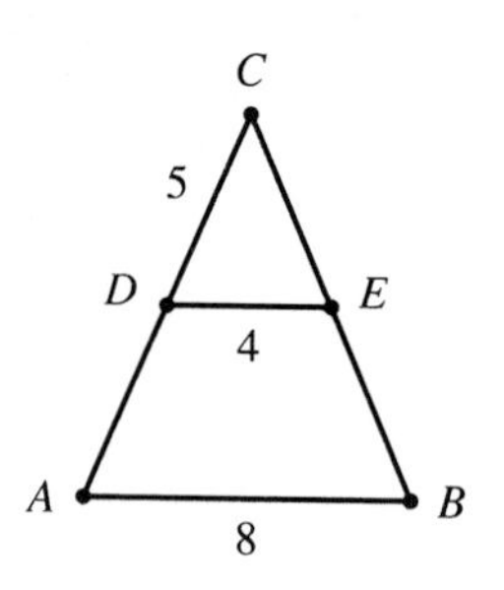

17. Given $\triangle ABC \cong \triangle DCB$, find $\angle DCB$ and $\angle CBD$.

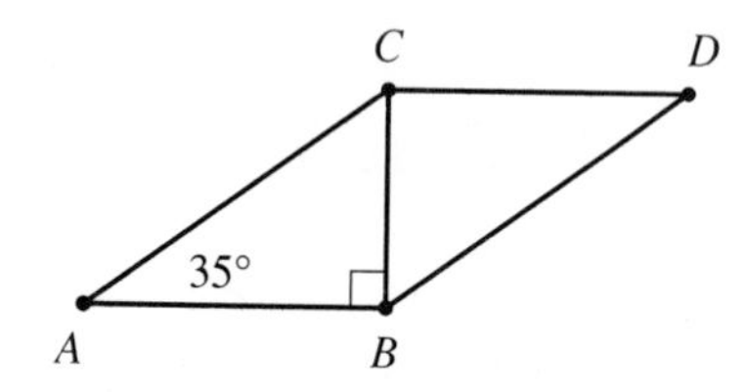

18. Given $\angle BAD = \angle CAD$ and $AB = AC$, identify the congruent triangles and name the property used.

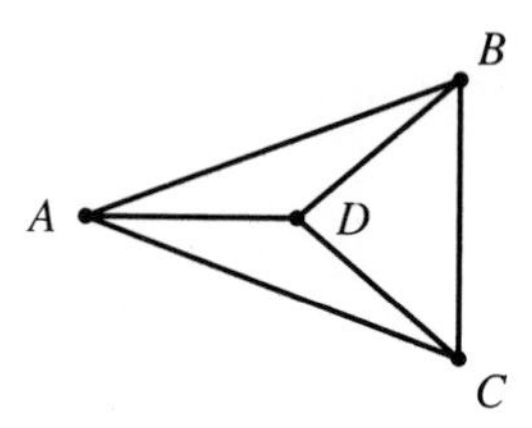

Cumulative Review Exercises for Chapters 1–16

In Exercises 1–13, perform the indicated operations. Reduce all fractions to lowest terms, and change any improper fractions to mixed numbers.

1. $27\frac{4}{9} - 12\frac{5}{6}$

2. $5\frac{5}{6} \div 2\frac{1}{2}$

3. $94.3 - 7.16$

4. 8.7×4.69

5. $8 - 2 \cdot 6 + (3 - 7)$

6. $(5x^2 - 3x - 9) - (2x^2 + x - 6)$

7. $(3x - 4y)(5x + 2y)$

8. $(4x - 3)^2$

9. $\dfrac{x - 2}{x^2 + x - 6} \cdot \dfrac{x^2 + 3x}{3x^3}$

10. $\dfrac{12x}{x^2 - 8x + 12} \div \dfrac{2x + 4}{x^2 - 4}$

11. $\dfrac{x^2}{x - 3} + \dfrac{9}{3 - x}$

12. $\dfrac{x - 3}{x + 4} - \dfrac{2}{x}$

13. $\dfrac{x - \frac{25}{x}}{1 + \frac{5}{x}}$

In Exercises 14–17, solve each equation.

14. $5 - 3(x - 1) = 2x - (x + 4)$

15. $3x^2 + x - 10 = 0$

16. $\dfrac{x + 7}{3} - \dfrac{x + 4}{6} = 2$

17. $\sqrt{3x + 1} = 4$

18. Solve by addition.

$2x + 3y = 4$

$5x + 4y = 3$

19. Solve by substitution.

$x + 2y = 1$

$2x - 3y = 16$

20. Solve the inequality and graph the solution on the number line.
$4x - 2 > 6x + 8$

21. Solve $P = 2L + 2W$ for L.

In Exercises 22–24, simplify and write answers using only positive exponents.

22. $(3x^3y^{-4})^2$

23. $(a^{-2}b^5)(a^{-3}b^{-4})$

24. $\dfrac{8xy^3}{4x^4y^{-2}}$

In Exercises 25–27, perform the indicated operations and simplify.

25. $\sqrt{12x^4y^3}$

26. $\sqrt{75} + 3\sqrt{20} - \sqrt{27}$

27. $(5\sqrt{2} - 3)(\sqrt{2} + 4)$

28. Divide and round off to two decimal places: $2.4 \div 0.46$

29. Divide using long division.
$(2x^3 - 7x^2 + 10x - 8) \div (x - 2)$

30. Change 0.076 to a fraction in lowest terms.

31. Evaluate $x^2 + xy - y^2$ if $x = -3$ and $y = 5$.

32. Perform the indicated operation and write the answer in decimal form.

$$7.9 + \frac{5}{8}$$

33. Perform the indicated operation and write the answer in scientific notation.

$$\frac{0.0072}{12{,}000}$$

34. Graph: $5x - 2y = 10$

35. Graph: $y = -4$

36. Find the equation of the line through the points (3, 6) and (5, −2).

37. Find the slope and the y-intercept of the line $3x + 2y = 8$.

38. Terry worked $6\frac{1}{2}$ hr on Monday, $4\frac{1}{4}$ hr on Tuesday, $3\frac{3}{4}$ hr on Wednesday, $2\frac{3}{4}$ hr on Thursday, and 4 hr on Friday. Find the average number of hours worked.

39. If 3 pounds of fertilizer will cover 400 square feet of lawn, how many pounds of fertilizer must be used on a 1,000 square foot lawn?

40. 34 is what percent of 40?

41. 65% of the students enrolled at the college are women. If there are 5,200 women students, how many students are enrolled at the college?

42. It took Ken 3 hours to drive to the mountains. Due to traffic it took him 4 hours to drive home. If his speed driving to the mountains was 15 mph faster than his speed coming home, find his speed coming home.

43. Nancy has 21 coins consisting of dimes and quarters. If the total value is $3.00, how many dimes does she have?

44. y varies directly with x. If $y = 9$ when $x = 6$, find y when $x = 10$.

45. A rectangle has a perimeter of 28 feet and a width of 5 feet. Find the area of the rectangle.

46. A circle has an area of 9π. Find its circumference.

47. In the right $\triangle ABC$, find AC.

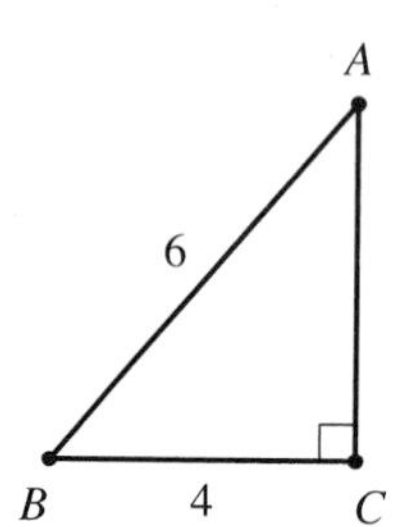

48. Given right $\triangle ABC$, $\overline{AB} \parallel \overline{DE}$, and $\angle A = 42°$. Find $\angle CED$.

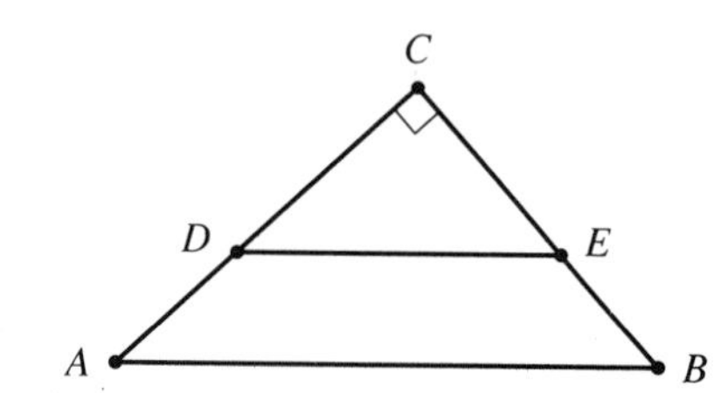

In Exercises 49 and 50, use the graph in Figure A.

49. Between which two years did the sales increase the most for store A?

50. How much more did store A sell than store B in 1986?

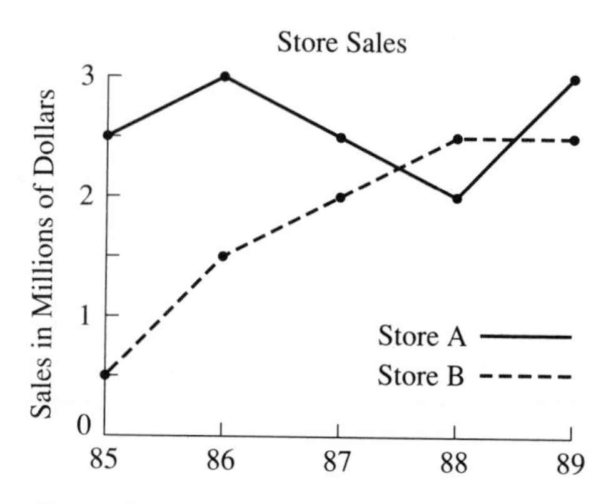

Figure A

APPENDIX I

Addition and Multiplication Facts Drill

Addition

Addition Facts

Table of Basic Addition Facts

+	*0*	*1*	*2*	*3*	*4*	*5*	*6*	*7*	*8*	*9*
0	0	1	2	3	4	5	6	7	8	9
1	1	2	3	4	5	6	7	8	9	10
2	2	3	4	5	6	7	8	9	10	11
3	3	4	5	6	7	8	9	10	11	12
4	4	5	6	7	8	9	10	11	12	13
5	5	6	7	8	9	10	11	12	13	14
6	6	7	8	9	10	11	12	13	14	15
7	7	8	9	10	11	12	13	14	15	16
8	8	9	10	11	12	13	14	15	16	17
9	9	10	11	12	13	14	15	16	17	18

To show how to use the table of addition facts, we give an example. To find the sum $7 + 8 = 15$, move right from 7 of the left column until you are directly below 8 of the top row. See the circled numbers in the table.

Addition Drill

The following addition combinations include every possible pair of digits that can be added together. If a student practices these daily, his or her addition skills will improve. A convenient way of practicing is to place a paper over the answers, then write or say the answers as quickly as possible. For variety in your practice, work from left to right across the page, then from right to left. Also add down one time, then up the next time.

2	0	8	5	3	9	6	5	1	6
2	6	2	6	3	5	4	4	9	9
4	6	10	11	6	14	10	9	10	15
6	2	9	8	7	3	5	8	2	7
5	4	4	3	7	6	3	9	7	4
11	6	13	11	14	9	8	17	9	11
4	8	2	6	5	2	7	8	3	6
4	1	3	8	5	9	0	5	8	6
8	9	5	14	10	11	7	13	11	12
4	7	3	9	2	7	8	7	4	9
6	8	7	9	8	1	4	6	5	6
10	15	10	18	10	8	12	13	9	15
8	4	2	7	9	6	3	5	4	8
8	3	5	5	7	1	3	8	4	7
16	7	7	12	16	7	6	13	8	15

5	3	0	7	3	4	7	3	2	6
2	2	8	7	9	8	3	5	6	3
7	5	8	14	12	12	10	8	8	9

7	5	3	5	9	4	9	1	8	7
2	9	4	7	8	7	3	5	8	9
9	14	7	12	17	11	12	6	16	16

9	5	4	6	9	8	6	4	6	9
2	5	2	2	9	6	7	9	6	0
11	10	6	8	18	14	13	13	12	9

Multiplication

Multiplication Facts

Table of Basic Multiplication Facts

×	*0*	*1*	*2*	*3*	*4*	*5*	*6*	*7*	*(8)*	*9*
0	0	0	0	0	0	0	0	0	0	0
1	0	1	2	3	4	5	6	7	8	9
2	0	2	4	6	8	10	12	14	16	18
3	0	3	6	9	12	15	18	21	24	27
4	0	4	8	12	16	20	24	28	32	36
5	0	5	10	15	20	25	30	35	40	45
6	0	6	12	18	24	30	36	42	48	54
(7)	0	7	14	21	28	35	42	49	(56)	63
8	0	8	16	24	32	40	48	56	64	72
9	0	9	18	27	36	45	54	63	72	81

To show how to use the table of multiplication facts, we give an example. To find the product, $7 \times 8 = 56$, move right from 7 of the left column until you are directly below 8 of the top row. See the circled numbers in the table. Therefore, 7 and 8 are factors of 56.

Multiplication Drill

The following multiplication combinations include every possible pair of digits that can be multiplied together. If a student practices these daily, his or her multiplication skills will improve. A convenient way of practicing is to place a paper over the answers, then write or say the answers as quickly as possible. To get variety in your practice, work from left to right across the page, then from right to left.

1	5	9	3	7	6	4	5	7	9
9	3	2	4	7	3	2	6	2	4
9	15	18	12	49	18	8	30	14	36

8	4	6	5	3	9	3	7	4	8
7	6	6	7	2	5	7	8	5	6
56	24	36	35	6	45	21	56	20	48

6	7	3	0	8	4	5	3	9	6
7	4	8	6	8	9	2	3	8	5
42	28	24	0	64	36	10	9	72	30
9	7	3	8	4	6	0	4	7	5
9	0	9	5	4	9	8	3	6	5
81	0	27	40	16	54	0	12	42	25
7	3	6	3	7	2	1	8	4	3
9	3	4	5	7	8	7	9	8	6
63	9	24	15	49	16	7	72	32	18
5	4	9	2	5	9	6	8	2	6
9	7	7	2	4	0	6	4	3	8
45	28	63	4	20	0	36	32	6	48
2	8	6	0	8	4	7	2	9	5
5	8	2	5	3	4	5	6	9	8
10	64	12	0	24	16	35	12	81	40
9	2	7	5	2	9	8	2	8	1
6	7	3	5	4	3	1	9	2	6
54	14	21	25	8	27	8	18	16	6

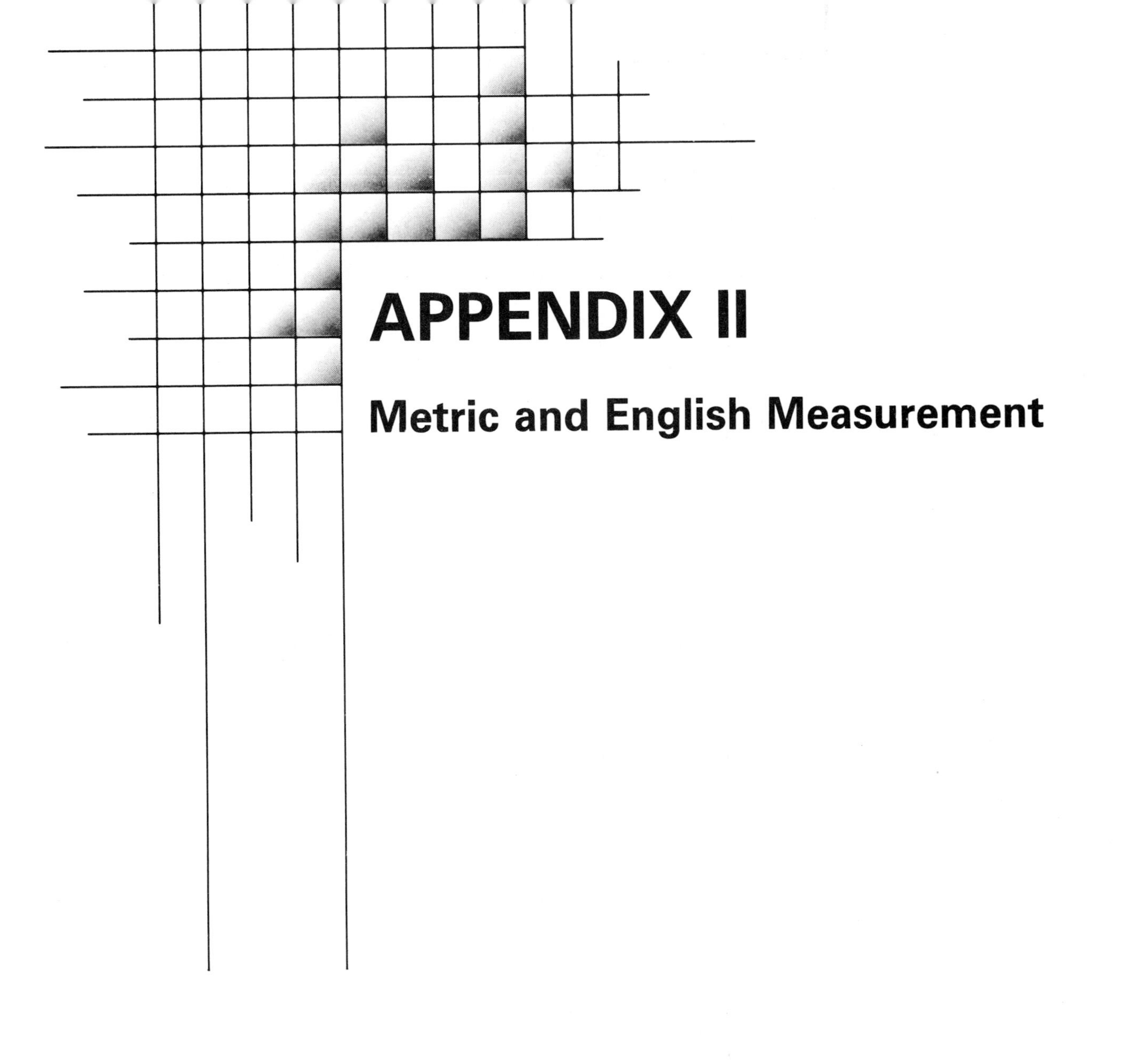

APPENDIX II

Metric and English Measurement

The metric system is a decimal system of measurement that originated in France during the French Revolution. Today, most of the world uses the metric system* or will use it in the near future. The metric system is already used in our country in medicine, nursing, the pharmaceutical industry, the National Aeronautics and Space Administration, photography, the military, and sports. For example, you are probably familiar with the 100-meter race, 35-millimeter film and slides, the liter bottle, the 105-millimeter gun, and so on. See Figure II.1.1.

FIGURE II.1.1
Metric Uses in the United States

Astronauts on the moon use meters to describe their positions. In many countries—Mexico, for example—metric units are used for speeds and distances on road signs, weight and volume, clothing sizes, and so forth. The United States is committed to changing to the metric system in the near future, so the metric system is a necessary part of your education.

II.1 Basic Metric Units of Measurement

We consider four of the original basic units used in the metric system.

The *meter* (m). Used to measure *length*.

---

*The metric system used most widely is the *International System of Units* (SI System).

The *liter* (ℓ). Used to measure *volume.*

The *gram* (g). Used to measure *weight.**

The *Celsius degree* (°C). Used to measure *temperature.*
(Also called Centigrade degree.)

The Meter

The basic unit for measuring length in the metric system is the **meter** (m). It corresponds roughly to the yard in the English system of measurement. (See Figure II.1.2.)

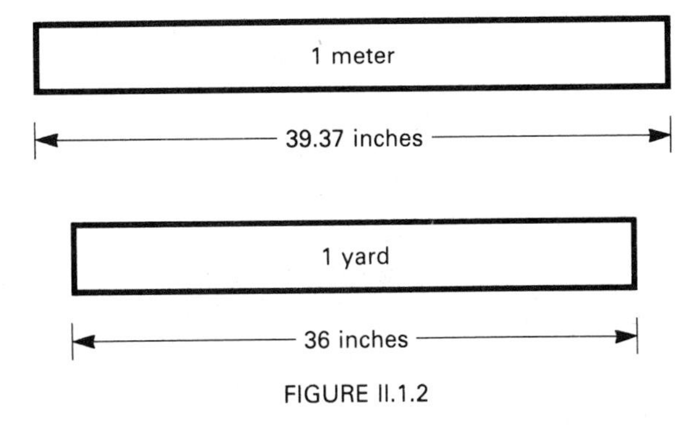

FIGURE II.1.2

The Liter

The basic unit for measuring volume (capacity) in the metric system is the **liter** (ℓ). It corresponds roughly to the quart in the English system of measurement. (See Figure II.1.3.)

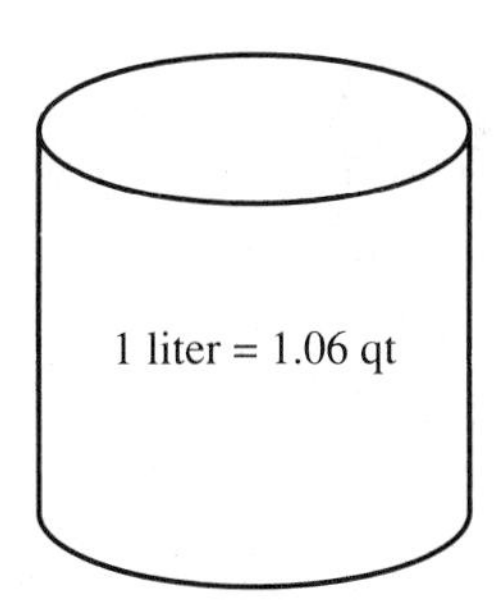

FIGURE II.1.3

The Gram

The original basic unit for measuring *weight* in the metric system is the **gram** (g). One gram is a small quantity of weight. A paper clip weighs about 1 gram. It takes approximately 454 grams to equal one pound. (See Figure II.1.4.)

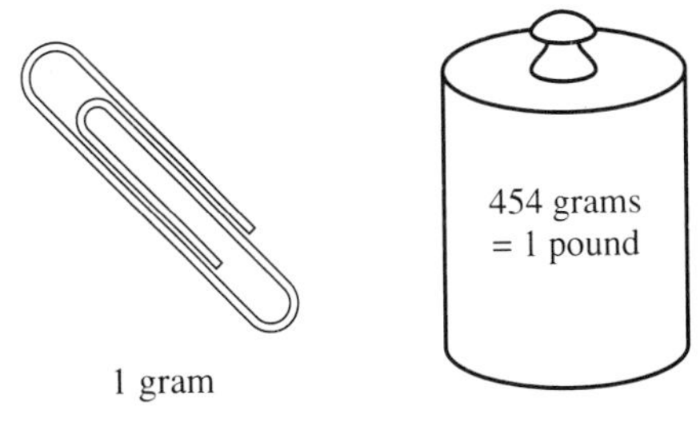

FIGURE II.1.4

* To be exact, the gram measures *mass* and not weight. However, in everyday, nonscientific use, the gram is used for *weight.*
For example:
a. the *weight* limit on luggage for overseas flights is given as 20 kilo*grams* per person.
b. A bottle of tomato catsup is marked "NET *WEIGHT* 14 oz. 397 *GRAMS.*"

The Celsius Degree

The basic unit for measuring temperature in the metric system is the **Celsius degree** (often called Centigrade degree).

SOME EQUIVALENT CELSIUS–FAHRENHEIT TEMPERATURES	
Boiling point of water	100°C = 212°F
Normal body temperature	37°C = 98.6°F
Freezing point of water	0°C = 32°F

II.2 Changing Units within the Metric System

Prefixes Basic units can be changed to larger or smaller units by means of *prefixes.* We consider the three most commonly used prefixes:

1. *Kilo* means 1,000.

   Therefore, 1 *kilo*meter (km) = 1,000 meters (m)
   1 *kilo*liter (kℓ) = 1,000 liters (ℓ)
   1 *kilo*gram (kg) = 1,000 grams (g)

2. *Centi* means $\frac{1}{100}$. $\left(\text{Remember, 1 cent} = \frac{1}{100}\text{ dollars}\right)$.

   Therefore, 1 *centi*meter (cm) $= \frac{1}{100}$ meter *or* 100 cm = 1 m
   1 *centi*liter (cℓ) $= \frac{1}{100}$ liter *or* 100 cℓ = 1 ℓ
   1 *centi*gram (cg) $= \frac{1}{100}$ gram *or* 100 cg = 1 g

3. *Milli* means $\frac{1}{1{,}000}$.

   Therefore, 1 *milli*meter (mm) $= \frac{1}{1{,}000}$ meter
   or 1,000 mm = 1 m
   1 *milli*liter (mℓ) $= \frac{1}{1{,}000}$ liter
   or 1,000 mℓ = 1 ℓ
   1 *milli*gram (mg) $= \frac{1}{1{,}000}$ gram
   or 1,000 mg = 1 g

Changing Units All the prefixes involve either a multiplication or a division by a power of ten. Multiplying or dividing a number by a power of ten can be carried out just by moving the decimal point. Therefore, *to change to larger or smaller units in the* *metric system, it is only necessary to move the decimal point.* This is one of the main advantages to using the metric system.

We now study each prefix in more detail.

II.2A Kilo (Means 1,000)

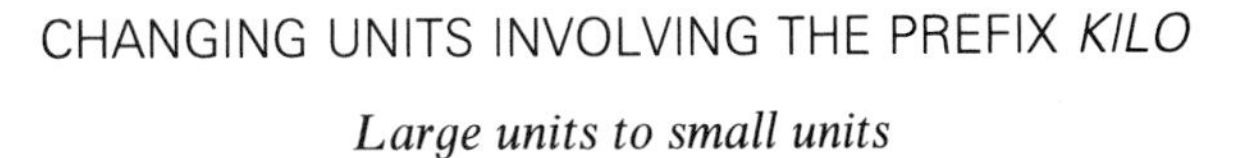

CHANGING UNITS INVOLVING THE PREFIX *KILO*

Large units to small units

To change {kilometers to meters, kiloliters to liters, kilograms to grams} move the decimal point 3 places to the *right*. (See Example 1.)

Small units to large units

To change {meters to kilometers, liters to kiloliters, grams to kilograms} move the decimal point 3 places to the *left*. (See Example 2.)

Example 1 Changing from large units to small units involving *kilo*

a. $0.42 \text{ km} = 0_\wedge 420. \text{ m} = 420 \text{ m}$ (arrow: 3)

b. $6.039 \text{ k}\ell = 6_\wedge 039. \, \ell = 6{,}039 \, \ell$ (arrow: 3)

c. $8.7 \text{ kg} = 8_\wedge 700. \text{ g} = 8{,}700 \text{ g}$ (arrow: 3)

Move decimal three places to the *right* ■

Example 2 Changing from small units to large units involving *kilo*

a. $9{,}025 \text{ m} = 9.025_\wedge \text{ km} = 9.025 \text{ km}$ (arrow: −3)

b. $640 \, \ell = .640_\wedge \text{ k}\ell = 0.640 \text{ k}\ell$ (arrow: −3)

c. $62{,}300 \text{ g} = 62.300_\wedge \text{ kg} = 62.3 \text{ kg}$ (arrow: −3)

Move decimal three places to the *left* (The −3 means a movement of 3 places to the left) ■

EXERCISES II.2A

1. 1.8 km = ___?___ m
2. 34 kℓ = ___?___ ℓ
3. 0.249 kg = ___?___ g
4. 5.71 km = ___?___ m
5. 60.5 ℓ = ___?___ kℓ
6. 322 g = ___?___ kg
7. 275 g = ___?___ kg
8. 56.4 ℓ = ___?___ kℓ
9. 0.78 km = ___?___ m
10. 9.3 kg = ___?___ g
11. 72,350 g = ___?___ kg
12. 2,365 m = ___?___ km

13. A pharmaceutical house's orders for hydrogen peroxide average 125 liters per month. How many kiloliters is this per month?

14. A rectangular alfalfa field measures 0.90 km by 0.20 km. Find the area of this field in square meters (m^2).

II.2B Centi (Means 1/100)

CHANGING UNITS INVOLVING THE PREFIX *CENTI*

Small units to large units

To change {centimeters to meters, centiliters to liters, centigrams to grams} move the decimal point 2 places to the *left*. (See Example 3.)

Large units to small units

To change {meters to centimeters, liters to centiliters, grams to centigrams} move the decimal point 2 places to the *right*. (See Example 4.)

Example 3 Changing from small units to large units involving *centi*

a. $155 \text{ cm} = 1.55_{\wedge} \text{ m} = 1.55 \text{ m}$ (← −2)

b. $76 \text{ c}\ell = .76_{\wedge}\ \ell = 0.76\ \ell$ (← −2)

c. $4.9 \text{ cg} = .04_{\wedge}9 \text{ g} = 0.049 \text{ g}$ (← −2)

Move decimal two places to the *left*

■

Example 4 Changing from large units to small units involving *centi*

a. $6.8 \text{ m} = 6_{\wedge}80. \text{ cm} = 680 \text{ cm}$ (2 →)

b. $0.47\ \ell = 0_{\wedge}47. \text{ c}\ell = 47 \text{ c}\ell$ (2 →)

c. $5.873 \text{ g} = 5_{\wedge}87.3 \text{ cg} = 587.3 \text{ cg}$ (2 →)

Move decimal two places to the *right*

■

EXERCISES II.2B

1. 279 cm = __?__ m

2. 54 cℓ = __?__ ℓ

3. 8.3 cg = __?__ g

4. 4,090 cm = __?__ m

5. 2.5 m = __?__ cm

6. 0.72 ℓ = __?__ cℓ

7. 3.906 g = __?__ cg

8. 0.842 m = __?__ cm

9. 632 cℓ = __?__ ℓ

10. 58.1 cg = __?__ g

11. 0.0263 g = __?__ cg

12. 0.092 ℓ = __?__ cℓ

13. Bob's height is 1.82 meters. Express his height in centimeters.

14. Hilda's gift weighed 1,430 grams. Express the weight of the gift in centigrams.

II.2C Milli (Means 1/1,000)

CHANGING UNITS INVOLVING THE PREFIX *MILLI*

Small units to large units

To change {millimeters to meters, milliliters to liters, milligrams to grams} move the decimal point 3 places to the *left*. (See Example 5.)

Large units to small units

To change {meters to millimeters, liters to milliliters, grams to milligrams} move the decimal point 3 places to the *right*. (See Example 6.)

Example 5 Changing from small units to large units involving *milli*

a. 56 mm = .056$_\wedge$ m = 0.056 m ($\xleftarrow{-3}$)

b. 4,800 mℓ = 4.800$_\wedge$ ℓ = 4.8ℓ ($\xleftarrow{-3}$)

c. 250 mg = .250$_\wedge$ g = 0.25 g ($\xleftarrow{-3}$)

Move decimal three places to the *left* ■

Example 6 Changing from large units to small units involving *milli*

a. 1.4 m = 1$_\wedge$400. mm = 1,400 mm ($\xrightarrow{3}$)

b. 0.68 ℓ = 0$_\wedge$680. mℓ = 680 mℓ ($\xrightarrow{3}$)

c. 0.2050 g = 0$_\wedge$205.0 mg = 205.0 mg ($\xrightarrow{3}$)

Move decimal three places to the *right* ■

Another commonly used metric unit is the *cubic centimeter.*

1 cubic centimeter (cc) = 1 milliliter (mℓ)

EXERCISES II.2C

1. 91 mm = ___?___ m
2. 5,600 mℓ = ___?___ ℓ
3. 470 mg = ___?___ g
4. 4,300 mm = ___?___ m
5. 2.6 m = ___?___ mm
6. 0.39 ℓ = ___?___ mℓ
7. 0.1080 g = ___?___ mg
8. 0.0827 m = ___?___ mm
9. 230 mℓ = ___?___ ℓ
10. 9,160 mg = ___?___ g
11. 7.04 g = ___?___ mg
12. 21.6 ℓ = ___?___ mℓ
13. 2 ℓ = ___?___ cc
14. 3.55 ℓ = ___?___ cc
15. 175 cc = ___?___ ℓ
16. 2,500 cc = ___?___ ℓ

17. A doctor recommends that Jackie take 1.5 g of vitamin C a day. How many milligrams would she take in a day?

18. Find the volume in cubic centimeters of a container that holds 1.75 liters of water. (1 cubic centimeter [cc] = 1 mℓ).

II.2D Additional Metric Prefixes

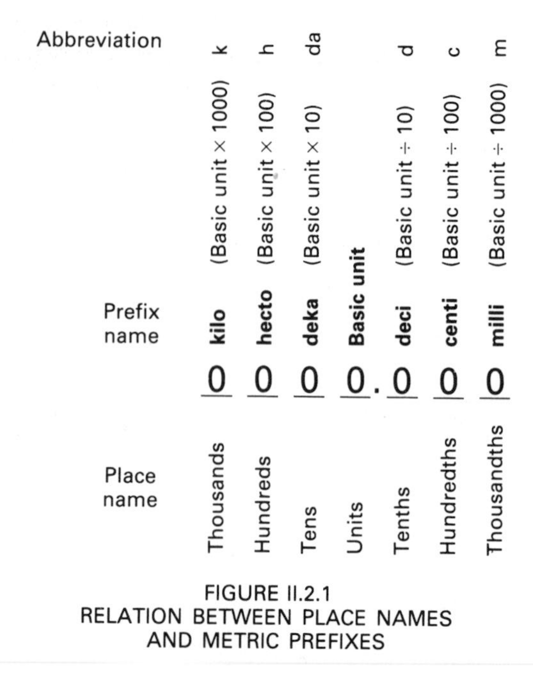

FIGURE II.2.1
RELATION BETWEEN PLACE NAMES
AND METRIC PREFIXES

Figure II.2.1 includes more metric prefixes than we have previously discussed. In the following box we show how Figure II.2.1 can be used when changing from one of these metric units to another.

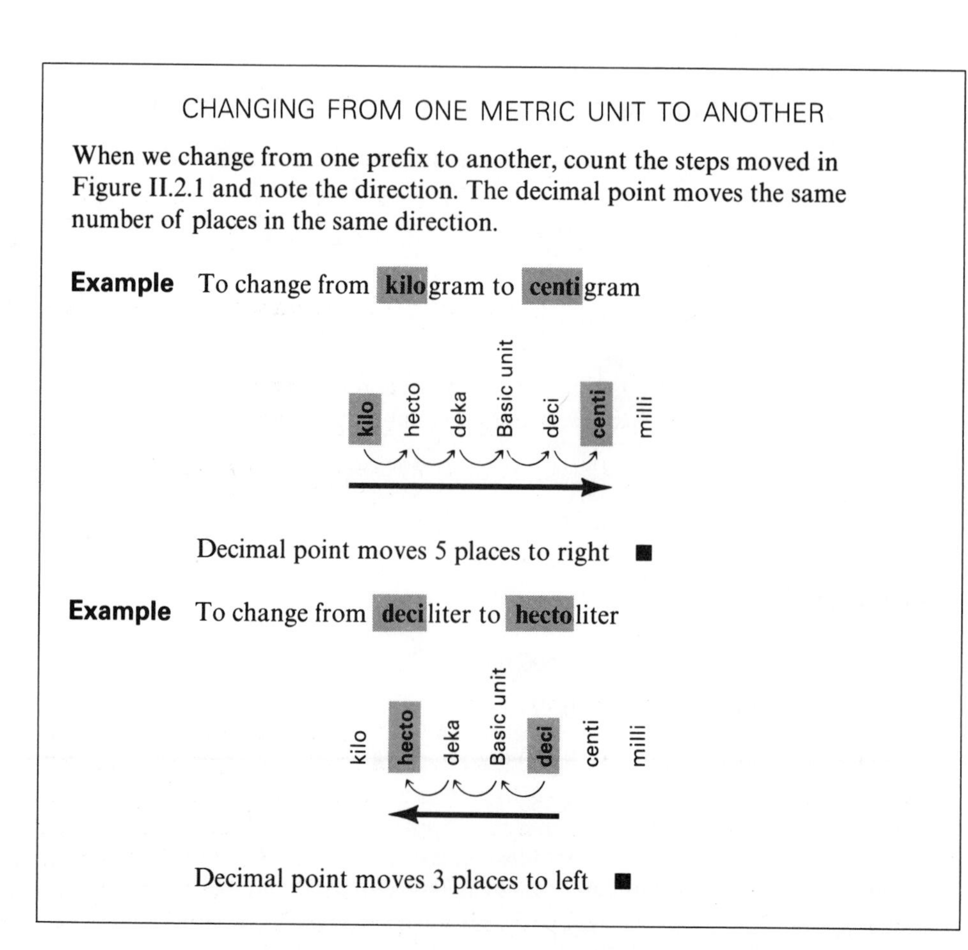

CHANGING FROM ONE METRIC UNIT TO ANOTHER

When we change from one prefix to another, count the steps moved in Figure II.2.1 and note the direction. The decimal point moves the same number of places in the same direction.

Example To change from **kilo**gram to **centi**gram

Decimal point moves 5 places to right ■

Example To change from **deci**liter to **hecto**liter

Decimal point moves 3 places to left ■

Example 7 Changing metric units

a. Change 573 *milli*liters to *deci*liters.
To change from milli to deci, we move *2 steps to the left* in Figure II.2.1.
Therefore, move decimal point *2 places to the left.*

$$573 \text{ m}\ell = \underset{\xleftarrow{-2}}{5.73_{\wedge}} \text{ d}\ell = 5.73 \text{ d}\ell$$

b. Change 0.094 *kilo*meters to *centi*meters.
To change from kilo to centi, we move *5 steps to the right* in Figure II.2.1.
Therefore, move decimal point *5 places to the right.*

$$0.094 \text{ km} = \underset{\xrightarrow{5}}{0_{\wedge}09400.} \text{ cm} = 9{,}400 \text{ cm}$$

■

EXERCISES II.2D

Use Figure II.2.1 on page 528 to help you solve the following exercises.

1. 3.54 km = __?__ m
2. 275 cm = __?__ m
3. 47 $k\ell$ = __?__ ℓ
4. 144 $c\ell$ = __?__ ℓ
5. 2,546 g = __?__ kg
6. 386 g = __?__ kg
7. 3.4 $d\ell$ = __?__ ℓ
8. 784 $m\ell$ = __?__ $d\ell$
9. 0.0516 km = __?__ dm
10. 0.074 dm = __?__ cm
11. 89.5 ℓ = __?__ $h\ell$
12. 607 ℓ = __?__ $k\ell$
13. 78.4 dam = __?__ km
14. 35.6 dm = __?__ hm
15. 456 hg = __?__ dg
16. 0.064 kg = __?__ cg
17. 3,402 mg = __?__ dag
18. 4,860 cg = __?__ kg
19. 5,614 $m\ell$ = __?__ $da\ell$
20. 956 $c\ell$ = __?__ $h\ell$
21. 3.75 m = __?__ mm

II.2E Temperature

When planning what clothing to wear or what activity to engage in, we usually check the temperature first. When traveling in countries that use the metric system, this can be confusing unless you can change Celsius (centigrade) to Fahrenheit. For example, if you hear that the temperature will be 30°C tomorrow, do you plan on going to the beach or going ice skating?

Changing Celsius to Fahrenheit This can be done by using the following formula.

TO CHANGE CELSIUS TO FAHRENHEIT

$$F = \frac{9}{5}C + 32$$

where F = number of °F
and C = number of °C

Example 8 30°C = __?__ °F (Round answer to nearest degree.)

Solution

$$F = \frac{9}{\cancel{5}_{1}}(\cancel{30}^{6}) + 32 = 54 + 32 = 86$$

Therefore, 30°C = 86°F. ■

Example 9 7°C = __?__ °F (Round answer to nearest degree.)

Solution
$$F = \frac{9}{5}(7) + 32 = \frac{63}{5} + 32 = 12.6 + 32 = 44.6 \approx 45$$

Therefore, 7°C ≈ 45°F. ■

Changing Fahrenheit to Celsius This can be done by using the following formula.

> TO CHANGE FAHRENHEIT TO CELSIUS
>
> $$C = \frac{5}{9}(F - 32)$$
>
> where C = number of °C
> and F = number of °F

Example 10 68°F = __?__ °C (Round answer to nearest degree.)

Solution
$$C = \frac{5}{9}(68 - 32) = \frac{5}{\cancel{9}_1}(\cancel{36}^{4}) = 20$$

Therefore, 68°F = 20°C. ■

Example 11 97°F = __?__ °C (Round answer to nearest degree.)

Solution
$$C = \frac{5}{9}(97 - 32) = \frac{5}{9}(65) = 36.1 \approx 36$$

Therefore, 97°F ≈ 36°C. ■

EXERCISES II.2E

Consider given numbers to be exact. Round off answers to the nearest degree.

1. 20°C = __?__ °F

2. 15°C = __?__ °F

3. 50°F = __?__ °C

4. 59°F = __?__ °C

5. 8°C = __?__ °F

6. 17°C = __?__ °F

7. 72°F = __?__ °C

8. 85°F = __?__ °C

9. A Frenchman traveling in the United States notes that a Fahrenheit thermometer reads 41°. What is the equivalent Celsius reading?

10. Is 10°C warmer or colder than 48°F? By how much?

II.3 The English System of Measurement

The English system that we use every day includes such units as inches, feet, yards, miles, pounds, and gallons. The relationships between some of these commonly used English system units are given in the following table.

SOME COMMON EQUIVALENT ENGLISH UNITS OF MEASURE	
Volume Units:	1 tablespoon (tbsp) = 3 teaspoons (tsp)
	1 cup = 16 tablespoons (tbsp)
	1 cup = 8 ounces (oz)
	1 pint (pt) = 2 cups
	1 pint (pt) = 16 ounces (oz)
	1 quart (qt) = 2 pints (pt)
	1 gallon (gal) = 4 quarts (qt)
	1 gallon (gal) = 231 cubic inches (cu in.)
Time Units:	1 minute (min) = 60 seconds (sec)
	1 hour (hr) = 60 minutes (min)
	1 day (da) = 24 hours (hr)
	1 week (wk) = 7 days (da)
	1year (yr) = 52 weeks (wk)
	1 year (yr) = 365 days (da)
Length Units:	1 foot (ft) = 12 inches (in.)
	1 yard (yd) = 3 feet (ft)
	1 mile (mi) = 5,280 feet (ft)
Weight Units:	1 pound (lb) = 16 ounces (oz)
	1 ton = 2,000 pounds (lb)

Figures II.3.1 and II.3.2 show how two English units were derived.

FIGURE 11.3.1
KING HENRY I DECREED:
"A YARD SHOULD BE THE DISTANCE FROM THE TIP OF MY NOSE TO THE END OF MY THUMB."

FIGURE 11.3.2
KING EDWARD I DECREED:
"ONE INCH SHOULD EQUAL THREE BARLEY CORNS LAID END TO END."

II.4 Changing Units within the English System

Unit Fraction

It is often necessary to change from one unit of measure to another. This can be done by means of ratios called **unit fractions**. We now show the unit fraction method most commonly used in science. We think you will find it helpful.

You can change from one unit of measure to another without using unit fractions, but students are sometimes confused as to whether they should divide or multiply in making the conversion. The method of unit fractions minimizes this confusion.

Example 1 Change 27 yards to feet.

Solution

$$1 \text{ yd} = 3 \text{ ft}$$

unit fraction

If we form the ratio $\frac{3 \text{ ft}}{1 \text{ yd}}$, this ratio must equal 1 because the numerator and denominator are equal.

Therefore, $27 \text{ yd} = 27 \text{ yd} \times (1) = \frac{27 \cancel{\text{yd}}}{1}\left(\frac{3 \text{ ft}}{1 \cancel{\text{yd}}}\right) = (27 \times 3) \text{ ft} = 81 \text{ ft}$

This is a unit fraction because 1 yd = 3 ft

Notice that units can be canceled as well as numbers.

Therefore, 27 yd = 81 ft. ■

You know that the value of the unit fraction must be 1. We now explain how to select the correct unit fraction.

TO SELECT THE CORRECT UNIT FRACTION

1. The unit fraction must have two units:
   a. the unit you want in your answer
   b. the unit you want to get rid of
2. The unit fraction must be written so that the unit of measure you want to get rid of can be canceled.
   a. *If the unit you want to get rid of is in the numerator* of the given expression, that same unit must be in the denominator of the unit fraction chosen. (See Example 2.)
   b. *If the unit you want to get rid of is in the denominator* of the given expression, that same unit must be in the numerator of the unit fraction chosen. (See Example 7.)

Example 2 Change 7 inches to feet.

Solution

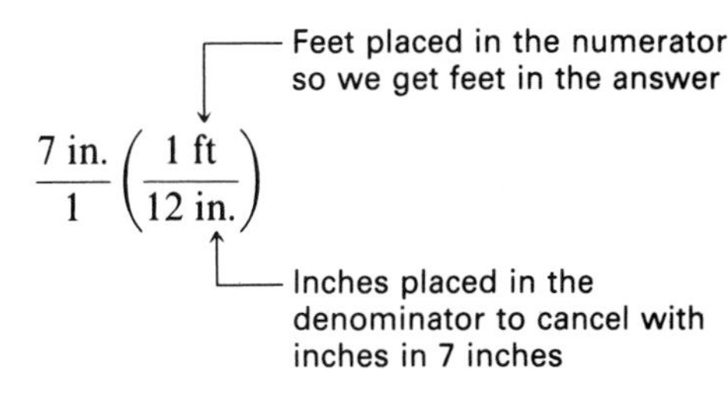

$$\frac{7\text{ in.}}{1}\left(\frac{1\text{ ft}}{12\text{ in.}}\right)$$

$$\frac{7\text{ }\cancel{\text{in.}}}{1}\left(\frac{1\text{ ft}}{12\text{ }\cancel{\text{in.}}}\right) = \frac{7}{12}\text{ ft}$$

Therefore, 7 in. $= \frac{7}{12}$ ft

If we try to use the ratio the other way, it won't work.

$$\frac{7\text{ in.}}{1}\left(\frac{12\text{ in.}}{1\text{ ft}}\right)$$

Here we see that the inches will not cancel, so we will not be left with just feet in the answer ■

Example 3 84 hr = __?__ da

Solution

Because 1 da = 24 hr

$$\frac{84\text{ }\cancel{\text{hr}}}{1}\left(\frac{1\text{ da}}{24\text{ }\cancel{\text{hr}}}\right) = \frac{84}{24}\text{ da} = \frac{7}{2}\text{ da} = 3\frac{1}{2}\text{ da}$$

Therefore, 84 hr $= 3\frac{1}{2}$ da. ■

Example 4 $7\frac{1}{2}$ gal = __?__ qt

Solution

Because 4 qt = 1 gal

$$\frac{7\frac{1}{2}\text{ }\cancel{\text{gal}}}{1}\left(\frac{4\text{ qt}}{1\text{ }\cancel{\text{gal}}}\right) = \left(7\frac{1}{2} \times 4\right)\text{qt} = \left(\frac{15}{\cancel{2}_1} \times \frac{\cancel{4}^2}{1}\right)\text{qt} = 30\text{ qt}$$

Therefore, $7\frac{1}{2}$ gal = 30 qt. ■

NOTE Sometimes it is necessary to use two or more unit fractions to change to the required units. (See Example 5.) ☑

Example 5 2 hr = __?__ sec

Solution

Step 1 Change hours to minutes.

$$\frac{2\text{ }\cancel{\text{hr}}}{1}\left(\frac{60\text{ min}}{1\text{ }\cancel{\text{hr}}}\right) = (2 \times 60)\text{ min} = 120\text{ min}$$

Step 2 Change minutes to seconds.

$$\frac{120\ \cancel{\text{min}}}{1}\left(\frac{60\text{ sec}}{1\ \cancel{\text{min}}}\right) = (120 \times 60)\text{ sec} = 7{,}200\text{ sec}$$

Therefore, 2 hr = 7,200 sec.

This problem could be worked in one step as follows:

$$\frac{2\ \cancel{\text{hr}}}{1}\left(\frac{60\ \cancel{\text{min}}}{1\ \cancel{\text{hr}}}\right)\left(\frac{60\text{ sec}}{1\ \cancel{\text{min}}}\right) = (2 \times 60 \times 60)\text{ sec} = 7{,}200\text{ sec}$$ ■

A WORD OF CAUTION The unit fraction must be written so that the unit of measure we want to get rid of can be canceled. ☑

Sometimes the unit we want to get rid of is in the numerator (Example 6).

Example 6 26 mi = __?__ ft

Solution *To get rid of miles* and be left with feet:

Correct choice	*Incorrect choice*
because miles cancel	because miles do *not* cancel
$\frac{26\ \cancel{\text{mi}}}{1}\left(\frac{5280\text{ ft}}{1\ \cancel{\text{mi}}}\right)$	$\frac{26\text{ mi}}{1}\left(\frac{1\text{ mi}}{5280\text{ ft}}\right)$

Therefore, 26 mi = (26 × 5280) ft = 137,280 ft. ■

Sometimes the unit you want to get rid of is in the denominator (Example 7).

Example 7 $2000\text{ mi per hr} = 2000\frac{\text{mi}}{\text{hr}} = \underline{\ ?\ }\frac{\text{mi}}{\text{min}} = \underline{\ ?\ }\text{ mi per min}$

Solution *To get rid of hours* and be left with minutes:

Correct choice	*Incorrect choice*
because hours cancel	because hours do *not* cancel
$\frac{2000\text{ mi}}{\cancel{\text{hr}}}\left(\frac{1\ \cancel{\text{hr}}}{60\text{ min}}\right)$	$\frac{2000\text{ mi}}{\text{hr}}\left(\frac{60\text{ min}}{1\text{ hr}}\right)$

Therefore, $2000\text{ mi per hr} = \left(\frac{2000}{60}\right)\frac{\text{mi}}{\text{min}} \approx 33.33\text{ mi per min.}$ ■

Example 8 Fred drives 10 miles in 20 minutes. What is his average speed in miles per hour?

Solution

$$\frac{10\text{ mi}}{\underset{1}{\cancel{20}}\ \cancel{\text{min}}}\left(\frac{\overset{3}{\cancel{60}}\ \cancel{\text{min}}}{1\text{ hr}}\right) = 30\frac{\text{mi}}{\text{hr}} = 30\text{ mph}$$ ■

EXERCISES II.4

Find the missing numbers.

1. 5 yd = __?__ ft

2. 3 ft = __?__ in.

3. $3\frac{1}{3}$ yd = __?__ ft

4. $5\frac{2}{3}$ ft = __?__ in.

5. 8 gal = __?__ qt

6. $2\frac{3}{4}$ gal = __?__ qt

7. 7 qt = __?__ pt

8. 2.5 qt = __?__ pt

9. $1\frac{1}{2}$ hr = __?__ min

10. 7.75 min = __?__ sec

11. $1\frac{3}{4}$ mi = __?__ ft

12. $3\frac{1}{4}$ da = __?__ hr

13. 10 yd = __?__ ft = __?__ in.

14. 8 gal = __?__ qt = __?__ pt

15. 24 in. = __?__ ft

16. 84 in. = __?__ ft

17. 90 sec = __?__ min

18. 150 min = __?__ hr

19. 5,280 ft = __?__ mi

20. 730 da = __?__ yr

21. 104 wk = __?__ yr

22. $\frac{1}{4}$ cup = __?__ oz

23. $\frac{3}{8}$ cup = __?__ tbsp

24. 1 gal = __?__ oz

25. $1\frac{1}{3}$ tbsp = __?__ tsp

26. 2 gal = __?__ cu. in.

27. $2\frac{1}{2}$ lb = __?__ oz

28. 2,200 lb = __?__ tons

29. $2\frac{1}{2}$ tons = __?__ lb

30. 48 oz = __?__ lb

31. Change 1.25 miles to feet.

32. Change 8,820 feet to miles.

33. Change 2 weeks to hours.

34. Change 5 miles to yards.

35. Change 2,160 minutes to days.

36. Change 60 ounces to pounds.

37. Louise drove 8 miles in 12 minutes. What was her average speed in miles per hour?

38. Abe walks 2 miles in 50 minutes. What is his average walking speed in miles per hour?

39. Sound travels about 1,100 feet per second. What is the speed of sound in miles per hour?

40. A certain glacier moves 100 yards per year. What is its speed in feet per month?

II.5 Simplifying Denominate Numbers

If you wish to describe how tall you are, how much you weigh, how old you are, or how much money you have in your pocket, you use certain *standard units of measure.* For example, you may be 69 *inches* tall, weigh 160 *pounds*, be 19 *years* old, and have 15 *dollars* in your pocket. Here the standard units of measure are inches, pounds, years, and dollars.

Numbers expressed in standard units of measure, such as inches, pounds, years, and dollars, are called *denominate numbers.* A denominate number is a number with a name (*nom*en means name in Latin). When we say 3 feet, the *3 feet* is a denominate number. When we say 5 hours, the *5 hours* is a denominate number. Numbers such as 5, 25, and $\frac{3}{4}$, which are not given with units, are called *abstract* numbers. When we say $35 + 10 = 45$, we are using abstract numbers, but when we say 35 feet + 10 feet = 45 feet, we are using denominate numbers.

When we say *like numbers*, we mean denominate numbers expressed in the same units.

Example 1 Examples of like numbers

a. 7 feet and 20 feet

b. 15 pounds and 10 pounds ■

Example 2 Examples of unlike numbers

a. 5 feet and 10 inches

b. 2 hours and 15 minutes ■

To simplify a denominate number having two or more different units, change a quantity in smaller units to larger units whenever possible.

Example 3 Simplify 2 feet 18 in.

Solution

$$\begin{aligned} & 2 \text{ ft} + \overbrace{18 \text{ in.}} \\ &= \underbrace{2 \text{ ft} + 1 \text{ ft}} + 6 \text{ in.} \\ &= 3 \text{ ft} + 6 \text{ in.} \\ &= 3 \text{ ft } 6 \text{ in.} \end{aligned}$$

Another way of writing the solution

feet	*inches*	
2	~~18~~	
1	6	18 in. = 1 ft 6 in.
3 ft	6 in. ■	

Example 4 Simplify 14 hr 126 min.

Solution

$$\begin{aligned} & 14 \text{ hr} + \overbrace{126 \text{ min}} \\ &= \underbrace{14 \text{ hr} + 2 \text{ hr}} + 6 \text{ min} \\ &= 16 \text{ hr} + 6 \text{ min} \\ &= 16 \text{ hr } 6 \text{ min} \end{aligned}$$

Another way of writing the solution

hours	*minutes*	
14	~~126~~	
2	6	126 min = 2 hr 6 min
16 hr	6 min	

Therefore, 14 hr 126 min = 16 hr 6 min. ■

Example 5 Simplify 2 da 23 hr 150 min.

Solution

days	*hours*	*minutes*	
2	23	~~150~~	
	2	30	150 min = 2 hr 30 min
	~~25~~		
1	1		25 hr = 1 da 1 hr
3 da	1 hr	30 min	

Therefore, 2 da 23 hr 150 min = 3 da 1 hr 30 min. ■

Example 6 Simplify 4 gal 11 qt 5 pt.

Solution

gallons	*quarts*	*pints*	
4	11	~~5~~	
	2	1	5 pt = 2 qt 1 pt
	~~13~~		
3	1		13 qt = 3 gal 1 qt
7 gal	1 qt	1 pt ■	

EXERCISES II.5

Simplify.

1. 4 ft 15 in.
2. 7 yd 5 ft
3. 2 wk 9 da
4. 2 da 36 hr
5. 3 gal 15 qt
6. 7 qt 11 pt
7. 5 yd 4 ft 27 in.
8. 10 yd 2 ft 34 in.
9. 2 hr 73 min 110 sec
10. 3 hr 82 min 125 sec
11. 3 gal 7 qt 5 pt
12. 4 gal 5 qt 6 pt
13. 2 mi 6,000 ft
14. 3 mi 6,400 ft
15. 2 yr 48 wk 75 da
16. 3 yr 51 wk 55 da
17. 2 tons 3,500 lb
18. 3 tons 2,250 lb
19. 4 lb 20 oz
20. 5 lb 56 oz
21. 2 da 23 hr 75 min

II.6 Adding Denominate Numbers

Only like numbers can be added.

5 dollars + 3 dollars = 8 dollars

2 feet + 4 feet = 6 feet

6 days + 11 days = 17 days

> TO ADD DENOMINATE NUMBERS
>
> 1. Write the denominate numbers under one another with like units in the same vertical line.
> 2. Add the numbers in each vertical line.
> 3. Simplify the denominate number found in (2).

Example 1 Add 7 ft 3 in., 3 ft 5 in., and 2 ft 7 in.

Solution

7 ft	3 in.	
3	5	
2	7	
12	~~15~~	
1	3	15 in. = 1 ft 3 in.
13 ft	3 in. ■	

Example 2 Add 3 da 18 hr 42 min, 1 da 9 hr 29 min, and 1 da 14 hr 51 min.

Solution

da	hr	min	
3 da	18 hr	42 min	
1	9	29	
1	14	51	
5	41	~~122~~	
	2	2	122 min = 2 hr 2 min
	~~43~~		
1	19		43 hr = 1 da 19 hr
6 da	19 hr	2 min ■	

EXERCISES II.6

Add and simplify.

1. 4 ft 5 in.
3 ft 6 in.
5 ft 2 in.

2. 2 yd 2 ft
3 yd 2 ft
4 yd 1 ft

3. 3 hr 15 min
2 hr 50 min
7 hr 24 min

4. 35 min 54 sec
48 min 27 sec

5. 3 gal 2 qt 1 pt
5 gal 3 qt 1 pt
8 gal 1 qt 1 pt

6. 3 gal 2 qt 1 pt
5 gal 3 qt 1 pt
4 gal 1 qt 1 pt

7. 3 yd 2 ft 10 in.
1 yd 1 ft 9 in.
8 yd 2 ft 7 in.

8. 1 mi 4,000 ft
2 mi 3,800 ft

9. 1 da 12 hr 15 min
5 da 23 hr 54 min
2 da 18 hr 47 min

10. 2 yr 41 wk 5 da
1 yr 18 wk 4 da
3 yr 27 wk 3 da

11. 3 tons 1,500 lb
5 tons 450 lb
7 tons 1,850 lb

12. 3 lb 5 oz
8 lb 15 oz
13 lb 9 oz

13. 7 lb 3 oz
15 lb 9 oz
22 lb 17 oz

14. 5 hr 17 min 35 sec
3 hr 44 min 47 sec
2 hr 53 min 24 sec

15. Add 2 yd 8 in., 3 yd 2 ft 5 in., and 4 yd 1 ft.

16. Add 3 hr 40 min, 2 hr 30 sec, and 5 hr 25 min 45 sec.

II.7 Subtracting Denominate Numbers

Only like numbers can be subtracted.

> TO SUBTRACT DENOMINATE NUMBERS
>
> 1. Write the number being subtracted under the number it is being subtracted from, writing like units in the same vertical line.
> 2. Subtract the numbers in each vertical line, borrowing when necessary from the first nonzero number to the left.
> 3. Simplify your answer.

Example 1 Subtract 2 ft 4 in. from 10 ft 6 in.

Solution

$$\begin{array}{rr} 10 \text{ ft} & 6 \text{ in.} \\ -\ 2 \phantom{\text{ ft}} & \underline{4} \phantom{\text{ in.}} \\ \hline 8 \text{ ft} & 2 \text{ in.} \end{array} \quad \blacksquare$$

Example 2 Subtract 4 gal 3 qt from 7 gal 1 qt.

Solution

$$\begin{array}{rr} 6 \phantom{\text{ gal}} & 5 \phantom{\text{ qt}} \\ \not{7} \text{ gal} & \not{1} \text{ qt} \\ -\ 4 \phantom{\text{ gal}} & 3 \phantom{\text{ qt}} \\ \hline 2 \text{ gal} & 2 \text{ qt} \end{array}$$

The 1 gallon *borrowed* makes 4 quarts, which, when added to 1 quart already there, gives 5 quarts ■

Example 3 Subtract 1 hr 50 min 40 sec from 3 hr 21 min.

Solution

$$\begin{array}{rrr} & 80 & \\ 2 \phantom{\text{ hr}} & \not{20} \phantom{\text{ min}} & 60 \phantom{\text{ sec}} \\ \not{3} \text{ hr} & \not{21} \text{ min} & \not{0} \text{ sec} \\ -\ 1 \phantom{\text{ hr}} & 50 \phantom{\text{ min}} & 40 \phantom{\text{ sec}} \\ \hline 1 \text{ hr} & 30 \text{ min} & 20 \text{ sec} \end{array} \quad \blacksquare$$

Example 4 Subtract 1 yd 2 ft 10 in. from 3 yd 4 in.

Solution

$$\begin{array}{rrr} & 2 & \\ 2 \phantom{\text{ yd}} & \not{3} \phantom{\text{ ft}} & 16 \phantom{\text{ in.}} \\ \not{3} \text{ yd} & \not{0} \text{ ft} & \not{4} \text{ in.} \\ -\ 1 \phantom{\text{ yd}} & 2 \phantom{\text{ ft}} & 10 \phantom{\text{ in.}} \\ \hline 1 \text{ yd} & 0 \text{ ft} & 6 \text{ in.} \end{array} = 1 \text{ yd } 6 \text{ in.} \quad \blacksquare$$

EXERCISES II.7

Subtract and simplify.

1. 8 ft 10 in.
− 3 ft 4 in.

2. 9 ft 6 in.
− 7 ft 2 in.

3. 13 yd 2 ft
− 7 yd 1 ft

4. 7 yd 2 ft
− 3 yd 1 ft

5. 8 gal 2 qt
− 3 gal 3 qt 1 pt

6. 8 hr 15 min
− 3 hr 50 min

7. 5 lb 3 oz
− 2 lb 8 oz

8. 3 lb 7 oz
− 1 lb 10 oz

9. 3 tons 700 lb
− 1 ton 1,200 lb

10. 4 tons 500 lb
− 2 tons 1,600 lb

11. 3 da 5 hr
− 1 da 15 hr

12. 5 gal 1pt
− 3 gal 3 qt

13. 5 yd 6 in.
− 2 yd 2 ft 9 in.

14. 3 yd 9 in.
− 1 yd 2 ft 10 in.

15. 4 mi 3,000 ft
− 1 mi 4,700 ft

16. 3 mi 2,000 ft
− 2 mi 2,350 ft

17. 35 min 40 sec
− 20 min 55 sec

18. 27 min 20 sec
− 15 min 32 sec

19. 5 da 13 hr 22 min
− 2 da 18 hr 45 min

20. 5 yr 43 wk 3 da
− 2 yr 50 wk 5 da

21. Subtract 2 hr 4 min 29 sec from 5 hr 14 sec.

22. Subtract 4 yd 2 ft 8 in. from 6 yd 6 in.

II.8 Multiplying Denominate Numbers

TO MULTIPLY A DENOMINATE NUMBER BY AN ABSTRACT NUMBER

1. Multiply each part of the denominate number by the abstract number.
2. Simplify the product.

Example 1 Multiply 4 × (2 ft 8 in.).

Solution

```
   2 ft    8 in.
         × 4
   8 ft   32 in. ←┐
   2       8      ├─ Simplifying the product
  10 ft    8 in. ←┘  ■
```

Example 2 Multiply 5 × (4 gal 3 qt 1 pt).

Solution

```
  4 gal  3 qt 1 pt
            × 5
 20 gal 15 qt 5 pt ←┐
         2    1     │
        17          ├─ Simplifying the product
  4      1          │
 24 gal  1 qt 1 pt ←┘  ■
```

Multiplying a Denominate Number by a Denominate Number Sometimes it is possible to multiply a denominate number by another denominate number.

Example 3 Find the area of a rectangle that is 3 yards long and 2 yards wide.

Solution

Area = length × width
= 3 yards × 2 yards
= 6 square yards (sq. yd)

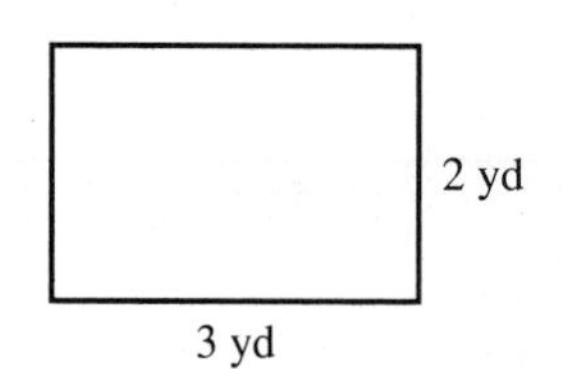

The reason we can multiply these two denominate numbers is that the following has meaning.

1 yd × 1 yd = 1 sq. yd ■

Example 4 Find the number of man-hours worked if a crew of 10 men worked for 8 hours.

Solution 10 men × 8 hours = 80 man-hours

The reason we can multiply these two denominate numbers is that the following has meaning.

1 man × 1 hour = 1 man-hour = work done by 1 person in 1 hour ■

Example 5 5 apples × 3 chairs

Solution 5 apples × 3 chairs = 15 apple-chairs

Because apple-chair has no meaning that we can think of, multiplying these denominate numbers does not make sense. ■

MULTIPLYING DENOMINATE NUMBERS

1. A denominate number can always be multiplied by an abstract number. (See Examples 1 and 2.)
2. A denominate number can be multiplied by another denominate number only when the product of their units has meaning.

EXERCISES II.8

Multiply and simplify.

1. 4 × (3 wk 5 da)
2. 3 × (2 yr 225 da)
3. 6 × (5 mi 2,850 ft)
4. 5 × (3 mi 1,550 ft)
5. 5 × (2 yd 1 ft 3 in.)
6. 3 × (5 gal 2 qt 1 pt)
7. 4 × (1 hr 25 min 11 sec)
8. 7 × (2 yd 2 ft 6 in.)
9. 8 × (3 gal 3 qt 1 pt)
10. 6 × (2 hr 15 min 21 sec)
11. Find the area of a rectangle that is 176 feet long and 48 feet wide.
12. Find the area of a rectangle that is 28 inches by 17 inches.
13. A baseball diamond is a square 90 feet on each side. Find the area enclosed in a baseball diamond.
14. A football field measures 100 yards between goal lines and measures 55 yards between side lines. Find the area enclosed by these lines.
15. A construction crew of 17 men worked 12 eight-hour days to construct a building. Find the total man-hours used to construct the building. Find the cost to construct the building if the cost per man-hour is $6.50.
16. In designing a new vacuum cleaner, 956.4 work-hours were used. Find the cost of designing this machine if the average cost of a work-hour was $14.23.

II.9 Dividing Denominate Numbers

To divide a denominate number expressed in unlike units by an abstract number, we divide the largest unit first, then the next largest unit, and so on until the division is completed. Our first example is simple and has no remainder.

Example 1 Divide (10 yd 2 ft 8 in.) by 2.

Solution

$$\begin{array}{r|l} & \text{5 yd 1 ft 4 in.} \\ \hline 2 & \text{10 yd 2 ft 8 in.} \end{array}$$ ■

> TO DIVIDE A DENOMINATE NUMBER BY AN ABSTRACT NUMBER
>
> 1. Divide the largest unit by the abstract number.
> 2. If a remainder is left from (1), change it to the next smaller unit and add it to those units already there.
> 3. Divide the sum found in (2) by the abstract number.
> 4. Repeat steps (2) and (3) until the division is complete.

Example 2 Divide (11 yd 1 ft 6 in) by 3.

Solution

$$\begin{array}{rllll}
 & \text{3 yd} & & \text{2 ft} & \text{6 in.} \\
3 \,| & \text{11 yd} & & \text{1 ft} & \text{6 in.} \\
 & 9 & & & \\
 & \text{2 yd} & = & \text{6 ft} & \\
 & & & \text{7 ft} & \\
 & & & 6 & \\
 & & & \text{1 ft} = & \text{12 in.} \\
 & & & & \text{18 in.} \\
 & & & & 18 \\
 & & & & 0
\end{array}$$ ■

Example 3 Divide (3 gal 2 qt 1 pt) by 4.

Solution

$$\begin{array}{rllll}
 & \text{0 gal} & \text{3 qt} & 1\frac{1}{4} & \text{pt} \\
4 \,| & \text{3 gal} & \text{2 qt} & 1 & \text{pt} \\
 & 0 & & & \\
 & \text{3 gal} = & \text{12 qt} & & \\
 & & \text{14 qt} & & \\
 & & 12 & & \\
 & & \text{2 qt} = & 4 & \text{pt} \\
 & & & 5 & \text{pt} \\
 & & & 4 & \text{pt} \\
 & & & 1 & \text{pt} \longrightarrow \dfrac{1 \text{ pt}}{4} = \dfrac{1}{4} \text{ pt}
\end{array}$$

This 1 pint has not been divided by 4. When it is divided by 4, we get $\frac{1}{4}$ pint. ■

Dividing a Denominate Number by a Denominate Number Sometimes it is possible to divide a denominate number by another denominate number.

Example 4 How many 3-foot shelves can be cut from a 12-foot board?

Solution

$$\frac{12\text{ ft}}{3\text{ ft}} = \frac{\overset{4}{\cancel{12}\,\cancel{\text{ft}}}}{\underset{1}{\cancel{3}\,\cancel{\text{ft}}}} = 4$$

Notice that the answer is an abstract number ■

Example 5 If a sponsor buys enough TV time to permit a total of one hour for commercials, how many 3-minute commercials can he put on?

Solution

$$\frac{1\text{ hour}}{3\text{ min}} = \frac{\overset{20}{\cancel{60}\,\cancel{\text{min}}}}{\underset{1}{\cancel{3}\,\cancel{\text{min}}}} = 20$$ ■

EXERCISES II.9

Divide and simplify.

1. (2 ft 6 in.) ÷ 2

2. (3 qt 1 pt) ÷ 3

3. (2 qt 1 pt) ÷ 4

4. (6 ft 8 in.) ÷ 5

5. (5 hr 30 min) ÷ 3

6. (6 hr 40 min) ÷ 4

7. (4 gal 3 qt 1 pt) ÷ 3

8. (5 yd 2 ft 6 in.) ÷ 3

9. (8 yd 2 ft 10 in.) ÷ 5

10. (5 gal 3 qt 1 pt) ÷ 6

11. (5 lb 8 oz) ÷ 7

12. (4 lb 10 oz) ÷ 8

13. (13 wk 5 da 15 hr) ÷ 3

14. (12 wk 4 da 10 hr) ÷ 5

15. (8 mi 4,500 ft) ÷ 6

16. (4 mi 4,000 ft) ÷ 12

17. How many 2-foot fence posts can be cut from a 16-foot board?

18. How many $1\frac{3}{4}$-foot stakes can be cut from a 14-foot board?

19. How many 15¢ postcards can $1.75 buy?

20. How many special-addressed envelopes costing 18 cents each can be bought for $17.50?

21. How many 2-minute radio commercials can be fitted into 1 hour and 30 minutes available just for commercials?

22. It takes 32 minutes to machine a special fitting. How many of these fittings can be made in an 8-hour work day?

II.10 Changing English System Units to Metric Units (and Vice Versa)

Have you ever had the experience of trying to use an American-made wrench on a bolt on a foreign-made car? The American-made wrench does not fit the metric-made bolt. Because over 90 percent of the people in the world today use the metric system, we must learn how to change the English system units that we use into metric units, and vice versa.

Here is a short table listing commonly used conversions between the metric and the English systems of measurement.

COMMON ENGLISH–METRIC CONVERSIONS

(These conversion factors have 3-significant-digit accuracy.)

1 inch (in.) = 2.54 centimeters (cm)
39.4 inches (in.) ≈ 1 meter (m)
0.621 miles (mi) ≈ 1 kilometer (km)
1 mile (mi) ≈ 1.61 kilometers (km)
1 pound (lb) ≈ 454 grams (g)
2.20 pounds (lb) ≈ 1 kilogram (kg)
1.06 quarts (qt) ≈ 1 liter (ℓ)
2.47 acres ≈ 1 hectare (ha)

TO CONVERT ENGLISH UNITS TO METRIC UNITS (AND VICE VERSA)

1. Select the conversion factor from the table that relates the units given in the problem.
2. If you cannot determine whether to multiply or divide by the conversion factor, use unit fractions.
3. Round off your answer to the allowable accuracy determined by the accuracy of the given numbers and the conversion factors used.

It is possible to get slightly different answers if conversion factors of different accuracy are used.

Example 1 Mrs. Peralta weighs 62 kg. What is her weight in pounds?

Solution 1 kg = 2.20 lb

$$62 \text{ kg} = \frac{62 \cancel{\text{kg}}}{1}\left(\frac{2.20 \text{ lb}}{1 \cancel{\text{kg}}}\right) = 62 \times 2.20 \text{ lb} \approx 140 \text{ lb}$$

1 lb

1 kg = 2.2 lb

Therefore, 62 kg ≅ 140 lb (rounded to same accuracy as 62 kg). ■

Example 2 A road sign in Mexico reads *85 kilometers to Ensenada.* How far is this in miles?

Solution 1 km = 0.621 mi

$$85 \text{ km} = \frac{85 \cancel{\text{km}}}{1}\left(\frac{0.621 \text{ mi}}{1 \cancel{\text{km}}}\right) = 85 \times 0.621 \text{ mi} \approx 53 \text{ mi}$$

Therefore, 85 km ≈ 53 mi (rounded to same accuracy as 85 km). ■

Example 3 Change 104 grams to ounces.

Solution

$$104 \text{ g} = \frac{104 \cancel{g}}{1}\left(\frac{1 \cancel{\text{lb}}}{454 \cancel{g}}\right)\left(\frac{16 \text{ oz}}{1 \cancel{\text{lb}}}\right) = \frac{104 \times 16}{454} \text{ oz} \approx 3.67 \text{ oz}$$

This is 1 because 1 lb = 454 g

This is 1 because 1 lb = 16 oz

Therefore, 104 g ≈ 3.67 oz (rounded to same accuracy as 104 g). ■

Example 4 Change 228 centimeters to feet.

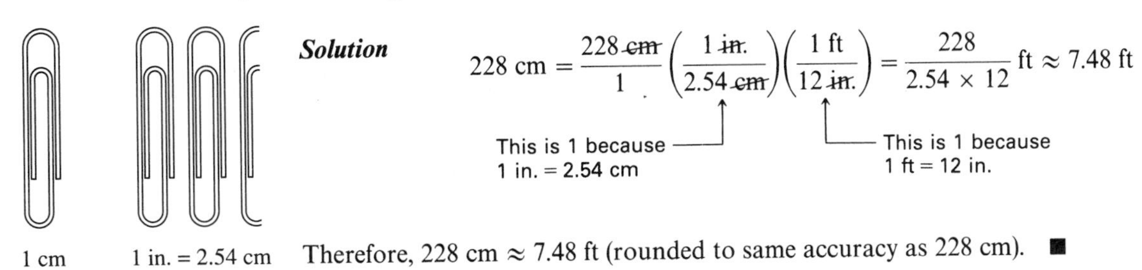

1 cm   1 in. = 2.54 cm

Solution

$$228 \text{ cm} = \frac{228 \cancel{\text{cm}}}{1}\left(\frac{1 \cancel{\text{in.}}}{2.54 \cancel{\text{cm}}}\right)\left(\frac{1 \text{ ft}}{12 \cancel{\text{in.}}}\right) = \frac{228}{2.54 \times 12} \text{ ft} \approx 7.48 \text{ ft}$$

This is 1 because 1 in. = 2.54 cm

This is 1 because 1 ft = 12 in.

Therefore, 228 cm ≈ 7.48 ft (rounded to same accuracy as 228 cm). ■

Example 5 Change 750 feet per second to kilometers per minute.

Solution

$$\frac{750 \cancel{\text{ft}}}{\cancel{\text{sec}}}\left(\frac{1 \cancel{\text{mi}}}{5{,}280 \cancel{\text{ft}}}\right)\left(\frac{1.61 \text{ km}}{1 \cancel{\text{mi}}}\right)\left(\frac{60 \cancel{\text{sec}}}{1 \text{ min}}\right) \approx 14 \text{ km per min}$$

This is 1 because 1 mi = 5,280 ft

This is 1 because 1 min = 60 sec

This is 1 because 1 mi = 1.61 km

Therefore, 750 ft per sec ≈ 14 km per min (rounded to same accuracy as 750 fps). ■

EXERCISES II.10

Round off your answers to the allowable accuracy determined by the accuracy of the given numbers and the conversion factors used.

1. 15 in. = __?__ cm

2. 18 in. = __?__ cm

3. 2.12 qt = __?__ ℓ

4. 3.18 qt = __?__ ℓ

5. 0.55 kg = __?__ lb

6. 5.1 kg = __?__ lb

7. 82 km = __?__ mi

8. 140 km = __?__ mi

9. 12 ℓ = __?__ qt

10. 17 ℓ = __?__ qt

11. 33 mi = __?__ km

12. 165 mi = __?__ km

13. 20.6 lb = __?__ kg

14. 28.4 lb = __?__ kg

15. 150 ha = __?__ acres

16. 65 acres = __?__ ha

17. 66 in. = __?__ m

18. 74 in. = __?__ m

19. 2 m = __?__ in.

20. 5 m = __?__ in.

21. 908 g = __?__ lb

22. 1,362 g = __?__ lb

23. 0.75 lb = __?__ g

24. 1.25 lb = __?__ g

25. 1.5 yd = __?__ cm

26. 100 cm = __?__ yd

27. 227 g = __?__ oz

28. 12 oz = __?__ g

29. A road sign in France reads *120 kilometers to Paris.* Find this distance in miles.

30. A speed control sign reads 30 kilometers per hour. What is this in miles per hour?

31. Mr. Dubois steps on a scale in Orly Airport. The scale reads 85.7 kilograms. What is his weight in pounds?

32. A crate of transistor radios arrives from Japan marked *67.5 kilograms net weight.* Find this weight in pounds.

33. In the Olympics, there is a 1,500-meter race which is about the same distance as our 1-mile race. Which race is longer and by how much? (Express the difference in feet.)

34. In the Olympics, there is a 400-meter race and a quarter-mile race. Which race is longer and by how much? (Express the difference in feet.)

35. A Howitzer muzzle measures 175 millimeters. What is this measurement in inches?

36. What is the width in inches of a 35-millimeter roll of film?

37. How many gallons does a 20-liter container hold?

38. A moon lander is descending at the rate of 1,200 miles per hour. Express this rate of descent in meters per second.

II.11 Converting Units Using a Calculator

Conversion of English system units to metric units (and vice versa) can be easily done using a calculator.

CONVERTING UNITS USING THE CALCULATOR

(These conversion factors have 6-significant-digit accuracy.)

English to Metric					*Metric to English*				
Length					*Length*				
Inches	×	2.54	=	Centimeters	Centimeters	×	.393701	=	Inches
Feet	×	30.48	=	Centimeters	Centimeters	×	.0328084	=	Feet
Yards	×	.9144	=	Meters	Meters	×	1.09361	=	Yards
Miles	×	1.60934	=	Kilometers	Kilometers	×	.621371	=	Miles
Volume (U.S.)					*Volume*				
Pints	×	.473176	=	Liters	Liters	×	2.11338	=	Pints
Quarts	×	.946353	=	Liters	Liters	×	1.05669	=	Quarts
Gallons	×	3.78541	=	Liters	Liters	×	.264172	=	Gallons
Weight					*Weight*				
Ounces	×	28.3495	=	Grams	Grams	×	.0352740	=	Ounces
Pounds	×	453.592	=	Grams	Grams	×	.00220462	=	Pounds

Temperature

Changing Fahrenheit to Celsius

°F [−] 32 [=] [×] 5 [÷] 9 [=] °C

Changing Celsius to Fahrenheit

°C [×] 9 [÷] 5 [+] 32 [=] °F

Example 1 Converting units using the calculator

a. Change 976 yards to meters.

976 [×] .9144 [=] 892.4544 $\approx$ 892 m

b. Change 15.9 gallons to liters.

15.9 [×] 3.78541 [=] 60.188019 ≈ 60.2 ℓ

c. Change 8.423 ounces to grams.

8.423 [×] 28.3495 [=] 238.7878385 ≈ 238.8 g

d. Change 52.8 centimeters to inches.

52.8 [×] .393701 [=] 20.7874128 ≈ 20.8 in.

e. Change 15.76 liters to quarts.

15.76 [×] 1.05669 [=] 16.6534344 ≈ 16.65 qt

f. Change 19.5°C into °F.

19.5 [×] 9 [÷] 5 [+] 32 [=] 67.1°F ■

EXERCISES II.11

Use calculator conversions and round off answers to allowable accuracy.

1. 2.65 ft = __?__ cm

2. 34.5 in. = __?__ cm

3. 419 qt = __?__ ℓ

4. 97 pt = __?__ ℓ

5. 14.75 oz = __?__ g

6. 5.7 lb = __?__ g

7. 82°F = __?__ °C

8. 23.5°C = __?__ °F

9. 1,500 m = __?__ yd

10. 528 cm = __?__ ft

11. 22 ℓ = __?__ gal

12. 183 ℓ = __?__ qt

13. 546.7 g = __?__ lb

14. 325.8 g = __?__ oz

15. 98.6°F = __?__ °C

APPENDIX III

Quadratic Formula

In this appendix we discuss methods for solving quadratic equations. We have already solved quadratic equations by factoring in Section 11.7.

General Form of a Quadratic Equation Any quadratic equation can be arranged as follows:

General Form (Standard Form)

THE GENERAL FORM OF A QUADRATIC EQUATION

$$ax^2 + bx + c = 0$$

where a, b, and c are real numbers ($a \neq 0$).

In this text when we write the general form of a quadratic equation, all coefficients will be integers. It is also helpful to write the general form in such a way that a is positive.

TO CHANGE A QUADRATIC EQUATION INTO GENERAL FORM

1. Remove fractions by multiplying each term by the LCD.
2. Remove grouping symbols.
3. Combine like terms.
4. Arrange all nonzero terms in descending powers on one side, leaving only zero on the other side.

Example 1 Change the quadratic equations into general form and identify a, b, and c.

a.
$$7x = 5 - 2x^2$$
$$2x^2 + 7x - 5 = 0 \quad \text{General form} \quad \begin{cases} a = 2 \\ b = 7 \\ c = -5 \end{cases}$$

b.
$$5x^2 = 3$$
$$5x^2 + 0x - 3 = 0 \quad \text{General form} \quad \begin{cases} a = 5 \\ b = 0 \\ c = -3 \end{cases}$$

c.
$$\frac{2}{3}x^2 - 5x = \frac{1}{2}$$
$$\frac{6}{1} \cdot \frac{2}{3}x^2 + \frac{6}{1} \cdot (-5x) = \frac{6}{1} \cdot \frac{1}{2}$$
Multiplied each *term* by the LCD, 6
$$4x^2 - 30x = 3$$
$$4x^2 - 30x - 3 = 0 \quad \text{General form} \quad \begin{cases} a = 4 \\ b = -30 \\ c = -3 \end{cases}$$

d.
$$x(x - 2) = 5$$
$$x^2 - 2x = 5$$
$$x^2 - 2x - 5 = 0 \quad \text{General form} \quad \begin{cases} a = 1 \\ b = -2 \\ c = -5 \end{cases}$$ ■

The ± Symbol

The symbol $\pm$ is read *plus or minus.*
± 2 is read *plus or minus 2.*
$x = \pm 2$ is read *x equals plus or minus 2.*
This mean $x = +2$ or $x = -2$.

A positive real number N has two square roots: a positive root called the *principal square root*, written $\sqrt{N}$, and a negative square root, written $-\sqrt{N}$. We can represent these two square roots by the symbol $\pm\sqrt{N}$.

Example 2 Solve $x^2 - 4 = 0$ (Here, $b = 0$).

Solution

$$x^2 - 4 = 0$$
$$x^2 = 4$$
$$x = \pm 2 \quad \text{Took square root of both sides}$$

Justification for the $\pm$ sign: This equation can be solved by factoring.

$$x^2 - 4 = 0$$
$$(x + 2)(x - 2) = 0$$
$$x + 2 = 0 \quad \Big| \quad x - 2 = 0$$
$$x = -2 \quad \Big| \quad x = 2$$

This shows why we must use $\pm$. ■

The Quadratic Formula The methods we have shown in previous sections can only be used to solve *some* quadratic equations. The method we show in this section can be used to solve *all* quadratic equations.

Completing the Square

In Example 3 we use a method called *completing the square* to solve a quadratic equation.

Example 3 Solve $x^2 - 4x + 1 = 0$.

Solution

$$x^2 - 4x = -1$$

Take $\frac{1}{2}(-4) = -2$

Then $(-2)^2 = 4$

$$x^2 - 4x + 4 = -1 + 4 \quad \text{Added 4 to both sides; this made the left side a trinomial square}$$
$$(x - 2)^2 = 3 \quad \text{Factored the left side}$$
$$\sqrt{(x - 2)^2} = \pm\sqrt{3}$$
$$x - 2 = \pm\sqrt{3} \quad \text{Took the square root of both sides}$$
$$x = 2 \pm \sqrt{3} \quad \text{Added 2 to both sides}$$

There are two solutions: $x = 2 + \sqrt{3}$ and $x = 2 - \sqrt{3}$.

Check for $x = 2 + \sqrt{3}$

$$x^2 - 4x + 1 = 0$$
$$(2 + \sqrt{3})^2 - 4(2 + \sqrt{3}) + 1 \stackrel{?}{=} 0$$
$$4 + 4\sqrt{3} + 3 - 8 - 4\sqrt{3} + 1 \stackrel{?}{=} 0$$
$$0 = 0$$

We leave the check for $x = 2 - \sqrt{3}$ for the student. ■

Deriving the Quadratic Formula

The method of completing the square can be used to solve *any* quadratic equation. We now use it to solve the general form of the quadratic equation and in this way derive the *quadratic formula.*

$$ax^2 + bx + c = 0$$ General form

$$ax^2 + bx = 0 - c$$ Subtracted c from both sides

$$x^2 + \frac{b}{a}x = -\frac{c}{a}$$ Divided both sides by a

Take $\frac{1}{2}\left(\frac{b}{a}\right) = \frac{b}{2a}$

Then $\left(\frac{b}{2a}\right)^2 = \frac{b^2}{4a^2}$

$$x^2 + \frac{b}{a}x + \frac{b^2}{4a^2} = \frac{b^2}{4a^2} - \frac{c}{a} = \frac{b^2}{4a^2} - \frac{4ac}{4a^2}$$ Added $\frac{b^2}{4a^2}$ to both sides

This made the left side a trinomial square

$$\left(x + \frac{b}{2a}\right)^2 = \frac{b^2 - 4ac}{4a^2}$$ Factored the left side and added the fractions on the right side

$$x + \frac{b}{2a} = \pm\sqrt{\frac{b^2 - 4ac}{4a^2}}$$ Took the square root of both sides

$$x + \frac{b}{2a} = \pm\frac{\sqrt{b^2 - 4ac}}{\sqrt{4a^2}}$$ Simplified radicals

$$x + \frac{b}{2a} = \pm\frac{\sqrt{b^2 - 4ac}}{2a}$$

$$x = -\frac{b}{2a} \pm \frac{\sqrt{b^2 - 4ac}}{2a}$$ Added $-\frac{b}{2a}$ to both sides

Therefore, $$x = \frac{-b \pm \sqrt{b^2 - 4ac}}{2a}$$ Quadratic formula

The procedure for using the quadratic formula can be summarized as follows:

TO SOLVE A QUADRATIC EQUATION BY FORMULA

1. Arrange the equation in general form.

$$ax^2 + bx + c = 0$$

2. Substitute the values of a, b, and c into the *quadratic formula.*

$$x = \frac{-b \pm \sqrt{b^2 - 4ac}}{2a} \quad (a \neq 0)$$

3. Simplify your answers.

Check your answers by substituting them in the *original* equation.

Example 4 Solve $x^2 - 5x + 6 = 0$ by formula.

Solution Substitute $\begin{cases} a = 1 \\ b = -5 \\ c = 6 \end{cases}$ in the formula $x = \dfrac{-b \pm \sqrt{b^2 - 4ac}}{2a}$.

$$x = \frac{-(-5) \pm \sqrt{(-5)^2 - 4(1)(6)}}{2(1)}$$

$$x = \frac{5 \pm \sqrt{25 - 24}}{2} = \frac{5 \pm \sqrt{1}}{2}$$

$$x = \frac{5 \pm 1}{2} = \begin{cases} \dfrac{5+1}{2} = \dfrac{6}{2} = 3 \\ \dfrac{5-1}{2} = \dfrac{4}{2} = 2 \end{cases}$$

This equation can also be solved by factoring.

$$x^2 - 5x + 6 = 0$$

$$(x - 2)(x - 3) = 0$$

$$x - 2 = 0 \quad | \quad x - 3 = 0$$

$$x = 2 \quad | \quad x = 3 \quad \blacksquare$$

Solving a quadratic equation by factoring is ordinarily shorter than using the formula. Therefore, first check to see if the equation can be solved by factoring. If it cannot, use the formula.

Example 5 Solve $x^2 - 6x - 3 = 0$.

Solution Substitute $\begin{cases} a = 1 \\ b = -6 \\ c = -3 \end{cases}$ in the formula $x = \dfrac{-b \pm \sqrt{b^2 - 4ac}}{2a}$.

$$x = \frac{-(-6) \pm \sqrt{(-6)^2 - 4(1)(-3)}}{2(1)}$$

$$= \frac{6 \pm \sqrt{36 + 12}}{2} = \frac{6 \pm \sqrt{48}}{2}$$

$$= \frac{6 \pm 4\sqrt{3}}{2} = \frac{\cancel{2}(3 \pm 2\sqrt{3})}{\cancel{2}} = 3 \pm 2\sqrt{3} \quad \blacksquare$$

Example 6 Solve $\dfrac{1}{4}x^2 = 1 - x$.

Solution

$$\frac{4}{1} \cdot \frac{1}{4}x^2 = 4 \cdot 1 - 4 \cdot x$$ Multiplied each term by the LCD, 4

$$x^2 = 4 - 4x$$

$$x^2 + 4x - 4 = 0$$ Changed the equation to general form

Substitute $\begin{cases} a = 1 \\ b = 4 \\ c = -4 \end{cases}$ into $x = \dfrac{-b \pm \sqrt{b^2 - 4ac}}{2a}$.

$$x = \frac{-(4) \pm \sqrt{(4)^2 - 4(1)(-4)}}{2(1)}$$

$$= \frac{-4 \pm \sqrt{16 + 16}}{2} = \frac{-4 \pm \sqrt{32}}{2}$$

$$= \frac{-4 \pm 4\sqrt{2}}{2} = \frac{\not{2}(-2 \pm 2\sqrt{2})}{\not{2}} = -2 \pm 2\sqrt{2}$$ ■

Example 7 Solve $4x^2 - 5x + 2 = 0$.

Solution Substitute $\begin{Bmatrix} a = 4 \\ b = -5 \\ c = 2 \end{Bmatrix}$ into $x = \dfrac{-b \pm \sqrt{b^2 - 4ac}}{2a}$.

$$x = \frac{-(-5) \pm \sqrt{(-5)^2 - 4(4)(2)}}{2(4)}$$

$$= \frac{5 \pm \sqrt{25 - 32}}{8} = \frac{5 \pm \sqrt{-7}}{8}$$ ← Solution is not a real number because radicand is negative ■

We can use Table I or a calculator to evaluate the square root and express the answers as approximate decimals.

Example 8 Solve $x^2 - 5x + 3 = 0$

Solution Express the answers as decimals correct to two decimal places.

Substitute $\begin{Bmatrix} a = 1 \\ b = -5 \\ c = 3 \end{Bmatrix}$ into $x = \dfrac{-b \pm \sqrt{b^2 - 4ac}}{2a}$.

$$x = \frac{-(-5) \pm \sqrt{(-5)^2 - 4(1)(3)}}{2(1)}$$

$$= \frac{5 \pm \sqrt{25 - 12}}{2} = \frac{5 \pm \sqrt{13}}{2}$$ $\sqrt{13} \approx 3.606$ from Table I.

$$\approx \frac{5 \pm 3.606}{2} = \begin{cases} \dfrac{5 + 3.606}{2} = \dfrac{8.606}{2} = 4.303 \approx 4.30 \\ \dfrac{5 - 3.606}{2} = \dfrac{1.394}{2} = 0.697 \approx 0.70 \end{cases}$$

We suggest you use a calculator to check these answers in the original equation. ■

EXERCISES III

Use the quadratic formula to solve each of the following equations.

1. $3x^2 - x - 2 = 0$ **2.** $2x^2 + 3x - 2 = 0$ **3.** $x^2 - 4x + 1 = 0$

4. $x^2 - 4x - 1 = 0$ **5.** $x^2 - 4x + 2 = 0$ **6.** $4x^2 = 12x - 7$

7. $2x^2 = 8x - 5$ **8.** $3x^2 = 6x - 2$ **9.** $3x^2 + 2x + 1 = 0$

10. $4x^2 + 3x + 2 = 0$ **11.** $x(x - 2) = 3$ **12.** $(x + 1)(x + 2) = 12$

13. $\frac{x}{2} + \frac{2}{x} = \frac{5}{2}$

14. $\frac{x}{3} + \frac{2}{x} = \frac{7}{3}$

15. $x^2 = \frac{3 - 5x}{2}$

16. $\frac{x}{3} + \frac{1}{x} = \frac{7}{6}$

17. $\frac{x^2}{2} = x + 1$

18. $\frac{4}{x + 6} + x = 0$

19. The difference between a number and its reciprocal is $\frac{2}{3}$. Find the number. $\left(\text{The reciprocal of a number } n \text{ is } \frac{1}{n}.\right)$

20. One leg of a right triangle is 2 units longer than the other. The hypotenuse is $2\sqrt{2}$. Find the length of each leg.

Answers

Exercises 1.1 (page 4)

1. 4 **3.** 0 **5.** 1 **7.** Cannot be found **9.** 99
11. 100 **13.** No **15.** Yes **17.** 1, 2, 3, 4, 5 **19.** <
21. > **23.** <

Exercises 1.2 (page 7)

1. a. 6 units **b.** 7 tens = 70 units **c.** 5 hundreds = 500 units
3. a. 4 tens **b.** 4 tens = 40 units **c.** 3 hundreds **d.** 30 tens
5. Sixteen thousand, three hundred forty-six
7. One million, seventy thousand, two hundred
9. Forty-nine million, seventy
11. Eight billion, seven hundred twenty-five thousand
13. 16,601 Sixteen thousand, six hundred one
15. 71,040 Seventy-one thousand, forty
17. 710,004,000 Seven hundred ten million, four thousand
19. 5,206,000,000 Five billion, two hundred six million
21. 20,005 **23.** 8,008,808 **25.** 7,000,007 **27.** 10,000,100,010
29. 100,029,006 **31.** 5,880,000,000,000

Exercises 1.3 (page 9)

1. 4,700 **3.** 930 **5.** 800 **7.** 63,000 **9.** 800
11. 20,000,000 **13.** 52,000,000 **15.** 3,500 **17.** 45,430,000
19. 1,600,000

Exercises 1.4 (page 12)

1. 637 **3.** 1,702 **5.** 1,794 **7.** 1,024 **9.** 1,818
11. 907 **13.** 9,081 **15.** 70,150 **17.** 257
19. 1,014,924,913 **21.** 1,049,125,745 **23.** 892,351,182
25. 392,008,000 **27.**

```
          30,006
      75,000,100
   2,000,000,500
 +    50,100,010
   -------------
   2,125,130,616
```

29. 45 **31.** 65,226 ft

33.

```
John        $75
Jim         $18
           ----
            $93   Amount John and Jim have together
          + 12
Marie      $105
Then, John   75
Jim          18
           ----
           $198   Total amount the three have together
```

Exercises 1.5 (page 15)

1. 8,205 **3.** 3,005 **5.** 94,200 **7.** 328 **9.** 129
11. 212 **13.** 2,321 **15.** 49,893 **17.** 337,603 **19.** 230,792
21. 2,886 **23.** 92,650,000 mi **25.** 3,448 mi **27.** $210
29.

```
Jim     75
Joe     57
Jack    92 = 75 + 17
Mike   119 = 75 + 57 − 13
      ----
      $343
```

Exercises 1.6 (page 20)

1. 2,552 **3.** 6,734 **5.** 51,976 **7.** 69,318 **9.** 24,966
11. 501,504 **13.** 8,576,220 **15.** 97,760 **17.** 4,521,606
19. 22,594,496 **21.** 72,624 **23.** 2,697,036 **25.** 7,872,218
27. 540,822,290 **29.** 2,250,795,063 **31.** 940,000
33. 29,016,800 **35.** 1,875,500 **37.** 2,963,400 **39.** 74,983,500
41. 50,176,000 **43.** 713,500 sheets
45.

```
 16 | 000
 18 |
----
128 |
 16 |
---------
288 | 000¢ = $2,880
```

47. $24,500

Exercises 1.7 (page 23)

1. 8 **3.** 25 **5.** 27 **7.** 36 **9.** 1 **11.** 100
13. 10,000 **15.** 1,000,000 **17.** 0 **19.** 1 **21.** 16 **23.** 1
25. 248,832 **27.** 531,441 **29.** 14,706,125 **31.** 1

Exercises 1.8 (page 24)

1. 300 **3.** 300 **5.** 160 **7.** 8,000 **9.** 100,000 **11.** 4
13. 2,000 **15.** 80,000 **17.** 9,000 **19.** 10,300,000

Exercises 1.9 (page 30)

1. 154 R 3 **3.** 1,050 R 4 **5.** 7,306 **7.** 27 R 10
9. 20 R 25 **11.** 53 R 66 **13.** 286 **15.** 207
17. 370 R 5 **19.** 270 **21.** 602 R 2 **23.** 1,070 R 44
25. 1,407 R 8 **27.** 2,084 R 27 **29.** 351 R 123
31. 1,400 **33.** 790 R 240 **35.** 9,003 **37.** 62 hours
39. a. 7 bars **b.** 3¢ **41. a.** $1,560 **b.** $360
43. 6,025 boxes **45.** 12 + 4 = 16 boxes
47. All remainders are zero.

Exercises 1.10 (page 31)

1. 5 **3.** 8 **5.** 9 **7.** 4 **9.** 1 **11.** 12 **13.** 10
15. 100 **17.** 23 **19.** 75

Exercises 1.11 (page 33)

1. 25 **3.** 8 **5.** 26 **7.** 17 **9.** 9

11.
$$\begin{aligned} &5^2 - 2^3 + \sqrt{9} = \\ &25 - 8 + 3 = \\ &17 + 3 = 20 \end{aligned}$$

13. 23 **15.** 42 **17.** 20 **19.** 12 **21.** 5

23.
$$\begin{aligned} &20 - (2 + 4 \cdot 3) = \\ &20 - (2 + 12) = \\ &20 - 14 = 6 \end{aligned}$$

25. 19 **27.** 98

29.
$$\begin{aligned} &5 \cdot 8 + 2\sqrt{100} = \\ &5 \cdot 8 + 2 \cdot 10 = \\ &40 + 20 = 60 \end{aligned}$$

31.
$$\begin{aligned} &24 - 8 \div 4 \cdot 2 + 2^3 = \\ &24 - 8 \div 4 \cdot 2 + 8 = \\ &24 - 2 \cdot 2 + 8 = \\ &24 - 4 + 8 = \\ &20 + 8 = 28 \end{aligned}$$

33. 31 **35.** 52 **37.** 205 **39.** 606 **41.** 3,500
43. 684 **45.** 5,909 **47.** 96

Exercises 1.12 (page 35)

1. 6 **3.** 7 **5.** 26 **7.** 79 **9.** 79 **11.** 77 in. **13.** 75

15.
Group A's average 84
Group B's average − 79
Group A is higher by 5

17. 9 in.

Exercises 1.13A (page 37)

1. 40 ft **3.** 42 in. **5.** 18 ft

7.

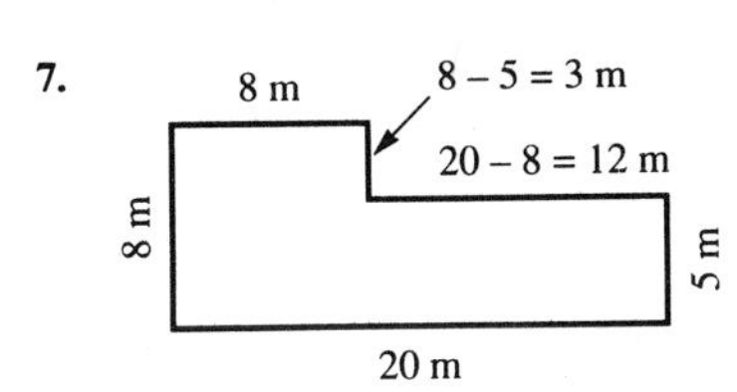

$P = 8 + 8 + 20 + 5 + 12 + 3 = 56$ m

9. 60 in. **11.** $72

Exercises 1.13B (page 39)

1. 120 sq. in. **3.** 36 sq. in. **5.** 64 sq. ft
7. a. 30 sq. yd **b.** $450 **9.** $42

Exercises 1.14 (page 40)

1. 19,000 **3.** 70,000 **5.** 2,100,000 **7.** 200 **9.** 50,000
11. a. 2(400) + 200 + 300 = 1,300 calories **b.** 1,292 calories
13. a. 9,000 ft **b.** 8,938 ft

Review Exercises 1.15 (page 43)

1. Three billion, seventy-five million, six hundred thousand, eight.
2. 5,072,006 **3.** 0, 1, 2, 3, 4 **4.** 10
5. a. 17 > 12 **b.** 0 < 19 **6.** subtraction and division
7. factors or divisors **8.** subtraction **9.** trial divisor
10. inverse

11.
8 minuend
− 2 subtrahend
6 difference

12.
14 ← quotient
divisor → 17)243 ← dividend
17
73
68
5 ← remainder

13. a. 280,000 **b.** 9,700 **c.** 43,000,000
14. a. undefined **b.** 0 **c.** 4 **d.** undefined
15. 1,941 **16.** 1,210 **17.** 86,314 **18.** 7,889 **19.** 5,117
20. 679,158 **21.** 27,510 **22.** 6,481,728 **23.** 3,439,389
24. 4,086 **25.** 57 R 14 **26.** 480 **27.** 25 **28.** 16
29. 1 **30.** 6 **31.** 9 **32.** 7,500 **33.** 80,000 **34.** 30,000

35.
$$\begin{aligned} &10 + 20 \div 2 \cdot 5 = \\ &10 + 10 \cdot 5 = \\ &10 + 50 = 60 \end{aligned}$$

36.
$$\begin{aligned} &36 - 6 + 3 \cdot 2^3 = \\ &36 - 6 + 3 \cdot 8 = \\ &36 - 6 + 24 = \\ &30 + 24 = 54 \end{aligned}$$

37.
$$\begin{aligned} &2(3)^2 - 4(3) + 5 = \\ &2 \cdot 9 - 4(3) + 5 = \\ &18 - 12 + 5 = \\ &6 + 5 = 11 \end{aligned}$$

38.
$$\begin{aligned} &10 - (8 - 3 \cdot 2) - 6 = \\ &10 - (8 - 6) - 6 = \\ &10 - 2 - 6 = \\ &8 - 6 = 2 \end{aligned}$$

39. $758 **40.** $1,792 **41.** 2,054,691 **42.** 5 bottles
43. $127 **44.** $48 **45.** 280,000 words

46.
John 45
Harry 23
Bill 80 = 45 + 23 + 12
$148

47.
$ 82
× 23
246
164
$1886 Paid

$1,950
− 1,886
$64 Final Payment

48. 1100 people **49.** $140 a month **50.** 871,200 times

51.
67
× 35
335
201
$2345 Paid

2365
− 2345
$20 Left

52. 352 miles

53. a. $19 a month
24)456
24
216
216

b.
$19
× 17
133
19
$323 Paid

$456
− 323
$133 Still owed

54.
53,731
− 53,408
323 mi

17 mi per gal
19)323
19
133
133
0

55. 26 hr **56.** 75
57. a. 40 in. **b.** 96 sq. in. **58. a.** 36 ft **b.** 81 sq. ft
59. a. 1,800,000 **b.** 200 **60.** $390

Chapter 1 Diagnostic Test (page 47)

Following each problem number is the textbook section reference (in parentheses) in which that kind of problem is discussed.

1. (1.2) Five trillion, eight hundred seventy-nine billion, two hundred million

2. (1.2) 54,007,506,080

3. (1.4) $$\begin{array}{r} 5843 \\ 209 \\ +\ 6027 \\ \hline 12{,}079 \end{array}$$

4. (1.4) $$\begin{array}{r} 946 \\ 7328 \\ 407 \\ +\ 24 \\ \hline 8705 \end{array}$$

5. (1.5) $$\begin{array}{r} {\scriptstyle 2\ 14\ 16} \\ 3564 \\ -\ 782 \\ \hline 2{,}782 \end{array}$$

6. (1.5) $$\begin{array}{r} {\scriptstyle 9} \\ {\scriptstyle 4\ 10\ 3\ 10\ 16} \\ 50{,}406 \\ -\ 35{,}008 \\ \hline 15{,}398 \end{array}$$

7. (1.6) $$\begin{array}{r} 576 \\ \times\ 89 \\ \hline 5184 \\ 4608 \\ \hline 51{,}264 \end{array}$$

8. (1.6) $$\begin{array}{r} 3084 \\ \times\ 706 \\ \hline 18504 \\ 21588 \\ \hline 2{,}177{,}304 \end{array}$$

9. (1.9) $$\begin{array}{r} 80 \text{ R } 15 \\ 63\overline{)5055} \\ 504 \\ \hline 15 \\ 0 \\ \hline 15 \end{array}$$

10. (1.9) $$\begin{array}{r} 706 \\ 495\overline{)349{,}470} \\ 3465 \\ \hline 297 \\ 0 \\ \hline 2970 \\ 2970 \\ \hline \end{array}$$

11. (1.7) **a.** $2^3 = 2 \cdot 2 \cdot 2 = 8$ **b.** $8^2 = 8 \cdot 8 = 64$

12. (1.10) **a.** $\sqrt{16} = 4$ **b.** $\sqrt{100} = 10$

13. (1.8) **a.** $40 \cdot 100 = 4{,}000$ **b.** $16 \cdot 10^4 = 160{,}000$

14. (1.3) **a.** 79,000 **b.** 3,700

15. (1.5) $$\begin{array}{r} 71{,}304 \\ -\ 67{,}856 \\ \hline 3{,}448 \text{ mi} \end{array}$$

16. (1.9) $$\begin{array}{r} \$56 \\ 36\overline{)2016} \\ 180 \\ \hline 216 \\ 216 \\ \hline \end{array}$$

17. (1.6) $$\begin{array}{r} 168 \\ \times\ 52 \\ \hline 336 \\ 840 \\ \hline \$8{,}736 \end{array}$$

18. (1.6) $$\begin{array}{r} 58 \\ \times\ 23 \\ \hline 174 \\ 116 \\ \hline \$1334 \end{array} \qquad \begin{array}{r} \$1{,}350 \\ -\ 1{,}334 \\ \hline \$16 \end{array}$$ Final Payment

19. (1.13) $$\begin{aligned} P &= 2\ell + 2w \\ P &= 2 \cdot 18 + 2 \cdot 12 \\ P &= 36 + 24 \\ P &= 60 \text{ in.} \end{aligned}$$

20. (1.13) $$\begin{aligned} A &= s^2 \\ A &= 10^2 \\ A &= 100 \text{ sq. cm} \end{aligned}$$

21. (1.14) $80{,}000 \cdot 200 = 16{,}000{,}000$

22. (1.12) $$\begin{array}{r} 76 \\ 84 \\ 92 \\ 63 \\ +\ 70 \\ \hline 385 \end{array} \qquad \begin{array}{r} 77 \text{ average} \\ 5\overline{)385} \\ 35 \\ \hline 35 \\ 35 \\ \hline \end{array}$$

23. (1.11) $$\begin{aligned} &6 + 18 \div 3 \cdot 2 = \\ &6 + 6 \cdot 2 = \\ &6 + 12 = 18 \end{aligned}$$

24. (1.11) $$\begin{aligned} &15 - 3^2 + 4 = \\ &15 - 9 + 4 = \\ &6 + 4 = 10 \end{aligned}$$

25. (1.11) $$\begin{aligned} &5 + 2(8 - 4) = \\ &5 + 2 \cdot 4 = \\ &5 + 8 = 13 \end{aligned}$$

Exercises 2.1 (page 52)

1. $\frac{3}{8}$ **3.** $\frac{1}{6}$ **5.** $\frac{3}{7}$ **7.** $\frac{5}{11}, \frac{17}{22}, \frac{1}{2}, \frac{98}{107}, \frac{1}{31}$ **9.** 5

11. $\frac{8}{8}$ and $\frac{4}{4}$ **13.** 4 **15.** 4 **17.** 1 **19.** 9 **21.** 20

Exercises 2.2 (page 54)

1. $\frac{10}{21}$ **3.** $\frac{21}{32}$ **5.** $\frac{12}{25}$ **7.** $\frac{8}{27}$ **9.** $\frac{9}{16}$ **11.** $\frac{25}{96}$

13. $\frac{33}{64}$ **15.** $\frac{77}{104}$ **17.** $\frac{3}{40}$ **19.** $\frac{6}{175}$

21. $2 \cdot \frac{1}{3} = \frac{2}{1} \cdot \frac{1}{3} = \frac{2 \cdot 1}{1 \cdot 3} = \frac{2}{3}$ **23.** $\frac{4}{5}$ **25.** $\frac{15}{2}$

27. $\frac{49}{12}$ **29.** $\frac{25}{48}$

Exercises 2.3 (page 56)

1. $1\frac{2}{3}$ **3.** $1\frac{4}{5}$ **5.** $2\frac{3}{5}$ **7.** $3\frac{3}{4}$ **9.** $3\frac{5}{6}$ **11.** $1\frac{3}{13}$

13. $2\frac{6}{7}$ **15.** $$\begin{array}{r} 10 \text{ R } 17 \\ 19\overline{)207} \\ 19 \\ \hline 17 \end{array} = 10\frac{17}{19}$$ **17.** $3\frac{3}{17}$

19. $5\frac{3}{4}$ **21.** $14\frac{2}{5}$

23.

Number of students		*Score*		*Points*
3	×	5	=	15
5	×	4	=	20
2	×	2	=	4
10 students				39 points

$\text{Average} = \frac{39}{10} = 3\frac{9}{10}$

Exercises 2.4 (page 57)

1. $\frac{3}{2}$ **3.** $\frac{13}{4}$ **5.** $\frac{29}{6}$ **7.** $\frac{37}{10}$ **9.** $\frac{67}{12}$ **11.** $\frac{38}{3}$

13. $\frac{27}{4}$ **15.** $\frac{52}{15}$ **17.** $\frac{63}{4}$ **19.** $\frac{27}{10}$

Exercises 2.5 (page 60)

1. $\frac{5}{10}$ **3.** $\frac{1}{2}$ **5.** $\frac{4}{6}$ **7.** $\frac{3}{5}$ **9.** $\frac{9}{15}$ **11.** $\frac{7}{10}$ **13.** $\frac{36}{52}$

15. $\frac{6}{9}$ **17.** $\frac{40}{55}$ **19.** $\frac{5}{2}$

Exercises 2.6 (page 62)

1. $14 = 2 \cdot 7$ **3.** $21 = 3 \cdot 7$ **5.** $16 = 2^4$ **7.** 29 is prime.

9. 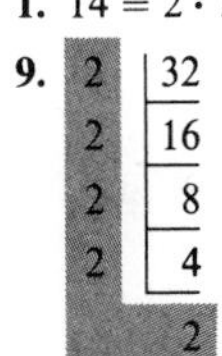

$32 = 2^5$

11. 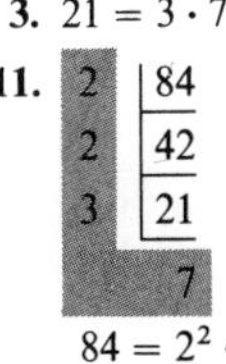

$84 = 2^2 \cdot 3 \cdot 7$

13. 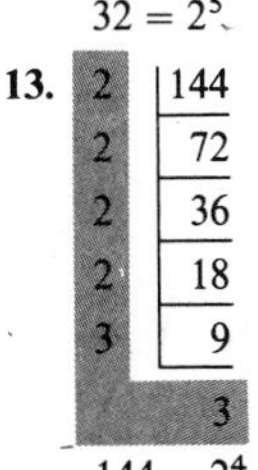

$144 = 2^4 \cdot 3^2$

15. 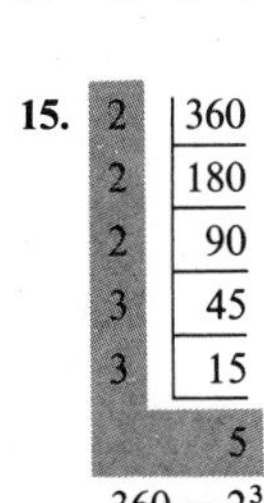

$360 = 2^3 \cdot 3^2 \cdot 5$

17. 5 is prime. **19.** 13 is prime. **21.** C; $12 = 2^2 \cdot 3$

23. C; $21 = 3 \cdot 7$ **25.** 41 is prime. **27.** C; $49 = 7^2$

29. C; $51 = 3 \cdot 17$

Exercises 2.7A (page 65)

1. $\frac{2}{3}$ **3.** $\frac{3}{4}$ **5.** $\frac{3}{4}$ **7.** $\frac{3}{4}$ **9.** $\frac{5}{9}$ **11.** $\frac{4}{5}$ **13.** $\frac{3}{5}$

15. $\frac{4}{5}$ **17.** $\frac{3}{5}$ **19.** $\frac{3}{5}$ **21.** $\frac{39}{51} = \frac{\not{3} \cdot 13}{\not{3} \cdot 17} = \frac{13}{17}$

23. $\frac{4}{9}$ **25.** $\frac{5}{7}$

Exercises 2.7B (page 66)

1. $\frac{4}{7}$ **3.** $\frac{1}{3}$ **5.** 1 **7.** $\frac{5}{32}$ **9.** $\frac{2}{15}$ **11.** $\frac{1}{5}$ **13.** $\frac{3}{7}$

15. $\frac{1}{5}$ **17.** 9 **19.** 240 **21.** 31 lb

23. $\frac{3}{8} \cdot 24 = \frac{3}{\not{8}_1} \cdot \frac{\not{24}^3}{1} = 9$ gal left

$24 - 9 = 15$ gal used

Exercises 2.8 (page 68)

1. $\frac{2}{3}$ **3.** 1 **5.** $\frac{1}{5}$ **7.** $\frac{1}{3}$ **9.** $1\frac{1}{2}$ **11.** $\frac{5}{8}$ **13.** $\frac{3}{5}$

15. $\frac{17}{27}$ **17.** $1\frac{1}{2}$ **19.** $\frac{2}{3}$

Exercises 2.9 (page 71)

1. 12 **3.** 8 **5.** 15 **7.** 70 **9.** 35 **11.** 72

13. 600 **15.** 420 **17.** 78 **19.** 132

Exercises 2.10 (page 74)

1. $1\frac{1}{4}$ **3.** $\frac{9}{10}$ **5.** $\frac{1}{2}$ **7.** $\frac{1}{4}$ **9.** $\frac{11}{12}$ **11.** $\frac{5}{6}$ **13.** $\frac{11}{21}$

15. $\frac{1}{5}$ **17.** $\frac{1}{32}$

19.
$$\frac{56}{64} = \frac{7}{8} = \frac{21}{24}$$
$$-\frac{14}{24} = -\frac{7}{12} = -\frac{14}{24}$$
$$\frac{7}{24}$$

$8 = 2^3$
$12 = 2^2 \cdot 3$
$\text{LCD} = 2^3 \cdot 3 = 24$

21. $\frac{11}{40}$ **23.** $1\frac{11}{18}$ **25.** $\frac{17}{60}$ **27.** $\frac{13}{70}$ **29.** 2 **31.** $1\frac{2}{5}$

33. $1\frac{11}{24}$

35.
$$\frac{4}{6} = \frac{2}{3} = \frac{14}{21}$$
$$\frac{6}{14} = \frac{3}{7} = \frac{9}{21}$$
$$+\frac{2}{3} = \frac{2}{3} = \frac{14}{21}$$
$$\frac{37}{21} = 1\frac{16}{21}$$

37. $1\frac{7}{24}$ **39.** $1\frac{11}{28}$

Exercises 2.11 (page 76)

1. $3\frac{7}{10}$ **3.** $5\frac{5}{6}$ **5.** $7\frac{5}{6}$ **7.** $4\frac{1}{8}$ **9.** $11\frac{1}{2}$ **11.** $5\frac{5}{8}$

13. $7\frac{1}{12}$ **15.** $10\frac{5}{8}$ **17.** $37\frac{7}{12}$ **19.** $49\frac{9}{20}$ **21.** $247\frac{11}{12}$

23. 61 **25.** $120\frac{7}{16}$

27.
$$117\frac{5}{6} = 117 + \frac{15}{18}$$
$$28\frac{1}{2} = 28 + \frac{9}{18}$$
$$+\ 232\frac{7}{9} = 232 + \frac{14}{18}$$
$$377 + \frac{38}{18}$$
$$377 + 2\frac{2}{18} = 379\frac{2}{18} = 379\frac{1}{9}$$

29.
$$7\frac{3}{4} = 7 + \frac{60}{80}$$
$$5\frac{7}{8} = 5 + \frac{70}{80}$$
$$8\frac{5}{16} = 8 + \frac{25}{80}$$
$$+\ 10\frac{2}{5} = 10 + \frac{32}{80}$$
$$30 + \frac{187}{80} = 30 + 2\frac{27}{80} = 32\frac{27}{80} \text{ oz}$$

$4 = 2^2$
$8 = 2^3$
$16 = 2^4$
$5 = 5$
$\text{LCD} = 2^4 \cdot 5 = 80$

31.
$$10\frac{1}{5} = 10 + \frac{4}{20}$$
$$5\frac{3}{10} = 5 + \frac{6}{20}$$
$$12\frac{3}{4} = 12 + \frac{15}{20}$$
$$+\ 6\frac{1}{2} = 6 + \frac{10}{20}$$
$$33 + \frac{35}{20} = 33 + 1\frac{15}{20} = 34\frac{3}{4} \text{ m}$$

Exercises 2.12 (page 78)

1. $4\frac{1}{2}$ **3.** $12\frac{5}{8}$ **5.** $\begin{aligned} 8 &= 7 + \frac{2}{2} = & 7\frac{2}{2} \\ -4\frac{1}{2} &= & -4\frac{1}{2} \\ & & 3\frac{1}{2} \end{aligned}$ **7.** $3\frac{2}{5}$

9. $\begin{aligned} 4\frac{1}{4} &= 3 + 1 + \frac{1}{4} = 3 + \frac{4}{4} + \frac{1}{4} = & 3\frac{5}{4} \\ -1\frac{3}{4} &= & -1\frac{3}{4} \\ & & 2\frac{2}{4} = 2\frac{1}{2} \end{aligned}$ **11.** $5\frac{1}{10}$

13. $\begin{aligned} 3\frac{1}{12} &= 2 + 1 + \frac{1}{12} = 2 + \frac{12}{12} + \frac{1}{12} = & 2\frac{13}{12} \\ -1\frac{1}{6} &= & -1\frac{2}{12} \\ & & 1\frac{11}{12} \end{aligned}$ **15.** $6\frac{1}{3}$

17. $\begin{aligned} 68\frac{5}{16} &= 67 + 1 + \frac{5}{16} = 67 + \frac{16}{16} + \frac{5}{16} = & 67\frac{21}{16} \\ -53\frac{3}{4} &= & -53\frac{12}{16} \\ & & 14\frac{9}{16} \end{aligned}$

19. $\begin{aligned} 234\frac{5}{14} &= 233 + \frac{14}{14} + \frac{5}{14} = & 233\frac{19}{24} \\ -157\frac{3}{7} &= & -157\frac{6}{14} \\ & & 76\frac{13}{14} \end{aligned}$ **21.** $\frac{3}{8}$ lb

23. $1\frac{1}{4}$ sq. yd

Exercises 2.13 (page 79)

1. $4\frac{1}{6}$ **3.** 6 **5.** 30 **7.** $10\frac{1}{2}$ **9.** $1\frac{19}{20}$ **11.** $2\frac{2}{3}$

13. $3\frac{3}{10}\cdot\frac{6}{11}\cdot 1\frac{2}{3} = \frac{\overset{3}{\cancel{33}}}{\underset{1}{\underset{\cancel{2}}{\cancel{10}}}}\cdot\frac{\overset{1}{\overset{\cancel{2}}{\cancel{6}}}}{\underset{1}{\cancel{11}}}\cdot\frac{\overset{1}{\cancel{5}}}{\underset{1}{\cancel{3}}} = 3$

15. $3\frac{3}{4}\cdot 4\frac{2}{5}\cdot 3\frac{1}{2} = \frac{\overset{3}{\cancel{15}}}{\underset{2}{\cancel{4}}}\cdot\frac{\overset{11}{\cancel{22}}}{\underset{1}{\cancel{5}}}\cdot\frac{7}{2} = \frac{231}{4} = 57\frac{3}{4}$ **17.** $19\frac{1}{2}$ lb

19. $8\frac{2}{5}$ tons **21.** $10\frac{4}{5}$ m^2

Exercises 2.14 (page 81)

1. $1\frac{1}{2}$ **3.** $\frac{3}{4}$ **5.** $\frac{1}{10}$ **7.** 4 **9.** 2 **11.** 2 **13.** $\frac{5}{12}$

15. 112 **17.** $\frac{35}{16}\div\frac{42}{22} = \frac{\overset{5}{\cancel{35}}}{16}\cdot\frac{\overset{11}{\cancel{22}}}{\underset{3}{\underset{\cancel{21}}{\cancel{42}}}} = \frac{55}{48} = 1\frac{7}{48}$ **19.** 45 **21.** $\frac{9}{28}$

23. $\frac{14}{24}\div 210 = \frac{\overset{1}{\overset{\cancel{2}}{\cancel{14}}}}{\underset{12}{\cancel{24}}}\cdot\frac{1}{\underset{30}{\cancel{210}}} = \frac{1}{360}$ **25.** 12

27. $1\frac{7}{9}\div 2\frac{2}{3} = \frac{16}{9}\div\frac{8}{3} = \frac{\overset{2}{\cancel{16}}}{\underset{3}{\cancel{9}}}\cdot\frac{\overset{1}{\cancel{3}}}{\underset{1}{\cancel{8}}} = \frac{2}{3}$ **29.** $2\frac{1}{2}$ **31.** $\frac{1}{5}$ **33.** $3\frac{1}{2}$

35. $1\frac{7}{8}$ **37.** $4\frac{1}{2}\div 3 = \frac{9}{2}\div\frac{3}{1} = \frac{\overset{3}{\cancel{9}}}{2}\cdot\frac{1}{\underset{1}{\cancel{3}}} = \frac{3}{2} = 1\frac{1}{2}$ tablets

39. $36\div 1\frac{1}{2} = \frac{36}{1}\div\frac{3}{2} = \frac{\overset{12}{\cancel{36}}}{1}\cdot\frac{2}{\underset{1}{\cancel{3}}} = 24$ books

Exercises 2.15 (page 83)

1. $4\frac{1}{2}$ **3.** $1\frac{1}{3}$ **5.** 2 **7.** $\frac{9}{10}$ **9.** 9 **11.** $8\frac{3}{4}$ **13.** $\frac{1}{6}$

15. $\frac{1}{15}$ **17.** $\frac{2}{5}$ **19.** $3\frac{9}{10}$

21. $\dfrac{\frac{1}{8}+\frac{3}{4}}{\frac{1}{2}-\frac{1}{3}} = \dfrac{\frac{1}{8}+\frac{6}{8}}{\frac{3}{6}-\frac{2}{6}} = \dfrac{\frac{7}{8}}{\frac{1}{6}} = \frac{7}{8}\div\frac{1}{6} = \frac{7}{\underset{4}{\cancel{8}}}\cdot\frac{\overset{3}{\cancel{6}}}{1} = \frac{21}{4} = 5\frac{1}{4}$

23. $\dfrac{\frac{1}{7}+\frac{9}{28}}{\frac{13}{14}-\frac{3}{7}} = \dfrac{\frac{4}{28}+\frac{9}{28}}{\frac{13}{14}-\frac{6}{14}} = \dfrac{\frac{13}{28}}{\frac{7}{14}} = \frac{13}{28}\div\frac{7}{14} = \frac{13}{\underset{2}{\cancel{28}}}\cdot\frac{\overset{1}{\cancel{14}}}{7} = \frac{13}{14}$

25. $\dfrac{\frac{13}{18}-\frac{11}{24}}{\frac{5}{12}-\frac{7}{36}} = \dfrac{\frac{52}{72}-\frac{33}{72}}{\frac{15}{36}-\frac{7}{36}} = \dfrac{\frac{19}{72}}{\frac{8}{36}} = \frac{19}{72}\div\frac{8}{36} = \frac{19}{\underset{2}{\cancel{72}}}\cdot\frac{\overset{1}{\cancel{36}}}{8} = \frac{19}{16} = 1\frac{3}{16}$

27. $\dfrac{11\frac{1}{4}+12\frac{1}{2}+11\frac{5}{8}+9\frac{7}{8}+11}{5}$

$= \dfrac{11\frac{2}{8}+12\frac{4}{8}+11\frac{5}{8}+9\frac{7}{8}+11}{5}$

$= \dfrac{54\frac{18}{8}}{5} = 56\frac{1}{4}\div 5 = \frac{225}{4}\div\frac{5}{1} = \frac{\overset{45}{\cancel{225}}}{4}\cdot\frac{1}{\underset{1}{\cancel{5}}} = \frac{45}{4} = 11\frac{1}{4}$

Exercises 2.16 (page 86)

1. $3\frac{1}{3}$ **3.** $4\frac{5}{6}$ **5.** $1\frac{4}{5}$ **7.** $1\frac{1}{2}$ **9.** $\frac{7}{15}$

11. $\left(\frac{2}{3}\right)^2 + 1\frac{2}{3}\cdot\frac{1}{10} =$

$\frac{2}{3}\cdot\frac{2}{3} + \frac{\overset{1}{\cancel{5}}}{3}\cdot\frac{1}{\underset{2}{\cancel{10}}} =$

$\frac{4}{9} + \frac{1}{6} =$

$\frac{8}{18} + \frac{3}{18} = \frac{11}{18}$

13. $19\frac{3}{4}$ **15.** $\frac{8}{9}$ **17.** 17

19. $P = 2\ell + 2w$

$$P = 2 \cdot 5\frac{1}{3} + 2 \cdot 3\frac{1}{4}$$

$$P = \frac{2}{1} \cdot \frac{16}{3} + \frac{\cancel{2}^{1}}{1} \cdot \frac{13}{\cancel{4}_{2}}$$

$$P = \frac{32}{3} + \frac{13}{2}$$

$$P = \frac{64}{6} + \frac{39}{6}$$

$$P = \frac{103}{6} = 17\frac{1}{6} \text{ yd}$$

$$\text{cost} = 17\frac{1}{6} \cdot 12 = \frac{103}{\cancel{6}_{1}} \cdot \frac{\cancel{12}^{2}}{1} = \$206$$

Exercises 2.17 (page 87)

1. $\frac{5}{6} > \frac{3}{4} > \frac{2}{3}$ **3.** $\frac{3}{4} > \frac{11}{16} > \frac{5}{8}$ **5.** $\frac{3}{4} > \frac{5}{7} > \frac{9}{14}$

7. $\frac{3}{10} > \frac{1}{6} > \frac{2}{15}$ **9.** $\frac{1}{2} = \frac{12}{24}$ Because $\frac{12}{24} < \frac{13}{24}$

$$\frac{1}{2} < \frac{13}{24}$$

$$5\frac{1}{2} < 5\frac{13}{24}$$

11. > **13.** $\frac{1}{20}$ of 15 ? $\frac{5}{16}$ of 2

$$\frac{1}{\cancel{20}_{4}} \cdot \frac{\cancel{15}^{3}}{1} \;?\; \frac{5}{\cancel{16}_{8}} \cdot \frac{\cancel{2}^{1}}{1}$$

$$\frac{3}{4} \;?\; \frac{5}{8}$$

$$\frac{6}{8} > \frac{5}{8}$$

15. Up by $\frac{1}{8}$ **17.** 45 in.

Review Exercises 2.18 (page 90)

1. $2\frac{1}{3}$ **2.** $1\frac{3}{8}$ **3.** $5\frac{3}{5}$ **4.** $2\frac{5}{12}$ **5.** $\frac{13}{5}$ **6.** $\frac{31}{8}$

7. $\frac{101}{11}$ **8.** $\frac{83}{6}$ **9.** $\frac{7}{9}$ **10.** $\frac{3}{7}$ **11.** $\frac{28}{57}$ **12.** $\frac{4}{5}$

13. $1\frac{7}{9}$ **14.** $4\frac{4}{15}$ **15.** $4\frac{15}{16}$ **16.** $289\frac{3}{20}$ **17.** $2\frac{1}{10}$

18.
$$\begin{aligned} 5\frac{1}{3} &= 5\frac{4}{12} = 4\frac{16}{12} \\ -2\frac{3}{4} &= -2\frac{9}{12} = -2\frac{9}{12} \\ \hline & \qquad\qquad 2\frac{7}{12} \end{aligned}$$

19. $10\frac{1}{8}$ **20.** $78\frac{8}{9}$ **21.** $10\frac{1}{2}$ **22.** $24\frac{1}{5}$ **23.** 6

24. $12\frac{3}{4}$ **25.** $2\frac{1}{4}$ **26.** $\frac{1}{4}$ **27.** 26 **28.** $\frac{5}{32}$

29.
$$\frac{7}{12} + 2\frac{2}{3} \cdot \frac{1}{4} =$$
$$\frac{7}{12} + \frac{\cancel{8}^{2}}{3} \cdot \frac{1}{\cancel{4}_{1}} =$$
$$\frac{7}{12} + \frac{2}{3} =$$
$$\frac{7}{12} + \frac{8}{12} =$$
$$\frac{15}{12} = 1\frac{3}{12} = 1\frac{1}{4}$$

30.
$$\frac{2}{3} \div 1\frac{1}{4} \cdot \left(\frac{3}{4}\right)^2 =$$
$$\frac{2}{3} \div \frac{5}{4} \cdot \frac{3}{4} \cdot \frac{3}{4} =$$
$$\frac{\cancel{2}^{1}}{\cancel{3}_{1}} \cdot \frac{\cancel{4}^{1}}{5} \cdot \frac{\cancel{3}^{1}}{\cancel{4}_{1}} \cdot \frac{3}{\cancel{4}_{2}} = \frac{3}{10}$$

31. $\frac{3}{4}$ **32.** 22

33.
$$\frac{1\frac{2}{3}}{\frac{3}{4} + \frac{1}{2}} = \frac{\frac{5}{3}}{\frac{3}{4} + \frac{2}{4}} = \frac{\frac{5}{3}}{\frac{5}{4}} = \frac{5}{3} \div \frac{5}{4} = \frac{\cancel{5}^{1}}{3} \cdot \frac{4}{\cancel{5}_{1}} = \frac{4}{3} = 1\frac{1}{3}$$

34.
$$\frac{\frac{2}{3} + \frac{1}{2}}{\frac{8}{9} - \frac{1}{3}} = \frac{\frac{4}{6} + \frac{3}{6}}{\frac{8}{9} - \frac{3}{9}} = \frac{\frac{7}{6}}{\frac{5}{9}} = \frac{7}{6} \div \frac{5}{9} = \frac{7}{\cancel{6}_{2}} \cdot \frac{\cancel{9}^{3}}{5} = \frac{21}{10} = 2\frac{1}{10}$$

35. $\frac{2}{5} > \frac{3}{10} > \frac{1}{4}$ **36.** $\frac{5}{6} > \frac{2}{3} > \frac{5}{8}$ **37.** $\frac{8}{24} = \frac{1}{3}$

38. 56 books **39.** $40\frac{3}{4}$ hr **40. a.** $19\frac{1}{4}$ sq. yd **b.** $154

41. 1 week = $7 \cdot 24 = 168$ hours

$$168 \div 2\frac{1}{3} = \frac{168}{1} \div \frac{7}{3} = \frac{\cancel{168}^{24}}{1} \cdot \frac{3}{\cancel{7}_{1}} = 72 \text{ trips}$$

42. $24\frac{1}{4}$ in.

43. a. $16 \div 3\frac{1}{2} = \frac{16}{1} \div \frac{7}{2} = \frac{16}{1} \cdot \frac{2}{7} = \frac{32}{7} = 4\frac{4}{7}$

Therefore, four shelves can be cut from the 16-foot board.

b. $4 \times 3\frac{1}{2} = \frac{\cancel{4}^{2}}{1} \times \frac{7}{\cancel{2}_{1}} = 14$ ft for shelves

Four cuts $= 4 \times \frac{1}{8} = \frac{4}{8} = \frac{1}{2}$ inch for saw cuts

Part left $= 16 \text{ ft} - 14 \text{ ft} - \frac{1}{2}$ in.

$= 2 \text{ ft} - \frac{1}{2} \text{ in.} = 24 \text{ in.} - \frac{1}{2}$ in.

$= 23\frac{1}{2}$ in.

44.
$$\begin{aligned} 1\frac{1}{4} &= 1\frac{2}{8} \\ +\ \frac{5}{8} &= \frac{5}{8} \\ \hline & \ \ 1\frac{7}{8} \end{aligned}$$
yards for each girl

$$26 \cdot 1\frac{7}{8} = \frac{\cancel{26}^{13}}{1} \cdot \frac{15}{\cancel{8}_{4}} = \frac{195}{4}$$
$$= 48\frac{3}{4} \text{ total yards}$$

45. 8 tons

46.
$$\begin{array}{rl} 3\frac{1}{2} = & 3\frac{4}{8} \\ 2\frac{3}{4} = & 2\frac{6}{8} \\ +\ 1\frac{5}{8} = & 1\frac{5}{8} \\ \hline & 6\frac{15}{8} = 7\frac{7}{8} \text{ lb lost} \end{array}
\qquad
\begin{array}{rl} 200 = & 199\frac{8}{8} \\ -\ 7\frac{7}{8} = & 7\frac{7}{8} \\ \hline & 192\frac{1}{8} \text{ lb} \end{array}$$

47. $2\frac{1}{8}$ lb

48.
$$\text{Spends} = \left(\frac{1}{3}+\frac{1}{4}+\frac{1}{3}\right) \text{ of } \$18{,}000$$
$$= \left(\frac{4}{12}+\frac{3}{12}+\frac{4}{12}\right) \text{ of } \$18{,}000$$
$$= \frac{11}{\cancel{12}_{2}} \cdot \frac{\cancel{18{,}000}^{3{,}000}}{1}$$
$$= \frac{33{,}000}{2} = \$16{,}500$$

$$\text{Left over} = \begin{array}{r} \$18{,}000 \\ -\ 16{,}500 \\ \hline \$1{,}500 \end{array}$$

49. a. 5 ft **b.** $1\frac{9}{16}$ sq. ft **50. a.** $14\frac{1}{2}$ m **b.** $12\frac{3}{8}$ sq. m

Chapter 2 Diagnostic Test (page 93)

Following each problem number is the textbook section reference (in parentheses) in which that kind of problem is discussed.

1. (2.3) $\frac{69}{8} = 8\overline{)69}$ → 8 R 5; 64, 5; $= 8\frac{5}{8}$

2. (2.4) $5\frac{7}{12} = \frac{5 \cdot 12 + 7}{12} = \frac{67}{12}$

3. (2.7A) $\frac{18\cancel{0}}{54\cancel{0}} = \frac{\cancel{18}^{2}}{\cancel{54}_{6}} = \frac{2}{6} = \frac{1}{3}$

4. (2.7B) $\frac{7}{8}$ of 120 = $\frac{7}{\cancel{8}_{1}} \cdot \frac{\cancel{120}^{15}}{1} = 105$

5. (2.10)
$$\begin{array}{rl} \frac{5}{6} = & \frac{20}{24} \\ +\ \frac{7}{8} = & \frac{21}{24} \\ \hline & \frac{41}{24} = 1\frac{17}{24} \end{array}$$

6. (2.11)
$$\begin{array}{rl} 4\frac{2}{3} = & 4 + \frac{4}{6} \\ +\ 3\frac{1}{2} = & 3 + \frac{3}{6} \\ \hline & 7 + \frac{7}{6} = \\ & 7 + 1\frac{1}{6} = 8\frac{1}{6} \end{array}$$

7. (2.7B) $\frac{\cancel{5}^{1}}{\cancel{6}_{2}} \cdot \frac{\cancel{3}^{1}}{\cancel{20}_{4}} = \frac{1}{8}$

8. (2.13) $1\frac{1}{3} \cdot 42 = \frac{4}{\cancel{3}_{1}} \cdot \frac{\cancel{42}^{14}}{1} = 56$

9. (2.14) $\frac{3}{8} \div \frac{9}{16} = \frac{\cancel{3}^{1}}{\cancel{8}_{1}} \cdot \frac{\cancel{16}^{2}}{\cancel{9}_{3}} = \frac{2}{3}$

10. (2.10)
$$\begin{array}{rl} \frac{7}{10} = & \frac{21}{30} \\ -\ \frac{1}{6} = & \frac{5}{30} \\ \hline & \frac{16}{30} = \frac{8}{15} \end{array}$$

11. (2.12)
$$\begin{array}{rl} 8 = & 7\frac{9}{9} \\ -\ 3\frac{4}{9} = & 3\frac{4}{9} \\ \hline & 4\frac{5}{9} \end{array}$$

12. (2.14) $2\frac{2}{9} \div 3\frac{1}{3} = \frac{20}{9} \div \frac{10}{3} = \frac{\cancel{20}^{2}}{\cancel{9}_{3}} \cdot \frac{\cancel{3}^{1}}{\cancel{10}_{1}} = \frac{2}{3}$

13. (2.13) $4\frac{1}{5} \cdot 2\frac{1}{7} = \frac{\cancel{21}^{3}}{\cancel{5}_{1}} \cdot \frac{\cancel{15}^{3}}{\cancel{7}_{1}} = 9$

14. (2.12)
$$\begin{array}{rll} 5\frac{1}{4} = & 5\frac{3}{12} = & 4\frac{15}{12} \\ -\ 2\frac{5}{6} = & 2\frac{10}{12} = & 2\frac{10}{12} \\ \hline & & 2\frac{5}{12} \end{array}$$

15. (2.12)
$$\begin{array}{rll} 124\frac{2}{3} = & 124\frac{10}{15} = & 123\frac{25}{15} \\ -\ 17\frac{4}{5} = & -\ 17\frac{12}{15} = & -\ 17\frac{12}{15} \\ \hline & & 106\frac{13}{15} \end{array}$$

16. (2.10)
$$\begin{array}{rl} \frac{5}{12} = & \frac{10}{24} \\ \frac{3}{8} = & \frac{9}{24} \\ +\ \frac{5}{6} = & \frac{20}{24} \\ \hline & \frac{39}{24} = 1\frac{15}{24} = 1\frac{5}{8} \end{array}$$

17. (2.15) $\dfrac{\frac{5}{8}}{\frac{15}{16}} = \frac{5}{8} \div \frac{15}{16} = \frac{\cancel{5}^{1}}{\cancel{8}_{1}} \cdot \frac{\cancel{16}^{2}}{\cancel{15}_{3}} = \frac{2}{3}$

18. (2.15) $\dfrac{\frac{3}{8}+\frac{3}{4}}{7\frac{1}{2}} = \dfrac{\frac{3}{8}+\frac{6}{8}}{\frac{15}{2}} = \dfrac{\frac{9}{8}}{\frac{15}{2}} = \frac{9}{8} \div \frac{15}{2} = \frac{\cancel{9}^{3}}{\cancel{8}_{4}} \cdot \frac{\cancel{2}^{1}}{\cancel{15}_{5}} = \frac{3}{20}$

19. (2.16)
$$\frac{4}{5} \div 2\frac{2}{3} \cdot \frac{5}{6} =$$
$$\frac{4}{5} \div \frac{8}{3} \cdot \frac{5}{6} =$$
$$\frac{\cancel{4}^{1}}{\cancel{5}_{1}} \cdot \frac{\cancel{3}^{1}}{\cancel{8}_{2}} \cdot \frac{\cancel{5}^{1}}{\cancel{6}_{2}} = \frac{1}{4}$$

20. (2.16)
$$\frac{3}{4} + \frac{1}{4} \cdot 1\frac{3}{5}$$
$$\frac{3}{4} + \frac{1}{\cancel{4}_{1}} \cdot \frac{\cancel{8}^{2}}{5} =$$
$$\frac{3}{4} + \frac{2}{5} =$$
$$\frac{15}{20} + \frac{8}{20} =$$
$$\frac{23}{20} = 1\frac{3}{20}$$

21. (2.12)
$$\begin{array}{rl} 1\frac{1}{8} = & \frac{9}{8} \\ -\ \frac{3}{4} = & -\ \frac{6}{8} \\ \hline & \frac{3}{8} \text{ lb} \end{array}$$

22. (2.14) $7\frac{1}{2} \div 3 = \frac{15}{2} \div \frac{3}{1} = \frac{\overset{5}{\cancel{15}}}{2} \cdot \frac{1}{\underset{1}{\cancel{3}}} = \frac{5}{2} = 2\frac{1}{2}$

23. (2.13) $A = \ell w$

$$A = 5\frac{1}{4} \cdot 3\frac{1}{3}$$

$$A = \frac{\overset{7}{\cancel{21}}}{\underset{2}{\cancel{4}}} \cdot \frac{\overset{5}{\cancel{10}}}{\underset{1}{\cancel{3}}}$$

$$A = \frac{35}{2}$$

$$A = 17\frac{1}{2} \text{ sq. ft}$$

24. (2.11)

$$\begin{aligned} 23\frac{1}{2} &= 23\frac{4}{8} \\ 16\frac{1}{4} &= 16\frac{2}{8} \\ +\ 37\frac{7}{8} &= 37\frac{7}{8} \\ \hline & 76\frac{13}{8} = 76 + 1\frac{5}{8} = 77\frac{5}{8} \text{ lb} \end{aligned}$$

25. (2.17) $\frac{8}{15} = \frac{16}{30}, \frac{7}{10} = \frac{21}{30}, \frac{3}{5} = \frac{18}{30}$

$$\frac{21}{30} > \frac{18}{30} > \frac{16}{30}$$

$$\frac{7}{10} > \frac{3}{5} > \frac{8}{15}$$

Exercises 3.1 (page 99)

1. Thirty-five hundredths **3.** Three and sixteen thousandths
5. Four ten-thousandths
7. Twenty and nine hundred thousandths
9. Nine thousand and fifty hundredths
11. Five thousand six ten-thousandths **13.** 0.09 **15.** 2.003
17. 0.0400 **19.** 60.08 **21.** 0.720 **23.** 3,000.55
25. 100.0004

Exercises 3.2 (page 102)

1. 7.2 **3.** 9.03 **5.** 400 **7.** 0.0603 **9.** 3.210 **11.** 107
13. 490 **15.** 0.00982 **17.** 0.10

Exercises 3.3 (page 103)

1. 374.988 **3.** $267.48 **5.** 258.9556 **7.** 11,036.075

9. a.

$$\begin{array}{r} \$5 \\ 7 \\ 4 \\ 6 \\ +\ 5 \\ \hline \$27 \end{array}$$

b. $27.62

11.

$$\begin{array}{r} \overset{1}{3},050.37 \\ 5.00002 \\ +\ \ 70.0150 \\ \hline 3,125.38502 \end{array}$$

Exercises 3.4 (page 105)

1. 4.41 **3.** 201.34 **5.** 299.855 **7.** 78,499.76 **9.** 4.8185
11. 177.2

13.

Beginning balance plus deposits	*Checks written*	*Computing balance*	
254.39	27.15	750.50	
183.50	86.94	− 566.09	
233.75	123.47	$184.41	Ending balance
78.86	167.66		
$750.50	122.20		
	38.67		
	$566.09		

Exercises 3.5 (page 107)

1. 0.024 **3.** 0.0024 **5.** 12.6 **7.** 0.136 **9.** 2.7
11. 0.3922 **13.** 395 **15.** 8,170 **17.** $0.9 \times 0.4 = 0.36$
19. $0.7 \times 40 = 28$ **21. a.** $6 × 30 = $180 **b.** $159.90 **23. a.** 90 sq. m **b.** 107.5 sq. m **c.** $2,203.75

25. $357.50 **27.**

$$\begin{array}{r} 875 \\ \times\ .15 \\ \hline 4375 \\ 875\ \ \\ \hline \$131.25 \end{array} \qquad \begin{array}{r} 12 \\ \times\ 7 \\ \hline \$84 \end{array} \qquad \begin{array}{r} 131.25 \\ +\ 84.00 \\ \hline \$215.25 \end{array} \text{ total cost}$$

Exercises 3.6A (page 108)

1. 10.87 **3.** 15.6 **5.** 1.05 **7.** 0.175 **9.** 0.079
11. ≈1.22 **13.** ≈47.038 **15.** ≈1.4 **17.** ≈0.051
19. ≈0.37 **21.** 8.65 **23.** ≈2.4 in.

25.

$$\begin{aligned} 3 \times \$8.50 &= \ \ \$25.50 \\ +\ 2 \times \$6.95 &= +\ 13.90 \\ \hline 5 \text{ tapes for } & \ \ \$39.40 \end{aligned}$$

$$\text{Average} = \frac{\$39.40}{5} = \$7.88$$

Exercises 3.6B (page 111)

1. 3.56 **3.** 78.4 **5.** 0.064 **7.** 0.014 **9.** 0.906
11. 35.7 **13.** 820 **15.** 2.85 **17.** ≈2.65 **19.** ≈30.8
21. ≈6.406 **23.** ≈0.0310 **25.** ≈904.1 **27.** ≈$68.53

29.

$$\begin{array}{r} 14.\ \text{pieces} \\ 3.5_\wedge \overline{)\,50.0_\wedge} \\ 35\ \ \\ \hline 150 \\ 140 \\ \hline 10 \end{array}$$

Exercises 3.7 (page 114)

1. 9.56 **3.** 5.73 **5.** 2,780 **7.** 209.4 **9.** 7.502
11. 984.6 **13.** 0.1 **15.** 274,000

17. $\frac{146.35}{10} = 14.635 \approx \$14.64$ **19.** $537 \times 100 = \$53,700$

Exercises 3.8 (page 115)

1. 4.44 **3.** 15.82 **5.** 7.95 **7.** 5.742 **9.** 40.74

11. $2.3 + 1.6 + 1.2 \div 3 =$
$2.3 + 1.6 + 0.4 = 4.3$ **13.** $4 + 7 - 4 = 7$

15. $2(9 + 5) = 28$ **17.** $\dfrac{2 \times 6}{0.3} = \dfrac{12}{0.3} = 40$ **19.** 10.7802

21. 1,816.71 **23.** 12.464 **25.** $P = 2\ell + 2w$
$P = 2(8.65) + 2(5.3)$
$P = 17.3 + 10.6$
$P = 27.9$ m

27. Fare $= 2 + 0.75 \times 5 = \$5.75$

29. **a.** 9963 [−] 9876 [=] 87 therms used
[−] 65 [=] 22 therms at higher rate
cost = 65 [×] .365 [+] 22 [×] 785 [=]
$= 40.995 \approx \$41.00$

b. 87 [÷] 30 [=] 2.9 therms

c. 41 [÷] 30 [=] \$1.37

Exercises 3.9 (page 118)

1. 0.75 **3.** 2.5 **5.** 0.125 **7.** 0.666 . . . or $0.\overline{6}$
9. 0.272727 . . . or $0.\overline{27}$ **11.** 3.0666 . . . or $3.0\overline{6}$ **13.** 0.04
15. 0.3125 **17.** 6.05 **19.** ≈ 0.64 **21.** ≈ 0.22 **23.** ≈ 0.087
25. ≈ 6.467 **27.** 0.5625 **29.** ≈ 2.2917 **31.** ≈ 9.3846

Exercises 3.10A (page 120)

1. $\frac{3}{5}$ **3.** $\frac{1}{20}$ **5.** $\frac{3}{40}$ **7.** $\frac{7}{8}$ **9.** $2\frac{1}{2}$ **11.** $5\frac{9}{10}$ **13.** $\frac{1}{16}$

15. $37\frac{1}{2}$ **17.** $65\frac{5}{8}$ **19.** $\frac{7}{8}$ in.

Exercises 3.10B (page 121)

1. $\frac{3}{8}$ **3.** $\frac{1}{3}$ **5.** $\frac{23}{40}$

7. $1.0\frac{1}{5} \to \dfrac{10\frac{1}{5}}{10} = \dfrac{\frac{51}{5}}{10} = \dfrac{51}{5} \div \dfrac{10}{1} = \dfrac{51}{5} \cdot \dfrac{1}{10} = \dfrac{51}{50} = 1\dfrac{1}{50}$

9. $0.00\frac{5}{12} \to \dfrac{\frac{5}{12}}{100} = \dfrac{5}{12} \div \dfrac{100}{1} = \dfrac{\cancel{5}^{1}}{12} \cdot \dfrac{1}{\cancel{100}_{20}} = \dfrac{1}{240}$

11. $0.001\frac{1}{6} \to \dfrac{1\frac{1}{6}}{1{,}000} = \dfrac{\frac{7}{6}}{1{,}000} = \dfrac{7}{6} \div \dfrac{1{,}000}{1}$
$= \dfrac{7}{6} \cdot \dfrac{1}{1{,}000} = \dfrac{7}{6{,}000}$

13. $\frac{5}{6}$ **15.** $\frac{2}{15}$ **17.** $2\frac{3}{28}$

Exercises 3.11 (page 123)

1. 5.32 **3.** 0.55 **5.** 0.375 **7.** 25

9. $0.35 = \dfrac{35}{100} = \dfrac{7}{20}$

$\dfrac{7}{8} + 0.35 =$

$\dfrac{7}{8} + \dfrac{7}{20} = \quad \dfrac{7}{8} = \dfrac{35}{40}$
$+\dfrac{7}{20} = \dfrac{14}{40}$
$\dfrac{49}{40} = 1\dfrac{9}{40}$

11. $0.75 = \dfrac{75}{100} = \dfrac{3}{4}$ **13.** $\frac{7}{15}$ **15.** $3\frac{1}{3}$

$6.75 - 2\dfrac{5}{6} =$

$6\dfrac{3}{4} - 2\dfrac{5}{6} = \quad 6\dfrac{3}{4} = 6\dfrac{9}{12} = 5\dfrac{21}{12}$
$-2\dfrac{5}{6} = 2\dfrac{10}{12} = 2\dfrac{10}{12}$
$3\dfrac{11}{12}$

17. **a.** $9 + 5 = 14$
b. 8 [+] 5 [÷] 6 [=] [+] 4.797 [=] ≈ 13.630

19. **a.** $7 \times 0.1 = 0.7$
b. 7 [+] 2 [÷] 11 [=] [×] .1238 [=] ≈ 0.889

21. \$9.38 **23.** \$48.38 **25.** $18{,}000 \div 22\frac{1}{2} =$
$18{,}000 \div 22.5 =$ 800. gal
$22.5_\wedge \overline{)\,18000.0_\wedge}$
1800
000

Then, 1.05
× 800
840.00 = \$840 a year

Exercises 3.12 (page 124)

1. $0.49 > 0.41 > 0.409 > 0.4$ **3.** $3.1 > 3.075 > 3.05 > 3.009$
5. $0.07501 > 0.075 > 0.0749 > 0.07$
7. $5.5 > 5.0501 > 5.05 > 5.0496 > 5$ **9.** $<$ **11.** $>$

13. $\dfrac{5}{16} = 0.3125$
No, 0.325-inch pin > 0.3125-inch hole

15. Star: $\dfrac{1.02}{7} \approx 14.6$¢ per oz
Whale: $\dfrac{1.89}{12} \approx 15.8$¢ per oz
Therefore, Star Fish is best buy.
$15.8 - 14.6 = 1.2$¢ per oz difference.

Review Exercises 3.13 (page 126)

1. One hundred forty-five thousandths
2. Two hundred fifty and six hundredths **3.** 0.0016
4. 500.75 **5.** 0.837 **6.** 0.60 **7.** 24.7 **8.** 190
9. 35,201.8505 **10.** 36.7666 **11.** 4.992 **12.** 0.086

13. 468.255 **14.** 7,850 **15.** 0.0064 **16.** 0.00826 **17.** 2.05

18. 56 **19.** 436.6 **20.** 0.096 **21.** ≈10.42 **22.** ≈3.05

23. $12.5 + 5.6 \times 10^2 =$
$12.5 + 560 = 572.5$ **24.** 59.2

25. $48 \div 0.8 \times 0.3 =$
$60 \times 0.3 = 18$ **26.** 9.2 **27.** 0.875

28. $5.0555\ldots$ or $5.0\overline{5}$ **29.** ≈0.83 **30.** ≈3.64 **31.** $\frac{17}{25}$

32. $4\frac{1}{40}$ **33.** $0.16\frac{2}{3} = \frac{16\frac{2}{3}}{100} = \frac{50}{3} \div \frac{100}{1} = \frac{\cancel{50}^{1}}{3} \cdot \frac{1}{\cancel{100}_{2}} = \frac{1}{6}$

34. $0.4\frac{1}{6} = \frac{4\frac{1}{6}}{10} = \frac{25}{6} \div \frac{10}{1} = \frac{\cancel{25}^{5}}{6} \cdot \frac{1}{\cancel{10}_{2}} = \frac{5}{12}$ **35.** 6.435 **36.** 4.5

37. $0.24 = \frac{24}{100} = \frac{6}{25}$

$\frac{9}{10} + 0.24 =$

$\frac{9}{10} + \frac{6}{25} = \quad \frac{9}{10} = \frac{45}{50}$

$+\frac{6}{25} = \frac{12}{50}$

$\frac{57}{50} = 1\frac{7}{50}$

38. $0.025 \times 3\frac{1}{3} =$

$\frac{\cancel{25}^{1}}{\cancel{1000}_{4}} \times \frac{\cancel{10}^{1}}{3} = \frac{1}{12}$

39. $0.3(7 + 4) = 0.3(11) = 3.3$ **40.** $\frac{0.5 \times 0.5}{8} = \frac{0.25}{8} \approx \frac{0.24}{8} = 0.03$

41. $0.603 > 0.6 > 0.063 > 0.06$

42. $3.908 > 3.9 > 3.89 > 3.098 > 3$ **43.** 7.5 **44.** 0.55

45. **a.** 21.1 m
b. 26.875 sq. m

46. **a.** 13.6 cm
b. 11.56 sq. cm

47. 40 pieces

48. Yes, 0.625-inch pin < 0.6875-inch hole **49.** $121.55

50.

Check written	*Deposits made*		
$15.98	$250	Beginning Balance	$275.38
46.75	+ 350	Deposits	+ 600.00
87.45	$600		875.38
135.46		Checks	− 353.64
+ 68.00		Ending Balance	$521.74
$353.64			

51. ≈ $19.72 **52.** ≈ $4.50

53. ≈ $0.065 per mi which is $6\frac{1}{2}$¢ per mi

54. 14 gal used

$14\overline{)196}$
14
56
56

$1.329
× 14
5316
1329
$18.606 ≈ $18.61

Chapter 3 Diagnostic Test (page 129)

Following each problem number is the textbook section reference (in parentheses) in which that kind of problem is discussed.

1. (3.1) Nine and fifteen thousandths

2. (3.1) 420.05

3. (3.3)

```
  1 1
    7.8
   56.
    0.017
 + 500.94
  -------
  564.757
```

4. (3.4)

```
  3 10 5 10
  40.60   (crossed: 4→3, 0→10, 6→5, 0→10)
 − 3.54
 ------
  37.06
```

5. (3.4)

```
  8 10
  9.073   (crossed: 9→8, 0→10)
 −0.87
 -----
  8.203
```

6. (3.5)

```
     3.7
 × 0.058
 -------
     296
    185
 -------
  0.2146
```

7. (3.6B)

```
          805.
 .078^)62.790^
       624
       ---
         390
         390
```

8. (3.8) $26 + 1.3 \times 0.8 =$
$26 + 1.04 = 27.04$

9. (3.7) $0.46 \times 10^3 = 460$ **10.** (3.7) $6.9 \div 100 = 0.069$

11. (3.6B)

```
          2 0.5 7 ≈ 20.6
 0.3 5^)7.2 0^0 0
        7 0
        ---
          2 0
            0
          ---
          2 0 0
          1 7 5
          -----
            2 5 0
            2 4 5
```

12. (3.9) $2\frac{3}{16} = \frac{35}{16} =$

```
     2.1875
 16)35.0000
    32
    --
     30
     16
     --
     140
     128
     ---
      120
      112
      ---
        80
        80
```

13. (3.10A) $0.78 = \frac{78}{100} = \frac{39}{50}$

14. (3.10B) $0.1\frac{9}{11} = \frac{1\frac{9}{11}}{10} = \frac{20}{11} \div \frac{10}{1} = \frac{\cancel{20}^{2}}{11} \cdot \frac{1}{\cancel{10}_{1}} = \frac{2}{11}$

15. (3.11) $1\frac{4}{5} = \frac{9}{5} =$

```
    1.8
 5)9.0
   5
   -
   40
   40
```

$1\frac{4}{5} \times 0.65 =$

```
  0.65
 × 1.8
 -----
   520
   65
 -----
 1.170
```

16. (3.8) $\frac{10 \times 2}{0.4} = \frac{20}{0.4} = 50$

17. (3.12) $7.48\boxed{0} > 7.408 > 7.4\boxed{00} > 7.084$
$7.48 > 7.408 > 7.4 > 7.084$

18. (3.6A) $\dfrac{45.60 + 42.76 + 44.02}{3} = \dfrac{132.38}{3}$

$44.126 \approx 44.13$ sec

$$\begin{array}{r} 44.126 \\ 3\overline{)132.380} \\ \underline{12} \\ 12 \\ \underline{12} \\ 03 \\ \underline{3} \\ 08 \\ \underline{6} \\ 20 \\ \underline{18} \end{array}$$

19. (3.6) $79.722 \approx \$79.72$

$$\begin{array}{r} 79.722 \\ 36\overline{)2870.000} \\ \underline{252} \\ 350 \\ \underline{324} \\ 260 \\ \underline{252} \\ 80 \\ \underline{72} \\ 80 \\ \underline{72} \end{array}$$

20. (3.4) Checks:

Checks
$17.75
64.57
91.35
135.46
+ 186.40
$495.53

Beginning balance	$346.52
Deposit	+ 325.00
	671.52
Checks	− 495.53
Ending balance	$175.99

Exercises 4.1 (page 134)

1. $\frac{2}{3}$ **3.** $\frac{9}{4}$ **5.** $\frac{5}{1}$ **7.** $\frac{17}{3}$ **9.** $\frac{1}{5}$ **11.** $\frac{3}{4}$ **13.** $\frac{3}{1}$

15. $\frac{9}{1}$ **17.** $\frac{1}{8}$ **19.** $\frac{1}{2}$ **21.** $\frac{3}{4}$ **23.** $\frac{1}{4}$

25. a. $\frac{24}{30} = \frac{4}{5}$ **b.** $\frac{24}{6} = \frac{4}{1}$ **c.** $\frac{6}{24} = \frac{1}{4}$

27. a. $\frac{12}{8} = \frac{3}{2}$ **b.** $\frac{12}{4} = \frac{3}{1}$
c. total pies = 12 + 8 + 4 = 24
$\frac{8}{24} = \frac{1}{3}$

29. $\dfrac{2 \text{ hr } 40 \text{ min}}{4 \text{ hr}} = \dfrac{160 \text{ min}}{240 \text{ min}} = \dfrac{2}{3}$

31. $\dfrac{4\frac{1}{2}\text{-ft shadow}}{15\text{-ft pole}} = 4\frac{1}{2} \div 15$

$= \dfrac{\overset{3}{\cancel{9}}}{2} \cdot \dfrac{1}{\underset{5}{\cancel{15}}}$

$= \dfrac{3}{10}$

Exercises 4.2 (page 140)

1. a. 8 **b.** 14 **c.** 16 **d.** 28 **e.** 14 and 16 **f.** 8 and 28

3. Yes **5.** No **7.** Yes, $\dfrac{2\frac{1}{4}}{27} = \dfrac{3\frac{1}{3}}{40}$

$2\frac{1}{4} \cdot 40 = 27 \cdot 3\frac{1}{3}$

$\dfrac{9}{\underset{1}{\cancel{4}}} \cdot \dfrac{\overset{10}{\cancel{40}}}{1} = \dfrac{\overset{9}{\cancel{27}}}{1} \cdot \dfrac{10}{\underset{1}{\cancel{3}}}$

$90 = 90$

9. 12 **11.** 8 **13.** 75 **15.** $2\frac{2}{3}$ **17.** $2\frac{2}{3}$ **19.** 8

21. $\frac{7}{20}$ or 0.35 **23.** 2 **25.** 6

27. $\dfrac{\frac{5}{6}}{1\frac{1}{4}} = \dfrac{24}{x}$ **29.** 0.208 **31.** 320

$\frac{5}{6} \cdot x = \dfrac{5}{\underset{1}{\cancel{4}}} \cdot \dfrac{\overset{6}{\cancel{24}}}{1}$

$\dfrac{\frac{\cancel{5}}{\cancel{6}} \cdot x}{\frac{\cancel{5}}{\cancel{6}}} = \dfrac{30}{\frac{5}{6}}$

$x = 30 \div \frac{5}{6}$

$x = \dfrac{\overset{6}{\cancel{30}}}{1} \cdot \dfrac{6}{\underset{1}{\cancel{5}}}$

$x = 36$

Exercises 4.3 (page 143)

1. 30 gal **3.** 30 ft **5.** $9,000 **7.** $750 **9.** 25 lb

11. 630 lb **13.** $3.96 **15.** $\dfrac{\frac{1}{2}\text{ in.}}{12 \text{ mi}} = \dfrac{2\frac{1}{4}\text{ in.}}{x \text{ mi}}$

$\frac{1}{2} \cdot x = \dfrac{\overset{3}{\cancel{12}}}{1} \cdot \dfrac{9}{\underset{1}{\cancel{4}}}$

$\dfrac{\frac{\cancel{1}}{\cancel{2}} \cdot x}{\frac{\cancel{1}}{\cancel{2}}} = \dfrac{27}{\frac{1}{2}}$

$x = 27 \div \frac{1}{2}$

$x = \frac{27}{1} \cdot \frac{2}{1} = 54$ mi

17. 1,400 mi **19.** 30 men

21. Let ℓ = length of room
w = width of room

$$\frac{1 \text{ in.}}{8 \text{ ft}} = \frac{3 \text{ in.}}{\ell \text{ ft}} \qquad \frac{1 \text{ in.}}{8 \text{ ft}} = \frac{2\frac{1}{2} \text{ in.}}{w \text{ ft}}$$

$$1 \cdot \ell = 8 \cdot 3 \qquad 1 \cdot w = \frac{\overset{4}{\cancel{8}}}{1} \cdot \frac{5}{\underset{1}{\cancel{2}}}$$

$$\ell = 24 \text{ ft} \qquad w = 20 \text{ ft}$$

Exercises 4.4A (page 145)

1. 27% **3.** 6% **5.** 140% **7.** 18.6% **9.** 7.5%
11. 290% **13.** 200.5% **15.** 136% **17.** 400%

Exercises 4.4B (page 146)

1. 0.45 **3.** 1.25 **5.** 0.065 **7.** 0.09

9. $2\frac{1}{2}\% = 2.5\% = 0.025$ **11.** $3\frac{1}{4}\% = 3.25\% = 0.0325$

13. 0.1005 **15.** $\frac{3}{4}\% = .75\% = 0.0075$

17. $66\frac{2}{3}\% \approx 66.67\% = 0.6667$

Exercises 4.4C (page 147)

1. 50% **3.** 40% **5.** 37.5% **7.** 90% **9.** 16%

11. $\frac{1}{3} \approx .3333 = 33.33\%$ **13.** $\frac{5}{6} \approx .8333 = 83.33\%$ **15.** 43.75%

17. $2\frac{5}{8} = 2.625 = 262.5\%$

Exercises 4.4D (page 148)

1. $\frac{3}{4}$ **3.** $\frac{1}{10}$ **5.** $\frac{7}{20}$ **7.** $\frac{4}{5}$ **9.** $\frac{1}{20}$

11. $250\% = \frac{250}{100} = \frac{5}{2} = 2\frac{1}{2}$

13. $\frac{1}{2}\% = \frac{\frac{1}{2}}{100} = \frac{1}{2} \div 100 = \frac{1}{2} \cdot \frac{1}{100} = \frac{1}{200}$

15. $2\frac{1}{2}\% = \frac{2\frac{1}{2}}{100} = 2\frac{1}{2} \div 100 = \frac{\overset{1}{\cancel{5}}}{2} \cdot \frac{1}{\underset{20}{\cancel{100}}} = \frac{1}{40}$

17. $33\frac{1}{3}\% = \frac{33\frac{1}{3}}{100} = 33\frac{1}{3} \div 100 = \frac{\cancel{100}}{3} \cdot \frac{1}{\cancel{100}} = \frac{1}{3}$

	Fraction	Decimal	Percent
19.	$\frac{1}{2}$	0.5	50%
21.	$\frac{3}{5}$	0.6	60%
23.	$\frac{1}{10}$	0.1	10%
25.	$\frac{3}{4}$	0.75	75%
27.	$\frac{3}{50}$	0.06	6%
29.	$\frac{12}{25}$	0.48	48%
31.	$1\frac{1}{8}$	1.125	112.5%
33.	$\frac{11}{25}$	0.44	44%
35.	$\frac{1}{40}$	0.025	2.5%
37.	$3\frac{1}{2}$	3.50	350%
39.	$6\frac{1}{4}$	6.25	625%
41.	$\frac{21}{400}$	0.0525	$5\frac{1}{4}\%$
43.	$\frac{3}{400}$	0.0075	$\frac{3}{4}\%$

45. a. $\frac{10}{40} = \frac{1}{4}$ **b.** $\frac{1}{4} = 4\overline{)1.00}$ = 0.25 (8, 20, 20, 0) **c.** $0.25_{\wedge} = 25\%$

47. $100\% - 20\% = 80\% = \frac{80}{100} = \frac{4}{5}$

49. $25\% + 20\% = 45\% = \frac{45}{100} = \frac{9}{20}$

Exercises 4.5 (page 151)

1. 40 **3.** 12 **5.** 9 **7.** 30 **9.** $2\frac{2}{5}$ **11.** 1.5

13. $13\frac{1}{3}\% = \dfrac{13\frac{1}{3}}{100} = 13\frac{1}{3} \div 100$

$= \dfrac{\overset{2}{\cancel{40}}}{3} \cdot \dfrac{1}{\underset{5}{\cancel{100}}} = \dfrac{2}{15}$

Then $13\frac{1}{3}\%$ of $600 = \dfrac{2}{\underset{1}{\cancel{15}}} \cdot \dfrac{\overset{40}{\cancel{600}}}{1} = 80$

15. $1\frac{1}{2}$ or 1.5 **17. a.** \$101.25 **b.** \$273.75

19. 8 days **21.** \$922.50 **23.** 56 hr **25.** \$2,116.50

27. 7.25 **29.** 101.7 **31.** 238.96 **33.** 168.75 **35.** $\approx$\$45.72

Exercises 4.6 (page 154)

1. 50 **3.** 46% **5.** 10 **7.** 850 **9.** 225% **11.** 31.2

13. 600 **15.** 125% **17.** 24 **19.** 250

21. $P = 66\frac{2}{3}$, $A = 42$, B is unknown.

$\dfrac{42}{B} = \dfrac{66\frac{2}{3}}{100}$

$66\frac{2}{3} \cdot B = 4{,}200$

$B = \dfrac{4{,}200}{66\frac{2}{3}} = 4{,}200 \div 66\frac{2}{3}$

$= \dfrac{4{,}200}{1} \div \dfrac{200}{3}$

$= \dfrac{\overset{21}{\cancel{4{,}200}}}{1} \cdot \dfrac{3}{\underset{1}{\cancel{200}}} = 63$

Exercises 4.7 (page 159)

1. 85 games **3.** $16\frac{2}{3}\%$ **5.** 25% **7.** \$6,500

9. (What) is (24%) of (1,800)?
A P B

$\dfrac{A}{1{,}800} = \dfrac{24}{100}$

$\dfrac{100A}{100} = \dfrac{43{,}200}{100}$

$A = 432$

Then, $432 - 256 = 176$ more students

11. 2,303 students **13.** Yes; 40% of 50 = 20 **15.** 8.6%

17. 72% **19. a.** \$960 **b.** \$12,960 **21.** 12,220 people

23. a. \$34.50 **b.** 75%

25. \$8,750 − 2,100 = \$6,650 decrease

(\$6,650) is (what %) of (\$8,750)? ← original cost
A P B

$\dfrac{6{,}650}{8{,}750} = \dfrac{P}{100}$

$8{,}750P = 665{,}000$

$\dfrac{8{,}750P}{8{,}750} = \dfrac{665{,}000}{8{,}750}$

$P = 76\%$

Review Exercises 4.8 (page 161)

1. $\frac{9}{4}$ **2.** $\frac{1}{9}$ **3.** $\frac{6}{1}$ **4.** Yes **5.** No **6.** Yes

7. 36 **8.** 63 **9.** $11\frac{1}{4}$ or 11.25

10. $\dfrac{\frac{x}{2}}{3} = \dfrac{\frac{9}{16}}{\frac{5}{6}}$

$\dfrac{5}{6} \cdot x = \dfrac{\overset{1}{\cancel{2}}}{\underset{1}{\cancel{3}}} \cdot \dfrac{\overset{3}{\cancel{9}}}{\underset{8}{\cancel{16}}}$

$\dfrac{\frac{\cancel{5}}{\cancel{6}} \cdot x}{\frac{\cancel{5}}{\cancel{6}}} = \dfrac{\frac{3}{8}}{\frac{5}{6}}$

$x = \dfrac{3}{8} \div \dfrac{5}{6}$

$x = \dfrac{3}{\underset{4}{\cancel{8}}} \cdot \dfrac{\overset{3}{\cancel{6}}}{5} = \dfrac{9}{20}$

11. 3.5 **12.** $\frac{3}{4}$ **13.** \$1.50 **14.** 8 quarts **15.** 315 students

16. 72 ft **17.** 70 mi **18.** 256 parts **19. a.** 70% **b.** 125%

20. a. 0.04 **b.** 0.65 **21. a.** 60% **b.** 87.5%

22. a. $\frac{6}{25}$ **b.** $\frac{3}{8}$ **23.** 20 **24.** 48 **25.** 135 **26.** 2.4

27. 125% **28.** 240

29. $\dfrac{A}{3{,}560} = \dfrac{5}{100}$

$\dfrac{100A}{100} = \dfrac{17{,}800}{100}$

$A = 178$

Fall enrollment = 3,560 + 178 = 3,738

30. 75% **31.** \$76 **32.** \$450 **33.** 15% **34.** \$1,700

35. \$10,070 **36.** Increase = \$1,015 − \$875 = \$140

$\dfrac{140}{875} = \dfrac{P}{100}$

$\dfrac{875P}{875} = \dfrac{14{,}000}{875}$

$P = 16\%$

37. $\approx$\$6.46

38.

$$\text{Sale price} = 250 \boxed{\times} \underset{\uparrow\,(100\% - 25\%)}{75} \boxed{\%}$$
$$= \$187.50$$

$$\text{Employee pays} = 187.5 \boxed{\times} \underset{\uparrow\,(100\% - 15\%)}{85} \boxed{\%}$$
$$= 159.375 \approx \$159.38$$

Chapter 4 Diagnostic Test (page 163)

Following each problem number is the textbook section reference (in parentheses) in which that kind of problem is discussed.

1. (4.1) **a.** $\frac{15}{24} = \frac{5}{8}$ **b.** $\frac{15}{9} = \frac{5}{3}$

2. (4.1) $\frac{9 \text{ in.}}{2 \text{ ft}} = \frac{9 \text{ in.}}{24 \text{ in.}} = \frac{3}{8}$

3. (4.2)
$$\frac{x}{12} = \frac{\overset{2}{\cancel{40}}}{\underset{3}{\cancel{60}}}$$
$$\frac{x}{12} = \frac{2}{3}$$
$$3 \cdot x = 24$$
$$\frac{\cancel{3} \cdot x}{\cancel{3}} = \frac{24}{3}$$
$$x = 8$$

4. (4.2)
$$\frac{\overset{4}{\cancel{20}}}{\underset{15}{\cancel{75}}} = \frac{x}{300}$$
$$\frac{4}{15} = \frac{x}{300}$$
$$15 \cdot x = 1200$$
$$\frac{\cancel{15} \cdot x}{\cancel{15}} = \frac{1200}{15}$$
$$x = 80$$

5. (4.2)
$$\frac{3\frac{1}{2}}{x} = \frac{21}{40}$$
$$21 \cdot x = \frac{7}{\underset{1}{\cancel{2}}} \cdot \frac{\overset{20}{\cancel{40}}}{1}$$
$$\frac{\cancel{21} \cdot x}{\cancel{21}} = \frac{140}{21}$$
$$x = \frac{20}{3} = 6\frac{2}{3}$$

6. (4.4A) $0.80_{\wedge} = 80\%$ (decimal point moved $\rightarrow$ two places right)

7. (4.4B) ${}_{\wedge}12.5\% = 0.125$ (decimal point moved $\leftarrow$ two places left)

8. (4.4C) $\frac{3}{40} = \; 40\overline{)3.000}$ $\; .075 = 7.5\%$

9. (4.4D) $76\% = \frac{76}{100} = \frac{19}{25}$

10. (4.5) $\frac{5}{8}$ of $72 = \frac{5}{\underset{1}{\cancel{8}}} \cdot \frac{\overset{9}{\cancel{72}}}{1} = 45$

11. (4.5) 0.7 of $35 = 0.7 \times 35 = 24.5$

12. (4.6)
$$\frac{A}{250} = \frac{36}{100}$$
$$100 \cdot A = 250 \cdot 36$$
$$\frac{\cancel{100} \cdot A}{\cancel{100}} = \frac{9000}{100}$$
$$A = 90$$

13. (4.6)
$$\frac{15}{B} = \frac{\overset{5}{\cancel{125}}}{\underset{4}{\cancel{100}}}$$
$$\frac{15}{B} = \frac{5}{4}$$
$$5 \cdot B = 15 \cdot 4$$
$$\frac{\cancel{5} \cdot B}{\cancel{5}} = \frac{60}{5}$$
$$B = 12$$

14. (4.6)
$$\frac{12}{30} = \frac{P}{100}$$
$$30 \cdot P = 12 \cdot 100$$
$$\frac{\cancel{30} \cdot P}{\cancel{30}} = \frac{120\cancel{0}}{3\cancel{0}}$$
$$P = 40\%$$

15. (4.3) Let x = cost of $7\frac{1}{2}$ lb
$$\frac{5 \text{ lb}}{\$4} = \frac{7\frac{1}{2} \text{ lb}}{\$x}$$
$$5 \cdot x = \frac{\overset{2}{\cancel{4}}}{1} \cdot \frac{15}{\underset{1}{\cancel{2}}}$$
$$\frac{\cancel{5} \cdot x}{\cancel{5}} = \frac{30}{5}$$
$$x = \$6$$

16. (4.3) Let x = height of pole
$$\frac{\overset{3}{\cancel{6}}\text{-ft man}}{\underset{2}{\cancel{4}}\text{-ft shadow}} = \frac{x\text{-ft pole}}{18\text{-ft shadow}}$$
$$2 \cdot x = 3 \cdot 18$$
$$\frac{\cancel{2} \cdot x}{\cancel{2}} = \frac{54}{2}$$
$$x = 27 \text{ ft}$$

17. (4.3) Let x = quarts of oil for 6,000 mi
$$\frac{2 \text{ qt}}{1{,}500 \text{ mi}} = \frac{x \text{ qt}}{6{,}000 \text{ mi}}$$
$$1{,}500 \cdot x = 2 \cdot 6{,}000$$
$$\frac{\cancel{1{,}500} \cdot x}{\cancel{1{,}500}} = \frac{12{,}0\cancel{00}}{1{,}5\cancel{00}}$$
$$x = 8 \text{ qt}$$

18. (4.7) (34 / A) is (85% / P) of (what / B)?
$$\frac{34}{B} = \frac{85}{100}$$
$$85 \cdot B = 34 \cdot 100$$
$$\frac{\cancel{85} \cdot B}{\cancel{85}} = \frac{3{,}400}{85}$$
$$B = 40 \text{ problems}$$

19. (4.7) (Discount / A) is (25% / P) of (96 / B)
$$\frac{A}{96} = \frac{\overset{1}{\cancel{25}}}{\underset{4}{\cancel{100}}}$$
$$4 \cdot A = 96$$
$$\frac{\cancel{4} \cdot A}{\cancel{4}} = \frac{96}{4}$$
$$A = 24$$
Sale price = \$96 − \$24 = \$72

20. (4.7) Increase = 4,600 − 4,000 = 600

600 is what % of 4,000 ?
A P B

$$\frac{600}{4{,}000} = \frac{P}{100}$$
$$4{,}000 \cdot P = 600 \cdot 100$$
$$\frac{\cancel{4{,}000} \cdot P}{\cancel{4{,}000}} = \frac{60,\cancel{000}}{4,\cancel{000}}$$
$$P = 15\%$$

Exercises 5.1 (page 168)

1. 20,000,000 people **3.** New York **5.** California
7. 2,000,000 more people **9.** 2 times **11.** $40,000
13. February **15.** $15,000 **17.** January to February
19. 17,000,000 sq. mi **21.** 9,500,000 + 7,000,000 = 16,500,000 sq. mi

23. 9,500,000 − 7,000,000 = 2,500,000 sq. mi

Exercises 5.2 (page 171)

1. 20,000 homes **3.** Increase **5.** 20,000 more homes
7. 3,500 students **9.** 3,500 + 3,500 = 7,000 students
11. 3,500 + 6,000 = 9,500 students **13.** 20,000 students

15. $\frac{7{,}000}{20{,}000} = \frac{P}{100}$
$$\frac{7}{20} = \frac{P}{100}$$
$$20P = 700$$
$$P = 35\%$$

Exercises 5.3 (page 174)

1. 10,000,000 people **3.** 1960 **5.** 15,000,000 people
7. 1940 to 1980 **9.** $10 **11.** $25 in 1986 **13.** Stock A
15. $5 more **17.** 1988 to 1989

Exercises 5.4 (page 176)

1. 34% **3.** 10% + 5% + 10% = 25% **5.** 21–25 years old
7. Under 21 years old **9.** 5,000 students **11.** 7,500 students
13. 40% **15.** 20% **17.** Husband's salary **19.** 15%
21. $15,000 **23.** $1,000

Exercises 5.5 (page 180)

1. a. 5 **b.** 4 **c.** 4 **d.** 7
3. a. 25 **b.** 24 **c.** None **d.** 12
5. a. 8.1 mpg **b.** 19.8 mpg **c.** 20 mpg
7. a. 173 **b.** 176 **c.** 54
9. a. 7.5 **b.** 7 **c.** 6 **d.** 4
11. a. ≈740 hr **b.** 726 hr **c.** 147 hr

Review Exercises 5.6 (page 181)

1. 60°F **2.** 70°F **3.** February **4.** August
5. February to August **6.** August to February
7. February to March **8.** September to October
9. 250 ft **10.** 175 ft **11.** 20 mph **12.** 40 mph
13. 80 ft **14.** 130 ft **15.** 30% **16.** 24% **17.** 96%
18. 58% **19.** 420 people **20.** 42,000 people **21.** Saturday
22. Wednesday **23.** $8,000 **24.** $7,500 **25.** Monday
26. Wednesday **27.** $5,500 **28.** $50,000 **29.** $\frac{10{,}000}{50{,}000} = \frac{P}{100}$
$$\frac{1}{5} = \frac{P}{100}$$
$$5P = 100$$
$$P = 20\%$$

30. 9% **31.** 5 **32.** 5.5 **33.** 6 **34.** 6 **35.** 67°F
36. 67.5°F **37.** 60°F and 80°F **38.** 40°F

Chapter 5 Diagnostic Test (page 185)

Following each problem number is the textbook section reference (in parentheses) in which that kind of problem is discussed.

1. (5.3) 60°F **2.** (5.3) 2 P.M.
3. (5.3) 70°F − 65°F = 5°F **4.** (5.3) 6 A.M. to 2 P.M.
5. (5.3) 2 P.M. **6.** (5.1) Cheerios **7.** (5.1) 3 grams
8. (5.1) Raisin Bran **9.** (5.1) 4 times
10. (5.1) 11 − 3 = 8 grams **11.** (5.2) 50 cars
12. (5.2) Carson **13.** (5.2) 40 cars
14. (5.2) 10 × 20 = 200 cars

15. (5.2) $\frac{60}{200} = \frac{P}{100}$
$$200P = 6{,}000$$
$$P = \frac{6{,}000}{200} = 30\%$$

16. (5.4) 10% **17.** (5.4) 35% + 25% + 10% = 70%

18. (5.4) C **19.** (5.4) $\frac{A}{48} = \frac{25}{100}$
$$\frac{A}{48} = \frac{1}{4}$$
$$4A = 48$$
$$A = 12 \text{ students}$$

20. (5.5) $\frac{87 + 72 + 82 + 93 + 71}{5} = \frac{405}{5} = 81$

21. (5.5) 71, 72, 82, 87, 93
↑ 82 is the median

22. (5.5) 71°F − 53°F = 18°F

23. (5.5) $\frac{53 + 66 + 71 + 62}{4} = \frac{252}{4} = 63°\text{F}$

24. (5.5) 2, 2, 4, 5, 6, 6, 6, 9
$$\frac{5 + 6}{2} = \frac{11}{2} = 5.5$$

25. (5.5) 6

Cumulative Review Exercises for Chapters 1–5 (page 187)

1. 5,700.09 **2.** $1\frac{5}{9}$ **3.** $12\frac{3}{4}$ **4.** $4\frac{3}{8}$ **5.** $1\frac{5}{9}$

6. 49.12 **7.** 4.608 **8.** 350 **9.** 16 **10.** $\frac{5}{6}, \frac{3}{4}, \frac{7}{12}$

11. 6 **12.** $\frac{3}{125}$ **13.** 40% **14.** $\frac{9}{20}$ **15.** 80 **16.** 60 mi

17. 56 cm **18.** No **19.** \$45 **20.** 6.1 in. **21.** 19 yd

22. \$96 **23.** 15% **24.** 12,000,000 accounts

25. 1985 to 1986

Exercises 6.1 (page 193)

1. Negative seventy-five **3.** -54 **5.** -62 **7.** -2

9. -10 **11.** -1 **13.** Cannot be found **15.** *F* **17.** *D*

19. *A* **21.** *E* **23.** > **25.** < **27.** < **29.** >

Exercises 6.2 (page 197)

1. 9 **3.** -7 **5.** -1 **7.** 4 **9.** -11 **11.** 3

13. 11 **15.** -3 **17.** -9 **19.** -7 **21.** 0 **23.** 7

25. -12 **27.** -9 **29.** -6 **31.** 3 **33.** -9 **35.** 2

37. -21 **39.** 7 **41.** 10 **43.** -40 **45.** 41 **47.** -32

49.
$$\begin{array}{rcr} \left(-1\frac{1}{2}\right) & = & \left(-1\frac{5}{10}\right) \\ +\left(-3\frac{2}{5}\right) & = & \left(-3\frac{4}{10}\right) \\ \hline & & -4\frac{9}{10} \end{array}$$

51.
$$\begin{array}{rcr} \left(4\frac{5}{6}\right) & = & \left(4\frac{5}{6}\right) \\ +\left(-1\frac{1}{3}\right) & = & \left(-1\frac{2}{6}\right) \\ \hline & & 3\frac{3}{6} = 3\frac{1}{2} \end{array}$$

53. -12.78 **55.** 0.084 **57.** $(-35) + (53) = 18°F$

59. 8 [+/−] [+] 3 [=] -5

61. 12 [+/−] [+] 26 [+/−] [=] -38

Exercises 6.3 (page 199)

1. 6 **3.** -1 **5.** -8 **7.** 14 **9.** 9 **11.** 2

13. -7 **15.** -4 **17.** 3 **19.** -12 **21.** -7

23. -6 **25.** 14 **27.** 7 **29.** -12 **31.** -3 **33.** -4

35. 5 **37.** -25 **39.** -55 **41.** -59 **43.** 663

45.
$$\begin{array}{rcr} \left(-4\frac{2}{3}\right) & = & \left(-4\frac{4}{6}\right) \\ -\left(2\frac{1}{6}\right) & = & +\left(-2\frac{1}{6}\right) \\ \hline & & -6\frac{5}{6} \end{array}$$

47.
$$\begin{array}{rcr} \left(-3\frac{3}{4}\right) & = & \left(-3\frac{9}{12}\right) \\ -\left(-2\frac{1}{6}\right) & = & +\left(+2\frac{2}{12}\right) \\ \hline & & -1\frac{7}{12} \end{array}$$

49. $(-10) - (-7) =$
$(-10) + (7) \quad = -3$

51. -9.36 **53.** -47.67

55.
$$\begin{array}{rcr} \left(-5\frac{3}{10}\right) & = & \left(-5\frac{6}{20}\right) \\ -\left(+3\frac{1}{4}\right) & = & +\left(-3\frac{5}{20}\right) \\ \hline & & -8\frac{11}{20} \end{array}$$

57. $(42) - (-7) =$
$(42) + (7) \quad = 49°F$ Rise

59. 2,022 ft

61. $(-141) - (68) \quad =$
$(-141) + (-68) = -209$ ft

63. 8 [−] 2 [+/−] [=] 10

65. 17 [+/−] [−] 6 [+/−] [=] -11

Exercises 6.4 (page 203)

1. -6 **3.** -10 **5.** 16 **7.** -32 **9.** -18 **11.** -54

13. 28 **15.** -30 **17.** -63 **19.** 100 **21.** -56

23. -260 **25.** 200 **27.** $-1{,}125$ **29.** -150 **31.** 140

33. -40 **35.** 56

37. $\left(2\frac{1}{4}\right)\left(-1\frac{1}{3}\right) = -\left(\frac{\cancel{9}^{3}}{\cancel{4}_{1}} \cdot \frac{\cancel{4}^{1}}{\cancel{3}_{1}}\right) = -3$

39. $\left(-1\frac{7}{8}\right)\left(-2\frac{4}{5}\right) = +\left(\frac{\cancel{15}^{3}}{\cancel{8}_{4}} \cdot \frac{\cancel{14}^{7}}{\cancel{5}_{1}}\right) = \frac{21}{4} = 5\frac{1}{4}$

41. -274 **43.** 38.18

45. 9 [+/−] [×] 8 [=] -72

47. 15 [+/−] [×] 25 [+/−] [=] 375

Exercises 6.5 (page 205)

1. 2 **3.** -4 **5.** -5 **7.** 2 **9.** -5 **11.** -4

13. 3 **15.** -3 **17.** 9 **19.** -15 **21.** -3 **23.** -3

25. $\frac{-15}{6} = \frac{-5}{2} = -2\frac{1}{2}$ or -2.5

27. -15 **29.** 7 **31.** -3.67

33. $\frac{2\frac{1}{2}}{-5} = \frac{\frac{5}{2}}{-\frac{5}{1}} = \frac{5}{2} \div \left(-\frac{5}{1}\right) = -\left(\frac{\cancel{5}^{1}}{2} \cdot \frac{1}{\cancel{5}_{1}}\right) = -\frac{1}{2}$

35. $\frac{-4\frac{1}{2}}{-1\frac{7}{8}} = \frac{-\frac{9}{2}}{-\frac{15}{8}} = \left(-\frac{9}{2}\right) \div \left(-\frac{15}{8}\right) = +\left(\frac{\cancel{9}^{3}}{\cancel{2}_{1}} \cdot \frac{\cancel{8}^{4}}{\cancel{15}_{5}}\right)$
$= \frac{12}{5} = 2\frac{2}{5}$

37. 84 [+/−] [÷] 7 [=] -12

39. 132 [+/−] [÷] 12 [+/−] [=] 11

Exercises 6.6 (page 208)

1. True. Because of the commutative property of addition (order of numbers changed).

3. True. Because of the associative property of addition (grouping changed).

5. $6 - 2 \stackrel{?}{=} 2 - 6$
$4 \neq -4$
False. Commutative property does not hold for subtraction.

7. True. Because of the associative property of multiplication.

9. $8 \div 4 \stackrel{?}{=} 4 \div 8$
   $2 \neq \frac{1}{2}$
   False. Commutative property does not hold for division.

11. True. Because of the commutative property of multiplication.

13. False. $(4)(-5) \stackrel{?}{=} (-5) + (4)$
    $-20 \neq -1$

15. True. Because of the commutative property of addition.

17. True. Because of the commutative property of addition.

19. True. Because of the associative and commutative properties of addition.

21. False. Subtraction is not commutative.

23. True. Because of the associative property of multiplication (grouping changed).

25. True. Because of the commutative property of addition (order changed).

27. True. Because of the commutative property of multiplication.

29. True. Because of the commutative property of multiplication.

Exercises 6.7 (page 210)

1. 0 **3.** 4 **5.** -6 **7.** 0 **9.** 11 **11.** Undefined
13. 0 **15.** Undefined **17.** -789 **19.** Undefined

21. 9 ÷ 0 = E ← The E in the display indicates that an error has been made. In this case, division by 0 is undefined.

23. Undefined **25.** 0

Exercises 6.8 (page 212)

1. 27 **3.** $(-5)^2 = (-5)(-5) = 25$ **5.** 49 **7.** 0
9. -10 **11.** 1,000 **13.** $-100{,}000$ **15.** 4
17. $-(-3)^2 = -(-3)(-3) = -9$ **19.** 625 **21.** 64,000
23. $-1{,}728$ **25.** -1 **27.** $-2^2 = -(2 \cdot 2) = -4$
29. -1 **31.** 1 **33.** $\frac{9}{16}$ **35.** $-\frac{1}{1{,}000}$
37. 59,049 **39.** $-16{,}384$ **41.** 5.0625

Exercises 6.9 (page 214)

1. 4 because $4^2 = 16$ **3.** $-\sqrt{4} = -(\sqrt{4}) = -2$
5. 9 **7.** 10 **9.** -7 **11.** 8 **13.** -10 **15.** $\frac{4}{5}$
17. ≈ 3.606 **19.** ≈ 2.646 **21.** ≈ 7.071 **23.** ≈ 13.565

Review Exercises 6.10 (page 217)

1. 8, 9 **2.** 0, 1, 2 **3.** 9 **4.** 1 **5.** Cannot be done.
6. -1 **7. a.** < **b.** < **c.** > **d.** > **e.** <

8. a. True. Because of the associative property of multiplication.
b. True. Because of the commutative property of addition.
c. False. $5 - (-2) = 5 + 2 = 7$
$(-2) - 5 = (-2) + (-5) = -7$
d. True. Because of the associative property of addition.
e. True. Because of the commutative property of multiplication.
f. True. Because of the commutative property of addition.

9. 3 **10.** $(-5) - (-3) =$ $(-5) + (3) = -2$ **11.** 3 **12.** 12 **13.** -11

14. -72 **15.** 4 **16.** -6 **17.** -9 **18.** -6 **19.** -42
20. -10 **21.** $-3^2 = -(3 \cdot 3) = -9$ **22.** -4 **23.** 3
24. -3 **25.** 0 **26.** $(-5)^2 = (-5)(-5) = 25$ **27.** -12
28. -24 **29.** 5 **30.** 0 **31.** 6
32. $-2^4 = -(2 \cdot 2 \cdot 2 \cdot 2) = -16$ **33.** 13 **34.** -5 **35.** 48
36. 9 **37.** -7 **38.** 0 **39.** -14 **40.** 5 **41.** -11
42. -9 **43.** 1 **44.** -7 **45.** -126
46. $(-3)^3 = (-3)(-3)(-3) = -27$ **47.** 12 **48.** -5
49. -3 **50.** 5 **51.** 4 **52.** 80 **53.** -17 **54.** 10,000
55. -11 **56.** -16 **57.** Undefined **58.** $-\frac{4}{3} = -1\frac{1}{3}$
59. $2\frac{1}{6}$ **60.** $-1\frac{13}{20}$ **61.** $-12\frac{1}{2}$ **62.** Undefined
63. $1\frac{5}{22}$ **64.** $-5\frac{11}{12}$ **65.** $(-7) - (9) =$
$(-7) + (-9) = -16$
66. -8 **67.** 7.280 **68.** 9.592 **69.** 12.369 **70.** 13.601

Chapter 6 Diagnostic Test (page 219)

Following each problem number is the textbook section reference (in parentheses) in which that kind of problem is discussed.

1. (6.1) < **2.** (6.1) > **3.** (6.1) < **4.** (6.1) 0, 1, 2
5. (6.1) 1 **6.** (6.3) $(38) - (-20) =$
$(38) + (+20) = 58°F$
7. (6.1) Negative thirty-five
8. (6.6) True. Because of the commutative property of addition.
9. (6.6) True. Because of the associative property of addition.
10. (6.6) True. Because of the commutative property of multiplication.
11. (6.6) False. Subtraction is not commutative.
$7 - 4 \stackrel{?}{=} 4 - 7$
$3 \neq -3$
12. (6.2) -4 **13.** (6.3) $20 - (-10) =$ $20 + (+10) = 30$ **14.** (6.4) 50
15. (6.5) -3 **16.** (6.8) 36 **17.** (6.2) 3 **18.** (6.4) -36
19. (6.3) $(-10) - (+5) =$ $(-10) + (-5) = -15$ **20.** (6.7) Undefined
21. (6.4) -40 **22.** (6.5) $-\frac{5}{6}$ **23.** (6.8) -64
24. (6.2) 7 **25.** (6.3) $(-9) - (-7) =$ $(-9) + (\ 7) = -2$ **26.** (6.4) -42
27. (6.3) $(5) - (\ 12) =$ $(5) + (-12) = -7$ **28.** (6.3) $(-2) - (-6) =$ $(-2) + (\ 6) = 4$
29. (6.2) -11 **30.** (6.8) 1 **31.** (6.4) -54 **32.** (6.5) 4
33. (6.3) $(6) - (\ 10) =$ $(6) + (-10) = -4$ **34.** (6.2) -9 **35.** (6.4) -24
36. (6.2) 5 **37.** (6.8) 0
38. (6.3) $(-4) - (\ 8) =$ $(-4) + (-8) = -12$ **39.** (6.2) -7

40. (6.4) $\left(-1\frac{7}{8}\right)\left(2\frac{2}{5}\right) = -\left(\frac{\overset{3}{\cancel{15}}}{\underset{2}{\cancel{8}}} \cdot \frac{\overset{3}{\cancel{12}}}{\underset{1}{\cancel{5}}}\right) = -\frac{9}{2} = -4\frac{1}{2}$

41. (6.2)

$$\begin{array}{r} \left(-2\frac{3}{4}\right) = \left(-2\frac{6}{8}\right) \\ +\left(-1\frac{5}{8}\right) = \left(-1\frac{5}{8}\right) \\ \hline -3\frac{11}{8} = -4\frac{3}{8} \end{array}$$

42. (6.5) -3

43. (6.8) $(-2)^4 = (-2)(-2)(-2)(-2) = 16$ **44.** (6.7) 0

45. (6.7) 0 **46.** (6.3) $(9) - (-5) =$ $(9) + (\ \ 5) = 14$

47. (6.2) -5 **48.** (6.9) 7 **49.** (6.9) 9 **50.** (6.9) -8

Exercises 7.1 (page 223)

1. -2 **3.** -14 **5.** 15 **7.** 15 **9.** 25 **11.** -1

13. 18 **15.** 36 **17.** 0 **19.** 50 **21.** 84

23.
$$\begin{aligned} &-3^2 - 4^2 = \\ &-9 - 16 = \\ &-9 + (-16) = -25 \end{aligned}$$

25. -6 **27.** 36

29.
$$\begin{aligned} &2^3 + (3 - 2 \cdot 8) = \\ &2^3 + (3 - 16) = \\ &2^3 + (-13) = \\ &8 + (-13) = -5 \end{aligned}$$

31. -24

33.
$$\begin{aligned} &-6 - 8 \div 2 \cdot 4 = \\ &-6 - 4 \cdot 4 = \\ &-6 - 16 = -22 \end{aligned}$$

35. -28

37.
$$\begin{aligned} &2 \cdot \frac{7}{16} + \frac{9}{20} \div \frac{3}{5} = \\ &\frac{\overset{1}{\cancel{2}}}{1} \cdot \frac{7}{\underset{8}{\cancel{16}}} + \frac{\overset{3}{\cancel{9}}}{\underset{4}{\cancel{20}}} \cdot \frac{\overset{1}{\cancel{5}}}{\underset{1}{\cancel{3}}} = \\ &\frac{7}{8} + \frac{3}{4} = \\ &\frac{7}{8} + \frac{6}{8} = \frac{13}{8} = 1\frac{5}{8} \end{aligned}$$

39.
$$\begin{aligned} &\left(\frac{2}{3}\right)^2 + 3\frac{1}{3} \cdot \frac{1}{4} = \\ &\frac{4}{9} + \frac{\overset{5}{\cancel{10}}}{3} \cdot \frac{1}{\underset{2}{\cancel{4}}} = \\ &\frac{4}{9} + \frac{5}{6} = \\ &\frac{8}{18} + \frac{15}{18} = \frac{23}{18} = 1\frac{5}{18} \end{aligned}$$

41.
$$\begin{aligned} &2.3 + 5(3.7) \div 100 = \\ &2.3 + 18.5 \div 100 = \\ &2.3 + .185 = 2.485 \end{aligned}$$

43. -98 **45.** -34 **47.** 43 **49.** -72 **51.** 132

Exercises 7.2 (page 225)

1. -27 **3.** 12 **5.** 15 **7.** -26 **9.** 12 **11.** 5

13. -1 **15.** 1

17.
$$\begin{aligned} &\frac{3^2 + 5}{2} - \frac{(-4)^2}{8} = \\ &\frac{9 + 5}{2} - \frac{16}{8} = \\ &\frac{14}{2} - \frac{16}{8} = \\ &7 - 2 = 5 \end{aligned}$$

19.
$$\begin{aligned} &6 - 3\{8 - [4 - (-1)]\} = \\ &6 - 3\{8 - [\ 4 + 1\ ]\} = \\ &6 - 3\{8 - 5\} = \\ &6 - 3\{3\} = \\ &6 - 9 = -3 \end{aligned}$$

21. 5 **23.** -2

25.
$$\begin{aligned} &15 - \{4 - [2 - 3(6 - 4)]\} = \\ &15 - \{4 - [2 - 3(2)]\} = \\ &15 - \{4 - [2 - 6]\} = \\ &15 - \{4 - [-4]\} = \\ &15 - \{8\} = 7 \end{aligned}$$

Exercises 7.3 (page 227)

1. -20 **3.** -4 **5.** 25 **7.** 21 **9.** 21

11.
$$\begin{aligned} &3b - ab + xy = \\ &3(-5) - (3)(-5) + (4)(-7) = \\ &-15 + 15 - 28 = -28 \end{aligned}$$

13. -33 **15.** -5 **17.** 19 **19.** 13

21.
$$\begin{aligned} &a^2 - 2ab + b^2 = \\ &(3)^2 - 2(3)(-5) + (-5)^2 = \\ &9 - 2(3)(-5) + 25 = \\ &9 + 30 + 25 = 64 \end{aligned}$$

23. -1 **25.** 0 **27.** 11

29. $-\frac{2}{3}$

31.
$$\begin{aligned} &\frac{(1 + G)^2 - 1}{H} = \\ &\frac{[1 + (-5)]^2 - 1}{-4} = \\ &\frac{[-4]^2 - 1}{-4} = \\ &\frac{16 - 1}{-4} = \\ &\frac{15}{-4} = -3\frac{3}{4} \end{aligned}$$

33. -1

35.
$$\begin{aligned} &G - \sqrt{G - 4EH} = \\ &(-5) - \sqrt{(-5)^2 - 4(-1)(-4)} = \\ &(-5) - \sqrt{25 - 16} = \\ &(-5) - \sqrt{9} = \\ &(-5) - 3 = -8 \end{aligned}$$

37.
$$\frac{3x^2}{z} = \frac{3\left(\frac{1}{2}\right)^2}{\left(\frac{3}{8}\right)} = \frac{\frac{3}{1} \cdot \frac{1}{4}}{\frac{3}{8}} = \frac{\frac{3}{4}}{\frac{3}{8}} = \frac{3}{4} \div \frac{3}{8} = \frac{\overset{1}{\cancel{3}}}{\underset{1}{\cancel{4}}} \cdot \frac{\overset{2}{\cancel{8}}}{\underset{1}{\cancel{3}}} = 2$$

39.
$$\begin{aligned} &(x + y + z)^2 = \\ &\left[\left(\frac{1}{2}\right) + \left(-\frac{3}{4}\right) + \left(\frac{3}{8}\right)\right]^2 = \\ &\left[\frac{4}{8} + \left(-\frac{6}{8}\right) + \frac{3}{8}\right]^2 = \\ &\left[\frac{1}{8}\right]^2 = \frac{1}{64} \end{aligned}$$

Exercises 7.4 (page 229)

1. 105 **3.** 5 **5.** 243 **7.** 77 **9.** 314 **11.** 144

13. 400

15.
$$\begin{aligned} &\sigma = \sqrt{npq} \\ &\sigma = \sqrt{(100)(.9)(.1)} \\ &\sigma = \sqrt{9} = 3 \end{aligned}$$

17. -25

19.
$$\begin{aligned} &C = \frac{a}{a + 12} \cdot A \\ &C = \frac{6}{6 + 12} \cdot 30 \\ &C = \frac{\overset{1}{\cancel{6}}}{\underset{\underset{1}{\cancel{3}}}{\cancel{18}}} \cdot \frac{\overset{10}{\cancel{30}}}{1} = 10 \end{aligned}$$

21. 1,628.89 **23.** 9.4

Review Exercises 7.5 (page 232)

1. 9

2.
$$\begin{aligned} &11 - 7 \cdot 3 = \\ &11 - 21 = -10 \end{aligned}$$

3. 9 **4.** 2

5. $6-[8-(3-4)]=$
$6-[8-\ (-1)]\ =$
$6-\ \ [9]\ =-3$

6. 16 **7.** 2

8. 10 **9.** -5 **10.** 45 **11.** 21 **12.** -17

13. $9-\{6-[4-(7-8)]\}=$
$9-\{6-[4-\ (-1)\ ]\}=$
$9-\{6-\ \ 5\ \ \}\ =$
$9-\ \ 1\ =8$

14. $\left(\frac{3}{4}-\frac{1}{3}\right)\div\left(\frac{4}{9}+\frac{2}{3}\right)=$

$\left(\frac{9}{12}-\frac{4}{12}\right)\div\left(\frac{4}{9}+\frac{6}{9}\right)=$

$\frac{5}{12}\div\frac{10}{9}=$

$\frac{\cancel{5}^{1}}{\cancel{12}_{4}}\cdot\frac{\cancel{9}^{3}}{\cancel{10}_{2}}=\frac{3}{8}$

15. -13

16. $6-x[y-z]\ =$
$6-(-2)[3-(-4)]=$
$6-(-2)[\ \ 7\ \ ]=$
$6-(-14)\ =$
$6+\ \ 14\ =20$

17. 36

18. $y^2\ -\ 2yz\ +\ z^2\ =$
$(3)^2-2(3)(-4)+(-4)^2=$
$9\ -2(3)(-4)+\ 16\ =$
$9\ -\ \ 6(-4)+\ 16\ =$
$9\ +\ \ 24\ +\ 16\ =49$

19. $x-2(y-xz)\ =$
$-2-2[3-(-2)(-4)]\ =$
$-2-2[3-\ \ 8\ \ ]=$
$-2-2[\ \ -5\ \ ]=$
$-2-(-10)\ =$
$-2+\ \ 10\ =8$

20. $\frac{(x+y)^2-z^2}{x-2y}=$

$\frac{(-2+3)^2-(-4)^2}{-2-2(3)}=$

$\frac{1-16}{-2-6}=$

$\frac{-15}{-8}=1\frac{7}{8}$

21. $(\ x^2\ +\ y^2\ )(\ x^2\ -\ y^2\ )=$
$[(-2)^2+(3)^2][(-2)^2-(3)^2]=$
$[\ \ 4\ +9\ \ [\ \ 4\ -9\ \ ]=$
$[13]\ \ [-5]\ =-65$

22. $(\ x\ -\ y\ )(\ x^2\ +\ xy\ +\ y^2\ )=$
$[(-2)-(3)][(-2)^2+(-2)(3)+(3)^2]=$
$[-5]\ \ [\ \ 4\ +\ (-6)\ +\ 9\ \ ]=$
$[-5]\ \ [7]\ =-35$

23. 14 **24.** 31.5

25. $C=\frac{5}{9}(F-32)$

$C=\frac{5}{9}\left(15\frac{1}{2}-32\right)$

$C=\frac{5}{9}\left(\frac{31}{2}-\frac{64}{2}\right)$

$C=\frac{5}{\cancel{9}_{3}}\left(-\frac{\cancel{33}^{11}}{2}\right)$

$C=-\frac{55}{6}=-9\frac{1}{6}$

26. 5 **27.** 1,300

28. $S=\frac{1}{2}gt^2$

$S=\frac{1}{2}(32)\left(\frac{3}{2}\right)^2$

$S=\frac{1}{\cancel{2}_{1}}\cdot\frac{\cancel{32}^{\cancel{8}\,4}}{1}\cdot\frac{9}{\cancel{4}_{1}}=36$

29. 5 **30.** 628

Chapter 7 Diagnostic Test (page 235)

Following each problem number is the textbook section reference (in parentheses) in which that kind of problem is discussed.

1. (7.1) $17-9-6+11=$
$8\ \ -6+11=$
$2\ \ +11=13$

2. (7.1) $5+2\cdot3=$
$5+\ \ 6\ =11$

3. (7.1) $-12\div2\cdot3=$
$-6\cdot3=-18$

4. (7.1) $-5^2+(-4)^2=$
$-25+\ \ 16\ =-9$

5. (7.1) $2\cdot3^2-4=$
$2\cdot9\ -4=$
$18\ \ -4=14$

6. (7.1) $3\sqrt{25}-5(-4)=$
$3\cdot5-5(-4)=$
$15\ +\ \ 20\ =35$

7. (7.2) $\frac{8-12}{-6+2}=\frac{-4}{-4}=1$

8. (7.1) $(2^3-7)(5^2+4^2)=$
$(8-7)(25+16)=$
$(1)\ \ (41)\ =41$

9. (7.2) $10-[6-(5-7)]=$
$10-[6-\ (-2)\ ]=$
$10-[6+2]\ =$
$10-\ \ [8]\ =2$

10. (7.1) $2+(6-3\cdot4)=$
$2+(6-\ \ 12\ )=$
$2+\ \ (-6)\ =-4$

11. (7.2) $\sqrt{10^2-6^2}=$
$\sqrt{100-36}=$
$\sqrt{64}=8$

12. (7.2) $5-2[4-(6-9)]=$
$5-2[4-\ (-3)\ ]=$
$5-2[4+\ \ 3\ \ ]=$
$5-2\ \ [7]\ =$
$5-\ \ 14\ =-9$

13. (7.3) $3a+\ 6x\ -\ \ cy\ =$
$3(-2)+4(5)-(-3)(-6)=$
$-6\ +\ 20\ -\ \ 18\ =$
$14\ \ -\ \ 18\ =-4$

14. (7.3) $4x\ -[\ a\ -\ (3c-b)]\ =$
$4(5)-[-2-(3(-3)-4)]=$
$4(5)-[-2-\ \ (-9\ -4)]=$
$4(5)-[-2-\ \ (-13)\ ]=$
$4(5)-\ \ [11]\ =$
$20\ -\ \ 11\ =9$

15. (7.3) $x^2+\ \ 2xy\ -\ \ y^2\ =$
$(5)^2+2(5)(-6)-(-6)^2=$
$25+2(5)(-6)\ -\ 36\ =$
$25+\ (-60)\ -\ 36\ =$
$-35\ -\ 36\ =-71$

16. (7.4) $C=\frac{5}{9}(F-32)$ $F=-4$

$C=\frac{5}{9}(-4-32)$

$C=\frac{5}{\cancel{9}_{1}}\left(-\frac{\cancel{36}^{4}}{1}\right)$

$C=-20$

17. (7.4) $A=\pi R^2$ $\begin{cases}\pi\approx3.14\\ R=20\end{cases}$

$A\approx(3.14)(20)^2$
$A\approx3.14(400)$
$A\approx1{,}256$

18. (7.4) $C=\frac{a}{a+12}\cdot A$ $\begin{cases}a=7\\ A=38\end{cases}$

$C=\frac{7}{7+12}(38)$

$C=\frac{7}{19}\left(\frac{38}{1}\right)$

$C=14$

19. (7.4)
$$\begin{aligned} A &= P(1 + rt) \\ A &= 600(1 + 0.10(2.5)) \\ A &= 600(1 + 0.25) \\ A &= 600(1.25) \\ A &= 750 \end{aligned} \qquad \begin{cases} P = 600 \\ r = 0.10 \\ t = 2.5 \end{cases}$$

20. (7.4)
$$\begin{aligned} V &= C - Crt \\ V &= 600 - 600(0.04)(10) \\ V &= 600 - 240 \\ V &= 360 \end{aligned} \qquad \begin{cases} C = 600 \\ r = 0.04 \\ t = 10 \end{cases}$$

Exercises 8.1 (page 241)

1. a. One term **b.** No second term

3. a. Three terms **b.** $-5F$ **5. a.** Three terms **b.** $\dfrac{2x + y}{3xy}$

7. a. 4th **b.** -3 **9. a.** 2nd **b.** 2 **11. a.** 3rd **b.** 1

13. a. 1st **b.** $-\dfrac{1}{3}$ **15.** 2nd **17.** 5th **19.** 4th

21. Not a polynomial **23.** $8x^5 + 7x^3 - 4x - 5$

25. $xy^3 + 8xy^2 - 4x^2y$

Exercises 8.2 (page 245)

1. x^{13} **3.** y^{10} **5.** x^5 **7.** a^5 **9.** x^{28} **11.** a^4

13. z **15.** 10,000,000 **17.** 1,000,000 **19.** 10,000,000,000

21. x^2y^3 cannot be simplified because the bases are different.

23. $\dfrac{6x^2}{2x} = \dfrac{\overset{3}{\cancel{6}}}{\cancel{2}} \cdot \dfrac{x^2}{x} = \dfrac{3}{1} \cdot \dfrac{x}{1} = 3x$

25. $\dfrac{a^3}{b^2}$ cannot be simplified because the bases are different.

27. $\dfrac{10x^4}{5x^3} = \dfrac{\overset{2}{\cancel{10}}}{\cancel{5}} \cdot \dfrac{x^4}{x^3} = \dfrac{2}{1} \cdot \dfrac{x}{1} = 2x$

29. $\dfrac{12h^4k^3}{8h^2k} = \dfrac{12}{8} \cdot \dfrac{h^4}{h^2} \cdot \dfrac{k^3}{k} = \dfrac{3}{2} \cdot \dfrac{h^2}{1} \cdot \dfrac{k^2}{1} = \dfrac{3h^2k^2}{2}$ **31.** 5^{u+v}

33. Same form. See Word of Caution, page 244. **35.** y^6

37. a^2 **39.** z^8 **41.** y^2 **43.** 729 **45.** hk^2 **47.** $\dfrac{m^4}{n^2}$

49. $\dfrac{5a^2b}{3}$ **51.** $2 \cdot 2^3 \cdot 2^2 = 2^{1+3+2} = 2^6$ or 64

53. $x \cdot x^3 \cdot x^4 = x^{1+3+4} = x^8$ **55.** $x^2y^3x^5 = x^{2+5}y^3 = x^7y^3$

57. $ab^3a^5 = a^{1+5}b^3 = a^6b^3$ **59.** $3^{a \cdot b}$

Exercises 8.3 (page 248)

1. $(-2a)(4a^2) = -(2 \cdot 4)(a \cdot a^2) = -8a^3$ **3.** $-35x^2y$ **5.** $64x^6$

7. $30x^6$ **9.** $-16m^6n^5$

11. $(5 \cdot 2)(x^2 \cdot x)(y \cdot y)(z^3 \cdot z) = 10x^3y^2z^4$ **13.** $5a + 30$

15. $3m - 12$ **17.** $-2x + 6$ **19.** $-6x^2 + 12x - 15$

21. $12x^3 - 24x$ **23.** $-10x^3 - 6x^2 + 8x$ **25.** $-x^2 + y^2$

27. $-x - 3$ **29.** $-2x^2 - 4x + 7$ **31.** $6x - 24$

33. $7y^2 - 28y + 21$ **35.** $8x^3 - 12x^2 + 20x$

37. $x^2y - 3x$ **39.** $3a^2b - 6a^3$ **41.** $6xy - 12x^2y^2$

43.
$$\begin{aligned} &-2xy(x^2y - y^2x - y - 5) \\ &= (-2xy)(x^2y) + (-2xy)(-y^2x) + (-2xy)(-y) + (-2xy)(-5) \\ &= -2x^3y^2 + 2x^2y^3 + 2xy^2 + 10xy \end{aligned}$$

45.
$$\begin{aligned} &(3x^3 - 2x^2y + y^3)(-2xy) \\ &= (3x^3)(-2xy) + (-2x^2y)(-2xy) + (y^3)(-2xy) \\ &= -6x^4y + 4x^3y^2 - 2xy^4 \end{aligned}$$

47. $12a^5b^4 - 8a^2b^3 - 20ab^3$

49. $-12x^3y^3z^2 + 6xy^3z^3 + 8x^2y^2z^4$

Exercises 8.4 (page 250)

1. $12x$ **3.** $-7a$ **5.** $4x^2$ **7.** $8x^3$ **9.** $-7xy$

11. $4x^2 + 3x$ **13.** $3a$ **15.** x **17.** $2y$ **19.** 0

21. $-2xy$ **23.** $-2a^2b$ **25.** $3x^2 + 5x - 2$

27. $-2x^3 - 2x^2 - 4x$ **29.** $x^3 - 5x^2 - x + 3$

31. $5x^2 + 5xy - 6y^2$ **33.** $3x^2y - 3xy^2$

35.
$$\begin{aligned} &\frac{2}{3}y^2 - \frac{1}{2}y^2 + \frac{5}{6}y^2 \\ &= \left(\frac{4}{6} - \frac{3}{6} + \frac{5}{6}\right)y^2 = \frac{6}{6}y^2 = y^2 \end{aligned}$$

37. 15.02 sec

Exercises 8.5 (page 253)

1. $5m^2 - 1$ **3.** $-2x^2 + 7x - 17$ **5.** $13b^2 - 2b - 22$

7. $2x^3 + 4x^2 - x + 3$ **9.** $-a^2 + 3a + 12$ **11.** $a^2 - 8a + 10$

13. $4x^2 - 10x - 1$ **15.** $-2x$ **17.** $2y^2 - 7y + 4$

19. $17a^3 + 8a^2 - 2a$ **21.** $12x^2y^3 - 9xy^2$

23. $7x^3 - 4x^2 - 9x + 17$ **25.** $3a^3 + 4a^2 + 3a + 3$

27. $v^3 + v^2 + 8v + 13$ **29.** $4x^2y^2 - 10x^2y + 6xy + 3$

31.
$$\begin{aligned} &(5x - 3) - [(x + 2) + (2x - 6)] \\ &= 5x - 3 - [3x - 4] \\ &= 5x - 3 - 3x + 4 = 2x + 1 \end{aligned}$$

33. $3x^2 + 10$

35. $5x^2 + 4x - 7$ **37.** $-x^2 + 8x - 10$

39. $-16a^3 + 11a^2 + 34a - 26$

41.
$$\begin{aligned} &[(5 + xy^2 + x^3y) + (-6 - 3xy^2 + 4x^3y)] \\ &\quad - [(x^3y + 3xy^2 - 4) + (2x^3y - xy^2 + 5)] \\ &= [5 + xy^2 + x^3y - 6 - 3xy^2 + 4x^3y] \\ &\quad - [x^3y + 3xy^2 - 4 + 2x^3y - xy^2 + 5] \\ &= [-1 - 2xy^2 + 5x^3y] - [3x^3y + 2xy^2 + 1] \\ &= -1 - 2xy^2 + 5x^3y - 3x^3y - 2xy^2 - 1 \\ &= -2 - 4xy^2 + 2x^3y \end{aligned}$$

43. $4x^2 + 3x + 8$ **45.** $4.4x^2 + 4.1x + 11.8$

47. $-4.78x^2 - 20.85x + 2.46$

Exercises 8.6 (page 256)

1. $-15z^5 + 12z^4 - 6z^3 + 24z^2$ **3.** $6x^3y^2 - 2x^2y^3 + 8xy^4$

5. $15a^3b^2 - 30a^2b^3 + 25ab^4$ **7.** $3x^5y^4z^4 + 12x^4y^4z^6 - 6x^2y^3z^8$

9. $x^2 + x - 6$ **11.** $2y^2 - 3y - 20$

13. $(x + 4)^2 = (x + 4)(x + 4)$
$$\begin{array}{r} x + 4 \\ x + 4 \\ \hline 4x + 16 \\ x^2 + 4x \phantom{{}+16} \\ \hline x^2 + 8x + 16 \end{array}$$

15. $2x^3 + 4x^2 - 5x - 10$

17. $x^3 - x^2 - 10x + 6$ **19.** $z^3 - 8z^2 + 32z - 64$

21. $6x^3 + 17x^2 - 3x - 20$ **23.** $4x^3 + 25x^2 - 45x + 18$
25. $-2x^3 + 11x^2 - 14x + 45$ **27.** $2x^4 - 8x^3 + 3x^2 - 5x - 28$
29. $2y^4 - 17y^3 + 23y^2 - 11y + 12$
31. $x^4 + 2x^3 - x^2 + 22x - 24$ **33.** $x^4 + 4x^3 + 10x^2 + 12x + 9$
35. $x^3 + 6x^2 + 12x + 8$ **37.** $(x + y)^2(x - y)^2$

$$\begin{array}{r} (x^2 + 2xy + y^2)(x^2 - 2xy + y^2) \\ x^2 + 2xy + y^2 \\ \underline{x^2 - 2xy + y^2} \\ x^2y^2 + 2xy^3 + y^4 \\ -2x^3y - 4x^2y^2 - 2xy^3 \\ \underline{x^4 + 2x^3y + x^2y^2} \\ x^4 \qquad - 2x^2y^2 \qquad + y^4 \end{array}$$

Exercises 8.7 (page 260)

1. $x^2 + 5 + 4$ **3.** $a^2 + 7a + 10$ **5.** $m^2 - 2m - 8$
7. $72 + y - y^2$ **9.** $x^2 - 4$ **11.** $h^2 - 36$ **13.** $a^2 + 8a + 16$
15. $x^2 + 6x + 9$ **17.** $16 - 8b + b^2$ **19.** $6x^2 - 7x - 20$
21. $4x^2 - 9y^2$ **23.** $2a^2 + 7ab + 5b^2$ **25.** $8x^2 + 26xy - 7y^2$
27. $49x^2 - 140xy + 100y^2$ **29.** $(3x + 4)^2 = (3x)^2 + 2(3x)(4) + (4)^2 = 9x^2 + 24x + 16$
31. $25m^2 - 20mn + 4n^2$ **33.** $2x^4 + 5x^2y^2 + 3y^4$
35. $4m^4 - 10m^2n^2 - 6n^4$ **37.** $x^4 - y^4$ **39.** $x^4 + 2x^2y^2 + y^4$

41. $\left(\frac{2x}{3} + 1\right)\left(\frac{2x}{3} - 1\right) = \left(\frac{2x}{3}\right)^2 - (1)^2 = \frac{4x^2}{9} - 1$

43. $\left(2 - \frac{x}{3}\right)^2 = 2^2 + 2(2)\left(-\frac{x}{3}\right) + \left(-\frac{x}{3}\right)^2 = 4 - \frac{4x}{3} + \frac{x^2}{9}$

Exercises 8.8A (page 261)

1. $x + 2$ **3.** $1 + 2x$ **5.** $-3x + 4y$ **7.** $2x + 3$
9. $3a - 1$ **11.** $3z^2 - 4z + 2$ **13.** $x - \frac{4}{5} + \frac{2}{x^2}$
15. $3xyz + 6$ **17.** $x - 2y + 3$ **19.** $5x^4y^3 + 9xy + 1$

Exercises 8.8B (page 264)

1. $x + 3$ **3.** $x - 5$ **5.** $2x + 3$ **7.** $3v + 8$ R 66
9. $4a + b$ **11.** $3a - 2b$ R $7b^2$
13.

$$\begin{array}{r} x^2 - 4x + 6 \\ x + 4 \overline{)\, x^3 + 0x^2 - 10x + 24} \\ \underline{x^3 + 4x^2 \qquad\qquad} \\ -4x^2 - 10x \qquad \\ \underline{-4x^2 - 16x \qquad} \\ 6x + 24 \\ \underline{6x + 24} \\ 0 \end{array}$$

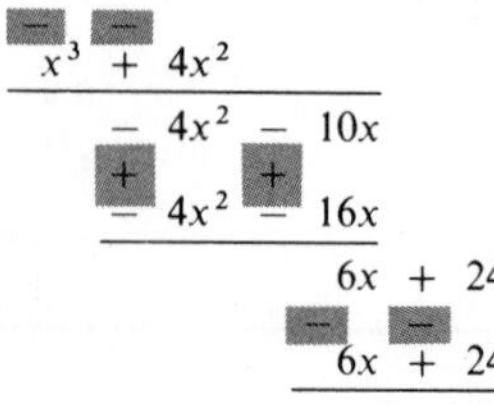

15. $4x^2 + 8x + 8$ R -6 **17.** $a^2 + 2a + 4$ **19.** $2x^2 - 3x + 4$
21. $3x^2 - 2x - 1$ **23.** $x^2 + x - 1$ R 3

Review Exercises 8.9 (page 265)

1. Three terms. Second term: $2ab$
2. Two terms. Second term: $-2(x^2 + y^2)$
3. Two terms. Second term: $-4x$
4. $3x^4 + 7x^2 - x - 6$
5. Degree of first term: 2nd; Degree of polynomial: 3rd
6. x^4 **7.** m^9 **8.** m^{18}
9. m^3 **10.** $6x^5$ **11.** $4x^6$ **12.** $8x^2y + 5xy^2$
13. $9a^3 - 3ab^2 + 5$ **14.** $9x^2y + xy^2 - 12 + 2y^2$
15. $4a^2b + 9ab^2 + 4a - 8$ **16.** $6a^2 - 11a + 2$

17.
$$\begin{aligned} &(2x + 9) - [(3x - 5) - (2 - x)] \\ &= (2x + 9) - [3x - 5 - 2 + x] \\ &= (2x + 9) - [4x - 7] \\ &= 2x + 9 - 4x + 7 \\ &= -2x + 16 \end{aligned}$$

18.
$$\begin{aligned} &(5y^2 - 6y - 4) - (3y^2 - 5y + 1) \\ &= 5y^2 - 6y - 4 - 3y^2 + 5y - 1 \\ &= 2y^2 - y - 5 \end{aligned}$$

19. $2a^3 - 5a^2 - 11a - 4$
20. $-2x^2y^2 - 5x^2y + 5xy$ **21.** $-63x^{12}y^8$ **22.** $20a^{10}b^6$
23. $6x^3 - 8x^2$ **24.** $15x^3y^3 + 20x^3y - 10x^2yz$
25. $6a^2 - 2a - 20$ **26.** $9x^2 + 24x + 16$ **27.** $x^2 - 25y^2$
28. $4m^2 - 18m + 18$ **29.** $15u^2 + 11u + 2$ **30.** $49x^2 - 16$
31. $x^4 - 12x^2 + 36$ **32.** $35x^2 - 18xy - 8y^2$
33. $2x^4 - x^2y^2 - 3y^4$ **34.** $\left(\frac{2x}{3} + 4\right)^2 = \left(\frac{2x}{3}\right)^2 + 2\left(\frac{2x}{3}\right)(4) + (4)^2 = \frac{4x^2}{9} + \frac{16x}{3} + 16$
35. $x^3 - 8$ **36.** $x - 2y$ **37.** $3ab^2 - \frac{4b}{5} + 2$

38.
$$\begin{array}{r} 2x - 1 \\ 2x + 1 \overline{)\, 4x^2 + 0x - 1} \\ \underline{4x^2 + 2x \qquad} \\ -2x - 1 \\ \underline{-2x - 1} \\ 0 \end{array}$$

39. $3x + 3$ R 25 **40.** $2a + 5b$

41.
$$\begin{array}{r} a^2 \qquad\qquad - 1 \\ 2a^2 - a + 3 \overline{)\, 2a^4 - a^3 + a^2 + a - 3} \\ \underline{2a^4 - a^3 + 3a^2 \qquad\qquad} \\ -2a^2 + a - 3 \\ \underline{-2a^2 + a - 3} \\ 0 \end{array}$$

Chapter 8 Diagnostic Test (page 267)

Following each problem number is the textbook section reference (in parentheses) in which that kind of problem is discussed.

1. (8.1) $x^2 - 4xy^2 + 5$
- **a.** The first term is 2nd degree.
- **b.** The polynomial is 3rd degree.
- **c.** The numerical coefficient of the 2nd term is -4.

2. (8.5)
$$\begin{aligned} &(-4z^2 - 5z + 10) - (10 - z + 2z^2) \\ &= -4z^2 - 5z + 10 - 10 + z - 2z^2 \\ &= -6z^2 - 4z \end{aligned}$$

3. (8.5)
$$\begin{array}{r} -5x^3 + 2x^2 \qquad - 5 \\ 7x^3 \qquad + 5x - 8 \\ 3x^2 - 6x + 10 \\ \hline 2x^3 + 5x^2 - x - 3 \end{array}$$

4. (8.5)
$$\begin{array}{r} 9a^2b - 2ab^2 - 3ab \\ -(4a^2b + 6ab^2 - 7ab) \\ \hline \end{array} \Rightarrow \begin{array}{r} 9a^2b - 2ab^2 - 3ab \\ -4a^2b - 6ab^2 + 7ab \\ \hline 5a^2b - 8ab^2 + 4ab \end{array}$$

5. (8.5) $(6x^2 - 5 - 11x) + (15x - 4x^2 + 7)$
$= 6x^2 - 5 - 11x + 15x - 4x^2 + 7$
$= 2x^2 + 4x + 2$

6. (8.5) $(3xy^2 - 4xy) - (6xy - 3xy^2) + (4xy + x^3)$
$= 3xy^2 - 4xy - 6xy + 3xy^2 + 4xy + x^3$
$= 6xy^2 - 6xy + x^3$

7. (8.5) $(2m^2 - 5) - [(7 - m^2) - (4m^2 - 3)]$
$= (2m^2 - 5) - [7 - m^2 - 4m^2 + 3]$
$= 2m^2 - 5 - 7 + m^2 + 4m^2 - 3$
$= 7m^2 - 15$

8. (8.5) $(2x - 8 - 7x^2) + (12 - 2x^2 + 11x) - (5x^2 - 9x + 6)$
$= 2x - 8 - 7x^2 + 12 - 2x^2 + 11x - 5x^2 + 9x - 6$
$= -14x^2 + 22x - 2$

9. (8.2) $10^2 \cdot 10^3 = 10^5 = 100{,}000$ **10.** (8.2) $(a^3)^5 = a^{15}$

11. (8.2) $\dfrac{12x^6}{2x^2} = 6x^4$ **12.** (8.2) $x^3 \cdot x^5 = x^8$

13. (8.2) $\dfrac{2^8}{2^5} = 2^3 = 8$

14. (8.3) $(-3xy)(5x^3y)(-2xy^4)$
Because two factors are negative, the answer is positive.
$(3 \cdot 5 \cdot 2)(xx^3x)(yyy^4) = 30x^5y^6$

15. (8.3) $4x(3x^2 - 2) = (4x)(3x^2) + (4x)(-2)$
$= 12x^3 - 8x$

16. (8.3) $-2ab(5a^2 - 3ab^2 + 4b)$
$= (-2ab)(5a^2) + (-2ab)(-3ab^2) + (-2ab)(4b)$
$= -10a^3b + 6a^2b^3 - 8ab^2$

17. (8.7) $(2x - 5)(3x + 4) = 6x^2 - 7x - 20$

18. (8.7) $(2y - 3)^2 = (2y)^2 + 2(2y)(-3) + (-3)^2$
$= 4y^2 - 12y + 9$

19. (8.7) $(m + 3)(2m + 6) = 2m^2 + 12m + 18$

20. (8.7) $(5a + 2)(5a - 2) = (5a)^2 - (2)^2$
$= 25a^2 - 4$

21. (8.7) $(4x - 2y)(3x - 5y) = 12x^2 - 26xy + 10y^2$

22. (8.6)
$$\begin{array}{r} w^2 + 3w + 8 \\ w - 3 \\ \hline -3w^2 - 9w - 24 \\ w^3 + 3w^2 + 8w \qquad \\ \hline w^3 \qquad - w - 24 \end{array}$$

23. (8.8A) $\dfrac{6x^3 - 4x^2 + 8x}{2x} = \dfrac{6x^3}{2x} + \dfrac{-4x^2}{2x} + \dfrac{8x}{2x}$
$= 3x^2 - 2x + 4$

24. (8.8B) $5x + 3 \quad R\ -2$ or $5x + 3 - \dfrac{2}{2x - 1}$
$$\begin{array}{r} 5x + 3 \\ 2x - 1 \overline{)\,10x^2 + x - 5} \\ \boxed{-}10x^2 \boxed{+} - 5x \quad \\ \hline 6x - 5 \\ \boxed{-}6x \boxed{+} - 3 \\ \hline -2 \end{array}$$

25. (8.8B)
$$\begin{array}{r} 2x^2 - 6x - 2 \\ x + 3 \overline{)\,2x^3 + 0x^2 - 20x - 6} \\ \boxed{-}2x^3 \boxed{-}+ 6x^2 \qquad\qquad \\ \hline -6x^2 - 20x \quad \\ \boxed{+}-6x^2 \boxed{+}- 18x \quad \\ \hline -2x - 6 \\ \boxed{+}-2x \boxed{+}- 6 \\ \hline 0 \end{array}$$

Exercises 9.1 (page 274)

1. Yes **3.** No **5.** Yes **7.** 3 **9.** 7 **11.** −7 **13.** 4 **15.** 3 **17.** −3 **19.** 42 **21.** 4 **23.** 23 **25.** 12 **27.** −10 **29.** 28

31.
$$\begin{array}{rcl} -28 &=& -15 + x \\ +\ 15 & & +\ 15 \\ \hline -13 &=& x \end{array}$$
Check: $-28 = -15 + x$
$-28 \stackrel{?}{=} -15 + (-13)$
$-28 = -28$

33. 45

35.
$$\begin{array}{rcl} 5.6 + x &=& 2.8 \\ -5.6 & & -5.6 \\ \hline x &=& -2.8 \end{array}$$
Check: $5.6 + x = 2.8$
$5.6 + (-2.8) \stackrel{?}{=} 2.8$
$2.8 = 2.8$

37. 7 **39.** −3 **41.** −117 **43.** 31.4 **45.** 8 **47.** $\frac{9}{10}$ **49.** $\frac{5}{12}$

Exercises 9.2 (page 276)

1. 2 **3.** 2 **5.** 13 **7.** −4 **9.** 3 **11.** 4 **13.** $6\frac{3}{4}$ **15.** $10\frac{1}{2}$ **17.** $5\frac{2}{3}$

19.
$$\begin{array}{rcl} -73 &=& 24x + 31 \\ -31 & & -31 \\ \hline -104 &=& 24x \end{array}$$
$\dfrac{-104}{24} = \dfrac{24x}{24}$
$-4\frac{1}{3} = x$

Check: $-73 = 24x + 31$
$-73 \stackrel{?}{=} 24\left(-4\frac{1}{3}\right) + 31$
$-73 \stackrel{?}{=} \overset{8}{\cancel{24}}\left(-\dfrac{13}{\underset{1}{\cancel{3}}}\right) + 31$
$-73 \stackrel{?}{=} -104 + 31$
$-73 = -73$

21. 0.6 **23.** $4\frac{1}{2}$ **25.** $-5\frac{1}{3}$

27.
$$\begin{array}{rcl} -2x + 7 &=& -13 \\ -7 & & -7 \\ \hline -2x &=& -20 \end{array}$$
$\dfrac{-2x}{-2} = \dfrac{-20}{-2}$
$x = 10$

29. −10.8

Exercises 9.3 (page 277)

1. 3 **3.** 3 **5.** -2 **7.** 7 **9.** 5 **11.** 9 **13.** 2

15. 1 **17.** 4 **19.** 3 **21.**

$$\begin{aligned} 7 - 9x - 12 &= 3x + 5 - 8x \\ -9x - 5 &= -5x + 5 \\ +9x \quad & \quad +9x \\ -5 &= 4x + 5 \\ -5 \quad & \quad -5 \\ -10 &= 4x \\ \frac{-10}{4} &= \frac{4x}{4} \\ -\frac{5}{2} &= x \quad \text{or} \quad x = -2\frac{1}{2} \end{aligned}$$

23. 2 **25.** $\frac{3}{4}$

Exercises 9.4 (page 279)

1. -3 **3.** 2 **5.** -5 **7.** 1 **9.** 4 **11.** 1

13.

$$\begin{aligned} 2(3x - 6) - 3(5x + 4) &= 5(7x - 8) \\ 6x - 12 - 15x - 12 &= 35x - 40 \\ -9x - 24 &= 35x - 40 \\ 16 &= 44x \\ \frac{16}{44} &= x \\ \frac{4}{11} &= x \end{aligned}$$

15. $-3\frac{9}{10}$ **17.** 2 **19.** -6 **21.** $7\frac{1}{5}$ **23.** 2

25.

$$\begin{aligned} 5(3 - 2x) - 10 &= 4x + [-(2x - 5) + 15] \\ 15 - 10x - 10 &= 4x + [-2x + 5 + 15] \\ 15 - 10x - 10 &= 4x - 2x + 20 \\ -15 &= 12x \\ -\frac{15}{12} &= x \\ -1\frac{1}{4} &= x \end{aligned}$$

27. -3

29.

$$10 - 6\left(\frac{1}{2}x - \frac{2}{3}\right) = 8\left(\frac{3}{4}x + \frac{5}{8}\right)$$

$$10 - \frac{\overset{3}{\cancel{6}}}{1} \cdot \frac{1}{\underset{1}{\cancel{2}}}x + \frac{\overset{2}{\cancel{6}}}{1} \cdot \frac{2}{\underset{1}{\cancel{3}}} = \frac{\overset{2}{\cancel{8}}}{1} \cdot \frac{3}{\underset{1}{\cancel{4}}}x + \frac{\overset{1}{\cancel{8}}}{1} \cdot \frac{5}{\underset{1}{\cancel{8}}}$$

$$\begin{aligned} 10 - 3x + 4 &= 6x + 5 \\ -3x + 14 &= 6x + 5 \\ +3x \quad &= +3x \\ 14 &= 9x + 5 \\ -5 \quad & \quad -5 \\ 9 &= 9x \\ 1 &= x \end{aligned}$$

Exercises 9.5 (page 281)

1. Conditional; 5 **3.** No solution **5.** Identity

7. Conditional; 2 **9.** Identity **11.** No solution

13.

$$\begin{aligned} 7(2 - 5x) - 32 &= 10x - 3(6 + 15x) \\ 14 - 35x - 32 &= 10x - 18 - 45x \\ -35x - 18 &= -35x - 18 \quad \text{Identity} \end{aligned}$$

15.

$$\begin{aligned} 2(2x - 5) - 3(4 - x) &= 7x - 20 \\ 4x - 10 - 12 + 3x &= 7x - 20 \\ -22 &\neq -20 \quad \text{No solution} \end{aligned}$$

17.

$$\begin{aligned} 2[3 - 4(5 - x)] &= 2(3x - 11) \\ 2[3 - 20 + 4x] &= 6x - 22 \\ 6 - 40 + 8x &= 6x - 22 \\ 2x &= 12 \\ x &= 6 \quad \text{Conditional} \end{aligned}$$

19. Identity

Exercises 9.6 (page 285)

1. $x < 7$

3. $x > -4$

5. $x \le -5$

7. $x \le 3$

9.

$$\begin{aligned} 6 - x &> 2 \\ -x &> -4 \\ (-1)(-x) &< (-4)(-1) \\ x &< 4 \end{aligned}$$

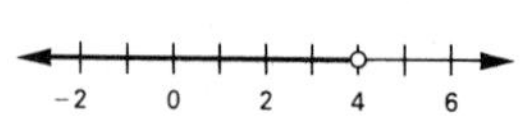

11. $x \ge -2$

13. $x < -1$

15.

$$\begin{aligned} 2x - 9 &> 3(x - 2) \\ 2x - 9 &> 3x - 6 \\ -x - 9 &> -6 \\ -x &> 3 \\ (-1)(-x) &< 3(-1) \\ x &< -3 \end{aligned}$$

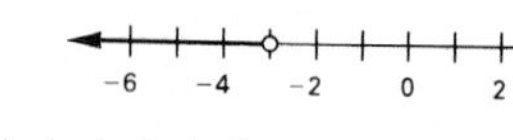

17. $x \ge 2$

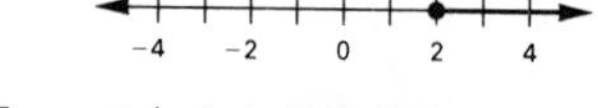

19. $x \ge -5$

21. $x \le 1$

23. $x < 3\frac{1}{8}$

25.

$$\begin{aligned} 2[3 - 5(x - 4)] &< 10 - 5x \\ 2[3 - 5x + 20] &< 10 - 5x \\ 2[23 - 5x] &< 10 - 5x \\ 46 - 10x &< 10 - 5x \\ 46 - 5x &< 10 \\ -5x &< -36 \\ \frac{-5x}{-5} &> \frac{-36}{-5} \\ x &> 7\frac{1}{5} \end{aligned}$$

27.

$$\begin{aligned} 4x - \{6 - [2x - (x + 3)]\} &\le 6 \\ 4x - \{6 - [2x - x - 3]\} &\le 6 \\ 4x - \{6 - [x - 3]\} &\le 6 \\ 4x - \{6 - x + 3\} &\le 6 \\ 4x - \{9 - x\} &\le 6 \\ 4x - 9 + x &\le 6 \\ 5x - 9 &\le 6 \\ 5x &\le 15 \\ x &\le 3 \end{aligned}$$

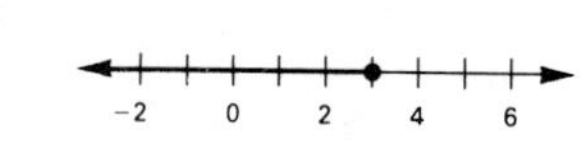

Review Exercises 9.7 (page 287)

1. Yes **2.** No **3.** 3 **4.** 3 **5.** 2 **6.** No solution

7. 75 **8.** 196 **9.** 5 **10.** Identity **11.** $\frac{5}{6}$ **12.** $-\frac{1}{2}$

13. $\frac{2}{3}$ **14.** 14 **15.** $\frac{2}{3}$ **16.** No solution **17.** -6.9

18. Identity

19.
$$\begin{aligned} 4[-24-6(3x-5)+22x] &= 0\\ 4[-24-18x+30+22x] &= 0\\ 4[6+4x] &= 0\\ 24+16x &= 0\\ 16x &= -24\\ x &= -1\tfrac{1}{2} \end{aligned}$$

20.
$$\begin{aligned} 2[-7y-3(5-4y)+10] &= 10y-12\\ 2[-7y-15+12y+10] &= 10y-12\\ 2[5y-5] &= 10y-12\\ 10y-10 &= 10y-12\\ -10 &\neq -12 \quad \text{No solution} \end{aligned}$$

21. $x < 3$

-2 0 2 4 6

22. $x \geq -6$

-8 -6 -4 -2 0

23. $x \geq -7$

-8 -6 -4 -2 0

24. $x < -2$

-4 -2 0 2 4

25.
$$\begin{aligned} 3[2-5(x-1)] &> 3-6x\\ 3[2-5x+5] &> 3-6x\\ 3[7-5x] &> 3-6x\\ 21-15x &> 3-6x\\ 21-9x &> 3\\ -9x &> -18\\ \frac{-9x}{-9} &< \frac{-18}{-9}\\ x &< 2 \end{aligned}$$

-4 -2 0 2 4

26.
$$\begin{aligned} 2(x+4)-13 &\leq 7x-5(2x-3)\\ 2x+8-13 &\leq 7x-10x+15\\ 2x-5 &\leq -3x+15\\ 5x-5 &\leq 15\\ 5x &\leq 20\\ x &\leq 4 \end{aligned}$$

-2 0 2 4 6

Chapter 9 Diagnostic Test (page 289)

Following each problem number is the textbook section reference (in parentheses) in which that kind of problem is discussed.

1. (9.1)
$$\begin{aligned} 4x-5 &= -7\\ 4(-3)-5 &\stackrel{?}{=} -7\\ -12-5 &\stackrel{?}{=} -7\\ -19 &\neq -7 \end{aligned}$$
Therefore, -3 is not a solution.

2. (9.1)
$$\begin{aligned} x-3 &= 7\\ +3 &\quad +3\\ x &= 10 \end{aligned}$$

3. (9.1)
$$\begin{aligned} 10 &= x+4\\ -4 &\quad -4\\ 6 &= x \end{aligned}$$

4. (9.2)
$$\begin{aligned} 3x+2 &= 14\\ -2 &\quad -2\\ 3x &= 12\\ \frac{3x}{3} &= \frac{12}{3}\\ x &= 4 \end{aligned}$$

5. (9.2)
$$\begin{aligned} 15-2x &= 7\\ -15 &\quad -15\\ -2x &= -8\\ \frac{-2x}{-2} &= \frac{-8}{-2}\\ x &= 4 \end{aligned}$$

6. (9.1)
$$\begin{aligned} -5 &= \frac{x}{7}\\ 7(-5) &= \frac{\cancel{7}}{1}\left(\frac{x}{\cancel{7}}\right)\\ -35 &= x \end{aligned}$$

7. (9.3)
$$\begin{aligned} 4x+5 &= 17-2x\\ 6x+5 &= 17\\ \frac{6x}{6} &= \frac{12}{6}\\ x &= 2 \end{aligned}$$

8. (9.3)
$$\begin{aligned} 6x-3 &= 5x+2\\ x-3 &= 2\\ x &= 5 \end{aligned}$$

9. (9.3)
$$\begin{aligned} 9+7x &= 5x-1\\ 9+2x &= -1\\ 2x &= -10\\ x &= -5 \end{aligned}$$

10. (9.3)
$$\begin{aligned} 2x-5 &= 8x-3\\ -6x-5 &= -3\\ \frac{-6x}{-6} &= \frac{2}{-6}\\ x &= -\frac{1}{3} \end{aligned}$$

11. (9.3)
$$\begin{aligned} 3z-21+5z &= 4-6z+17\\ 8z-21 &= 21-6z\\ 14z &= 42\\ \frac{14z}{14} &= \frac{42}{14}\\ z &= 3 \end{aligned}$$

12. (9.4)
$$\begin{aligned} 6k-3(4-5k) &= 9\\ 6k-12+15k &= 9\\ 21k-12 &= 9\\ 21k &= 21\\ \frac{21k}{21} &= \frac{21}{21}\\ k &= 1 \end{aligned}$$

13. (9.5)
$$\begin{aligned} 6x-2(3x-5) &= 10\\ 6x-6x+10 &= 10\\ 0 &= 0 \quad \text{Identity} \end{aligned}$$

14. (9.4)
$$\begin{aligned} 2m-4(3m-2) &= 5(6+m)-7\\ 2m-12m+8 &= 30+5m-7\\ -10m+8 &= 23+5m\\ -15m &= 15\\ \frac{-15m}{-15} &= \frac{15}{-15}\\ m &= -1 \end{aligned}$$

15. (9.5)
$$\begin{aligned} 2(3x+5) &= 14+3(2x-1)\\ 6x+10 &= 14+6x-3\\ 6x+10 &= 11+6x\\ 10 &\neq 11 \quad \text{No solution} \end{aligned}$$

16. (9.4)
$$\begin{aligned} 3[7-6(y-2)] &= -3+2y\\ 3[7-6y+12] &= -3+2y\\ 3[19-6y] &= -3+2y\\ 57-18y &= -3+2y\\ 60 &= 20y\\ \frac{60}{20} &= \frac{20y}{20}\\ 3 &= y \end{aligned}$$

17. (9.6)
$$\begin{array}{rcl} 5x-2 & > & 10-x \\ +\ x & & +x \\ \hline 6x-2 & > & 10 \\ +2 & & +2 \\ \hline 6x & > & 12 \\ \frac{6x}{6} & > & \frac{12}{6} \\ x & > & 2 \end{array}$$

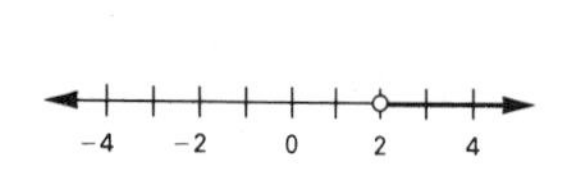

18. (9.6)
$$\begin{array}{rcl} 4x+5 & < & x-4 \\ 3x+5 & < & -4 \\ \frac{3x}{3} & < & \frac{-9}{3} \\ x & < & -3 \end{array}$$

19. (9.6)
$$\begin{array}{rcl} 2(x-4)-5 & \geq & 6+3(2x-1) \\ 2x-8-5 & \geq & 6+6x-3 \\ 2x-13 & \geq & 6x+3 \\ -6x & & -6x \\ \hline -4x-13 & \geq & 3 \\ +13 & & +13 \\ \hline -4x & \geq & 16 \\ \frac{-4x}{-4} & \leq & \frac{16}{-4} \\ x & \leq & -4 \end{array}$$

20. (9.6)
$$\begin{array}{rcl} 7x-2(5+4x) & \leq & -7 \\ 7x-10-8x & \leq & -7 \\ -10-x & \leq & -7 \\ +10 & & +10 \\ \hline -x & \leq & 3 \\ (-1)(-x) & \geq & 3(-1) \\ x & \geq & -3 \end{array}$$

Exercises 10.1A (page 293)

1. $x + 10$ **3.** $A - 5$ **5.** $6z$ **7.** $C - D$ **9.** $x + 10$

11. $x - UV$ **13.** $5x^2$ **15.** $(A + B)^2$ **17.** $\frac{x+7}{y}$

19. $\frac{A}{C+10}$ **21.** $15 + 2F$ **23.** $3x - 5y$ **25.** $x(y - 6)$

27. $3(A + 5)$ **29.** $2(6 - B)$

Exercises 10.1B (page 295)

1. Let S = Fred's salary.
Then $S + 75$ is Fred's salary plus seventy-five dollars.

3. Let N = number of children.
Then $N - 2$ is two less than the number of children.

5. Let x = Joyce's age.
Then $4x$ is four times Joyce's age.

7. Let c = cost of record.
Then $20c + 89$ is twenty times the cost of a record increased by eighty-nine cents.

9. Let c = cost of hamburger.
Then $\frac{1}{5}c$ or $\frac{c}{5}$ is one-fifth the cost of a hamburger.

11. Let x = speed of car.
Then $5x + 100$ is five times the speed of the car, plus one hundred miles per hour.

13. Let x = the unknown.
Then x^2 = the square of the unknown, and $5x^2 - 10$ is ten less than five times the square of the unknown.

15. Let x = the unknown number.
Then $\frac{8+x}{x^2}$ represents the statement.

17. Let C = Centigrade temperature.
Then $32 + \frac{9}{5}C$ represents the statement.

19. Let x = unknown.
Then $2(5 - x)$ represents the statement.

21. There are two unknowns.
First method:
Let w = width
Then $w + 12$ = length.
Second method:
Let ℓ = length
Then $\ell - 12$ = width.

23. Let x = one number
$60 - x$ = other number

25. Let B = brother's age
$B - 5$ = Henry's age

27. Let L = Linda's money
$L + 139$ = Tom's money

29. Let x = first consecutive integer
$x + 1$ = second consecutive integer
Then the product is $x(x + 1)$.

31. Let x = first odd
$x + 2$ = second odd
$x + 4$ = third odd
Then the sum is $x + (x + 2) + (x + 4)$.

33. Let w = width
$2w - 4$ = length

Exercises 10.2 (page 298)

1. a. Let x = unknown number
b. $13 + 2x = 25$
$x = 6$
c. Unknown number is 6.

3. a. Let x = unknown number
b. $5x - 8 = 22$
$x = 6$
c. Unknown number is 6.

5. a. Let x = unknown number
b. $\frac{x}{12} = 6$
$x = 72$
c. Unknown number is 72.

7. a. Let x = unknown number
b. $7 - x = x + 1$
$x = 3$
c. Unknown number is 3.

9. a. Let x = unknown number
b. $x + 7 = 2x$
$x = 7$
c. Unknown number is 7.

11. a. Let x = unknown number
b. $2(5 + x) = 26$
$x = 8$
c. Unknown number is 8.

13. a. Let x = unknown number
b. $2(4 + x) = x - 10$
$x = -18$
c. Unknown number is -18.

15. a. Let x = first integer
$x + 1$ = second integer
$x + 2$ = third integer
b. $x + (x + 1) + (x + 2) = 63$
$x = 20$
c. The integers are 20, 21, and 22.

17. a. Let x = first even
$x + 2$ = second even
$x + 4$ = third even
b. $x + (x + 2) - (x + 4) = 4$
$x = 6$
c. The integers are 6, 8, and 10.

19. a. Let x = first odd
$x + 2$ = second odd
$x + 4$ = third odd
b. $x + (x + 2) = 4(x + 4)$
$-7 = x$
c. The integers are -7, -5, and -3.

21. a. Let w = width
b. $36 = 2(12) + 2w$
$6 = w$
c. The width = 6 in.

23. a. Let w = width
$w + 3$ = length
b. $50 = 2(w + 3) + 2w$
$11 = w$
c. The width = 11 ft
and length = 14 ft

25. a. Let h = height
b. $24 = \frac{1}{2}(8)(h)$
$6 = h$
c. The height = 6 m

Exercises 10.3 (page 302)

1. a. Let x = lb of peanuts
$12 - x$ = lb of cashews
b. $4x + 8(12 - x) = 5(12)$
$x = 9$
c. 9 lb of peanuts
3 lb of cashews

3. a. Let x = lb of caramels
b. $1.25(15) + 1.50x = 1.35(x + 15)$
$x = 10$
c. 10 lb of caramels

5. a. Let x = lb of nuts
$10 - x$ = lb of raisins
b. $3.40x + 1.90(10 - x) = 25.00$
$x = 4$
c. 4 lb of nuts
6 lb of raisins

7. a. Let x = \$ per bushel of corn
Note: 100 bu mixture
− 30 bu soybeans
70 bu corn
b. $8.00(30) + x(70) = 4.85(100)$
$x = 3.5$
c. \$3.50 per bushel of corn

9. a. Let x = number of nickels
$13 - x$ = number of dimes
b. $5x + 10(13 - x) = 95$
$x = 7$
c. 7 nickels and 6 dimes

11. a. Let x = number of nickels
$12 - x$ = number of quarters
b. $5x + 25(12 - x) = 220$
$x = 4$
c. 4 nickels and 8 quarters

13. a. Let x = number of nickels
$x + 4$ = number of quarters
$3x$ = number of dimes
b. $5x + 25(x + 4) + 10(3x) = 400$
$x = 5$
c. 5 nickels, 9 quarters, and 15 dimes

Exercises 10.4 (page 304)

1. a. Let x = cc of 20% solution
b. $0.20x + 0.50(100) = 0.25(x + 100)$
$x = 500$
c. 20% solution x = 500 cc

3. a. Let x = gal of 30% solution
$100 - x$ = gal of 90% solution
b. $0.30x + 0.90(100 - x) = 0.75(100)$
$x = 25$
c. 30% solution x = 25 gal
90% solution $100 - x$ = 75 gal

5. a. Let x = mℓ of water
b. $0.40(500) + 0(x) = 0.25(500 + x)$
$x = 300$
c. water x = 300 mℓ

7. a. Let x = ounces of 10% solution
b. $0.10x + 0.25(20) = 0.16(x + 20)$
$x = 30$
c. 10% solution x = 30 oz

9. a. Let x = mℓ of 50% solution
$50 - x$ = mℓ of 25% solution
b. $0.50x + 0.25(50 - x) = 0.30(50)$
$x = 10$
c. 50% solution x = 10 mℓ
25% solution $50 - x$ = 40 mℓ

Exercises 10.5 (page 307)

1. a. Let x = speed of Malone's car
$x + 9$ = speed of King's car
$6x = 5(x + 9)$
$x = 45$
b. Malone's car x = 45 mph
King's car $x + 9$ = 54 mph
c. Distance = $6x = 6(45) = 270$ mi

3. a. Let x = time to lake
$x - 3$ = time returning
$2x = 5(x - 3)$
$x = 5$
b. Time to lake = x = 5 hr
c. Distance = $2x = 2(5) = 10$ mi

5. a. Let $x =$ Fran's speed
$x + 2 =$ Ron's speed
b. $3x + 3(x + 2) = 54$
$x = 8$
c. Fran's speed $x = 8$ mph
Ron's speed $x + 2 = 10$ mph

7. a. Let $x =$ speed in still water
Then $x + 2 =$ speed downstream
$x - 2 =$ speed upstream
b. $3(x + 2) = 5(x - 2)$
$x = 8$
c. Speed in still water $x = 8$ mph
Distance downstream $= 3(x + 2) = 3(8 + 2)$
$= 30$ miles

9. a. Let $x =$ time going
$10 - x =$ time returning
b. $80x = 120(10 - x)$
$x = 6$
c. Distance $80x = 80(6) = 480$ mi

Exercises 10.6 (page 311)

1. 28 **3.** 12 **5.** 3 **7.** -2 **9.** -125 **11.** 1,000

13. $\frac{3}{4}$ **15.** $\frac{3}{2}$ **17.** $s = kt^2$, $64 = k(2)^2$, $64 = 4k$, $16 = k$; $s = kt^2$, $s = 16(3)^2$, $s = 16 \cdot 9$, $s = 144$ ft

19. $I = \dfrac{k}{d^2}$, $15 = \dfrac{k}{(10)^2}$, $15 = \dfrac{k}{100}$, $1500 = k$; $I = \dfrac{k}{d^2}$, $I = \dfrac{1500}{(15)^2}$, $I = \dfrac{1500}{225}$, $I = \dfrac{20}{3} = 6\frac{2}{3}$ cd

21. $1\frac{1}{5}$ in.

23. 1,000 cc **25.** $17\frac{7}{9}$ lb

Review Exercises 10.7 (page 313)

1. 2 **2.** 9 **3.** -8, -7, and -6 **4.** 11 and 13

5. Length = 20 ft, Width = 15 ft **6.** Height = 10 cm **7.** 11 nickels, 16 dimes

8. Let $x =$ number of nickels.
Then $x + 4 =$ number of quarters.
And $4x =$ number of dimes.

Amount of money in nickels	+	amount in quarters	+	amount in dimes	=	240¢
$5x$	+	$25(x + 4)$	+	$10(4x)$	=	240

$5x + 25(x + 4) + 10(4x) = 240$
$5x + 25x + 100 + 40x = 240$
$70x = 140$
$x = 2$ nickels
$x + 4 = 6$ quarters
$4x = 8$ dimes

2 nickels, 6 quarters, and 8 dimes

9. 3 lb of almonds, 7 lb of walnuts **10.** 30 lb of Colombian coffee, 70 lb of Brazilian coffee

11. 50 cc of 50% solution **12.** 15 mℓ of water

13. First car's speed = 40 mph, Second car's speed = 50 mph **14.** Plane A's speed = 200 mph, Plane B's speed = 150 mph

15. $x = 4$ **16.** $y = -12$ **17.** $y = 36$ **18.** $y = \frac{9}{2}$ or $4\frac{1}{2}$

19. $R = \dfrac{k}{d^2}$, $9 = \dfrac{k}{(0.02)^2}$, $9 = \dfrac{k}{0.0004}$, $0.0036 = k$; $R = \dfrac{k}{d^2}$, $R = \dfrac{0.0036}{(0.03)^2}$, $R = \dfrac{0.0036}{0.0009}$, $R = 4$ ohms

20. $R = kn$, $\dfrac{12{,}000}{800} = \dfrac{k(800)}{800}$, $15 = k$; $R = kn$, $15{,}000 = 15n$, $1{,}000 = n$

21. 35¢ phone call uses 2 coins (1 quarter and 1 dime).
Therefore, after the phone call he had $10 - 2 = 8$ coins.
Let $x =$ number of dimes
$8 - x =$ number of quarters
$10x + 25(8 - x) = 110$
$10x + 200 - 25x = 110$
$-15x = -90$
$x = 6$ dimes

22. Let $c =$ speed of current
$20 + c =$ speed downstream
$20 - c =$ speed upstream
$6(20 - c) = 4(20 + c)$
$120 - 6c = 80 + 4c$
$40 = 10c$
$4 = c$
Speed of current = 4 mph

Chapter 10 Diagnostic Test (page 315)

Following each problem number is the textbook section reference (in parentheses) in which that kind of problem is discussed.

1. (10.2) Let $x =$ unknown number

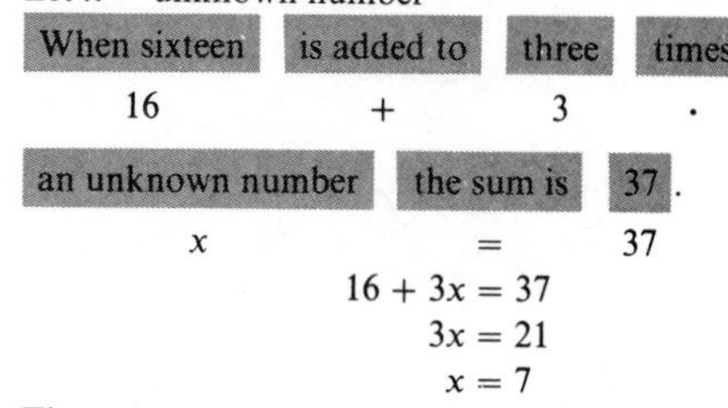

$16 + 3x = 37$
$3x = 21$
$x = 7$
The unknown number is 7.

2. (10.2) Let $x =$ first even
$x + 2 =$ second even
$x + 4 =$ third even
$x + (x + 2) + (x + 4) = 54$
$3x + 6 = 54$
$3x = 48$
$x = 16$
$x + 2 = 18$
$x + 4 = 20$
The integers are 16, 18, and 20.

3. (10.3) Let x = number of dimes
$15 - x$ = number of quarters

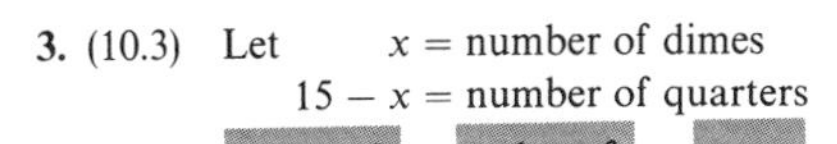

$$\begin{aligned}10x \quad + 25(15 - x) &= 300\\ 10x + 375 - 25x &= 300\\ 375 - 15x &= 300\\ -15x &= -75\\ x &= 5\\ 15 - x &= 10\end{aligned}$$

5 dimes and 10 quarters

4. (10.2) Let x = width
$x + 10$ = length

$$\begin{aligned}2(x + 10) + 2x &= 60\\ 2x + 20 + 2x &= 60\\ 20 + 4x &= 60\\ 4x &= 40\\ x &= 10\end{aligned}$$

Width = x = 10 in.
Length = $x + 10$ = 20 in.

5. (10.2) Let x = unknown number

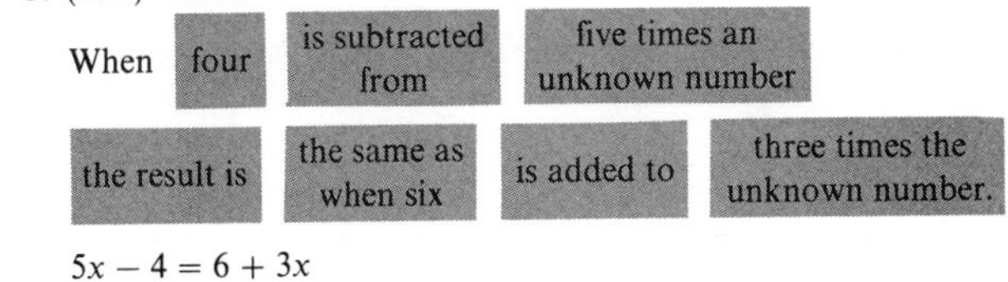

$$\begin{aligned}5x - 4 &= 6 + 3x\\ 2x &= 10\\ x &= 5\end{aligned}$$

The unknown number is 5.

6. (10.3) Let x = lb of Brand A
$10 - x$ = lb of Brand B

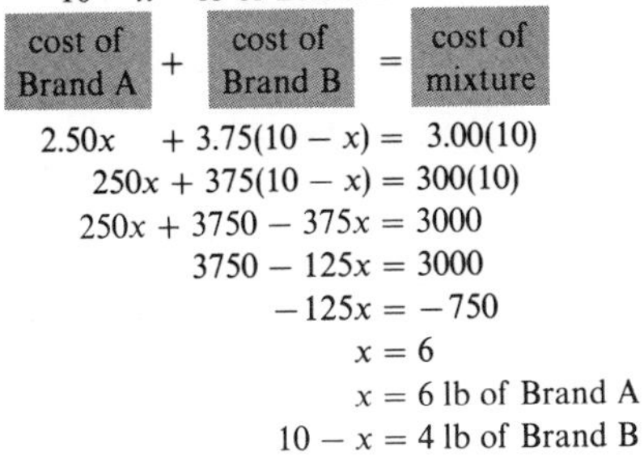

$$\begin{aligned}2.50x \quad + 3.75(10 - x) &= 3.00(10)\\ 250x + 375(10 - x) &= 300(10)\\ 250x + 3750 - 375x &= 3000\\ 3750 - 125x &= 3000\\ -125x &= -750\\ x &= 6\\ x &= 6 \text{ lb of Brand A}\\ 10 - x &= 4 \text{ lb of Brand B}\end{aligned}$$

7. (10.6)

$$y = \frac{k}{x} \qquad y = \frac{k}{x}$$

$$3 = \frac{k}{-4} \qquad y = \frac{-12}{\frac{4}{5}}$$

$$-12 = k \qquad y = \frac{-12}{1} \div \frac{4}{5}$$

$$y = \frac{-12}{1} \cdot \frac{5}{4}$$

$$y = -15$$

8. (10.4) Let x = pt of 2% solution

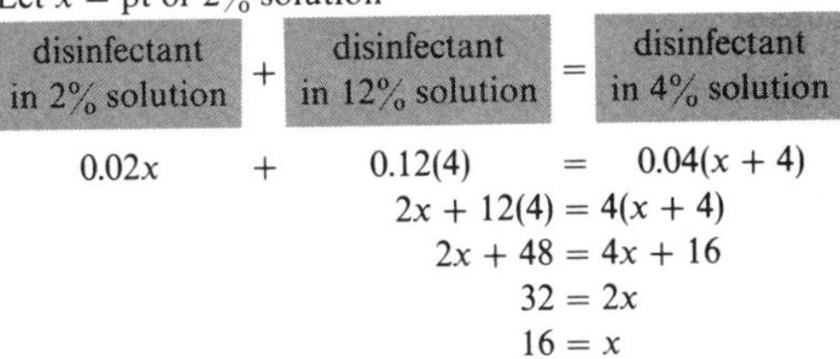

$$\begin{aligned}0.02x \quad + \quad 0.12(4) &= 0.04(x + 4)\\ 2x + 12(4) &= 4(x + 4)\\ 2x + 48 &= 4x + 16\\ 32 &= 2x\\ 16 &= x\end{aligned}$$

16 pt of the 2% solution

9. (10.5) Let x = speed going home
$x + 12$ = speed to mountains

	r ·	t =	d
To mountains	$x + 12$	4	$4(x + 12)$
Returning home	x	5	$5x$

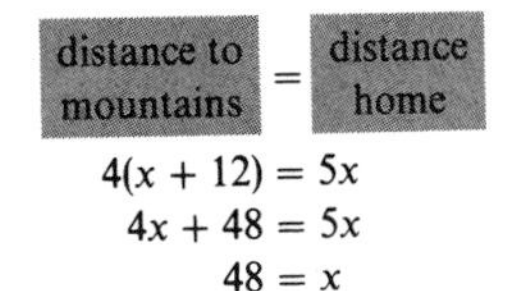

$$\begin{aligned}4(x + 12) &= 5x\\ 4x + 48 &= 5x\\ 48 &= x\end{aligned}$$

distance = $5x = 5(48) = 240$ mi

10. (10.6)

$$p = kd \qquad p = kd$$

$$\frac{17.32}{40} = \frac{k(40)}{40} \qquad p = 0.433(70)$$

$$0.433 = k \qquad p = 30.31 \text{ psi}$$

Cumulative Review Exercises for Chapters 1–10 (page 317)

1. $17\frac{1}{10}$ **2.** $3\frac{1}{3}$ **3.** 16.725 **4.** 850 **5.** -4 **6.** -7
7. $2x^2 - 2x - 7$ **8.** $4x^2 - 20x + 25$ **9.** $y^3 - 5y^2 + 10y - 8$
10. $2x - 4$ **11.** $x = -2$ **12.** $x = -10$ **13.** $m = 2$
14. $x \le -5$ **15.** $\frac{1}{12}$ **16.** 125% **17.** ≈ \$3.28
18. 15 qt **19.** Number is 8. **20.** $y = 75$ **21.** 4 oz
22. Boat A's speed = 50 mph
Boat B's speed = 30 mph **23.** 42 **24.** 66 in. **25.** 65 in.

Exercises 11.1 (page 324)

1. $\{1, 5\} \Rightarrow P$ **3.** $\{1, 13\} \Rightarrow P$ **5.** $\{1, 2, 3, 4, 6, 12\} \Rightarrow C$
7. $\{1, 3, 7, 21\} \Rightarrow C$ **9.** $\{1, 5, 11, 55\} \Rightarrow C$ **11.** $\{1, 7, 49\} \Rightarrow C$
13. $\{1, 3, 17, 51\} \Rightarrow C$ **15.** $\{1, 3, 37, 111\} \Rightarrow C$ **17.** $14 = 2 \cdot 7$
19. $21 = 3 \cdot 7$ **21.** $26 = 2 \cdot 13$ **23.** 29 is prime **25.** $32 = 2^5$
27. $34 = 2 \cdot 17$ **29.** **31.**

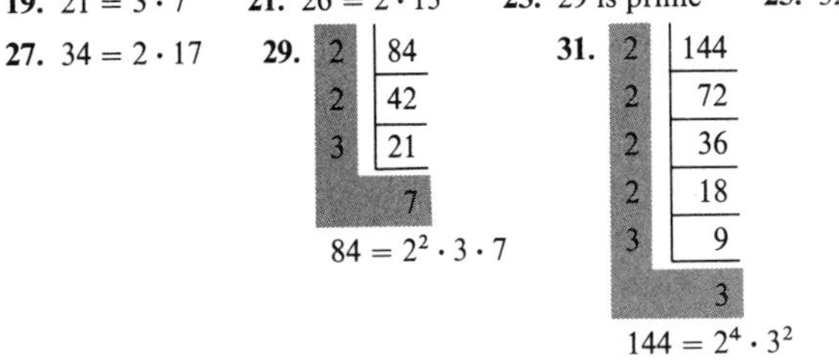

$84 = 2^2 \cdot 3 \cdot 7$

$144 = 2^4 \cdot 3^2$

33. $2(x - 4)$ **35.** $5(a + 2)$ **37.** $3(2y - 1)$ **39.** $3x(3x + 1)$
41. $5a^2(2a - 5)$ **43.** Cannot be factored **45.** $2ab(a + 2b)$
47. $6c^2d^2(2c - 3d)$ **49.** $4x(x^2 - 3 - 6x)$ **51.** $8(3a^4 + a^2 - 5)$
53. $14xy^3(-x^7y^6 + 3x^4y - 2)$ **55.** Cannot be factored
57. $6x^5y^2(3x^5y^3 + 4x^2y^2 - 2)$ **59.** $8m^3n^6(4m^2n - 3m^5n^3 - 5)$
61. $(a + b)(m + n)$ **63.** $(y + 1)(x - 1)$ **65.** $(a - 2b)(3a + 2)$

Exercises 11.2 (page 326)

1. $(a+b)(m+n)$ **3.** $(m-n)(x-y)$ **5.** $(y+1)(x-1)$
7. $(a-2b)(3a+2)$ **9.** $(3e-f)(2e-3)$ **11.** $(x+y)(a+b)$
13. Cannot be factored **15.** $10xy-15y+8x-12$
$= 5y(2x-3)+4(2x-3)$
$= (2x-3)(5y+4)$

17. x^3+3x^2-2x-6
$= x^2(x+3)-2(x+3)$
$= (x+3)(x^2-2)$

Exercises 11.3 (page 328)

1. $(x+3)(x-3)$ **3.** $(b+1)(b-1)$ **5.** $(m+n)(m-n)$
7. $(2c+5)(2c-5)$ **9.** $(5a+2b)(5a-2b)$
11. $(8a+7b)(8a-7b)$ **13.** $2x^2-8$
$= 2(x^2-4)$
$= 2(x+2)(x-2)$
15. $5x(y+a)(y-a)$
17. $3x(x+5)(x-5)$ **19.** Cannot be factored
21. $(x^3+a^2)(x^3-a^2)$ **23.** $(7u^2+6v^2)(7u^2-6v^2)$
25. $(ab+cd)(ab-cd)$ **27.** $(7+5wz)(7-5wz)$
29. $3x^4-27x^2$
$= 3x^2(x^2-9)$
$= 3x^2(x+3)(x-3)$
31. $ab(2a-3b)(2a+3b)$

Exercises 11.4 (page 331)

1. $(x+2)(x+4)$ **3.** $(x+1)(x+4)$ **5.** $(k+1)(k+6)$
7. $(u+2)(u+5)$ **9.** Cannot be factored **11.** $(b-2)(b-7)$
13. $(z-4)(z-5)$ **15.** $(x-2)(x-9)$ **17.** $(x-1)(x+10)$
19. $(z+2)(z-3)$ **21.** Cannot be factored
23. $(u^2-8)(u^2-8)=(u^2-8)^2$ **25.** $(v-4)(v-4)=(v-4)^2$
27. $(b+4d)(b-15d)$ **29.** $(r+3s)(r-16s)$ **31.** $(w-4)(w+6)$
33. $(n+2)(n-12)$ **35.** $(w+4z)(w-12z)$

Exercises 11.5 (page 335)

1. $(x+2)(3x+1)$ **3.** $(x+1)(5x+2)$ **5.** $(x+1)(4x+3)$
7. Cannot be factored **9.** $(a-3)(5a-1)$ **11.** $(b-7)(3b-1)$
13. $(z-7)(5z-1)$ **15.** $(n+5)(3n-1)$ **17.** $(k-7)(5k+1)$
19. $(x+3y)(7x+2y)$ **21.** $(h-k)(7h-4k)$
23. Cannot be factored **25.** $(v-3)(5v-2)$
27. $(2e^2-5)(3e^2+4)$ **29.** $(5k-1)(7k-1)$
31. $(7x-5)(x+1)$ **33.** $(2m-7n)(2m-n)$
35. $17u-12+5u^2$
$= 5u^2+17u-12$ Descending powers
$= (5u-3)(u+4)$

Exercises 11.6 (page 336)

1. $2(x+2y)(x-2y)$ **3.** $5(a^2+2b)(a^2-2b)$
5. $(x^2+y^2)(x+y)(x-y)$ **7.** $2(v+4)(2v-1)$
9. $4(z-2)(2z+1)$ **11.** $2(2x-1)(3x+4)$ **13.** $a(b-1)^2$
15. $3(a+5b)(a-5b)$ **17.** $(m^2+1)(m+1)(m-1)$
19. Cannot be factored **21.** $k(h-2)^2$
23. $mn^3(2m+n)(2m-n)$ **25.** $5w(z+2w)(z-w)$
27. $(x^2+9)(x+3)(x-3)$ **29.** $a^3b^2(a+2b)(a-2b)$
31. $2a(x+2ay)(x-2ay)$ **33.** $2u(u+3v)(u-2v)$
35. $8h^3-20h^2k+12hk^2$
$= 4h(2h^2-5hk+3k^2)$
$= 4h(2h-3k)(h-k)$
37. $12+4x-3x^2-x^3$
$= 4(3+x)-x^2(3+x)$
$= (3+x)(4-x^2)$
$= (3+x)(2+x)(2-x)$
39. $3(2my-3nz+5mz)$ **41.** $6(a+b)(c-d)$
43. $3(4x-1)(2x+3)$ **45.** $5a(9a-4)(a-1)$
47. $x^3-4x^2-4x+16$
$= x^2(x-4)-4(x-4)$
$= (x-4)(x^2-4)$
$= (x-4)(x+2)(x-2)$
49. $3(3a+2b)^2$
51. x^4-2x^2+1
$= (x^2-1)(x^2-1)$
$= (x+1)(x-1)(x+1)(x-1)$
$= (x+1)^2(x-1)^2$
53. $4xy^2(2x+3y)(x-y)$

Exercises 11.7 (page 340)

1. $5, -4$ **3.** $0, 4$ **5.** $\frac{3}{2}, -10$

7. $3(x+2)(x-1)=0$

$3 \neq 0$	$x+2=0$	$x-1=0$
	$-2 \quad -2$	$+1 \quad +1$
	$x = -2$	$x = 1$

9. $-1, -8$ **11.** $4, -3$ **13.** $9, -2$ **15.** $0, \frac{5}{3}$ **17.** $0, 6$

19. $3, \frac{1}{5}$ **21.** $-1, \frac{5}{3}$ **23.** $3, -6$ **25.** $0, 4$ **27.** $3, -\frac{2}{3}$

29. $0, \frac{5}{2}, -\frac{4}{3}$ **31.**
$(y-3)(y-6)=-2$
$y^2-9y+18=-2$
$y^2-9y+20=0$
$(y-4)(y-5)=0$
$y-4=0 \quad | \quad y-5=0$
$y=4 \quad | \quad y=5$
33. $1, 4$

35. $6, -2$ **37.** $0, \frac{1}{2}, -\frac{7}{3}$

39.
$2x^3+x^2=3x$
$2x^3+x^2-3x=0$
$x(2x^2+x-3)=0$
$x(1x-1)(2x+3)=0$
$x=0 \quad | \quad x-1=0 \quad | \quad 2x+3=0$
$x=1 \quad | \quad 2x=-3$
$x=-\frac{3}{2}$

41. $0, 5$ **43.**
$\frac{x-1}{x+4}=\frac{3}{x}$
$x(x-1)=3(x+4)$
$x^2-x=3x+12$
$x^2-4x-12=0$
$(x-6)(x+2)=0$
$x-6=0 \quad | \quad x+2=0$
$x=6 \quad | \quad x=-2$

45. $0, 3$

Exercises 11.8 (page 342)

1. a. Let x = smaller number
$x+5$ = larger number
b. $x(x+5)=14$
$x=2$ or $x=-7$
c. Two answers: $\{2, 7\}$ and $\{-7, -2\}$

3. a. Let x = one number
$12 - x$ = other number
b. $x(12 - x) = 35$
$x = 5$ or $x = 7$
c. Numbers are 5 and 7.

5. a. Let x = first consecutive odd
$x + 2$ = second consecutive odd
b. $x(x + 2) = 63$
$x = 7$ or $x = -9$
c. Two answers: $\{7, 9\}$ and $\{-9, -7\}$

7. a. Let x = first consecutive
$x + 1$ = second consecutive
b. $x(x + 1) = 11 + x + x + 1$
$x = 4$ or $x = -3$
c. Two answers: $\{4, 5\}$ and $\{-3, -2\}$

9. a. Let x = first consecutive
$x + 1$ = second consecutive
$x + 2$ = third consecutive
b. $x(x + 1) = 7 + (x + 2)$
$x = 3$ or $x = -3$
c. Two answers: $\{3, 4, 5\}$ and $\{-3, -2, -1\}$

11. a. Let x = first consecutive
$x + 1$ = second consecutive
b. $x^2 + (x + 1)^2 = 25$
$x = -4$ or $x = 3$
c. Two answers: $\{-4, -3\}$ and $\{3, 4\}$

13. a. Let x = one number
$3x$ = other number
b. $x^2 + (3x)^2 = 40$
$x = 2$ or $x = -2$
c. Two answers: $\{2, 6\}$ and $\{-2, -6\}$

15. a. Let w = width
$w + 5$ = length
b. $w(w + 5) = 84$
$w = 7$ or $w \neq -12$
c. Width = 7 ft
Length = 12 ft

17. a. Let x = height
then $x + 3$ = base

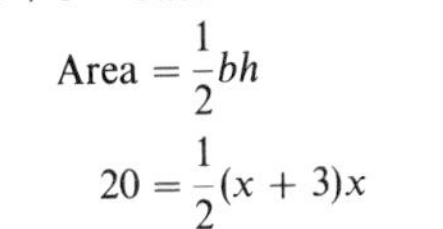

$$\text{Area} = \frac{1}{2}bh$$

b.
$$20 = \frac{1}{2}(x + 3)x$$

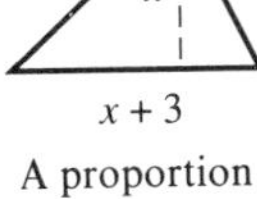

A proportion

$$\frac{20}{1} = \frac{(x + 3)x}{2}$$
$$1 \cdot (x + 3)x = 20(2)$$
$$x^2 + 3x = 40$$
$$x^2 + 3x - 40 = 0$$
$$(x + 8)(x - 5) = 0$$

$x = -8$	$x = 5$ height
No meaning	$x + 3 = 8$ base

c. Height = 5 in.
Base = 8 in.

19. a. Let ℓ = length
$\ell - 4$ = width
b. $\ell(\ell - 4) = 17 + 2\ell + 2(\ell - 4)$
$\ell = 9$ or $\ell \neq -1$
c. Length = 9 yd
Width = 5 yd

21. a. Let x = side of large square
$x - 3$ = side of small square

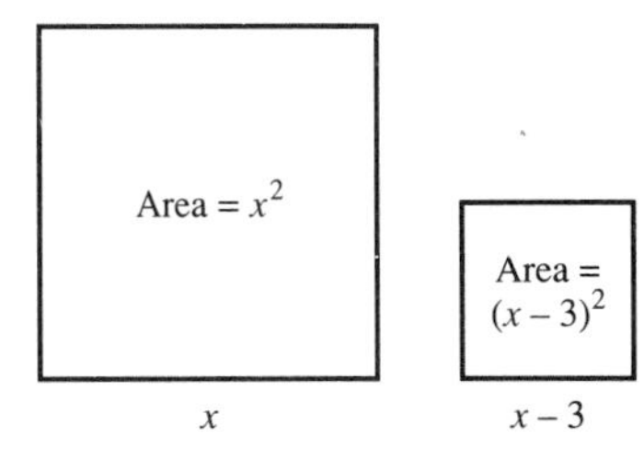

b.

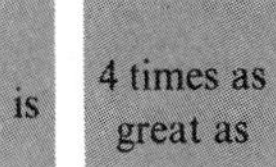

The area of the larger square	is	4 times as great as	the area of the smaller square.
x^2	$=$	$4 \cdot$	$(x - 3)^2$

$$x^2 = 4(x - 3)^2$$
$$x^2 = 4(x^2 - 6x + 9)$$
$$x^2 = 4x^2 - 24x + 36$$
$$0 = 3x^2 - 24x + 36$$
$$0 = 3(x^2 - 8x + 12)$$
$$0 = 3(x - 2)(x - 6)$$

$x = 2$	$x = 6$ large sq.
$x - 3 = -1$	$x - 3 = 3$ small sq.
Not possible	

c. Side of large square is 6 cm
Side of small square is 3 cm

23. a. Let x = first term
$x + 7$ = fourth term
b. $\frac{x}{3} = \frac{6}{x + 7}$
$x = 2$ or $x = -9$
c. The proportions are: $\frac{2}{3} = \frac{6}{9}$ or $\frac{-9}{3} = \frac{6}{-2}$

Review Exercises 11.9 (page 343)

1. $12 = 2^2 \cdot 3$ **2.** $36 = 2^2 \cdot 3^2$ **3.** 31 is prime

4. $42 = 2 \cdot 3 \cdot 7$ **5.**

3	210
5	105
3	21
	7

$210 = 2 \cdot 3 \cdot 5 \cdot 7$

6. $4(2x - 1)$

7. $(m + 2)(m - 2)$ **8.** $(x + 3)(x + 7)$ **9.** $2u(u + 2)$

10. $(z + 2)(z - 9)$ **11.** $(4x - 1)(x - 6)$ **12.** $9k^2 - 144$
$= 9(k^2 - 16)$
$= 9(k - 4)(k + 4)$

13. $8 - 2a^2$
$= 2(4 - a^2)$
$= 2(2 - a)(2 + a)$

14. $ab + 2b - a - 2$
$= b(a + 2) - 1(a + 2)$
$= (a + 2)(b - 1)$

15. $(y - 4)(x + 3)$ **16.** $(3x - 2)(x + 2)$ **17.** $3uv(5u - 1)$

18. $(a^2 + b^2)(a + b)(a - b)$ **19.** $(x^4 + 1)(x^2 + 1)(x + 1)(x - 1)$

20. $x^4 - 5x^2 + 4$
$= (x^2 - 4)(x^2 - 1)$
$= (x + 2)(x - 2)(x + 1)(x - 1)$

21. $4xy(x - 2y + 1)$

22. $15a^2 + 15ab - 30b^2$
$= 15(a^2 + ab - 2b^2)$
$= 15(a + 2b)(a - b)$

23. $3uv(2u^2v - 3v^2 - 4)$

24. 5, −3 **25.** 7, −2 **26.** 6, −3 **27.** 3, −3

28. 0, 4 **29.** $\frac{5}{2}, -\frac{1}{3}$ **30.** 0, −6, 2

31.
$$6x^3 = 9x^2 + 6x$$
$$6x^3 - 9x^2 - 6x = 0$$
$$3x(2x^2 - 3x - 2) = 0$$
$$3x(2x + 1)(x - 2) = 0$$

$3x = 0$	$2x + 1 = 0$	$x - 2 = 0$
$x = 0$	$2x = -1$	$x = 2$
	$x = -\frac{1}{2}$	

32.
$$\frac{x-2}{9} = \frac{4}{x+7}$$
$$(x - 2)(x + 7) = 36$$
$$x^2 + 5x - 14 = 36$$
$$x^2 + 5x - 50 = 0$$
$$(x + 10)(x - 5) = 0$$

$x + 10 = 0$	$x - 5 = 0$
$x = -10$	$x = 5$

33. Let x = smaller number
$x + 3$ = large number
$x(x + 3) = 28$
$x = 4$ or $x = -7$
There are two answers:
$\{4, 7\}$ and $\{-7, -4\}$

34. Let x = first consecutive
$x + 1$ = second consecutive
$x + 2$ = third consecutive
$$x(x + 1) - (x + 2) = 7$$
$$x^2 + x - x - 2 = 7$$
$$x^2 - 2 = 7$$
$$x^2 - 9 = 0$$
$$(x + 3)(x - 3) = 0$$

$x + 3 = 0$	$x - 3 = 0$
$x = -3$	$x = 3$
$x + 1 = -2$	$x + 1 = 4$
$x + 2 = -1$	$x + 2 = 5$

There are two answers:
$\{-3, -2, -1\}$ and $\{3, 4, 5\}$

35. Let ℓ = length
$\ell - 3$ = width
$\ell(\ell - 3) = 40$
$\ell = 8$ or $\ell \neq -5$
Length $= \ell = 8$
Width $= \ell - 3 = 5$

36. Let x = side of small square
$x + 6$ = side of large square
$$(x + 6)^2 = 16x^2$$
$$x^2 + 12x + 36 = 16x^2$$
$$0 = 15x^2 - 12x - 36$$
$$0 = 3(5x^2 - 4x - 12)$$
$$0 = 3(5x + 6)(x - 2)$$

$3 \neq 0$	$5x + 6 = 0$	$x - 2 = 0$
	$5x = -6$	$x = 2$
	$x = \frac{-6}{5}$	$x + 6 = 8$
	Not possible	

The side of the small square is 2, and the side of the large square is 8.

37. Let x = side of square
$x^2 = 3(4x)$
$x = 0$ or $x = 12$
The side of square is 12.

38. Let w = width
$w + 6$ = length
$w(w + 6) = 12 + 2w + 2(w + 6)$
$w = 4$ or $w \neq -6$
Width $= w = 4$ yd
Length $= w + 6 = 10$ yd

Chapter 11 Diagnostic Test (page 345)

Following each problem number is the textbook section reference (in parentheses) in which that kind of problem is discussed.

1. (11.1) **a.** $18 = 2 \cdot 9 = 3 \cdot 6$
18 is composite because it has factors other than itself and 1.
b. $21 = 3 \cdot 7 \Rightarrow$ composite
c. $31 = 1 \cdot 31 \Rightarrow$ prime
31 has no factors other than itself and 1.

2. (11.1)

5	45
3	9
	3

$45 = 3^2 \cdot 5$

3. (11.1)

2	160
2	80
2	40
2	20
2	10
	5

$160 = 2^5 \cdot 5$

4. (11.1) $5x + 10 = 5(x + 2)$ **5.** (11.1) $3x^2 - 6x = 3x(x - 2)$

6. (11.3) $16x^2 - 49y^2 = (4x + 7y)(4x - 7y)$

7. (11.6) $2z^2 - 8 = 2(z^2 - 4) = 2(z + 2)(z - 2)$

8. (11.4) $z^2 + 6z + 8 = (z + 2)(z + 4)$

9. (11.4) $m^2 + m - 6 = (m + 3)(m - 2)$

10. (11.5) $5w^2 - 12w + 7 = (5w - 7)(w - 1)$

11. (11.5) $3v^2 + 14v - 5 = (3v - 1)(v + 5)$

12. (11.2) $5n - mn - 5 + m = n(5 - m) - 1(5 - m) = (5 - m)(n - 1)$

13. (11.6) $6h^2k - 8hk^2 + 2k^3 = 2k(3h^2 - 4hk + k^2) = 2k(3h - k)(h - k)$

14. (11.6) $x^4 - 8x^2 - 9 = (x^2 + 1)(x^2 - 9) = (x^2 + 1)(x + 3)(x - 3)$

15. (11.6) $24a^2 + 2a - 12 = 2(12a^2 + a - 6) = 2(3a - 2)(4a + 3)$

16. (11.7)
$$x^2 - 12x + 20 = 0$$
$$(x - 2)(x - 10) = 0$$

$x - 2 = 0$	$x - 10 = 0$
$x = 2$	$x = 10$

17. (11.7)
$$3x^2 = 12x$$
$$3x^2 - 12x = 0$$
$$3x(x - 4) = 0$$

$3x = 0$	$x - 4 = 0$
$x = 0$	$x = 4$

18. (11.7)
$$36y = 18y^2 + 18$$
$$0 = 18y^2 - 36y + 18$$
$$0 = 18(y^2 - 2y + 1)$$
$$0 = 18(y - 1)(y - 1)$$

$18 \neq 0$	$y - 1 = 0$	$y - 1 = 0$
	$y = 1$	$y = 1$

19. (11.8) Let $w = \text{width}$

$w + 3 = \text{length}$

$w(w + 3) = 28$

$w^2 + 3w = 28$

$w^2 + 3w - 28 = 0$

$(w + 7)(w - 4) = 0$

$w + 7 = 0$	$w - 4 = 0$
$w = -7$	$w = 4$
Not possible	$w + 3 = 7$

Width $= 4$ ft

Length $= 7$ ft

20. (11.8) Let $x = \text{first odd}$

$x + 2 = \text{second odd}$

$x(x + 2) = 7 + x + x + 2$

$x^2 + 2x = 2x + 9$

$x^2 - 9 = 0$

$(x + 3)(x - 3) = 0$

$x + 3 = 0$	$x - 3 = 0$
$x = -3$	$x = 3$
$x + 2 = -1$	$x + 2 = 5$

Two answers: $\{-3, -1\}$ and $\{3, 5\}$

Exercises 12.1 (page 351)

1. 2 **3.** None **5.** 1 and -2 **7.** None **9.** 2 and -1

11. 0 and -3 **13.** -5 **15.** y **17.** -7

19. -5, the sign of the numerator was changed, $y - 6 = -(6 - y)$. Therefore, the sign of 5 must be changed to -5.

21. $\frac{3}{4}$ **23.** $2b$ **25.** $2x$ **27.** 5 **29.** -1

31. $\frac{x+6}{2(x-4)}$ **33.** $\frac{2+4}{4} = \frac{6}{4} = \frac{3}{2}$ **35.** Cannot be reduced

37. $\frac{y}{2x+y}$ **39.** $x - 1$ **41.** $\frac{3x-12}{4-x} = \frac{3\overset{-1}{\cancel{(x-4)}}}{\underset{1}{\cancel{4-x}}} = -3$

43. $\frac{x-6}{x-4}$ **45.** $\frac{x+4}{x+3}$ **47.** $\frac{3x-2}{5x-1}$ **49.** $\frac{x-y}{x+y}$

51. $\frac{(a-2b)\overset{-1}{\cancel{(b-a)}}}{(2a+b)\underset{1}{\cancel{(a-b)}}} = \frac{-1(a-2b)}{2a+b} = \frac{2b-a}{2a+b}$

53. $\frac{2y^2+xy-6x^2}{3x^2+xy-2y^2} = \frac{\overset{-1}{\cancel{(2y-3x)}}(y+2x)}{\underset{1}{\cancel{(3x-2y)}}(x+y)}$

$= \frac{-(y+2x)}{x+y}$ or $-\frac{y+2x}{x+y}$

55. $\frac{8x^2-2y^2}{2ax-ay+2bx-by} = \frac{2(4x^2-y^2)}{a(2x-y)+b(2x-y)}$

$= \frac{2(2x+y)\cancel{(2x-y)}}{\cancel{(2x-y)}(a+b)}$

$= \frac{2(2x+y)}{a+b}$

Exercises 12.2 (page 354)

1. $\frac{1}{2}$ **3.** $\frac{a}{b}$ **5.** $\frac{3x}{2}$ **7.** $\frac{5}{2}$ **9.** $\frac{x}{y}$ **11.** $\frac{5}{x^2}$

13. $\frac{b}{2}$ **15.** $\frac{a}{a-2}$ **17.** 1 **19.** $\frac{5}{z-4}$ **21.** $-\frac{3}{2}$

23. 1 **25.** $\frac{4a}{3b}$ **27.** $\frac{3u^2}{2}$ **29.** $\frac{1}{4}$ **31.** $\frac{x^2}{x+3}$ **33.** -1

35. $\frac{x-y}{9x+9y} \div \frac{x^2-y^2}{3x^2+6xy+3y^2}$

$= \frac{\cancel{(x-y)}}{\underset{3}{\cancel{9}}\cancel{(x+y)}} \cdot \frac{\cancel{3}\cancel{(x+y)}\cancel{(x+y)}}{\cancel{(x+y)}\cancel{(x-y)}} = \frac{1}{3}$

37. $\frac{2x^3y+2x^2y^2}{6x} \div \frac{x^2y^2-xy^3}{y-x}$

$= \frac{\cancel{2}x^2y(x+y)}{\underset{3}{\cancel{6}}x} \cdot \frac{\overset{-1}{\cancel{y-x}}}{xy^2\cancel{(x-y)}} = \frac{-(x+y)}{3y}$ or $-\frac{x+y}{3y}$

39. $\frac{x^2-8x-9}{x^3-3x^2-x+3} \cdot \frac{x^2-9}{x^2-9x}$

$= \frac{(x-9)(x+1)}{x^2(x-3)-1(x-3)} \cdot \frac{(x+3)(x-3)}{x(x-9)}$

$= \frac{(x-9)(x+1)}{(x-3)(x^2-1)} \cdot \frac{(x+3)(x-3)}{x(x-9)}$

$= \frac{\cancel{(x-9)}\cancel{(x+1)}}{\cancel{(x-3)}\cancel{(x+1)}(x-1)} \cdot \frac{(x+3)\cancel{(x-3)}}{x\cancel{(x-9)}}$

$= \frac{x+3}{x(x-1)}$

Exercises 12.3 (page 357)

1. $\frac{9}{a}$ **3.** $\frac{2}{a}$ **5.** $\frac{4}{x}$ **7.** $\frac{4}{x-y}$ **9.** 2 **11.** 1 **13.** 1

15. 3 **17.** -2 **19.** 3 **21.** $\frac{-8}{y-2}$ or $\frac{8}{2-y}$

23. $\frac{a+2}{2a+1} - \frac{1-a}{2a+1} = \frac{a+2-1+a}{2a+1}$

$= \frac{2a+1}{2a+1} = 1$

25. -1

27. 3 **29.** $\frac{7z}{8z-4} + \frac{6-5z}{4-8z}$

$= \frac{7z}{8z-4} + \frac{5z-6}{8z-4}$

$= \frac{7z+5z-6}{8z-4} = \frac{12z-6}{8z-4} = \frac{\overset{3}{\cancel{6}}\cancel{(2z-1)}}{\underset{2}{\cancel{4}}\cancel{(2z-1)}}$

$= \frac{3}{2}$

Exercises 12.4 (page 359)

1. 10 **3.** $2y$ **5.** 4 **7.** $6x$ **9.** x^3 **11.** $12y^2$

13. $36u^3v^3$ **15.** $a(a + 3)$

17. (1) $2(x + 2)$, 2^2x (denominators in factored form)

(2) $2, x, (x + 2)$ are the different factors.

(3) $2^2, x, (x + 2)$ are the highest powers of each factor.

(4) LCD $= 4x(x + 2)$

19. $(x + 2)^2$

21. (1) $(x + 1)(x + 2), (x + 1)^2$

(2) $(x + 1), (x + 2)$

(3) $(x + 1)^2, (x + 2)$

(4) LCD $= (x + 1)^2(x + 2)$

23. $(x + y)^2(x - y)$

25. (1) $2^2 \cdot 3 \cdot x^2(x+2), (x-2)^2, (x+2)(x-2)$
(2) $2, 3, x, (x+2), (x-2)$
(3) $2^2, 3, x^2, (x+2), (x-2)^2$
(4) $\text{LCD} = 12x^2(x+2)(x-2)^2$

Exercises 12.5 (page 363)

1. $\frac{3a+2}{a^3}$ **3.** $\frac{3x+4}{x^2}$ **5.** $\frac{x^2+6x-10}{2x^2}$ **7.** $\frac{2-3x}{xy}$

9. $\frac{5-3y}{xy}$ **11.** $\frac{3y+4}{y}$ **13.** $\frac{20x+21y}{24x^2y^2}$

15. First reduce $\frac{9}{6y}$ to $\frac{3}{2y}$. **17.** $\frac{3x^2-3}{x}$

LCD $= 4y^2$

$$\frac{5}{4y^2} + \frac{3}{2y} \cdot \frac{2y}{2y} = \frac{5+6y}{4y^2}$$

19. $\frac{a^2+3a+9}{a(a+3)}$ **21.** $\frac{2x^2-5x-10}{4x(x+2)}$

23. LCD $= x(x-2)$

$$\frac{x}{1} + \frac{2}{x} - \frac{3}{x-2}$$
$$= \frac{x}{1} \cdot \frac{x(x-2)}{x(x-2)} + \frac{2}{x} \cdot \frac{(x-2)}{(x-2)} - \frac{3}{x-2} \cdot \frac{x}{x}$$
$$= \frac{x^2(x-2)}{x(x-2)} + \frac{2(x-2)}{x(x-2)} - \frac{3x}{x(x-2)}$$
$$= \frac{x^3 - 2x^2 + 2x - 4 - 3x}{x(x-2)} = \frac{x^3 - 2x^2 - x - 4}{x(x-2)}$$

25. $\frac{a^2}{b(a-b)}$ **27.** $\frac{5}{6(e-1)}$ **29.** $\frac{8}{m-2}$ **31.** $\frac{4x}{x^2-1}$

33. $\frac{x^2+2x+4}{x^2+2x}$ **35.** $\frac{a^3-2a^2+a-6}{a^2-2a}$ **37.** $\frac{a^2}{ab+b^2}$

39. $\frac{3}{(x+1)(x+2)}$

41. LCD $= xy(x-y)$

$$\frac{y}{x(x-y)} \cdot \frac{y}{y} + \frac{-x}{y(x-y)} \cdot \frac{x}{x}$$
$$= \frac{y^2-x^2}{xy(x-y)} = \frac{(y+x)\overset{-1}{\cancel{(y-x)}}}{xy\underset{1}{\cancel{(x-y)}}} = \frac{-(y+x)}{xy} \text{ or } -\frac{y+x}{xy}$$

43. LCD $= (x-3)(x+3)$

$$\frac{2x}{x-3} \cdot \frac{x+3}{x+3} + \frac{-2x}{x+3} \cdot \frac{x-3}{x-3} + \frac{36}{(x+3)(x-3)}$$
$$= \frac{2x^2 + 6x - 2x^2 + 6x + 36}{(x-3)(x+3)} = \frac{12x+36}{(x-3)(x+3)}$$
$$= \frac{12\cancel{(x+3)}}{(x-3)\cancel{(x+3)}} = \frac{12}{x-3}$$

45. $\frac{2}{x-1}$ **47.** $\frac{2x+2}{(x+2)^2}$ **49.** $\frac{x}{x+4}$

51. $\frac{3}{(x-1)(x+2)}$ **53.** $\frac{a}{5}$

55.
$$\frac{3x}{(x-5)(x+4)} \cdot \frac{x-2}{x-2} - \frac{5}{(x-5)(x-2)} \cdot \frac{x+4}{x+4}$$
$$= \frac{3x^2-6x}{(x-5)(x+4)(x-2)} - \frac{5x+20}{(x-5)(x-2)(x+4)}$$
$$= \frac{3x^2 - 6x - (5x+20)}{(x-5)(x+4)(x-2)} = \frac{3x^2 - 6x - 5x - 20}{(x-5)(x+4)(x-2)}$$
$$= \frac{3x^2 - 11x - 20}{(x-5)(x+4)(x-2)} = \frac{(3x+4)\cancel{(x-5)}}{\cancel{(x-5)}(x+4)(x-2)}$$
$$= \frac{3x+4}{(x+4)(x-2)}$$

57. LCD $= (x+1)^2(x+2)$

$$\frac{x-4}{x^2+3x+2} + \frac{3x+1}{x^2+2x+1}$$
$$= \frac{x-4}{(x+1)(x+2)} + \frac{3x+1}{(x+1)^2}$$
$$= \frac{(x-4)}{(x+1)(x+2)} \cdot \frac{x+1}{x+1} + \frac{3x+1}{(x+1)^2} \cdot \frac{x+2}{x+2}$$
$$= \frac{(x-4)(x+1)}{(x+1)^2(x+2)} + \frac{(3x+1)(x+2)}{(x+1)^2(x+2)}$$
$$= \frac{x^2 - 3x - 4 + 3x^2 + 7x + 2}{(x+1)^2(x+2)} = \frac{4x^2+4x-2}{(x+1)^2(x+2)}$$

59. $\frac{2x(2x-y)}{4x^2+y^2}$

61. LCD $= (x+2)(x-2)$

$$\frac{2x}{x+2} \cdot \frac{x-2}{x-2} + \frac{x}{x-2} \cdot \frac{x+2}{x+2} - \frac{3x+2}{(x+2)(x-2)}$$
$$= \frac{2x^2-4x}{(x+2)(x-2)} + \frac{x^2+2x}{(x-2)(x+2)} - \frac{3x+2}{(x+2)(x-2)}$$
$$= \frac{2x^2 - 4x + (x^2+2x) - (3x+2)}{(x+2)(x-2)}$$
$$= \frac{2x^2 - 4x + x^2 + 2x - 3x - 2}{(x+2)(x-2)} = \frac{3x^2-5x-2}{(x+2)(x-2)}$$
$$= \frac{(3x+1)\cancel{(x-2)}}{(x+2)\cancel{(x-2)}} = \frac{3x+1}{x+2}$$

63. $\frac{2}{x-1}$

Exercises 12.6 (page 366)

1. $\frac{2}{7}$ **3.** $2\frac{2}{25}$ **5.** $\frac{3x^2}{2y^3}$

7. LCD $= 15a^3b^2$

$$\frac{15a^3b^2}{15a^3b^2} \cdot \frac{\frac{18cd^2}{5a^3b}}{\frac{12cd^2}{15ab^2}} = \frac{\frac{15a^3b^2}{1} \cdot \frac{18cd^2}{5a^3b}}{\frac{15a^3b^2}{1} \cdot \frac{12cd^2}{15ab^2}} = \frac{\overset{9}{\cancel{54}}bcd^2}{\underset{2}{\cancel{12}}a^2cd^2}$$
$$= \frac{9b}{2a^2}$$

9. ab **11.** 1 **13.** $\frac{a+b}{a-b}$ **15.** $\frac{d}{c-2d}$

17. The LCD of the secondary fractions is x.

$$\frac{x}{x} \cdot \frac{1}{1-\frac{1}{x}} = \frac{x(1)}{x(1) - \frac{x}{1}\left(\frac{1}{x}\right)} = \frac{x}{x-1}$$

19. $\frac{xy+1}{xy}$ **21.** $\frac{1+x^2}{1-x^2}$ **23.** $\frac{1}{3}$

25. LCD $= a^2$

$$\frac{a^2}{a^2} \cdot \frac{1 - \frac{1}{a^2}}{\frac{1}{a} - \frac{1}{a^2}} = \frac{\frac{a^2}{1} \cdot 1 - \frac{a^2}{1} \cdot \frac{1}{a^2}}{\frac{a^2}{1} \cdot \frac{1}{a} - \frac{a^2}{1} \cdot \frac{1}{a^2}} = \frac{a^2 - 1}{a - 1}$$

$$= \frac{(a+1)\cancel{(a-1)}}{\cancel{a-1}} = a + 1$$

27. $\frac{x - 2y}{y}$

29. LCD $= a^2b^2$

$$\frac{a^2b^2}{a^2b^2} \cdot \frac{\frac{1}{a^2} - \frac{1}{b^2}}{\frac{1}{a} + \frac{1}{b}} = \frac{\frac{a^2b^2}{1} \cdot \frac{1}{a^2} - \frac{a^2b^2}{1} \cdot \frac{1}{b^2}}{\frac{a^2b^2}{1} \cdot \frac{1}{a} + \frac{a^2b^2}{1} \cdot \frac{1}{b}}$$

$$= \frac{b^2 - a^2}{ab^2 + a^2b} = \frac{\cancel{(b+a)}(b-a)}{ab\cancel{(b+a)}}$$

$$= \frac{b-a}{ab}$$

Exercises 12.7 (page 370)

1. 12 **3.** 20 **5.** 10 **7.** $\frac{3}{2}$ **9.** $-4, 4$ **11.** 2

13. $\frac{9}{8}$ **15.** 3 **17.** 5 **19.** 1 **21.** $3, -3$ **23.** $0, -2$

25. $0, -1$ **27.** -2 **29.** 2 **31.** $\frac{14}{17}$

33. LCD $= x - 2$

$$\left(\frac{\cancel{x-2}}{1}\right)\frac{x}{\cancel{x-2}} = \left(\frac{\cancel{x-2}}{1}\right)\frac{2}{\cancel{x-2}} + \left(\frac{x-2}{1}\right)(5)$$

$$x = 2 + 5x - 10$$
$$8 = 4x$$
$$x = 2 \quad \text{(Not a solution)}$$

Therefore, this equation has no solution.

35. LCD $= 5(x+2)(x-2)$

$$5(x+2)\cancel{(x-2)}\frac{1}{\cancel{x-2}} - 5\cancel{(x+2)}(x-2)\frac{4}{\cancel{x+2}} = \cancel{5}(x+2)(x-2)\frac{1}{\cancel{5}}$$

$$5(x+2) - 20(x-2) = (x+2)(x-2)$$
$$5x + 10 - 20x + 40 = x^2 - 4$$
$$0 = x^2 + 15x - 54$$
$$0 = (x+18)(x-3)$$

$x + 18 = 0$ | $x - 3 = 0$
$x = -18$ | $x = 3$

37. LCD $= 3(x+1)(x-1)$

$$3(x+1)\cancel{(x-1)}\frac{1}{\cancel{x-1}} + 3\cancel{(x+1)}(x-1)\frac{2}{\cancel{x+1}} = \cancel{3}(x+1)(x-1)\frac{5}{\cancel{3}}$$

$$3(x+1) + 6(x-1) = 5(x^2 - 1)$$
$$3x + 3 + 6x - 6 = 5x^2 - 5$$
$$0 = 5x^2 - 9x - 2$$
$$0 = (5x+1)(x-2)$$

$5x + 1 = 0$ | $x - 2 = 0$
$x = -\frac{1}{5}$ | $x = 2$

39. LCD $= 4(2x+5)$

$$\frac{4\cancel{(2x+5)}}{1} \cdot \frac{3}{\cancel{2x+5}} + \frac{\cancel{4}(2x+5)}{1} \cdot \frac{x}{\cancel{4}} = \frac{\cancel{4}(2x+5)}{1} \cdot \frac{3}{\cancel{4}}$$

$$4(3) + x(2x+5) = 3(2x+5)$$
$$12 + 2x^2 + 5x = 6x + 15$$
$$2x^2 - x - 3 = 0$$
$$(1x+1)(2x-3) = 0$$

$x + 1 = 0$ | $2x - 3 = 0$
$x = -1$ | $2x = 3$
| $x = \frac{3}{2}$

Exercises 12.8 (page 373)

1. $x = \frac{4 - y}{2}$ **3.** $z = y + 8$ **5.** $y = 2x + 4$ **7.** $x = \frac{3y + 6}{2}$

9. $x = 5 - 2y$ **11.** $x = \frac{y - 4}{2}$ **13.** $x = 6y + 4$

15. $V = \frac{k}{P}$ **17.** $x = \frac{10z}{3y}$ **19.** $x = \frac{y - b}{m}$ **21.** $r = \frac{A - P}{Pt}$

23.
$$\frac{S}{1} = \frac{a}{1 - r} \quad \text{This is a proportion}$$
$$S(1 - r) = a \quad \text{Product of means = product of extremes}$$
$$S - Sr = a$$
$$-Sr = a - S$$
$$\frac{-Sr}{-S} = \frac{a - S}{-S}$$
$$r = \frac{S - a}{S}$$

25. $x = \frac{c}{a + b}$ **27.** $R = \frac{zr}{r - z}$

29. LCD $= Fuv$

$$\frac{\cancel{F}uv}{1} \cdot \frac{1}{\cancel{F}} = \frac{F\cancel{u}v}{1} \cdot \frac{1}{\cancel{u}} + \frac{Fu\cancel{v}}{1} \cdot \frac{1}{\cancel{v}}$$

$$uv = Fv + Fu$$
$$uv - Fu = Fv$$
$$u(v - F) = Fv$$
$$\frac{u(v - F)}{(v - F)} = \frac{Fv}{(v - F)}$$
$$u = \frac{Fv}{v - F}$$

31. $F = \frac{9C + 160}{5}$

33. Multiply by LCD $= x$

$$\frac{\cancel{x}}{1}\left(\frac{m + n}{\cancel{x}}\right) + \frac{x}{1}\left(\frac{-a}{1}\right) = \frac{\cancel{x}}{1}\left(\frac{m - n}{\cancel{x}}\right) + \frac{x}{1}\left(\frac{c}{1}\right)$$

$$m + n - ax = m - n + cx$$
$$2n = ax + cx \quad \text{Collecting terms}$$
$$2n = x(a + c) \quad \text{Factoring } x \text{ from } ax + cx$$
$$\frac{2n}{a + c} = \frac{x\cancel{(a + c)}}{\cancel{a + c}} \quad \text{Dividing both sides by } a + c$$
$$\frac{2n}{a + c} = x$$

Exercises 12.9 (page 378)

1. a. Let x = the number

b. $x + \frac{1}{x} = \frac{25}{12}$

$x = \frac{3}{4}$ or $x = \frac{4}{3}$

c. Two answers: $\frac{3}{4}$ and $\frac{4}{3}$

3. a. Let x = the number

b. $x - \frac{1}{x} = \frac{8}{3}$

$x = -\frac{1}{3}$ or $x = 3$

c. Two answers: $-\frac{1}{3}$ and 3

5. a. Let x = numerator

$x + 2$ = denominator

b. $\frac{x - 1}{(x + 2) + 5} = \frac{1}{5}$

$x = 3$

c. Fraction is $\frac{3}{5}$.

7. a. Let x = numerator

$2x$ = denominator

b. $\frac{x + 2}{2x + 8} = \frac{2}{5}$

$x = 6$

c. Fraction is $\frac{6}{12}$.

9. a. Let x = time each works

b. $\frac{x}{2} + \frac{x}{6} = 1$

$x = \frac{3}{2}$

c. $1\frac{1}{2}$ hr

11. a. Let x = time all pipes are on

b. $\frac{x}{3} + \frac{x}{2} + \frac{x}{6} = 1$

$x = 1$

c. 1 hr

13. a. Let x = time each works

b. $\frac{x}{30} + \frac{x}{50} = 4$

$x = 75$

c. 75 min, or 1 hr 15 min

15. a. Let x = time to sort mail

b. $\frac{1}{4} + \frac{x}{3} = 1$

$x = \frac{9}{4}$

c. $2\frac{1}{4}$ hr

17. a. Let x = Joe's speed

$\frac{4}{5}x$ = Ron's speed

b. $3x + 3\left(\frac{4x}{5}\right) = 54$

$5 \cdot 3x + \cancel{5} \cdot \frac{12x}{\cancel{5}} = 54 \cdot 5$

$15x + 12x = 270$

$27x = 270$

$x = 10$

c. Joe's speed = x = 10 mph

Ron's speed $= \frac{4}{5}x = \frac{4(10)}{5} = 8$ mph

19. a. Let x = score on fourth exam

b. $\frac{70 + 85 + 83 + x}{4} = 80$

$\frac{238 + x}{4} = 80$

$238 + x = 320$

$x = 82$

c. Must score at least 82

Review Exercises 12.10 (page 381)

1. -4 **2.** 0

3. $\frac{x - 1}{x^2 - 3x - 10} = \frac{x - 1}{(x - 5)(x + 2)}$

5 and -2 must be excluded because either makes the denominator zero.

4. 5 **5.** $x - 2$ **6.** 2 **7.** $\frac{2x^2}{y}$ **8.** $1 + 2m$ **9.** $a - 2$

10. $\frac{1}{x - 4}$ **11.** $\frac{y - x}{y + x}$

12. $\frac{a - b}{ax + ay - bx - by}$

$= \frac{a - b}{a(x + y) - b(x + y)}$

$= \frac{\cancel{a - b}}{(x + y)\cancel{(a - b)}} = \frac{1}{x + y}$

13. $\frac{5}{z}$ **14.** $\frac{10x - 3}{2x}$ **15.** 1 **16.** $-\frac{3ab}{2}$ **17.** $\frac{x^2}{4}$

18. LCD = 15

$\frac{3}{3} \cdot \frac{x + 4}{5} - \frac{5}{5} \cdot \frac{x - 2}{3}$

$= \frac{3x + 12 - (5x - 10)}{15} = \frac{3x + 12 - 5x + 10}{15} = \frac{-2x + 22}{15}$

19. LCD = $(a - 1)(a + 2)$

$\frac{a + 2}{a + 2} \cdot \frac{a - 2}{a - 1} + \frac{a - 1}{a - 1} \cdot \frac{a + 1}{a + 2}$

$= \frac{a^2 - 4 + a^2 - 1}{(a + 2)(a - 1)}$

$= \frac{2a^2 - 5}{(a + 2)(a - 1)}$

20. $8(x - 3) = 8x - 24$ **21.** 3 **22.** $\frac{x^2 - y^2}{x^2y^2}$

23. LCD = $(x + 4)(x + 1)(x + 2)$

$= \frac{3}{(x + 4)(x + 1)} \cdot \frac{x + 2}{x + 2} - \frac{2}{(x + 4)(x + 2)} \cdot \frac{x + 1}{x + 1}$

$= \frac{3x + 6}{(x + 4)(x + 1)(x + 2)} - \frac{2x + 2}{(x + 4)(x + 2)(x + 1)}$

$= \frac{3x + 6 - (2x + 2)}{(x + 4)(x + 1)(x + 2)} = \frac{3x + 6 - 2x - 2}{(x + 4)(x + 1)(x + 2)}$

$= \frac{\overset{1}{\cancel{x + 4}}}{\underset{1}{\cancel{(x + 4)}}(x + 1)(x + 2)} = \frac{1}{(x + 1)(x + 2)}$

24. $\frac{3k}{2m}$ **25.** $\frac{x + 3y}{x - 3y}$ **26.** y **27.** $\frac{6}{5}$ **28.** -3 **29.** 40

30. $\frac{14}{11}$

31. LCD = $2z$

$(2z)\frac{4}{2z} + (2z)\frac{2}{z} = (2z)1$

$4 + 4 = 2z$

$8 = 2z$

$z = 4$

32. LCD = $4x^2$

$\frac{4x^2}{1}\left(\frac{3}{x}\right) - \frac{4x^2}{1}\left(\frac{8}{x^2}\right) = \frac{4x^2}{1}\left(\frac{1}{4}\right)$

$12x - 32 = x^2$

$0 = x^2 - 12x + 32$

$0 = (x - 4)(x - 8)$

$x - 4 = 0 \quad | \quad x - 8 = 0$

$x = 4 \quad | \quad x = 8$

33. $3x - 4y = 12$

$3x = 4y + 12$

$x = \frac{4y + 12}{3}$

34. LCD = n

$(n)\frac{2m}{n} = (n)P$

$2m = nP$

$\frac{2m}{P} = n$ or $n = \frac{2m}{P}$

35. $H = \dfrac{V}{LW}$ **36.** $m = \dfrac{Egr}{v^2}$

37. $\dfrac{F - 32}{C} = \dfrac{9}{5}$ This is a proportion

$5(F - 32) = 9C$ Product of means = product of extremes

$5F - 160 = 9C$

$\dfrac{5F - 160}{9} = C$

38. $x = \dfrac{b}{c - a}$ **39.** $\dfrac{29}{49}$ **40.** Two answers: $\dfrac{1}{2}$ and 3

41. 200 min = 3 hr 20 min **42.** Debbie's speed = 8 mph
Wendy's speed = 6 mph

Chapter 12 Diagnostic Test (page 383)

Following each problem number is the textbook section reference (in parentheses) in which that kind of problem is discussed.

1. (12.1) **a.** $\dfrac{3x}{x - 4}$; x cannot be 4 because that would make the denominator zero

(12.1) **b.** $\dfrac{5x + 4}{x^2 + 2x} = \dfrac{5x + 4}{x(x + 2)}$; x cannot be 0 or -2 because either value makes the denominator zero

2. (12.1) **a.** $-\dfrac{-4}{5} = +\dfrac{+4}{5}$

b. $\dfrac{-3}{x - y} = \dfrac{+3}{-(x - y)} = \dfrac{+3}{y - x}$

3. (12.1) $\dfrac{\overset{2}{\cancel{6}}x^3y}{\underset{3}{\cancel{9}}x^2y^2} = \dfrac{2x}{3y}$

4. (12.1) $\dfrac{x^2 + 8x + 16}{x^2 - 16} = \dfrac{\cancel{(x + 4)}(x + 4)}{\cancel{(x + 4)}(x - 4)} = \dfrac{x + 4}{x - 4}$

5. (12.1) $\dfrac{6a^2 + 11ab - 10b^2}{6a^2b - 4ab^2} = \dfrac{(2a + 5b)\cancel{(3a - 2b)}}{2ab\cancel{(3a - 2b)}}$

$= \dfrac{2a + 5b}{2ab}$

6. (12.2) $\dfrac{a}{a + 2} \cdot \dfrac{4a + 8}{6a^2} = \dfrac{a}{\cancel{(a + 2)}} \cdot \dfrac{\overset{2}{\cancel{4}}\cancel{(a + 2)}}{\underset{3}{\cancel{6}}a^2} = \dfrac{2}{3a}$

7. (12.3) $\dfrac{-15}{4y - 5} - \dfrac{12y}{5 - 4y} = \dfrac{15}{5 - 4y} - \dfrac{12y}{5 - 4y}$

$= \dfrac{15 - 12y}{5 - 4y}$

$= \dfrac{3\cancel{(5 - 4y)}}{\cancel{5 - 4y}} = 3$

8. (15.5) $4 + \dfrac{2}{x} = \dfrac{4}{1} + \dfrac{2}{x} = \dfrac{4}{1} \cdot \boxed{\dfrac{x}{x}} + \dfrac{2}{x} = \dfrac{4x}{x} + \dfrac{2}{x} = \dfrac{4x + 2}{x}$

9. (12.5) LCD $= b(b - 1)$;

$\dfrac{b + 1}{b} - \dfrac{b}{b - 1} = \dfrac{b + 1}{b} \cdot \boxed{\dfrac{b - 1}{b - 1}} - \dfrac{b}{b - 1} \cdot \boxed{\dfrac{b}{b}}$

$= \dfrac{(b + 1)(b - 1)}{b(b - 1)} - \dfrac{b^2}{b(b - 1)}$

$= \dfrac{b^2 - 1}{b(b - 1)} - \dfrac{b^2}{b(b - 1)}$

$= \dfrac{b^2 - 1 - b^2}{b(b - 1)}$

$= \dfrac{-1}{b(b - 1)}$ or $\dfrac{1}{b(1 - b)}$

10. (12.2) $\dfrac{2x}{x^2 - 9} \div \dfrac{4x^2}{x - 3} = \dfrac{\cancel{2}x}{(x + 3)\cancel{(x - 3)}} \cdot \dfrac{\cancel{(x - 3)}}{\underset{2}{\cancel{4}}x^2}$

$= \dfrac{1}{2x(x + 3)}$

11. (12.5) LCD $= (x + 3)(x + 1)(x + 2)$

$\dfrac{2}{(x + 3)(x + 1)} \cdot \boxed{\dfrac{x + 2}{x + 2}} - \dfrac{1}{(x + 3)(x + 2)} \cdot \boxed{\dfrac{x + 1}{x + 1}}$

$= \dfrac{2x + 4}{(x + 3)(x + 1)(x + 2)} - \dfrac{x + 1}{(x + 3)(x + 2)(x + 1)}$

$= \dfrac{2x + 4 - (x + 1)}{(x + 3)(x + 1)(x + 2)} = \dfrac{2x + 4 - x - 1}{(x + 3)(x + 1)(x + 2)}$

$= \dfrac{\overset{1}{\cancel{x + 3}}}{\underset{1}{\cancel{(x + 3)}}(x + 1)(x + 2)} = \dfrac{1}{(x + 1)(x + 2)}$

12. (12.6) $\dfrac{\dfrac{9x^5}{10y}}{\dfrac{3x^2}{20y^3}} = \dfrac{9x^5}{10y} \div \dfrac{3x^2}{20y^3} = \dfrac{\overset{3}{\cancel{9}}x^5}{\cancel{10}y} \cdot \dfrac{\overset{2}{\cancel{20}}y^3}{\cancel{3}x^2} = 6x^3y^2$ (Method 2)

13. (12.6) LCD $= a^2$

$\boxed{\dfrac{a^2}{a^2}} \cdot \dfrac{1 + \dfrac{2}{a}}{1 - \dfrac{4}{a^2}} = \dfrac{\boxed{\dfrac{a^2}{1}} \cdot 1 + \boxed{\dfrac{a^2}{1}} \cdot \dfrac{2}{a}}{\boxed{\dfrac{a^2}{1}} \cdot 1 - \boxed{\dfrac{a^2}{1}} \cdot \dfrac{4}{a^2}}$

$= \dfrac{a^2 + 2a}{a^2 - 4} = \dfrac{a\cancel{(a + 2)}}{\cancel{(a + 2)}(a - 2)}$

$= \dfrac{a}{a - 2}$

14. (12.7) $\dfrac{y}{3} - \dfrac{y}{4} = 1$; LCD $= 3 \cdot 4 = 12$

$\boxed{\dfrac{\overset{4}{\cancel{12}}}{1}} \cdot \left(\dfrac{y}{\cancel{3}}\right) - \boxed{\dfrac{\overset{3}{\cancel{12}}}{1}} \cdot \left(\dfrac{y}{\cancel{4}}\right) = \boxed{12}(1)$

$4y - 3y = 12$

$y = 12$

15. (12.7) $\dfrac{x - 2}{5} = \dfrac{x + 1}{2} + \dfrac{3}{5}$; LCD $= 2 \cdot 5 = 10$

$\boxed{\dfrac{\overset{2}{\cancel{10}}}{1}} \cdot \dfrac{x - 2}{\cancel{5}} = \boxed{\dfrac{\overset{5}{\cancel{10}}}{1}} \cdot \dfrac{x + 1}{\cancel{2}} + \boxed{\dfrac{\overset{2}{\cancel{10}}}{1}} \cdot \dfrac{3}{\cancel{5}}$

$2(x - 2) = 5(x + 1) + 2 \cdot 3$

$2x - 4 = 5x + 5 + 6$

$-3x = 15$

$x = -5$

16. (12.7) $\dfrac{3}{a + 4} = \dfrac{5}{a}$ This is a proportion

$5a + 20 = 3a$ Product of means = product of extremes

$2a = -20$

$a = -10$

17. (12.7) $\dfrac{2x - 5}{x} = \dfrac{x - 2}{3}$ This is a proportion

$6x - 15 = x^2 - 2x$ Product of means = product of extremes

$0 = x^2 - 8x + 15$

$0 = (x - 3)(x - 5)$

$x = 3 \quad x = 5$

18. (12.8)
$$3x - 4y = 9$$
$$-4y = 9 - 3x$$
$$y = \frac{9 - 3x}{-4}$$
$$y = \frac{3x - 9}{4}$$

19. (12.8)
$$PM + Q = PN$$
$$PM - PN = -Q$$
$$P(M - N) = -Q$$
$$\frac{P(\cancel{M - N})}{\cancel{M - N}} = \frac{-Q}{M - N}$$
$$P = \frac{Q}{N - M}$$

20. (12.9) Let x = time each works

	Rate ·	*Time* =	*Amount of work*
Sid	$\frac{1}{20}$	x	$\frac{x}{20}$
George	$\frac{1}{30}$	x	$\frac{x}{30}$

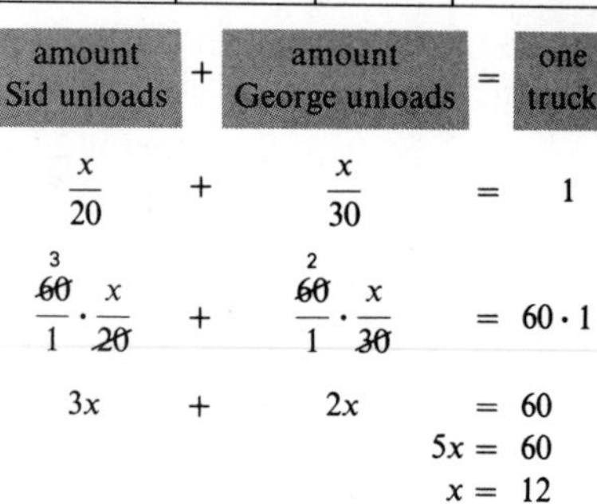

$$\frac{x}{20} + \frac{x}{30} = 1$$
$$\frac{\overset{3}{\cancel{60}}}{1} \cdot \frac{x}{\cancel{20}} + \frac{\overset{2}{\cancel{60}}}{1} \cdot \frac{x}{\cancel{30}} = 60 \cdot 1$$
$$3x + 2x = 60$$
$$5x = 60$$
$$x = 12$$

It will take 12 min.

Exercises 13.1 (page 388)

1.

3. a. (3, 0)
b. (0, 5)
c. (−5, 2)
d. (−4, −3)

5. a. 5 because A is 5 units to the right.
b. −3 because C is 3 units to the left.
c. 3 because E is 3 units to the right.
d. 1 because F is 1 unit to the right.

7. a. 1 **b.** 2 **9.** Origin **11.** Abscissa, horizontal coordinate

13.

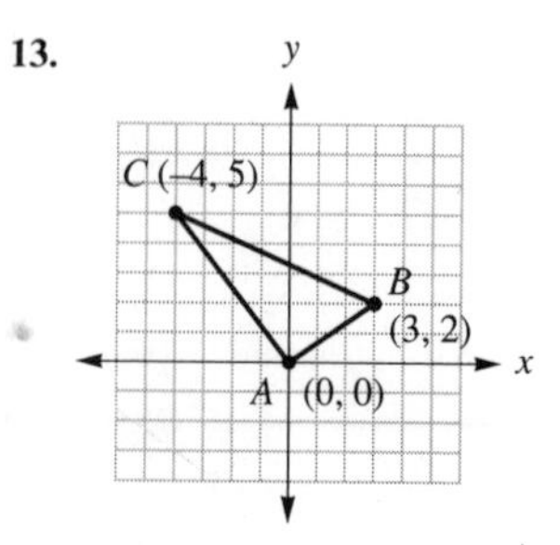

Exercises 13.2 (page 394)

1.

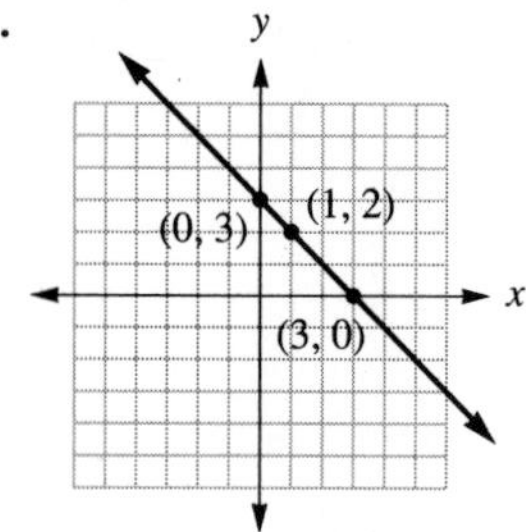

3.

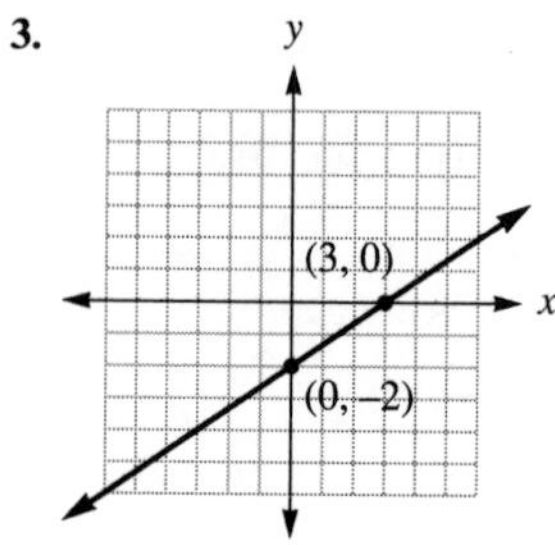

5. x can be any number but y is always 8.

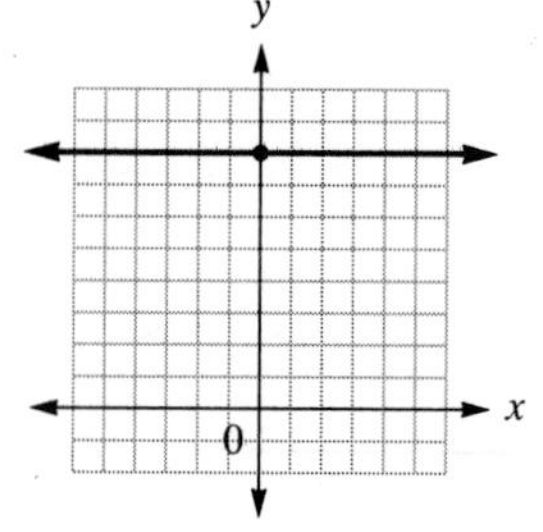

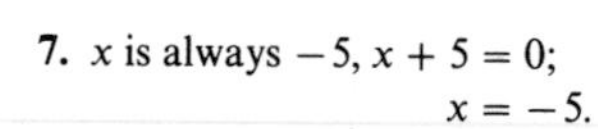
7. x is always -5, $x + 5 = 0$; $x = -5$.

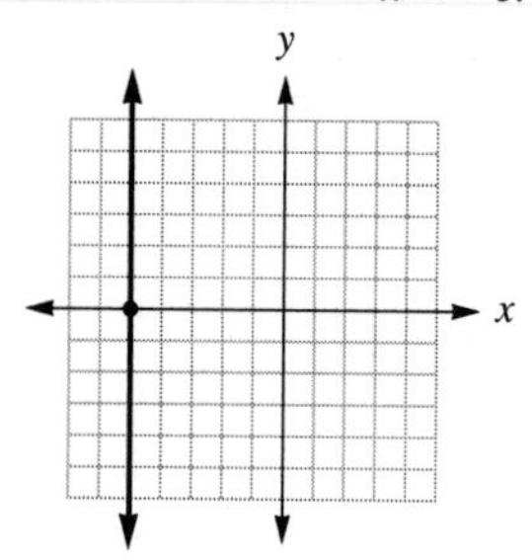

9.

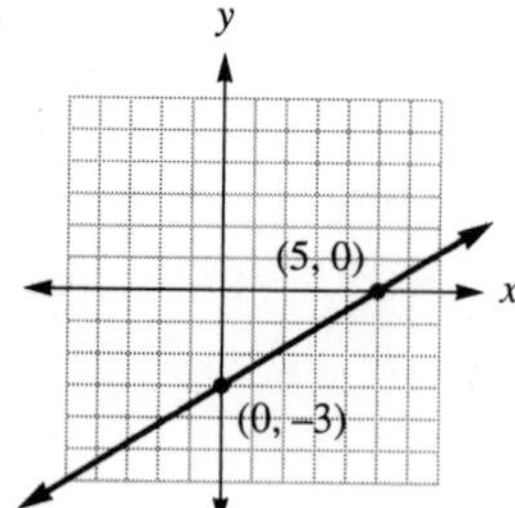

11.

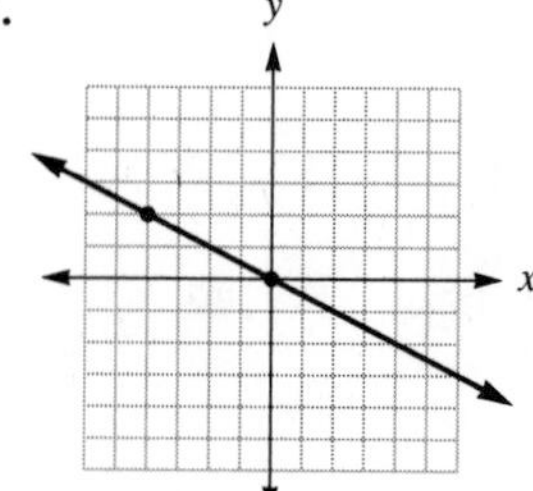

13.

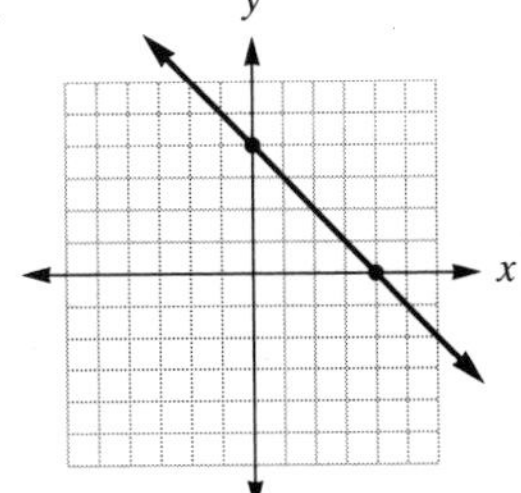

15.

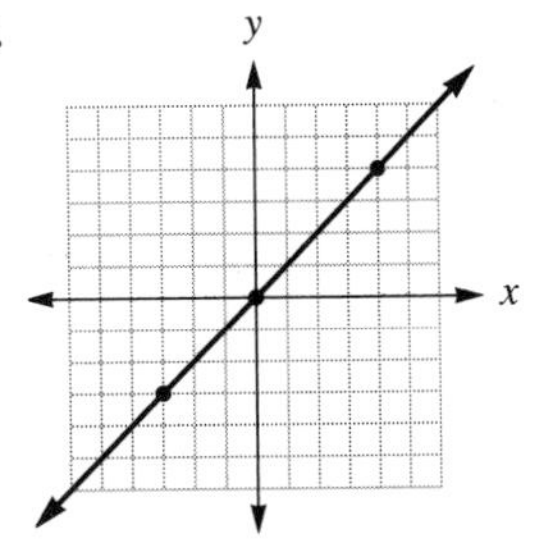

17.

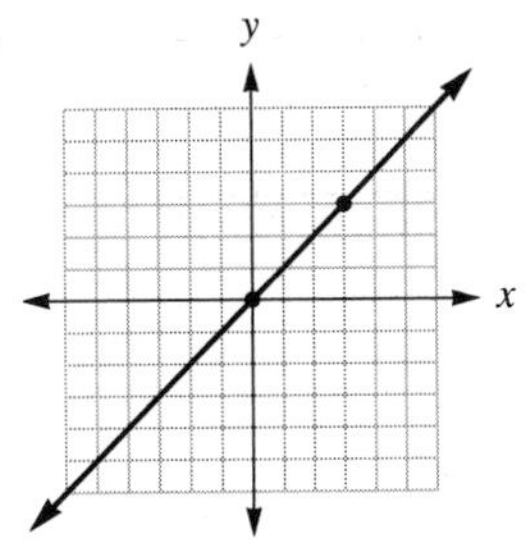

19.

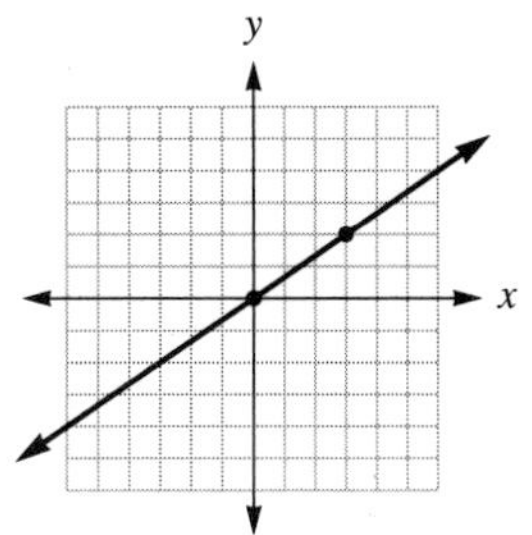

21.

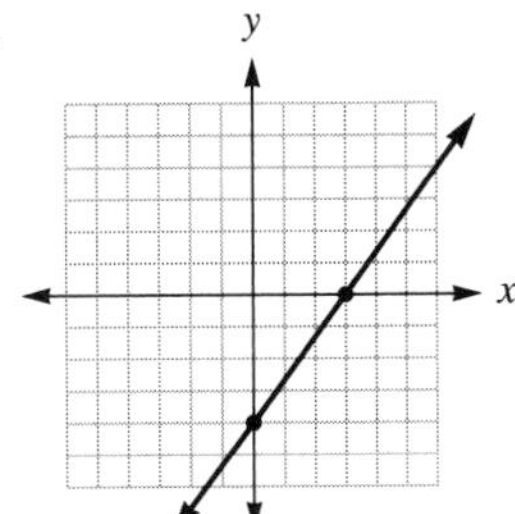

23.

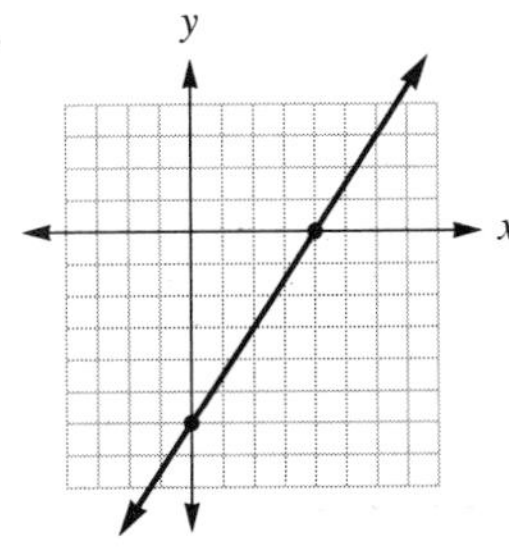

25.

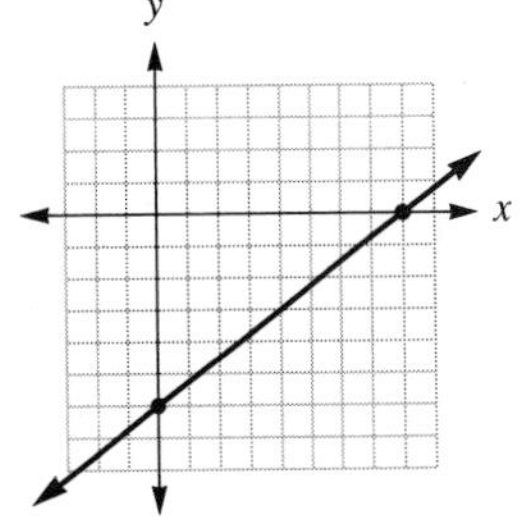

27.

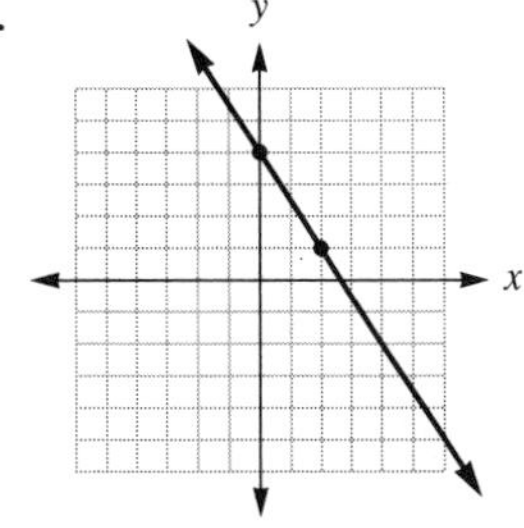

29. a. $x - y = 5$

x	y
0	-5
5	0

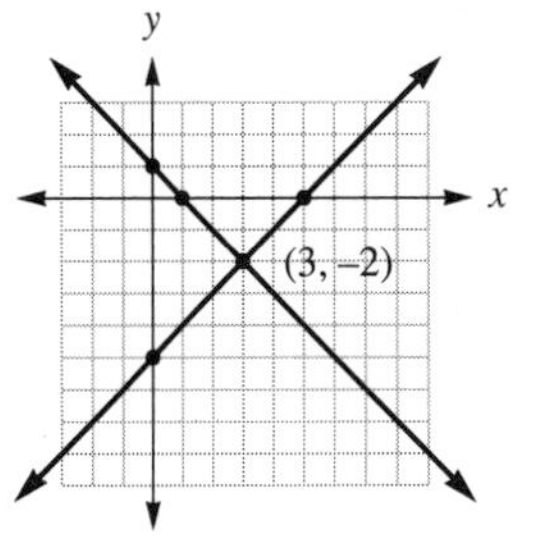

b. $x + y = 1$

x	y
0	1
1	0

Exercises 13.3 (page 400)

1. $-\frac{17}{11}$ **3.** $-\frac{1}{3}$ **5.** 0 **7.** $-\frac{8}{15}$ **9.** 1

11. No slope **13.** $-\frac{2}{3}$ **15.** $\frac{2}{9}$ **17.** $-\frac{5}{3}$

19. If $x = 0$: $2x + 3y = 6$

$2(0) + 3y = 6$

$y = 2$

Therefore, $P_1(0, 2)$

If $y = 0$: $2x - 3y = 6$

$2x - 3(0) = 6$

$x = 3$

Therefore, $P_2(3, 0)$

$$m = \frac{y_2 - y_1}{x_2 - x_1} = \frac{0 - 2}{3 - 0} = \frac{-2}{3} = -\frac{2}{3}$$

21. $-\frac{4}{5}$

23. Slope $= \frac{2}{3}$
y-intercept $= 4$

25. Slope $= -1$
y-intercept $= 0$

27. $3x + 4y = 12$
$$4y = -3x + 12$$
$$\frac{4y}{4} = \frac{-3x}{4} + \frac{12}{4}$$
$$y = -\frac{3}{4}x + 3$$
Slope $= -\frac{3}{4}$
y-intercept $= 3$

29. Slope $= \frac{2}{3}$
y-intercept $= -2$

31.

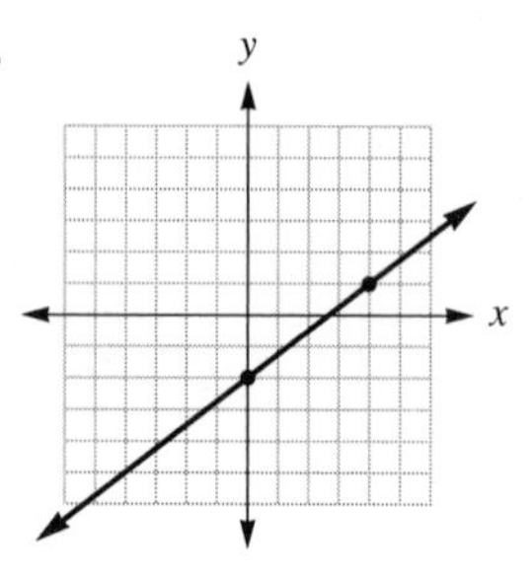

33.

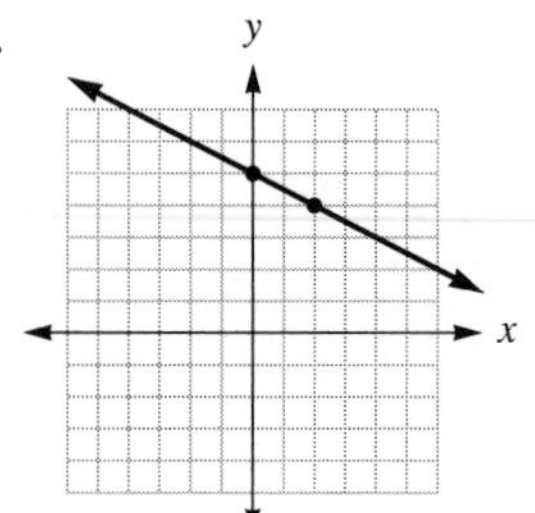

35.

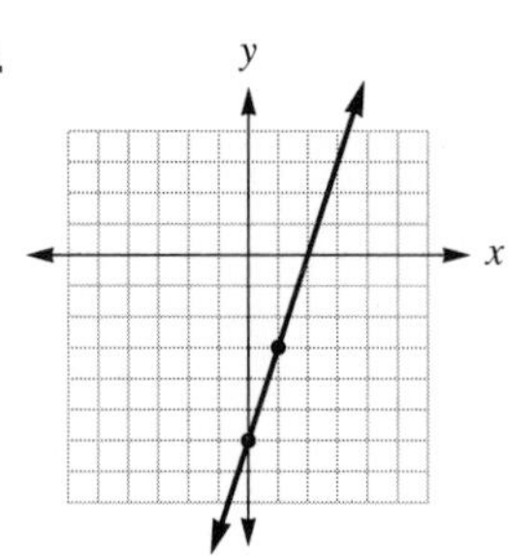

37.

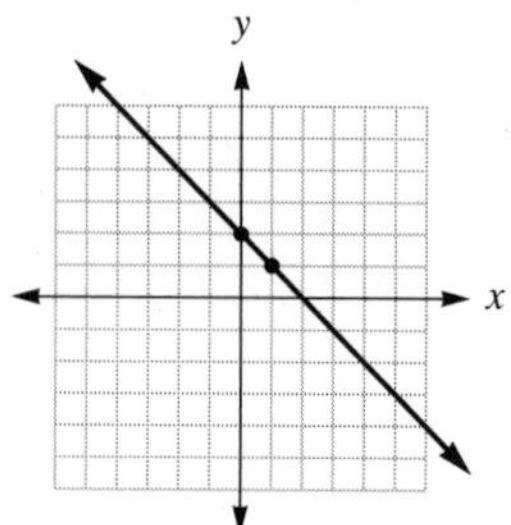

39. Slope $= -\frac{3}{2}$
y-intercept $= 6$

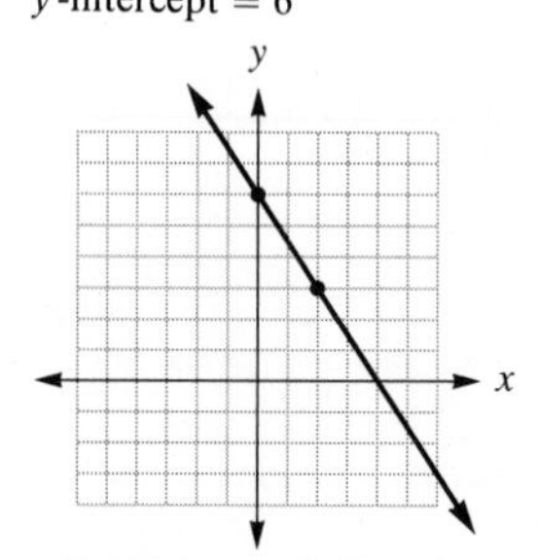

41. Slope $= \frac{3}{5}$
y-intercept $= 3$

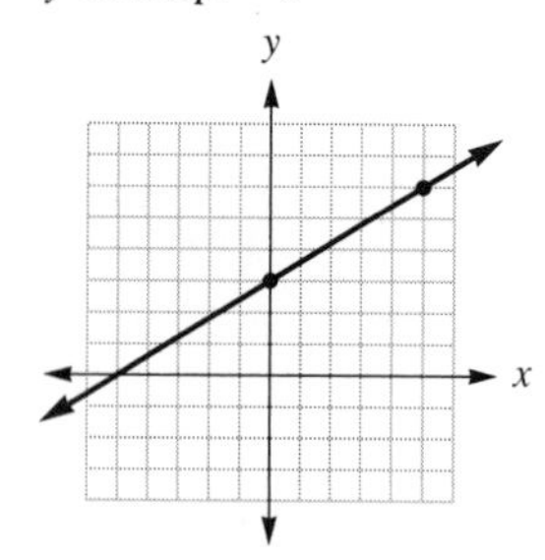

Exercises 13.4 (page 403)

1. $3x - 2y = -4$ **3.** $3x + 4y = -8$ **5.** $4x - 6y = 1$

7. $y - 2 = \frac{3}{4}(x + 1)$ **9.** $x - 2y = -5$
$$4(y - 2) = 4 \cdot \frac{3}{4}(x + 1)$$
$$4y - 8 = 3x + 3$$
$$-3x + 4y = 11$$
$$3x - 4y = -11$$

11. $2x + 3y = -8$ **13.** $x + 2y = 0$ **15.** $3x - 4y = 12$

17. $2x - 7y = 14$ **19.** $4x + 10y = 5$

21. Use the two points to find the slope, then use m and one point to find the equation of the line.
$$m = \frac{y_2 - y_1}{x_2 - x_1} = \frac{4 - (-1)}{2 - 4} = -\frac{5}{2}$$
$$y - y_1 = m(x - x_1)$$
$$y - 4 = -\frac{5}{2}(x - 2)$$
$$2y - 8 = -5x + 10$$
$$5x + 2y = 18$$

23. $4x - 3y = 0$ **25.** $x = 4$ **27.** $3x + 4y = 7$ **29.** $y = 5$

31. $x = -6$

Exercises 13.5 (page 406)

1.

x	y
-3	9
-2	4
-1	1
0	0
1	1
2	4
3	9

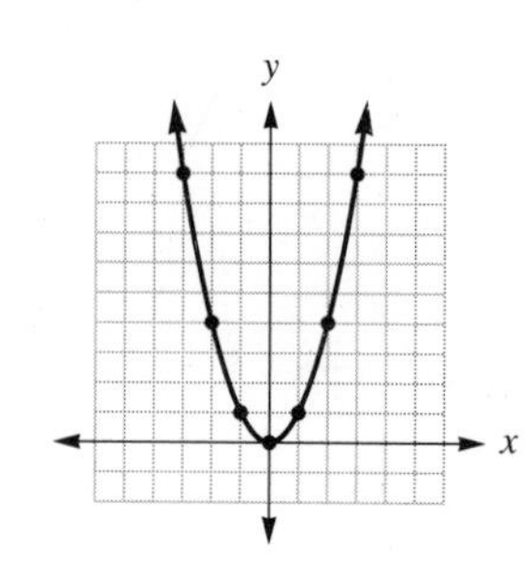

3.

x	y
-3	$4\frac{1}{2}$
-2	2
-1	$\frac{1}{2}$
0	0
1	$\frac{1}{2}$
2	2
3	$4\frac{1}{2}$

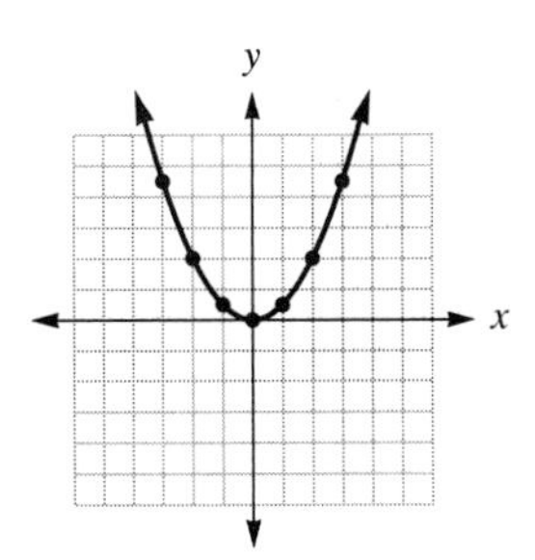

5.

x	y
-2	8
-1	3
0	0
1	-1
2	0
3	3
4	8

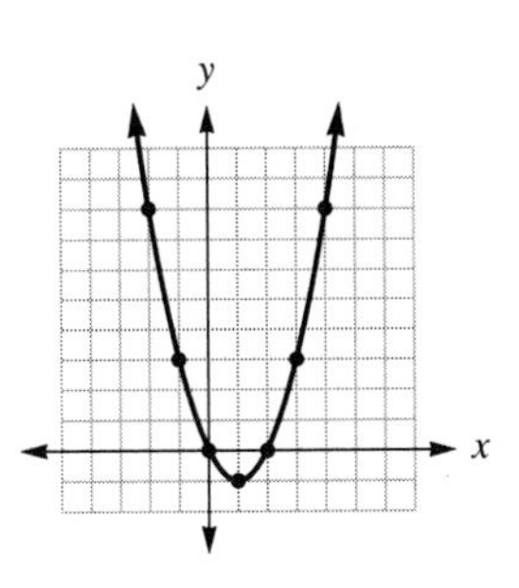

7.

x	y
-2	-8
-1	-3
0	0
1	1
2	0
3	-3
4	-8

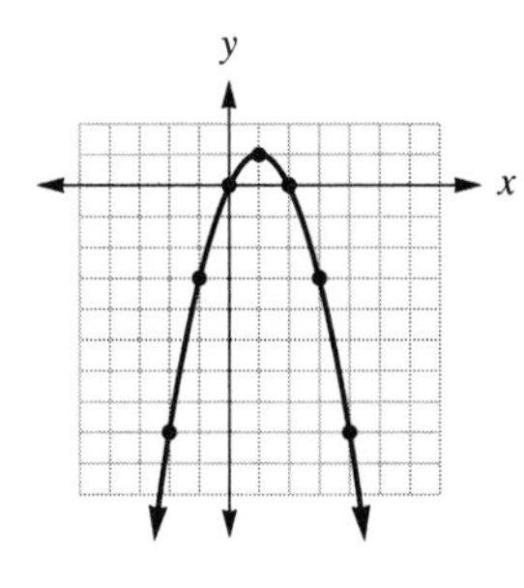

9.

x	y
-2	-8
-1	-1
0	0
1	1
2	8

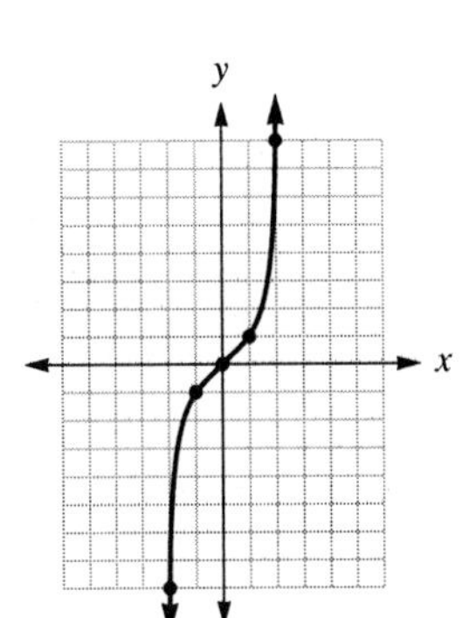

11.

x	y
-2	2
-1	6
0	4
1	2
2	6

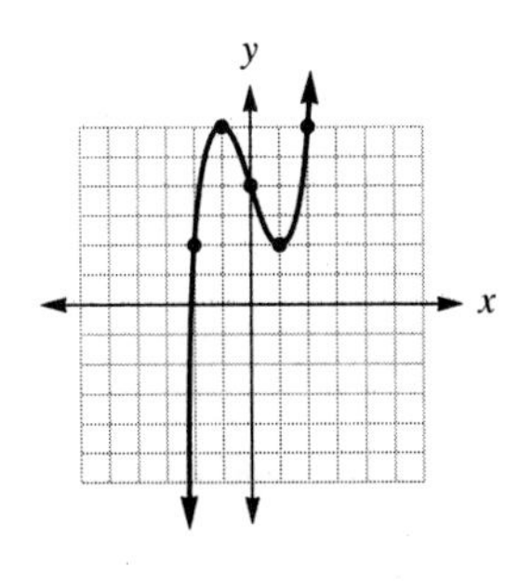

Exercises 13.6 (page 411)

1.

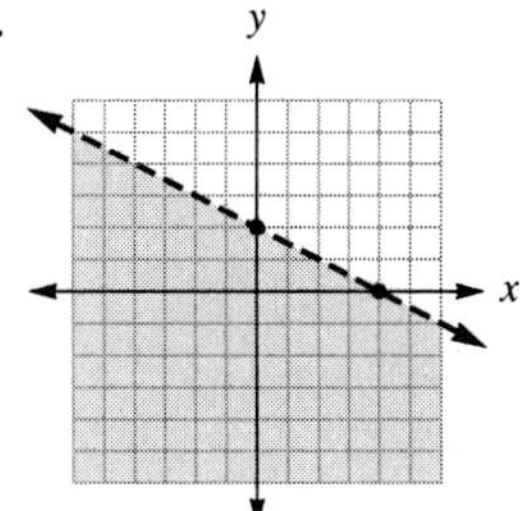

3.

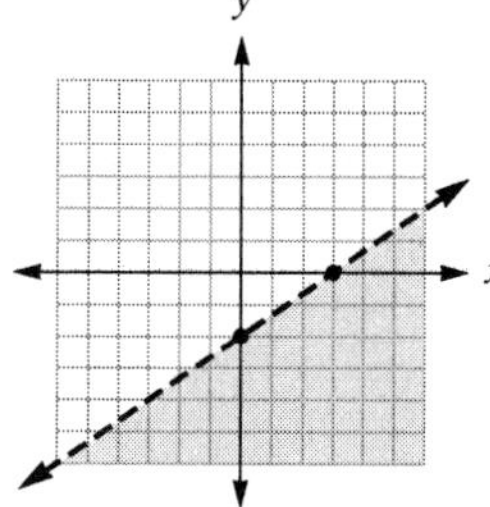

5. $x \geq -2$
All points to the right and including the line $x = -2$

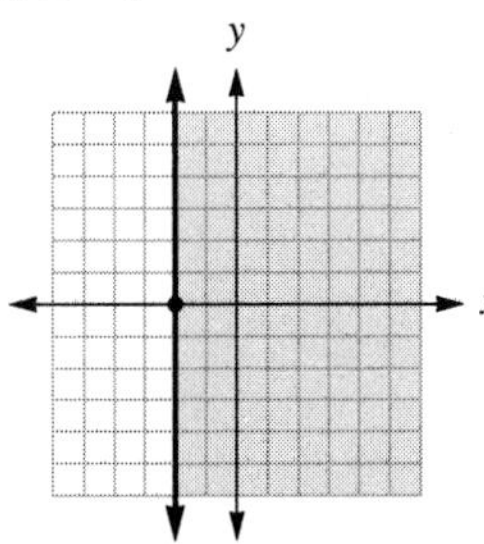

7.

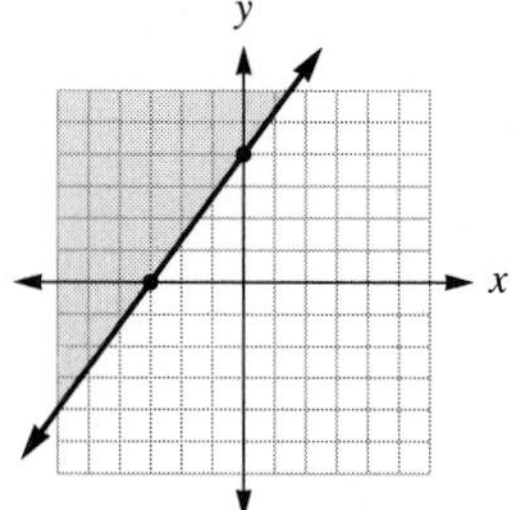

9.

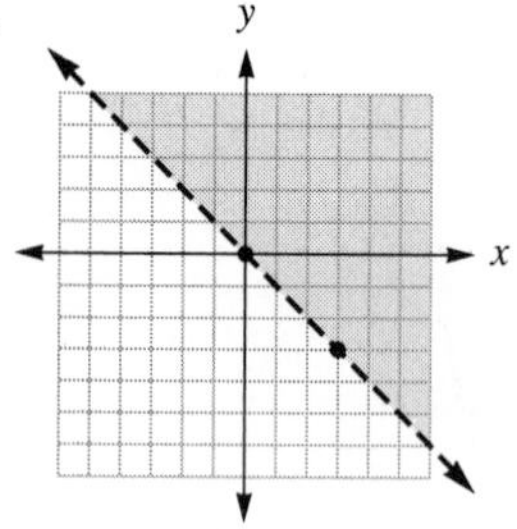

11. $3x - 4y \geq 10$
Boundary line:
$3x - 4y = 10$

$x = 0,\ y = \dfrac{10}{-4} = -2\dfrac{1}{2}$

$y = 0,\ x = \dfrac{10}{3} = 3\dfrac{1}{3}$

The corrrect half-plane does not include the origin

x	y
0	$-2\frac{1}{2}$
$3\frac{1}{3}$	0

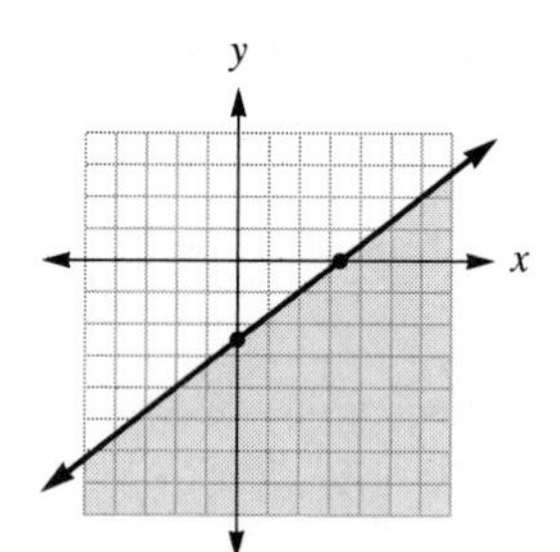

13.

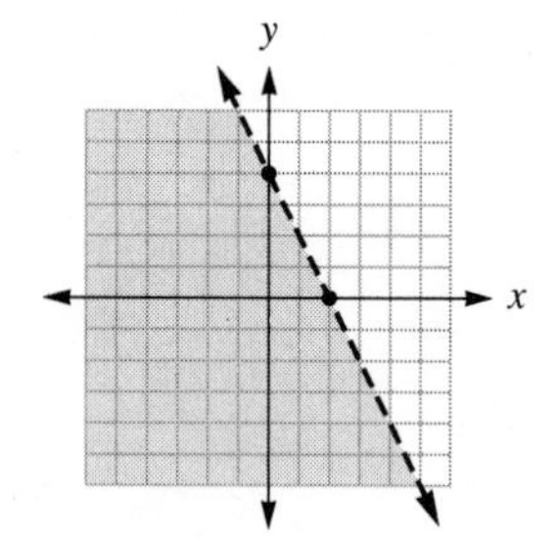

15.

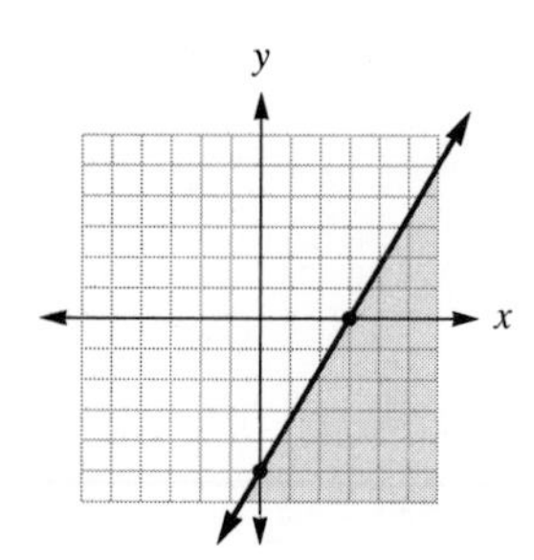

17.

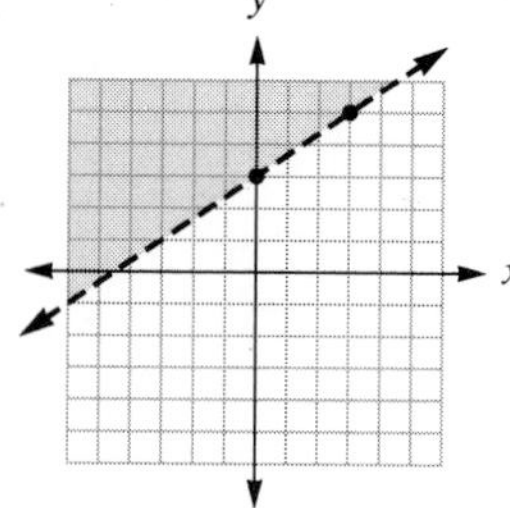

19.

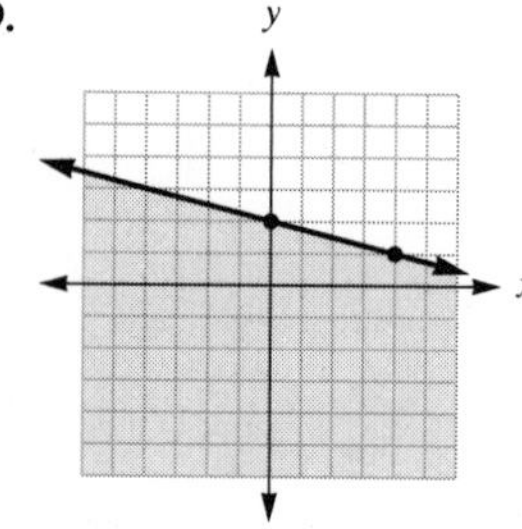

21.

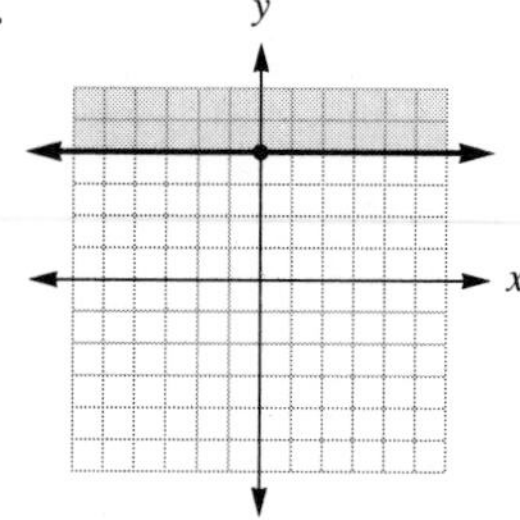

23.

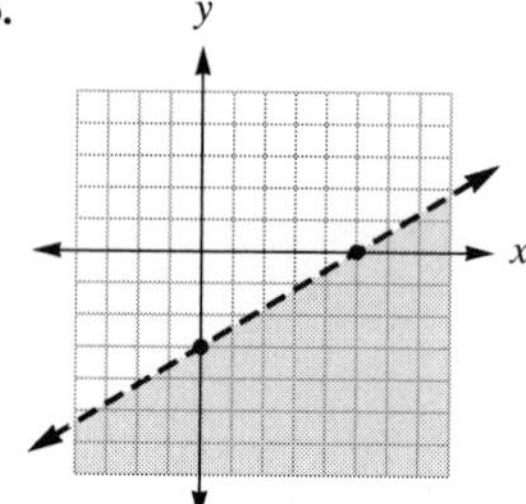

Review Exercises 13.7 (page 413)

1.

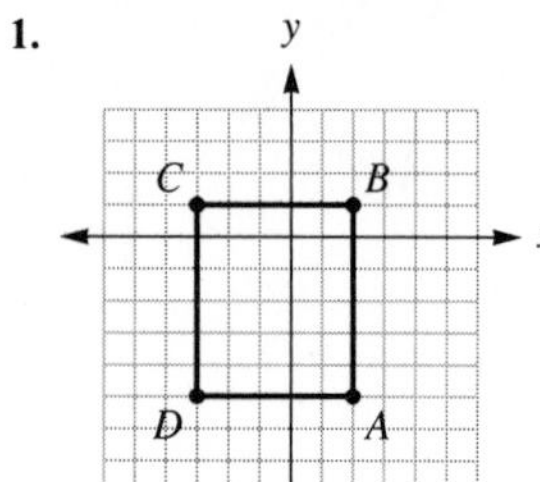

2. a. 5
b. -3
c. 5

3.

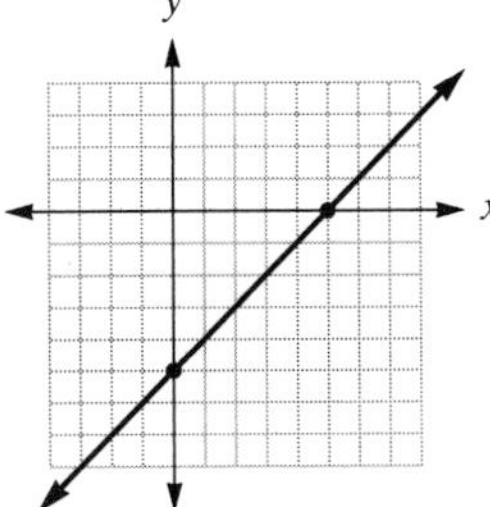

4.

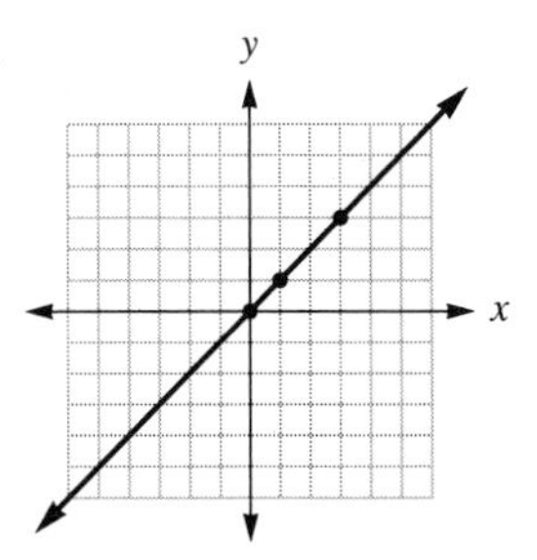

5.

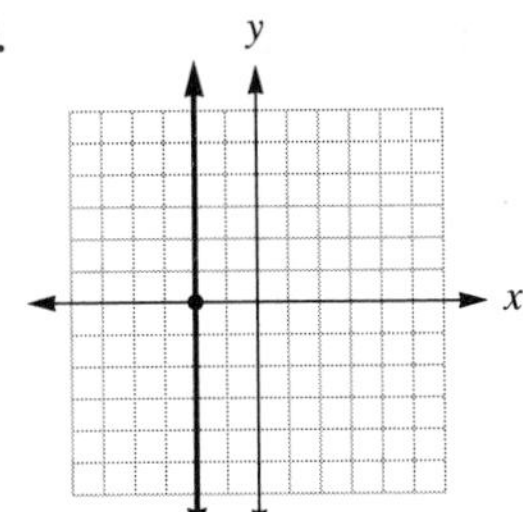

6.

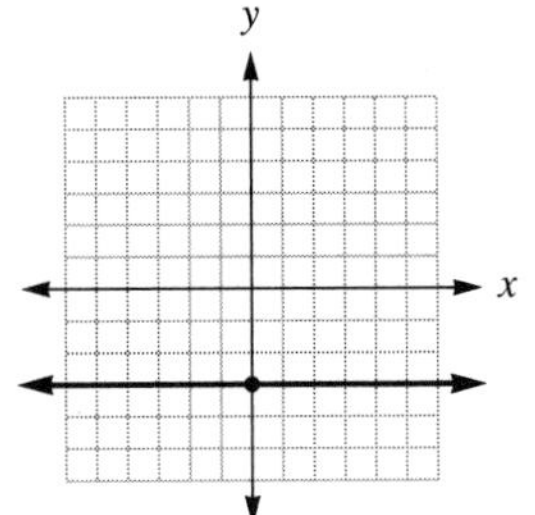

7.

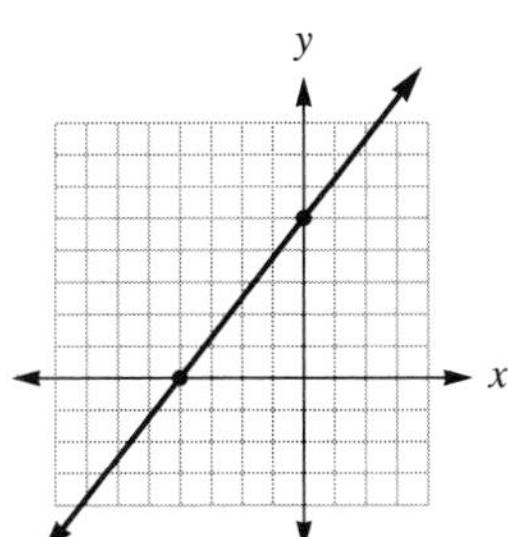

8.

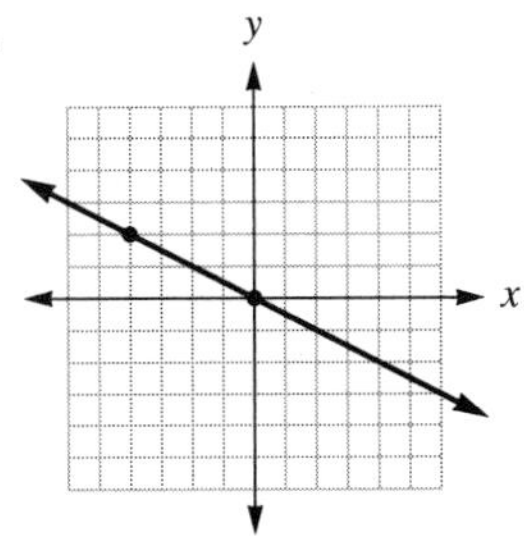

9.

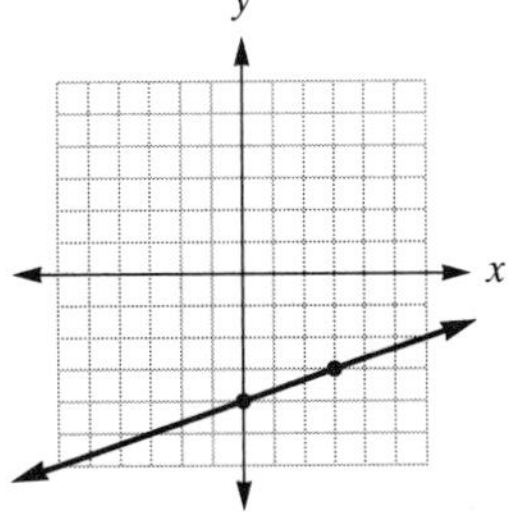

10.

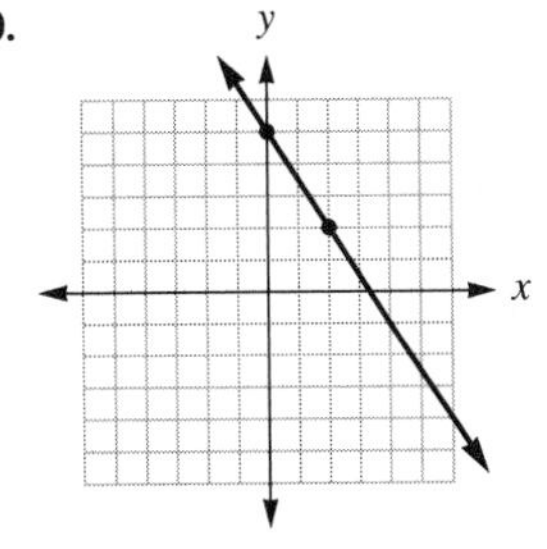

11. $y = \dfrac{x^2}{2}$

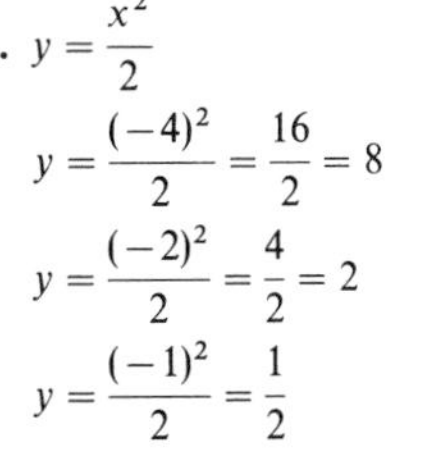

$$y = \frac{(-4)^2}{2} = \frac{16}{2} = 8$$
$$y = \frac{(-2)^2}{2} = \frac{4}{2} = 2$$
$$y = \frac{(-1)^2}{2} = \frac{1}{2}$$
$$y = \frac{0^2}{2} = 0$$
$$y = \frac{1^2}{2} = \frac{1}{2}$$
$$y = \frac{2^2}{2} = \frac{4}{2} = 2$$
$$y = \frac{4^2}{2} = \frac{16}{2} = 8$$

x	y
−4	8
−2	2
−1	$\frac{1}{2}$
0	0
1	$\frac{1}{2}$
2	2
4	8

12. $y = x^3 - 3x$

$y = (-3)^3 - (3)(-3)$
$= -27 + 9 = -18$
$y = (-2)^3 - 3(-2)$
$= -8 + 6 = -2$
$y = (-1)^3 - 3(-1)$
$= -1 + 3 = 2$
$y = 0^3 - 3(0) = 0$
$y = 1^3 - 3(1) = -2$
$y = 2^3 - 3(2) = 2$
$y = 3^3 - 3(3) = 18$

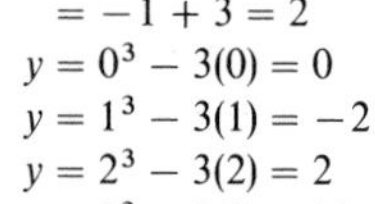

x	y
-3	-18
-2	-2
-1	2
0	0
1	-2
2	2
3	18

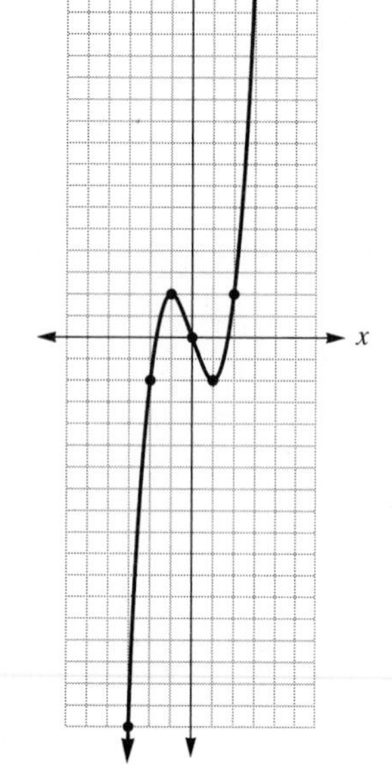

13. $-\frac{11}{5}$ **14.** Slope $= \frac{4}{15}$, y-intercept $= -2$

15. $x + 2y = 12$ **16.** $2x + 3y = -9$

17. First find m; then use m and one point to find the equation.

$$m = \frac{y_2 - y_1}{x_2 - x_1} = \frac{-2 - 4}{1 - (-3)} = \frac{-6}{4} = -\frac{3}{2}$$

$$y - y_1 = m(x - x_1)$$

$$y - (-2) = -\frac{3}{2}(x - 1)$$

$$2(y + 2) = -3(x - 1)$$

$$2y + 4 = -3x + 3$$

$$3x + 2y = -1$$

18. $x = -7$

19.

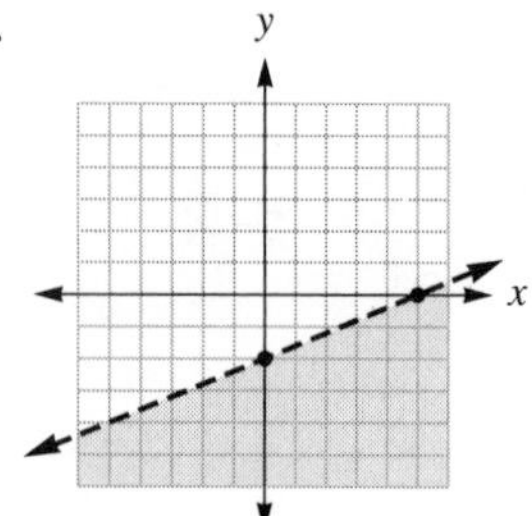

20.

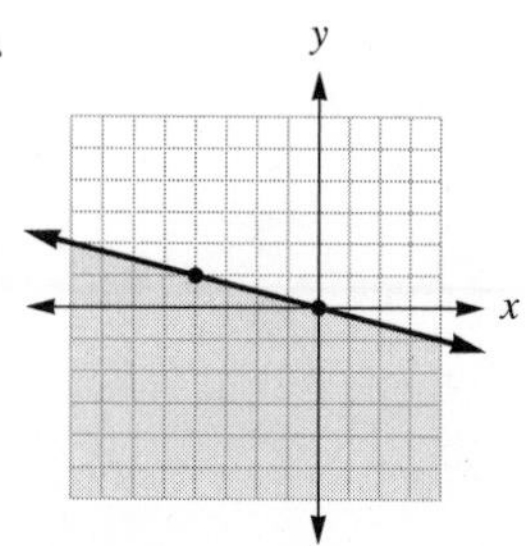

21.

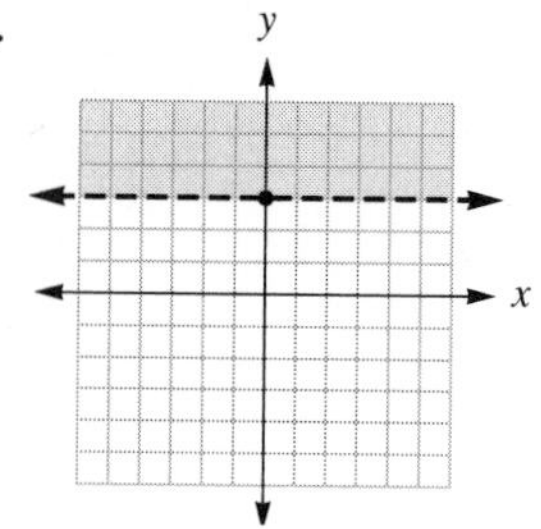

22.

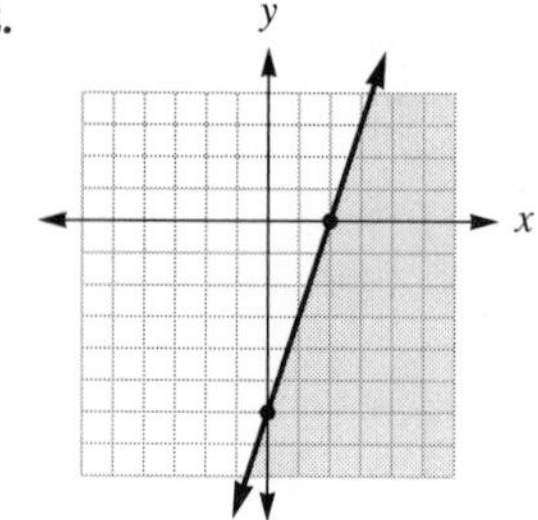

Chapter 13 Diagnostic Test (page 415)

Following each problem number is the textbook section reference (in parentheses) in which that kind of problem is discussed.

1. (13.1)

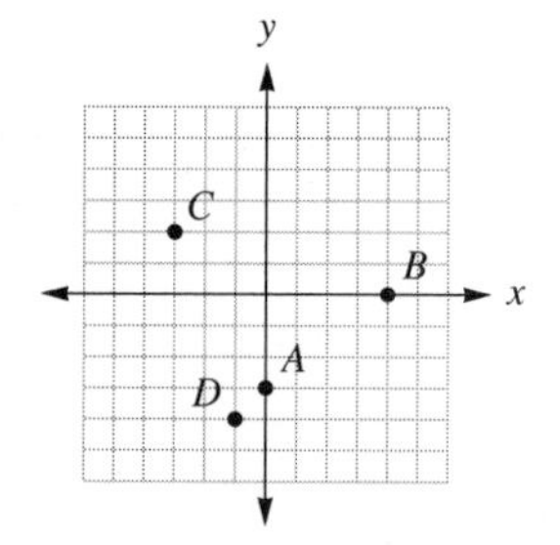

2. (13.3) $5x + 2y = 8$

$$2y = -5x + 8$$

$$\frac{2y}{2} = \frac{-5x}{2} + \frac{8}{2}$$

$$y = -\frac{5}{2}x + 4$$

Slope $= -\frac{5}{2}$, y-intercept $= 4$

3. (13.2)

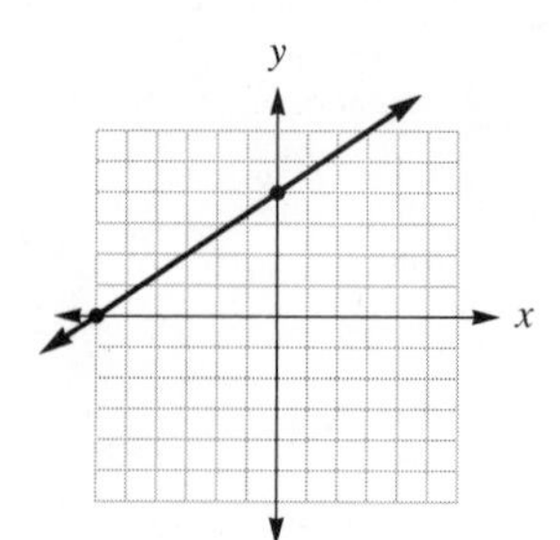

4. (13.3)

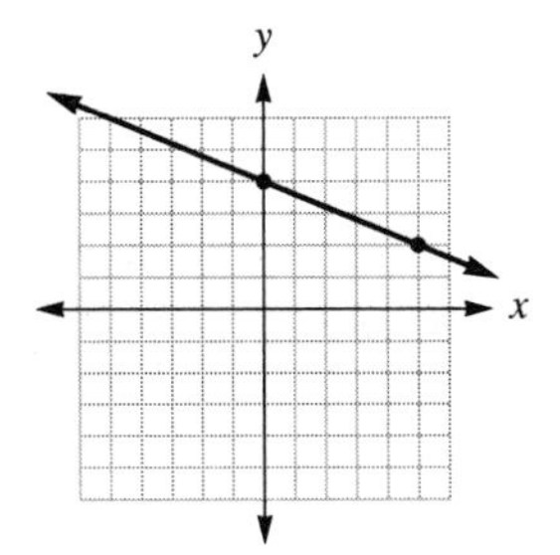

5. (13.2)

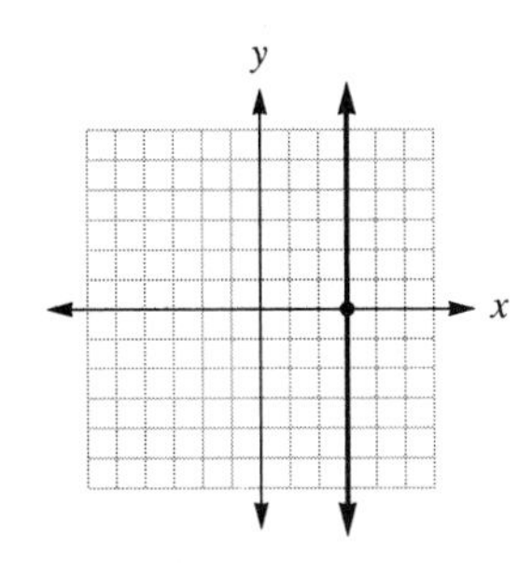

6. (13.5) $y = x^2 + x - 6$
$y = (-4)^2 + (-4) - 6 = 6$
$y = (-3)^2 + (-3) - 6 = 0$
$y = (-2)^2 + (-2) - 6 = -4$
$y = (-1)^2 + (-1) - 6 = -6$
$y = (0)^2 + (0) - 6 = -6$
$y = (1)^2 + (1) - 6 = -4$
$y = (2)^2 + (2) - 6 = 0$
$y = (3)^2 + (3) - 6 = 6$

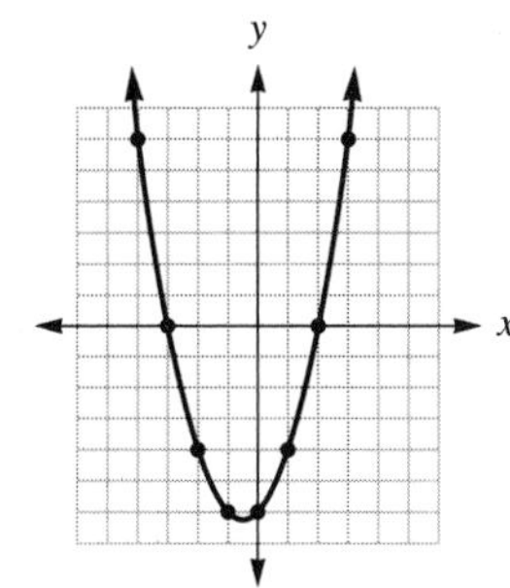

x	y
−4	6
−3	0
−2	−4
−1	−6
0	−6
1	−4
2	0
3	6

7. (13.6) Boundary line: $3x - 2y = 6$
If $x = 0$, $3(0) - 2y = 6 \Rightarrow y = -3$
If $y = 0$, $3x - 2(0) = 6 \Rightarrow x = 2$
Boundary line solid, because equality included.
$3x - 2y \leq 6$
Correct half-plane includes (0, 0), because
$3x - 2y \leq 6$
$3(0) - 2(0) \leq 6$
$0 \leq 6$ is true.

x	y
0	−3
2	0

8. (13.3) $m = \dfrac{y_2 - y_1}{x_2 - x_1}$

$m = \dfrac{-1 - 2}{1 - (-5)}$

$m = \dfrac{-3}{6} = -\dfrac{1}{2}$

9. (13.4) A horizontal line has an equation in the form $y = b$. Because the y-coordinate of (2, 5) is 5, the equation is $y = 5$.

10. (13.4)
$$y = mx + b$$
$$y = -\frac{2}{3}x + (-4)$$
$$3y = -2x - 12$$
$$2x + 3y = -12$$

11. (13.4)
$$y - y_1 = m(x - x_1)$$
$$y - 5 = \frac{1}{4}(x - (-2))$$
$$4(y - 5) = 1(x + 2)$$
$$4y - 20 = x + 2$$
$$-x + 4y = 22$$
$$x - 4y = -22$$

12. (13.4) $P_1(-4, 2)$, $P_2(1, -3)$

a. $m = \dfrac{-3 - 2}{1 - (-4)} = \dfrac{-5}{5} = -1$

b.
$$y - y_1 = m(x - x_1)$$
$$y - (-3) = -1(x - 1)$$
$$y + 3 = -x + 1$$
$$x + y = -2$$

Exercises 14.1 (page 422)

1. Is not a solution **3.** Is a solution

5.

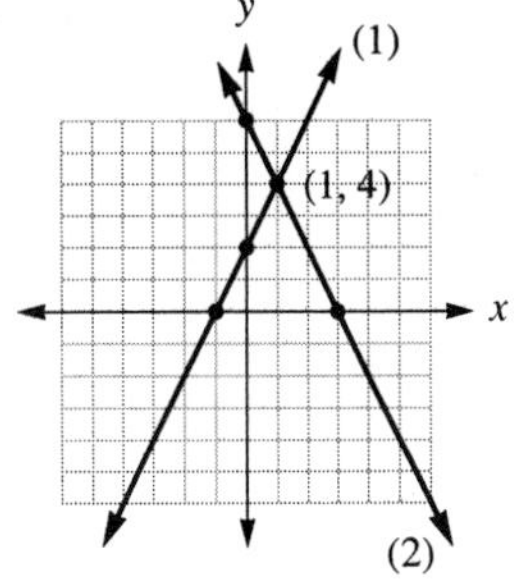

7.

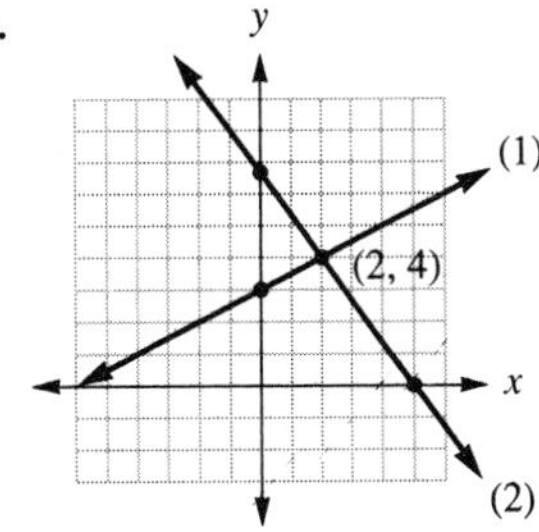

9. 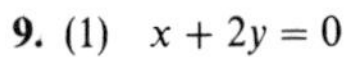
(1) $x + 2y = 0$
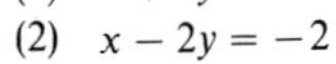
(2) $x - 2y = -2$

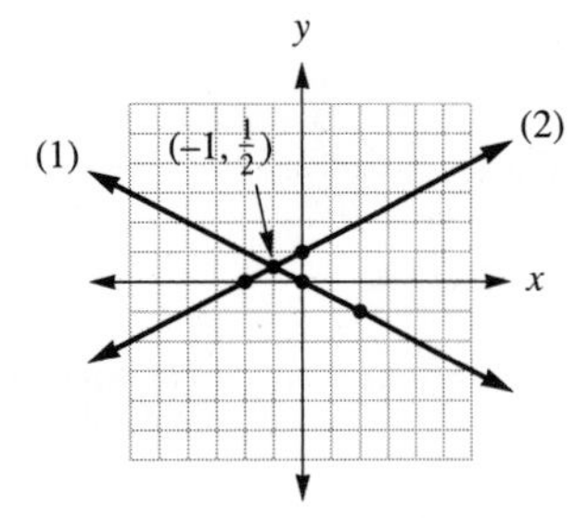

(1)

x	y
0	0
2	−1

(2)

x	y
0	1
−2	0

11.

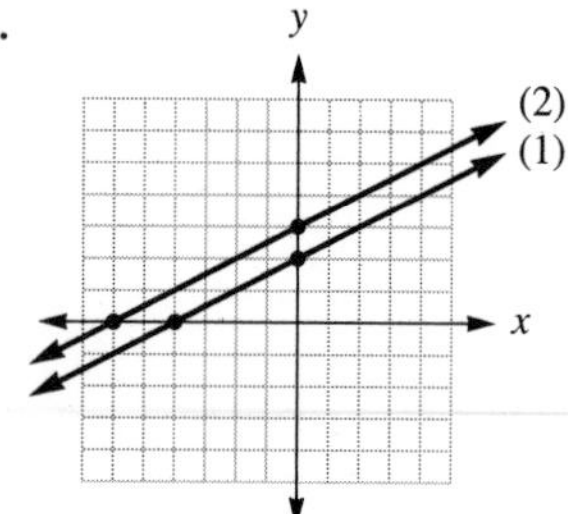

Inconsistent (no solution; parallel lines)

13. (1) $10x - 4y = 20$
(2) $6y - 15x = -30$
(1) Intercepts: (2, 0), (0, −5)
(2) Intercepts: (2, 0), (0, −5)
Because both lines have the same intercepts, they are the same line. Any point on the line is a solution.

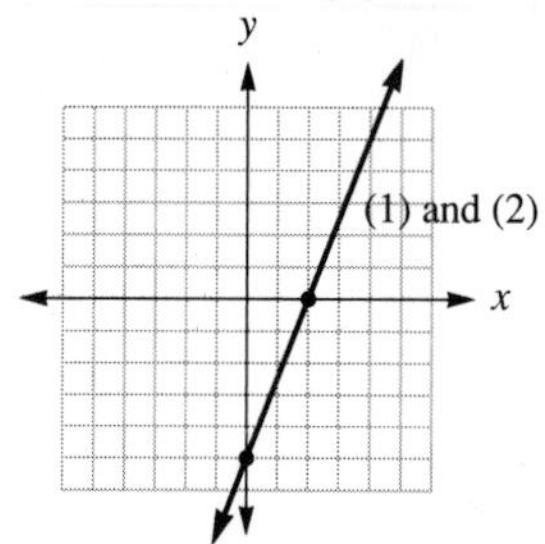

Dependent (many solutions; same line)

15.

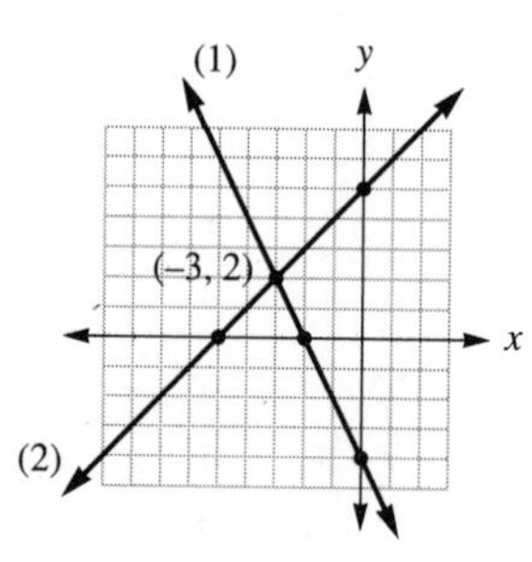

17.

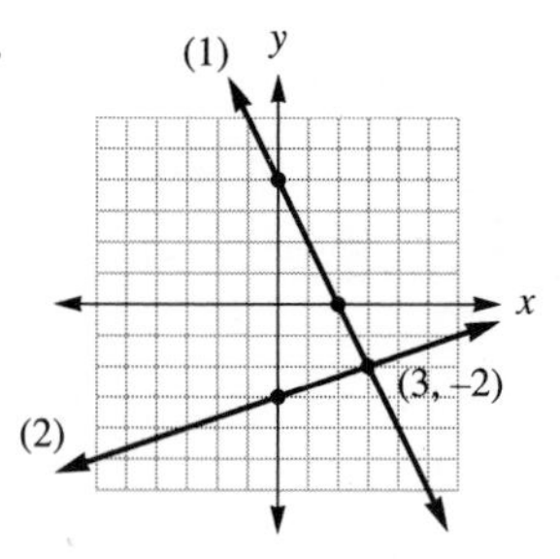

19.

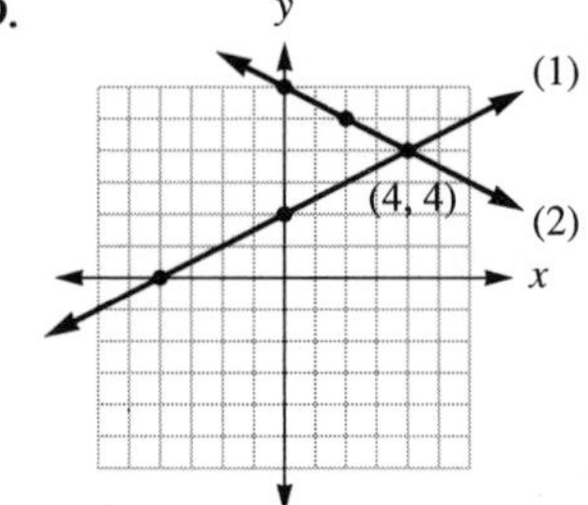

Exercises 14.2 (page 427)

1. $(-2, 0)$ **3.** $(6, -2)$ **5.** $(3, -1)$ **7.** $(-1, 3)$ **9.** $(0, 0)$

11. 3] −4] | 4 3

$$\begin{aligned} 4x + 3y &= 2 \Rightarrow & 12x + 9y &= 6 \\ 3x + 5y &= -4 \Rightarrow & -12x - 20y &= 16 \\ & & -11y &= 22 \\ & & y &= -2 \end{aligned}$$

Substituting $y = -2$ into equation (1),

$$\begin{aligned} (1) \quad 4x + 3y &= 2 \\ 4x + 3(-2) &= 2 \\ 4x - 6 &= 2 \\ 4x &= 8 \\ x &= 2 \end{aligned}$$

Solution: $(2, -2)$

13. 9] −6] | 3] −2] | 6 9

$$\begin{aligned} 6x - 10y &= 6 \Rightarrow & 18x - 30y &= 18 \\ 9x - 15y &= -4 \Rightarrow & -18x + 30y &= 8 \\ & & 0 &= 26 \quad \text{False} \end{aligned}$$

Inconsistent (no solution)

15. 2] 1]

$$\begin{aligned} 3x - 5y &= -2 \Rightarrow & 6x - 10y &= -4 \\ -6x + 10y &= 4 \Rightarrow & -6x + 10y &= 4 \\ & & 0 &= 0 \quad \text{True} \end{aligned}$$

Dependent (many solutions)

17. $(-2, 3)$ **19.** $\left(3, -\frac{3}{2}\right)$ **21.** Dependent (many solutions)

23. $(-3, -2)$

Exercises 14.3 (page 430)

1. $(5, 3)$ **3.** $(2, -1)$ **5.** $(1, -2)$

7. Dependent (many solutions) **9.** $(2, 1)$ **11.** $(5, -1)$

13. Inconsistent (no solution) **15.** $(-2, -6)$ **17.** $\left(\frac{1}{2}, 1\right)$

19. (1, 1)

21. $8\left(\frac{x}{2}\right) + 8\left(\frac{y}{8}\right) = 8(1)$

(1) $4x + y = 8 \Rightarrow y = 8 - 4x$

$10\left(\frac{x}{2}\right) + 10\left(\frac{y}{5}\right) = 10(4)$

(2) $5x + 2y = 40$

Substitute $8 - 4x$ in (2) for y:

$$5x + 2(8 - 4x) = 40$$
$$5x + 16 - 8x = 40$$
$$-3x = 24$$
$$x = -8$$

Substitute $x = -8$ in $y = 8 - 4x$

$y = 8 - 4(-8)$
$y = 8 + 32$
$y = 40$

Therefore, the solution is $(-8, 40)$

23. $4\left(\frac{x}{4}\right) + \overset{2}{4}\left(\frac{y+1}{2}\right) = 4(2)$

(1) $x + 2y + 2 = 8$

$x = 6 - 2y$

$\overset{3}{6}\left(\frac{x}{2}\right) + \overset{2}{6}\left(\frac{y-2}{3}\right) = 6(1)$

(2) $3x + 2y - 4 = 6$

$3x + 2y = 10$

Substitute $6 - 2y$ in (2) for x:

$$3(6 - 2y) + 2y = 10$$
$$18 - 6y + 2y = 10$$
$$-4y = -8$$
$$y = 2$$

Substitute $y = 2$ in $x = 6 - 2y$

$x = 6 - 2(2)$
$x = 6 - 4$
$x = 2$

Therefore, the solution is $(2, 2)$

Exercises 14.4 (page 433)

1. a. Let x = one number
y = other number
b. $x + y = 30$
$x - y = 12$
c. Numbers are 9 and 21

3. a. Let x = one angle
y = other angle
b. $x + y = 90$
$x - y = 40$
c. Angles are 25° and 65°

5. a. Let x = smaller number
y = larger number
b. $2x + 3y = 34$
$5x - 2y = 9$
c. Smaller is 5
Larger is 8

7. a. Let x = lb of almonds
y = lb of hazelnuts
b. $x + y = 20$
$85x + 140y = 1975$
c. 15 lb almonds
5 lb hazelnuts

9. a. Let ℓ = length
w = width

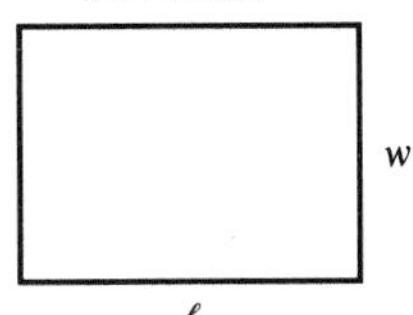

$2w + 2\ell$ = Perimeter

b. (1) $2w + 2\ell = 19$
(2) $\ell = w + 1.5$

Substituting $w + 1.5$ in equation (1) for ℓ, we get

$$2w + 2\ell = 19$$
$$2w + 2(w + 1.5) = 19$$
$$2w + 2w + 3 = 19$$
$$4w = 16$$
$$w = 4$$

Substituting $w = 4$ in equation (2), we get

$\ell = w + 1.5$
$\ell = (4) + 1.5$
$\ell = 5.5 = 5$ ft 6 in.

c. Width is 4 ft
Length is 5 ft 6 in.

11. a. Let x = number of 10¢ stamps
y = number of 25¢ stamps
b. $x + y = 22$
$10x + 25y = 340$
c. Number of 10¢ stamps is 14.
Number of 25¢ stamps is 8.

13. a. Let n = numerator
d = denominator

b. (1) $\frac{n}{d} = \frac{2}{3}$ and (2) $\frac{n+4}{d-2} = \frac{6}{7}$

$2d = 3n$ $\qquad 6(d - 2) = 7(n + 4)$

$d = \frac{3n}{2}$ $\qquad 6d - 12 = 7n + 28$

(3) $6d = 7n + 40$

Substitute $\frac{3n}{2}$ for d in (3) $\Rightarrow 6\left(\frac{3n}{2}\right) = 7n + 40$

$$9n = 7n + 40$$
$$2n = 40$$
$$n = 20$$

Substitute $n = 20$ in $d = \frac{3n}{2}$

$$d = \frac{3(20)}{2}$$
$$d = 30$$

c. Therefore, the original fraction is $\frac{20}{30}$.

15. a. Let x = Number of adult tickets
y = Number of children tickets
b. $x + y = 8$
$1.95x + .95y = 10.60$
c. Adult tickets is 3
Children tickets is 5

17. a. Let x = number of \$15 rolls
y = number of \$20 rolls
b. $x + y = 40$
$15x + 20y = 725$
c. 15 – \$15 rolls
25 – \$20 rolls

19. a. Let x = hours as tutor
y = hours as cashier
b. $x + y = 30$
$4x + 5y = 140$
c. 10 hours as tutor
20 hours as cashier

21. a. Let x = cost of tie
y = cost of pin
b. $x + y = 1.10$
$x = 1.00 + y$
c. Pin costs 5¢
Tie costs \$1.05

Review Exercises 14.5 (page 435)

1. 

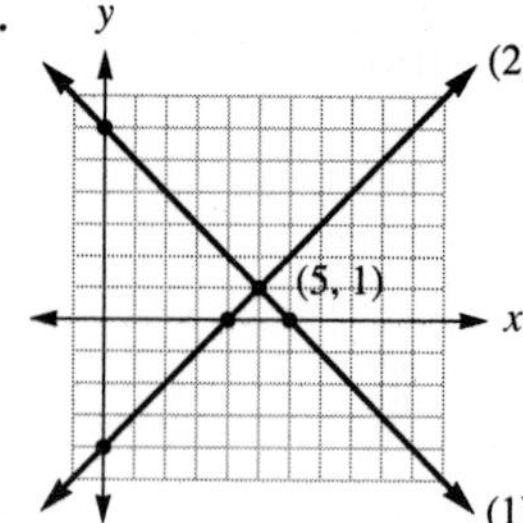

2.

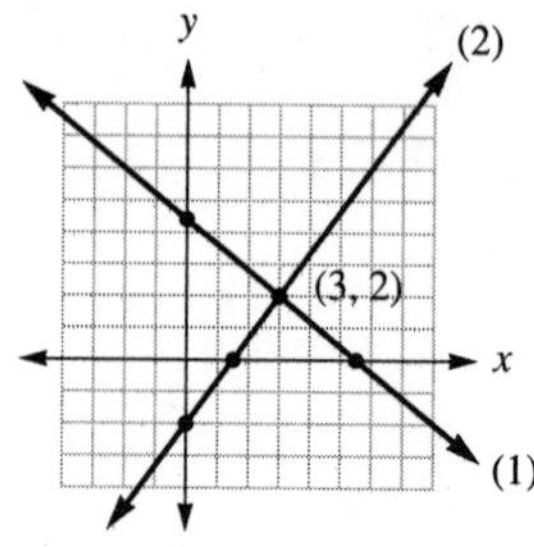

3. (1) $2x - 3y = 3$
(2) $3y - 2x = 6$

(1) Intercepts: $\left(1\frac{1}{2}, 0\right)$, $(0, -1)$

(2) Intercepts: $(-3, 0)$, $(0, 2)$

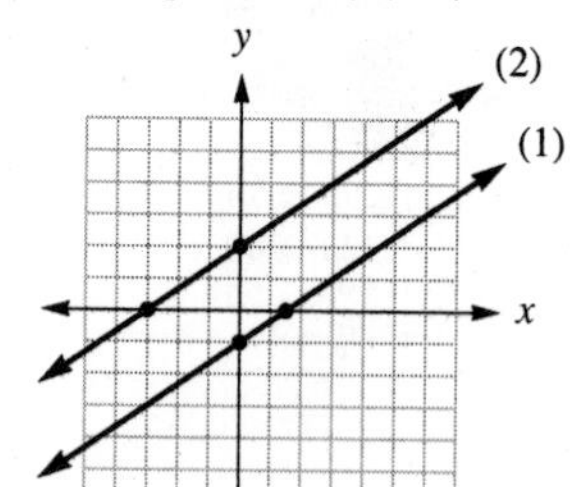

Inconsistent (no solution)

4.

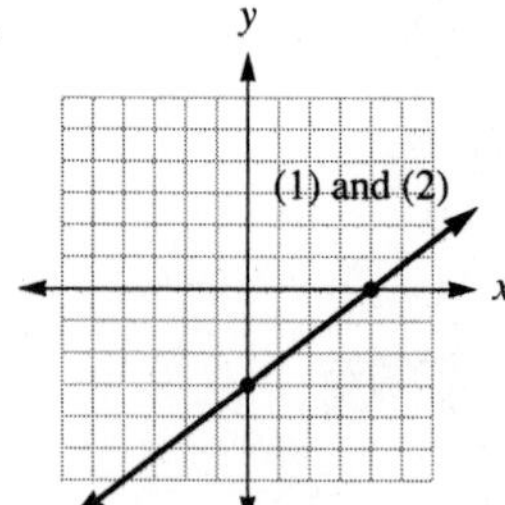

Dependent (many solutions)

5. (1, 2) **6.** Dependent (many solutions)

7. Inconsistent (no solution)

8. $4 \cdot \frac{x}{2} - 4 \cdot \frac{y}{4} = 4 \cdot 4 \Rightarrow 3]\quad 2x - y = 16$

$12 \cdot \frac{x}{3} + 12 \cdot \frac{y}{4} = 12 \cdot 1 \Rightarrow \quad 4x + 3y = 12$

$$\begin{aligned} 6x - 3y &= 48 \\ 4x + 3y &= 12 \\ \hline 10x \quad &= 60 \\ x &= 6 \end{aligned}$$

Substitute $x = 6$ into (1)

$2x - y = 16$
$2(6) - y = 16$
$12 - y = 16$
$-y = 4$
$y = -4$

Solution: $(6, -4)$

9. (7, 5) **10.** $(-4, -11)$ **11.** (0, 2)

12. $6 \cdot \frac{x}{3} + 6 \cdot \frac{y}{6} = 6(-1) \Rightarrow 2x + y = -6$

$12 \cdot \frac{x}{4} + 12 \cdot \frac{y}{12} = 12(-1) \Rightarrow 3x + y = -12$

$y = -3x - 12$

Substitute $-3x - 12$ for y in (1).

$2x + y = -6$
$2x + (-3x - 12) = -6$
$-x - 12 = -6$
$-x = 6$
$x = -6$

Substitute $x = -6$ in $y = -3x - 12$

$y = -3(-6) - 12$
$y = 18 - 12$
$y = 6$

Solution: $(-6, 6)$

13. (5, −4) **14.** Inconsistent **15.** (−3, −2)

16. $\overset{3}{\cancel{18}} \cdot \frac{x - 1}{\cancel{6}} + \overset{2}{\cancel{18}} \cdot \frac{y + 5}{\cancel{9}} = 18 \cdot 2$

$3x - 3 + 2y + 10 = 36$

(1) $3x + 2y = 29$

$\overset{3}{\cancel{12}} \cdot \frac{x + 1}{\cancel{4}} - \overset{2}{\cancel{12}} \cdot \frac{y + 2}{\cancel{6}} = 12 \cdot 1$

$3x + 3 - 2y - 4 = 12$

(2) $3x - 2y = 13$

$$\begin{aligned} (1)\quad 3x + 2y &= 29 \\ (2)\quad 3x - 2y &= 13 \\ \hline 6x \quad &= 42 \\ x &= 7 \end{aligned}$$

Substitute $x = 7$ in (1)

(1) $3x + 2y = 29$
$3(7) + 2y = 29$
$21 + 2y = 29$
$2y = 8$
$y = 4$

Solution: (7, 4)

17. 53 and 31 **18.** 8 and 5 **19.** Length is 11 feet
Width is 7 feet

20. x = hours as tutor
y = hours as waiter

(1) 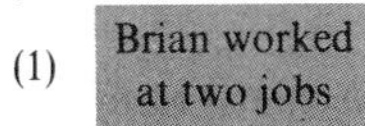

for a total of | 32 hours

$x + y = 32$

(2) He received a total of \$216.

$8x + 6y = \$216$

(1) $x + y = 32 \Rightarrow x = 32 - y$

Substitute $32 - y$ in (2) for x:

$$
\begin{aligned}
(2) \quad 8x + 6y &= 216 \\
8(32 - y) + 6y &= 216 \\
256 - 8y + 6y &= 216 \\
-2y &= -40 \\
y &= 20 \text{ hours}
\end{aligned}
$$

Substitute $y = 20$ in $x = 32 - y$

$x = 32 - 20 = 12$ hours

Solution: 12 hours as a tutor
20 hours as a waiter

21. Let x = number of \$10 rolls
y = number of \$8 rolls

(1) The total number of rolls is 80.

$x + y = 80$

(2) The total amount paid for stamps is \$730.

$10x + 8y = 730$

(1) $x + y = 80 \Rightarrow x = 80 - y$

Substitute $80 - y$ in (2) for x:

$$
\begin{aligned}
(2) \quad 10x + 8y &= 730 \\
10(80 - y) + 8y &= 730 \\
800 - 10y + 8y &= 730 \\
-2y &= -70 \\
y &= 35 \text{ (\$8 rolls)}
\end{aligned}
$$

Substitute $y = 35$ in $x = 80 - y$:

$x = 80 - 35 = 45$ (\$10 rolls)

Solution: 45 \$10 rolls and 35 \$8 rolls

22. 4 boxes legal size
16 boxes letter size

Chapter 14 Diagnostic Test (page 437)

Following each problem number is the textbook section reference (in parentheses) in which that kind of problem is discussed.

1. (14.1) (1) $3x + 2y = 2$
(2) $2x - 3y = 10$

(1) Intercepts: $\left(\frac{2}{3}, 0\right)$ (0, 1)

Check point: $(-2, 4)$

(2) Intercepts: $(5, 0), \left(0, -3\frac{1}{3}\right)$

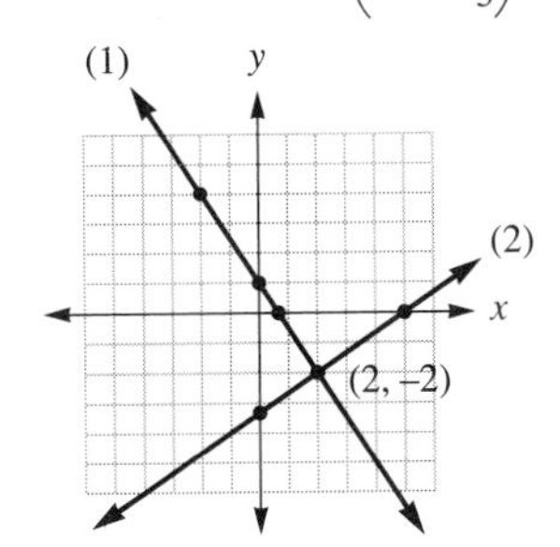

2. (14.2)

$$
\begin{aligned}
-3]\quad 3x - 4y = 1 &\Rightarrow -9x + 12y = -3 \\
4]\quad 5x - 3y = 9 &\Rightarrow \underline{20x - 12y = 36} \\
&\qquad 11x = 33 \\
&\qquad x = 3
\end{aligned}
$$

Substitute $x = 3$ into (1):

$$
\begin{aligned}
(1) \quad 3x - 4y &= 1 \\
3(3) - 4y &= 1 \\
9 - 4y &= 1 \\
-4y &= -8 \\
y &= 2
\end{aligned}
$$

Solution: (3, 2)

3. (14.2)

$$
\begin{aligned}
-6]\quad -3]\quad 3x - 4y = 1 &\Rightarrow -9x + 12y = -3 \\
4]\quad 2]\quad 5x - 6y = 5 &\Rightarrow \underline{10x - 12y = 10} \\
&\qquad x = 7
\end{aligned}
$$

Substitute $x = 7$ into (1):

$$
\begin{aligned}
(1) \quad 3x - 4y &= 1 \\
3(7) - 4y &= 1 \\
21 - 4y &= 1 \\
-4y &= -20 \\
y &= 5
\end{aligned}
$$

Solution: (7, 5)

4. (14.3) (1) $3x - 5y = 14$
(2) $x = y + 2$

Substitute $y + 2$ in place of x in (1):

$$
\begin{aligned}
(1) \quad 3x - 5y &= 14 \\
3(y + 2) - 5y &= 14 \\
3y + 6 - 5y &= 14 \\
-2y &= 8 \\
y &= -4
\end{aligned}
$$

Substitute $y = -4$ in $x = y + 2$:

$x = -4 + 2 = -2$

Solution: $(-2, -4)$

5. (14.3) (1) $4x + y = 2 \Rightarrow y = 2 - 4x$
(2) $2x - 3y = 8$

Substitute $2 - 4x$ for y in (2):

$$
\begin{aligned}
(2) \quad 2x - 3y &= 8 \\
2x - 3(2 - 4x) &= 8 \\
2x - 6 + 12x &= 8 \\
14x &= 14 \\
x &= 1
\end{aligned}
$$

Substitute $x = 1$ in $y = 2 - 4x$:

$y = 2 - 4(1) = 2 - 4 = -2$

Solution: $(1, -2)$

6. (14.3) (1) $4x - 3y = 7$
(2) $x - 2y = -2 \Rightarrow x = 2y - 2$

Substitute $2y - 2$ for x in (1)

$$
\begin{aligned}
(1) \quad 4x - 3y &= 7 \\
4(2y - 2) - 3y &= 7 \\
8y - 8 - 3y &= 7 \\
5y &= 15 \\
y &= 3
\end{aligned}
$$

Substitute $y = 3$ in $x = 2y - 2$

$x = 2(3) - 2 = 6 - 2 = 4$

Solution: (4, 3)

7. (14.2)

$$
\begin{aligned}
10]\quad 5]\quad 6x - 9y = 2 &\Rightarrow 30x - 45y = 10 \\
6]\quad 3]\quad -10x + 15y = -5 &\Rightarrow \underline{-30x + 45y = -15} \\
&\qquad 0 = -5 \quad \text{False}
\end{aligned}
$$

Inconsistent (no solution)

8. (14.4) Let x = larger number
y = smaller number

(1) The sum of two numbers is 18.
$$x + y = 18$$
(2) Their difference is 42.
$$x - y = 42$$
Using addition:
(1) $x + y = 18$
(2) $x - y = 42$
$2x = 60$
$x = 30$ (larger number)
Substitute $x = 30$ in (1):
(1) $x + y = 18$
$30 + y = 18$
$y = -12$ (smaller number)
Solution: 30 is the larger number
-12 is the smaller number

9. (14.4) Let x = number of classical records
y = number of pop records

(1)

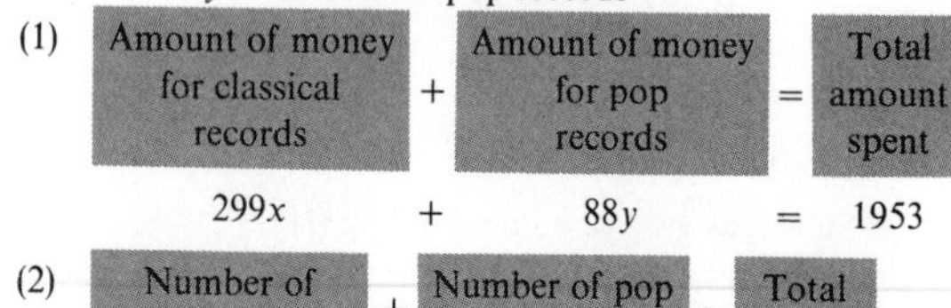

$$299x + 88y = 1953$$

(2) Number of classical records + Number of pop records = Total records
$$x + y = 15$$
Using substitution:
(1) $229x + 88y = 1953$
(2) $x + y = 15 \Rightarrow y = 15 - x$
Substitute $15 - x$ in place of y in (1):
$299x + 88(15 - x) = 1953$
$299x + 1320 - 88x = 1953$
$221x + 1320 = 1953$
$211x = 633$
$x = 3$
Number of classical records $= x = 3$
Number of pop records $= y = 15 - x = 15 - 3 = 12$

10. (14.4) Let ℓ = length
w = width

w

ℓ

Length is 3 cm more than width.

(1) $\ell = 3 + w$
$2\ell + 2w$ = perimeter
(2) $2\ell + 2w = 102$
Substituting $3 + w$ for ℓ in equation (2), we get
$2\ell + 2w = 102$
$2(3 + w) + 2w = 102$
$6 + 2w + 2w = 102$
$4w = 96$
$w = 24$
Substituting $w = 24$ in $\ell = 3 + w$
$\ell = 3 + 24$
$\ell = 27$
Therefore, the dimensions are
Length = 27 cm
Width = 24 cm

Exercises 15.1 (page 444)

1. $\frac{1}{x^4}$ **3.** $\frac{4}{a^3}$ **5.** a^4 **7.** $\frac{s}{r^4t^2}$ **9.** $\frac{y^3}{x^2}$ **11.** $\frac{s}{r^2t^4}$

13. xy^3 **15.** h^2k^4 **17.** $\frac{3}{x^4y}$ **19.** $\frac{a}{b^2}$ **21.** x **23.** 10

25. $\frac{1}{x^8}$ **27.** a^6 **29.** $\frac{1}{y^7}$ **31.** $\frac{m}{p^4}$ **33.** $\frac{1}{10}$ **35.** $\frac{1}{z^6}$

37. 10^5 **39.** 10^7 **41.** 1 **43.** x^{-2} **45.** hk^{-1}

47. $x^2y^{-1}z^{-5}$ **49.** 100 **51.** $\frac{1}{10{,}000}$ **53.** 4

55. 25 **57.** 64 **59.** 25 **61.** $\frac{1}{100{,}000{,}000}$

63. $\frac{10^{-3} \cdot 10^2}{10^5} = 10^{-3+2-5} = 10^{-6} = \frac{1}{10^6} = \frac{1}{1{,}000{,}000}$ **65.** -1

67. a^3c^2 **69.** $\frac{p^4t^2}{r}$ **71.** $\frac{2}{3x^4}$ **73.** $\frac{4h^2}{7}$ **75.** $\frac{y^3}{2x^3}$

77. $\frac{15m^0n^{-2}}{5m^{-3}n^4} = \frac{\overset{3}{\cancel{15}}}{\underset{1}{\cancel{5}}} \cdot \frac{m^0}{m^{-3}} \cdot \frac{n^{-2}}{n^4} = \frac{3}{1} \cdot \frac{m^{0-(-3)}}{1} \cdot \frac{n^{-2-4}}{1}$

$= \frac{3}{1} \cdot \frac{m^3}{1} \cdot \frac{n^{-6}}{1} = \frac{3}{1} \cdot \frac{m^3}{1} \cdot \frac{1}{n^6} = \frac{3m^3}{n^6}$

79. y^2z^5 **81.** $\frac{8}{5h^3}$ **83.** $\frac{2b^4}{3a}$ **85.** x^n **87.** y^{7n}

89. $x^{3m} \cdot x^{-m} = x^{3m-m} = x^{2m}$ **91.** $(x^{3b})^{-2} = x^{3b(-2)} = x^{-6b} = \frac{1}{x^{6b}}$

93. $\frac{x^{2a}}{x^{-5a}} = x^{2a-(-5a)} = x^{7a}$ **95.** $\frac{x + \frac{1}{y}}{y} \cdot \frac{y}{y} = \frac{xy + 1}{y^2}$

Exercises 15.2 (page 447)

1. x^5y^5 **3.** $\frac{16}{x^2}$ **5.** $\frac{1}{5x}$ **7.** $\frac{x^2}{25}$ **9.** a^4b^6

11. $4z^6$ **13.** $\frac{n^4}{m^8}$ **15.** $\frac{x^8}{y^{12}}$ **17.** $\frac{a^6}{27}$ **19.** $\frac{r^3s^9}{t^{12}}$

21. $R^{12}S^6$ **23.** $\frac{1}{M^8N^{12}}$ **25.** a^4b^2

27. $\left(\frac{10^2 \cdot 10^{-1}}{10^{-2}}\right)^2 = (10^{2-1+2})^2$
$= (10^3)^2 = 10^{3 \cdot 2} = 10^6$
29. m^4n^2 **31.** $\frac{1}{x^5y^2}$

33. $\left(\frac{\overset{2}{\cancel{8}}s^{-3}}{\underset{1}{\cancel{4}}st^2}\right)^{-2} = (2^1s^{-3-1}t^{-2})^{-2} = 2^{1(-2)}s^{-4(-2)}t^{-2(-2)}$
$= 2^{-2}s^8t^4 = \frac{s^8t^4}{2^2} = \frac{s^8t^4}{4}$
35. k^8

37. $(x^3 + y^4)^5$
Cannot be simplified by rules of exponents. Rule 2 or Rule 8 *cannot* be used here because the + sign means that x^3 and y^4 are *not* factors.

39. 1 **41.** $\frac{m^2}{6p}$ **43.** 1,000,000 **45.** $\frac{1}{x^5y^2}$ **47.** x^4y^{10}

49. $\frac{z}{8y}$

Exercises 15.3 (page 451)

	Common notation	*Scientific notation*
1.	$7_\wedge48.$	7.48×10^2
3.	$0.06_\wedge3$	6.3×10^{-2}
5.	$0.001_\wedge732$	1.732×10^{-3}
7.	$3_\wedge470{,}000.$	3.47×10^6
9.	$0.000001_\wedge91$	1.91×10^{-6}
11.	$0.00005_\wedge63$	5.63×10^{-5}
13.	$0.00005_\wedge8$	5.8×10^{-5}
15.	$0.000000000000000000000000006_\wedge547$	$= 6.547 \times 10^{-27}$
17.	$0.0000000004_\wedge77$	4.77×10^{-10}
19.	$8_\wedge3{,}600{,}000.$	8.36×10^7

21. 8.5×10^{-3} **23.** 3×10^5 **25.** 8.12×10 **27.** 1.5×10^{-2} **29.** 6×10^4 **31.** 3.6×10^{10} **33.** 3.75×10^{-12} **35.** 2.28×10^{12}

Exercises 15.4 (page 453)

1. A rational number is any number that can be expressed in the form $\frac{a}{b}$, where a and b are integers and b is not zero.

Therefore, $-5 = \frac{-5}{1}$; $\frac{3}{4}$; $2\frac{1}{2} = \frac{5}{2}$; $3.5 = \frac{7}{2}$; $\sqrt{4} = \frac{2}{1}$; and $3 = \frac{3}{1}$ are rational numbers.

3. a. 17 **b.** $x + 1$ **5.** 6 **7.** 1 **9.** 9 **11.** -2 **13.** $\frac{4}{7}$

Exercises 15.5A (page 457)

1. z **3.** x^2 **5.** $4x^3$ **7.** $2\sqrt{3}$ **9.** $3\sqrt{2}$ **11.** $2\sqrt{2}$ **13.** $2\sqrt{6}$ **15.** $x\sqrt{x}$ **17.** $m^3\sqrt{m}$ **19.** $hk\sqrt{h}$ **21.** $6\sqrt{7}$ **23.** $20\sqrt{2}$ **25.** $5x\sqrt{3x}$ **27.** $2x^2y^3\sqrt{5}$ **29.** $6x^{18}$ **31.** $2xy\sqrt{5y}$ **33.** $4x^2y\sqrt{xy}$ **35.** $3ab^2\sqrt{3a}$ **37.** $5x^2y\sqrt{10y}$ **39.** 13 **41.** 3 **43.** x **45.** $5x$ **47.** 5 **49.** $6x$

Exercises 15.5B (page 459)

1. $\frac{2}{9}$ **3.** $\frac{3}{7}$ **5.** $\frac{a}{b^3}$ **7.** $\frac{x^3}{v^4}$ **9.** $\frac{2x}{3}$ **11.** $\frac{m}{3}$ **13.** $\frac{x}{5}$ **15.** $\frac{2\sqrt{3}}{5}$ **17.** $\frac{5\sqrt{2}}{9}$ **19.** $\frac{x\sqrt{x}}{y^2}$ **21.** $\frac{\sqrt{2}}{2}$ **23.** $\frac{\sqrt{a}}{a}$ **25.** $\frac{2\sqrt{5}}{5}$ **27.** $\frac{8}{\sqrt{2}} \cdot \frac{\sqrt{2}}{\sqrt{2}} = \frac{8\sqrt{2}}{2} = 4\sqrt{2}$ **29.** $2\sqrt{x}$

Exercises 15.6 (page 460)

1. $7\sqrt{3}$ **3.** $4\sqrt{5}$ **5.** $9\sqrt{x}$ **7.** $6\sqrt{7}$ **9.** $8\sqrt{2} - 4\sqrt{3}$ **11.** $8\sqrt{x} - 2\sqrt{y}$ **13.** $7\sqrt{2}$ **15.** $\sqrt{2}$ **17.** $15\sqrt{3}$ **19.** $-4\sqrt{3}$ **21.** $7\sqrt{x}$ **23.** $5 - 2\sqrt{5}$ **25.** $7\sqrt{3}$ **27.** $5\sqrt{5} - 6\sqrt{3}$ **29.** $8 - 7\sqrt{2}$

Exercises 15.7 (page 462)

1. 6 **3.** $3x$ **5.** $6a^2$ **7.** $2\sqrt{3}$ **9.** $3x\sqrt{2}$ **11.** $5a\sqrt{2a}$ **13.** 15 **15.** 24

17.
$$\begin{aligned}\sqrt{12}\cdot 5\sqrt{2} &= 5\sqrt{12\cdot 2}\\ &= 5\sqrt{24}\\ &= 5\sqrt{4\cdot 6}\\ &= 5\cdot 2\sqrt{6}\\ &= 10\sqrt{6}\end{aligned}$$

19. $12\sqrt{10}$ **21.** $2 + \sqrt{2}$ **23.** $10 + 3\sqrt{5}$ **25.** $x - 2\sqrt{x}$ **27.** $6 - 2\sqrt{3}$ **29.** $13 + 5\sqrt{7}$ **31.** $22 - 11\sqrt{5}$

33.
$$\begin{aligned}&(5\sqrt{3} - 2)(2\sqrt{3} + 4)\\ &= 5\sqrt{3}\cdot 2\sqrt{3} + 4\cdot 5\sqrt{3} - 2\cdot 2\sqrt{3} - 2\cdot 4\\ &= 30 + 20\sqrt{3} - 4\sqrt{3} - 8\\ &= 22 + 16\sqrt{3}\end{aligned}$$

35. 7 **37.** 2 **39.** $19 + 8\sqrt{3}$ **41.** $9 - 6\sqrt{x} + x$ **43.** $8 + 2\sqrt{15}$

Exercises 15.8 (page 464)

1. 2 **3.** $\frac{1}{2}$ **5.** x **7.** $\frac{1}{a^2}$ **9.** $3x^2$ **11.** $2\sqrt{2}$ **13.** $2\sqrt{3}$ **15.** $\frac{\sqrt{3}}{3}$ **17.** $\frac{\sqrt{15}}{5}$ **19.** $\frac{\sqrt{6}}{3}$

21. $\frac{2}{\sqrt{8}} \cdot \frac{\sqrt{2}}{\sqrt{2}} = \frac{2\sqrt{2}}{\sqrt{16}} = \frac{2\sqrt{2}}{4} = \frac{\sqrt{2}}{2}$

23. $\frac{\sqrt{3}}{2}$ **25.** $\sqrt{2} + 1$ **27.** $\sqrt{5} - \sqrt{2}$ **29.** $3\sqrt{2} + 3$

31.
$$\begin{aligned}\frac{6}{\sqrt{5}+\sqrt{2}} \cdot \frac{\sqrt{5}-\sqrt{2}}{\sqrt{5}-\sqrt{2}} &= \frac{6(\sqrt{5}-\sqrt{2})}{5-2}\\ &= \frac{\overset{2}{\cancel{6}}(\sqrt{5}-\sqrt{2})}{\underset{1}{\cancel{3}}}\\ &= 2\sqrt{5} - 2\sqrt{2}\end{aligned}$$

33. $4 - 2\sqrt{2}$ **35.** $4\sqrt{6} + 8$ **37.** $\sqrt{x} - 2$

Exercises 15.9 (page 467)

1. 25 **3.** 64 **5.** 8 **7.** 7 **9.** 15 **11.** 40 **13.** 8 **15.** 6 **17.** 8 **19.** 2, 1 **21.** 5 **23.** $\frac{1}{2}$

25.
$$\begin{aligned}\sqrt{x-3} + 5 &= x\\ (\sqrt{x-3})^2 &= (x-5)^2\\ x - 3 &= x^2 - 10x + 25\\ 0 &= x^2 - 11x + 28\\ 0 &= (x-4)(x-7)\end{aligned}$$

$x - 4 = 0$, $x = 4$ | $x - 7 = 0$, $x = 7$

Check:

For $x = 7$:	For $x = 4$:
$\sqrt{x-3} + 5 = x$	$\sqrt{x-3} + 5 = x$
$\sqrt{7-3} + 5 \overset{?}{=} 7$	$\sqrt{4-3} + 5 \overset{?}{=} 4$
$\sqrt{4} + 5 \overset{?}{=} 7$	$\sqrt{1} + 5 \overset{?}{=} 4$
$2 + 5 \overset{?}{=} 7$	$1 + 5 \neq 4$
$7 = 7$	
Therefore, 7 is a solution.	Therefore, 4 is *not* a solution.

27. $x = 3$ **29.** $x = 6$

Review Exercises 15.10 (page 469)

1. x^{11} **2.** a^2 **3.** $\frac{1}{c^5}$ **4.** p^3 **5.** $\frac{1}{x^9}$ **6.** m^3

7. $10^4 = 10{,}000$ **8.** $10^7 = 10{,}000{,}000$ **9.** x^{10} **10.** $\frac{1}{p^{15}}$

11. m^8 **12.** 1 **13.** x^8y^{12} **14.** $\frac{p^2}{r^6}$ **15.** $\frac{2}{a^3}$ **16.** $9x^8$

17. $\frac{1}{16b^6}$ **18.** $\frac{1}{100}$ **19.** $\frac{x^{10}y^{15}}{z^{20}}$ **20.** $x^{12}z^8$

21. $\frac{w^{12}}{u^{15}v^6}$ **22.** $(5a^3b^{-4})^{-2} = 5^{-2}a^{-6}b^8 = \frac{b^8}{25a^6}$

23. $\left(\frac{4h^2}{ij^{-2}}\right)^{-3} = \frac{4^{-3}h^{-6}}{i^{-3}j^6} = \frac{i^3}{4^3h^6j^6} = \frac{i^3}{64h^6j^6}$

24. $\left(\frac{x^{10}y^5}{x^5y}\right)^3 = (x^5y^4)^3 = x^{15}y^{12}$

25. 1 **26.** x^{4d} **27.** $\frac{1}{x^{8a}}$ **28.** 6^{3x} **29.** x^2y^{-3}

30. m^2n^3 **31.** $a^3b^{-2}c^{-5}$ **32.** $\frac{1}{16}$ **33.** $\frac{1}{27}$ **34.** 8

35. $\frac{1}{100}$ **36.** $\frac{(-8)^2}{-8^2} = \frac{(-8)(-8)}{-(8)(8)} = \frac{64}{-64} = -1$

37. $\left(\frac{10^{-4}\cdot 10}{10^{-2}}\right)^5 = (10^{-4+1+2})^5 = (10^{-1})^5 = 10^{-5} = \frac{1}{10^5}$

38. 2.25×10^{-4} **39.** 9.6×10^5 **40.** $\frac{1}{200} = .005 = 5 \times 10^{-3}$

41. 7,800 **42.** 0.0000406 **43.** 0.01207 **44.** 9.6×10^{-2}

45. 3×10^6 **46.** 7×10^3

47. a. $\sqrt{3}, 2\sqrt{5}, \sqrt{5}$, because they cannot be expressed as a fraction.
b. $2\sqrt{5}$ and $\sqrt{5}$

48. a. $9x$ **b.** $\frac{1}{2}$ **c.** $x - 5$ **49.** 9 **50.** $4\sqrt{3}$ **51.** 8

52. a^3 **53.** $x\sqrt{x}$ **54.** $4xy^2$ **55.** $ab^2\sqrt{ab}$

56. $30\sqrt{2}$ **57.** $4\sqrt{5}$ **58.** $\frac{3}{x}$ **59.** $\sqrt{2}$ **60.** $\frac{\sqrt{10}}{5}$

61. $2\sqrt{3}$ **62.** $\sqrt{3}$ **63.** $10 + 3\sqrt{2}$ **64.** -4 **65.** $7 + 4\sqrt{3}$

66. $(3\sqrt{2} + 1)(2\sqrt{2} - 1)$
$= 3\sqrt{2}\cdot 2\sqrt{2} - 1\cdot 3\sqrt{2} + 1\cdot 2\sqrt{2} - 1\cdot 1$
$= 12 - 3\sqrt{2} + 2\sqrt{2} - 1$
$= 11 - \sqrt{2}$

67. $\frac{8}{\sqrt{3}-2}\cdot\frac{\sqrt{3}+2}{\sqrt{3}+2} = \frac{8(\sqrt{3}+2)}{3-4} = -8(\sqrt{3}+2)$
$= -8\sqrt{3} - 16$

68. $\frac{10}{\sqrt{6}-2}\cdot\frac{\sqrt{6}+2}{\sqrt{6}+2} = \frac{10(\sqrt{6}+2)}{6-4} = \frac{\overset{5}{\cancel{10}}(\sqrt{6}+2)}{\cancel{2}}$
$= 5\sqrt{6} + 10$

69. $x = 16$ **70.** $a = 12$ **71.** $x = 13$ **72.** $a = 3$

73. $(\sqrt{7x-6})^2 = (x)^2$
$7x - 6 = x^2$
$0 = x^2 - 7x + 6$
$0 = (x-6)(x-1)$
$x - 6 = 0 \mid x - 1 = 0$
$x = 6 \mid x = 1$

74. $\sqrt{2x-1} + 2 = x$
$(\sqrt{2x-1})^2 = (x-2)^2$
$2x - 1 = x^2 - 4x + 4$
$0 = x^2 - 6x + 5$
$0 = (x-5)(x-1)$
$x - 5 = 0 \mid x - 1 = 0$
$x = 5 \mid x = 1$

Check: $x = 5$
$\sqrt{2(5)-1} + 2 \overset{?}{=} 5$
$\sqrt{9} + 2 \overset{?}{=} 5$
$3 + 2 = 5$

Check: $x = 1$
$\sqrt{2(1)-1} + 2 \overset{?}{=} 1$
$\sqrt{1} + 2 \overset{?}{=} 1$
$1 + 2 \neq 1$

Therefore, $x = 5$ is the only solution.

Chapter 15 Diagnostic Test (page 471)

Following each problem number is the textbook section reference (in parentheses) in which that kind of problem is discussed.

1. (15.1) $a^{-6}\cdot a^2 = a^{-6+2} = a^{-4} = \frac{1}{a^4}$

2. (15.2) $(a^{-3}b)^2 = a^{-3\cdot 2}b^{1\cdot 2} = a^{-6}b^2 = \frac{b^2}{a^6}$

3. (15.1) $x^{3a}\cdot x^{2a} = x^{3a+2a} = x^{5a}$

4. (15.1) $\frac{x^5}{x^{-2}} = x^{5-(-2)} = x^7$

5. (15.2) $\left(\frac{3x^{-4}}{y^2}\right)^2 = \frac{3^2x^{-8}}{y^4} = \frac{9}{x^8y^4}$

6. (15.2) $\left(\frac{x^5y}{x^2y^3}\right)^3 = (x^{5-2}y^{1-3})^3$
$= (x^3y^{-2})^3$
$= x^9y^{-6} = \frac{x^9}{y^6}$

7. (15.1) $(10^{-3})^2 = 10^{-6} = \frac{1}{10^6} = \frac{1}{1{,}000{,}000}$

8. (15.1) $\frac{2^{-2}\cdot 2^3}{2^{-4}} = 2^{-2+3-(-4)} = 2^5 = 32$

9. (15.1) $\frac{-4^2}{(-4)^2} = \frac{-16}{16} = -1$

10. (15.1) $\frac{h^{-2}}{k^{-3}h^{-4}} = k^3h^{-2-(-4)} = k^3h^2$

11. (15.3) **a.** $723{,}000 = 7.23 \times 10^5$
b. $0.0048 = 4.8 \times 10^{-3}$

12. (15.3) $\frac{4200}{0.00014} = \frac{4.2 \times 10^3}{1.4 \times 10^{-4}}$
$= 3 \times 10^{3-(-4)}$
$= 3 \times 10^7$

13. (15.7) $\sqrt{2}\sqrt{18x^2} = \sqrt{2\cdot 18x^2} = \sqrt{36x^2} = 6x$

14. (15.7) $\sqrt{3}(2\sqrt{3} - 5) = (\sqrt{3})(2\sqrt{3}) + (\sqrt{3})(-5) = 6 - 5\sqrt{3}$

15. (15.5A) $\sqrt{72} = \sqrt{36\cdot 2} = 6\sqrt{2}$

16. (15.5B) $\sqrt{\frac{18}{2m^2}} = \sqrt{\frac{9}{m^2}} = \frac{3}{m}$

17. (15.8) $\frac{\sqrt{16}}{\sqrt{50}} = \sqrt{\frac{16}{50}} = \sqrt{\frac{8}{25}} = \frac{\sqrt{4 \cdot 2}}{\sqrt{25}} = \frac{2\sqrt{2}}{5}$

18. (15.8) $\frac{4}{\sqrt{3}+1} \cdot \frac{\sqrt{3}-1}{\sqrt{3}-1} = \frac{4(\sqrt{3}-1)}{3-1}$

$= \frac{\overset{2}{\cancel{4}}(\sqrt{3}-1)}{\underset{1}{\cancel{2}}}$

$= 2\sqrt{3} - 2$

19. (15.5A) $\sqrt{12x^4y^3} = \sqrt{4 \cdot 3 \cdot x^4 \cdot y^2 \cdot y}$
$= 2x^2y\sqrt{3y}$

20. (15.6) $2\sqrt{20} + \sqrt{45} = 2\sqrt{4 \cdot 5} + \sqrt{9 \cdot 5}$
$= 2 \cdot 2\sqrt{5} + 3\sqrt{5}$
$= 4\sqrt{5} + 3\sqrt{5}$
$= 7\sqrt{5}$

21. (15.7) $(3\sqrt{2} + 5)(2\sqrt{2} + 1)$
$= 3\sqrt{2} \cdot 2\sqrt{2} + 1 \cdot 3\sqrt{2} + 5 \cdot 2\sqrt{2} + 5 \cdot 1$
$= 12 + 3\sqrt{2} + 10\sqrt{2} + 5$
$= 17 + 13\sqrt{2}$

22. (15.5B) $\sqrt{\frac{4}{5}} = \frac{\sqrt{4}}{\sqrt{5}} = \frac{2}{\sqrt{5}} \cdot \frac{\sqrt{5}}{\sqrt{5}} = \frac{2\sqrt{5}}{5}$

23. (15.6) $\sqrt{28} + \sqrt{75} - \sqrt{27} =$
$\sqrt{4 \cdot 7} + \sqrt{25 \cdot 3} - \sqrt{9 \cdot 3} =$
$2\sqrt{7} + 5\sqrt{3} - 3\sqrt{3} = 2\sqrt{7} + 2\sqrt{3}$

24. (15.9) $(\sqrt{4x+5})^2 = (5)^2$
$4x + 5 = 25$
$4x = 20$
$x = 5$

Check: $\sqrt{4x+5} = 5$
$\sqrt{4(5)+5} \stackrel{?}{=} 5$
$\sqrt{25} \stackrel{?}{=} 5$
$5 = 5$

25. (15.9) $(\sqrt{5x-6})^2 = (x)^2$
$5x - 6 = x^2$
$x^2 - 5x + 6 = 0$
$(x-3)(x-2) = 0$
$x - 3 = 0 \quad | \quad x - 2 = 0$
$x = 3 \quad | \quad x = 2$

Check: $x = 3$
$\sqrt{5x-6} = x$
$\sqrt{5(3)-6} \stackrel{?}{=} 3$
$\sqrt{9} \stackrel{?}{=} 3$
$3 = 3$

Check: $x = 2$
$\sqrt{5x-6} = x$
$\sqrt{5(2)-6} \stackrel{?}{=} 2$
$\sqrt{4} \stackrel{?}{=} 2$
$2 = 2$

Exercises 16.1 (page 479)

1. **a.** Obtuse **b.** Straight **c.** Right **d.** Acute

3. 65° 5. 107° 7. 45° and 135° 9. 30° and 60°

11. 110° 13. $\angle 1 = 65°$, $\angle 2 = 25°$ 15. $\angle CBE = 35°$, $\angle BED = 35°$, $\angle DEF = 145°$

17. $\angle 1 = 60°$
$\angle 3 = 45°$
$\angle 4 = 120°$
$\angle 5 = 60°$
$\angle 7 = 135°$

19. **a.** Quadrilateral
b. Pentagon
c. Triangle
d. Hexagon

Exercises 16.2 (page 484)

1. **a.** Equilateral
b. Isosceles
c. Scalene
d. Isosceles Right

3. 34° 5. 99°

7. $x + x + x = 180$
$3x = 180$
$x = 60°$

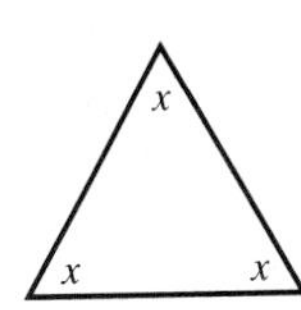

9. 52° 11. 30° and 60°

13. Let $3x = \angle A$
$x = \angle B$
$x - 10 = \angle C$
$3x + x + x - 10 = 180$
$5x - 10 = 180$
$5x = 190$
$x = 38$

Then $\angle A = 3x = 3(38) = 114°$
$\angle B = x = 38°$
$\angle C = x - 10 = 38 - 10 = 28°$

15. $\angle A = 58°$
$\angle ACD = 32°$
$\angle BCD = 58°$

17. $\angle 2 = 70°$
$\angle 3 = 110°$
$\angle 4 = 80°$
$\angle 6 = 70°$
$\angle 7 = 30°$

19. $5^2 + 12^2 \stackrel{?}{=} 13^2$
$25 + 144 \stackrel{?}{=} 169$
$169 = 169$
Is a right triangle.

21. Is not a right triangle.

23. $x = 20$ 25. $x = 3$ 27. $x = 3\sqrt{2}$

29. $(x+1)^2 + (\sqrt{20})^2 = (x+3)^2$
$x^2 + 2x + 1 + 20 = x^2 + 6x + 9$
$x^2 + 2x + 21 = x^2 + 6x + 9$
$21 = 4x + 9$
$12 = 4x$
$3 = x$

31. 6 and 8

33. $x^2 + b^2 = (2x)^2$
$x^2 + b^2 = 4x^2$
$b^2 = 3x^2$
$\sqrt{b^2} = \sqrt{3x^2}$
$b = x\sqrt{3}$

35. **a.**

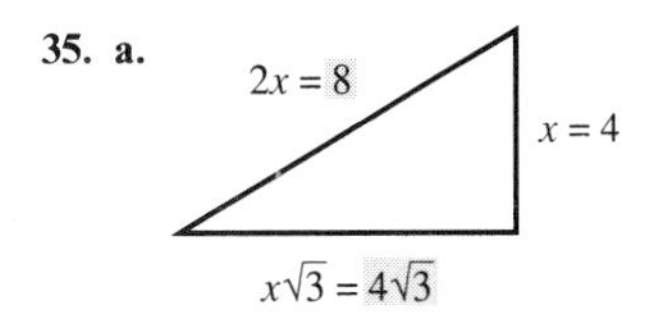

b.

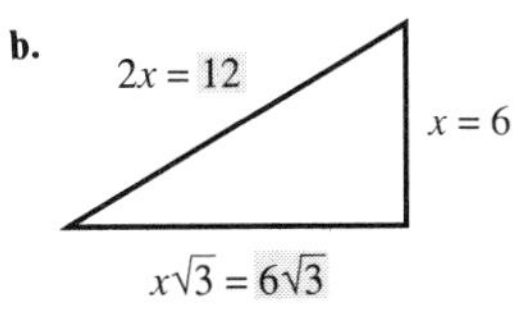

c.

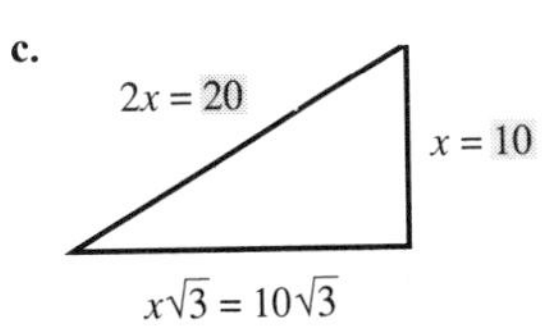

d.

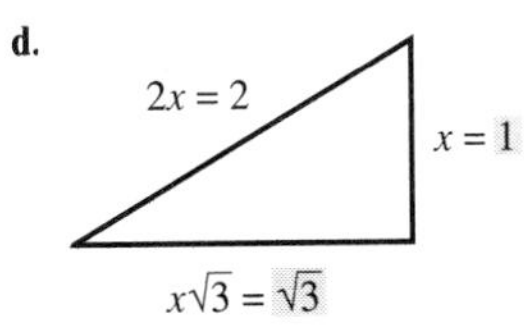

Exercises 16.3 (page 489)

1. $\angle ABD = \angle CDB$
$\angle ADB = \angle CBD$
$\angle A = \angle C$
$DB = DB$
$AB = DC$
$AD = BC$

3. $\angle A = \angle B$
$\angle ADF = \angle BEF$
$\angle DFA = \angle EFB$
$AF = BF$
$AD = BE$
$DF = EF$

5. $\angle ACD = \angle CAB$
$\angle ADC = \angle CBA$
$\angle DAC = \angle BCA$
$AC = AC$
$AD = CB$
$DC = AB$

7. $\angle F = 50°$
$\angle FBE = 40°$

9. $\angle D = 60°$
$\angle ECB = 55°$

11. $x = 6$
$y = 10$

13. $x = 5$
$y = 6$

15. $\triangle ACD \cong \triangle ACB$ by SSS **17.** $\triangle ABC \cong \triangle DCB$ by ASA

19. $\triangle ADC \cong \triangle BDC$ by SAS **21.** $\triangle AFD \cong \triangle BFE$ by ASA

Exercises 16.4 (page 493)

1. $DF = 9$
$EF = 6$

3. $AC = 16$
$EF = 8$

5. $AE = 14$ **7.** $BC = 12$

9. $\frac{6}{9} = \frac{4}{4 + BD}$
$36 = 24 + 6BD$
$12 = 6BD$
$2 = BD$

Exercises 16.5 (page 496)

1. 22 ft **3.** 50 in. **5.** 4π in. **7.** 16 in. **9.** 6 ft

11. $(6\pi + 12)$ ft **13.** $(4\pi + 24)$ cm

15.

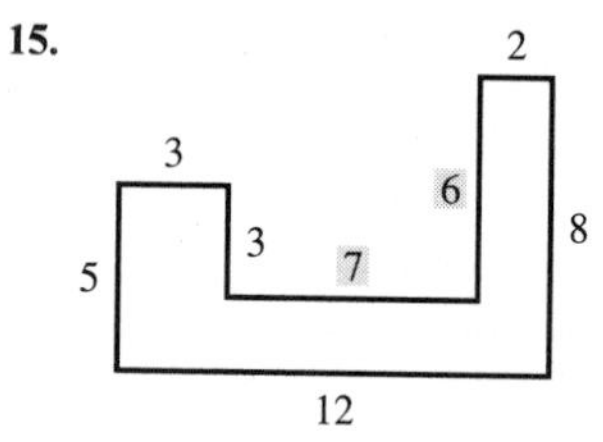

$P = 3 + 3 + 7 + 6 + 2 + 8 + 12 + 5 = 46$ yd

17. 30 cm **19.** 46 in.

21. Let $x =$ width
$2x - 5 =$ length
$2\ell + 2w = P$
$2(2x - 5) + 2x = 38$
$4x - 10 + 2x = 38$
$6x = 48$
$x = 8$
Width $= x = 8$ yd
Length $= 2x - 5 = 2(8) - 5 = 11$ yd

x
$2x - 5$

Exercises 16.6 (page 501)

1. 96 sq. ft **3.** $7\frac{1}{2}$ sq. in. **5.** 64π sq. ft **7.** 25 sq. in.

9. 18π sq. ft **11.** $(8\pi + 80)$ m^2 **13.** 15 cm^2

15. $(9\pi + 72)$ sq. in.

17. $A = \frac{1}{2}bh$
$30 = \frac{1}{2}b(12)$
$30 = 6b$
$5 = b$
$a^2 + b^2 = c^2$
$12^2 + 5^2 = c^2$
$144 + 25 = c^2$
$169 = c^2$
$13 = c$
Leg is 5 in.
Hypotenuse is 13 in.

c
$a = 12$
b

19. 8 m **21.** 9 sq. ft **23.** 196 sq. in.

25. $C = 2\pi r$
$\frac{10\pi}{2\pi} = \frac{2\pi r}{2\pi}$
$5 = r$
$A = \pi r^2$
$A = \pi 5^2$
$A = 25\pi$ sq. ft

27. 26 in. **29.** 9 times

Exercises 16.7 (page 505)

1. 900 cu. ft **3.** 640π cu. in. **5.** 288π cu. in. **7.** 112 cm^3

9. $(24\pi + 200)$ cu. ft **11.** 1,728 cu. in. **13.** 8 in.

15. a. 3600 in.^{3}
b. ≈ 130 lb

17. a. $V = \pi r^2 h$
$V = \pi(10)^2(50)$
$V = 5000\pi$ cu. in.
b. $\frac{5000(3.14)}{231} \approx 68$ gal

19. 27 times

Review Exercises 16.8 (page 509)

1. 15° and 75° **2.** 70° and 110°

3. $\angle 1 = 85°$
$\angle 2 = 30°$
$\angle 3 = 65°$
$\angle 4 = 150°$
$\angle 5 = 115°$

4. $\angle 2 = 40°$
$\angle 3 = 105°$
$\angle 4 = 75°$
$\angle 5 = 75°$
$\angle 6 = 105°$

5. 128° **6.** 30°

7. 12 **8.** 4 **9.** $2\sqrt{3}$ **10.** $2\sqrt{5}$ cm **11.** $8\sqrt{2}$ in.

12. $\angle C = 50°$
$\angle DBE = 40°$

13. $\angle DAE = 37°$
$AE = 4$

14. $\triangle ADE \cong \triangle BCE$ by ASA

15. $\triangle ADB \cong \triangle BCA$ by SAS

16. DE = 6 **17.** Let $x = EC$
$\frac{8}{6} = \frac{3 + x}{x}$
$8x = 18 + 6x$
$2x = 18$
$x = 9 = EC$

18. $P = 60$ cm
$A = 150$ cm^2

19. $P = 28$ in.
$A = 24$ in.^{2}

20. $P = 36$ ft
$A = 60$ ft^2

21. $P = (5\pi + 10)$ m
$A = 12\frac{1}{2}\pi$ m^2

22. $P = 42$ yd
$A = 57$ sq. yd

23. $P = (4\pi + 24)$ ft
$A = (8\pi + 64)$ sq. ft

24. Area of shaded = Area of Rectangle − Area of Square

$= \ell \cdot w - s^2$
$= 8 \cdot 6 - 4^2$
$= 48 - 16$
$= 32$ sq. in.

25. 5π mm^2 26. $(144 - 36\pi)$ sq. ft 27. 28 m^2
28. 24 cm 29. 16π ft 30. 24 in. 31. 4 in.
32. 28 m 33. 64 cu. ft 34. 3 cm

35. $V = \frac{1}{2}\left(\frac{4}{3}\pi r^3\right)$
$V = \frac{1}{2} \cdot \frac{4}{3} \cdot \pi \cdot 6^3$
$V = \frac{1}{\not{2}} \cdot \frac{4}{\not{3}} \cdot \pi \cdot \not{216}$ (216 → 108 → 36)
$V = 144\pi$ m^3

36. $\frac{3}{4}V = \frac{3}{4}(\pi r^2 h)$ (Tank is $\frac{3}{4}$ full)
$= \frac{3}{4}\pi(4)^2(8)$
$= \frac{3}{\not{4}} \cdot \pi \cdot \not{16} \cdot 8$ (16 → 4)
$= 96\pi$ cu. ft

Chapter 16 Diagnostic Test (page 513)

Following each problem number is the textbook section reference (in parentheses) in which that kind of problem is discussed.

1. (16.1) $\angle 2 = 75°$
$\angle 3 = 80°$
$\angle 4 = 155°$
$\angle 5 = 25°$
$\angle 7 = 100°$

2. (16.2) $\angle A + \angle B + \angle ACB = 180°$
$30 + \angle B + 115 = 180$
$\angle B = 35°$
$\angle B + \angle BDC + \angle BCD = 180°$
$35 + 90 + \angle BCD = 180$
$\angle BCD = 55°$

3. (16.2) $x^2 + 1^2 = 3^2$
$x^2 + 1 = 9$
$x^2 = 8$
$\sqrt{x^2} = \sqrt{8}$
$x = 2\sqrt{2}$

4. (16.2) $2^2 + (\sqrt{5})^2 = x^2$
$4 + 5 = x^2$
$9 = x^2$
$3 = x$

5. (16.1) Let x = one angle
$x + 30$ = other angle
$x + x + 30 = 180$
$2x = 150$
$x = 75$
$x + 30 = 105$
Angles are 75° and 105°

6. (16.5) $9^2 + 12^2 = c^2$
$81 + 144 = c^2$
$225 = c^2$
$15 = c$
$P = a + b + c$
$P = 9 + 12 + 15$
$P = 36$ in.

7. (16.5) $P = 4s$
$20 = 4s$
$5 = s$
$5^2 + 5^2 = d^2$
$25 + 25 = d^2$
$50 = d^2$
$\sqrt{50} = \sqrt{d^2}$
$5\sqrt{2}$ cm $= d$

8. (16.6) $A = b \cdot h$
$A = 8 \cdot 4$
$A = 32$ sq. ft

9. (16.5)
$P = 3 + 1 + 2 + 2 + 5 + 3 = 16$ m

10. (16.6) Area of Triangle $= A_1 = \frac{1}{2}bh$
$= \frac{1}{2}(6)(2)$
$= 6$ sq. in.
Area of Square $= A_2 = s^2$
$= 6^2$
$= 36$ sq. in.
Total Area $= A_1 + A_2$
$= 6 + 36$
$= 42$ sq. in.

11. (16.6) $A = \ell w$
$54 = \ell \cdot 6$
$9 = \ell$
$P = 2\ell + 2w$
$P = 2(9) + 2(6)$
$P = 18 + 12$
$P = 30$ ft

12. (16.6) $C = 2\pi r$
$\frac{12\pi}{2\pi} = \frac{2\pi r}{2\pi}$
$6 = r$
$A = \pi r^2$
$A = \pi 6^2$
$A = 36\pi$ sq. ft

13. (16.7) $V = \ell wh$
$V = 12 \cdot 3 \cdot 5$
$V = 180$ cu. in.

14. (16.7) $d = 2r$
$6 = 2r$
$3 = r$
$V = \frac{4}{3}\pi r^3$
$V = \frac{4}{3}\pi(3)^3$
$V = \frac{4}{\not{3}} \cdot \pi \cdot \not{27}$ (27 → 9)
$V = 36\pi$ m^3

15. (16.4) $\frac{AB}{4} = \frac{12}{6}$
$6AB = 48$
$AB = 8$

16. (16.4) Let $x = AD$
then $x + 5 = AC$
$\frac{4}{8} = \frac{5}{x + 5}$
$4x + 20 = 40$
$4x = 20$
$x = 5$

17. (16.3) $\angle DCB = \angle ABC = 90°$
$\angle DCB + \angle D + \angle CBD = 180$
$90 + 35 + \angle CBD = 180$
$\angle CBD = 55°$

18. (16.3) $AD = AD$ common side
$\triangle ABD \cong \triangle ACD$ by SAS

Cumulative Review Exercises: Chapters 1–16 (page 515)

1. $14\frac{11}{18}$ **2.** $2\frac{1}{3}$ **3.** 87.14 **4.** 40.803 **5.** -8

6. $3x^2 - 4x - 3$ **7.** $15x^2 - 14xy - 8y^2$ **8.** $16x^2 - 24x + 9$

9. $\frac{1}{3x^2}$ **10.** $\frac{6x}{x-6}$ **11.** $x + 3$ **12.** $\frac{x^2 - 5x - 8}{x(x+4)}$

13. $x - 5$ **14.** $x = 3$ **15.** $x = \frac{5}{3}$ or $x = -2$

16. $x = 2$ **17.** $x = 5$ **18.** $(-1, 2)$ **19.** $(5, -2)$

20. $x < -5$

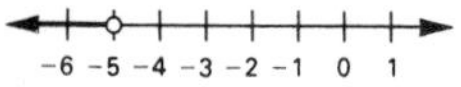

−6 −5 −4 −3 −2 −1 0 1

21. $L = \frac{P - 2W}{2}$ **22.** $\frac{9x^6}{y^8}$ **23.** $\frac{b}{a^5}$ **24.** $\frac{2y^5}{x^3}$

25. $2x^2y\sqrt{3y}$ **26.** $2\sqrt{3} + 6\sqrt{5}$ **27.** $-2 + 17\sqrt{2}$

28. ≈ 5.22 **29.** $2x^2 - 3x + 4$ **30.** $\frac{19}{250}$ **31.** -31

32. 8.525 **33.** 6×10^{-7}

34.

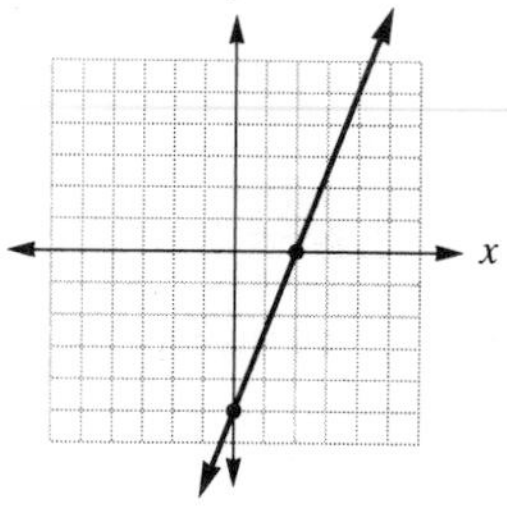

y

x

35.

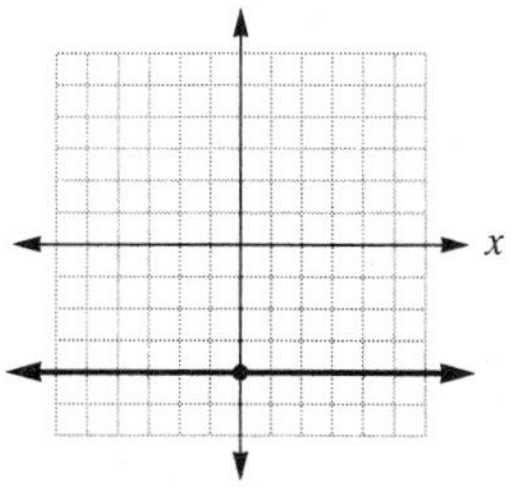

y

x

36. $4x + y = 18$ **37.** slope $= -\frac{3}{2}$, y-intercept $= 4$ **38.** $4\frac{1}{4}$ hr

39. $7\frac{1}{2}$ lb **40.** 85% **41.** 8,000 students **42.** 45 mph

43. 15 dimes **44.** $y = 15$ **45.** 45 sq. ft **46.** 6π

47. $AC = 2\sqrt{5}$ **48.** $\angle CED = 48°$

49. Between 1988 and 1989 **50.** $1,500,000

Exercises II.2A (page 525)

1. 1,800 m **3.** 249 g **5.** 0.0605 $k\ell$ **7.** 0.275 kg

9. 780 m **11.** 72.35 kg **13.** 0.125 $k\ell$

Exercises II.2B (page 526)

1. 2.79 m **3.** 0.083 g **5.** 250 cm **7.** 390.6 cg

9. 6.32 ℓ **11.** 2.63 cg **13.** 182 cm

Exercises II.2C (page 527)

1. 0.091 m **3.** 0.47 g **5.** 2,600 mm **7.** 108 mg

9. 0.23 ℓ **11.** 7,040 mg **13.** 2,000 cc

15. 0.175 ℓ **17.** 1,500 mg

Exercises II.2D (page 529)

1. 3,540 m **3.** 47,000 ℓ **5.** 2.546 kg **7.** 0.34 ℓ

9. 516 dm **11.** 0.895 $h\ell$ **13.** 0.784 km **15.** 456,000 dg

17. 0.3402 dag **19.** 0.5614 $da\ell$ **21.** 3,750 mm

Exercises II.2E (page 530)

1. 68°F **3.** 10°C **5.** ≈46°F **7.** ≈22°C **9.** 5°C

Exercises II.4 (page 534)

1. 15 ft **3.** $\frac{3\frac{1}{3}\,\cancel{\text{yd}}}{1}\left(\frac{3\text{ ft}}{1\,\cancel{\text{yd}}}\right) = \frac{10}{3} \times \frac{3}{1}\text{ ft} = 10\text{ ft}$

5. 32 qt **7.** 14 pt **9.** 90 min **11.** 9,240 ft

13. 30 ft, 360 in. **15.** 2 ft **17.** $1\frac{1}{2}$ min **19.** 1 mi **21.** 2 yr

23. 6 tbsp **25.** 4 tsp **27.** 40 oz **29.** 5,000 lb **31.** 6,600 ft

33. $\frac{2\,\cancel{\text{wk}}}{1}\left(\frac{7\,\cancel{\text{da}}}{1\,\cancel{\text{wk}}}\right)\left(\frac{24\text{ hr}}{1\,\cancel{\text{da}}}\right) = 2 \times 7 \times 24\text{ hr} = 336\text{ hr}$

35. $\frac{2{,}160\,\cancel{\text{min}}}{1}\left(\frac{1\,\cancel{\text{hr}}}{60\,\cancel{\text{min}}}\right)\left(\frac{1\text{ day}}{24\,\cancel{\text{hr}}}\right) = \frac{2{,}160}{60 \times 24}\text{ day} = 1.5\text{ day}$

37. 40 mph **39.** $\frac{1{,}100\,\cancel{\text{ft}}}{1\,\cancel{\text{sec}}}\left(\frac{60\,\cancel{\text{sec}}}{1\,\cancel{\text{min}}}\right)\left(\frac{60\,\cancel{\text{min}}}{1\text{ hr}}\right)\left(\frac{1\text{ mi}}{5{,}280\,\cancel{\text{ft}}}\right) = \frac{1{,}100 \times 60 \times 60\text{ mi}}{5280\text{ hr}} = 750\text{ mph}$

Exercises II.5 (page 537)

1. 5 ft 3 in. **3.** 3 wk 2 day **5.** 6 gal 3 qt **7.** 7 yd 3 in.

9. 3 hr 14 min 50 sec **11.** 5 gal 1 qt 1 pt **13.** 3 mi 720 ft

15. 3 yr 6 wk 5 day **17.** 3 tons 1,500 lb **19.** 5 lb 4 oz

21. 3 da 15 min

Exercises II.6 (page 538)

1. 13 ft 1 in. **3.** 13 hr 29 min **5.** 17 gal 3 qt 1 pt

7. 14 yd 1 ft 2 in. **9.** 10 day 6 hr 56 min **11.** 16 tons 1,800 lb

13. 45 lb 13 oz **15.** 10 yd 1 ft 1 in.

Exercises II.7 (page 539)

1. 5 ft 6 in. **3.** 6 yd 1 ft **5.**

$$\begin{array}{r} \overset{7}{\cancel{8}}\text{ gal }\overset{\overset{5}{\cancel{1}}}{\cancel{2}}\text{ qt }\overset{2}{\cancel{0}}\text{ pt} \\ -\,3\qquad 3\qquad 1\quad \\ \hline 4\text{ gal }2\text{ qt }1\text{ pt} \end{array}$$

7. 2 lb 11 oz **9.** 1 ton 1,500 lb **11.** 1 da 14 hr
13. 2 yd 9 in. **15.** 2 mi 3,580 ft **17.** 14 min 45 sec
19. 2 da 18 hr 37 min **21.** 2 hr 55 min 45 sec

Exercises II.8 (page 541)

1. 14 wk 6 da **3.** 33 mi 1,260 ft **5.** 12 yd 3 in.

7. 5 hr 40 min 44 sec **9.**

```
   3 gal   3 qt  1 pt
×                 8
24 gal  24 qt  8 pt
```

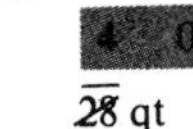

28 qt

31 gal

11. 8,448 sq. ft **13.** 8,100 sq. ft

15. Man-hours = 17 × 12 × 8 = 1,632
Cost = \$6.50 × 1,632 = \$10,608

Exercises II.9 (page 543)

1. 1 ft 3 in.

3. 4 ⟌ 2 qt 1 pt, quotient: 0 qt $1\frac{1}{4}$ pt

```
2 qt   1 pt
  └──→ 4
       5 pt
       4
       1 pt → 1 pt/4 = 1/4 pt
```

5. 1 hr 50 min **7.** 1 gal 2 qt 1 pt

9. 5 ⟌ 8 yd 2 ft 10 in., quotient: 1 yd 2 ft $4\frac{2}{5}$ in.

```
5 ⟌ 8 yd   2 ft   10 in.
    5
    3 yd =  9 ft
           11 ft
           10 ft
            1 ft = 12 in.
                   22 in.
                   20
                    2 in. → 2 in./5 = 2/5 in.
```

11. $12\frac{4}{7}$ oz **13.** 4 wk 4 da 5 hr **15.** 1 mi 2,510 ft **17.** 8

19. 11 postcards with 10¢ left over **21.** 45

Exercises II.10 (page 545)

1. ≈38 cm **3.** ≈2.00 ℓ **5.** ≈1.2 lb **7.** ≈51 mi
9. ≈13 qt **11.** ≈53 km **13.** ≈9.36 kg **15.** ≈370 acres
17. ≈1.7 m **19.** ≈80 in. **21.** ≈2.00 lb **23.** ≈340 g

25. $\frac{1.5\,\text{yd}}{1}\left(\frac{36\,\text{in.}}{1\,\text{yd}}\right)\left(\frac{2.54\text{ cm}}{1\,\text{in.}}\right) = 137.16\text{ cm} \approx 140\text{ cm}$

27. $\frac{227\,\text{g}}{1}\left(\frac{1\,\text{lb}}{454\,\text{g}}\right)\left(\frac{16\text{ oz}}{1\,\text{lb}}\right) \approx 8.00\text{ oz}$ **29.** ≈75 mi

31. ≈189 lb

33. $\frac{1{,}500\,\text{m}}{1}\left(\frac{39.4\,\text{in.}}{1\,\text{m}}\right)\left(\frac{1\text{ ft}}{12\,\text{in.}}\right) = 4{,}925\text{ ft}$

```
  5,280
– 4,925
    355
```

1 mile = 5,280 ft.
Therefore, the 1,500 m race is 355 ft shorter than the mile race.

35. $\frac{175\,\text{mm}}{1}\left(\frac{1\text{ cm}}{10\,\text{mm}}\right)\left(\frac{1\text{ in.}}{2.54\text{ cm}}\right) = 6.88976378\text{ in.} \approx 6.89\text{ in.}$

37. $\frac{20\,\ell}{1}\left(\frac{1.06\,\text{qt}}{1\,\ell}\right)\left(\frac{1\text{ gal}}{4\,\text{qt}}\right) = 5.30\text{ gal} \approx 5\text{ gal}$

Exercises II.11 (page 547)

1. ≈80.8 cm **3.** ≈397 ℓ **5.** ≈418.2 g **7.** ≈28°C
9. ≈1,600 yd **11.** ≈5.8 gal **13.** ≈1.205 lb **15.** 37°C

Exercises III (page 554)

1. $1, -\frac{2}{3}$ **3.** $2 \pm \sqrt{3}$ **5.** $2 \pm \sqrt{2}$ **7.** $\frac{4 \pm \sqrt{6}}{2}$

9. $3x^2 + 2x + 1 = 0$ $\begin{cases} a = 3 \\ b = 2 \\ c = 1 \end{cases}$

$$x = \frac{-(2) \pm \sqrt{(2)^2 - 4(3)(1)}}{2(3)}$$

$$x = \frac{-2 \pm \sqrt{4 - 12}}{6} = \frac{-2 \pm \sqrt{-8}}{6}$$

Solution is not a real number because radicand is negative.

11. $3, -1$ **13.** $4, 1$ **15.** $\frac{1}{2}, -3$ **17.** $1 \pm \sqrt{3}$

19. Let x = the number

then $x - \frac{1}{x} = \frac{2}{3}$ LCD = $3x$

$$\frac{3x}{1} \cdot \frac{x}{1} - \frac{3x}{1} \cdot \frac{1}{x} = \frac{3x}{1} \cdot \frac{2}{3}$$

$$3x^2 - 3 = 2x$$
$$3x^2 - 2x - 3 = 0$$ $\begin{cases} a = 3 \\ b = -2 \\ c = -3 \end{cases}$

$$x = \frac{-(-2) \pm \sqrt{(-2)^2 - 4(3)(-3)}}{2(3)}$$

$$x = \frac{2 \pm \sqrt{4 + 36}}{6} = \frac{2 \pm \sqrt{40}}{6}$$

$$x = \frac{2 \pm 2\sqrt{10}}{6} = \frac{1 \pm \sqrt{10}}{3}$$

Two answers: $\frac{1 + \sqrt{10}}{3}$ and $\frac{1 - \sqrt{10}}{3}$

Index

COMMON METRIC MEASURES

Basic Units	*Prefixes*
Meter measures length.	*Kilo* means 1,000.
Liter measures volume.	*Milli* means $\frac{1}{1{,}000}$.
Gram measures weight.	*Centi* means $\frac{1}{100}$.

1 kilometer (km) = 1,000 meters (m)
1 meter (m) = 100 centimeters (cm)
1 meter (m) = 1,000 millimeters (mm)
1 centimeter (cm) = 10 millimeters (mm)
1 kilogram (kg) = 1,000 grams (g)
1 gram (g) = 1,000 milligrams (mg)
1 liter (ℓ) = 1,000 milliliters (mℓ)
1 cubic centimeter (cc) = 1 milliliter (mℓ)
1 liter (ℓ) = 1,000 cubic centimeters (cc)

COMMON ENGLISH MEASURES

Length

1 foot (ft) = 12 inches (in.)
1 yard (yd) = 3 feet (ft)
1 mile (mi) = 5,280 feet (ft)

Volume

1 tablespoon (tbsp) = 3 teaspoons (tsp)
1 cup = 8 ounces (oz) = 16 tablespoons
1 pint (pt) = 2 cups = 16 ounces (oz)
1 quart (qt) = 2 pints (pt)
1 gallon (gal) = 4 quarts (qt)
1 gallon (gal) = 231 cubic inches (cu. in.)

Weight

1 pound (lb) = 16 ounces (oz)
1 ton = 2,000 pounds (lb)

Time

1 minute (min) = 60 seconds (sec)
1 hour (hr) = 60 minutes (min)
1 day (da) = 24 hours (hr)
1 week (wk) = 7 days (da)
1 year (yr) = 52 weeks (wk) = 365 days (da)

GEOMETRIC FORMULAS (See also page 508)

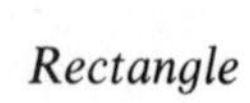

Rectangle
$A = \ell \times w$
$P = 2\ell + 2w$

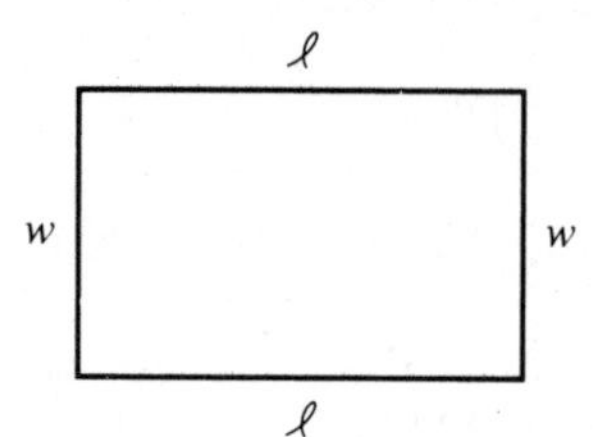

Triangle
$A = \frac{1}{2}bh$
$P = a + b + c$

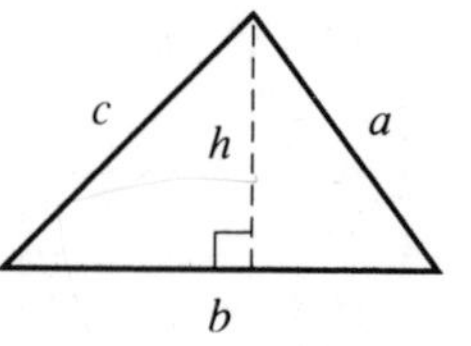

Circle
$A = \pi r^2$
$C = \pi d = 2\pi r$

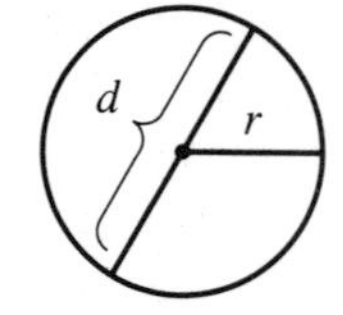

Rectangular Box
$V = \ell \times w \times h$

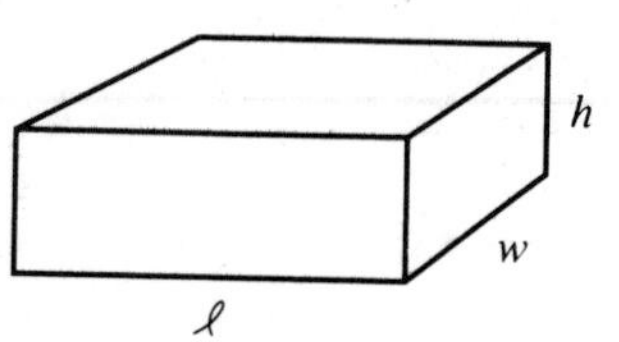

Cylinder
$V = \pi r^2 h$

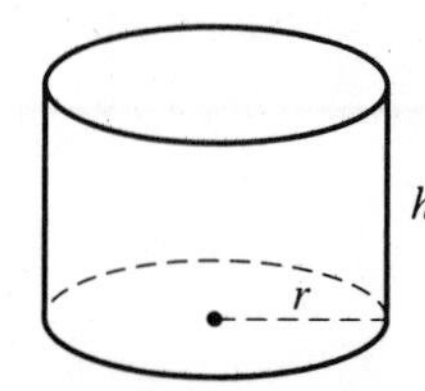

Sphere
$V = \frac{4}{3}\pi r^3$

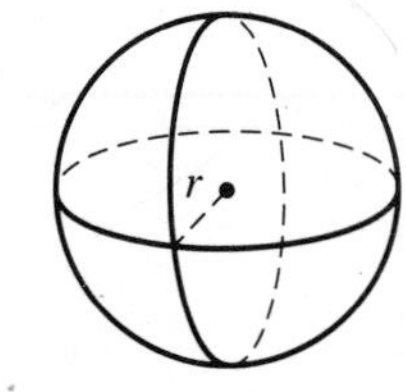